Jan de Vries
Topological Dynamical Systems

De Gruyter Studies in Mathematics

Volume 59

Jan de Vries

Topological Dynamical Systems

An Introduction to the Dynamics of Continuous Mappings

DE GRUYTER

Mathematics Subject Classification 2010
Primary: 37-01, 54H20; Secondary: 37B10, 37B20, 37B25 37B40, 37D45

Author
Dr. Jan de Vries
Ontginningsweg 1
9865 XA Opende
Netherlands
jandv3@xs4all.nl

Retired from CWI, Centrum voor Wiskunde & Informatica (Center for Mathematics and Computer Science), Amsterdam, the Netherlands

ISBN 978-3-11-034073-0
e-ISBN 978-3-11-034240-6
Set-ISBN 978-3-11-034241-3
ISSN 0179-0986

Library of Congress Cataloging-in-Publication Data
A CIP catalog record for this book has been applied for at the Library of Congress.

Bibliographic information published by the Deutsche Nationalbibliothek
The Deutsche Nationalbibliothek lists this publication in the Deutsche Nationalbibliografie; detailed bibliographic data are available in the Internet at http://dnb.dnb.de.

Typesetting: le-tex publishing services GmbH, Leipzig
Printing and binding: CPI buch bücher.de GmbH, Birkach
♾Printed on acid-free paper
Printed in Germany

MIX
Papier aus verantwortungsvollen Quellen
FSC® C003147

www.degruyter.com

Preface

Thus the superior man understands the transitory
in the light of the eternity of the end.
After: I Ching, Hexagram 54.

This book is just what its title says: an introduction. It is addressed primarily to graduate students who want to learn the basic ideas of topological dynamics. Thus, the fundamental notions of (topological) dynamical systems are defined and their elementary properties are discussed. Students who have mastered this book will have a firm basis to start research related with the topics discussed here, even though we have not included all most recent results. Unfortunately, in order to keep this book reasonably sized many important topics could not be included (or even mentioned). The choice of which topics to include is to a great extend determined by my wish to concentrate on the purely topological aspects of the theory, in the spirit of the work by Birkhoff.

When I started writing [deV][1], '(topological) dynamics' was a relatively unknown topic. At present the situation is completely different and one might say that this topic is quite popular, to the extend that introductory courses in differential equations are sometimes called courses in dynamical systems. On the other hand, the purely topological approach to dynamical systems theory gets less attention than it deserves. Therefore I decided to publish the lecture notes of a course in 'applied topology' that I gave at the Free University in Amsterdam from 1995 until my retirement in 2002.

The process of transforming my Dutch lecture notes into an English book was more complicated and took much more time than anticipated. Not only had the material to be reorganized, streamlined and expanded, but also a severe illness prevented me from working during a couple of years.

The book you have before you treats systems consisting of a topological space (the phase space) and iterations of a single continuous mapping of this space into itself (the phase mapping). Except in Section 2.1 no use is made of differentiability of the phase mapping. The assumption that the phase space is metrizable is avoided as much as possible.

The book is organized as follows: In the Introduction the 'dynamical systems approach' is explained: the philosophy behind the material in the book and the red thread through the subsequent chapters. Though most of the Introduction contains no specific results needed in the rest of the book, the reader is strongly advised to have a look at it. Moreover, the examples at the end of the Introduction are heavily used later on (but reading them can be postponed until they are needed).

The Chapters 1 and 2 together with the examples in the Introduction are used throughout the remainder of the book. In Chapter 1 the basic notions of the theory are

1 Items like [deV] refer to the literature.

defined, the elementary properties of dynamical systems are discussed and illustrated by examples. Chapter 2 treats dynamical systems on intervals in $\mathbb{R}$; it culminates in a proof of Šarkovskij's Theorem.

The Chapters 3 and 4 are about stability: stability of invariant sets and variants of Poisson-stability (which we call 'recurrence'), including almost periodic and non-wandering points. Chapter 3 also discusses attraction, though we refrain from giving a formal definition of an 'attractor'. See, however, the Notes at the end of that chapter. Chapter 3 ends with a discussion of the space of components of a transitive (asymptotically) stable set in a locally connected locally compact phase space – providing full proofs of statements that have unconvincing proofs elsewhere. The discussion of recurrence and almost periodicity in Chapter 4 is restricted to a bare minimum because there is much other literature about these topics: see [GH] and [deV] to get an idea.

Next, in Chapter 5 we discuss shift systems (spaces of sequences with the shift operator) and in Chapter 6 we investigate how such systems can be used to represent other systems by means of a suitable coding (symbolic dynamics). The study of shift systems has a strong algebraic/combinatorial flavour and it has many applications in and points of contact with other parts of mathematics, from artificial languages to coding theory and from automata theory to probability theory. We discuss none of these applications; the interested reader is referred to D. Lind & B. Marcus [1995] or B. Kitchens [1998]. We give only some applications of symbolic dynamics to 1-dimensional systems; these are used in Chapter 8 to compute the topological entropy of those systems.

Chapter 7 deals with notions of chaos. There are many definitions of this notion, all about 'erratic behaviour'. We concentrate on sensitive dependence on initial conditions and the existence of large so-called 'scrambled' sets. In Chapter 8 we discuss the notion of topological entropy. We include a proof of the fact that, for maps of an interval into itself, positive entropy is equivalent to the existence of a point with odd primitive period greater than 1 which, by the results of Chapter 7, implies chaos.

The results of the Chapters 3 and 4 are not needed for a good understanding of the later chapters, though in Chapter 5 some examples are given that illustrate notions and results from these previous chapters. Similarly, the Chapters 5 and 6 are not needed for an understanding of the Chapters 7 and 8, though also here examples in the latter chapters are taken from the former. So possible courses can be be based on the Chapters 1 and 2 followed by either the Chapters 3 and/or 4, or 5 and 6, or 7 and 8. Of course, other selections are possible, depending on the available time.

Every chapter concludes with a set of exercises. Most of them are routine applications of the material in the chapter, others deal with extensions of the theory. For the more challenging ones one may find hints (or if you prefer: telegramm-style answers) at the end of the book. The exercises are followed by a section of Notes in which also references to the literature are given. These Notes are rather sketchy and the references are far from complete. In particular, they are not meant as complete historical introductions; rather, they tell how the results came to me. In point of fact, many results included in the book are common knowledge and are known already for decennia. My knowledge of dynamical systems grew over a rather long period and often I lost track of where I read or heard the various results. Consequently, for many results in this book references to the original sources are missing. But I can safely state that I learned the 'classical' results from the book [GH]. I started reading [GH] around 1970 (I was interested in almost periodic functions and transformation groups) but after a couple of hours I threw the book aside in despair (some outstanding mathematicians told me they had the same experience) – try it, and you will understand why. A couple of months later I tried again, then a year later.... Over the years I learned how to extract information from that book and about 1985 I could reasonably find my way in it. It really contains almost everything that was know about topological dynamics in, say, 1950. Of course, much in the present book was discovered later, and I have tried to give credit to whom it deserves.

Finally, a few words about terminology. In the literature the objects studied in this book are often called *semi*-dynamical systems. Moreover, what we call orbits, or limit sets, or invariant sets, are usually called *positive semi-orbits*, *positive (or omega-)limit sets*, *positively (or forward) invariant sets*, etc. As we pay no special attention to invertible systems and, consequently, no negative semi-orbits, or negative (or alpha-)limit sets, or backward invariant sets, etc., are discussed, it would seem somewhat redundant to prefix all notions with 'semi-', 'positive(ly)' or 'forward'. So I decided to omit all those prefixes. This has the disadvantage that the terminology in this book does not always agree with the usual one in the literature. I have tried to obviate difficulties caused by conflicting terminology by adding remarks on 'invertible vs. non-invertible systems' at the end of the relevant chapters. These notes do not cover all differences: I tried to restrict myself to the discrepancies which I believe are likely to cause confusion. The reader who gets confused by seemingly conflicting results in this book on the one hand and the literature on the other is advised to have a look at these notes.

The prerequisites for understanding this book are are rather modest. A reader who has mastered a course in General Topology and has a working knowledge of Calculus should be able to follow all arguments. For a good understanding of the Introduction some familiarity with the theory of differential equations is useful. For easy reference there are two Appendices at the end of the book: one with the preliminaries from general topology and a second one about the Cantor set.

The internal reference system is rather straightforward. The numbering of sections starts anew in each chapter; thus, 'Section $k.m$' refers to the m-th section in Chapter k. Within each section the items are numbered consecutively: '$k.m.n$' refers to item n in Section $k.m$. Equations are numbered separately with the a similar system, but a slightly different style is used: '($k.m$-n)' refers to the n-th formula in Section $k.m$. Figures and Exercises are numbered as follows: 'Figure $k.n$' refers to the n-th figure in Chapter k and 'Exercise $k.n$' refers to the n-th exercise at the end of Chapter k; moreover, 'Exercise $k.n$ (i)' refers to part (i) of Exercise $k.n$.

Finally, I want to express my indebtedness to the people who contributed to my (mathematical) education and my knowledge of dynamical systems. First, I mention the late prof. J. F. Koksma. With a course on 'Almost Periodic Functions' (covering most of Maak's book) he roused my interest in almost periodicity. Most of my later research was in some sense related to this topic. Next, prof. P. C. Baayen (Cor) played a decisive role in my mathematical life. He was my Ph. D. thesis supervisor and, as head of the Pure Mathematics Department of the 'Mathematisch Centrum' in Amsterdam – presently called CWI – he invited me to work there after my graduation. Later, as Scientific Director of the CWI (1980–1994) he directly contributed to the ideal research environment that the CWI was at that time, and in this way he indirectly contributed to my research. His broad knowledge of mathematics (not to mention history, Tolkien, science fiction, ...) has always inspired me. In the 1980's I supervised one of his Ph. D. students, Jaap van der Woude, in his research in topological dynamics – or rather, Jaap took me in tow and I had to work quite hard to keep pace with him. This cooperation eventually gave rise to the book [deV]. I also want to mention Jan Aarts; my lectures at the Free University were based on the lecture notes of a course he gave at the Technical University in Delft. Finally, I owe much to Mike Keane and the guest speakers at his seminars in Delft and in Amsterdam.

My last words here are for my wife Liet. This month we celebrated the fact that we first met exactly 50 years ago. Since, her caring love has sustained me. It kept me, literally, in life during the preparation of this book.

Opende, november 2013 *Jan de Vries*

Notation

The best notation is no notation...
Paul Halmos, How to write Mathematics (1970)

The purpose of the list of symbols below is twofold. First we establish some general notation concerning sets and mappings. With a minor exception, this notation is not defined in this book and it will be used without further reference. Also some notation about topological and metric spaces is collected. Most of it is defined in Appendix A, but the descriptions below should be sufficiently clear and no references to particular page numbers are given. Secondly, in this list we collect the notation about dynamical systems developed in this book. Here the page numbers where the respective notions are defined are added between square brackets.

Sets and mappings

- $\left.\begin{matrix} P := Q \\ Q =: P \end{matrix}\right\}$: P is defined as Q.
- $A \subseteq B$: A is a subset of B (possibly $A = B$).
- $A \subset B$: $A \subseteq B$ and $A \neq B$.
- $\mathbb{R}$: the real line.
- $\mathbb{Z}$: the integers.
- $\mathbb{Q}$: the set of rational numbers.
- $\mathbb{R}^+ := \{s \in \mathbb{R} \,\vdots\, s \geq 0\}$.
- $\mathbb{Z}^+ := \mathbb{Z} \cap \mathbb{R}^+$.
- $\mathbb{N} := \mathbb{Z}^+ \setminus \{0\}$.
- $\mathbb{S} := \{z \in \mathbb{C} \,\vdots\, |z| = 1\}$ (the circle).
- $sA + tB := \{sa + tb \,\vdots\, a \in A \text{ and } b \in B\}$ for $A, B \subseteq \mathbb{R}$ and $s, t \in \mathbb{R}$.
- $[a; b] := \{s \in \mathbb{R} \,\vdots\, a \leq s \leq b\}$ (closed interval).
- $(a; b) := \{s \in \mathbb{R} \,\vdots\, a < s < b\}$ (open interval)
- $(a; b] := \{s \in \mathbb{R} \,\vdots\, a < s \leq b\}$ (left open, right closed interval).
- $[a; b) := \{s \in \mathbb{R} \,\vdots\, a \leq s < b\}$ (left closed, right open interval).

If J is a bounded interval then $|J|$ will denote the length of J. So if J is one of the above intervals then $|J| = b - a$.

- $[t] := e^{2\pi i t} \in \mathbb{S}$ for $t \in \mathbb{R}$.
- $d_c([s], [t]) := 2\pi \min\{|s - t| (\text{mod } 1), 1 - |s - t| (\text{mod } 1)\}$ for $s, t \in \mathbb{R}$ (metric in $\mathbb{S}$).

If X is a set then $\text{id}_X : x \mapsto x : X \to X$ is the identity mapping. If $n \in \mathbb{N}$ then $X^n := X \times \cdots \times X$ (n times) and $\Delta_X := \{(x, \ldots, x) \in X^n \,\vdots\, x \in X\}$ (used almost exclusively in the case that $n = 2$).

Let $f: X \to Y$ be a mapping, $A \subseteq X$, $B \subseteq Y$ and $y \in Y$. Then:
- $f[A] := \{f(x) \,\vdots\, x \in A\}$.
- $f^{\leftarrow}[B] := \{x \in X \,\vdots\, f(x) \in B\}$.
- $f^{\leftarrow}[y] := f^{\leftarrow}[\{y\}]$.

If f is a bijection then the *inverse* of f is denoted f^{-1}. In that case:
- $f^{\leftarrow}[B] = f^{-1}[B]$ and $f^{\leftarrow}[y] = \{f^{-1}(y)\}$.

If $I \subseteq X$ and $J \subseteq Y$ then:
- $f: I \overset{\circ}{\to} J$ means: $f[I] \supseteq J$ (f maps I over J) [80],
- $f: I \twoheadrightarrow J$ means: $f[I] = J$ (f maps I onto J) [80].

Topological and metric spaces

Let X be a topological space, $x \in X$ and $A \subseteq X$. Then:
- $\mathcal{N}_x$: the set of all neighbourhoods of the point x.
- $\mathcal{N}_A$: the set of all neighbourhoods of the set A.
- $\overline{A}$, A^-, $\mathrm{cl}_X(A)$ (or $\mathrm{cl}(A)$ if X is understood) the closure of A in X.
- A°, $\mathrm{int}_X(A)$ (or $\mathrm{int}(A)$ if X is understood) the interior of A.
- $C(X, X)$ the set of all continuous mappings from X into itself.

Let (X, d) be a metric space, $x \in X$, $A \subseteq X$ and $\varepsilon > 0$. Then:
- $B_\varepsilon(x) := \{x' \in X \,\vdots\, d(x, x') < \varepsilon\}$ (open ball about x with radius ε).
- $S_\varepsilon(x) := \{x' \in X \,\vdots\, d(x, x') \leq \varepsilon\}$ (closed ball about x with radius ε).
- $d(x, A) := \inf\{d(x, y) \,\vdots\, y \in A\}$ (distance between x and A).
- $B_\varepsilon(A) := \{x' \in X \,\vdots\, d(x', A) < \varepsilon\} = \bigcup\{B_\varepsilon(y) \,\vdots\, y \in A\}$ (open ε-neighbourhood of A).
- $C_c(X, Y) : C(X, Y)$ endowed with the compact-open topology.
- $C_u(X, Y) : C(X, Y)$ endowed with the topology of uniform convergence (Y a metric space).

Continuous functions defining dynamical systems

- $f_\mu : x \mapsto \mu x(1 - x) : \mathbb{R} \to \mathbb{R}$ for $\mu > 0$: the quadratic or logistic family [9].
- $\varphi_a : [t] \mapsto [a + t] : \mathbb{S} \to \mathbb{S}$ for $a \in \mathbb{R}$: the rigid rotation of the circle [11].
- $\psi : [t] \mapsto [2t] : \mathbb{S} \to \mathbb{S}$: the argument-doubling transformation [12].
- $T : x \mapsto \min\{2x, 2(1 - x)\} : [0; 1] \to [0; 1]$: the tent map [9].
- $T_\lambda : x \mapsto \min\{T(x), \lambda\} : [0; 1] \to [0; 1]$: the truncated tent map [84].
- $T_s : x \mapsto \frac{s}{2}(1 - |2x - 1|) : [0; 1] \to [0; 1]$: the generalized tent map [314].
- $\sigma : (x_n)_{n \in \mathbb{Z}^+} \mapsto (x_{n+1})_{n \in \mathbb{Z}^+} : \Omega_S \to \Omega_S$: the shift (shift map) [223].
- σ_X: the shift map restricted to a subshift X [226].

Dynamical notions

Let (X, f) be a dynamical system, $x \in X$ and $A, B \subseteq X$. Then:

- $\mathcal{O}_f(x) := \{f^n(x) : n \in \mathbb{Z}^+\}$: the orbit of x under f [7,17].
- $\omega_f(x) := \bigcap_{n=0}^{\infty} \overline{\mathcal{O}_f(f^n(x))}$: the (positive) limit set of x [33].
- $D(x, B) := \{n \in \mathbb{Z}^+ : f^n(x) \in B\}$: the dwelling set of x in B [28].
- $D(A, B) := \{n \in \mathbb{Z}^+ : f^n[A] \cap B \neq \emptyset\} = \{n \in \mathbb{Z}^+ : A \cap (f^n)^{\leftarrow}[B] \neq \emptyset\}$: the dwelling set of A in B [31].
- $\mathcal{B}_f(A) := \{x \in X : \emptyset \neq \omega_f(x) \subseteq A\}$: the basin of attraction of A (A not empty, closed and invariant) [120].
- $R(X, f)$: the set of recurrent points of (X, f) [165].
- $Z(X, f) := \overline{R(X, f)}$: the centre of (X, f) [178].
- $\Omega(X, f)$: the non-wandering set of (X, f) [175].
- $CR(X, f)$: the chain recurrent set of (X, f) [185].
- $\mathbf{Trans}(X, f)$: the set of all transitive points in (X, f) [66].
- $\mathbf{Eq}(X, f)$: the set of all points x such that the system is equicontinuous (that is, stable) at x [326].

If the phase space X is a metric space:

- $d_n^f(x, y) := \max_{0 \leq i \leq n-1} d\big(f^i(x), f^i(y)\big)$ $(x, y \in X, n \in \mathbb{N})$ [380].
- $h(K, f)$: the topological entropy of K under f ($K \subseteq X$ compact) [382].
- $h(f)$: the topological entropy of f [382].

If the phase space X is a any space and $\mathcal{A}$ and $\mathcal{B}$ are covers of X then:

- $\mathcal{A} < \mathcal{B}$: $\mathcal{B}$ is *finer than* $\mathcal{A}$ (or: $\mathcal{A}$ is *coarser than* $\mathcal{B}$) [395].
- $\mathcal{A} \vee \mathcal{B}$: the *join* of the covers $\mathcal{A}$ and $\mathcal{B}$ [395].
- $\mathcal{A}^{f,n} := \mathcal{A} \vee f^{\leftarrow}\mathcal{A} \vee \cdots \vee (f^{n-1})^{\leftarrow}\mathcal{A}$ ($\mathcal{A}$ a cover, $n \in \mathbb{N}$) [395]
- $N(\mathcal{A})$: the minimal cardinality of a finite subcover of a special cover $\mathcal{A}$ [396].
- $H(\mathcal{A}) := \log N(\mathcal{A})$ [396].
- $h(f, \mathcal{A})$: the entropy of f with respect to the special cover $\mathcal{A}$ [397].

Shift spaces and symbolic representations

Let $\mathbf{S}$ be a finite set with at least two elements.

- $\mathbf{S}^* := \bigcup_{n=0}^{\infty} \mathbf{S}^n$: the langue – or alphabet – over the symbol set $\mathbf{S}$ [219].
- $\square$: the unique element of $\mathbf{S}^0$, i.e., the empty word [219].
- $\Omega_{\mathbf{S}} := \mathbf{S}^{\mathbb{Z}^+}$: the (full) shift space over the symbol set $\mathbf{S}$ [218].
- $\widetilde{B}_k(x) := \{y \in \Omega_{\mathbf{S}} : y_i = x_i \text{ for } 0 \leq i \leq k-1\}$: the cylinder about the point $x \in \Omega_{\mathbf{S}}$, based on the initial block of x with length $k \geq 1$ [221].
- $\mathfrak{X}(\mathcal{B}) := \{x \in \Omega_{\mathbf{S}} : \text{no member of } \mathcal{B} \text{ occurs in } x\}$ ($\mathcal{B} \subseteq \mathbf{S}^* \setminus \{\square\}$) [228].

- $\mathcal{A}(X) := \{b \in \mathbf{S}^* : b$ does not occur in any point of $X\}$ (X a shift space): the words absent from X [227].
- $\mathcal{L}(X) := \mathbf{S}^* \setminus \mathcal{V}(X)$ (X a shift space): the language of a shift space X, that is, the set of X-present words [230].
- $\mathcal{A}_k(X) := \{b \in \mathbf{S}^k : b$ does not occur in any point of $X\}$ (X a shift space, $k \in \mathbb{N}$): the words of length k absent from X [234].
- $\mathcal{L}_k(X) := \mathbf{S}^k \setminus \mathcal{V}_k(X)$ (X a shift space, $k \in \mathbb{N}$): the X-present words of length k [234].
- $\mathfrak{M}_v(G)$: the SFT defined by the faithfully vertex-labelled graph G [245].
- $\mathfrak{W}_v(G)$: the shift space defined by the infinite walks on a vertex-labelled graph G [249].
- $\mathfrak{W}_e(G)$: the sofic shift space defined by the infinite walks on an edge-labelled graph G [250].

If $\mathcal{P} := \{P_0, \ldots, P_{s-1}\}$ is a topological partition of X and $\mathbf{S} := \{0, \ldots, s-1\}$ then
- $\iota(x) \in \Omega_{\mathbf{S}}$, the point with coordinates $\iota(x)_n$ such that $f^n(x) \in \mathcal{P}_{\iota(x)_n}$ for $n \in \mathbb{Z}^+$: the itinerary of a point x with respect to the topological partition $\mathcal{P}$ [283].
- $X^* := X^*(\mathcal{P}, f)$: the set of points $x \in X$ with a full itinerary $\iota(x)$ [283].
- $Z := Z(\mathcal{P}, f)$: the symbolic model of X generated by $\mathcal{P}$ (provided $\mathcal{P}$ is f-adapted) [284/285].
- $D_k(b) := \bigcap_{n=0}^{k-1} (f^n)^{\leftarrow}[P_{b_n}]$: ($\{P_0, \ldots, P_{s-1}\}$ an indexed family of subsets of X, $b = b_0 \cdots b_{k-1}$ a k-tuple of symbols from the set $\{0, \ldots, s-1\}$) [284, also 407].
- $\psi := \psi_{\mathcal{P},f} : (Z, \sigma_Z) \to (X, f)$: the symbolic representation of (X, f) (provided $\mathcal{P}$ is a pseudo-Markov partition) [287].

Contents

I'm writing a book. I've got the page numbers done.
Unknown author.

0 Introduction

Abstract. By a *dynamical system* we understand an ordered pair (X, f), where X is a topological Hausdorff space and $f: X \to X$ is a continuous mapping. We consider the study of such objects as a part of topology, a part with its own problems and methods, which distinguishes it from other topics in topology. In this introduction we explain which problems are characteristic for the theory of dynamical systems (or 'topological dynamics', as this theory is often called).

0.1 Definition and a (very brief) historical overview

The theory of (discrete) dynamical systems is about the 'behaviour' of points of a topological space X under the application of the iterates of a given continuous mapping $f: X \to X$. As usual, the iterates of f are defined inductively, as follows: $f^0 := \mathrm{id}_X$ and, for every $n \in \mathbb{Z}^+$, $f^{n+1} := f \circ f^n$. A simple, yet crucial, property of the iterates of a mapping $f: X \to X$ is

$$f^0 = \mathrm{id}_X \quad \text{and} \quad f^k \circ f^n = f^{k+n} \quad \text{for all } k, n \in \mathbb{Z}^+ . \tag{0.1-1}$$

If f is a homeomorphism it makes also sense to define $f^{-n} := (f^{-1})^n$ for every $n \in \mathbb{N}$. So in that case, f^n is defined for all $n \in \mathbb{Z}$. It is easy to see that for a bijection $f: X \to X$ equation (0.1-1) holds for all $k, n \in \mathbb{Z}$.

Iterative procedures have been in use in mathematics long before the theory of dynamical systems emerged. As a reminder, we mention some examples. These have little or nothing to do with the theory of dynamical systems.

Recurrent sequences. Let $D \subseteq \mathbb{R}$ and let $f: D \to D$ be a function. Each choice of an initial value $x_0 \in D$ determines a unique sequence $(x_n)_{n \in \mathbb{Z}^+}$ in D, defined inductively by $x_n := f(x_{n-1})$ $(n \in \mathbb{N})$. Example: Let $x_0 > 0$ and let $x_n := \sqrt{x_{n-1}}$ for $n \in \mathbb{N}$. Then $\lim_{n \to \infty} 2^n(x_n - 1) = \ln(x_0)$. This might be used as the definition of the natural logarithm of x_0. In particular, one may try to derive from this definition some of the well-known properties of the (natural) logarithm.

Newton's method. Consider a differentiable function $f: J \to \mathbb{R}$ (J an interval), select an initial value $x_0 \in J$ and for each $n \in \mathbb{Z}^+$, let $x_{n+1} := x_n - f(x_n)/f'(x_n)$. If certain conditions are fulfilled then the sequence so obtained converges in $\mathbb{R}$ to a root of the equation $f(x) = 0$.

Perhaps the best way to describe the 'special way' the theory of dynamical systems looks at iterations is to give a (very) brief sketch of the historical background of the theory.

At the end of the 19th century it was discovered that no explicit expression can be given for the solutions of the differential equations for celestial mechanics (the N-body problem). This discovery gave rise to what is now called the *qualitative the-*

ory of differential equations: try to say something about the existence and stability of equilibrium states and periodic solutions or, more generally, study the behaviour of solutions in the long run, *without first solving the equations*. The French mathematician Henri Poincaré attacked this problem by considering the geometric picture of the solution curves of a differential equation (the *phase portrait*) – of which one can often get an idea without solving the equations – and trying to interpret outstanding geometric features of this picture in terms of significant physical phenomena (see, for example, the explanation of a gradient flow in Section 0.2 below). As a result of this qualitative approach the attention in the study of differential equations focussed more and more on the geometry (topology) of the phase portrait. The American mathematician G. D. Birkhoff made the transition from the qualitative theory of differential equations to topology most explicit by studying possible phase portraits without making any reference at all to the fact that they might be defined by differential equations. The present book stands in this tradition.

In its most general form a dynamical system consists of a set X provided with an additional structure like a topology, a metric or a differential structure, and a set $\{\pi_t\}_{t\in T}$ of mappings of X into itself compatible with the additional structure of X. The space X is viewed as the space of all possible *states*[1] of some fictitious 'physical' system. It is called the *phase space* of the system (in physics, 'phase' is an often used synonym for 'state'). The set T is the collection of all time values that are relevant for the system. Usually one takes for T the set $\mathbb{R}$ or $\mathbb{R}^+$ (continuous time) or else $\mathbb{Z}$ or $\mathbb{Z}^+$ (discrete time). The mappings π_t for $t \in T$ describe all possible developments of the system (depending on the 'physical' laws by which it is governed), as follows: if at time $t = 0$ the system is in state $x \in X$ (the *initial state*) then $\pi_t(x)$ will denote the state of the system at time t. This means that we consider *deterministic* systems: at any moment $t \in T$ the state $\pi_t(x)$ is completely determined by the initial state $x \in X$ (and the value of t, of course).

In this book we consider only the special case that X is a topological Hausdorff space and that each mapping $\pi_t : X \to X$ is continuous. In addition, we shall always assume that our system is *stationary*, that is, the rules that govern the system do not change with time (when such a system is described by differential equations then these equations would be *autonomous*): the special moment $t = 0$ can be anytime. This means that the state of the system after a particular time interval depends only on its initial state at the beginning of the time interval and on the length of that time interval, but not on the moment that the time interval actually begins. Consequently, the

1 Classically, it is natural to think of the state of a system as the sum of its dispositions to respond to the range of circumstances it might encounter in the future.In any specific type of physical system one has to make clear what 'state' means. For example, in the description of a moving particle its state is given by its position and its momentum, but in the description of the evolution of a population the state may be its size.

expression $\pi_t(x)$ for $x \in X$ and $t \in T$ denotes the state which the system has reached after a time-interval of length t when it starts at state x. A moment's reflection then shows that $\pi_0(x) = x$ and $\pi_s(\pi_t(x)) = \pi_{s+t}(x)$ for all $s, t \in T$ and every $x \in X$, that is,

$$\pi_0 = \mathrm{id}_X \quad \text{and} \quad \pi_s \circ \pi_t = \pi_{s+t} \,. \tag{0.1-2}$$

In the case that $T = \mathbb{Z}$ or $\mathbb{R}$, each of the mappings π_t with $t \in T$ is a bijection with inverse π_{-t}. However, if $T = \mathbb{Z}^+$ or $\mathbb{R}^+$ then the mappings π_t are not necessarily injective or surjective. It is easy to show that in the case that $T = \mathbb{Z}$ or $\mathbb{Z}^+$ the set $\{\pi_t\}_{t\in T}$ is equal to the set of functions consisting of $f := \pi_1 : X \to X$ and its iterates f^n with $n \in \mathbb{Z}$ or $\mathbb{Z}^+$, respectively. In this case, the equations in (0.1-2) take the form of (0.1-1).

Thus, if (X, f) is a dynamical system then X is considered as the set of all possible states of a fictitious 'physical' system. Every point $x \in X$ will denote a possible state of that system at any moment and $f(x)$ is the state that this system will have one unit of time later.

0.2 Continuous vs. discrete time

We briefly sketch what the 'geometric method' means for autonomous differential equations. This may seem a bit awkward, as differential equations typically define systems with continuous time, while this book is about systems with discrete time. But conceptually it is easier to deal with a system defined by differential equations, as in that case the orbits (the definition of an orbit will be given later) are the solution curves.

So consider a 'physical' system and assume that its evolution according to the 'physical' laws by which it is governed is given by the autonomous differential equation $\dot{x} = F(x)$ in an open and connected subset G of $\mathbb{R}^n$. We shall describe how this equation defines a dynamical system with phase space G. For convenience, we assume that this equation has unique solutions that are defined on all of $\mathbb{R}$. For every point $x \in G$ there is a unique solution curve $t \mapsto \pi_t(x) \colon \mathbb{R} \to G$ with $\pi_0(x) = x$. Then x can be considered as the initial state of the system and the past and future states of the system are found as the points $\pi_t(x)$ of the solution curve through x in G. It is a straightforward consequence of the unicity of solutions that the conditions of (0.1-2) are fulfilled and that solution curves cannot intersect without coinciding. It follows that the solution curves form a partition of the set G.

Looking at the phase space in this way gives the idea of a *flow*: when t increases, the points $\pi_t(x)$ of the phase space move along the solution curves which act as flow lines. In general, such 'flow lines' are called *orbits*. The velocity of the flow at any point x of G given by the vector $F(x)$.

The most simple type of flow is a *gradient flow*. Imagine a horizontal plane: the phase space of the system. Above this plane there is a hilly landscape: the graph of a real valued smooth function Φ. Now assume that the surface of this landscape is

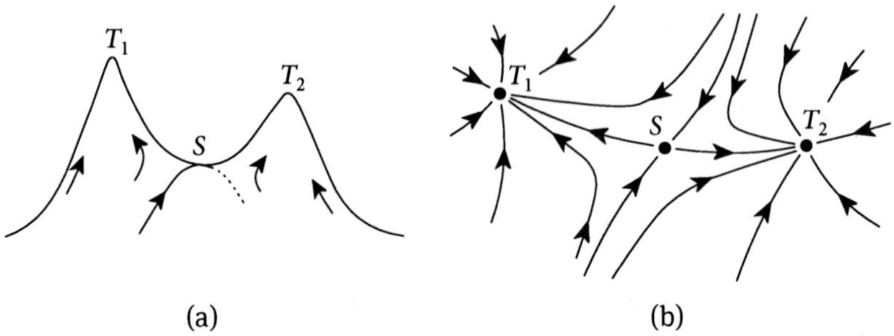

Fig. 1. A gradient flow: (b) shows the solution curves of the differential equation $\dot{x}(t) = \nabla\Phi(x(t))$; Φ is the function describing the hilly landscape in (a). The symbol ∇ – pronounced *nabla* – denotes the operator $(\partial/\partial x_1, \partial/\partial x_2)$. It is well-known that in every point z of G the tangent vector $\nabla\Phi(z)$ in z to the solution curve through z – hence the velocity of the flow – is perpendicular to the level curve $\{ x \in G \vdots \Phi(x) = C \}$ (C is a constant) through that point.

covered by fine sand and that all grains have the tendency to move upwards as steep as possible (of course, downwards would be more natural, but we want to conform to usage). The projection of this movement on the flat horizontal plane is the gradient flow. See Figure 1 (a) for the hilly landscape and Figure 1 (b) for the corresponding flow in the phase space. In the local extremes and in the saddle points of Φ there will be *equilibrium points* (also called *invariant points*). Some of these points (the tops T_1 and T_2) are *attracting* all neighbouring points in the sense that those points flow towards these particular equilibrium points. In physics, such an equilibrium is called 'stable'. Continuity of Φ implies that the flow will slow down close to an equilibrium point. Therefore, a neighbouring point will never actually reach an attracting equilibrium point: only in the limit it will end up there (in models of real-life 'physical' systems this means that after a certain moment it will be indistinguishable from it). The (open) set of all states that approach a particular attracting equilibrium point is called the *basin of attraction* of that equilibrium point. Valleys (which do not occur in our picture) would correspond to *repelling* equilibrium points. Saddle points (like the point S) and the flow lines to and from these points play a particular role: these curves act as *separatrices* of the various basins. In Figure 1 (b) one can see immediately what the evolution of the system will be for every possible initial state and which 'final' state eventually will be approached. It is important to note that such a picture can be sketched if one knows the function Φ: the geometrical structure of the collection of solution curves – the *phase portrait* – is known without solving the differential equation (which in many cases would be difficult or impossible). Thus, significant properties of the solutions of the differential equation can be found without solving it[2] .

2 This is not meant to suggest that the construction and analysis of the phase portrait is always an easy task.

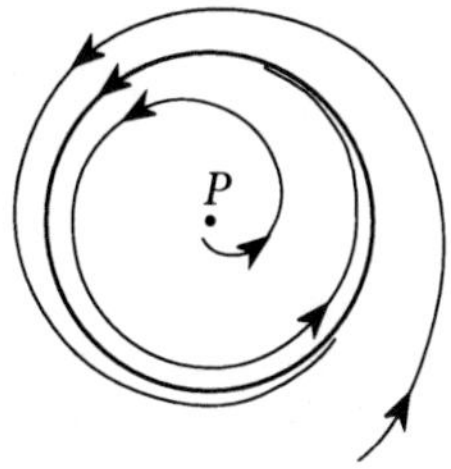

Fig. 2. An attracting periodic solution curve around a repelling equilibrium point P: phase portrait of the differential equations $\dot{r} = r(1 - r)$, $\dot{\theta} = 1$ (in polar coordinates).

An example of a gradient flow in $\mathbb{R}$ is the flow defined by the differential equation $\dot{x} = Mx(1 - x)$ with $M > 0$. Its phase portrait is rather simple; it looks like the following picture: 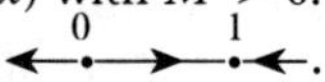

. Every positive initial state tends to the equilibrium 1 and every negative state tends to $-\infty$. The qualitative behaviour of the system is completely described by these observations. However, the pictures tell us nothing about the quantitative behaviour of the system (e.g., how much time does it take for the system to approach, from a given initial state, the equilibrium 1 up to a given error).

Other types of equations can have *periodic* solutions: solution curves that recur into itself after a finite amount of time (this cannot happen in a gradient flow, except in a world where Escher's "Waterfall" can be realized). In that case the solution curve is, topologically, a circle. It is possible that such a periodic solution attracts nearby solutions, in the sense that they spiral towards it, or that it repels them. See Figure 2.

All sorts of additional complications can occur. Often the phase space is not simply an open subset of $\mathbb{R}^n$ but a more complicated manifold. There are flows in which a certain non-periodic initial state can approach itself infinitely often and arbitrarily close.

It can be shown that such a flow cannot exist in $\mathbb{R}^2$, but it can exist on a torus. Such a flow on a torus may be embedded in a larger flow in such a way that it attracts (or repels) nearby orbits. But in many cases, attracting sets are not simply points (equilibrium points), closed curves (periodic motions) or simple surfaces like tori: sometimes their geometry is so complicated that they have a non-integer dimension: *strange attractors*. There are also flows, or attracting subsets in flows, whose behaviour is so chaotic that it cannot be distinguished from stochastic behaviour (*chaotic* systems). The more examples of flows became known during the 20th century, the more one got removed from the 'paradise' of the gradient flows. But the idea has remained the same: try to find an interpretation for outstanding geometrical/topological features of the phase portrait and try to deduce from this what the behaviour of the system will be in the long run (given any initial state).

From the preceding discussion one may, correctly, infer that the study of dynamical systems was initiated with continuous-time systems in mind. So why would one study systems with discrete time, as we do in this book? We mention a few reasons.

The first one is: for simplicity. For a dynamical system with $T = \mathbb{R}$ it may have advantages to study only the mapping $f := \pi_1$ and its iterates. It follows from formula

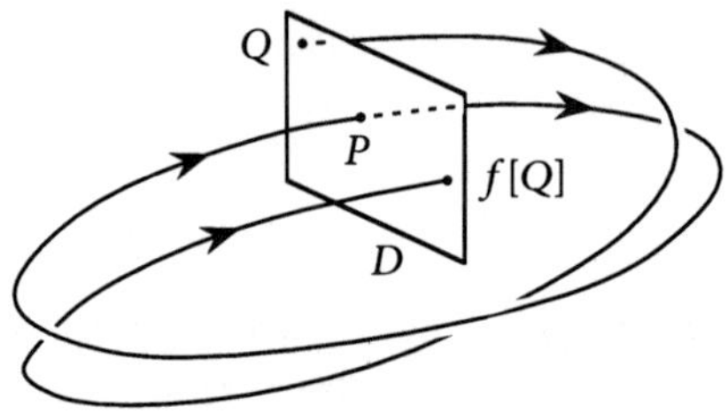

Fig. 3. A cross section D to a flow with continuous time in a periodic point P. Any point of D close to P will return in D after some time.

(0.1-2) that $f^n = \pi_n$ for all $n \in \mathbb{Z}$, so that, in fact, we study just π_t for $t \in \mathbb{Z}$. The discrete dynamical system (X, π_1) is called the *time-1 discrete system for* $\{\pi_t\}_{t\in\mathbb{R}}$.

Another reason to consider systems with discrete time are the so-called *first return maps*. Suppose a system with continuous time in an open subset of $\mathbb{R}^n$ has a periodic orbit. In order to study the behaviour of neighbouring orbits, consider a transversal hyperplane D through a point P of the periodic orbit. Let Q be a point of D that is sufficiently close to P and follow this point along its orbit. Then that point will stay close to the periodic orbit of P for a long time and it will return into D again. Let $f(Q)$ be the point of first return in D. In this way we obtain a mapping f of a part of D into itself. See Figure 3. The behaviour of the points in D under this 'first return map' often gives useful information about the behaviour of the full orbits near the original periodic orbit.

Finally, there are 'physical' systems for which discrete time is more natural than continuous time. This is, for example, the case when one studies the annual growth of a biological population. In such a case one can model the 'physical' system straight away as a dynamical system with discrete time. Often, this leads to surprisingly new insights: when one makes a mathematical model with discrete time it can behave rather differently from the time-1 discrete system of a continuous-time model of the same 'physical' system, even if both models are based on the same principles.

Example. Consider the population of some species with the property that at any moment t the velocity $\dot{z}(t)$ of the growth of its size $z(t)$ is proportional to the size $z(t)$ itself and to the difference $C - z(t)$ for a certain constant $C > 0$ (this is to account for a slowing down of the growth when there is danger of overpopulation). This leads to the equation $\dot{z} = cz(C - z)$ for some constant $c > 0$. Put $x := z/C$ and get the differential equation $\dot{x} = Mx(1 - x)$ with $M > 0$ (the *Verhulst* model). The general behaviour of its solutions has been indicated above: all positive initial states evolve towards the equilibrium state 1 (negative initial states make no sense here).

If one is only interested in the yearly growth of the population then, as a first approximation, one might consider the time -1 discrete system obtained from the above continuous-time model. It is clear how the points of $\mathbb{R}^+$ move in this system: instead of flowing continuously towards the equilibrium point 1, they 'hop' to it.

But we can also make a model of this population by starting right away with discrete time and assuming that the annual growth $x_{n+1} - x_n$ in year n is proportional to the average size x_n in year n and to a factor $C - x_n$ for some constant $C > 0$. After a suitable rescaling we get $x_{n+1} = \mu x_n(1 - x_n)$ with $\mu > 0$. So now we consider the behaviour of points of $\mathbb{R}^+$ under iterations of the mapping $f_\mu : x \mapsto \mu x(1 - x)$. We shall yet study the dynamical system $(\mathbb{R}, f_\mu)$ in detail for various positive values of μ. Its behaviour will turn out to be completely different from the simple behaviour of the time-1 mapping of the system with continuous time.

The dynamical systems obtained from autonomous differential equations (with continuous time or the time-1 discrete systems) are invertible: the phase mappings are homeomorphisms. In such systems the past and the future appear symmetrically. In this book the stress is on systems that have only a future: the phase mapping f is not invertible and, consequently, only iterates f^n with $n \geq 0$ can be considered. For example in the system $(\mathbb{R}, f_\mu)$ considered above the mapping f_μ is not surjective and, more importantly, not injective. For some results this makes no difference, but often similar-looking definitions have different meanings for invertible and non-invertible systems. More about the reason to study non-invertible systems rather than invertible ones is in Note 8 at the end of this chapter.

0.3 The dynamical systems point of view

In this book we study dynamical systems (X, f) from the point of view of pure (non-applied) mathematics. Thus, we pay no attention to problems related to the building of mathematical models of real-life systems (like the discrepancy related to the Verhulst model mentioned above), nor shall we discus applications. On the other hand, the problems that we do discuss are ultimately motivated by applications: how will the fictitious 'physical' system of which (X, f) is meant to be the mathematical model evolve in time? If $x \in X$ then we are not particularly interested in the value $f(x)$, but in the question: what is the 'behaviour' of $f^n(x)$ for large n? In fact, we are interested in the sequence

$$x,\ f(x),\ f^2(x),\ \dots,\ f^n(x),\ \dots$$

which will be called the *orbit* of x, denoted by $\mathcal{O}(x)$. The vague notion of 'behaviour' mentioned above can now be specified more accurately by means of the following questions[3] : Are there initial states x in X for which this sequence is constant (an invariant point or equilibrium state) or a finite set which is traversed cyclically (a periodic orbit, each point of which is a periodic state)? It goes without saying that states with periodic behaviour are as important in real-life systems as equilibria: after the

3 We shall often ask for the behaviour of a particular state x when we actually mean the behaviour of $f^n(x)$ for large n.

evolution of such a state has been observed for a sufficiently long time the future development is completely predictable. Apparently, not in every system all states are periodic, but can the set of periodic states be dense in the phase space? This would mean that every orbit can be approximated with preselected precision and for a preselected stretch of time by a periodic orbit.

Opposed to the regular behaviour of periodic points is the erratic behaviour of a point x_0 whose orbit is dense in the phase space. Historically, the question whether in a given system such a point exists is related to Boltzmann's 'ergodic hypothesis'. Every possible state will be approximated if we start in such a point x_0, and there seems to be no predictability at all. How can such systems be characterized? Is it possible that such a system has a dense set of periodic points, a seemingly contradictory situation? These, and related, questions are studied in Chapter 1, and in Chapter 2 in the context of dynamical systems with an interval as phase space.

Another question is whether for a certain (or for every) initial state the orbit has a limit (which has to be an equilibrium point). If not, what does it mean that the orbit of a point x approaches a periodic orbit? More generally, does there exist a 'decent' subset A of X such that for (almost) all points x in a neighbourhood of A the points $f^n(x)$ approach A (in a sense that must yet be specified)? This is related with the question of stability of equilibria. See Chapter 3 for the theory concerning these questions.

Yet another question is whether there exist 'approximately periodic' states (we shall use the term 'recurrent'), i.e., states $x \in X$ for which future states $f^n(x)$ approximate the initial state x infinitely often arbitrarily close (as was hypothesized by Poisson for our solar system)? This type of behaviour is treated in Chapter 4.

Some (deterministic) systems behave so erratically that they seem to be completely random: future states cannot be predicted from the initial state. This seems to contradict the fact that those systems are deterministic! However, in applications the real question is not whether the state of the system can be completely determined for all future times, but: how accurately can it be predicted over what length of time, given a certain amount of initial information. The reason a deterministic system can be difficult to predict is that what happens in the future can depend very sensitively on its current state: the slightest variation in the latter causes considerable deviations in the future states. This type of instability will be discussed in Chapter 7.

In general, the study of real-life dynamical systems relies heavily on additional structures on the phase space; often differentiability of the phase mapping is needed, or the existence of an invariant measure. In this book we consider only topological structures, and all problems and their solutions will be formulated in topological terms. A large part of the theory is devoted to the development of a theoretical framework that will enable us to discuss questions like those posed above.

0.4 Examples

For easy reference we describe here a few examples which will often recur in our text. As we have not yet developed much theory, we cannot discuss these examples here at full length (at the end of Chapter 1 we shall revisit them). In these examples we illustrate simple notions like *invariant point* and *periodic orbit* as defined above. Recapitulating: a point x in the phase space is said to be *invariant under the phase mapping* f whenever $f(x) = x$ (or, equivalently, $f^n(x) = x$ for all $n \in \mathbb{N}$), and it is said to be *periodic under* f *with period* p whenever $f^p(x) = x$ for some $p \geq 1$ (which is easily seen to imply $f^n(x) = f^{n(\mathrm{mod}\ p)}$ for all $n \in \mathbb{N}$); in this case the orbit of x is called a *periodic orbit* and p is called a *period* of the point x; we also often say that the orbit of x has period p. Obviously, a point is periodic with period p iff it is an invariant point for f^p.

0.4.1 (The quadratic family). The *quadratic family* (also called *logistic family*) is the set of functions $\{f_\mu\}_{\mu>0}$, defined as follows: for every $\mu > 0$ the mapping f_μ is given by

$$f_\mu : x \mapsto \mu x(1 - x) : \mathbb{R} \to \mathbb{R}\,.$$

Each member of this family has one *critical point* (that is, a point where the derivative is zero), situated at $1/2$. Clearly, it has a maximum there; its maximal value is $f_\mu(1/2) = \mu/4$. So for $0 < \mu \leq 4$, f_μ maps the unit interval $[0; 1]$ into itself. Moreover, the equation $f_\mu(x) = x$ has two solutions, namely, $x = 0$ and $x = p_\mu := 1 - 1/\mu$. This means that the points 0 and p_μ are invariant under f_μ. Note that $p_\mu \notin [0; 1]$ if $\mu < 1$.

The mappings in this family have been studied extensively. An important aspect is the phenomenon of *bifurcation*: if μ increases from 1 to 4 then the phase portrait of the system changes significantly when μ passes certain values. We shall give here a brief description of this phenomenon, but not much attention will be paid to it later on. Numerical experiments show that there is a strictly increasing sequence $(\mu_n)_{n\in\mathbb{N}}$ of real numbers such that for μ between μ_n and μ_{n+1} there is an attracting periodic orbit with period 2^{n-1}. This manifests itself as follows: up to some initial terms the orbit of the point $1/2$ virtually coincides with that periodic orbit. If μ increases and passes the value μ_{n+1}, that attracting periodic orbit becomes repelling and near each point of this (now repelling) periodic orbit appear two points that are part of an attracting periodic orbit with period 2^n. This phenomenon is called *period doubling*. In Chapter 2 we shall compute the first few terms of this sequence: we shall see that $\mu_1 = 1$, $\mu_2 = 3$ and $\mu_3 = 1 + \sqrt{6}$. The sequence $(\mu_n)_{n\in\mathbb{N}}$ converges with limit $\mu_\infty := 3.57\ldots$ (often called the *Feigenbaum point*). When μ has this value the dynamical system $([0; 1], f_\mu)$ seems to be completely random.

0.4.2 (The tent map). The *tent map* $T : [0; 1] \to [0; 1]$ – so called because of the shape of its graph – is defined by

$$T(x) := 1 - |2x - 1| = \begin{cases} 2x & \text{for } 0 \leq x \leq 1/2\,, \\ 2(1 - x) & \text{for } 1/2 \leq x \leq 1\,. \end{cases}$$

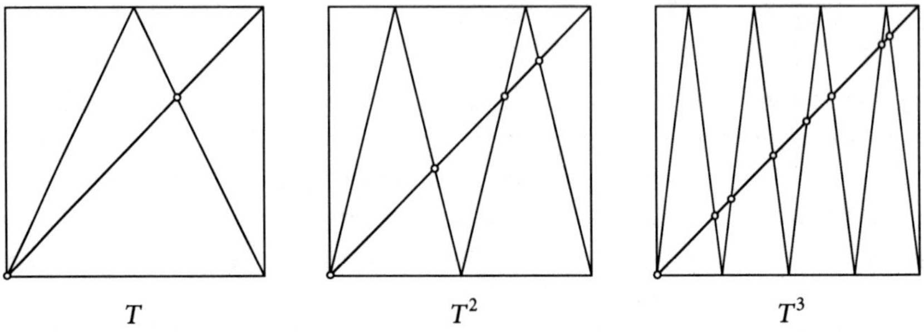

Fig. 4. The graphs of T, T^2 and T^3.

In Figure 4 the graphs of T, T^2 and T^3 are sketched. The tent map has two invariant points: one in 0 and one in $2/3$. Both of them are repelling: for a point close to one of these invariant points the distance to that invariant point is doubled by T.

By induction one easily shows that, for every $n \in \mathbb{N}$, the graph of the mapping T^n consists of 2^{n-1} 'tents'. Consequently, the graph of T^n intersects the diagonal of the unit square in 2^n points. Hence there are 2^n points with period n. Moreover, since each of the intervals $[k2^{-n}; (k+1)2^{-n}]$ with $k = 0, \ldots, 2^n - 1$ contains such a periodic point it follows that, by taking n sufficiently large, every subinterval of $[0; 1]$ contains a periodic point. *Consequently, the periodic points are dense in the interval* $[0; 1]$.

Our final examples are on the circle. First, we introduce some notation.

Notation. The unit circle in the complex plane will be denoted by $\mathbb{S}$, so, by definition, $\mathbb{S} = \{ z \in \mathbb{C} : |z| = 1 \}$ or, alternatively, $\mathbb{S} = \{ e^{2\pi i t} : t \in \mathbb{R} \}$. Let

$$[\,t\,] := e^{2\pi i t} \quad \text{for } t \in \mathbb{R}\,.$$

The mapping $t \mapsto [t] : \mathbb{R} \to \mathbb{S}$ is a continuous surjection of the real line onto the circle. It maps each interval of the form $[a; a+1)$ with $a \in \mathbb{R}$ bijectively onto $\mathbb{S}$. Moreover, if $s, t \in \mathbb{R}$ then $[s] = [t]$ iff $s = t \pmod 1$, iff $s - t \in \mathbb{Z}$.

We consider $\mathbb{S}$ as a topological space with the relative topology of the complex plane. Being a bounded closed subset of the complex plane, $\mathbb{S}$ is a compact metric space. However, we shall not use the metric that $\mathbb{S}$ inherits from $\mathbb{C}$ (the Euclidean metric). It will be more convenient to use as the distance of two points $[s]$ and $[t]$ of $\mathbb{S}$ the length of the shortest arc in $\mathbb{S}$ that connects the two points: if $s, t \in \mathbb{R}$ then

$$d_c([s], [t]) := 2\pi \min\{ |s-t| \bmod 1 \,,\; 1 - |s-t| \bmod 1 \}\,.$$

If $|s - t| < 1/2$ then $d_c([s], [t]) = 2\pi\,|s-t|$; in this case the length $d_c([s], [t])$ of the arc between $[s]$ and $[t]$ is related in the following way to the euclidean distance $d_{\text{eucl}}([s], [t])$ of the points $[s]$ and $[t]$ in $\mathbb{C}$:

$$d_{\text{eucl}}([s], [t]) = 2\,\sin \pi\,|s-t| = 2\,\sin \frac{d_c([s], [t])}{2}\,.$$

This implies that the two metrics d_c and d_{eucl} on $\mathbb{S}$ are equivalent. So without limitation of generality we may use the metric d_c on $\mathbb{S}$.

0.4.3 (The rigid rotation of the circle). For every $a \in \mathbb{R}$, $a \neq 0$, we consider the dynamical system $(\mathbb{S}, \varphi_a)$, where the mapping $\varphi_a : \mathbb{S} \to \mathbb{S}$ is unambiguously defined by

$$\varphi_a([s]) := [a + s] \quad \text{for } [s] \in \mathbb{S}\,.$$

We call φ_a the *(rigid) rotation* of the circle over a. One can visualize the effect of the mapping φ_a by considering $\mathbb{S}$ as a rigid body that rotates around the origin over an angle of $2\pi a$ radians. For every $a \in \mathbb{R}$ the mapping $\varphi_a : \mathbb{S} \to \mathbb{S}$ is continuous. In fact, it is an isometry:

$$d_c(\varphi_a([s]), \varphi_a([t])) = d_c([a + s], [a + t]) = d_c([s], [t])$$

for all $[s], [t] \in \mathbb{S}$. In particular, the mapping φ_a is a homeomorphism of $\mathbb{S}$ onto itself. It is obvious that $\varphi_0 = \mathrm{id}_{\mathbb{S}}$. More generally, $\varphi_a = \mathrm{id}_{\mathbb{S}}$ iff $a \in \mathbb{Z}$. Moreover, $\varphi_a = \varphi_{a \,(\mathrm{mod}\ 1)}$ for all $a \in \mathbb{R}$, so without limitation of generality we may always assume that $0 \leq a < 1$.

As to the existence of periodic points, there are two mutually exclusive situations, depending on the value of $a \in \mathbb{R}$.

Case 1. *If $a \in \mathbb{Q}$ then all points of $\mathbb{S}$ are periodic under φ_a and all points have the same period.*

Proof. Let $a = p/q$ with $p \in \mathbb{Z}$ and $q \in \mathbb{N}$ such that $\gcd(p, q) = 1$. If $s \in \mathbb{R}$ then $(\varphi_a)^q([s]) = [p + s] = [s]$, so the point $[s] \in \mathbb{S}$ is periodic under φ_a with period q, independent of the choice of $[s]$. We show that q is the smallest period of the point $[s]$ (also independent of the choice of $[s]$), as follows: Let k be any period of $[s]$, i.e., $(\varphi_a)^k([s]) = [s]$. Then $[kp/q + s] = (\varphi_a)^k[s] = [s]$, so we have $kp/q \in \mathbb{Z}$. Because $\gcd(p, q) = 1$ this implies that k is an integer multiple of q. □

Case 2. *If $a \notin \mathbb{Q}$ then for every $s \in \mathbb{R}$ the orbit $\mathcal{O}([s])$ of $[s]$ under φ_a is dense in $\mathbb{S}$. In particular, no orbit is periodic.*

Proof. We shall show that $\mathcal{O}([0])$ is dense in $\mathbb{S}$. Once this has been proved, the following simple argument shows that every orbit is dense: if $[s] \in \mathbb{S}$ then the homeomorphism $\varphi_s : [t] \mapsto [t + s] : \mathbb{S} \to \mathbb{S}$ is easily seen to map the orbit of the point $[0]$ onto the orbit of the point $[s]$: for every $n \in \mathbb{Z}^+$ we have

$$\varphi_s(\varphi_a^n([0]) = [(na + 0) + s] = [na + s] = \varphi_a^n([s])\,.$$

As a homeomorphism maps dense sets onto dense sets, this completes the proof that every orbit is dense if the orbit of $[0]$ is dense.

In order to show that the orbit $\mathcal{O}([0])$ of $[0]$ is dense in $\mathbb{S}$ it is sufficient to prove that, for every $\varepsilon > 0$, every arc in $\mathbb{S}$ with length ε contains a point of $\mathcal{O}([0])$, that is, $\mathcal{O}([0])$ is ε-dense in $\mathbb{S}$. So let $\varepsilon > 0$ be arbitrary, let $k \in \mathbb{N}$ be so large that $2\pi/k < \varepsilon$

and divide the circle in k equal arcs. As $a \notin \mathbb{Q}$ we have $\varphi_a^l[0] = [la] \neq [ma] = \varphi_a^m[0]$ for all $l, m \in \mathbb{Z}^+$ with $l \neq m$. This implies that the orbit of the point $[0]$ is an infinite set. Therefore, at least one of the k arcs into which $\mathbb{S}$ is divided contains at least two different points of that orbit: there are $l, m \in \mathbb{Z}^+$ with $0 \leq l < m$ such that the points $[la]$ and $[ma]$ have a distance less than ε. Because the mapping φ_{-la} is an isometry, it follows that the points $[0] = \varphi_{-la}[la]$ and $[(m-l)a] = \varphi_{-la}[ma]$ have a distance less than ε as well. Hence also the successive points $[n(m-l)a]$ for $n \in \mathbb{Z}^+$ have a distance less than ε: the isometry $\varphi_{n(m-l)a}$ maps the arc with end points $[0]$ and $[(m-l)a]$ onto the arc with end points $[n(m-l)a]$ and $[(n+1)(m-l)a]$. Consequently, every arc of length ε contains at least one of the points $[n(m-l)a]$ with $n \in \mathbb{Z}^+$. As all of these points belong to the orbit $\mathcal{O}([0])$ this completes the proof. □

0.4.4 (The argument-doubling transformation). This is the mapping $\psi : \mathbb{S} \to \mathbb{S}$, defined by $\psi([s]) := [2s]$ for $s \in \mathbb{R}$. This mapping is well-defined, for if $s, t \in \mathbb{R}$ and $[s] = [t]$ then $[2s] = [2t]$. It is surjective but clearly not injective: for example, $\psi([0]) = \psi([1/2]) = [0]$. It is easy to see that $[0]$ is the unique invariant point under ψ. Indeed, the equation $[2s] = [s]$ for $s \in \mathbb{R}$ is equivalent to the equation $2s - s \in \mathbb{Z}$, hence with $[s] = 0$. This invariant point is 'repelling' in nature: every sufficiently small neighbourhood J of $[0]$ is expanding and each of its points, except $[0]$ itself, will eventually leave J. For example, take for J the arc $\{ [s] \vdots s \in \mathbb{R}, |s| < 1/4 \}$. In fact, for all pairs of points that are close to each other the transformation ψ doubles the distance. This type of instability occurs everywhere: points that are close to each other will eventually get a large mutual distance. This type of instability – which also occurs under the tent map in 2 above – will be investigated in Chapter 7.

A point $[s]$ in $\mathbb{S}$ is periodic under ψ with period $p \geq 2$ iff $2^p s - s \in \mathbb{Z}$, iff there exists $k \in \mathbb{Z}$ such that $s = k/(2^p - 1)$. Next, observe that for all choices of $k, l \in \mathbb{Z}$ we have $[k/(2^p - 1)] = [l/(2^p - 1)]$ iff $k - l$ is an integer multiple of $2^p - 1$. Conclusion: the points $[k/(2^p - 1)]$ for $k = 0, \dots, 2^p - 2$ are mutually different and together they form the set of all periodic points with period p ($p \in \mathbb{N}$, $p \geq 2$).

Thus, the periodic points with period p are evenly distributed around the circle with a distance of $2\pi/(2^p - 1)$ between two successive points. By taking p sufficiently large it follows that, for every $\varepsilon > 0$, every arc of length ε contains a periodic point. This shows: *the periodic points form a dense subset of* $\mathbb{S}$.

Exercises

0.1.

(1) Let $\{\pi_t\}_{t \in T}$ be a collection of self-maps of a space X with $T = \mathbb{Z}$ or $\mathbb{Z}^+$. Show that this collection satisfies the conditions of formula (0.1-2) iff there is a function $f : X \to X$ such that $\pi_n = f^n$ for all $n \in T$.

(2) Let $T = \mathbb{R}$ or $\mathbb{Z}$ and let $\{\pi_t\}_{t\in T}$ be a set of self-maps of a set X for which formula (0.1-2) holds. Show that each of the mappings π_t with $t \in T$ is a bijection. Show that, consequently, the full orbits $\{\pi_t(x) : t \in T\}$ for $x \in X$ form a partition of X. If X is a topological space and all maps π_t for $t \in T$ are continuous, then every mapping π_t with $t \in T$ is a homeomorphism.

0.2. The *phase portrait* of a mapping $f: X \to X$ can be visualized as a picture of the space X in which for 'typical' points $x \in X$ an arrow is drown from the point x to the point $f(x)$.

(1) For each of the following mappings, find the invariant points and give a sketch of the phase portrait:
 (a) $f(x) = \frac{1}{2}x$ for $-\infty < x < \infty$,
 (b) $f(x) = -3x$ for $-\infty < x < \infty$,
 (c) $f(x) = x^2$ for $0 \le x \le 1$,
 (d) $f(x) = \sqrt{x}$ for $0 \le x \le 1$,
 (e) $f(x) = \frac{\pi}{2}\sin x$ for $0 \le x \le \pi$.

(2) Describe the phase portrait of the mapping $x \mapsto ax: \mathbb{R} \to \mathbb{R}$ with $a \in \mathbb{R}$, $a \neq 0$.

0.3.

(1) Prove: if $\lim_{n\rightsquigarrow\infty} f^n(x) =: z$ exists then z is an invariant point.

(2) Call an invariant point z in a dynamical system (X, f) *topologically attracting* whenever there is a neighbourhood U of z in X such that $\lim_{n\rightsquigarrow\infty} f^n(x) = z$ for all $x \in U$. Which of the invariant points in Exercise 0.2 is topologically attracting?

0.4. Let $f: \mathbb{R}^2 \to \mathbb{R}^2$ be the linear mapping with matrix A. Describe the phase portrait of the dynamical system $(\mathbb{R}^2, f)$ in the following cases:

$$\text{(a)}\quad A = \begin{pmatrix} 2 & 0 \\ 0 & 3 \end{pmatrix}, \qquad \text{(c)}\quad A = \begin{pmatrix} 2 & 0 \\ 1 & 2 \end{pmatrix},$$

$$\text{(b)}\quad A = \begin{pmatrix} 1/2 & 0 \\ 0 & 1/3 \end{pmatrix}, \qquad \text{(d)}\quad A = \begin{pmatrix} 0 & 2 \\ 3 & 0 \end{pmatrix}.$$

0.5. Assume that the phase space X of a dynamical system (X, f) is metrizable with metric d and that it has a dense subset of points that are periodic under the phase mapping f. Show that for every point $x \in X$, for every $\varepsilon > 0$ and for every $n \in \mathbb{N}$ there is a periodic point $z \in X$ such that $d(f^n(x), f^n(z)) < \varepsilon$ for $n = 0, \dots, N$.

NB. See Note 5 below.

Notes

1 Another example of an iterative procedure is the following method of approximation of the solution of a differential equation (due to Picard): let G be an open subset of $\mathbb{R}^n$ and let $F: G \to G$ be a Lipschitz function (which means that there is a constant $M > 0$ such that $\|F(x_1) - F(x_2)\| \le M\|x_1 - x_2\|$ for all $x_1, x_2 \in G$). Consider the autonomous differential equation $\dot{x} = F(x)$ in G (actually, a system of n

coupled equations in n variables). Given $x_0 \in G$ and $t_0 \in \mathbb{R}$ one can find a $\delta > 0$ and, subsequently, one can define an operator T on the space $\mathcal{X}_{t_0,\delta}$ of all differentiable G-valued functions on the interval $[t_0 - \delta; t_0 + \delta]$ by

$$T(\varphi)(t) := x_0 + \int_{t_0}^{t} F(\varphi(s))\,ds \quad (\varphi \in \mathcal{X}_{t_0,\delta} \text{ and } |t - t_0| \le \delta)$$

(coordinate-wise integration of a vector-valued function). For each choice of $\varphi \in \mathcal{X}_{t_0,\delta}$ the sequence $(T^n(\varphi))_{n\in\mathbb{N}}$ turns out to converge uniformly on the interval $[t_0 - \delta; t_0 + \delta]$ to an element f in $\mathcal{X}_{t_0,\delta}$ which is a fixed point of T, that is, $T(f) = f$; this follows from the Banach Fixed Point Theorem: see Appendix A.7.9. This fixed point of T is the unique solution on the interval $[t_0 - \delta; t_0 + \delta]$ of the differential equation under consideration satisfying the additional initial condition $f(t_0) = x_0$.

2 Almost simultaneously with Poincaré, the Russian mathematician A. M. Lyapunov developed his theory of stability of solutions of differential equations, also using methods that did not require the equations under consideration to be solved. Also A. A. Markov and H. Whitney contributed to the 'geometrical theory' of differential equations.

3 I know only Dutch books about Escher, e.g., J. L. Locher (ed.) [1971]. The picture "Waterfall" was inspired by a paper by L. S. Penrose and R. Penrose in the February 1958 issue of the British Journal of Psychology. In fact, "Waterfall" is the artists view on the well-known drawing (by R. Penrose in the same Journal) reproduced below.

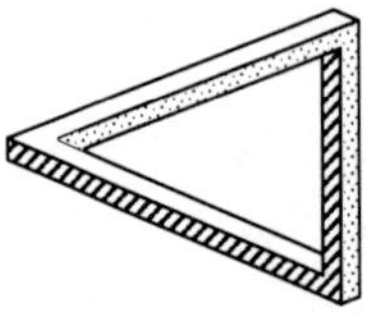

4 Time-1 maps can be used to define a discrete dynamical system with certain desired properties. We do not need this in the present book.

5 The importance of the existence of a dense set of periodic points was already observed by Poincaré, as the following citation illustrates:

> *Étant données des équations ... et une solution particulière quelconque de ces équations, on peut toujours trouver une solution périodique (dont la période peut, il est vrai, être très longue), telle que la différence entre les deux solutions soit aussi petite qu'on le veut, pendant un temps aussi long qu'on le veut. D'ailleurs, ce qui nous rend ces solutions périodiques si précieuses, c'est quelles sont, pour ansi dire, la seule brèche par où nous puissions esseyer de pénétrer dans une place jusqu'ici réputée inabordable.*
>
> – H. Poincaré, Les méthodes nouvelles de la méchanique céleste.

6 There are several definitions of 'chaotic behaviour', but they always include some instability condition for orbits: a slight difference in initial states produces a large variation in the orbits . The possibility of such behaviour was already noticed by Maxwell 1873. Henri Poincaré said in his 1908 book 'Science et Méthode' in a discussion on fortuitous events, after the remark that very small causes that escape our attention can have considerable consequences which are impossible to neglect:

> *Si nous connaissions exactement les lois de la nature et la situation de l'univers à l'instant initial, nous pourrions prédire exactement la situation de ce même univers à un instant ultérieur. Mais, lors même que les lois naturelles n'auraient plus de secret pour nous, nous ne pourrons connaître la*

*situation initiale qu'*approximativement. *Si cela nous permet de prévoir la situation ultérieure* avec la meme appoximation, *c'est tout ce qu'il nous faut, nous disons que le phénomène a été prévu, qu'il est régi par des lois; mais il n'en est pas toujours ainsi, il peut arriver que de petites différences dans les conditions initiales en engendrent de très grandes dans les phénomènes finaux; une petite erreur sur les premières produirait une erreur énorme sur les derniers. La prédiction devient impossible et nous avons le phénomène fortuit.*

– H. Poincaré, Science et méthode.

7 The logistic map was popularized in the paper R. M. May [1976], in part as a discrete-time demographic model analogous to the logistic equation $\dot{x} = rx(1 - x)$ published by Pierre François Verhulst in P. F. Verhulst [1838)]. There is an abundant literature on the logistic map (see "Logistic Map" in Wikipedia) and much more is known than we can attempt to cover here.

8 Invertible vs. non-invertible systems. Much of the existing literature about discrete dynamical systems (in fact, most of the literature before, say, 1960) is about *invertible systems*, i.e., dynamical systems (X, f) with f a homeomorphism. Often the difference between invertible and non-invertible systems (where the phase mapping is just a continuous function) is negligible, though sometimes the proofs for non-invertible systems are more involved than for invertible ones (this is the case, for example, in Section 3.3). But the difference goes deeper. There are questions which have only trivial answers for invertible systems but which have interesting answers for non-invertible ones. To give a simple example, we shall see later that a homeomorphism on an interval admits no points with a dense orbit, or no periodic points with period greater than 2. The tent map shows that this is not longer true for non-invertible continuous mappings (for the existence of a dense orbit, see Example 1 after Theorem 1.3.5 below). As a less trivial example, in Chapter 8 we shall see that a homeomorphism on an interval has topological entropy zero ('entropy' is a measure of the complexity of a system), whereas a non-invertible mapping can have positive entropy. There are more differences between invertible systems and non-invertible ones, but at this point we cannot yet explain them. Some of those differences will be indicated in the Notes to the subsequent chapters.

In addition, there is a profound difference in the applicability of invertible systems and the non-invertible ones. For invertible systems it makes sense to ask for the past of any state. Formulated differently, in such a system one can reverse time and consider the behaviour of points under iterates of f^{-1}. No doubt, the study of reversible systems was motivated by the fact that the basic laws of physics are time-symmetric (as far as I know, the behaviour of the neutral kaon K^0 is the only exception). This means that if a given process is permitted by physical laws then so too is the reverse process – what we would see if a film of the original process were shown in reverse. Thus, if one replaces the time t by $-t$ in a solution of the equations for such a process then one gets, again, a solution (be it with different initial conditions). For the related time-1 discrete dynamical system this means exactly that the phase mapping is a homeomorphism.

Notwithstanding this, many observed (macroscopic) processes are irreversible. In particular, this holds for biological processes (to my knowledge, the only person living backwards in time is (was?) Merlin in T. H. White's *The Once and Future King*; Benjamin Button is a totally different case). For this reason, non-invertible dynamical systems had to be studied as well. For systems with discrete time this means: phase mappings that have no inverse.

Though this explanation sounds rather trivial, it is connected with a deep, partly philosophical, problem, namely: how can it be explained that backwards solutions for many processes are never observed even though they are equally consistent with the laws of mechanics as the forward solutions? For example, buildings collapse into rubble, but rubble does not 'uncollapse' into buildings. Similarly, after a stone has been dropped into a pond one observes concentrically outgoing waves, but concentrically focussing waves ejecting a stone from the water are never observed. The physical part of the problem is to explain why these backwards solutions are not observed. The philosophical part of the problem is to investigate how this relates to the question what ‘time’ actually is, and what the expression ‘the arrow of time’ means (if it is meaningful at all). For a very readable account on these topics we refer to H. Price [1996].

1 Basic notions

Abstract. In this chapter we define and discuss the basic notions of the theory of (discrete) dynamical systems which will be used in the remainder of this book.

Notation and terminology. Unless stated otherwise, (X, f) will always denote a dynamical system. Thus, X is a (non-empty) topological Hausdorff space, the *phase space* of the system, and $f: X \to X$ is a continuous mapping, the *phase mapping*. Metrizability of X is not automatically assumed. We shall sometimes refer to a point of X as a *state* of the system.

1.1 Invariant and periodic points

The *orbit* of a point x in X *under* f is the set

$$\mathcal{O}_f(x) := \{ f^n(x) : n \in \mathbb{Z}^+ \} = \{x, f(x), f^2(x), \ldots\}.$$

When f is understood we call this just 'the orbit of x' and we shall simply write $\mathcal{O}(x)$ instead of $\mathcal{O}_f(x)$. In this connection, the point x is called the *initial state* and the points $f^n(x)$ with $n > 0$ the *future states* (of this orbit). We also say that the point x of X has position $f^n(x)$ at time n and that it *moves* at time n from $f^n(x)$ to $f^{n+1}(x)$. It is often convenient to consider an orbit as a *sequence*, so that it makes sense to consider subsequences of an orbit, or to ask for its limit points or even its limit.

Though in non-metric spaces sequences are not that useful, the notions of convergence, of the limit and a of limit point of a sequence can be defined as in metric spaces. See also Appendix A.7.4.

It will always be clear from the context if an orbit is considered as a set or as a sequence.

It is important to keep in mind that for every $x \in X$ and every $k, n \in \mathbb{Z}^+$ we have $f^n(f^k(x)) = f^{k+n}(x)$. Hence for every $k \in \mathbb{Z}^+$ the orbit of $f^k(x)$ is a subset – even a subsequence – of the orbit of x. In fact,

$$\mathcal{O}(f^k(x)) = \{ f^{k+n}(x) : n \in \mathbb{Z}^+ \} = \{ f^m(x) : m \in \mathbb{Z}^+, m \geq k \}.$$

If $x_0 \in X$ then a sequence $(x_{-n})_{n\in\mathbb{N}}$ in X such that $f(x_{-n}) = x_{-n+1}$ for all $n \in \mathbb{N}$ is called a *complete past* of x_0. A complete past may not be unique. E.g., the point [0] in the argument-doubling system has countably many complete pasts. If x_0 has a complete past then $x_0 \in \bigcap_{i=0}^{\infty} f^i[X]$. The converse is not generally true, but it is true if X is compact; see Exercise 1.4. An *orbit closure* is the (topological) closure of an orbit in the phase space under consideration. The points of the orbit of x are the states that will actually be reached in the future when starting at state x, the orbit closure of x also includes the points that can be approximated arbitrarily close by future states.

A point x_0 in X is said to be *invariant* (under f), or it is called a *rest point* or an *equilibrium point*, whenever $f(x_0) = x_0$. If x_0 is an invariant point then (by induction) $f^n(x_0) = x_0$ for every $n \in \mathbb{Z}^+$. So the point x_0 is invariant iff $\mathcal{O}(x_0) = \{x_0\}$. A point $x_0 \in X$ is said to be *eventually invariant* whenever its orbit contains an invariant point: there exists $n \in \mathbb{Z}^+$ such that the point $f^n(x_0)$ is invariant. Then $f^m(x_0) = f^{m-n}(f^n(x_0)) = f^n(x_0)$ for all $m \geq n$, which implies that $\mathcal{O}(x_0) = \{x_0, \ldots, f^n(x_0)\}$, a finite set.

A point x_0 is said to be *periodic* (under f) and $\mathcal{O}(x_0)$ is called a *periodic orbit* whenever there exists $p \in \mathbb{N}$ (so $p \neq 0$) such that $f^p(x_0) = x_0$. In that case, the number p is called *a period*[1] of the point x_0. Clearly, an invariant point is a periodic point with period 1. Note also that a point is periodic under f with period p iff it is invariant under f^p. Moreover, if p is a period of the periodic point x_0, then for every $k \in \mathbb{N}$ the number kp is a period of x_0 as well. Obviously, every periodic point has a smallest period, which is called the *primitive period* of that point. We shall see in a moment that the set of all periods of x_0 is the set of all integer multiples of the primitive period (see Case (c) in Lemma 1.1.2 below).

Finally, a point x_0 of X is called *eventually periodic* whenever its orbit contains a periodic point: there exists $n \in \mathbb{Z}^+$ such that the point $f^n(x_0)$ is periodic. It is easily seen that an eventually periodic point has a finite orbit, a straightforward consequence of the fact that a periodic orbit has a finite orbit (see also Corollary 1.1.3 ahead). Below we shall see that eventually periodic points are characterized by this property: see Theorem 1.1.5.

Examples.

(1) Let $X := [0; 1]$ and let $f(x) := x^2$ for $x \in X$. The points 0 and 1 are invariant, and if x is strictly between 0 and 1 then $\mathcal{O}(x) = \{x, x^2, x^4, x^8, \ldots\}$ is a sequence converging to the invariant point 0.

(2) Consider the rigid rotation $(\mathbb{S}, \varphi_a)$ with $a \in \mathbb{Q}$, say, $a = p/q$ with $p, q \in \mathbb{N}$ and $\gcd(p, q) = 1$. It follows from the proof of Case 1 of Example 0.4.3 in the Introduction that all points of $\mathbb{S}$ are periodic under φ_a with primitive period q.

(3) In the Examples 0.4.2 and 0.4.4 we determined for every $n \in \mathbb{N}$ the number of periodic points with period n for the tent map and for the argument-doubling transformation. In Exercise 1.2 (1) an iterative procedure is suggested for determining how many points there are with a given primitive period.

(4) Let $X := \mathbb{R}$ and let f be defined as follows:

$$f(x) := \begin{cases} x + 2 & \text{for } x \leq 1\,, \\ -x & \text{for } -1 \leq x \leq 1\,, \\ x - 2 & \text{for } x \geq 1\,. \end{cases}$$

1 Since $f^0 = \mathrm{id}_X$, it makes no sense to consider periods equal to 0: every point would be periodic with period 0 under every mapping.

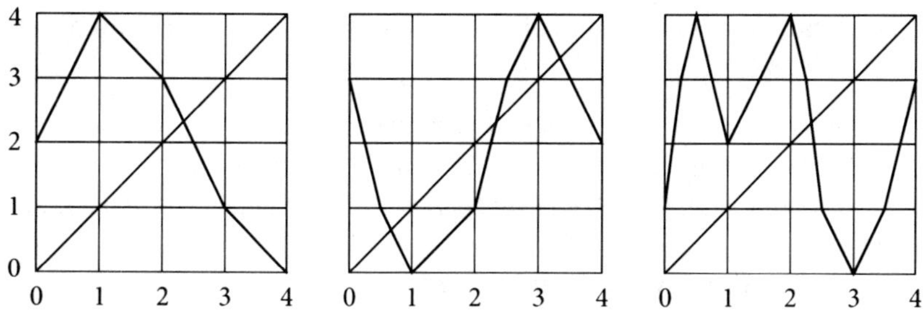

Fig. 1.1. The graphs of f, f^2 and f^3 of the mapping f of Example (5).

All points in the interval $[-1; 1]$ are periodic with period 2 and 0 is an invariant point. If $x > 1$ then there is a unique first value of $n \in \mathbb{N}$ such that $x_n := f^n(x) = x - 2n$ belongs to the interval $[-1; 1]$. Then $\mathcal{O}(x) = \{x, x-2, \dots, x_n, -x_n, x_n, \dots\}$ (as a sequence), so every point $x > 1$ is eventually periodic. Similarly, every point $x < -1$ is eventually periodic. All points of the set $\{2n \colon n \in \mathbb{Z}\}$ are eventually invariant.

(5) Let $f\colon [0; 4] \to [0; 4]$ be defined as follows: f is affine on the intervals with end points 0, 1, 2, 3 and 4, and in these points f is given by $f(0) := 2$, $f(1) := 4$, $f(2) := 3$, $f(3) := 1$ and $f(4) := 0$; see Figure 1.1. Clearly, the point 0 is periodic with primitive period 5 and $\mathcal{O}(0) = \{0, 1, 2, 3, 4\}$ (as a set). It can be shown that f has no periodic points with period 3; see Exercise 1.2 (2).

Proposition 1.1.1. *For fixed $p \in \mathbb{N}$, the set of all periodic points with period p is a closed subset of X. In particular, the set of all invariant points under f is closed in X.*

Proof. The set of periodic points with period p is equal to the set of points where the two continuous functions f^p and id_X from X to itself coincide. This set is closed because X is a Hausdorff space. □

Remarks.

(1) In this proposition, 'period p' cannot be replaced by 'primitive period p'. For example, the set of all points with primitive period 2 under the mapping $x \mapsto -x\colon \mathbb{R} \to \mathbb{R}$ is equal to the open set $\mathbb{R} \setminus \{0\}$. It follows that the function that assigns to each periodic point its primitive period is not continuous on the set of periodic points. See also Exercise 1.3.

(2) For every $n \in \mathbb{N}$, the set of all periodic points with period less than or equal to n is a closed set. However, the set of *all* periodic points need not be closed: both for the tent map and the argument-doubling transformation, the set of of all periodic points is a dense proper subset of the phase space: see the Examples 0.4.2 and 0.4.4 in the Introduction.

For every point $x \in X$ we denote by $D(x, x)$ the set of 'time-values' that x returns into itself:

$$D(x, x) := \{n \in \mathbb{Z}^+ \colon f^n(x) = x\}.$$

The sets $D(x,x)$ can be used to distinguish the various types of points introduced above.

Lemma 1.1.2. *For every $x \in X$ the set $D(x,x)$ is a subsemigroup of the additive semigroup $\mathbb{Z}^+$. The following, mutually exclusive, cases can be distinguished:*
(a) *$D(x,x) = \{0\}$. In this case, the point x is not periodic (in particular, the point x is not invariant).*
(b) *$D(x,x) = \mathbb{Z}^+$. In this case, x is an invariant point.*
(c) *There exists $p \in \mathbb{N}$, $p \geq 2$, such that*

$$D(x,x) = \mathbb{Z}^+ p = \{np : n \in \mathbb{Z}^+\}.$$

In this case, x is a non-invariant periodic point with primitive period p.

Proof. All statements in the lemma are either obvious or easy to prove, except perhaps that the three cases cover all possibilities. So suppose that neither of the cases (a) or (b) applies. Then $D(x,x)\setminus\{0\} \neq \emptyset$ and $1 \notin D(x,x)$. It follows that the smallest element p of the set $D(x,x) \setminus \{0\}$ is not equal to 1, so $p \geq 2$. We show that $\mathbb{Z}^+ p = D(x,x)$ for this choice of p.

"⊆": As $D(x,x)$ is a subsemigroup of the additive subgroup $\mathbb{Z}^+$, the fact that $p \in D(x,x)$ implies that $\mathbb{Z}^+ p \subseteq D(x,x)$.

"⊇": Let $m \in D(x,x)$. Write m as $m = kp + r$ with integers $k, r \in \mathbb{Z}^+$ such that $0 \leq r \leq p-1$. Then

$$f^m(x) = f^r(f^{kp}(x)) = f^r(x). \tag{1.1-1}$$

However, $f^m(x) = x$, hence $f^r(x) = x$ as well, so $r \in D(x,x)$. As p is the smallest non-zero element of $D(x,x)$, it follows that $r = 0$. □

Corollary 1.1.3. *Let x_0 be a non-invariant periodic point in X and let p be its primitive period (so $p \geq 2$). Then*

$$f^n(x_0) = f^{n \,(\mathrm{mod}\ p)}(x_0) \quad \text{for all } n \in \mathbb{Z}^+. \tag{1.1-2}$$

Consequently, $\mathcal{O}(x_0) = \{x_0, \dots, f^{p-1}(x_0)\}$, with $f^i(x_0) \neq f^j(x_0)$ for every pair i, j with $0 \leq i, j \leq p-1$ and $i \neq j$. So the orbit $\mathcal{O}(x_0)$ consists of exactly p distinct points. Each of those points is periodic with primitive period p and has the same orbit, namely, $\mathcal{O}(x_0)$.

Proof. The first statement follows immediately from the equality in formula (1.1-1) above. This implies that $\mathcal{O}(x_0) = \{f^i(x_0) : i = 0, \dots, p-1\}$.

Next, we show that the points $x_i := f^i(x_0)$ with $0 \leq i \leq p-1$ are mutually different. To this end, consider $i, j \in \mathbb{Z}^+$ with $0 \leq i \leq j \leq p-1$ such that $x_i = x_j$. Then

$$f^{p-j}(f^i(x_0)) = f^{p-j}(x_i) = f^{p-j}(x_j) = x_0.$$

So $p-(j-i)$ is a period of the point x_0. Since $1 \leq p-(j-i) \leq p$ the only possibility is that $p-(j-i) = p$, hence that $i = j$. Consequently, if $i \neq j$ then $x_i \neq x_j$. We can rephrase this property by saying that the number of points in the the orbit of a periodic point is equal to its primitive period.

Finally, we show that the points of $\mathcal{O}(x_0)$ are all periodic with the same primitive period p as x_0. To this end, consider any $i \in \{0, \ldots, p-1\}$. It is easy to see that the point x_i is periodic: $f^p(x_i) = f^{p+i}(x_0) = f^i(x_0) = x_i$. Moreover, since $x_i \in \mathcal{O}(x_0)$ it follows that $\mathcal{O}(x_i) \subseteq \mathcal{O}(x_0)$. On the other hand, the equality $f^{p-i}(x_i) = x_0$ implies that $x_0 \in \mathcal{O}(x_i)$, so $\mathcal{O}(x_0) \subseteq \mathcal{O}(x_i)$. This shows that $\mathcal{O}(x_i) = \mathcal{O}(x_0)$. Consequently, the orbit of the periodic point x_i consists of p points, hence its primitive period equals p. □

Remarks.

(1) This Corollary is trivially true (with $p = 1$) for an invariant point. This case was excluded in order to avoid trivialities in the formulation.

(2) A periodic orbit is always a finite set and equation (1.1-2) expresses the fact that that such an orbit is traversed cyclically by each of its points. This means that f maps the points of the periodic orbit $\mathcal{O}(x_0)$ onto each other in the following way: $x_0 \mapsto x_1 \mapsto \cdots \mapsto x_{p-1} \mapsto x_0 \mapsto \ldots$ (here, as in the proof above, $x_i := f^i(x_0)$ for $i = 0, \ldots, p-1$). Note that this observation, which is almost immediately clear from the definitions, is just a reformulation of the proof of Corollary 1.1.3 and that formula (1.1-2) is a straightforward consequence of this observation.

(3) As observed earlier, eventually periodic points have finite orbits. Conversely, if the orbit of a point x is finite then not all points $f^n(x)$ can be mutually different, so there are $k, l \in \mathbb{Z}^+, l \neq k$, such that $f^l(x) = f^k(x)$. Assuming that $l > k$, this implies that $f^{l-k}(f^k(x)) = f^l(x) = f^k(x)$, so the point $f^k(x)$ is periodic. Hence the point x is eventually periodic (even periodic if k happens to be 0). This shows that a point has a finite orbit iff it is eventually periodic.

(4) When a periodic point x_0 has period p and p is a prime number then it follows from Lemma 1.1.2 (c) that p is the primitive period of x_0; in point of fact, p is not an integer multiple of any number different from 1 or p. The converse need not be true: for example, the point $[1/15] \in \mathbb{S}$ has primitive period 4 under the argument-doubling transformation.

Next, we make a few remarks about the relationship between the primitive periods of a periodic point $x \in X$ under f and under f^k for $k \in \mathbb{N}$. First an example: if the point x has primitive period 4 under f then under f^6 it has period 4 as well, because $(f^6)^4(x) = f^{24}(x) = (f^4)^6(x) = x$. But 4 is not the primitive period of x under f^6, because x also has period 2 under f^6, as is easily computed. By Lemma 1.1.2 (c), the point x is not invariant under f^6 as 6 is not a multiple of 4. So the primitive period of x under f^6 is 2. This result is in agreement with statement 1 in the next proposition.

The converse problem, to derive the primitive period of a point x under f from the primitive period of x under f^k, usually has no unique solution. For example, if x has primitive period 2 under f^6 then the above shows that x may have primitive period 4 under f. However, a similar argument shows that if x has primitive period 12 under f then x also has primitive period 2 under f^6.

Proposition 1.1.4. *Let* (X, f) *be a dynamical system, let* $x \in X$ *and let* $k \in \mathbb{N}$.

(1) *If the point* x *is periodic under* f *with primitive period* p *then* x *is periodic under* f^k *with primitive period* $\text{lcm}(p, k)/k$.
(2) *If the point* x *is periodic under* f^k *with primitive period* m *then* x *is periodic under* f *and there exists* $b \in \mathbb{N}$ *such that* $\gcd(b, m) = 1$, $b|k$ *and the natural number* mk/b *is the primitive period of* x *under* f.

Proof. (1) Obviously, $(f^k)^p(x) = (f^p)^k(x) = x$, so x is periodic under f^k. Let m be any period of x under f^k. Then km is a period of x under f, so by Lemma 1.1.2 (c) it is a multiple of p. Obviously, km is a multiple of k, so km is a common multiple of k and p. Consequently, every period of x under f^k is equal to a common multiple of p and k, divided by k.

Now let $m_0 := \text{lcm}(p, k)/k$. Then $m_0 \in \mathbb{N}$ and m_0k is a period of x under f, because it is a multiple of p. It follows immediately that m_0 is a period of x under f^k. Taking into account the conclusion of the previous paragraph one easily sees that this completes the proof.

(2) Obviously, $f^{km}(x) = (f^k)^m(x) = x$, so the point x is periodic under f with period km. If we denote the primitive period of x under f by p then there exists $b \in \mathbb{N}$ such that $km = bp$. If $q := \gcd(m, b) > 1$ then $m = qm'$ and $b = qb'$ with $m', b' \in \mathbb{N}$. After division by q the equality $km = bp$ becomes $km' = b'p$ with $m', b' \in \mathbb{N}$. Hence km', being a multiple of p, is a period of x under f which, in turn, implies that m' is a period of x under f^k. Since $1 \leq m' < m$, this contradicts the choice of m as the primitive period of x under f^k. Consequently, $\gcd(m, b) = 1$, hence the equality $km = bp$ implies that that b divides k. □

Examples. Let $m = 2^n$ for some $n \in \mathbb{N}$ and assume that the point $x \in X$ is periodic under f^m.

(1) Assume that the point x has primitive period r under f^m, where $r \in \mathbb{N}$ is even. By statement 2 of the proposition, the primitive period of x under f is $p = mr/b$ with $\gcd(b, r) = 1$ and $b|m$. The former q condition implies that b is odd (for r has a factor 2), hence the latter implies that $b = 1$. So $p = mr = 2^n r$.
(2) Assume that the point x has primitive period r under f^m, where $r \in \mathbb{N}$ is odd. Again, x has primitive period $p = mr/b$ with $\gcd(b, r) = 1$ and $b|m$. We can draw no conclusion from the first condition, but the second condition implies that b is a power of 2. It follows that $p = 2^i r$ with $1 \leq i \leq n$.

In the literature a periodic orbit is sometimes called a 'closed orbit', meaning that, geometrically, the orbit is a closed curve. However, in this book a closed orbit is an orbit that is a closed subset of the phase space. Simple examples show that an orbit which is a closed subset need not be (eventually) periodic: consider e.g. the orbit of a point in $\mathbb{R}$ under the mapping $x \mapsto x+1$. On the other hand, compactness characterizes eventually periodic orbits:

Theorem 1.1.5. *Let $x_0 \in X$. The following conditions are equivalent:*
(i) *The point x_0 is eventually periodic.*
(ii) *The orbit of x_0 is finite.*
(iii) *The orbit of x_0 is compact.*

Proof. The implications (i)⇒(ii)⇒(iii) are obvious. For a direct and easy proof of the implication (ii)⇒(i), see Remark 3 after Corollary 1.1.3. It remains to prove the implication (iii)⇒(ii). So assume that $\mathcal{O}(x_0)$ is compact. Without limitation of generality, we may assume that $X = \mathcal{O}(x_0)$. For convenience, put $x_i := f^i(x_0)$ for $i = 0, 1, 2, \dots$.

First, note that $X = \bigcup_{i=0}^{\infty}\{x_i\}$, where each $\{x_i\}$ is a closed subset of X. As X is a Baire space, at least one of these closed sets has a non-empty interior. This means that there exists $k \in \mathbb{Z}^+$ such that the set $\{x_k\}$ is open, that is, x_k is an isolated point. Claim: all points of X are isolated. If this claim is true, then X is covered by open singleton sets. As X is compact, it is covered by finitely many of them: X is a finite set.

In order to prove the claim, assume the contrary. Then there are two possibilities: in the sequence $x_0, x_1, \dots$ either some non-isolated point precedes an isolated point, or the isolated points precede all non-isolated ones,

Assume that the first possibility holds: there are $i, k \in \mathbb{Z}^+$, $i < k$, such that x_i is not isolated in X and x_k is isolated. Because $f^{k-i}(x_i) = x_k$, the set $(f^{k-i})^{\leftarrow}[x_k]$ is an open neighbourhood of the non-isolated point x_i. Consequently, it contains a point x_j of X different from x_i, that is, there exists $j \in \mathbb{Z}^+$, $j \neq i$, with $f^{k-i}(x_j) = x_k$. Then $f^{k-i+j}(x_0) = f^k(x_0)$ and since $k \neq k-i+j$ it follows from the argument used in Remark 3 after Corollary 1.1.3 that the point x_0 is eventually periodic. But then X is finite and all points of X are isolated. This contradicts our assumptions.

In the second case there exists $k \in \mathbb{N}$ such that the points $x_0, \dots, x_{k-1}$ are isolated and all points x_i for $i \geq k$ are non-isolated. Clearly, now the subset $X' := X \setminus \{x_0, \dots, x_{k-1}\}$ of X is clopen, hence compact, and X' is the orbit of the point $x_0' := x_k$. Thus, we have a compact orbit, which contains (according to the first part of this proof) an isolated point, isolated in X', that is. However, X' is open in X, so the isolated point in X' is isolated in X as well, contradicting the assumption that none of the points x_i with $i > k$ is isolated in X. This contradiction completes the proof of our claim. □

1.2 Invariant sets

A subset A of X is said to be *invariant in* (X, f), or *invariant under* f, or f*-invariant*, whenever $f[A] \subseteq A$. If f is understood then an f-invariant set will be called just *invariant*. A *completely invariant* set is an invariant set which is mapped onto itself by f, i.e., a subset A of X such that $f[A] = A$.

If A is an invariant subset of X then $f^n[A] \subseteq A$ for every $n \in \mathbb{Z}^+$. In particular, a subset A of X is invariant iff $\mathcal{O}(x) \subseteq A$ for all $x \in A$. So a closed subset A of X is invariant iff $\overline{\mathcal{O}(x)} \subseteq A$ for all $x \in A$.

If A is an invariant subset of X then $f|_A$ is a continuous mapping of A into itself (assuming that A has the relative topology of X). So if A is *non-empty* we have another dynamical system, namely, $(A, f|_A)$. We call this the *subsystem* of (X, f) *on* (or: *defined by*) A. If no confusion is likely to arise we shall denote this subsystem simply by (A, f).

Examples.

(1) In any dynamical system (X, f), $\emptyset$ and X are invariant sets, the *trivial* invariant subsets. Moreover, the set $\bigcap_{n=0}^{\infty} f^n[X]$ includes every completely invariant subset of X and is itself completely invariant as well: it is the largest completely invariant subset of X.

(2) Obviously, every orbit in any dynamical system is invariant and every periodic orbit is completely invariant. The orbit of a non-periodic point x is never completely invariant because $x \notin f[\mathcal{O}(x)]$: if $x \in f[\mathcal{O}(x)] = \{f(x), f^2(x), \dots\}$ then x would be a periodic point. In addition, the set of all invariant points and the set of all periodic points are completely invariant.

(3) In the system considered in Example (1) in Section 1.1, every proper subinterval of the closed unit interval $[0; 1]$ with 0 as a left end point is invariant but not completely invariant. In the system considered in Example (2) in Section 1.1, every interval which includes the interval $[-1; 1]$ is invariant. A subinterval of the interval $[-1; 1]$ is invariant iff it is symmetric. In that case, it is completely invariant.

Proposition 1.2.1. *Let A be an invariant subset of X.*

(1) *Let $B \subseteq A$. Then B is invariant in the subsystem (A, f) iff B is invariant in the full system (X, f).*

(2) *For every $k \in \mathbb{N}$, the set $f^k[A]$ is invariant. Similarly, if A is completely invariant then so is $f^k[A]$.*

Proof. (1) Trivial.

(2) If A is invariant and $k \in \mathbb{N}$, then $f[f^k[A]] = f^k[f[A]] \subseteq f^k[A]$, which shows that the set $f^k[A]$ is invariant. In the case that A is completely invariant we have an equality here instead of an inclusion, which shows that the set $f^k[A]$ is completely invariant. □

Proposition 1.2.2. *The union of a collection of (completely) invariant sets is (completely) invariant. Moreover, the intersection of a collection of invariant sets is invariant.*

Proof. Straightforward. □

Examples.

(1) The complement of a (completely) invariant set need not be invariant. Consider the tent map $T: [0; 1] \to [0; 1]$. Then the interval $[0; 1/2]$ is completely invariant, but its complement, the interval $(1/2; 1]$, is not invariant.

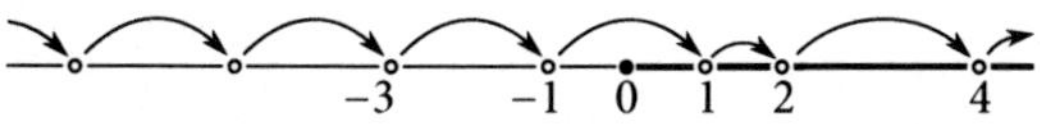

Fig. 1.2. The set A of white dots is completely invariant, as is the set B indicated by the heavy line, but their intersection is not completely invariant.

(2) The intersection of completely invariant sets, though invariant, need not be completely invariant. For example, let $f : \mathbb{R} \to \mathbb{R}$ be defined by

$$f(x) = \begin{cases} x + 2 & \text{for } x \le -1\,, \\ -x & \text{for } -1 \le x \le 0\,, \\ 2x & \text{for } x \ge 0\,. \end{cases}$$

The subsets $A := \{1 - 2n : n \in \mathbb{N}\} \cup \{2^n : n \in \mathbb{Z}^+\}$ and $B := \{x \in \mathbb{R} : x \ge 0\}$ are completely invariant. Their intersection, the set $C := \{n^2 : n \in \mathbb{N}\}$, is not completely invariant. See Figure 1.2.

NB. The intersection of compact completely invariant sets need not be completely invariant either. The easiest way to see this is to extend the above example to the two-point compactification $\{-\infty\} \cup \mathbb{R} \cup \{\infty\}$ of $\mathbb{R}$ by making the points $\pm\infty$ invariant under f. See also Exercise 3.10 (2).

Proposition 1.2.3. *The closure of an invariant set is invariant. If a set is completely invariant and its closure is compact, or f is a homeomorphism, then the closure is completely invariant as well.*

Proof. To prove the first statement, use the fact that continuity of f implies that for any subset A of X we have $f[\overline{A}] \subseteq \overline{f[A]}$; see formula (A.3-1) in Appendix A.3.1. So if A is invariant then $f[\overline{A}] \subseteq \overline{f[A]} \subseteq \overline{A}$.

Moreover, if $\overline{A}$ is compact then the equality $f[\overline{A}] = \overline{f[A]}$ holds. Obviously, this equality also holds if f is a homeomorphism. If, in addition, A is completely invariant then we get $f[\overline{A}] = \overline{f[A]} = \overline{A}$. □

Examples.

(1) The interior of an invariant or completely invariant set need not be invariant (let alone completely invariant). For example, the closed interval $[0; 1]$ is completely invariant under the mapping $f_4 : x \mapsto 4x(1 - x) : \mathbb{R} \to \mathbb{R}$, but its interior in $\mathbb{R}$, the interval $(0; 1)$, is not invariant: $f_4(1/2) = 1 \notin (0; 1)$.

(2) If the closure of a completely invariant set is not compact then that closure is not necessarily completely invariant (though it is invariant). Let the mapping $f : \mathbb{R} \to \mathbb{R}$ have a graph as sketched in Figure 1.3. It is easy to see that the subset $(0; \infty)$ is completely invariant, but its closure $[0; \infty)$ is not completely invariant.

Corollary 1.2.4. *Every orbit closure is invariant.*

Proof. An orbit is invariant, so its closure is invariant as well. □

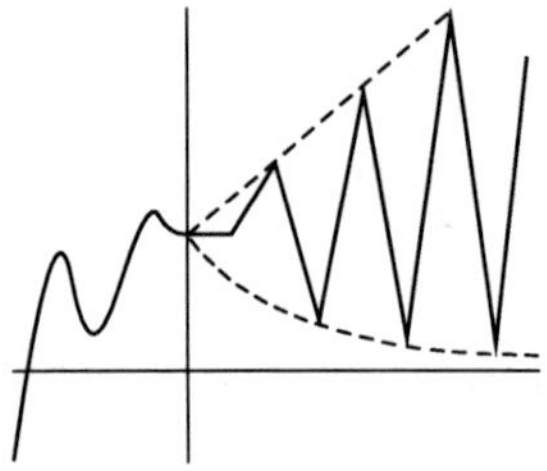

Fig. 1.3. Graph of a mapping $f : \mathbb{R} \to \mathbb{R}$ for which the set $A := (0; \infty)$ is completely invariant and $\overline{A}$ is not completely invariant.

Non-empty closed invariant subsets which have no proper closed invariant subsets have a special position. Such sets are called minimal: a subset A of X is said to be *minimal in* (X, f) (or: *under* f) whenever it has the following properties: it is non-empty, closed and invariant, and if $B \subseteq A$ is closed (in X) and invariant then either $B = \emptyset$ or $B = A$. So a subset A of X is minimal iff it is a minimal element in the partial order of the collection of all non-empty closed invariant subsets of X under inclusion. If A is a minimal subset of X and B is a non-empty closed invariant subset of X then $A \cap B$ is a closed invariant subset of A. Consequently, either $A \cap B = \emptyset$ or $A \subseteq B$. In particular, different minimal subsets of X are mutually disjoint.

A dynamical system is said to be *minimal* whenever its phase space is a minimal set. In that case, the system has no proper closed invariant subsets.

Proposition 1.2.5. *Let A be a non-empty closed invariant set in X. Then the set A is minimal in (X, f) iff the subsystem (A, f) is a minimal system.*

Proof. This follows easily from the observation that, because A is closed and invariant, the closed invariant sets in the subsystem (A, f) are just the closed and invariant sets of (X, f) that are included in A. □

Proposition 1.2.6. *Let A be a non-empty subset of X. Then A is minimal iff $A = \overline{\mathcal{O}(x)}$ for every $x \in A$. In particular, a system (X, f) is minimal iff every point has a dense orbit in X.*

Proof. “Only if”: Suppose A is a minimal set. Every orbit closure of a point of A is a non-empty closed invariant subset of A, hence all of A.

“If”: Assume the every orbit in A is dense in A. Let B be a closed non-empty invariant subset of A and let $x \in B$. Then $\overline{\mathcal{O}(x)} \subseteq B$. Since, by assumption, $\overline{\mathcal{O}(x)} = A$, this implies that $B = A$. □

Examples.

(1) An invariant point is a minimal set, as is every periodic orbit.

(2) The system $(\mathbb{S}, \varphi_a)$ with $a \notin \mathbb{Q}$ is minimal: in Case 2 of Example 0.4.3 in the Introduction it was shown that every point in this system has a dense orbit. For other examples, see 1.7.6, Section 4.2, the Propositions 5.6.3 and 5.6.7, and also 5.6.14 and 6.3.7 ahead.

In view of the property expressed in Proposition 1.2.6, a minimal subset is often called a *minimal orbit closure*.

Theorem 1.2.7. *Every non-empty compact invariant subset of X includes a minimal set. In particular, if the phase space itself is compact then the system under consideration has a minimal subset.*

Proof. This is a straightforward consequence of Zorn's Lemma. Let $\mathcal{F}$ be the collection of all non-empty compact invariant subsets of X, partially ordered by inclusion. Since every descending chain in $\mathcal{F}$ has a non-empty intersection (compactness), which is invariant by Proposition 1.2.2, such an intersection is again a member of $\mathcal{F}$, hence a lower bound in $\mathcal{F}$ for the chain. This means that the partial order in $\mathcal{F}$ is inductive, i.e., it satisfies the conditions for Zorn's Lemma. Therefore, every member of $\mathcal{F}$ includes a minimal member. By definition, such a minimal member is exactly what we have called a minimal set. □

Remark. This theorem has its main importance in theoretical considerations. In not too complicated examples, the minimal subsets often are just the invariant points and the periodic orbits.

Minimal sets under homeomorphism have been investigated at length. For some time it was not even obvious that there actually exist compact minimal systems with a non-invertible phase mappings. Presently, examples of such systems are known: see Remark 2 after Proposition 5.6.4 and also Proposition 5.6.7. In fact, these two examples are special cases of Proposition 5.3.2 ahead, which implies that every non-trivial minimal subsystem of the shift system – the system defined and studied in Chapter 5 – has a non-injective phase mapping. However, the phase mapping of a compact minimal system is 'almost' a homeomorphism:

Theorem 1.2.8. *Let (X, f) be a minimal dynamical system with a compact phase space X. Then f is an irreducible, hence semi-open, mapping. If, in addition, X is metrizable then f is also an almost 1,1-mapping.*

Proof. First, note that f is surjective: $f[X]$ is a compact, hence closed, invariant subset of X, so $f[X] = X$.

Next, consider a closed subset A of X such that $f[A] = X$. We have to show that $A = X$. For every $n \in \mathbb{Z}^+$, let $A_n := A \cap f^{\leftarrow}[A] \cap \cdots \cap (f^n)^{\leftarrow}[A]$. Then for every $n \in \mathbb{Z}^+$, A_n is a closed subset of X and we show by induction that $A_n \neq \emptyset$. For $n = 0$ we have $A_0 = A$, which is non-empty because otherwise $f[A]$ could not be equal to X. Suppose for certain $n \in \mathbb{Z}^+$ we have $A_n \neq \emptyset$. Now observe that

$$A_{n+1} = A \cap f^{\leftarrow}[A \cap \cdots \cap (f^n)^{\leftarrow}[A]] = A \cap f^{\leftarrow}[A_n]\,.$$

It follows that $f[A_{n+1}] = f[A] \cap A_n = X \cap A_n = A_n$, which is non-empty by assumption. Then A_{n+1} cannot be empty.

So $(A_n)_{n\in\mathbb{Z}^+}$ is a descending sequence of non-empty compact sets, which has, consequently, a non-empty intersection $\bigcap_{n=0}^{\infty}(f^n)^{\leftarrow}[A] =: Z$. Then Z is closed in X and it is easily seen that Z is invariant under f. As (X, f) is minimal, it follows that $Z = X$, whence $A = X$. This completes the proof that f is irreducible. The Corollary in Appendix A.9.2 now implies that f is a semi-open mapping.

For the last statement of the theorem, use Theorem A.9.7 in Appendix A. □

Combining this result with the Proposition in Appendix A.9.3 we get for a dynamical system (X, f) with a compact phase space:

(a) (X, f) is minimal $\Rightarrow f$ is semi-open;
(b) (X, f) is minimal and f is open $\Rightarrow f$ is a homeomorphism.

There are minimal systems (X, f) such that f is not a homeomorphism, so the conclusion in (a) cannot be improved to 'f is open' and in (b) the assumption that f is open cannot be weakened to 'f is semi-open'.

1.3 Transitivity

In many systems there are points that approach any other point in the system arbitrarily close under iterations of the phase mapping. Topologically, this means that such points have dense orbits. Actually, we have to be a little bit more precise, as the following examples show:

Example. Let $X := \{0\}\cup\{1/2^n : n \in \mathbb{Z}^+\}$ and $f(x) := x/2$ for $x \in X$. Under iterations of f the point 1 visits all points of $X \setminus \{0\}$ and it approaches the point 0 arbitrarily close. So the orbit of the point 1 is dense in X. Yet this is not what we have in mind: after at time n the initial point 1 has reached the point $1/2^n$ it will never come back in the vicinity of the point $1/2^n$.

A point x_0 in a dynamical system (X, f) is said to be *(topologically) transitive* whenever for every non-empty open subset U of X the 'dwelling set' of x_0 in U, that is, the set $D(x_0, U) := \{n \in \mathbb{Z}^+ : f^n(x_0) \in U\}$, is infinite. Equivalently, the point x_0 is transitive whenever for every non-empty open subset U of X and every $k \in \mathbb{Z}^+$ there exists $n \geq k$ with $f^n(x_0) \in U$. Thus, a transitive point visits every non-empty open set infinitely often. A dynamical system that has a transitive point will be called a *transitive* system. A non-empty subset of a dynamical system is said to be *transitive* whenever it is closed and invariant, and the subsystem on it is transitive.

A trivial example of a transitive system is a (finite) system consisting of one single periodic orbit. More generally, minimal systems will turn out to be transitive, as well as the (non-minimal) systems of the tent map and the argument-doubling transformation (see the end of this section). Additional examples can be found at the end of 1.7.5 (2) and in the examples after Proposition 5.3.7 ahead.

Proposition 1.3.1. *Let $x_0 \in X$. The following are equivalent:*
(i) *The point x_0 is transitive.*
(ii) *All points of $\mathcal{O}(x_0)$ are transitive.*
(iii) *All points of $\mathcal{O}(x_0)$ have a dense orbit.*
If these conditions are fulfilled then the set of all transitive points is a dense subset of X.

Proof. "(i)⇒(ii)": Assume (i), consider a point $x_k := f^k(x_0)$ of the orbit of x_0 ($k \in \mathbb{N}$) and let U be a non-empty open subset of X. By using the equality $f^n(x_k) = f^{n+k}(x_0)$ one easily shows that $D(x_k, U) = D(x_0, U) - k$. As the set $D(x_0, U)$ is infinite it follows that the set $D(x_k, U)$ is infinite as well. This means that x_k is a transitive point.

"(ii)⇒(iii)": Every transitive point has a dense orbit.

"(iii)⇒(i)": Assume that all points of $\mathcal{O}(x_0)$ have a dense orbit. Let U be a non-empty open set in X and let $k \in \mathbb{Z}^+$. Since the point $f^k(x_0)$ has a dense orbit, there exists $i \in \mathbb{Z}^+$ with $f^i(f^k(x_0)) \in U$, that is, $f^n(x_0) \in U$ with $n := k + i \geq k$. This shows that the point x_0 is transitive.

The final conclusion should be clear now: the orbit of a transitive point is dense and is included in the set of all transitive points. □

Remark. Proposition 1.2.6 and the implication (iii)⇒(i) above imply: *in a minimal system all points are transitive.* For conditions that transitivity implies minimality, see Exercises 1.5 (4) and 1.5 (5).

Proposition 1.3.2. *The following conditions are equivalent:*
(i) *The system (X, f) is transitive.*
(ii) *There is a point with a dense orbit and the set $f[X]$ is dense in X.*
In fact, if $f[X]$ is dense in X then then every point with a dense orbit is transitive.

Proof. "(i)⇒(ii)": If the system has a transitive point x_0 then it follows from the preceding proposition that the orbit of x_0 is dense. In addition, the orbit of the point $f(x_0)$ is dense as well. However, this orbit is included in the set $f[X]$, hence this set is dense.

"(ii)⇒(i)": Assume that the point $x_0 \in X$ has a dense orbit and that the set $f[X]$ is dense in X. Taking into account that a dense subset of a dense subset of X is dense in X, one easily shows by induction that the set $f^n[X]$ is dense in X for every $n \in \mathbb{N}$: if $f^n[X]$ is dense in X, then $f^{n+1}[X]$ is dense in $f[X]$, hence dense in X. Next, note that $f^n[\mathcal{O}(x_0)] = \mathcal{O}(f^n(x_0))$ for every $n \in \mathbb{N}$. As $\mathcal{O}(x_0)$ is dense in X it follows that $\mathcal{O}(f^n(x_0))$ is dense in the dense subset $f^n[X]$ of X, hence dense in X. Now an application of Proposition 1.3.1 (iii)⇒(i) completes the proof. □

Remarks.
(1) By the initial example in this section: if there is a point in X with a dense orbit under f then $f[X]$ is not necessarily dense. Another example: let $X := \mathbb{S} \cup \{x_0\}$, with $x_0 := 2 \in \mathbb{C}$ isolated in X, and let $f: X \to X$ be defined by $f|_{\mathbb{S}} := \varphi_a$ with $a \in \mathbb{R} \setminus \mathbb{Q}$ and $f(x_0) := [0]$. Then the point x_0 has a dense orbit in X, but the set $f[X] = \mathbb{S}$ is not dense in X. These examples are typical for the situation: see Exercise 1.6 (2).

(2) If the system (X, f) is transitive and X is compact (or if by some other reason the set $f[X]$ happens to be closed in X) then $f[X] = X$. In general, however, if the system (X, f) is transitive then $f[X]$ may be a proper dense subset of X, as the system in Exercise 1.9 shows.

Corollary 1.3.3. *Let $x_0 \in X$ and suppose that $\mathcal{O}(x_0)$ is dense in X. If either $x_0 \in f[X]$ or the point x_0 is not isolated in X then x_0 is a transitive point.*

Proof. In view of Proposition 1.3.2 it is sufficient to show that $f[X]$ is dense in X. If $\mathcal{O}(x_0)$ is dense in X then, obviously, the set $f[X] \cup \{x_0\}$ is dense as well. So if $x_0 \in f[X]$ then $f[X]$ is dense in X. If $x_0 \notin f[X]$ but x_0 is not isolated, proceed as follows: if U is a non-empty open subset of X then the set $U \setminus \{x_0\}$ is non-empty and open open as well, hence it includes a point of the dense set $f[X] \cup \{x_0\}$. Obviously, this cannot be the point x_0, hence U includes a point of $f[X]$. This shows that in this case the set $f[X]$ is dense. □

Remarks.
(1) The condition that x_0 is not isolated is 'almost' necessary: by Proposition 1.3.4 below an *infinite* transitive system has no isolated points.
(2) By Corollary 1.2.4, every orbit closure $A := \overline{\mathcal{O}_f(x_0)}$ in a dynamical system (X, f) defines a subsystem (A, f) in which the orbit of the point x_0 is dense. This subsystem is not necessarily transitive. For example, in the system $([0; 1], f)$, where $f(x) := x/2$ for $x \in [0; 1]$, the subsystem on the orbit closure of the point 1 is not transitive.

The existence of isolated points plays a determining role in the distinction between trivial and non-trivial transitive systems:

Proposition 1.3.4. *Suppose X has an isolated point. Then (X, f) is a transitive system iff X consists of a single periodic orbit.*

Proof. "If": Obvious.

"Only if": Let x_0 be a transitive point and let y_0 be an isolated point. Then there are $k, l \in \mathbb{N}$ such that $k < l$ and both $f^k(x_0)$ and $f^l(x_0)$ are in the open set $\{y_0\}$, i.e., are equal to y_0. It follows that $f^{l-k}(y_0) = y_0$, so y_0 is a periodic point. But y_0, being a point of the orbit of the transitive x_0, has a dense orbit in X. As this orbit is finite, hence closed, it is all of X. □

Remark. By Corollary 1.3.3 and Proposition 1.3.4: *If X is infinite then the system (X, f) is transitive iff it has a dense orbit and the space X has no isolated points. In particular, an infinite minimal set has no isolated points.*

The remark after Proposition 1.3.1 implies that in a minimal system all points are transitive, while in a non-minimal transitive system there is 'only' a dense set of transitive points (however, see Remark (1) after Theorem 1.3.5 below); if X is (locally) compact then this set has empty interior: see Exercise 1.5 (5). Yet another difference: in a minimal system a closed invariant set which is not all of the phase space is (by definition) empty, in a non-minimal transitive system it has 'only' an empty interior: if the interior were not empty then it would contain a (dense) 'tail' of the orbit of the transitive point, hence the set would be all of the phase space. This turns out to characterize an important property, namely, topological ergodicity.

A dynamical system (X, f) is said to be *topologically ergodic* whenever for every choice of two non-empty open subsets U and V of X there exists an $n \in \mathbb{Z}^+$ such that $f^n[U] \cap V \neq \emptyset$ or, equivalently, $U \cap (f^n)^{\leftarrow}[V] \neq \emptyset$ (we leave the trivial proof of this equivalence to the reader). The characterization of ergodicity mentioned above is formalized in Exercise 1.6 (3).

The following notation will be convenient in connection with this definition: if A and B are subsets of X, put

$$\begin{aligned} D_f(A, B) &:= \{n \in \mathbb{Z}^+ : A \cap (f^n)^{\leftarrow}[B] \neq \emptyset\} \\ &= \{n \in \mathbb{Z}^+ : f^n[A] \cap B \neq \emptyset\}. \end{aligned}$$

If the system (X, f) is understood we shall simply write $D(A, B)$ instead of $D_f(A, B)$. The elements of $D(A, B)$ are the times at which points of A dwell in B; for this reason, $D(A, B)$ is sometimes called a 'dwelling set' (also the term 'set of hitting times' is in use). Thus, the system (X, f) is topologically ergodic iff $D(U, V) \neq \emptyset$ for every choice of non-empty open sets U and V of X. If this is the case then $D(U, V)$ turns out to be always infinite: see Exercise 1.6 (6).

Note that *in a topologically ergodic system the set $f[X]$ is dense in X*. For if the set $X \setminus f[X]$ has an interior point y and x is any other point of X then the points x and y have disjoint neighbourhoods U and V, respectively, with $V \cap f[X] = \emptyset$. For every $n \in \mathbb{Z}^+$ the set $f^n[U]$ is either included in U (if $n = 0$) or in $f[X]$ (if $n \geq 1$), hence is disjoint from V, contradicting topological ergodicity of the system.

Theorem 1.3.5. *If the system (X, f) is transitive then it is topologically ergodic. The converse is true if X is a 2nd-countable Baire space.*

Proof. Assume that there is a transitive point x_0 in X and consider two non-empty open subsets U and V of X. Then there are $k, l \in \mathbb{Z}^+$ such that $f^k(x_0) \in U$, $f^l(x_0) \in V$ and $l > k$ (recall that there are infinitely many values of l such that $f^l(x_0) \in V$). Then $f^{l-k}(f^k(x_0)) = f^l(x_0) \in f^{l-k}[U] \cap V$. This completes the proof that (X, f) is topologically ergodic.

Conversely, assume that the system (X, f) is topologically ergodic. Moreover, let the topology of X have a countable base $\{B_i\}_{i \in \mathbb{N}}$. We may assume that $B_i \neq \emptyset$ for every $i \in \mathbb{N}$. Then for every $i \in \mathbb{N}$ the set

$$U_i := \bigcup_{n=0}^{\infty} (f^n)^{\leftarrow}[B_i]$$

is non-empty and open. Moreover, if U is any non-empty open set in X then by topological ergodicity there exists $n \in \mathbb{Z}^+$ such that $U \cap (f^n)^{\leftarrow}[B_i] \neq \emptyset$, hence $U \cap U_i \neq \emptyset$. It follows that the set U_i is dense in X ($i = 1, 2, \ldots$). Consequently, if X is a Baire space then the countable intersection

$$D := \bigcap_{i=1}^{\infty} U_i$$

is a dense subset of X as well. In particular, $D \neq \emptyset$. It is quite easy to show that every point of D has a dense orbit, as follows: Let $x \in D$ and consider a non-empty open set V in X. As $\{B_i\}_{i\in\mathbb{N}}$ is a base for the topology of X there is $j \in \mathbb{N}$ such that $B_j \subseteq V$. As $x \in D \subseteq U_j$, there exists $n \in \mathbb{Z}^+$ such that $x \in (f^n)^{\leftarrow}[B_j]$, hence $f^n(x) \in B_j \subseteq V$. This shows that the orbit of x is dense in X.

Since $f[X]$ is dense in X (see the observation preceding this theorem), and all points of D have a dense orbit, it follows from Proposition 1.3.2 that all points of D are transitive. □

Remarks.

(1) The proof of Theorem 1.3.5 shows that *in a transitive system on a 2nd-countable Baire space the set of transitive points includes a dense G_δ-set D, hence is a residual set.*

(2) We might also conclude the above proof by observing that X has no isolated points (Exercise 1.6 (4)), so that Corollary 1.3.3 can be applied.

Using Theorem 1.3.5 we can now easily give non-trivial examples of transitive systems.

Examples.

(1) Consider the tent map $T: [0;1] \to [0;1]$. We show that the system $([0;1], T)$ is topologically ergodic, hence transitive ($[0;1]$ is a 2nd countable Baire space).
Let U be a non-empty open subset of $[0;1]$ and let J be an open interval included in U. If a subinterval of $[0;1]$ does not include the point $1/2$ then the length of J is doubled by an application of T. So there must be $k \in \mathbb{N}$ such that $1/2 \in T^k[J]$, hence $0 \in T^n[J]$ for all $n \geq k+2$. It easily follows that $T^n[J] = [0;1]$ for sufficiently large n, (Alternative proof: J includes a complete 'tent' of T^n for sufficiently large n.) This implies that $T^n[U]$ meets every non-empty open subset V of $[0;1]$.

(2) Let $\psi: \mathbb{S} \to \mathbb{S}$ be the argument-doubling transformation. The system $(\mathbb{S}, \psi)$ is topologically ergodic, hence transitive ($\mathbb{S}$ is a 2nd countable Baire space). The proof is along the same lines as in the previous example. For let U be a non-empty open set in $\mathbb{S}$ and let J be an open arc in U. As the length of the arc is doubled under each application of ψ, there will be a natural number n such that $\psi^n[J] = \mathbb{S}$. This implies that $\psi^n[U]$ meets every non-empty open set V of $\mathbb{S}$.

(3) An example of a topologically ergodic system (on a non-Baire space) that is not transitive is given in Exercise 5.3 (2).

The tent map T and the argument-doubling transformation ψ show that in contrast to the phase mapping of a minimal system, a transitive mapping need not be irreducible or almost 1-to-1. However, transitive interval maps are semi-open: see Lemma 2.6.4 (2) ahead.

1.4 Limit sets

In this section we discuss only the most elementary properties of limit sets. Here is the formal definition. If $x \in X$ then the set

$$\omega_f(x) := \bigcap_{n=0}^{\infty} \overline{\mathcal{O}_f(f^n(x))} \tag{1.4-1}$$

is called the *limit set of the point x under f*. When f is understood we call this just 'the limit set of x' and we shall simply write $\omega(x)$ instead of $\omega_f(x)$.

It follows from the definition that $\omega(x) \subseteq \overline{\mathcal{O}(x)}$ for every $x \in X$. Being an intersection of closed sets, a limit set is always a (possibly empty) closed subset of X. The following will be used frequently and without further reference: if A is a *closed* invariant subset of X and $x \in A$ then the limit set of x in the subsystem $(A, f|_A)$ agrees with its limit set in the full system; consequently, the latter is included in A.

Example. Let the mapping $f: \mathbb{R}^+ \to \mathbb{R}^+$ be given by $f(x) := \sqrt{x}$ for $x \in \mathbb{R}^+$. Then, clearly, $\omega(0) = \{0\}$. If $x \neq 0$ and $n \in \mathbb{Z}^+$ then the points $f^n(x)$ for $n \geq 0$ form a monotonous sequence (constant if $x = 1$) converging to 1, and $\mathcal{O}(f^n(x))$ is included in the interval with end points $f^n(x)$ and 1. Hence $\omega(x) = \{1\}$.

Lemma 1.4.1. *Let $x \in X$.*
(1) *A point z of X belongs to $\omega(x)$ iff every neighbourhood of z contains points $f^k(x)$ for infinitely many values of $k \in \mathbb{Z}^+$.*
(2) *If X is a metric space then*

$$\omega(x) = \{\, z \in X : z \text{ is limit of a subsequence of } \mathcal{O}(x) \,\}.$$

(3) $\overline{\mathcal{O}(x)} = \mathcal{O}(x) \cup \omega(x)$. *Consequently, if $\omega(x) = \emptyset$ then $\mathcal{O}(x)$ is a closed set.*

Proof. (1) Use the following observation: If $z \in X$ and $n \in \mathbb{Z}^+$ then z belongs to the orbit closure $\overline{\mathcal{O}(f^n(x))}$ iff every neighbourhood U of z meets the set $\mathcal{O}(f^n(x))$, i.e., iff there exists $k \geq n$ such that $f^k(x) \in U$.
(2) Denote the set of all limits of subsequences of the orbit of x by $L(x)$. We want to show that $L(x) = \omega(x)$.
"$L(x) \subseteq \omega(x)$": Let $z \in L(x)$. Every neighbourhood of z contains almost all terms of a subsequence of the orbit of x, so the condition of part 1 of the lemma is met. Consequently, $z \in \omega(x)$.
"$\omega(x) \subseteq L(x)$": Let $z \in \omega(x)$. By part 1 of the lemma, for every $n \geq 0$ and for every neighbourhood U of z there exists $k \geq n$ such that $f^k(x) \in U$. Using this, one can inductively define a sequence $n_1 < n_2 < n_3 < \dots$ such that $f^{n_k}(x) \in B_{1/k}(z)$ for every $k \in \mathbb{N}$. Then $z = \lim_{k \rightsquigarrow \infty} f^{n_k}(x)$.
(3) "$\subseteq$": If $y \in X \setminus (\mathcal{O}(x) \cup \omega(x))$ then, by 1 above, y has a neighbourhood U containing only finitely many points of $\mathcal{O}(x)$, all different from y. Delete these points from U and retain a neighbourhood of y containing no points of $\mathcal{O}(x)$. Hence $y \notin \overline{\mathcal{O}(x)}$.
"$\supseteq$": Obvious. □

Proposition 1.4.2.
(1) *If x is a periodic point then $\omega(x) = \mathcal{O}(x)$. In particular, $x \in \omega(x)$.*
(2) *A point $x \in X$ is transitive iff $\omega(x) = X$.*
(3) *The system (X, f) is minimal iff $\omega(x) = X$ for every point $x \in X$.*

Proof. (1) As finite sets are closed, we have $\mathcal{O}(f^n(x)) = \mathcal{O}(x) = \overline{\mathcal{O}(f^n(x))}$ for all $n \geq 0$. Consequently, $\omega(x) = \mathcal{O}(x)$.
(2) Clear from the definition of a transitive point in Section 1.3 and the characterization of a limit set in Lemma 1.4.1 (1).
(3) By the Remark after Proposition 1.3.1, if X is minimal then every point of X is transitive. So by 2, $\omega(x) = X$ for every $x \in X$. Conversely, if $\omega(x) = X$ for every $x \in X$ then every point of X is transitive, hence has a dense orbit which, in view of Proposition 1.2.6, implies that X is minimal under f. □

Remark. If the phase space X is Čech complete, then the converse of the first part of statement 1 above holds: see Corollary 4.1.5 ahead. If X is just a Hausdorff space then this converse may not hold. For example, let x be a point in an infinite minimal system (Y, g), let $X := \mathcal{O}_g(x)$ and let $f := g|_X$. Then $\omega_f(x) = X = \mathcal{O}_f(x)$, but the point x is not periodic under f. As to the second part of statement 1, if $x \in \omega(x)$ then x is not necessarily periodic (e.g., x is a non-periodic transitive point). In Section 4.1 we shall pay more attention to points x such that $x \in \omega(x)$ (so-called *recurrent points*).

Proposition 1.4.3. *Let $x \in X$. Then:*
(1) *$\omega(f^n(x)) = \omega(x)$ for all $n \in \mathbb{N}$.*
(2) *$\omega(x)$ is a closed and invariant subset of X.*
(3) *If $y \in \omega(x)$ then $\omega(y) \subseteq \omega(x)$.*

Proof. The sets $\overline{\mathcal{O}(f^k(x))}$ for $k \in \mathbb{Z}^+$ form a decreasing nested sequence. This implies that, for every $n \in \mathbb{N}$,

$$\omega(x) = \bigcap_{k\geq 0} \overline{\mathcal{O}(f^k(x))} = \bigcap_{k \geq n} \overline{\mathcal{O}(f^k(x))} .$$

Moreover, $f^k(x) = f^{k-n}(f^n(x))$ for all $k \geq n$, so the right-hand side of this equality can be written as

$$\bigcap_{j\geq 0} \overline{\mathcal{O}(f^j(f^n(x)))} = \omega(f^n(x)) .$$

This completes the proof of 1. In order to prove 2, note that by Corollary 1.2.4 and Proposition 1.2.2, intersections of orbit closures are invariant, and closed, of course. Finally, 3 is clear from 2. In fact, if A is any closed invariant subset of X and $y \in A$, then $\omega(y) \subseteq \overline{\mathcal{O}(y)} \subseteq A$. □

Corollary 1.4.4. *A non-empty subset A of X is minimal iff $\omega(x) = A$ for every point $x \in A$.*

Proof. "Only if": If A is a minimal set in (X, f) then by Proposition 1.2.5, the system $(A, f|_A)$ is minimal. Hence Proposition 1.4.2 (3) implies that $A = \omega_{f|_A}(x)$ for every $x \in A$.

Since A is closed in X and invariant under f it is easily seen that $\omega_{f|_A}(x) = \omega_f(x)$ for every $x \in A$ (see also Exercise 3.1 (1) ahead).

"If": If A is not empty and $\omega(x) = A$ for every $x \in A$ Then Proposition 1.4.3 (2) implies that A is closed and invariant in X. Then by Proposition 1.4.2 (3) – again using that $\omega_{f|_A}(x) = \omega_f(x)$ for $x \in A$ – and Proposition 1.2.5, A is a minimal set in X. □

Theorem 1.4.5. *Let $x \in X$ and assume that its orbit closure $\overline{\mathcal{O}(x)}$ is compact. Then $\omega(x)$ is a non-empty compact invariant subset of X.*

Proof. The sets $\overline{\mathcal{O}(f^n(x))}$ for $n \in \mathbb{N}$ form a decreasing sequence of closed subsets of the compact set $\overline{\mathcal{O}(x)}$. Consequently, the intersection $\omega(x)$ of this sequence of sets is not empty and compact. That $\omega(x)$ is invariant was already established in Proposition 1.4.3 (2) above. □

Remark. Let A be a non-empty compact invariant subset of X. If $x \in A$ then $\overline{\mathcal{O}(x)} \subseteq A$, hence $\overline{\mathcal{O}(x)}$ is compact and the conclusions of the theorem hold for the point x.

Later on, in Chapter 3, we shall say more about limit sets. In particular, we shall discuss the role they play in several notions of stability.

1.5 Topological conjugacy and factor mappings

In this section we define the meaning (or rather: a possible meaning) of a phrase like: 'These two systems are essentially the same'. For example, in a dynamical system with phase space $\mathbb{R}^n$ the behaviour of points and the topological properties of orbits will not depend on the choice of the coordinate system in $\mathbb{R}^n$ which is used to describe the phase mapping. After a coordinate transformation the (description of the) phase mapping looks different: it looks like we have got another system. But the two systems are the same, only their descriptions differ. Thus, we get the same result if we first apply the phase mapping (with respect to the old coordinates) and then perform the coordinate transformation, or if we first perform the coordinate transformation and then apply the phase mapping with respect to the new coordinates.

In topological spaces the equivalence of a coordinate transformation is a homeomorphism. This leads us to the following definition: Two dynamical systems (X, f) and (Y, g) are said to be *conjugate* whenever there exists a homeomorphism $\varphi: X \to Y$ such that $\varphi \circ f = g \circ \varphi$, that is, such that the following diagram commutes:

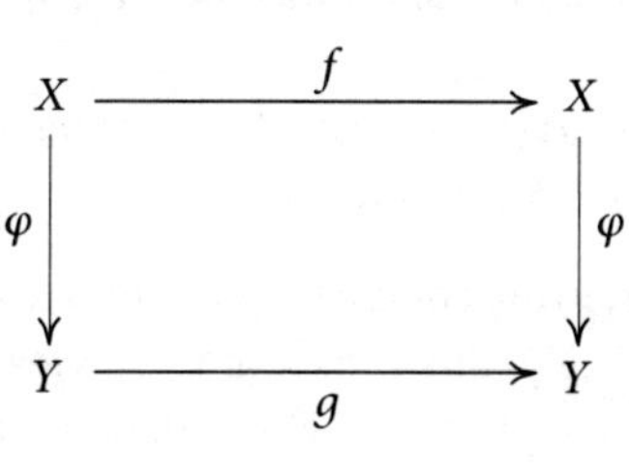

In that case, φ is called a *conjugation* of the systems (X, f) and (Y, g) (in this order). We also say that *the mappings* f and g are conjugate (*by* φ). Notation: $\varphi\colon (X, f) \xrightarrow{\simeq} (Y, g)$, and also $(X, f) \simeq (Y, g)$. A conjugation of a system with itself is also called an *automorphism* of the system.

Examples.

(1) The mappings $f\colon x \mapsto 2x\colon \mathbb{R} \to \mathbb{R}$ and $g\colon x \mapsto 3x\colon \mathbb{R} \to \mathbb{R}$ are conjugate by means of the conjugation $\varphi\colon (X, f) \xrightarrow{\simeq} (Y, g)$, given by

$$\varphi(x) := \begin{cases} \operatorname{sign}(x)\, x^{\kappa} & \text{for } x > 0\,, \\ 0 & \text{for } x = 0\,, \end{cases}$$

where $\kappa := \ln 3/\ln 2$. Observe that both f and g and even φ are differentiable, but that the inverse φ^{-1} is not. In this respect see Exercise 1.7 (3).

(2) The system $([0; 1], T)$ of the tent map and the quadratic system $([0; 1], f_4)$ are conjugate by means of the mapping

$$\varphi\colon x \mapsto \sin^2(\tfrac{1}{2}\pi x)\colon [0; 1] \to [0; 1]\,.$$

Indeed, φ is continuous, monotonous and surjective, hence a homeomorphism, and the following formulas show that $f_4 \circ \varphi = \varphi \circ T$: if $x \in [0; 1]$ then

$$\begin{aligned} f_4(\varphi(x)) &= 4\sin^2(\tfrac{1}{2}\pi x)\,(1 - \sin^2(\tfrac{1}{2}\pi x)) = \sin^2 \pi x\,, \\ \varphi(T(x)) &= \begin{cases} \sin^2 \pi x & \text{if } 0 \le x \le \tfrac{1}{2}\,, \\ \sin^2 \pi(1 - x) & \text{if } \tfrac{1}{2} \le x \le 1\,. \end{cases} \end{aligned}$$

The following lemma states that 'conjugation' is an equivalence relation on the collection of all dynamical systems: it is reflexive, symmetric and transitive. Hence the order in which the systems are mentioned in the definition of a conjugation is irrelevant and we can say that two dynamical systems are conjugate *to each other*. The proofs are left for the reader.

Lemma 1.5.1.

(1) *For every dynamical system* (X, f), $\mathrm{id}_X\colon X \to X$ *defines a conjugation of the system* (X, f) *with itself.*

(2) *If* $\varphi\colon (X, f) \xrightarrow{\simeq} (Y, g)$ *is a conjugation then* $\varphi^{-1}\colon (Y, g) \xrightarrow{\simeq} (X, f)$ *is a conjugation as well.*

(3) *If* $\varphi\colon (X, f) \xrightarrow{\simeq} (Y, g)$ *and* $\psi\colon (Y, g) \xrightarrow{\simeq} (Z, h)$ *are conjugations then their composition* $\psi \circ \varphi\colon (X, f) \xrightarrow{\simeq} (Z, h)$ *is also a conjugation.*

If two systems (X, f) and (Y, g) are conjugate then one expects that they have the same dynamical properties. Formally, this phrase has no meaning, for we did not yet define the notion of a 'dynamical property'. We shall do so now: a *dynamical property* is an equivalence class of mutually conjugate dynamical systems. So it is a tautology to say that a dynamical property is an invariant for conjugations: a system has a such a property (belongs to a conjugacy class) iff all conjugate systems have it (belong to that class). In general, one can say that properties that are formulated in terms of the topology of the phase space and the phase mapping are dynamical. In the next proposition we identify some dynamical properties.

Proposition 1.5.2. *Let $\varphi: (X, f) \simeq (Y, g)$ be a conjugation of dynamical systems, let $x \in X$ and let $A \subseteq X$. Then:*

(1) $\varphi[\mathcal{O}_f(x)] = \mathcal{O}_g(\varphi(x))$, *hence* $\varphi[\overline{\mathcal{O}_f(x)}] = \overline{\mathcal{O}_g(\varphi(x))}$ *and* $\varphi[\omega_f(x)] = \omega_g(\varphi(x))$.

(2) *The point x is periodic under f iff the point $\varphi(x)$ is periodic under g, in which case both points have the same primitive period.*

(3) *The set A is f-invariant iff the set $\varphi[A]$ is g-invariant. Similarly, A is completely invariant under f iff $\varphi[A]$ is completely invariant under g.*

(4) *The orbit of x under f converges in X to the point $z \in X$ (or has limit point z) iff the orbit of $\varphi(x)$ under g converges in Y to the point $\varphi(z) \in Y$ (has limit point $\varphi(z)$, respectively).*

(5) *The orbit of x is dense in X iff the orbit of $\varphi(x)$ is dense in Y. In particular, (X, f) is transitive iff (Y, G) is transitive.*

(6) *The set A is minimal in X under f iff the set $\varphi[A]$ is minimal in Y under g. In particular, (X, f) is minimal iff (Y, G) is minimal.*

(7) *(X, f) is topologically ergodic iff (Y, g) is topologically ergodic.*

Proof. First, observe that in all statements it is sufficient to prove the 'only if' part for φ: by applying this to the conjugation φ^{-1} one gets the 'if' part for φ. In order to prove the first equality in statement 1, first observe that it follows by induction from the equality $\varphi \circ f = g \circ \varphi$ that

$$\varphi \circ f^n = g^n \circ \varphi \quad \text{for all } n \in \mathbb{Z}^+ . \tag{1.5-1}$$

This immediately implies that $\varphi[\mathcal{O}_f(x)] = \mathcal{O}_g(\varphi(x))$. The other equalities in 1 follow from the first one, because the homeomorphism φ preserves closures and intersections. The statements about periodicity are straightforward consequences of formula (1.5-1) and the fact that φ is a bijection. Also, statements 3, 4 and 5 follow easily from 1; note that in 4, 5 and 6 the continuity of φ and its inverse are involved. In order to prove 7, consider non-empty open sets U and V in Y, let U' and V' be their preimages in X under φ, find $n \in \mathbb{Z}^+$ such that $f^n[U'] \cap V' \neq \emptyset$ and apply φ to this equality: using formula (1.5-1) one gets $g^n[U] \cap V \neq \emptyset$. □

Remark. If $\varphi : (X, f) \simeq (Y, g)$ is a conjugation then a set A is dense in X iff the set $\varphi[A]$ is dense in Y. So the system (X, f) has a dense subset of periodic points iff the system (Y, g) has the same property. Thus, having a dense set of periodic points is a dynamical property.

A topological conjugacy is, in fact, an *isomorphism* of dynamical systems. The corresponding notion of a *morphism* is obtained by relaxing the condition that φ is a homeomorphism: just require that φ is a continuous mapping. Thus, we get the following definition: let (X, f) and (Y, g) be dynamical systems. A *morphism (of dynamical systems)* is a continuous mapping $\varphi: X \to Y$ such that $\varphi \circ f = g \circ \varphi$. Notation: $\varphi: (X, f) \to (Y, g)$. For examples that are not conjugations, see Exercise 1.7 (4),(5). If φ is an embedding of X into Y then φ is also called an *embedding* of dynamical systems. In this case, the system (X, f) is conjugate to the subsystem $(\varphi[X], g|_{\varphi[X]})$ on the subset $\varphi[X]$ of Y (by Proposition 1.5.4 (3) below, $\varphi[X]$ is g-invariant). If φ is surjective then φ is also called a *factor mapping* and (Y, g) is called a *factor* of (X, f). In that case, (X, f) is often called an *extension* of (Y, g). Finally, if $(Y, g) = (X, f)$ then φ is also called an *endomorphism* of (X, f).

Lemma 1.5.3. *The composition of two morphisms is also a morphism. Consequently, the composition of two embeddings or two factor mappings of dynamical systems is again an embedding or a factor mapping, respectively.*

Proof. Straightforward. □

We say that a morphism $\varphi: (X, f) \to (Y, g)$ *preserves* a dynamical property whenever the fact that (X, f) has this property implies that (Y, g) has it. The following proposition lists some properties that are preserved by all morphisms:

Proposition 1.5.4. *Let $\varphi: (X, f) \to (Y, g)$ be a morphism of dynamical systems, let $x \in X$ and let $A \subseteq X$. Then:*
(1) *$\varphi[\mathcal{O}_f(x)] = \mathcal{O}_g(\varphi(x))$, hence $\varphi[\,\overline{\mathcal{O}_f(x)}\,] \subseteq \overline{\mathcal{O}_g(\varphi(x))}$, with equality if the orbit closure $\overline{\mathcal{O}_f(x)}$ of x is compact.*
(2) *If the point x is periodic under f then the point $\varphi(x)$ is periodic under g and the primitive period of $\varphi(x)$ is a divisor of the primitive period of x. In particular, if x is invariant under f then $\varphi(x)$ is invariant under g.*
(3) *If the set A is invariant under f then the set $\varphi[A]$ is invariant under g. Similarly, if A is completely invariant under f then so is $\varphi[A]$ under g.*
(4) *If the orbit of the point x under f converges in X to the point $z \in X$ (or has limit point z) then the orbit of the point $\varphi(x)$ under g converges in Y to the point $\varphi(z) \in Y$ (has limit point $\varphi(z)$, respectively).*
(5) *If the orbit of x is dense in X then the orbit of $\varphi(x)$ is dense in $\varphi[X]$. In particular, a factor of a transitive system is transitive and a factor of a minimal system is minimal.*
(6) *If the set A is minimal under f and $\varphi[A]$ is a closed subset of Y then the set $\varphi[A]$ is minimal under g.*
(7) *If $\varphi[X]$ is dense in Y and (X, f) is topologically ergodic then (Y, g) is topologically ergodic.*

Proof. The proofs of (1) through (5) are straightforward (use the proofs of the various statements in Proposition 1.5.2 as hints). As to (6), if A is minimal under f then $\varphi[A]$ is a non-empty g-invariant subset of Y, given to be closed. In order to show that $\varphi[A]$ is minimal, consider a non-empty closed g-invariant subset B of $\varphi[A]$. Then $\varphi^{\leftarrow}[B]$ is a closed and f-invariant subset of X. So $A \cap \varphi^{\leftarrow}[B]$ is a non-empty closed invariant subset of A. Hence $A \cap \varphi^{\leftarrow}[B] = A$, which implies that $B = \varphi[A]$. Finally, the proof of (7) is formally the same as the proof of Proposition 1.5.2 (7) (use that $\varphi[X]$ is dense in Y to show that U' and V' are non-empty). □

Remarks.

(1) In general, the inclusion $\varphi[\,\overline{\mathcal{O}_f(x)}\,] \subseteq \overline{\mathcal{O}_g(\varphi(x))}$ in statement 1 cannot be improved: see Example (2) below, or Exercise 1.9.

(2) In Proposition 3.1.2 ahead it will be shown that $\varphi[\omega_f(x)] \subseteq \omega_g(\varphi(x))$ and that, if $\overline{\mathcal{O}_f(x)}$ is compact we have an equality: $\varphi[\omega_f(x)] = \omega_g(\varphi(x))$.

(3) The additional condition in statement 6 that $\varphi[A]$ is closed in Y is certainly fulfilled if X is compact. If this condition does not hold then the conclusion of statement 6 may not be true. See Exercise 1.9.

(4) If $n \in \mathbb{N}$ then the mapping $f^n : X \to X$ defines an endomorphism. By the statements 1, 2, 3 and 4 of the proposition, f^n preserves orbits, periodic points, invariant sets and limits of orbits. This is in accordance with Proposition 1.2.1 (2).

In general the statements in the above proposition cannot be improved:

Examples.

(1) Let $X := \{0, 1, 2, 3\}$ and define $f : X \to X$ by $f(t) := t + 1 \pmod 4$ for $t \in X$. Moreover, let $Y := \{a, b\}$ and define $g : Y \to Y$ by $g(a) := b$, $g(b) := a$. So X and Y are periodic orbits with primitive periods 4 and 2, respectively. Finally, let $\varphi : X \to Y$ be given by $\varphi(0) = \varphi(2) := a$, $\varphi(1) = \varphi(3) := b$. Then φ is a morphism that maps periodic points with primitive period 4 onto periodic points with primitive period 2.

(2) Let $X := ([0; 1] \times \{0, 1\}) \setminus \{(0, 1)\}$, $Y := [0; 1]$ and let the mappings $f : X \to X$ and $g : Y \to Y$ be defined by $f(s, i) := (s^2, i)$ for $(s, i) \in X$ and $g(s) := s^2$ for $s \in Y$. Let $\varphi : X \to Y$ be defined by $\varphi(s, i) := s$ for $(s, i) \in X$. Then $\varphi : (X, f) \to (Y, g)$ is a factor map. For any $s \in (0; 1]$ the image under φ of the closure of $\mathcal{O}_f(s, 1)$ in X is equal to $\mathcal{O}_g(s)$, which is a proper subset of the orbit closure of s under g: this orbit closure includes the point 0. So the inclusion in Proposition 1.5.4 (1) may be proper.

The preceding proposition and examples are about *preservation* of dynamical properties by factor maps. The question of *lifting* dynamical properties by factor mappings is important as well: if the image under φ has a certain dynamical property, what does this imply for the original? In general, one needs additional conditions in order to guarantee lifting of certain properties. Roughly speaking, if φ identifies too many points of its domain (its fibres are too large) then information lifted into the domain from the range of φ is too coarse so as to be useful. Note also that if $\varphi : (X, f) \to (Y, g)$ is a morphism then, obviously, one cannot expect that properties concerning points and sets in $Y \setminus \varphi[X]$ can be lifted. For this reason we discuss only lifting for factor maps.

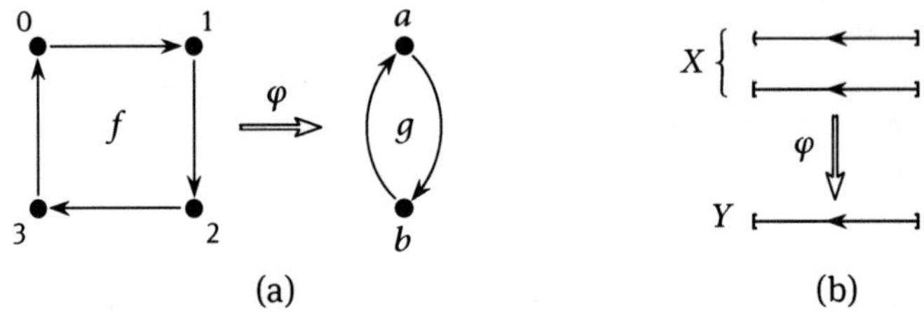

Fig. 1.4. (a): Illustrating Example (1). (b): Illustrating Example (2).

Remarks.

(1) In the statements 2 and 3 above no topology is involved, hence they would also hold if φ were not continuous. In particular, if φ is injective then they hold for the mapping $\varphi^{-1}: \varphi[X] \to X$ (with f and g interchanged). Thus, if for some point $x_0 \in X$ the point $\varphi(x_0) \in Y$ is periodic under g and φ is injective, then x_0 is periodic under f.

(2) The following example shows that periodic orbits and invariant or minimal sets are not lifted by non-injective factor maps: let $p \in \mathbb{N}$. Consider the factor mapping $\varphi: (X, f) \to (Y, g)$, where $X := \mathbb{Z}$ with $f(n) := n+1$ for $n \in \mathbb{Z}$ and $Y := \{0, \ldots, p-1\}$ with $g(i) := (i+1) \pmod{p}$ for $i \in Y$, and $\varphi(n) := n \pmod{p}$ for $n \in \mathbb{Z}$. In order to see that invariant points need not be lifted, take $p = 1$. This example also shows that transitivity is not lifted.

In the next proposition we discuss lifting of transitivity and minimality.

Proposition 1.5.5. *Let $\varphi: (X, f) \to (Y, g)$ be a factor mapping of compact dynamical systems, and let Y_0 be a closed invariant subset of Y.*

(1) *If Y_0 is transitive under g then there is a transitive closed invariant set X_0 in (X, f) such that $\varphi[X_0] = Y_0$. Moreover, if y_0 is a transitive point of Y_0 (under g) then $X_0 \cap \varphi^{\leftarrow}[y_0] \neq \emptyset$ and every point of $X_0 \cap \varphi^{\leftarrow}[y_0]$ is transitive in X_0 (under f).*

(2) *If Y_0 is minimal under g then there is a minimal subset X_0 of X such that $\varphi[X_0] = Y_0$.*

Proof. (1) Let $\mathcal{F}$ be the collection of all non-empty closed f-invariant subsets of X that are mapped onto Y_0 by φ. Note that the set $\varphi^{\leftarrow}[Y_0]$ is closed, f-invariant and that $\varphi[\varphi^{\leftarrow}[Y_0]] = Y_0 \cap \varphi[X] = Y_0$ because φ is surjective. This shows that $\varphi^{\leftarrow}[Y_0] \in \mathcal{F}$, so $\mathcal{F} \neq \emptyset$. It is standard to show that $\mathcal{F}$ is inductively ordered by inclusion (use Appendix A.3.5 to prove this; here compactness of X is involved), so Zorn's Lemma implies that $\mathcal{F}$ has a minimal element X_0.

Now let y_0 be any transitive point of Y_0. Since $\varphi[X_0] = Y_0$, there exists a point $x_0 \in X_0$ such that $\varphi(x_0) = y_0$. This shows that $X_0 \cap \varphi^{\leftarrow}[y_0] \neq \emptyset$. Now consider any point $x_0 \in X_0 \cap \varphi^{\leftarrow}[y_0]$. Then $\overline{\mathcal{O}_f(x_0)}$ is a closed invariant subset of X such that, by Proposition 1.5.4 (1) and compactness, $\varphi[\,\overline{\mathcal{O}_f(x_0)}\,] = \overline{\mathcal{O}_g(y_0)}$. The point y_0 is transitive in Y_0, so the right-hand side of this equality is equal to Y_0. It follows that $\overline{\mathcal{O}_f(x_0)} \in \mathcal{F}$ and therefore the choice of X_0 as a minimal element of $\mathcal{F}$ implies that $\overline{\mathcal{O}_f(x_0)} = X_0$. In view of Proposition 1.3.2 it remains to show that $f[X_0]$ is dense in X_0 (as $f[X_0]$ is compact, it will even be equal to X_0). In order to prove this, note that $f[X_0]$ is a closed invariant subset of X and that $\varphi[f[X_0]] = g[\varphi[X_0]] = g[Y_0] = Y_0$, where the final equality follows from the transitivity of (Y_0, g). It follows that $f[X_0] \in \mathcal{F}$, so by the choice of X_0 we get $f[X_0] = X_0$.

(2) Let X_0 be as in 1. Now every point of Y is transitive under g, so every point x of X_0 is mapped onto a transitive point of Y, hence is transitive in X_0. Consequently, X_0 is minimal under f. □

Remarks.
(1) For a related statement, see Exercise 1.8 (2).
(2) Theorem 1.2.7 can be obtained by applying statement 2 above to the factor map φ of (X, f) onto the trivial one-point system.

Corollary 1.5.6. *Let $\varphi: (X, f) \to (Y, g)$ be a factor mapping of compact dynamical systems. If $x_0 \in X$ is such that $\varphi^{\leftarrow}[\varphi(x_0)] = \{x_0\}$ and the orbit closure of the point $\varphi(x_0)$ in Y is transitive or minimal under g, then the orbit closure of x_0 in X is transitive or minimal under f, respectively.*

Proof. Let $y_0 := \varphi(x_0)$ and apply Proposition 1.5.5, taking into account that x_0 is the unique point of $\varphi^{\leftarrow}[y_0]$: as $X_0 \cap \varphi^{\leftarrow}[y_0] \neq \emptyset$ it necessarily follows that $x_0 \in X_0$, so that X_0 is the orbit closure of x_0. □

A factor map of compact minimal systems has a special topological property:

Theorem 1.5.7. *Let $\varphi: (X, f) \to (Y, g)$ be a factor mapping of compact minimal dynamical systems. Then φ is semi-open.*

Proof. Let U be a non-empty open subset of X and let V be a non-empty open subset V of U such that $\overline{V} \subseteq U$. By Proposition 1.2.6, minimality of the system (X, f) implies that $X = \bigcup_{n=0}^{\infty} (f^n)^{\leftarrow}[V] = \bigcup_{n=0}^{\infty} (f^n)^{\leftarrow}[\overline{V}]$. Since $\varphi[(f^n)^{\leftarrow}[\overline{V}]] \subseteq (g^n)^{\leftarrow}[\varphi[\overline{V}]]$ for every $n \in \mathbb{Z}^+$, it follows easily that $Y = \varphi[X] = \bigcup_{n=0}^{\infty} (g^n)^{\leftarrow}[\varphi[\overline{V}]]$. This represents the Baire space Y as a countable union of closed sets, so at least one of these sets has a non-empty interior: there exists $n \in \mathbb{Z}^+$ and there is an open subset W of Y such that $W \subseteq (g^n)^{\leftarrow}[\varphi[\overline{V}]] \subseteq (g^n)^{\leftarrow}[\varphi[U]]$. Consequently, $g^n[W] \subseteq \varphi[U]$. Since the system (Y, g) is compact and minimal (it is a factor of such a system), Theorem 1.2.8 and Lemma A.3.7 in Appendix A imply that the mapping g and its iterates are semi-open. So the set $g^n[W]$ has a non-empty interior as well. It follows that $\varphi[U]$ has a non-empty interior. □

Remark. Unlike their phase mappings – see Theorem 1.2.8 – factor maps of compact minimal systems are not necessarily irreducible. For example, in Example (1) after Proposition 1.5.4, φ maps the proper closed subset $\{0, 1\}$ of X onto Y.

1.5.8 (Application: a factor defined by a partition). Let (X, f) be a dynamical system and let R be any equivalence relation on X with the property that

$$\forall\, x, y \in X : \; xRy \Rightarrow f(x)\, R\, f(y) \tag{1.5-2}$$

or, equivalently, $f[R[x]] \subseteq R[f(x)]$ for every $x \in X$.

Equivalently, let $\mathcal{P}$ be a partition of X. This partition defines an equivalence relation R on X by the rule $\forall\, x, y \in X : \; xRy \Leftrightarrow \exists F \in \mathcal{P} : x, y \in F$. Obviously, this equivalence relation satisfies condition (1.5-2) iff for every $F \in \mathcal{P}$ there exists $F' \in \mathcal{P}$ such that $f[F] \subseteq F'$.

If we have an equivalence relation satisfying (1.5-2) then there exists an unambiguously defined mapping $\tilde{f} : X \to X/R$ such that the following diagram commutes:

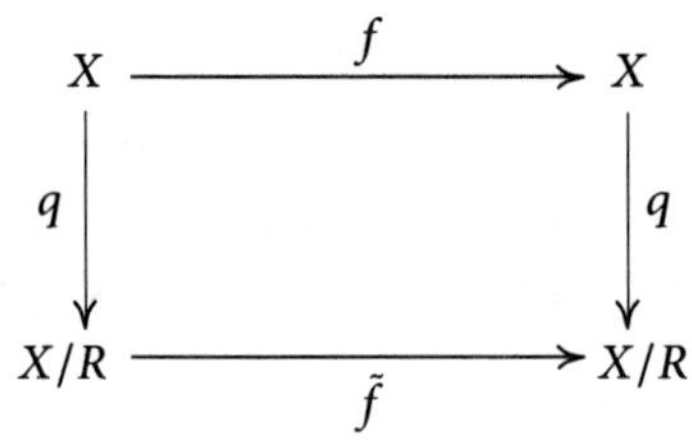

Here $q : X \to X/R$ is the quotient map. Since $\tilde{f} \circ q$ is continuous (it is equal to the continuous mapping $q \circ f$) it follows that $\tilde{f} : X/R \to X/R$ is a continuous mapping with respect to the quotient topology on X/R. Stated otherwise, we have a dynamical system $(X/R, \tilde{f})$ and a factor mapping $q : (X, f) \to (X/R, \tilde{f})$, *provided X/R with the quotient topology is a Hausdorff space.*

As an application of the above construction, consider the equivalence relation R on X defined by the partition of X into its connected components. Thus, if $x \in X$ then $R[x]$ is the component of x. For every $x \in X$ the set $f[R[x]]$ is connected and contains the point $f(x)$, so it is included in the component of $f(x)$, hence implication (1.5-2) holds, and we get a continuous mapping $\tilde{f} : X/R \to X/R$ as above. Recall from Theorem A.6.3 in Appendix A: *if X be a compact Hausdorff space then X/R is a 0-dimensional compact Hausdorff space.* So we really get a factor mapping $q : (X, f) \to (X/R, \tilde{f})$ of dynamical systems.

Theorem 1.5.9. *Let (X, f) be a transitive system on a compact Hausdorff space X. Then one of the following holds:*

(1) *X/R is a single periodic orbit under $\tilde{f}$.*
(2) *X/R is an infinite 0-dimensional compact Hausdorff space without isolated points, transitive under $\tilde{f}$.*

If, in addition, X is metrizable then in case (2) *X/R is a Cantor space.*

Proof. By Proposition 1.5.4 (5), the system $(X/R, \tilde{f})$ is transitive. If X/R has an isolated point then Proposition 1.3.4 implies that we have the situation described in (1). On the other hand, if X/R has no isolated points then we are in the situation of (2) above. In that case, if X is metrizable then it follows from the theorem in Appendix A.7.8 that X/R is metrizable as well. So by Brouwer's Theorem (see Appendix B.) X/R now is a Cantor space. □

Remark. If case (1) applies, and $X_0, \dots, X_{n-1}$ are the components of X, then the numbering of these sets can be chosen such that f maps X_i into $X_{i+1 \pmod n}$. We say in this case that *f permutes the components of X cyclically*. In this situation f maps the set X_i *onto* $X_{i+1 \pmod n}$: if $X_{i+1 \pmod n} \setminus f[X_i] \neq \emptyset$ for some i then none of its points can be image under f of any point of X, contradicting the fact that $f[X] = X$ by transitivity. Consequently, *if one of the components X_i consists of a single point then every component consists of a single point, in which case X is a single periodic orbit.*

By applying the above theorem to the subsystem on a compact transitive subset A of a dynamical system on a Hausdorff phase space we get:

Corollary 1.5.10. *Let A be a compact transitive subset of a dynamical system (X, f). Then there are the following, mutually exclusive, possibilities:*
(1a) *A consists of a single periodic orbit.*
(1b) *A is the union of finitely many*[2] *components, each of which has at least two points, and these components are cyclically permuted by f.*
(2) *A/R is an infinite 0-dimensional compact Hausdorff space, transitive under $\tilde{f}$ and without isolated points.*

Proof. By the Remark above, case (1) of Theorem 1.5.9 – applied to the subsystem (A, f) – splits in the two possibilities (1a) and (1b). □

For an application of this result to interval mappings we need a lemma which has some interest in its own.

Lemma 1.5.11. *Let X be a locally connected Hausdorff space and let A be a compact transitive subset of X. If A has non-empty interior A° then A has a finite number of components that are cyclically permuted by f.*

Proof. Let notation be as above. We want to show that A/R is a periodic orbit. As the set of transitive points is dense in A, there is a transitive point $x \in A^\circ$. Let U be a connected neighbourhood of x included in A°. Then $U \subseteq R[x]$, the connected component of the point x in A. Since the point x is transitive, there exists an $n \geq 1$ such that $f^n(x) \in U$. Consequently, $f^n[R[x]] \cap R[x] \neq \emptyset$, hence the set $f^n[R[x]] \cup R[x]$ is con-

2 Possibly, just one (namely, if X is connected). Example: f is the tent map on $[0; 1]$.

nected and therefore $f^n[R[x]] \subseteq R[x]$. By the definition of $\tilde{f}$ in 1.5.8, this implies that $\tilde{f}^n(q(x)) = q(x)$, that is, $q(x)$ is a periodic point in A/R.

Since x is a transitive point in A, it follows that $q(x)$ is a transitive point in A/R. Since it is also a periodic point, it follows that A/R consists of a single periodic orbit. □

Remark. If A is locally connected then the components of A are open, so by compactness A has only finitely many components, and Corollary 1.5.10 (1) applies. In the lemma above, local connectedness of A is replaced by local connectedness of the ambient space X and the condition that $A^\circ \neq \emptyset$.

Corollary 1.5.12. *Let (X, f) be a dynamical system on an interval in $\mathbb{R}$ and let A be a compact transitive subset of X. Then there are the following, mutually exclusive, possibilities:*
(1a) *A is a finite periodic orbit.*
(1b) *A is the union of finitely many closed non-degenerate intervals that are cyclically permuted by f.*
(2) *A is a Cantor space.*

Proof. Obviously, we can apply Corollary 1.5.10. Taking into account that the components of A are compact intervals in $\mathbb{R}$, it is easy to see that case (1b) of Corollary 1.5.10 implies the present possibility (1b).

Now assume that A/R is infinite, that is, case (2) of Corollary 1.5.10 applies. Then Lemma 1.5.11 implies that A has empty interior, so every component of A is a degenerate interval, i.e., a point. Therefore, the quotient mapping $q: A \to A/R$ is injective, hence a homeomorphism (recall that A is compact and that A/R is a Hausdorff space). Consequently, A is a compact 0-dimensional space without isolated points; moreover A, being a subset of $\mathbb{R}$, is a metric space. So by definition, it is a Cantor space. □

In Section 3.5 ahead more will be said about the component space of a compact transitive set. There we shall also present examples, showing that all possibilities mentioned in Theorem 1.5.9 – and Corollary 1.5.12 – actually occur.

1.6 Equicontinuity and weak mixing

At first sight, these notions have nothing to do with each other. But we shall see that (the absence of) equicontinuity can be used to characterize weak mixing, at least for compact minimal systems.

A dynamical system (X, f) is said to be *weakly mixing* whenever the product of the system with itself, i.e., the system $(X \times X, f \times f)$, is topologically ergodic. Here $(f \times f)(x_1, x_2) := (f(x_1), f(x_2))$ for $x_1, x_2 \in X$. Obviously, the system (X, f) is weakly mixing iff for every choice of two non-empty basic open sets $U_1 \times U_2$ and $V_1 \times V_2$ in $X \times X$ (where U_i and V_i are non-empty open subsets of X for $i = 1, 2$) there exists $n \in \mathbb{Z}^+$ such that $(f \times f)^n[U_1 \times U_2] \cap (V_1 \times V_2) \neq \emptyset$, that is, $f^n[U_1] \cap V_1 \neq \emptyset$ and $f^n[U_2] \cap V_2 \neq \emptyset$ (both for the same value of n). In that case, by Exercise 1.6 (6) , there infinitely many of such values of n.

Using the notion of dwelling sets, this definition can be reformulated as follows: the system (X, f) is weakly mixing iff $D(U_1, V_1) \cap D(U_2, V_2) \neq \emptyset$ for every choice of four non-empty open subsets U_1, U_2, V_1 and V_2 of X.

It is obvious that a trivial system (i.e., the phase space consists of a single point) is weakly mixing. However, it is easily seen that *a non-trivial weakly mixing system* (X, f) *can have no isolated points*: if x_0 is an isolated point then take in the definition above $U_1 = U_2 = V_1 = \{x_0\}$ and $V_2 := X \setminus \{x_0\}$.

A stronger notion is the following: a system (X, f) is said to be *strongly mixing* whenever $f^n[U] \cap V \neq \emptyset$ for almost all $n \in \mathbb{Z}^+$ for every choice of two non-empty open sets U and V in X. The following implications follow easily from the definitions:

$$\text{strongly mixing} \Rightarrow \text{weakly mixing} \Rightarrow \text{topologically ergodic.}$$

For the first implication, note that the intersection of two cofinal subsets of $\mathbb{Z}^+$ (i.e., sets with a finite complement) is cofinal in $\mathbb{Z}^+$, hence not empty. For the second implication, take $U_1 = U_2$ and $V_1 = V_2$ in the above. The examples (2) and 3 below show that, in general, none of these implications can be reversed.

Examples.

(1) In the Examples (1) and (2) after Theorem 1.3.5 we are dealing with systems (X, f) such that for every non-empty open subset of X one has $f^n[U] = X$ for almost all $n \in \mathbb{Z}^+$. So the systems under consideration – the tent map and the argument-doubling transformation – are strongly (hence weakly) mixing.

(2) In Exercise 5.13 (2) below is an example of a weakly but not strongly mixing system.

(3) A system consisting of one single periodic orbit with more than one point is topologically ergodic but not weakly mixing. Also, every rigid rotation $(\mathbb{S}, \varphi_a)$ with $a \in \mathbb{R} \setminus \mathbb{Q}$ is (minimal, hence) topologically ergodic and not weakly mixing: apply the definition with $U_1 = U_2$ a small arc and V_1 and V_2 small arcs that are far apart from each other.

Much of the remainder of this section is devoted to results that are needed for the construction of an example of a minimal weakly mixing system on a compact space. In Chapter 5 we shall present an example that does not need this construction, so the reader who is not interested in the particular construction presented here may skip most of this section. But the theory in this section touches many important topics that will not be discussed further on in this book, so the reader is advised to at least skim through the present section.

Recall that a *filter base* in a set X is a collection $\mathcal{B}$ of non-empty subsets of X with the property that the intersection of any two members of $\mathcal{B}$ includes a member of $\mathcal{B}$. By induction, it follows that the intersection of finitely many members of $\mathcal{B}$ includes a member of $\mathcal{B}$, hence is not empty.

Lemma 1.6.1. *A dynamical system (X, f) is weakly mixing iff the collection of all dwelling sets $D(U, V)$ with U and V non-empty open subsets of X is a filter base. Consequently, the intersection of a finite number of such dwelling sets includes a dwelling set, hence is not empty.*

Proof. "If": Clear from the definition.

"Only if": Let U_i and V_i $(i = 1, 2)$ be non-empty open subsets of X. Assuming that the system under consideration is weakly mixing, it is possible to select $k \in D(U_1, U_2) \cap D(V_1, V_2)$. Then $U := U_1 \cap (f^k)^{\leftarrow}[U_2]$ and $V := V_1 \cap (f^k)^{\leftarrow}[V_2]$ are non-empty open sets. Now it is sufficient to show that

$$D(U, V) \subseteq D(U_1, V_1) \cap D(U_2, V_2) .$$

If $n \in D(U, V)$ then the left-hand side of the equality

$$(U_1 \cap (f^k)^{\leftarrow}[U_2]) \cap (f^n)^{\leftarrow}[V_1 \cap (f^k)^{\leftarrow}[V_2]] = \\ (U_1 \cap (f^n)^{\leftarrow}[V_1]) \cap (f^k)^{\leftarrow}[U_2 \cap (f^n)^{\leftarrow}[V_2]]$$

is not empty, hence for $i = 1, 2$ the set $U_i \cap (f^n)^{\leftarrow}[V_i]$ is not empty (in the case that $i = 2$, note that if the pre-image of a set under f^k is not empty then the set itself is not empty), that is $n \in D(U_1, V_1) \cap D(U_2, V_2)$. □

Proposition 1.6.2. *If the system (X, f) is weakly mixing then for every choice of non-empty open subsets U and V of X the set $D(U, V)$ includes arbitrarily long segments of consecutive elements of $\mathbb{Z}^+$.*

Proof. Let U and V be non-empty open subsets of X and let $m \in \mathbb{N}$. By Lemma 1.6.1, the intersection of the sets $D(U, (f^i)^{\leftarrow}[V])$ for $i = 0, \dots m$ is not empty. If k is in this intersection then $k + i \in D(U, V)$ for $i = 0, \dots m$. □

A subset of $\mathbb{Z}$ that contains arbitrarily large segments of consecutive integers is called a *thick* subset of $\mathbb{Z}$. Thus, in a weakly mixing system every set of the form $D(U, V)$ for non-empty open subsets U and V of the phase space is thick. In a strongly mixing system every such a set is cofinal in $\mathbb{Z}^+$.

Corollary 1.6.3. *The product of a weakly mixing system with every compact minimal system is topologically ergodic.*

Proof. Let the system (X, f) be weakly mixing and let (Y, g) be an arbitrary minimal system with a compact phase space Y. Consider any pair of non-empty open subsets U_1 and V_1 of X and any pair of non-empty open subsets U and V of Y. We have to show that the set

$$D_{f\times g}(U_1 \times U, V_1 \times V) = D_f(U_1, V_1) \cap D_g(U, V)$$

is not empty.

As every point of Y has a dense orbit it is clear that $\bigcup_{i\in\mathbb{Z}^+}(g^i)^{\leftarrow}[V] = Y$. Each of the sets $(g^i)^{\leftarrow}[V]$ is open in Y and Y is compact, hence there exists $N \in \mathbb{N}$ such that

$Y = \bigcup_{i=0}^{N}(g^i)^{\leftarrow}[V]$. Fix an arbitrary point $y_0 \in U$. Then for every $n \in \mathbb{Z}^+$ there exists $i \in \{0, \dots, N\}$ such that $g^n(y_0) \in (g^i)^{\leftarrow}[V]$, that is, $n + i \in D_g(y_0, V) \subseteq D_g(U, V)$. This implies that every segment of N consecutive integers in $\mathbb{Z}^+$ meets $D_g(U, V)$.

This property is often expressed by saying that the sets $D_g(y_0, V)$ and $D_g(U, V)$ have 'bounded gaps' (bounded by N in this case). See also Theorem 4.2.2 (2) below.

Since the set $D_f(U_1, V_1)$ is thick, it is obvious that it has a non-empty intersection with the set $D_g(U, V)$. This completes the proof. □

There is a class of compact dynamical systems, called *scattering systems*, which are characterized by the property that their product with every compact minimal system is topologically ergodic. Every scattering system is topologically ergodic: it is conjugate to the product of itself with the trivial minimal system consisting of one invariant point. By the above corollary, *every compact weakly mixing system is scattering*. The converse is not generally true: in E. Akin & S. Glasner [2001] it is shown that there exists a non-weakly mixing system that is scattering; an explicit example is in W. Huang & X. Ye [2002b]. However, *a minimal scattering system with a compact phase space is weakly mixing*: by the characterization of scattering systems, the product of such a system with itself is topologically ergodic.

Proposition 1.6.4. *A factor of a weakly or strongly mixing system is weakly or strongly mixing, respectively.*

Proof. Use that if $\varphi: (X, f) \to (Y, g)$ is a factor mapping then for non-empty open subsets U and V of Y the sets $\varphi^{\leftarrow}[U]$ and $\varphi^{\leftarrow}[V]$ are non-empty and open in X and $D_f(\varphi^{\leftarrow}[U], \varphi^{\leftarrow}[V] \subseteq D_g(U, V)$. (Alternatively, for the weakly mixing case use Proposition 1.5.4 (7), taking into account that$(Y \times Y, g \times g)$ is a factor of $(X \times X, f \times f)$ by $\varphi \times \varphi$.) □

Let (X, f) be a dynamical system and assume that X is a metric space[3] with metric d. The system (X, f) is said to be *equicontinuous* whenever the collection of mappings $\{ f^n : n \in \mathbb{Z}^+ \}$ is equicontinuous on X (i.e., at every point of X): for every point $x \in X$ and for every $\varepsilon > 0$ there is a neighbourhood U of x in X such that

$$d(f^n(x), f^n(y)) < \varepsilon \quad \text{for all } y \in U \text{ and all } n \in \mathbb{Z}^+, \tag{1.6-1}$$

that is, $f^n[U] \subseteq B_\varepsilon(f^n(x))$ for all $n \in \mathbb{Z}^+$. In (1.6-1) it may be assumed that $U = B_\delta(x)$ for some $\delta > 0$. So an equivalent formulation is: for every point $x \in X$ and for every $\varepsilon > 0$ there exists $\delta > 0$ such that

$$f^n[B_\delta(x)] \subseteq B_\varepsilon(f^n(x)) \quad \text{for all } n \in \mathbb{Z}^+. \tag{1.6-1*}$$

Usually the δ here depends not only on ε but also on the choice of the point x. If for every $\varepsilon > 0$ the δ can be chosen independently of the choice of x then the system is

3 Most of the following can be done in arbitrary compact Hausdorff spaces, using their a unique uniform structures.

called *uniformly equicontinuous*. Thus, the system is uniformly equicontinuous iff for every $\varepsilon > 0$ there exists $\delta > 0$ such that for all $x, y \in X$ with $d(x, y) < \delta$ one has $d(f^n(x), f^n(y)) < \varepsilon$ for all $n \in \mathbb{Z}^+$. If for every $\varepsilon > 0$ we define a subset B_ε of $X \times X$ by

$$B_\varepsilon := \{(x, y) \in X \times X : d(x, y) < \varepsilon\}$$

then the definition of uniform equicontinuity can be reformulated as

$$\forall \varepsilon > 0 \exists \delta > 0 : (f \times f)^n[B_\delta] \subseteq B_\varepsilon \quad \text{for all } n \in \mathbb{Z}^+ . \tag{1.6-2}$$

The following is well-known: *If the system (X, f) with metric phase space X is equicontinuous and X is a compact then the system is uniformly equicontinuous.* See Appendix A.7.2. In the sequel, if the discussion is about an equicontinuous system on a compact metric space then we shall use without further reference that the system is uniformly equicontinuous.

Examples.
(1) Every system consisting of a single periodic orbit is equicontinuous with respect to the metric in which two different points have distance 1.
(2) For every $a \in \mathbb{R}$ the rigid rotation $(\mathbb{S}, \varphi_a)$ is equicontinuous.
(3) The system defined by the tent map is not equicontinuous. In point of fact, the set $\{T^n : n \in \mathbb{Z}^+\}$ is not equicontinuous at 0, because for every $n \in \mathbb{Z}^+$ there is a right neighbourhood of 0 on which T^n has direction coefficient 2^n. By a similar reasoning, the system defined by the argument-doubling transformation is not equicontinuous .

In view of Example (1) at the beginning of this section, Example (3) above is an illustration of:

Proposition 1.6.5. *A non-trivial equicontinuous system on a metric space is not weakly mixing.*

Proof. Let (X, f) be an equicontinuous system and assume that X has at least three[4] points x, y_1 and y_2 in X such that $y_1 \neq y_2$, Choose $\varepsilon > 0$ such that $d(y_1, y_2) = 5\varepsilon$ and select $\delta > 0$ according to (1.6-1*) at the point x. Let $U_1 = U_2 := B_\delta(x)$ and for $i = 1, 2$, let $V_i := B_\varepsilon(y_i)$. If $n \in D(U_1, V_1)$ then the set $f^n[U_1]$ meets the set $V_1 = B_\varepsilon(y_1)$. As the choice of δ implies that the diameter of the set $f^n[U_1]$ is at most 2ε it follows that $f^n[U_1] \subseteq B_{3\varepsilon}(y_1)$. But then the set $f^n[U_2] = f^n[U_1]$ cannot meet the set $V_2 = B_\varepsilon(y_2)$, for otherwise by the triangle inequality the distance of y_1 and y_2 would be less than $3\varepsilon + \varepsilon = 4\varepsilon$. So $D(U_1, V_1) \cap D(U_2, V_2) = \emptyset$.

Alternative proof: It is straightforward to show that, if the system (X, f) is equicontinuous then the system $(X \times X, f \times f)$ is equicontinuous as well. If it is also topologi-

4 Two would be sufficient, as there is no reason why x and y_1 should be different, except for clarity of the presentation of the proof.

cally ergodic then, by Exercise 1.5 (4) , it is minimal. As Δ_X is a closed invariant subset of $X \times X$ this implies that $\Delta_X = X$, so X consists of one point. □

Corollary 1.6.6. *A weakly mixing system has no non-trivial equicontinuous factor with a metric phase space.*

Proof. Clear from the above proposition and Proposition 1.6.4. □

Remark. In the above corollary the weakly mixing system itself need not be on a metric space. In the following theorem it is not necessary to assume that X is a metric space either:

Theorem 1.6.7. *Let (X, f) be a minimal dynamical system with X a compact space. The following statements are equivalent:*
(i) *(X, f) is weakly mixing.*
(ii) *(X, f) has no non-trivial equicontinuous factor with a metric phase space.*

Proof. In view of Corollary 1.6.6, only the implication (ii)⇒(i) needs yet a proof. Much of the proof falls outside the scope of this book. One of the ingredients needed is that the system admits an *invariant regular probability measure* μ. By minimality of the system such a measure has full support, i.e., every non-empty open set has positive measure.

Another component of the proof is the notion of a *continuous invariant pseudo-metric*: a continuous mapping $\rho\colon X \times X \to \mathbb{R}^+$ satisfying the conditions for a metric except the condition that $\rho(x, y) = 0$ implies $x = y$; that it is invariant means that $\rho(f(x), f(y)) = \rho(x, y)$ for all $x, y \in X$. If ρ is any continuous invariant pseudo-metric then by identifying points that have distance 0 under ρ one gets a factor of the system (X, f) on which ρ induces an invariant metric (i.e., the phase mapping is an isometry): an equicontinuous factor. By the assumption of (ii) this factor is trivial, which implies that $\rho(x, y) = 0$ for all $x, y \in X$.

Now the trick is to use the invariant measure μ to define for every non-empty closed $(f \times f)$-invariant subset N of $X \times X$ a continuous invariant pseudo-metric ρ_N, in the following way: first, define for every $x \in X$ the 'x-section' of N by $N(x) := \{z \in X : (x, z) \in N\}$, then note that $N(x)$ is a closed set and, finally, put $\rho_N(x, y) := \mu(N(x) \Delta N(y))$ for all $x, y \in X$, where Δ denotes the symmetric difference. The proof that ρ_N is an invariant pseudo-metric is straightforward (but tedious); that it is continuous is less straightforward and depends on the fact that the measure μ is regular. So by what was observed above, $\rho_N(x, y) = 0$ for all $x, y \in X$. It is easy to show that for every non-empty open subset U of X one has

$$\forall\, x, y \in X : \ U \subseteq N(x) \Leftrightarrow U \subseteq N(y)\,, \tag{1.6-3}$$

as follows: Let U be a non-empty open set such that $U \subseteq N(x)$. Then

$$\mu(U \setminus N(y)) \leq \mu(N(x) \setminus N(y)) \leq \mu(N(x) \Delta N(y)) = \rho_N(x, y)\,.$$

Here $U \setminus N(y)$ is an open set, so $\mu(U \setminus N(y)) > 0$ if it is not empty. Consequently, as $\rho_N(x, y) = 0$ it follows that $U \subseteq N(y)$. As x and y appear symmetrically in (1.6-3), the implication the other way round follows immediately.

Now let U_1, U_2, V_1 and V_2 be non-empty open subsets of X. We want to show that there exists $j \in \mathbb{Z}^+$ such that $(f \times f)^j[U_1 \times U_2] \cap (V_1 \times V_2) \neq \emptyset$. First, note that by minimality of (X, f) (or, rather, its ergodicity), there is an $n \in \mathbb{Z}^+$ such that $W := U_2 \cap (f^n)^{\leftarrow}[V_2] \neq \emptyset$. Now let $x \in U_1$. Then

$$\{x\} \times W \subseteq U_1 \times U_2 \subseteq \mathrm{cl}_{X \times X}\Big(\bigcup_{i=0}^{\infty} f^i[U_1 \times U_2]\Big) =: N\,.$$

The set N so defined is a non-empty closed invariant subset of $X \times X$, and, by definition, $W \subseteq N(x)$. Hence by (1.6-3), $W \subseteq N(y)$, that is, $\{y\} \times W \subseteq N$, for every $y \in X$. If we take $y \in (f^n)^{\leftarrow}[V_1]$ then we see that the set $(f^n)^{\leftarrow}[V_1] \times W$ has a non-empty intersection with N. As this set is open it follows that it has a non-empty intersection with the set $\bigcup_{i=0}^{\infty} f^i[U_1 \times U_2]$ of which N is the closure. Consequently, there exists $i \in \mathbb{Z}^+$ such that $(f^n)^{\leftarrow}[V_1] \cap f^i[U_1] \neq \emptyset$ and $(U_2 \cap (f^n)^{\leftarrow}[V_2]) \cap f^i[U_2] \neq \emptyset$, hence $(f^n)^{\leftarrow}[V_2] \cap f^i[U_2] \neq \emptyset$. It easily follows that $V_1 \cap f^{n+i}[U_1] \neq \emptyset$ and $V_2 \cap f^{n+i}[U_2] \neq \emptyset$ (even though f may be not injective; one doesn't even need that f is – by minimality – surjective). This completes the proof. □

Lemma 1.6.8. *Let (X, f) be an arbitrary dynamical system. Then the closure $E(X, f)$ of the set $\{f^n : n \in \mathbb{Z}^+\}$ in X^X is a subsemigroup of the semigroup X^X with composition of mappings as a semigroup operation.*

Proof. We have to show that if $\xi, \eta \in E(X, f)$ then also $\xi \circ \eta \in E(X, f)$. Let $F := \{f^n : n \in \mathbb{Z}^+\}$, so that $E(X, f)$ is the closure of F in X^X; needless to say, the topology in X^X is the product topology, i.e., the topology of pointwise convergence; see Appendix A.10.2 (1).

First, observe that for every $n \in \mathbb{Z}^+$ the mapping $\xi \mapsto f^n \circ \xi : X^X \to X^X$ is continuous, because for every point $x \in X$ its composition with the evaluation mapping at the point x is continuous: if $x \in X$ then $\xi \mapsto \xi(x) : X^X \to X$ is continuous, and since f^n is continuous, $\xi \mapsto f^n(\xi(x)) : X^X \to X$ is continuous as well. Obviously, the mapping $\xi \mapsto f^n \circ \xi$ maps F into F hence, by continuity, it maps the closure of F into the closure of F, i.e., it maps the set $E(X, f)$ into itself.

Stated otherwise, for every $\xi \in X^X$ the mapping $\eta \mapsto \eta \circ \xi$ maps F into $E(X, f)$. This mapping is continuous (again, its composition with every evaluation mapping is continuous), hence it maps the closure $E(X, f)$ of F into $E(X, f)$ as well. This completes the proof. □

The semigroup $E(X, f)$ defined above is called the *enveloping semigroup* of the dynamical system (X, f) (also: the *Ellis*[5] *semigroup* of the system). Note that, by Tychonov's

5 After Robert Ellis, who introduced enveloping semigroups in order to study the structure of compact minimal dynamical systems. See R. Ellis [1969].

Theorem, if X is compact then X^X is compact. In that case the enveloping semigroup $E(X, f)$, being a closed subset of X^X, is compact as well.

In the proof of the above lemma it is shown that all right translations $\rho_\xi : \eta \mapsto \eta \circ \xi : E(X, f) \to E(X, f)$ for $\xi \in E(X, f)$ are continuous and that all left translations $\lambda_\eta : \xi \mapsto \eta \circ \xi : E(X, f) \to E(X, f)$ for $\eta \in F$ are continuous. In point of fact, the same proof shows that λ_η is continuous for every *continuous* $\eta \in E(X, f)$.

The following result shows that, if (X, f) is equicontinuous and X is compact, then $E(X, f)$ is actually a compact Abelian topological group of homeomorphisms of X onto itself, homeomorphic with X: the proof, though quite lengthy, consists of a number of straightforward simple steps.

Theorem 1.6.9. *Let (X, f) be an equicontinuous minimal system with compact metric phase space X. Then $E(X, f)$ is a compact Abelian topological group of homeomorphisms of X, and for every $x \in X$ the evaluation mapping $\delta_x : E(X, f) \to X$ is a homeomorphism.*

Proof. By Tychonov's Theorem, X^X is compact; hence $E(X, f)$ is compact as well.

In order to show that $E(X, f)$ is a sub*group* of X^X it is, obviously, sufficient to show that $\mathrm{id}_X \in E(X, f)$ and that every $\eta \in E(X, f)$ is a homeomorphism whose inverse belongs to $E(X, f)$. It will be convenient to first show that the semigroup $E(X, f)$ is commutative, that is: $\xi \circ \eta = \eta \circ \xi$ for all $\xi, \eta \in E(X, f)$.

As in the proof of Lemma 1.6.8, let $F := \{f^n : n \in \mathbb{Z}^+\}$. Since F is equicontinuous, a standard result from topology implies that the set $E(X, f)$ is equicontinuous as well: see Appendix A.10.2 (3); in particular, all members of $E(X, f)$ are continuous mappings from X into itself. Hence not only all right translations $\rho_\xi : \eta \mapsto \eta \circ \xi : E(X, f) \to E(X, f)$ for $\xi \in E(X, f)$ are continuous, but all left translations $\lambda_\eta : \xi \mapsto \eta \circ \xi : E(X, f) \to E(X, f)$ for $\eta \in E(X, f)$ are continuous as well.

Now, if $\eta \in F$ then, obviously, the continuous mappings ρ_η and λ_η are equal to each other on the dense subset F of $E(X, f)$ (members if F commute), hence $\rho_\eta = \lambda_\eta$ on $E(X, f)$. Thus, $\xi \circ \eta = \eta \circ \xi$ for all $\eta \in F$ and $\xi \in E(X, f)$. This means that for every $\xi \in E(X, f)$ the continuous mappings λ_ξ and ρ_ξ agree on F. Consequently, they agree on $E(X, f)$. This completes the proof that $E(X, f)$ is commutative.

Next, note that for every point $x \in X$ the (continuous) evaluation mapping $\delta_x : \xi \mapsto \xi(x) : E(X, f) \to X$ maps the dense subset F of $E(X, f)$ onto the orbit of x under f, which is dense in X. Since $E(X, f)$ is compact, it follows that δ_x maps $E(X, f)$ onto all of X. Stated otherwise, for every $x \in X$ one has $\{\xi(x) : \xi \in E(X, f)\} = X$. Now fix a point $x_0 \in X$ and let $\eta \in E(X, f)$. The above observation can be applied to the point $x := \eta(x_0)$ to the effect that $\{\xi(\eta(x_0)) : \xi \in E(X, f)\} = X$. In particular, there exists $\xi \in E(X, f)$ such that $(\xi \circ \eta)(x_0) = x_0$. As $E(X, f)$ is commutative, it follows that

$$(\xi \circ \eta)(\zeta(x_0)) = (\xi \circ \eta \circ \zeta)(x_0) = (\zeta \circ \xi \circ \eta)(x_0) = \zeta(x_0)$$

for every $\zeta \in E(X, f)$. However, $\{\zeta(x_0) : \zeta \in E(X, f)\} = X$ so it follows that $\xi \circ \eta = \mathrm{id}_X$ and, by commutativity of $E(X, f)$, $\eta \circ \xi = \mathrm{id}_X$. Consequently, η is a continuous bijection, hence a homeomorphism (recall that X is a compact Hausdorff space), with inverse $\eta^{-1} := \xi \in E(X,f)$. Note that the fact that $E(X,f)$ is a semigroup and the fact that $\xi \circ \eta = \mathrm{id}_X$ together imply that $\mathrm{id}_X \in E(X,f)$. This completes the proof that $E(X,f)$ is a group.

For later reference, note that for every $\eta \in E(X,f)$ the continuous mappings ρ_η and $\rho_{\eta^{-1}}$ of $E(X,f)$ into itself are inverse to each other; consequently, ρ_η is a homeomorphism of $E(X,f)$ onto itself.

In order to show that δ_x is a homeomorphism for any $x \in X$ it remains to show that δ_x is injective, for $E(X,f)$ is compact, X is a Hausdorff space and δ_x is surjective. The proof is as follows: Let $\xi, \eta \in E(X,f)$ and assume that $\xi(x) = \eta(x)$ for some $x \in X$. Then for any $\zeta \in E(X,f)$ we have

$$\xi(\zeta(x)) = \zeta(\xi(x)) = \zeta(\eta(x)) = \eta(\zeta(x)),$$

because $E(X,f)$ is commutative. As $\{\zeta(x) : \zeta \in E(X,f)\} = X$ it follows that $\xi = \eta$. This completes the proof that δ_x is a homeomorphism of $E(X,f)$ onto X.

Finally, we show that $E(X,f)$ is a topological group. By the definition of a topological group, this means that we have to prove that the multiplication mapping $\mu: (\xi,\eta) \mapsto \xi \circ \eta: E(X,f) \times E(X,f) \to E(X,f)$ and the mapping $\iota: \xi \mapsto \xi^{-1}: E(X,f) \to E(X,f)$ are continuous. To this end, recall that $E(X,f)$, being the pointwise closure of an equicontinuous set, is equicontinuous on X. Consequently, the pointwise topology (in which $E(X,f)$ is compact) and the topology of uniform convergence coincide on $E(X,f)$; see Appendix A.10.2 (3). We shall show now that the above mappings are continuous with respect to the latter topology.

In order to show that μ is continuous on $E(X,f) \times E(X,f)$, consider any point $(\xi_0, \eta_0) \in E(X,f) \times E(X,f)$. Then for every $(\xi,\eta) \in E(X,f) \times E(X,f)$ and every point $x \in X$ we have

$$d((\xi \circ \eta)(x), (\xi_0 \circ \eta_0)(x)) \le d(\xi(\eta(x)), \xi(\eta_0(x))) + d(\xi(\eta_0(x)), \xi_0(\eta_0(x))).$$

If $\varepsilon > 0$ then, by (uniform) equicontinuity of $E(X,f)$, there exists $\delta > 0$ such that $d(\xi(y), \xi(y')) < \varepsilon$ for all $y, y' \in X$ with $d(y,y') < \delta$ and all $\xi \in E(X,f)$. So if $d(\eta(x), \eta_0(x)) < \delta$ for all $x \in X$ – this specifies a neighbourhood of η_0 with respect to the uniform topology – then the first term in the right-hand side of the above inequality is at most ε for all $x \in X$. Moreover, the condition that the second term of the right-hand side of this inequality is at most ε for all $x \in X$ specifies a neighbourhood of ξ_0 with respect to the uniform topology (recall that η_0 is a homeomorphism). Consequently, for all (ξ,η) in a neighbourhood of the point (ξ_0,η_0) in $E(X,f) \times E(X,f)$ the left-hand side of the above inequality is less than 2ε. This completes the proof that μ is continuous at the point (ξ_0,η_0) of $E(X,f) \times E(X,f)$.

Continuity of the mapping $\iota: E(X,f) \to E(X,f)$ is shown in two steps. First, if $\xi \in E(X,f)$ and $d(x, \xi(x)) < \varepsilon$ for all $x \in X$ – this specifies a neighbourhood of id_X in $E(X,f)$ with respect to the uniform topology – then, by replacing x by $\xi(y)$ for arbitrary

$y \in X$, we get $d(\xi^{-1}(y), y) < \varepsilon$ for all $y \in X$. Stated otherwise, ι maps a neighbourhood of id_X into itself. This shows that ι is continuous at the point id_X of $E(X, f)$. Then, by using that each right translation ρ_{ξ_0} is a homeomorphism of $E(X, f)$ onto itself, it is easily shown that ι is continuous at every point of $E(X, f)$ (this is standard topological group theory). □

Remark. In the situation of the theorem, the phase mapping f – being a member of $E(X, f)$ – is a homeomorphism. This consequence can also be proved by other means; see 7.1.11 ahead.

We have seen above that if (X, f) is a dynamical system then the mapping

$$\lambda_f : \xi \mapsto f \circ \xi : E(X, f) \to E(X, f)$$

is continuous. Thus, we have a 'natural' dynamical system $(E(X, f), \lambda_f)$. This system has a dense orbit: by definition, the orbit of id_X under λ_f is dense in $E(X, f)$ (it is the set F in the proof of Lemma 1.6.8). It is straightforward to check that, for every point $x \in X$, the continuous evaluation mapping δ_x commutes with λ_f and f:

$$(\delta_x \circ \lambda_f)(\xi) = (\lambda_f \xi)(x) = (f \circ \xi)(x) = f(\xi(x)) = (f \circ \delta_x)(\xi)$$

for every $\xi \in E(X, f)$. Thus, $\delta_x : (E(X, f), \lambda_f) \to (X, f)$ is a morphism of dynamical systems. In particular, if (X, f) is compact and x is a transitive point (e.g., because (X, f) is minimal) then δ_x is a factor mapping (recall that $E(X, f)$ is compact and is mapped onto the orbit closure of x by δ_x).

Corollary 1.6.10. *Let (X, f) be an equicontinuous minimal system with compact phase metric space X. Then X has the structure of a compact Abelian topological group such that f is the left (= right) translation over one of its elements.*

Proof. Clear from the above observation and Theorem 1.6.9. □

By $\mathbb{T}$ we denote the unit circle of the complex plane, considered as a group with multiplication of complex numbers as group operation.

Theorem 1.6.11. *Let (X, f) be a non-trivial equicontinuous minimal system with a compact metric phase space. Then there are a non-constant continuous function $\chi : X \to \mathbb{T}$ and a complex number $c \in \mathbb{T}$, $c \neq 1$, such that $\chi(f(x)) = c\,\chi(x)$ for all $x \in X$.*

Proof. By Corollary 1.6.10 we may assume that X is a compact Abelian topological group, say, with multiplication $(x, y) \mapsto x \cdot y : X \times X \to X$, such that for some $a \in X$ we have $f(x) = a \cdot x$ for all $x \in X$. As f has a dense orbit, a cannot be the unit element of the group X (otherwise we would have $f = \mathrm{id}_X$).

A fundamental result from Harmonic Analysis states that the points of a compact Abelian topological group are separated by the continuous group homomorphisms – also called characters – from X to $\mathbb{T}$. So there is a character $\chi : X \to \mathbb{T}$ such that

$\chi(a) \neq 1$. As χ is a group homomorphism, it is clear that $\chi(f(x)) = \chi(a \cdot x) = \chi(a)\chi(x)$ for all $x \in X$. This proves the theorem with $c := \chi(a)$. □

The result from Harmonic Analysis mentioned in the proof is the basic step in the proof of the Pontryagin Duality Theorem. Let G be a (locally) compact Abelian group and let $G^\frown$ denote the set of all characters of G. If $\delta_g : G^\frown \to \mathbb{T}$ is the evaluation mapping ($g \in G$), then the mapping $\underline{\delta} : g \mapsto \delta_g : G \to C(G^\frown, \mathbb{T})$ turns out to be injective: this is precisely the statement that $G^\frown$ separates the points of G. See Theorem (22.17) in E. Hewitt & K. A. Ross [1963]. The set $G^\frown$ is an Abelian group with pointwise multiplication as a group operation; it is a locally compact topological group when it is given the compact-open topology on G. The Pontryagin Duality Theorem states that $\underline{\delta}$ is a topological group isomorphism (= homeomorphism + isomorphism of groups) of G and $(G^\frown)^\frown$. See Theorem (24.8) in E. Hewitt & K. A. Ross [1963] (where $G^\frown$ is denoted by X and $(G^\frown)^\frown$ by ω).

Let (X, f) be a dynamical system. An *eigenfunction* of the system is a continuous mapping $\chi : X \to \mathbb{T}$ with the property that there exists $c \in \mathbb{T}$ such that $\chi(f(x)) = c\,\chi(x)$ for every $x \in X$; the constant c is called the *eigenvalue* of χ.

If (X, f) is transitive (so in particular, if (X, f) is minimal) and χ is an eigenfunction with eigenvalue c then the condition that χ is constant on X is equivalent to the condition that $c = 1$. For if $c \neq 1$ then, obviously, χ is not constant. Conversely, if $c = 1$ then $\chi(f^n(x)) = \chi(x)$ for all $n \in \mathbb{Z}^+$ and every point $x \in X$. If we select for x a point with a dense orbit and χ is continuous, then we get $\chi(x') = \chi(x)$ for all $x' \in X$. This means that χ is constant.

Theorem 1.6.12. *Let (X, f) be a minimal dynamical system with X a compact metric space. The following statements are equivalent:*
(i) *(X, f) is weakly mixing.*
(ii) *Every eigenfunction of (X, f) is constant.*

Proof. “(i)⇒(ii)”: Let $\chi : X \to \mathbb{T}$ be an eigenfunction of (X, f) with eigenvalue c. It is straightforward to check that the dynamical system $(\mathbb{T}, \lambda_c)$, with $\lambda_c : z \mapsto cz : \mathbb{T} \to \mathbb{T}$, is equicontinuous (in fact, λ_c is an isometry) and that $\chi \circ f = \lambda_c \circ \chi$. Thus, $\chi : (X, f) \to (\mathbb{T}, \lambda_c)$ is a morphism of dynamical systems. If χ is not constant then the subsystem of $(\mathbb{T}, \lambda_c)$ on $\chi[X]$ is a non-trivial equicontinuous system, so in that case (X, f) has a non-trivial equicontinuous factor on a compact metric space. By Corollary 1.6.6, this implies that (X, f) is not weakly mixing.

“(ii)⇒(i)”: Suppose (X, f) is not weakly mixing. Then by Theorem 1.6.7 the system (X, f) has a non-trivial equicontinuous factor $\varphi : (X, f) \to (Y, g)$ with Y a compact metric space. Being a factor of a minimal system, (Y, g) is minimal hence it has, by Theorem 1.6.11, a non-constant eigenfunction χ with eigenvalue $c \neq 1$. Now it is straightforward to check that $\chi \circ \varphi : X \to \mathbb{T}$ is a non-constant eigenfunction of (X, f) with eigenvalue c. □

Remarks.

(1) The implication (i)⇒(ii) is also true if the system under consideration is not minimal.

(2) The system $(\mathbb{T}, \lambda_c)$ introduced in the above proof is conjugate to the system $(\mathbb{S}, \varphi_a)$ with $a \in \mathbb{R}$ such that $c = [a]$ (recall that $[a] := \exp(2\pi a)$.

The implication (ii)⇒(i) in this theorem will enable us to prove that the construction we are about to present produces a weakly mixing minimal system on a compact space.

1.6.13. Let (X, f) be a dynamical system with compact Hausdorff phase space X and assume that $X = X_1 \cup X_2$, where X_1 and X_2 are mutually disjoint non-empty clopen subsets of X (not necessarily invariant under f). Let X_2' be a copy of X_2 and let $\varphi: X_2 \to X_2'$ be a homeomorphism. Finally, let X^* be the disjoint union of X and X_2' and define $f^*: X^* \to X^*$ by

$$f^*(x) := \begin{cases} f(x) & \text{if } x \in X_1 \\ \varphi(x) & \text{if } x \in X_2 \\ f(\varphi^{-1}(x)) & \text{if } x \in X_2' \end{cases}$$

See Figure 1.5. The following can easily be proved by induction: if $x \in X$ and $n \in \mathbb{N}$ then there exists $k \geq n$ such that $(f^*)^k(x) = f^n(x)$.

(1) *X^* is a compact Hausdorff space, $f^*: X^* \to X^*$ is continuous and if the system (X, f) is minimal then the system (X^*, f^*) is minimal as well.*

Proof. As X^* is the disjoint union of two compact Hausdorff spaces it is clear that X^* is a compact Hausdorff space as well. Moreover, f^* is continuous on each of the open(!) parts X_1, X_2 and X_2' of X^*, so f^* is continuous on X^*.

Now assume that (X, f) is minimal. In order to show that X^* is minimal under f^*, consider a point $z \in X^*$ and a non-empty open subset U of X^*. We want to show that there exists $m \in \mathbb{Z}^+$ such that $(f^*)^m(z) \in U$. Since the sets X_1, X_2 and X_2' are open in X^* we may assume that U is included in one of these sets. The point z and the open set U can independently of each other be situated in one of the sets X_1, X_2 or X_2', so we have to distinguish nine cases. We treat one of these cases, leaving the others to the reader. Assume that $z \in X_2'$ and $U \subseteq X_2'$. Then $\varphi^{-1}[U]$ is open in X_2, hence it is open in X, and non-empty. Let $x := f^*(z)$. By minimality of the system (X, f), there exists $n \in \mathbb{Z}^+$ such that $f^n(x) \in \varphi^{-1}[U]$. By the observation immediately after the definition

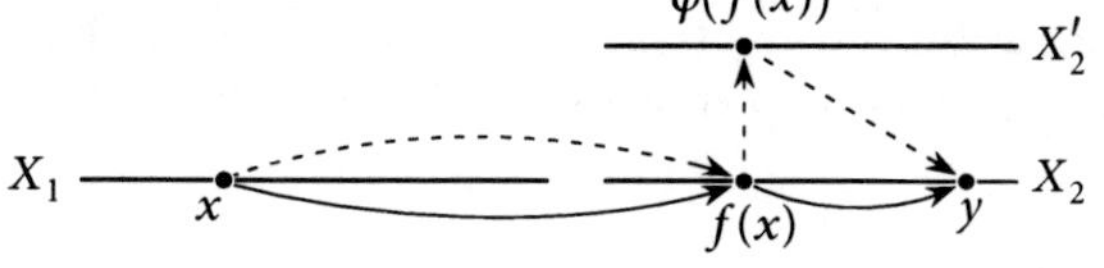

Fig. 1.5. The introduction of a delay: $y = f^2(x) = (f^*)^3(x)$. The solid arrows represent f on X, the dashed arrows represent f^* on X^*.

of f^*, there exists $k \geq n$ such that $(f^*)^k(x) = f^n(x)$. Then $(f^*)^k(x) \in \varphi^{-1}[U]$, hence $(f^*)^{k+2}(z) \in U$. □

(2) *If the system (X^*, f^*) has an eigenfunction with eigenvalue c then there exists a continuous function $\chi\colon X \to \mathbb{T}$ such that*

$$\chi(f(x)) = \begin{cases} c\,\chi(x) & \text{for } x \in X_1\,, \\ c^2\chi(x) & \text{for } x \in X_2\,. \end{cases} \tag{1.6-4}$$

Proof. If $\chi^*: X^* \to \mathbb{T}$ is an eigenfunction with eigenvalue c then it is easily checked that $\chi := \chi^*|_X$ satisfies the above conditions: for $x \in X_1$ this is trivial, and for $x \in X_2$, use that $f(x) = (f^*)^2(x)$. □

Remark. The converse is also true: if condition (1.6-4) holds for the continuous function $\chi\colon X \to \mathbb{T}$ then the function $\chi^*: X^* \to \mathbb{T}$ defined by $\chi^*|_X := \chi|_X$ and $\chi^*|_{X_2'} := c\,(\chi \circ \varphi^{-1})|_{X_2'}$ is an eigenfunction of the system (X^*, f^*) with eigenvalue c.

(3) *Assume that there are points $x_1, x_2 \in X$ such that*
  (a) *$x_1 \in X_1$ and $x_2 \in X_2$,*
  (b) *For every $n \geq 1$ the points $f^n(x_1)$ and $f^n(x_2)$ are both in X_1 or they are both in X_2 and $\overline{\mathcal{O}_{f\times f}(x_1, x_2)} \cap \Delta_X \neq \emptyset$,*
  (d) *The point (x_1, x_2) has a complete past $P = \{(x_{1,n}, x_{2,n}) : n \in \mathbb{N}\}$ under $f \times f$ such that for all $n \in \mathbb{N}$ the points $x_{1,n}$ and $x_{2,n}$ are both in X_1 or in X_2, and $\overline{P} \cap \Delta_X \neq \emptyset$.*

Then every eigenfunction of X^* has eigenvalue 1, hence is constant.

Proof. By 2 above, we have to show that if a function $\chi\colon X \to \mathbb{T}$ has the property expressed by (1.6-4) then $c = 1$. For $i = 1, 2$ and $n \in \mathbb{N}$ we have

$$\frac{\chi(f^n(x_i))}{\chi(x_i)} = \frac{\chi(f^n(x_i))}{\chi(f^{n-1}(x_i))} \cdot \frac{\chi(f^{n-1}(x_i))}{\chi(f^{n-2}(x_i))} \cdots \frac{\chi(f(x_i))}{\chi(x_i)}\,.$$

For $j = 1, \ldots, n-1$ the points $f^j(x_1)$ and $f^j(x_2)$ are both in X_1 or in X_2, so by (1.6-4) the fractions $\chi(f^{j+1}(x_1))/\chi(f^j(x_1))$ and $\chi(f^{j+1}(x_2))/\chi(f^j(x_2))$ are both equal to c or to c^2, hence equal to each other. Moreover, as $x_1 \in X_1$ and $x_2 \in X_2$ it is clear that $\chi(f(x_1))/\chi(x_1) = c$ and $\chi(f(x_2))/\chi(x_2) = c^2$. It follows that, for every $n \in \mathbb{N}$,

$$c = \frac{\chi(f^n(x_2))}{\chi(f^n(x_1))} \cdot \frac{\chi(x_1)}{\chi(x_2)}\,. \tag{1.6-5}$$

Next, let $z \in X$ be such that $(z, z) \in \overline{\mathcal{O}_{f\times f}(x_1, x_2)} \cap \Delta_X$, and for $k \in \mathbb{N}$ let V_k be a neighbourhood of the point z such that $|\chi(x) - \chi(z)| < 1/2k$ for all $x \in V_k$. Then there exists $n_k \in \mathbb{N}$ such that $(f^{n_k}(x_1), f^{n_k}(x_2)) \in V_k \times V_k$. It follows easily that

$$\lim_{k\rightsquigarrow\infty} \frac{\chi(f^{n_k}(x_2))}{\chi(f^{n_k}(x_1))} = 1\,.$$

By taking the limit for $k \rightsquigarrow \infty$ in (1.6-5) with $n = n_k$ we get $\chi(x_1) = c\chi(x_2)$.

By the definition of a complete past, the points of P satisfy the condition that, for $i = 1, 2$ and all $n \geq 2$, $f(x_{i,n}) = x_{i,n-1}$, and that $f(x_{i,1}) = x_i$. Similar to the above computation one finds

$$\frac{\chi(x_i)}{\chi(x_{i,n})} = \frac{\chi(x_i)}{\chi(x_{i,1})} \cdot \frac{\chi(x_{i,1})}{\chi(x_{i,2})} \cdots \frac{\chi(x_{i,n-1})}{\chi(x_{i,n})} .$$

According to assumption (d), the points $x_{1,n}$ and $x_{2,n}$ are for every $n \in \mathbb{N}$ both in X_1 or both in X_2, hence the right-hand side of the above equality does not depend on i. Consequently,

$$\frac{\chi(x_{2,n})}{\chi(x_{1,n})} = \frac{\chi(x_2)}{\chi(x_1)}$$

for all $n \in \mathbb{N}$. As before, one shows that the left-hand side of this equality tends to 1 along a subsequence of $\mathbb{N}$. Hence $\chi(x_1) = \chi(x_2)$. Above, we have seen that $\chi(x_1) = c\chi(x_2)$. It follows that $c = 1$. □

(4) **Conclusion.** *If (X, f) is a compact minimal dynamical system that satisfies the conditions of 3 above then the corresponding system (X^*, f^*) is a compact minimal weakly mixing system.*

Proof. Use Theorem 1.6.12, taking into account 1 and 3 above. □

It remains to provide an example of a dynamical system that satisfies the conditions of 3 above. This is not yet possible: it will come in 6.3.7 (3) below. Another example of a minimal weakly mixing system on a compact space, not using the above construction, will be presented in Theorem 5.6.12.

1.7 Miscellaneous examples

We recapitulate the properties of the four systems described in the Introduction and we discuss some additional examples.

1.7.1 **(The rigid rotation of the circle).** Recall that, for every $a \in \mathbb{R}$ the *rigid rotation of the circle* $\varphi_a : \mathbb{S} \to \mathbb{S}$ is defined by

$$\varphi_a([s]) := [a + s] \quad \text{for } [s] \in \mathbb{S} .$$

For each $a \in \mathbb{R}$ the mapping $\varphi_a : \mathbb{S} \to \mathbb{S}$ is an isometry with respect to the metric of $\mathbb{S}$. It follows that the system $(\mathbb{S}, \varphi_a)$ is uniformly equicontinuous. Moreover, φ_a is a homeomorphism of $\mathbb{S}$ onto itself; this is in agreement with Theorem 1.6.9.

For every $a, b \in \mathbb{R}$ it is easily shown that $\varphi_a \circ \varphi_b = \varphi_b \circ \varphi_a$. Hence φ_b is an automorphism of the system $(\mathbb{S}, \varphi_a)$, mapping the point $[0]$ onto the point $[b]$. If we select two arbitrary points in $\mathbb{S}$ then for a suitable choice of b the mapping φ_b sends the one point onto the other. Hence by Proposition 1.5.2: *If one point of* $\mathbb{S}$ *is periodic then all points of* $\mathbb{S}$ *are periodic with the same primitive period and, similarly, if one point has a minimal orbit closure, or is transitive, then all points have a minimal orbit closure or are transitive, respectively.* Actually, Theorem 1.2.7 implies that there exists a minimal orbit closure in $\mathbb{S}$. Consequently, *every orbit closure under* φ_a *is minimal*. This is in agreement with the two cases mentioned in the Introduction: if $a \in \mathbb{Q}$ then all points of $\mathbb{S}$ are periodic under φ_a and all points have the same primitive period, and if $a \notin \mathbb{Q}$ then the system $(\mathbb{S}, \varphi_a)$ is minimal.

1.7.2 (The argument-doubling transformation). Recall that this is the mapping $\psi: \mathbb{S} \to \mathbb{S}$ defined by $\psi([s]) := [2s]$ for $s \in \mathbb{R}$. In the Introduction, in Example (2) after Theorem 1.3.5 and in Example (1) in the beginning of Section 1.6 it has been shown that

- The periodic points form a dense subset of $\mathbb{S}$.
- The system $(\mathbb{S}, \psi)$ is topologically ergodic, hence transitive; it is even strongly mixing.

There is a graphical way to illustrate the above: see Exercise 1.11. The idea is to represent the circle as the unit interval $[0; 1]$ in which the end points 0 and 1 are identified (this is just what the mapping $s \mapsto [s] : [0; 1] \to \mathbb{S}$ does). One can then draw the 'graph' of the mapping ψ and of ψ^n ($n \in \mathbb{N}$). From these graphs it is easy to derive the above statements.

Recall that the proof that ψ is transitive on $\mathbb{S}$ is based on the observation that for every non-degenerate arc J in $\mathbb{S}$ there exists $k_J \in \mathbb{N}$ such that $\psi^k[J] = \mathbb{S}$ for all $k \geq k_J$. Obviously, for every $n \in \mathbb{N}$ the mapping ψ^n has the same property: $(\psi^n)^k[J] = \mathbb{S}$ for all $k \geq k_J/n$. Consequently, for every $n \in \mathbb{N}$ the mapping ψ^n is transitive on $\mathbb{S}$.

Finally, the argument used to show that the set of ψ-periodic points is dense in $\mathbb{S}$ can easily be adapted to show that ψ^n has a dense set of periodic points as well.

1.7.3 (The tent map). Recall that the *tent map* $T: [0; 1] \to [0; 1]$ is defined by the formula $T(x) := 1 - |2x - 1|$ for $0 \leq x \leq 1$. We have shown earlier:

- The periodic points form a dense subset of $[0; 1]$.
- The system $([0; 1], T)$ is topologically ergodic, hence transitive; it is even strongly mixing.

Similar to the argument used in 1.7.2 one shows that for every $n \in \mathbb{N}$ the mapping T^n is transitive on $[0; 1]$. We shall reconsider this special property of T in Section 2.6 in the next chapter.

The argument used to show that the periodic points for T form a dense set in $[0; 1]$ can easily be adapted to show that T^n has a dense set of periodic points in $[0; 1]$. By Theorem 2.6.2 ahead, this follows also from transitivity of T^n.

1.7.4 (The quadratic family). Recall that the *quadratic family* is the set of functions $\{f_\mu\}_{\mu>0}$ defined by $f_\mu : x \mapsto \mu x(1 - x) : \mathbb{R} \to \mathbb{R}$ for $\mu > 0$. In Example (2) before Lemma 1.5.1 above it was shown that the systems $([0; 1], T)$ and $([0; 1], f_4)$ are mutually conjugate. Consequently, the following statements hold for f_4, because they are true for the tent map: the periodic points of f_4 are dense in $[0;1]$, and the system $([0;1], f_4)$ is topologically ergodic, hence transitive. In 2.1.5 below we shall consider this family more closely.

The members of this family are conjugate to the members of another much-investigated family of mappings: Myrberg's family. See Exercise 1.7 (6).

1.7.5 (A system on the Cantor set). Let $f: \mathbb{R} \to \mathbb{R}$ be defined by

$$f(x) := \frac{3}{2}\Big(1 - |2x-1|\Big) = \min\{\,3x,\ 3(1-x)\,\} \quad \text{for } x \in \mathbb{R}\,.$$

We shall show that the Cantor set C is an f-invariant set. We could do this by applying f to the points of C represented by their ternary expansions as explained in Proposition B.1.1 of Appendix B.. But this would not enable us to obtain invariant Cantor sets for other functions: see (3) below.

The essential feature of f is that there are two mutually disjoint closed subintervals I_0^1 and I_1^1 of the interval $[0;1]$ whose images under f include both I_0^1 and I_1^1, namely, $I_0^1 := [0; \frac{1}{3}]$ and $I_1^1 := [\frac{2}{3}; 1]$. Both intervals are mapped homeomorphically onto $[0;1]$, i.e., f increases monotonously from 0 to 1 on I_0^1 and f decreases monotonously from 1 to 0 on I_1^1.

We shall define inductively a descending sequence of closed subsets Λ_n of $[0;1]$ ($n \in \mathbb{N}$) and a sequence $(d_n)_{n\in\mathbb{N}}$ of real numbers with limit 0 such that the following properties hold for every $n \in \mathbb{N}$:

(1_n) Λ_n is the union of 2^n mutually disjoint closed subintervals I_b^n of the unit interval, labelled by n-tuples $b = (b_0, \dots, b_{n-1})$ of 0's and 1's (so $b \in \{0,1\}^n$).

(2_n) $\forall\, b \in \{0,1\}^n$: f maps the interval I_b^n monotonously onto the interval $I_{b'}^{n-1}$, where $b' := (b_1, \dots, b_{n-1})$ (if $n = 1$ then read $[0;1]$ for $I_{b'}^{n-1}$).

(3_n) $\Lambda_{n+1} \subseteq \Lambda_n$, and the intervals of which Λ_{n+1} is the union are pairwise included in the intervals that form Λ_n: for all $b \in \{0,1\}^n$ the interval I_b^n includes the two intervals I_{b0}^{n+1} and I_{b1}^{n+1}.

(4_n) $\forall\, b \in \{0,1\}^n$: the intervals I_{b0}^{n+1} and I_{b1}^{n+1} are the (closed) extreme third parts of the interval I_b^n.

Put $d_n := \max\{|I_b^n| \,\vdots\, b \in \{0,1\}^n\}$, where $|I|$ denotes the length of an interval I. It follows by induction from the properties (4_n) for $n \in \mathbb{N}$ and the fact that the intervals I_0^1 and I_1^1 have length 3^{-1} that $d_n = 3^{-n}$ for all $n \in \mathbb{N}$, so that $d_n \leadsto 0$ if n tends to infinity. Moreover, comparison of the properties of the sets Λ_n with those of the sets C_n in Section B.1 of Appendix B. clearly shows that $\Lambda_n = C_n$ for every $n \in \mathbb{N}$. Consequently, $\Lambda := \bigcap_{n\in\mathbb{N}} \Lambda_n = C$, the Cantor set.

(1) The starting point of the construction is $\Lambda_1 := I_0^1 \cup I_1^1$. Then, obviously, the properties (1_1) and (2_1) are satisfied. In order to prove the properties (3_1) and (4_1) we first have to define the set Λ_2. This will be done in such a way that these properties are fulfilled automatically, as follows: Since f maps the interval I_i^1 for $i = 0, 1$ monotonously

onto the interval [0; 1] – which includes I_0^1 and I_1^1 – it is clear that there are two subintervals I_{i0}^2 and I_{i1}^2 of I_i^1 such that $f[I_{i0}^2] = I_0^1$ and $f[I_{i1}^2] = I_1^1$. See Figure 1.6 (a). So if we put $\Lambda_2 := \bigcup\{I_{ij}^2 : i, j \in \{0, 1\}\}$ then it is clear that property (3_1) holds. But f not just maps I_i^1 monotonously onto [0; 1], it uniformly stretches it by a factor 3. Consequently, I_i^1 is a shrunken (by a factor 1/3) copy of the unit interval, with I_{i0}^2 and I_{i1}^2 corresponding to I_0^1 and I_1^1, respectively. Hence the intervals I_{i0}^2 and I_{i1}^2 are the extreme third parts of I_i^1 $(i = 0, 1)$, so property (4_1) holds.

Though it will not play any role in the construction, observe that, because f is decreasing on I_1^1, the order in which the intervals I_{10}^2 and I_{11}^2 appear in I_1^1 is opposed to the order in which I_0^1 and I_1^1 appear in [0; 1]. On the other hand, the order in which the intervals I_{00}^2 and I_{01}^2 appear in I_1^0 is the same as the order in which I_0^1 and I_1^1 appear in [0; 1].

By the very definition of the intervals I_{ij}^2 for $(i, j) \in \{0, 1\}^2$ and of Λ_2 it is clear that the properties (1_2) and (2_2) hold. The properties (3_2) and (4_2) will follow from the way the set Λ_3 will be defined. For the construction of the intervals I_{ijk}^3 for $(i, j, k) \in \{0, 1\}^3$ we refer the reader to Figure 1.6 (b), where the construction is illustrated for the case that $i = 1$ (recall that f is decreasing on I_{1j}^2) and $j = 1$ (recall that the interval I_{11}^2 lies below the interval I_{10}^2). Put $\Lambda_3 := \bigcup\{I_{ijk}^3 : (i, j, k) \in \{0, 1\}^3\}$. Then, clearly, the properties (3_2) and (4_2) hold, as well as the properties (1_3) and (2_3). Thus, the principle of the construction is that an interval at level 2 is mapped onto an interval of level 1, which is already known to split into sets of level 2, and this induces a splitting of the given interval into two intervals of level 3.

Now assume that for some $n \in \mathbb{N}$, $n \geq 2$, we have sets $\Lambda_1 \supseteq \cdots \supseteq \Lambda_n$ such that the properties (1_k) and (2_k) hold for $k = 2, \ldots, n$ and that the properties (3_k) and (4_k) hold for $k = 2, \ldots, n-1$. By the properties (2_n) and (3_{n-1}), for every n-tuple $b = b_0 \ldots b_{n-1}$ the interval I_b^n is mapped monotonously onto the interval $I_{b_1 \ldots b_{n-1}}^{n-1} \supseteq I_{b_1 \ldots b_{n-1} 0}^n \cup I_{b_1 \ldots b_{n-1} 1}^n$.

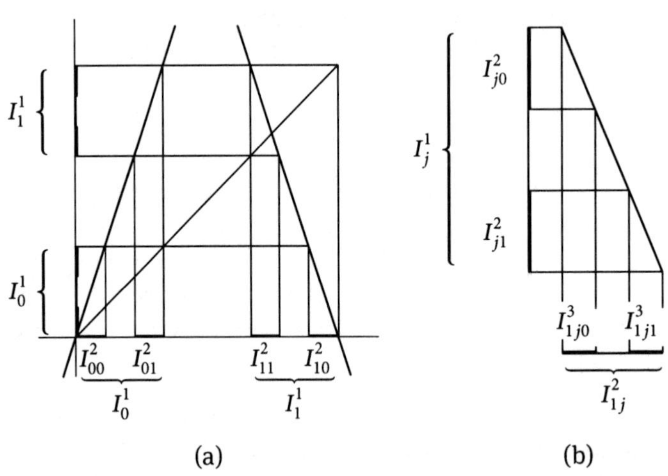

Fig. 1.6. (a) The mapping $x \mapsto \min\{3x, 3(1-x)\} : \mathbb{R} \to \mathbb{R}$ on the unit interval. (b) The construction of I_{ijk}^3 for $i = 1$, $j = 1$ and $k = 0, 1$.

Hence there are two mutually disjoint closed subintervals I_{b0}^{n+1} and I_{b1}^{n+1} of the interval I_b^n such that $f[I_{b0}^{n+1}] = I_{b_1\dots b_{n-1}0}^n$ and $f[I_{b1}^{n+1}] = I_{b_1\dots b_{n-1}1}^n$. So if we define the set Λ_{n+1} by

$$\begin{aligned}\Lambda_{n+1} &:= \bigcup\{\, I_{bj}^{n+1} : b \in \{0,1\}^n, j \in \{0,1\}\,\} \\ &= \bigcup\{\, I_b^{n+1} : b \in \{0,1\}^{n+1}\,\}\end{aligned}$$

then the properties (3_n), (1_{n+1}) and (2_{n+1}) hold. As f has constant derivative with absolute value 3 on I_b^n (note that $I_b^n \subseteq I_{b_0\dots b_{k-2}}^{n-1} \subseteq \dots \subseteq I_{b_0}^1$), the interval I_b^n is a scaled-down copy of the interval $I_{b_1\dots b_{n-1}}^{n-1}$ (possibly upside-down), hence it should be clear from property (4_{n-1}) that the two intervals I_{bj}^{n+1} for $j = 0, 1$ are the extreme third parts of the interval I_b^n. So properties (4_n) hold as well.

This completes the proof (by induction) that we have a sequence of non-empty closed subsets Λ_n of the unit interval and real numbers d_n with the required properties.

(2) *Λ is a non-empty closed completely invariant subset of* $[0;1]$.

Proof. By construction, $\Lambda_n \supseteq \Lambda_{n+1}$ for every $n \in \mathbb{N}$, so by compactness we may conclude that $\Lambda := \bigcap_{n=1}^{\infty} \Lambda_n$ is a non-empty closed subset of $[0;1]$.

In order to show that Λ is completely invariant under f, first note that for every $n \in \mathbb{N}$ and every n-tuple $b = (b_0 \dots b_{n-1})$ of 0's and 1's we have, by property (2_n), $f[I_b^n] = I_{b_1\dots b_{n-1}}^{n-1}$. As $b_1 \dots b_{n-1}$ runs through the set of all $(n-1)$-tuples of 0's and 1's if b runs through the set of all n-tuples, it follows easily that $f[\Lambda_n] = \Lambda_{n-1}$. Now Lemma A.3.3 in Appendix A implies that

$$f[\Lambda] = f\Big[\bigcap_{n=1}^{\infty} \Lambda_n\Big] = \bigcap_{n=1}^{\infty} f[\Lambda_n] = \bigcap_{n=1}^{\infty} \Lambda_{n-1} = \Lambda. \qquad \square$$

Conclusion. *As $\Lambda = C$, the Cantor set C is completely invariant under f.*

We shall show later that the subsystem (Λ, f) of $(\mathbb{R}, f)$ has a dense set of periodic points, is transitive and includes infinite minimal sets: see 6.3.6 (2) ahead (the system (Ω_2, σ) mentioned there has these properties).

(3) One can employ other functions $f: \mathbb{R} \to \mathbb{R}$ to obtain in a similar way invariant Cantor sets. In point of fact, if there are two mutually disjoint closed subintervals I_0^1 and I_1^1 of $[0;1]$ that are mapped homeomorphically onto the interval $[0;1]$ (i.e., on each of these intervals f increases monotonously from 0 to 1 or decreases monotonously from 1 to 0) then the above construction can be performed without any modifications, so the properties (1_n), (2_n) and (3_n) hold for all $n \in \mathbb{N}$. In the general case property (4_n) may fail completely, so that we do *not* necessarily have the equality $\Lambda_n = C_n$ for all $n \in \mathbb{N}$ (and Λ may be not *the* Cantor set C). Moreover, in order to get $d_n \rightsquigarrow 0$ for $n \rightsquigarrow \infty$ we have to make an additional assumption, for example: *f is differentiable on the set $\Lambda_1 = I_0^1 \cup I_1^1$ and there is a constant $c > 1$ such that $|f'(x)| \geq c$ for all $x \in \Lambda_1$.*

It is straightforward to check that in this case we are in the situation dealt with in the Example in Appendix B.1.3. We leave the details to the reader. Let it be sufficient

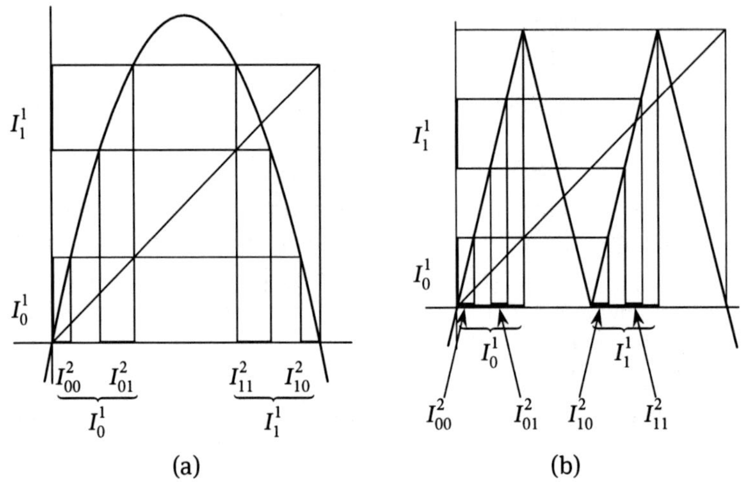

Fig. 1.7. (a) The mapping $x \mapsto \mu x(1-x)\colon \mathbb{R} \to \mathbb{R}$ for $\mu > 4$. (b) The mapping $T^2\colon \mathbb{R} \to \mathbb{R}$.

to observe that by the Mean Value Theorem we have, for all $k \in \mathbb{N}$ and all $b \in \{0,1\}^k$, $|I^{k-1}_{b_1\ldots b_{k-1}}| = |f[I^k_{b_0 b_1 \ldots b_{k-1}}]| \leq c|I^k_b|$ (recall that by $|I|$ we denote the length of an interval I), which implies (by induction) that $|I^n_b| \leq c^{-n}$ for all $n \in \mathbb{N}$.

Conclusion. *If f satisfies the conditions mentioned above then the completely invariant set Λ is a Cantor space.*

Example. The following mappings satisfy the conditions mentioned above:

- $f\colon x \mapsto \mu x(1-x)$ with $\mu > 2+\sqrt{5}$; see Figure 1.6 (b). A straightforward calculation shows that one must take $I_0^1 := [0; \frac{1}{2}(1-\sqrt{1-4/\mu})]$; then $|f'(x)|$ is minimal on this interval in its right end point with value greater than 1; by symmetry one has to take $I_1^1 := [\frac{1}{2}(1+\sqrt{1-4/\mu}); 1]$.
-
$$f\colon x \mapsto \begin{cases} a(x-x_1) & \text{for } x \leq x_0 \\ b(x_2-x) & \text{for } > x_0 \end{cases}$$
with $0 \leq x_1 < x_0 < x_2 \leq 1$ and $a, b > 1$ with $a(x_0-x_1) > 1$ and $b(x_2-x_0) > 1$ (a skew tent map with possibly a discontinuity at x_0).
- $f\colon x \mapsto T^2(x)$, where T is the tent map; as I_0^1 and I_1^1 one can take, for example, the intervals $[0; \frac{1}{4}]$ and $[\frac{1}{2}; \frac{3}{4}]$, respectively.

(4) In one respect the situation for function f considered above is different from that of the last example (i.e., for the mapping T^2), namely, for f we have $f^{\leftarrow}[0;1] = \Lambda_1$ (points not in Λ_1 are mapped outside of $[0;1]$; a similar remark holds for the mappings in the first two examples). We shall show by induction that this implies that

$$\forall n \in \mathbb{N}: f^{\leftarrow}[\Lambda_n] = \Lambda_{n+1}\,. \tag{1.7-1}$$

(We have already observed already that $f[\Lambda_{n+1}] = \Lambda_n$, i.e., $f^{\leftarrow}[\Lambda_n] \supseteq \Lambda_{n+1}$ for every $n \in \mathbb{N}$, but we shall not use this.)

For $n = 1$: first, note that $f^{\leftarrow}[\Lambda_1] \subseteq f^{\leftarrow}[0;1] = \Lambda_1$. However, the construction of Λ_2 implies that the only points of Λ_1 that are mapped into Λ_1 are those of Λ_2, that is, $\Lambda_2 = \Lambda_1 \cap f^{\leftarrow}[\Lambda_1]$. Hence $f^{\leftarrow}[\Lambda_1] = \Lambda_2$, i.e., the equality (1.7-1) holds for $n = 1$. Now suppose that (1.7-1) holds for a certain value of $n \in \mathbb{N}$. Then $f^{\leftarrow}[\Lambda_{n+1}] \subseteq f^{\leftarrow}[\Lambda_n] = \Lambda_{n+1}$ (the equality is the induction hypothesis). Recall that for every $b \in \{0,1\}^n$ and $i, j \in \{0,1\}$ one has $I^{n+2}_{ibj} = I^{n+1}_{ib} \cap f^{\leftarrow}[I^{n+1}_{bj}]$, which implies that $\Lambda_{n+2} = f^{\leftarrow}[\Lambda_{n+1}] \cap \Lambda_{n+1}$, and therefore $\Lambda_{n+2} = f^{\leftarrow}[\Lambda_{n+1}]$. This completes the proof of (1.7-1).

Since $f^{\leftarrow}[0;1] = \Lambda_1$ it follows immediately by induction from (1.7-1) that $\Lambda_n = (f^n)^{\leftarrow}[0;1]$. Hence

$$\Lambda = \bigcap_{n=1}^{\infty} (f^n)^{\leftarrow}[0;1] .$$

It is an easy exercise to show that this implies that Λ is the largest invariant subset of $\mathbb{R}$ under f.

(5) It is interesting to compare the construction of the intervals I^n_b for $n \in \mathbb{N}$ and $b \in \{0,1\}^n$ with the condtruction of the intervals J^n_b of Section B.1 of Appendix B.. For both constructions the intervals at any level are obtained from those at the previous level by omitting middle thirds. Consequently, for every $n \in \mathbb{N}$ the set of intervals I^n_b for $b \in \{0,1\}^n$ is the same as the set of intervals J^n_b for $b \in \{0,1\}^n$. Stated otherwise, for every $n \in \mathbb{N}$ there is a permutation $b \mapsto \kappa(n,b)$ of $\{0,1\}^n$ such that $I^n_b = J^n_{\kappa(n,b)}$ for all $b \in \{0,1\}^n$. For example, at level 2 the intervals from left to right are I^2_{00}, I^2_{01}, I^2_{11} and I^2_{10}, whereas in Appendix B. one has J^2_{00}, J^2_{01}, J^2_{10} and J^2_{11}. This means that $\kappa(2,0j) = 0j$ and $\kappa(2,1j) = 1\overline{j}$ $(j = 0,1)$, where $\overline{0} := 1$ and $\overline{1} := 0$.

In what follows we shall denote the permutation $\kappa(n,\cdot)$ of $\{0,1\}^n$ simply by κ. If one keeps in mind that $\kappa(b)$ is an n-tuple iff b is an n-tuple then this will cause no confusion. The following algorithm describes the action of κ: define $t\colon \{0,1\} \to \{0,1\}$ by $t(j) := \overline{j}$ for $j = 0,1$. Moreover, for $b \in \{0,1\}^n$ and $i = 0,\dots,n-1$, let $p(b,i)$ be the number of 1's in b preceding the coordinate b_i, that is, the number of 1's in the i-tuple $(b_0,\dots,b_{i-1})$. Then $\kappa(n,b)_i = t^{p(b,i)}(b_i)$ for $i = 0,\dots,n-1$. Thus, one applies κ by rewriting an n-tuple b from left to right, taking into account the number of 1's that one has passed. Note that $t^p = t^{p \pmod 2}$ for every $p \in \mathbb{Z}^+$, so only the parity of the number of 1's left of the coordinate one is dealing with is relevant. For example, $\kappa(01|1|01|001|1|0) = 01|0|01|110|1|1$. For details, see Exercise 1.13.

1.7.6 (Ellis' minimal system). Let us introduce the following notation: If x_1 and x_2 are distinct points in $\mathbb{S}$ then $(x_1;x_2)$ will denote the counter-clockwise open arc from x_1 to x_2. For example, if $x_1 = [t_1]$ and $x_2 = [t_2]$ with $0 < t_1 < t_2 < 1$ then $(x_1;x_2)$ equals the open arc $\{\,[s] \mathrel{\vdots} t_1 < s < t_2\,\}$, whereas $(x_2;x_1)$ is the arc $\{\,[s] \mathrel{\vdots} t_2 < s < t_1 + 1\,\}$. Moreover, let $[x_1;x_2) := \{x_1\} \cup (x_1;x_2)$ and $(x_1;x_2] := (x_1;x_2) \cup \{x_2\}$.

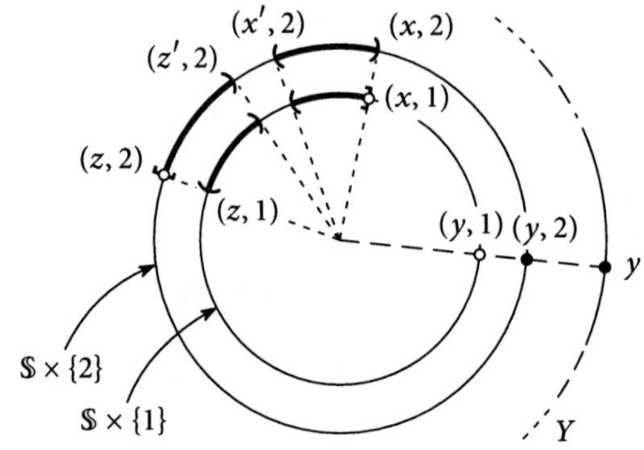

Fig. 1.8. Basic neighbourhoods of the points $(x, 1)$ and $(z, 2)$ in X. For every point $y \in Y = \mathbb{S}$ we have $\varphi^{\leftarrow}[y] = \{(y, 1), (y, 2)\}$.

Let $X := \mathbb{S} \times \{1, 2\}$. It is convenient to visualize X as the union of two concentric circles, $\mathbb{S} \times \{2\}$ slightly larger than $\mathbb{S} \times \{1\}$. We define a topology in X by specifying in every point a local base, as follows:

$$\text{at the point } (x, 1)\colon \text{ all sets } ([x; x') \times \{1\}) \cup ((x; x') \times \{2\})$$
$$\text{at the point } (z, 2)\colon \text{ all sets } ((z'; z) \times \{1\}) \cup ((z'; z] \times \{2\})$$

with $x' \neq x$ and $z' \neq z$; see Figure 1.8. Obviously, with this topology, X is a Hausdorff space. Moreover, the basic open sets defined above are clopen, so X is 0-dimensional. Next, we show that X is compact. First, note that X is countably compact. In point of fact, if a sequence in $\mathbb{S} \times \{1\}$ converges monotonously to a point $(x, 1)$ in the Euclidean topology, then it converges in X to the point $(x, 1)$ if it is decreasing, and it converges in X to the point $(x, 2)$ if it is increasing. A similar result holds for a sequence in $\mathbb{S} \times \{2\}$ that converges monotonously in the Euclidean topology. Since every sequence in X has a subsequence in $\mathbb{S} \times \{1\}$ or $\mathbb{S} \times \{2\}$, and that subsequence has, in turn, a subsequence that converges monotonously in the Euclidean topology, it follows that every sequence in X has a convergent subsequence. This means that X is countably compact (i.e., every countable open cover has a finite subcover). Next, observe that X is a Lindelöf space. This is due to the fact that both $\mathbb{S} \times \{1\}$ and $\mathbb{S} \times \{2\}$ with their relative topology in X are Lindelöf spaces (which means that every open cover has a countable subcover). The proof is virtually the same as the proof that the Sorgenfrey line is a Lindelöf space: see, for example, [Eng], Example 3.8.14. Since X is both countably compact and a Lindelöf space, it follows that every open cover of X has a finite subcover, that is, X is compact.

Select $a \in \mathbb{R} \setminus \mathbb{Q}$ and define $f\colon X \to X$ by $f(x, i) := (\varphi_a(x), i)$ for $x \in \mathbb{S}$ and $i = 1, 2$, where φ_a is the rigid rotation on $\mathbb{S}$ over a. Then f is easily seen to be continuous (it is even a homeomorphism). Moreover, the system (X, f) is minimal: this follows immediately from the the minimality of $(\mathbb{S}, \varphi_a)$ and the fact that every non-empty open subset of X meets both circles $\mathbb{S} \times \{1\}$ and $\mathbb{S} \times \{2\}$ in a set with a non-empty Euclidean interior. Thus, (X, f) is a minimal system on a 0-dimensional compact Hausdorff space; it is called the *Ellis minimal system*.

Finally, let the (continuous!) mapping $\varphi\colon X \to \mathbb{S}$ be given by $\varphi(x, i) := x$ for $x \in \mathbb{S}$ and $i = 1, 2$. Then $\varphi\colon (X, f) \to (\mathbb{S}, \varphi_a)$ is a factor map. Obviously, every (basic) open set of X includes a full fibre of φ, so φ is irreducible. (In view of Exercise 1.8 (2) , this accounts for the fact that (X, f) is minimal.) Since all fibres consist of two points, φ is not almost 1-to-1. So by Theorem A.9.7 in Appendix A, the space X is not metrizable. Finally, note that φ is semi-open – this is in accordance with Theorem 1.5.7 – but that it is not an open mapping.

Exercises

1.1. Let X be a locally compact or a completely metrizable Hausdorff space and let x be a point in X. Show that if $\mathcal{O}(x)$ is closed in X then all points of $\mathcal{O}(x)$ are isolated in $\mathcal{O}(x)$. Is the converse true?
NB. The result holds in every Čech-complete space.

1.2.
(1) For every $n \in \mathbb{N}$, let $p_n(X, f)$ denote the number of periodic points with primitive period n and let $P_n(X, f)$ be the number of all periodic points with period n. Then

$$P_n(X, f) = \sum_{k|n} p_k(X, f)\,.$$

for every $n \in \mathbb{N}$. Find recursively for k=1, 2, 3, 4, 5 and 6 the values of $p_k(X, f)$ for the argument-doubling system and for the tent map.
(2) Consider the dynamical system defined in Example (5) in Section 1.1. Show that there is no periodic point of primitive period 3 (i.e., the only point with period 3 is the invariant point in the interval $[2; 3]$).

1.3.
(1) Let $p_0 \in \mathbb{Z}^+$ and let x be a point in the closure of the set of periodic points with primitive period p_0. Then the point x is periodic with primitive period p such that $p \mid p_0$, that is, there exists $k \in \mathbb{N}$ such that $p_0 = kp$.
(2) Denote the primitive period of the periodic point $x \in X$ by $p(x)$. Show that for every $c \in \mathbb{R}^+$ the set $\{\, x \in X \,\vdots\, p(x) \le c \,\}$ is closed.
(3) Construct a dynamical system (X, f) where all points except one have primitive period 6, and where the exceptional point has primitive period 3.

1.4. Let X be a compact Hausdorff space and let $x \in X$. Then the point x has a complete past in X iff $x \in \bigcap_{i=0}^{\infty} f^i[X]$.
NB 1. Compactness of X can be replaced by the condition that each of the closed sets $(f^n)^{\leftarrow}[x]$ is compact. This is certainly the case if f is finite-to-one.
NB 2. If X is not compact then a point of $\bigcap_{i=0}^{\infty} f^i[X]$ does not necessarily have a complete past: see Figure 3.7 ahead.

1.5.

(1) A dynamical system on a non-degenerate interval in $\mathbb{R}$ is not minimal.

(2) Let A be a minimal set in a dynamical system (X, f). Show that $\overline{f[A]} = A$ and that $f[A] = A$ if A is compact.
NB. The system in Exercise 1.9 shows that $f[A]$ may be a proper subset of A.

(3) Let M be a minimal set in a dynamical system (X, f) such that[6] $f^{\leftarrow}[M] \subseteq M$ (equivalently, $f[X \setminus M] \subseteq X \setminus M$). Show that $\operatorname{int}_X(M)$ is either empty or equal to M, in which case M is clopen in X.

(4) Show that an equicontinuous system on a metric space is minimal iff it is transitive, iff it is topologically ergodic.

(5) Let (X, f) be a transitive dynamical system on a locally compact Hausdorff space X. Let $\mathbf{Trans}\,(X, f)$ denote the set of all transitive points in (X, f). If $\mathbf{Trans}\,(X, f)$ has a non-empty interior, then (X, f) is minimal and X is compact.

(6) If (X, f) is a minimal system with X locally compact then X is compact.

(7) If X is an infinite locally compact Hausdorff space and $\mathbf{Trans}\,(X, f) \neq \emptyset$ then either X is compact and $\mathbf{Trans}\,(X, f)$ equals all of X (i.e., X is minimal) or $\mathbf{Trans}\,(X, f)$ is a residual set with a dense complement (i.e., the set of intransitive points is dense).

1.6.

(1) Suppose there are two different points x_1 and x_2 in X which have a dense orbit. Show that the points x_1 and x_2 are transitive.

(2) Let $x_0 \in X$ be such that $\mathcal{O}(x_0)$ is dense in X, but assume that the set $f[X]$ is not dense in X. Show that the point x_0 is isolated in X and that $X \setminus \{x_0\} = \overline{\mathcal{O}(f(x_0))}$.

(3) A dynamical system (X, f) is topologically ergodic iff for every closed invariant subset A of X either $A^\circ = \emptyset$ or $A = X$.

(4) Let (X, f) be topologically ergodic. If X contains an isolated point then X consists of one single periodic orbit.

(5) If (X, f) is topologically ergodic then show that the collection of sets $D(U, U)$ with U a non-empty open subset of X is a filter base. In particular, the intersection of finitely many of such sets is not empty.

(6) Show that if (X, f) is topologically ergodic then for every two non-empty open subsets U and V of X the set $D(U, V)$ is infinite.

1.7.

(1) Show that no two of the following mappings are conjugate to each other:

$$\begin{aligned} &g_1 : x \mapsto x : \mathbb{R} \to \mathbb{R}, && g_4 : x \mapsto 4x : \mathbb{R} \to \mathbb{R}, \\ &g_2 : x \mapsto -x : \mathbb{R} \to \mathbb{R}, && g_5 : x \mapsto 4x(1-x) : \mathbb{R} \to \mathbb{R}, \\ &g_3 : x \mapsto x + 3 : \mathbb{R} \to \mathbb{R}, && g_6 : x \mapsto 6x(1-x) : \mathbb{R} \to \mathbb{R}. \end{aligned}$$

---

6 The condition $f^{\leftarrow}[M] \subseteq M$ cannot be deleted. Let $X := [-1; 1] \cup \{2\}$, $f(x) := 2$ if $x \in [-1; 1]$ and $f(2) := 0$. Then $\{0, 2\}$ is a non-clopen minimal set (a periodic orbit) with non-empty interior in X.

(2) The systems $(\mathbb{R}, f)$ and $(\mathbb{R}, f^{-1})$ with $f(x) := 3x$ are not conjugate.
(3) Show that there exists no conjugation φ of the systems of the first example in Section 1.5 such that both φ and φ^{-1} are differentiable.
(4) Show that $F\colon [t] \mapsto \sin^2 \pi t\colon \mathbb{S} \to [0;1]$ $(t \in \mathbb{R})$ defines a 2-to-1 factor map from the argument-doubling system $(\mathbb{S}, \psi)$ onto the logistic system $([0;1], f_4)$.
(5) The mapping $T^*\colon \mathbb{S} \to [0;1]$ given by $T^*([t]) := T(t)$ for $0 \leq t < 1$ defines a morphism $T^*\colon (\mathbb{S}, \psi) \to ([0;1], T)$ which is everywhere 2-to-1, except in the points $[0]$ and $[1/2]$.
(6) For $c \geq -1/4$, consider the set of mapping $g_c\colon x \mapsto x^2 - c\colon \mathbb{R} \to \mathbb{R}$ (Myrberg's family). Show that he members of this family are conjugate to the members of the quadratic family $\{f_\mu\}_{\mu \geq 1}$.

1.8. Let $\varphi\colon (X, f) \to (Y, g)$ be a factor mapping.
(1) If $B \subseteq Y$ is an g-invariant then $\varphi^{\leftarrow}[B]$ is f-invariant in X, closed if B is closed.
(2) Suppose φ is a closed and irreducible mapping. If the point $y \in Y$ is transitive under g then every point $x \in \varphi^{\leftarrow}[y]$ is transitive in X under f. In particular, if the system (Y, g) is minimal then so is (X, f).

1.9. Let $a \in \mathbb{R} \setminus \mathbb{Q}$, let X be the orbit of the point $[0]$ in $\mathbb{S}$ under φ_a and let $f := \varphi_a|_X$, where $\varphi\colon (X, f) \to (\mathbb{S}, \varphi_a)$ is the embedding mapping of X into $\mathbb{S}$. Show that the system (X, f) is minimal but $\varphi[X]$ is not a minimal set in $(\mathbb{S}, \varphi_a)$.

1.10. A minimal dynamical systems (X, f) is said to be *totally minimal* whenever for every $n \in \mathbb{N}$ the system (X, f^n) is minimal. A transitive system (X, f) is said to be *totally transitive* whenever for every $n \in \mathbb{N}$ the system (X, f^n) is transitive.
(1) Show that the rigid rotation $(\mathbb{S}, \varphi_a)$ with $a \in \mathbb{R} \setminus \mathbb{Q}$ is totally minimal. Show that the systems of the tent map and of the argument-doubling transformation are totally transitive.
(2) Show that the following construction can be used to get a minimal or transitive system (X, f) that is not totally transitive (hence not not totally minimal): let (Y, g) be any transitive or minimal system, let $X := Y \times \{1, 2\}$ and define the mapping f by $f(y, 1) := (g(y), 2)$, $f(y, 2) := (y, 1)$ for $y \in Y$.
(3) Let (X, f) be a minimal system and assume that the phase space X is compact. Then the following conditions are equivalent:
   (i) The system (X, f) is not totally minimal.
   (ii) There is a partition $X = A_0 \cup \cdots \cup A_{p-1}$ of X into $p \geq 2$ mutually disjoint clopen subsets such that $f[A_i] = A_{i+1 \pmod p}$.
   (iii) There exists a factor mapping $\varphi\colon (X, f) \to (Y, g)$ where (Y, g) is a non-trivial system consisting of one single periodic orbit.
(4) Use 3 to show that a compact minimal system is totally transitive iff it is totally minimal.
(5) Show that a weakly mixing system on a second countable Baire space is totally transitive.

(6) Show that a minimal weakly mixing system on a compact space is totally minimal.
(7) Show that a system (X, f) is totally transitive iff for every $n \in \mathbb{N}$ the product system $(X \times Y_n, f \times g_n)$ is transitive, where (Y_n, g_n) is the system consisting of a single periodic orbit with period n.

1.11. Define $F\colon [0;1] \to [0;1]$ by $F(x) := 2x \pmod 1$ for $0 \le x \le 1$. Obviously, F is continuous, except at the point $x = 1/2$. This mapping is sometimes called the *1-dimensional baker's transformation.*
(1) Show that for every $n \in \mathbb{N}$, the graph of F^n consists of 2^n line segments, increasing from 0 to 1 on an interval of the form $[k2^{-n}; (k+1)2^{-n}]$ $(k = 0, \dots, 2^n - 1)$.
(2) F has 2^n periodic points of period n and the set of all periodic points of F is dense in $[0; 1]$.
(3) For every non-degenerated interval J in $[0; 1]$ there exists $k \in \mathbb{N}$ such that $F^n[J] = [0; 1]$ for all $n \ge k$.
(4) Use the relationship $[\cdot] \circ F = \psi \circ [\cdot]$ to derive from 2 and 3 above that the argument-doubling transformation ψ has a dense set of periodic points and is topologically ergodic, hence transitive.

1.12. Let $\varphi\colon (X, f) \to (Y, g)$ be a factor mapping of dynamical systems on compact metric spaces X and Y.
(1) The following statements are equivalent
  (i) The system (X, f) is equicontinuous (i.e., $\{f^n : n \in \mathbb{Z}^+\}$ is pointwise equicontinuous on X).
  (ii) The set $\{f^n : n \in \mathbb{Z}^+\}$ is uniformly equicontinuous on X.
  (iii) For every $\varepsilon > 0$ there is a finite subset K of $\mathbb{Z}$ such that
$$\forall n \in \mathbb{Z}^+ \exists k \in K : d(f^n(x), f^k(x)) < \varepsilon \quad \text{for all } x \in X\,.$$
(2) If (X, f) is equicontinuous then so is (Y, f). Consequently, on a compact metrizable phase space equicontinuity is independent of the metric used.

1.13. Let notation be as in 1.7.5.
(1) For every $n \in \mathbb{N}$ and every n-tuple $b = (b_0, \dots, b_{n-1}) \in \{0, 1\}^n$ the intervals I_{b0}^{n+1} and I_{b1}^{n+1} (in this order) are the left and right third parts of the interval I_b^n iff the number of 1's among the coordinates of b is even.
(2) If the number of 1's among the coordinates of b is even (or odd, respectively) then $\kappa(bj) = \kappa(b)j$ (or $\kappa(bj) = \kappa(b)\bar{j}$, respectively) for $j = 0, 1$.
(3) Prove the algorithm for κ mentioned in 1.7.5 (5).

Notes

1 In the literature also the terms 'confining set' and 'trapping set' are used to denote invariant sets, though the latter term is also used in a more restricted sense, namely, for compact sets A with the property that $f[A] \subseteq \text{int}_X(A)$ (see 3.4.4 ahead for an application of this particular property).
2 Theorem 1.2.8 is taken from S. Kolyada, L. Snoha & S. Trofimchuk [2001].

3 The use of the notion of transitivity in the literature is slightly confusing: often a system is called transitive if it is topologically ergodic. According to Theorem 1.3.5, the notions coincide in second countable Baire spaces.

4 With a bit more effort it can be shown that the converse of Proposition 1.6.2 is also true: see E. Glasner [2003].

5 If a system on a metric space is equicontinuous at a point x_0 then the point x_0 is often called *stable in the sense of Lyapunov*. The reason that this is called 'stability' is, of course, that $f^n(x)$ will remain as close to $f^n(x_0)$ as one wants if $x \in X$ is sufficiently close to x_0. See also Note 4 at the end of Chapter 3, and Section 7.1.

6 In order to appreciate Corollary 1.6.6, note that every system has equicontinuous factors (for example, the trivial system is an equicontinuous factor of every system). It follows 'easily' that every system has a *largest equicontinuous factor*. The straightforward proof is, essentially, as follows: form the product of all equicontinuous factors and consider the induced mapping of the given system into that product; then the corestriction of that mapping to its range is the largest equicontinuous factor. But a lot of details have been omitted in this 'proof': one has to introduce a partial ordering in the collection of all equicontinuous factors, and one has to show that, essentially, there is only a *set* of equicontinuous factors (otherwise the product above is not defined). This is not particularly difficult, but we do not need these results, so we refer the interested reader to the literature for these details. See e.g., [deV], IV.4.40.

For the missing details of the proof of Theorem 1.6.7 we refer to [deV], V.1.13 (the second half of the proof) and [deV], V.1.15. The proofs there are given for a group T of homeomorphisms acting on X, but they remain valid if T is replaced by the semigroup $\{f^n : n \in \mathbb{Z}^+\}$ of continuous mappings, except the final part of the proof in [deV], V.1.15 which, in the final part of the above proof, we have adapted to our situation. For the existence of an invariant regular probability measure on X, see Section 6.2 in P. Walters [1982].

7 A popular phase space in the theory of dynamical systems is the torus, which is, topologically, the Cartesian product of a finite number of copies of the circle. One reason for this is that many systems in classical mechanics have models with a (higher dimensional) torus as a phase space. See also the Notes to Section 1 of Chapter III of [deV].

A well-known example is the following: let $\mathbb{S}^2 := \mathbb{S} \times \mathbb{S}$ be the 2-dimensional torus, a compact metrizable space and let, for any $a = (a_1, a_2) \in \mathbb{R}^2$, $\tau_a := \varphi_{a_1} \times \varphi_{a_2}$, the *translation* of $\mathbb{S}^2$ over a. So

$$\tau_a([t_1], [t_2]) := ([a_1 + t_1], [a_2 + t_2]) \quad \text{for } ([t_1], [t_2]) \in \mathbb{S}^2 .$$

Clearly, τ_a is a continuous mapping of $\mathbb{S}^2$ onto itself, even a homeomorphism; its inverse is the translation τ_{-a}; in point of fact, τ_a is an isometry.

For every $b \in \mathbb{R}^2$ the mapping τ_b is an automorphism of the dynamical system $(\mathbb{S}^2, \tau_a)$. If we select two arbitrary points in $\mathbb{S}^2$ then for suitable $b \in \mathbb{R}^2$ the automorphism τ_b maps the one point onto the other. So Proposition 1.5.2 implies that if one point of $\mathbb{S}^2$ is periodic then all points of $\mathbb{S}^2$ are periodic with the same primitive period and if one point has a minimal orbit closure then all points have a minimal orbit closure. Consequently, *every orbit closure under τ_a is minimal* as, by Theorem 1.2.7, there is at least one minimal orbit closure in $\mathbb{S}^2$. Quite similar to the situation of the rigid rotation on the circle, there are a two possibilities for the system $(\mathbb{S}^2, \tau_a)$:

Case 1. *If the real numbers a_1, a_2 and 1 are linearly dependent over $\mathbb{Q}$ then no orbit under τ_a is dense in the torus, so the system $(\mathbb{S}^2, \tau_a)$ is neither transitive nor minimal.*

Case 2. *If the real numbers a_1, a_2 and 1 are linearly independent over $\mathbb{Q}$ then every orbit under τ_a is dense in $\mathbb{T}$. In particular, the system $(\mathbb{T}, \tau_a)$ is minimal.*

For proofs, see e.g., [deV] III.1.13.

A well-known application of Case 2 is the following approximation theorem, originally due to Kronecker: if θ_1, θ_2 and 1 are linearly independent over $\mathbb{Q}$ then for every $\varepsilon > 0$ and every choice of $(\theta_1, \theta_2) \in \mathbb{R}^2$ the equations

$$\left| \theta_i - \frac{k_i}{n} \right| < \frac{\varepsilon}{n}, \; i = 1, 2$$

have infinitely many solutions with $n \in \mathbb{N}$ and $k_1, k_2 \in \mathbb{Z}$. This follows readily from approximation of the point $O := ([0], [0])$ by points $\tau_\theta^n(O)$ of its orbit.

The rigid rotation of the circle and the translation of the torus is a special case of the following construction: then for every $a \in G$ the *left translation* $\lambda_a : g \mapsto ag : G \to G$ is a continuous mapping (even a homeomorphism). For every $b \in G$ the *right translation* $\rho_b : g \mapsto gb : G \to G$ is a homeomorphism that commutes with λ_a, hence it is an automorphism of the dynamical system (G, λ_a). For any two points in G there is such an automorphism, mapping the first point onto the second, and because by Theorem 1.2.7 there is at least one point with a minimal orbit closure, all points have a minimal orbit closure.

8 In 1.7.5 (3) it was shown that for $\mu > 2 + \sqrt{5}$ the unit interval includes a Cantor set which is invariant under the quadratic mapping f_μ. Moreover, in 6.3.6 it will be shown that f_μ on this Cantor set is conjugate to a system – the full shift on two symbols – that behaves chaotically (see the Examples (2) and (3) in Section 7.2 ahead). It can be shown that these results hold for all $\mu > 4$. For a rather simple proof, see R. L. Kraft [1999].

9 Ellis' minimal system was introduced in R. Ellis & W. H. Gottschalk [1960]. It is a variation of an example by Alexandroff and Urysohn, cf. [Eng], Exercise 3.10.C. At the end of 1.7.6 we have seen that the phase space X of this system cannot be metrizable. This is well-known: as in [Eng], 1.2.2 one shows that X is not 2nd countable.

10 By Exercise 1.5 (1) there is no minimal system possible on a non-degenerate closed interval. On the circle minimal systems are possible (e.g., the rigid rotation), but it can be shown that the phase mapping always is a homeomorphism. Actually, a construction due to H. Furstenberg – see [deV], III(5.2) for references and a description – shows that every minimal system on the circle is conjugate to a rigid rotation. But on the torus $\mathbb{S}^2$ it is possible to construct a minimal system with a non-invertible phase mapping: see S. Kolyada, L. Snoha & S. Trofimchuk [2001].

11 By using the formula in Exercise 1.2, the numbers $p_k(X, f)$ can be computed inductively from the numbers $P_k(X, f)$. Readers with a good knowledge of number theory will recognize the possibility to apply the *Möbius inversion formula*:

$$p_n(X, f) = \sum_{d|n} \mu(d)\, P_{\frac{n}{d}}(X, f),$$

where $\mu : \mathbb{N} \to \{-1, 0, 1\}$ is the *Möbius function*, defined by

$$\mu(n) := \begin{cases} 1 & \text{if } n \text{ is square-free with an even number of different prime factors,} \\ -1 & \text{if } n \text{ is square-free with an odd number of different prime factors,} \\ 0 & \text{if } n \text{ is not square-free.} \end{cases}$$

The first few values of μ (starting with $\mu(1) = 1$) are: $1, -1, -1, 0, -1, 1, -1, 0, 0, \ldots$ For details and a proof, see G. H. Hardy & W. M. Wright [1979] or any other good book on number theory. For example, $p_6(X, f) = P_6(X, f) - P_3(X, f) - P_2(X, f) + P_1(X, f)$.

12 Invertible vs. non-invertible systems. Our results remain true for dynamical systems with a homeomorphism as phase mapping. In that case, some proofs are easier, because a homeomorphism preserves closures, interiors, complements and intersections of sets. Some results can even be slightly extended (e.g., the interior of an invariant set is invariant, as is the complement of a completely invariant set). Some results become trivial, like Theorem 1.2.8. Finally, some concepts are superfluous for invertible systems, for example the notions of eventually invariant points and eventually periodic points. However, it is not our intention to pay much attention to such relatively trivial matters. Instead, we want to warn the reader for conflicting terminology and definitions. In order to state clearly what I mean, let me define an *invertible system*, also called a *bilateral system*, as a dynamical system (X, f) where f is a homeomorphism of X onto itself which and where negative iterates of f are also taken into account (of course, $f^{-n} := (f^{-1})^n$ for all $n \in \mathbb{N}$). By way of contrast, the not necessarily invertible systems considered in this book is often called *semi-dynamical systems*. We shall mention a number of discrepancies between the theories of semi-dynamical systems and bilateral systems, but we will not go into details about all consequences.

First of all, in the theory of invertible systems the orbit of a point x under a homeomorphism f is defined as the set $\{ f^n(x) : n \in \mathbb{Z} \}$. We shall call this the *full orbit* of x. Then the set $\{ f^n(x) : n \in \mathbb{Z}^+ \}$ – which we have called the orbit – is usually called the *positive semi-orbit* of x and is denoted by $\mathcal{O}^+(x)$ (we leave the definition of the negative semi-orbit to the reader). It follows easily that full orbits form a partition of the phase space. Accordingly, a subset A is said to be invariant whenever $f[A] \subseteq A$ and $f^{-1}[A] \subseteq A$ or, equivalently, $f[A] = A$ (implying that $X \setminus A$ is invariant as well); let us call this notion *two-sided invariant* (in the literature also called *bilaterally invariant*). What we have termed invariant is then called *positively invariant*. It is easy to see that closures, interiors, complements and intersections of two-sided invariant sets are two-sided invariant as well and that full orbits and full orbit closures under a homeomorphism are two-sided invariant; compare these statements with Proposition 1.2.3, Corollary 1.2.4 and the Examples after Proposition 1.2.3.

It follows that our definition of minimality differs, at first sight, essentially from the definition of minimality in the theory of invertible systems. In fact, in the theory of invertible systems, a non-empty closed subset A of the phase space is called minimal whenever it is invariant and includes no proper subset with these properties – but 'invariant' means here 'two-sided invariant'. In the invertible theory our definition in Section 1.2 would read: a non-empty closed set A is *positively minimal* whenever it is positively invariant and includes no proper subset with these properties. Surprisingly, *if A is compact or if the ambient space is locally compact then the two definitions agree* (in the last case it follows that A is compact: see Exercise 1.5 (6)). For an outline of the proof of this result, originally due to W. H. Gottschalk, see [deV], III(9.6)1–3. The proof is based on Proposition 1.4.2 (3) and the observation that $\omega_f(x)$ is two-sided invariant if f is a homeomorphism, which can be proved following the lines of Proposition 3.1.2 (3) ahead – with f or f^{-1} instead of φ – and taking into account that Proposition 1.4.3 (1) now holds for all $n \in \mathbb{Z}$ (see also Exercise 3.2). More on limit sets in invertible systems is in Note 8 to Chapter 3.

Similarly, there is a discrepancy in the definition of transitivity. First of all, often a *system* is said to be transitive whenever it is topologically ergodic (for the definition in the invertible theory, see below). In the invertible theory, a *point* is called *positively (negatively) transitive* whenever its positive (negative) semi-orbit is dense. Moreover, a point that is both positively and negatively transitive is called *bilaterally transitive*. At first sight, these notions are weaker than the notion of transitivity defined in Section 1.3, which by Proposition 1.4.2 (2) is equivalent to saying that the limit set – which is now called the *positive limit set* – of the point under consideration equals the phase space. In the following discussion we shall refer to this notion as *ω-transitive*. Note that in an invertible system there is also the notion of α-transitivity: a point x is α-transitive whenever $\alpha_f(x) = X$. Here $\alpha_f(x)$ denotes the *negative limit set* of the point x, i.e., the set $\alpha_f(x) := \omega_{f^{-1}}(x)$.

Using this terminology, Proposition 1.3.1 can be rephrased as follows: a point is ω-transitive iff all points of its positive semi-orbit are positively transitive. Consequently, (the proof of) Proposition 1.3.2 can be adapted so as to show: *a point is ω-transitive iff it is positively transitive* (note that in an invertible system (X, f) always $f[X] = X$). So for homeomorphisms, positive transitivity is the same notion as defined in Section 1.3. Similarly, negative transitivity is the same as α-transitivity.

In general, density of a full orbit is a weaker property than density of a semi-orbit. In point of fact, the property of having a dense full orbit does neither imply positive nor negative transitivity (let alone, bilateral transitivity), as the following example shows (an invertible version of the initial example in Section 1.3):

> For $n \in \mathbb{N}$, let $t_n := 1 - \frac{1}{n}$ and $t_{-n} := -t_n$, and let $t_0 := 0$. Moreover, let X be the subset $\{-1, 1\} \cup \{t_n : n \in \mathbb{Z}\}$ of $\mathbb{R}$ with its relative topology in $\mathbb{R}$. Finally, define $f: X \to X$ by $f(t_n) := t_{n+1}$ for $n \in \mathbb{Z}$, $f(-1) := -1$ and $f(1) := 1$.

However, the following modification of Corollary 1.3.3 holds: *if X has no isolated points and the full orbit of the point x_0 is dense then for every point $x \in X$ and every neighbourhood U of x the set $D(x_0, U) := \{n \in \mathbb{Z} : f^n(x_0) \in U\}$ is infinite*: use that for any $k \in \mathbb{N}$ the set $U \setminus \{f^n(x_0) : -k \le n \le k\}$ is a neighbourhood of x as well. Equivalently: the set $D(x_0, U)$ is unbounded for every non-empty open subset U of X (but perhaps not for every U unbounded in the same direction).

In the invertible theory topological ergodicity (often also called transitivity) is defined as follows: for any two non-empty open sets U and V in the phase space there is $n \in \mathbb{Z}$ such that $f^n[U] \cap V \neq \emptyset$ (no guarantee that such values of n are all positive or are all negative; to appreciate the difference, note that in this setting Exercise 1.6 (4) may be false: see the example above). Then the first part of the proof of Theorem 1.3.5 shows that a system with a dense orbit is topologically ergodic. With a minor modification – let $U_i := \bigcup_{n=-\infty}^{\infty} f^n[B_i]$ – the second part of the proof of Theorem 1.3.5 shows that in a 2nd countable Baire space ergodicity implies that there is a dense G_δ-set of points with a dense full orbit. So we don't get positive or negative transitivity, and in this sense the definition in the invertible theory is weaker than our definition. See, however, Note 9 in Chapter 4. There we observe that *an invertible ergodic system without isolated points is ergodic according to our definition* in Section 1.3. So if the phase space is a 2nd countable Baire space without isolated points then there is a dense G_δ-set of positively transitive points. Similarly, there is a dense G_δ-set of negatively transitive points. Since in a Baire space the intersection of two dense G_δ-sets is a dense G_δ-set, it follows that there is a dense G_δ-set of bilaterally transitive points. Note also that on a space without isolated points the notions of weak mixing (i.e., ergodicity of the product of the system with itself) in the invertible theory and in our theory of semi-dynamical systems agree.

As a final remark on transitivity, let me observe that in 1.5.8 up to Corollary 1.5.12 the existence of a dense orbit is not enough: it is essential that 'transitive' means positively (or negatively) transitive, otherwise Proposition 1.3.4 cannot be applied.

We close these observations on conflicting definitions with a note on almost 1-to-1 factor mappings. In the theory of invertible dynamical systems one is used to say that a factor mapping $\varphi: (X, f) \to (Y, g)$ is almost 1-to-1 whenever there is a point $x_0 \in X$ with dense orbit under f such that $\varphi^{\leftarrow}[\varphi(x_0)] = \{x_0\}$, i.e., $x_0 \in S_\varphi$ (see, for example, Corollary 1.5.6, Theorem 1.2.7 and Corollary 7.2.8 ahead). If f and g are homeomorphisms then this implies that the underlying continuous mapping $\varphi: X \to Y$ is weakly almost 1-to-1 according to the definition in Section A.9 of Appendix A, because then the orbit of x_0 is included in S_φ. But there are examples showing that in general the notions are unrelated.

2 Dynamical systems on the real line

Abstract. In this chapter we concentrate on dynamical systems (X, f) where X is an interval in $\mathbb{R}$. In particular, we introduce methods for the study of such 1-dimensional systems (graphical iteration, the role of the absolute value of the derivative in a fixed point, though, strictly speaking, this falls outside of topology). We prove the Li and Yorke Theorem, which states that when a system on an interval has a point of period 3 then all periods are present. We also discuss Šarkovskij's Theorem about which combinations of periods can occur in a dynamical system on an interval and we show by examples that what is allowed by Šarkovskij's Theorem is also possible. Finally, we discuss transitive systems on intervals.

2.1 Graphical iteration

Graphical iteration is a method to construct initial segments of orbits under a mapping of an interval into itself. It often enables a quick understanding of the behaviour of various points under such a mapping.

In this section, we consider a mapping $f: X \to X$, where X is an interval in $\mathbb{R}$ (open, closed or half open, not necessarily bounded). Draw the graph of f, that is, plot all points $(x, f(x))$ in $X \times X$. In addition, draw the *diagonal* of $X \times X$: the set of points (x, x) with $x \in X$. Obviously, *the invariant points under f correspond to the points of intersection of the graph of f and the diagonal.*

In order to construct an initial segment of the orbit of a non-invariant point $x_0 \in X$, recall that one finds $f(x_0)$ (as a point of the vertical axis) by drawing a vertical line through the point $(x_0, 0)$ until it meets the graph of f in the point $(x_0, f(x_0))$. In order to find $f(f(x_0))$ one first needs the point $(f(x_0), 0)$ of the horizontal axis. This can be found by drawing a horizontal line through the point $(x_0, f(x_0))$ until it meets the diagonal in the point $(f(x_0), f(x_0))$. From there draw a vertical line until it meets the graph of f to find the point $(f(x_0), f^2(x_0))$. Etc.. Figure 2.1 shows how to construct an initial segment of the orbit of x_0 by zigzagging[1] between the diagonal and the graph of f. In point of fact, one identifies the horizontal axis with the diagonal by identifying the point $(x, 0)$ with the point (x, x) for every $x \in \mathbb{R}$. In order to get some experience with graphical iteration it is useful to sketch some graphs and apply this method for several different choices of the initial state. In the following observations we formulate some conclusions that force themselves upon the reader during such experiments.

1 If f is decreasing in the neighbourhood of an invariant point (as in Figure 2.3 below) the picture near that point looks more like a cobweb than a zigzag line. Therefore, these pictures are sometimes called *cobweb diagrams*.

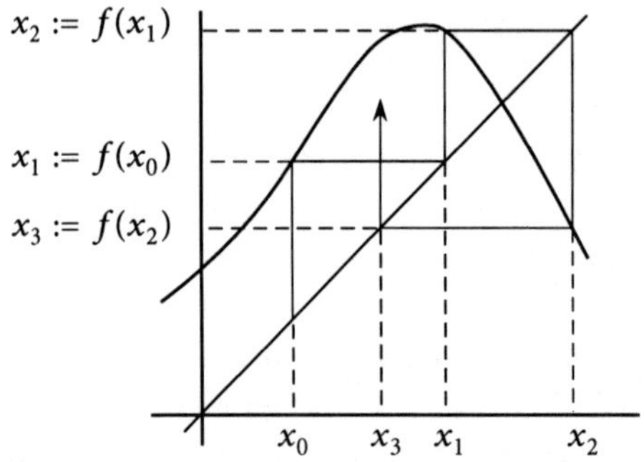

Fig. 2.1. Graphical iteration.

Observation 2.1.1. *Let J be an interval in X and assume that $f(x) \geq x$ for all $x \in J$. Consider a point $x_0 \in J$ such that $\mathcal{O}(x_0) \subseteq J$. Then the orbit of x_0 is either strictly monotonously increasing or eventually constant. In both cases it has a limit $z \in \mathrm{cl}_{\mathbb{R}}(J)$ (in the first case, possibly $z = \infty$). If $z \in X$ then it is an f-invariant point.*

Proof. First, show by induction that $f^{n+1}(x_0) \geq f^n(x_0)$ for all $n \in \mathbb{Z}^+$. Consequently, $\mathcal{O}(x_0)$ is a monotonously increasing sequence. If is is not strictly increasing, then $f^{k+1}(x_0) = f^k(x_0)$ for some $k \in \mathbb{Z}^+$. This means that $f^k(x_0)$ is an invariant point in J, hence $f^n(x_0) = f^k(x_0)$ for all $n \geq k$. If $\mathcal{O}(x_0)$ is strictly increasing and not bounded, then it has limit $z = \infty$. If it is bounded, then it converges to a finite limit $z \in \mathrm{cl}_{\mathbb{R}}(J)$. If $z \in X$ then

$$f(z) = f(\lim_{n\rightsquigarrow\infty} f^n(x_0)) = \lim_{n\rightsquigarrow\infty} f^{n+1}(x_0) = z$$

by continuity of f. This shows that z is an invariant point of f. □

Remark. The condition that $\mathcal{O}(x_0) \subseteq J$ is obviously fulfilled if J is an invariant subset of X. However, for some applications of the observation it would be unsuitable to require invariance of J.

Observation 2.1.2. *Let J be an interval in X, assume that $f(x) \leq x$ for all $x \in J$ and let $x_0 \in J$ such that $\mathcal{O}(x_0) \subseteq J$. Then the orbit of x_0 is either strictly monotonously decreasing or eventually constant. In both cases it has a limit $z \in \mathrm{cl}_{\mathbb{R}}(J)$ (in the first case, possibly $z = -\infty$). If $z \in X$ then it is an f-invariant point.*

Proof. The proof is an obvious modification of the proof of 2.1.1. □

Example. The quadratic mapping $f_\mu : x \mapsto \mu x(1-x) : \mathbb{R} \to \mathbb{R}$ has two invariant points, 0 and $p_\mu := 1 - 1/\mu$. Assume that $0 < \mu < 1$. Claim:

$$\lim_{n\rightsquigarrow\infty} f^n(x) = \begin{cases} -\infty & \text{for } x < p_\mu \text{ and for } x > \hat{p}_\mu\,, \\ 0 & \text{for } p_\mu < x < \hat{p}_\mu\,. \end{cases}$$

Here $\hat{p}_\mu := 1 - p_\mu = 1/\mu$, so that $f_\mu(\hat{p}_\mu) = p_\mu$.

In order to prove this, first observe that f_μ is increasing on the intervals $[p_\mu; 0]$ and $[0; 1/2]$. It follows that these intervals are invariant under f. On the first of these inter-

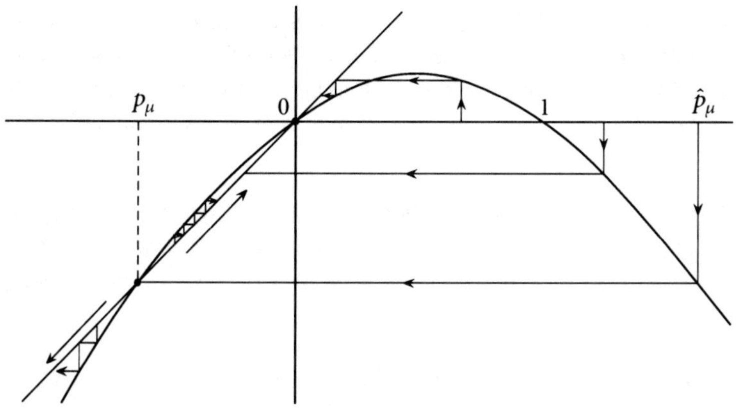

Fig. 2.2. Graphical iteration for $f_\mu : x \mapsto \mu x(1-x) : \mathbb{R} \to \mathbb{R}$ with $0 < \mu < 1$.

vals we have $f_\mu(x) \geq x$, on the second $f_\mu(x) \leq x$. Hence the Observations 2.1.1 and 2.1.2 can be applied: for a point x in the interior of one of these intervals, $f_\mu^n(x)$ tends to an invariant point (which can only be 0) for $n \rightsquigarrow \infty$. For a point x strictly between $1/2$ and $\hat{p}_\mu$, the point $f_\mu(x)$ lies in the interior of one of the intervals just mentioned, so $f_\mu^{n+1}(x) = f_\mu^n(f_\mu(x)) \rightsquigarrow 0$ for $n \rightsquigarrow \infty$. If $x < p_\mu$ then the orbit of x is decreasing but cannot converge to any invariant point, hence it tends to $-\infty$. Finally, if $x > \hat{p}_\mu$ then $f_\mu(x) < p_\mu$, hence $f_\mu^{n+1}(x) \rightsquigarrow -\infty$ for $n \rightsquigarrow \infty$.

For a related observation, see Exercise 2.1. These observations do not apply in the vicinity of an invariant point where f is decreasing; see Figure 2.3. From pictures like this, one may guess that iterations starting close to an invariant point converge to it if the graph of f is not too steep at that point. This guess is confirmed in the Propositions 2.1.3 and 2.1.4 below.

For the proofs of these propositions it is useful to recall the following: the closure $\text{cl}_X(J)$ in X of an open subinterval $J := (a; b) \subseteq X$ is equal to the set $X \cap \text{cl}_{\mathbb{R}}(J)$. Here

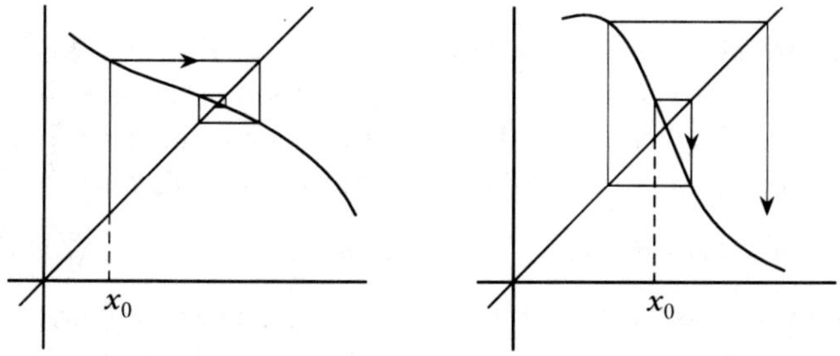

Fig. 2.3. The points $f^n(x_0)$ are situated alternately to the left and the right side of the invariant point. Depending on the steepness of the graph the orbit may or may not converge to this invariant point.

$\mathrm{cl}_{\mathbb{R}}(J) = [a;b]$ if a and b are finite, $\mathrm{cl}_{\mathbb{R}}(J) = [a;\infty)$ if only a is finite, and $\mathrm{cl}_{\mathbb{R}}(J) = (-\infty;b]$ if only b is finite (if both a and b are infinite we have $X = J = \mathbb{R}$). So in all cases $\mathrm{cl}_X(J)$ is obtained from J by adding to J the end points of J that belong to X. In particular, it is a subinterval of X. Consequently, points situated *strictly between* two points of $\mathrm{cl}_X(J)$ are always in J.

Proposition 2.1.3. *Let $f: X \to X$ be a differentiable function with a continuous derivative and let x_0 be an invariant point of f in X. Assume that $|f'(x_0)| < 1$. Then there exists a real number $h_0 > 0$ such that for every neighbourhood $J_h := X \cap (x_0 - h; x_0 + h)$ of x_0 in X with $0 < h \leq h_0$:*

$$f[\,\mathrm{cl}_X(J_h)\,] \subseteq J_h \quad \textit{and} \quad \bigcap_{n=0}^{\infty} f^n[\,\mathrm{cl}_X(J_h)\,] = \{x_0\}\,. \tag{2.1-1}$$

In particular,

$$\lim_{n\rightsquigarrow\infty} f^n(x) = x_0 \quad \textit{for all } x \in X \textit{ with } |x - x_0| \leq h_0\,. \tag{2.1-2}$$

Proof. Let c be a real number such that $|f'(x_0)| < c < 1$. Because of the continuity of f' on X there exists $h_0 > 0$ such that $|f'(x)| \leq c$ for every point $x \in X$ with $|x - x_0| < h_0$. Now let $h \in \mathbb{R}$, $0 < h \leq h_0$, and let J_h be defined as in the statement of the proposition. It follows from the mean value theorem that for every point x in $\mathrm{cl}_X(J_h)$ there is a point ξ strictly between x and x_0, hence in J_h, such that

$$|f(x) - x_0| = |f(x) - f(x_0)| = |f'(\xi)| \cdot |x - x_0| \overset{*}{\leq} ch < h\,, \tag{2.1-3}$$

where the inequality $\overset{*}{\leq}$ is justified by the fact that $\xi \in J_h \subseteq J_{h_0}$. It follows immediately from (2.1-3) that $f(x) \in (x_0 - h; x_0 + h)$ for all $x \in \mathrm{cl}_X(J_h)$; of course, also $f(x) \in X$, hence $f(x) \in J_h$ for these values of x. This proves the first part of (2.1-1). With induction it easily follows that

$$f^n(x) \in J_h \quad \text{for all } x \in \mathrm{cl}_X(J_h) \text{ and all } n \in \mathbb{Z}^+.$$

Next, consider again an arbitrary point $x \in \mathrm{cl}_X(J_h)$. We claim that

$$|f^n(x) - x_0| \leq c^n h \tag{2.1-4}$$

for every $n \in \mathbb{N}$. The proof is by induction. For $n = 1$ this inequality follows from (2.1-3). If the inequality (2.1-4) is true for some $n \in \mathbb{N}$ then an application of (2.1-3) with x replaced by $f^n(x)$ and h replaced by $c^n h$ establishes the desired inequality for $n + 1$. This completes the proof of the claim. Note that the claim can be reformulated as

$$f^n[\,\mathrm{cl}_X(J_h)\,] \subseteq [x_0 - c^n h; x_0 + c^n h] \quad \text{for all } n \in \mathbb{N}\,. \tag{2.1-5}$$

As $\lim_{n\rightsquigarrow\infty} c^n h = 0$ this clearly implies that $\bigcap_{n=0}^{\infty} f^n[\,\mathrm{cl}_X(J_h)\,] \subseteq \{x_0\}$. Since $x_0 = f^n(x_0) \in f^n[\mathrm{cl}_X(J_h)]$ for all values of n in $\mathbb{Z}^+$, this completes the proof of the second part of (2.1-1).

Finally, the condition that $x \in X$ and $|x - x_0| \leq h_0$ is equivalent to the condition that $x \in \mathrm{cl}_X(J_{h_0})$. Now apply formula (2.1-4), taking into account that $c^n h_0 \rightsquigarrow 0$ if $n \rightsquigarrow \infty$ (recall that $0 < c < 1$). This proves (2.1-2). □

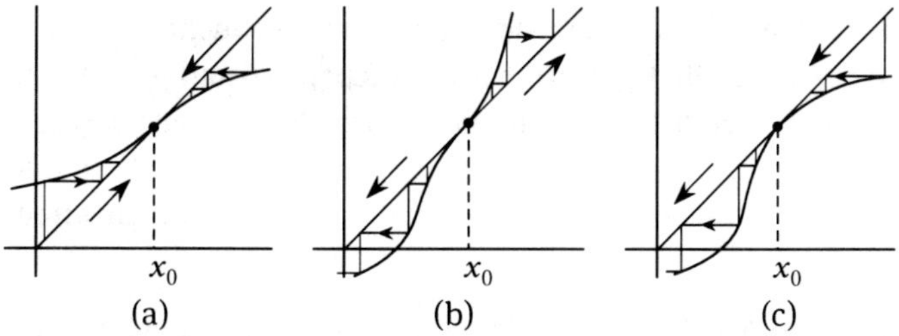

Fig. 2.4. (a) The point x_0 has an attracting character. (b) x_0 has a repelling character. (c) x_0 has a mixed character.

Proposition 2.1.4. *Let $f: X \to X$ be a differentiable function with a continuous derivative and let x_0 be an invariant point of f in X. Assume that $|f'(x_0)| > 1$. Then there exists a real number $h > 0$ such that for the neighbourhood $J_h := X \cap (x_0 - h; x_0 + h)$ of x_0 in X one has*

$$\forall\, x \in J_h \setminus \{x_0\}\ \exists n \geq 1 \,\vdots\, f^n(x) \notin \mathrm{cl}_X(J_h) \quad \textit{and} \quad f[J_h] \supseteq \mathrm{cl}_X(J_h)\,.$$

Proof. There are real numbers $c > 1$ and $h > 0$ such that $|f'(x)| \geq c$ for all $x \in \mathrm{cl}_X(J_h)$. Using this, a similar application of the mean value theorem as in the preceding proof shows the correctness of our proposition. □

In the situation of Proposition 2.1.3 we call the invariant point x_0 an *attracting invariant point*. From now on we shall use this term only for this situation: an invariant point of a continuously differentiable mapping where the derivative is in absolute value smaller than 1. Similarly, a *repelling* invariant point is an invariant point of a continuously differentiable mapping where the derivative has absolute value greater than 1.

If $f: X \to X$ is a continuously differentiable function and $x_0 \in X$ is an invariant point with $|f'(x_0)| = 1$, then x_0 may have an attracting character, a repelling character or it may have a mixed character. See Figure 2.4.

Remarks.

(1) It is obvious that if h is sufficiently small then $\mathrm{cl}_X(J_h) = \mathrm{cl}_{\mathbb{R}}(J_h)$ (also if x_0 is an end point of X, for then $x_0 \in X$), which is a closed and bounded interval, hence a compact set.

(2) If the first condition in formula (2.1-1) is fulfilled for a certain h then the sets $f^n[\,\mathrm{cl}_X(J_h)\,]$ for $n \in \mathbb{N}$ form a descending chain of sets. Suppose, in addition, that h is sufficiently small, so that these sets are compact. In that case the condition $\bigcap_{n=0}^{\infty} f^n[\,\mathrm{cl}_X(J_h)\,] = \{x_0\}$ is equivalent to the following:

$$\forall\, U \in \mathcal{N}_{x_0} \;:\; f^n[\,\mathrm{cl}_X(J_h)\,] \subseteq U \quad \text{for almost all } n$$

(that is: $f^n(x) \rightsquigarrow x_0$ *uniformly in* x on the compact interval $\mathrm{cl}_X(J_h)$). Obviously, because X is a Hausdorff space, the second statement implies the first. The converse follows easily from the compactness argument presented in Lemma A.2.2 in Appendix A.

2.1.5 (Application to the quadratic family). Consider the quadratic functions $f_\mu : x \mapsto \mu x(1-x) : \mathbb{R} \to \mathbb{R}$ with $\mu > 0$. Recall that f_μ has the invariant points 0 and $p_\mu := 1 - 1/\mu$. The values of the derivative of f in these invariant points are $f'(0) = \mu$ and $f'(p_\mu) = |2-\mu|$. We shall also have occasion to use the point $\hat{p}_\mu := 1 - p_\mu = 1/\mu$. Note that $f_\mu(\hat{p}_\mu) = f_\mu(p_\mu) = p_\mu$. Finally, observe that $f_\mu(1) = 0$, so 1 is an eventually invariant point.

$0 < \mu < 1$. In this case, $p_\mu < 0$. The invariant point p_μ is repelling and the invariant point 0 is attracting. Moreover, the conclusion of Proposition 2.1.3 holds for the attracting invariant point 0 with $h_0 = \min\{|p_\mu|, 1/2\}$. On the other hand, we have seen in the Example after 2.1.2 that $\lim_{n\rightsquigarrow\infty} f_\mu^n(x) = 0$ for every point x in the open interval $(p_\mu; \hat{p}_\mu)$. For all other points x, except the points p_μ and $\hat{p}_\mu$, we have $\lim_{n\rightsquigarrow\infty} f_\mu^n(x) = -\infty$

$\mu = 1$. Now $p_\mu = 0$, so 0 is the unique invariant point. Because $f_\mu'(0) = 1$, the Propositions 2.1.3 and 2.1.4 don't give a decisive answer. In fact, the point 0 turns out to be of 'mixed type':

$$\lim_{n\rightsquigarrow\infty} f_\mu^n(x) = \begin{cases} -\infty & \text{for } x < 0 \text{ and for } x > 1\,, \\ 0 & \text{for } 0 \le x \le 1\,. \end{cases}$$

The proof consists of a straightforward application of the methods of 2.1.1 and 2.1.2: the orbit of x is monotonously decreasing for $x \in (-\infty; 0)$ and for $x \in (0; 1/2]$. In the first case it cannot converge to 0 and since there are no other invariant points, it tends to $-\infty$. In the second case it is bounded, hence converges to an invariant point, which must be 0. Finally, for all points x with $1/2 \le x < 1$ or $x > 1$, use that $f_\mu(x)$ is in the one of the intervals mentioned above.

$1 < \mu \le 2$. Now $0 < p_\mu \le 1/2$, 0 is a repelling and p_μ is an attracting invariant point. As before, the observations in 2.1.1 and 2.1.2 show that the sequence $f_\mu^n(x)$ tends to $-\infty$ if $x < 0$ and also if $x > 1$ (the latter case reduces to the first because $f_\mu(x) < 0$ for $x > 1$). Moreover, $f_\mu^n(x)$ converges to the invariant point p_μ if x is in one of the (invariant) intervals $(0; p_\mu]$ and $[p_\mu; 1/2]$. This is also the case for every point x in the interval $(1/2; 1)$, because then the point $f_\mu(x)$ is in one of the intervals just mentioned. It follows that $\lim_{n\rightsquigarrow\infty} f_\mu^n(x) = p_\mu$ for all $x \in (0; 1)$.

$2 \le \mu \le 3$. Now $p_\mu \ge 1/2$, and as in the previous case, 0 is a repelling invariant point and, for $\mu \ne 3$, p_μ is an attracting invariant point. Again, it is clear that $f_\mu^n(x) \rightsquigarrow -\infty$ for $n \rightsquigarrow \infty$ for $x < 0$ and for $x > 1$. We shall show that the orbit of x converges to the attracting invariant point p_μ for all $x \in (0; 1)$, also if $\mu = 3$. This cannot be done by a straightforward application of the methods used above, because no intervals can be found on which $f_\mu(x) \ge x$ or $f_\mu(x) \le x$ and which are also invariant. Instead, we shall identify an interval J with the following properties: (a) $p_\mu \in J$ and for all $x \in J$ we have $\lim_{n\rightsquigarrow\infty} f_\mu^n(x) = p_\mu$, and (b): J 'catches' the orbits of all other points in the open unit interval, that is, for every point $x \in (0; 1)$ there exists $k_x \in \mathbb{Z}^+$ such that

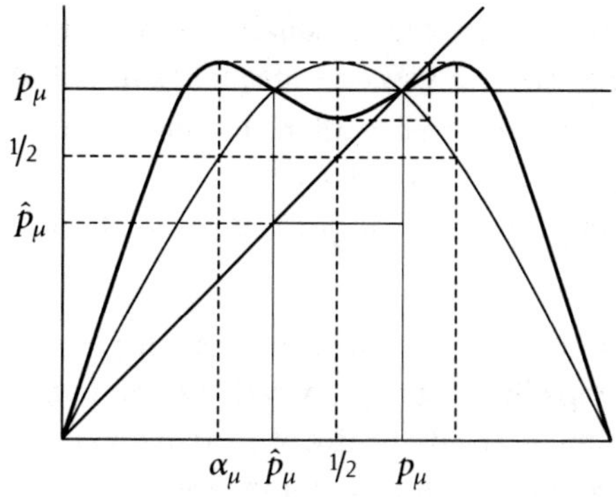

Fig. 2.5. The graph of f_μ^2 for $2 \le \mu \le 3$. Also the graph of f_μ is drawn. For the meaning of α_μ, see Exercise 2.2.

$f_\mu^{k_x}(x) \in J$. This is sufficient: in that case, it is clear that $\lim_{n\rightsquigarrow\infty} f_\mu^n(f_\mu^{k_x}(x)) = p_\mu$, hence $\lim_{n\rightsquigarrow\infty} f_\mu^n(x) = p_\mu$ as well.

To achieve our goal[2] we have to consider the mapping f_μ^2 and its iterates. See Figure 2.5 for a sketch of the graph of f_μ^2. In what follows we shall rely heavily on this graph, but we urge the reader to prove all statements analytically; see Exercise 2.2 for some hints.

It is straightforward to show that, for all $x \in [\hat{p}_\mu; p_\mu]$ we have $f_\mu^2(x) \ge x$ and $f_\mu^2(x) \in [1/2; p_\mu] \subseteq [\hat{p}_\mu; p_\mu]$, so 2.1.1 implies that for every point x in the interval $[\hat{p}_\mu; p_\mu]$ the orbit under f_μ^2 converges to the point p_μ. This takes care of the even itinerates of f_μ. For the odd itinerates of f_μ we get

$$\lim_{n\rightsquigarrow\infty} f_\mu^{2n+1}(x) = f_\mu(\lim_{n\rightsquigarrow\infty} f_\mu^{2n}(x)) = f_\mu(p_\mu) = p_\mu$$

for $x \in [\hat{p}_\mu; p_\mu]$. Since both sequences have the same limit, we get

$$\lim_{n\rightsquigarrow\infty} f_\mu^n(x) = p_\mu \quad \text{for all } x \in [\hat{p}_\mu; p_\mu]\,.$$

Thus, condition (a) above is fulfilled by the interval $[\hat{p}_\mu; p_\mu]$. In order to prove that this interval also satisfies condition (b), consider a point x in the interval $(0; \hat{p}_\mu)$ and suppose that $f_\mu^n(x) \in (0; \hat{p}_\mu)$ for all $n \in \mathbb{Z}^+$. Then Observation 2.1.1 would imply that the (increasing) sequence $\left(f_\mu^n(x)\right)_{n\in\mathbb{Z}^+}$ converges to an invariant point in the interval $[0; \hat{p}_\mu]$, which cannot be 0. Since $\hat{p}_\mu < p_\mu$, it cannot be p_μ either. Consequently, there is a first value of n such that $f_\mu^n(x) \notin (0; \hat{p}_\mu)$, that is, there exists $k \in \mathbb{N}$ such that

$$f_\mu^{k-1}(x) < \hat{p}_\mu \quad \text{and} \quad f_\mu^k(x) \ge \hat{p}_\mu\,. \tag{2.1-6}$$

2 In view of Proposition 2.1.3 there exists an interval J_{h_0} around p_μ satisfying condition (a), but this turns out to be too small to satisfy condition (b). Another approach would be as follows: the closure J of the open interval on which $|f_\mu'(x)| < 1$ is large enough to satisfy condition (b) and the mean value theorem would imply that J satisfies condition (a), *provided* it is invariant. Straightforward computations show that this is certainly not the case for $1 + \sqrt{3} < \mu \le 3$.

Because f_μ is increasing on the interval $(0; \hat{p}_\mu]$, the first inequality implies that $f_\mu^k(x) = f_\mu(f_\mu^{k-1}(x)) < f_\mu(\hat{p}_\mu) = p_\mu$. Together with the second inequality in (2.1-6) this means precisely that $f_\mu^k(x) \in [\hat{p}_\mu; p_\mu]$. As observed earlier, this implies that $\lim_{n\rightsquigarrow\infty} f_\mu^n(x) = p_\mu$. Finally, consider a point $x \in (p_\mu; 1)$. Then $f_\mu(x) \in (0; \hat{p}_\mu)$, so the orbit of the point $f_\mu(x)$ and hence the orbit of x, converges to p_μ. This completes the proof.

$\mu > 3$. In this case both invariant points 0 and p_μ are repelling. However, for μ only slightly larger than 3 there is, close to the repelling point p_μ, a periodic orbit with period 2. Heuristically, this can be seen as follows: The derivative of f_μ^2 at the point p_μ is $f_\mu'(f_\mu(p_\mu)) \cdot f_\mu'(p_\mu) = |2 - \mu|^2$, which is equal to 1 for $\mu = 3$ and larger than 1 if $\mu > 3$. Hence for μ slightly larger than 3 the graph of f_μ^2 intersects the diagonal close to p_μ in two points x_1^μ and x_2^μ, one at each side of p_μ. This suggests the existence of a periodic orbit with period 2. For details, see Exercise 2.2 (d). There the reader is also asked to show that for $\mu < 1 + \sqrt{6} \approx 2.45\ldots$ this periodic orbit is attracting. We have not yet defined the notion of an attracting periodic orbit, and here we give only the definition for the special case under consideration. The theoretical background will be given later, in Section 3.3.

The periodic orbit $\{x_1^\mu, x_2^\mu\}$ is said to be *attracting* whenever x_1^μ and x_2^μ are attracting invariant points of f_μ^2, that is, the derivative of f_μ^2 in both points has absolute value less than 1. If that is the case then for $i = 1, 2$ there exists, by Proposition 2.1.3, a neighbourhood J_i of x_i^μ such that $f_\mu^{2n}(x) \rightsquigarrow x_i^\mu$ for $x \in J_i$. Thus, if $n \rightsquigarrow \infty$ then

$$f_\mu^{2n}(x) \rightsquigarrow x_1^\mu \quad \text{and} \quad f_\mu^{2n+1}(x) \rightsquigarrow f_\mu(x_1^\mu) = x_2^\mu \quad \text{for } x \in J_1 \cap f_\mu^{\leftarrow}[J_2]$$

Similar formulas, with x_1^μ and x_2^μ interchanged, hold for $x \in J_2 \cap f_\mu^{\leftarrow}[J_1]$. Thus, for every point in the neighbourhood $(J_1 \cap f_\mu^{\leftarrow}[J_2]) \cup (J_2 \cap f_\mu^{\leftarrow}[J_1])$ of the periodic orbit the even iterates under f_μ converge to one of the periodic points and the odd iterates converge to the other, alternately hopping from the vicinity of x_1^μ to the vicinity of x_2^μ and back, approaching these points closer and closer in the process.

2.2 Existence of periodic orbits

In this section (X, f) is, again, a dynamical system on an interval X in $\mathbb{R}$ (open, closed or half-open, not necessarily bounded).

Notation. Let I and J be subsets of X and let g be any mapping whose domain includes I. The following notation will be used:
- $g : I \overset{\circ}{\rightarrow} J$ means: $g[I] \supseteq J$ (g maps I *over* J),
- $g : I \twoheadrightarrow J$ means: $g[I] = J$ (g maps I *onto* J).

When there is no danger of confusion about which mapping is considered we shall also employ notations like $g\colon I_1 \overset{\circ}{\rightarrow} I_2 \twoheadrightarrow I_3 \overset{\circ}{\rightarrow} I_4$; it should be clear that in this case we are talking about $g^3 : I_1 \overset{\circ}{\rightarrow} I_4$.

Lemma 2.2.1. *Let I, J and I_k for $k = 0, \dots, n$ be compact intervals included in X.*
(1) *If $f : I \overset{\circ}{\rightarrow} I$ then I contains an invariant point of f.*
(2) *If $f : I \overset{\circ}{\rightarrow} J$ then there is a closed subinterval I' of I such that $f : I' \twoheadrightarrow J$.*
(3) *If $I_n \overset{\circ}{\rightarrow} I_{n-1} \overset{\circ}{\rightarrow} \dots \overset{\circ}{\rightarrow} I_1 \overset{\circ}{\rightarrow} I_0$ then for $k = 1, \dots, n$ there is a closed subinterval J_k of I_k such that $J_n \twoheadrightarrow J_{n-1} \twoheadrightarrow \cdots \twoheadrightarrow J_1 \twoheadrightarrow I_0$.*

Proof. Let $I = [a; b]$ with $a < b$. As $a, b \in I \subseteq f[I]$, there are points x_a and x_b in I such that $f(x_a) = a$ and $f(x_b) = b$. Hence $x_a \geq a = f(x_a)$ and $x_b \leq b = f(x_b)$. So the continuous mapping $x \mapsto f(x) - x$ changes sign between x_a en x_b, hence assumes the value 0. This means that f has an invariant point in I.

Statement 3 follows easily from statement 2 by induction in n. In order to prove statement 2 it is sufficient to prove the following two claims:

(a) There exists a closed subinterval I' of I such that $f[I'] \supseteq J$ and which is minimal for this condition, i.e., if I'' is a closed subinterval of I' such that $f[I''] \supseteq J$, then $I'' = I'$.
(b) If a subinterval I' of I satisfies condition (a) then $f[I'] = J$.

Ad (a): This is a direct consequence of Zorn's Lemma. Let $\mathcal{L}$ be the collection of all closed subintervals I' of I with the property that $f[I'] \supseteq J$. Then $\mathcal{L}$ is partially ordered by inclusion and in order to apply Zorn's Lemma we have to show that this partial ordering is inductive. To this end, it is sufficient to show: if $\mathcal{K}$ is a chain in $\mathcal{L}$ then $\bigcap \mathcal{K}$ is a member of $\mathcal{L}$. Being an intersection of closed subintervals of I, $\bigcap \mathcal{K}$ is a closed subinterval of I, and

$$f[\bigcap \mathcal{K}] \overset{(1)}{=} \bigcap_{K \in \mathcal{K}} f[K] \overset{(2)}{\supseteq} J . \tag{2.2-1}$$

Here identity (1) follows from the lemma in Appendix A.3.5 and inclusion (2) is clear from the choice of $\mathcal{K}$ such that $f[K] \supseteq J$ for every $K \in \mathcal{K}$. Consequently, $\bigcap \mathcal{K} \in \mathcal{L}$. This completes the proof of (a).

Ad (b): Consider a subinterval $I' = [a'; b']$ of I such that $f[I'] \supseteq J$ and which is minimal with respect to this property. Let p and q be the left and right end point of J, respectively, so that $J = [p; q]$ with $p < q$. There are points a'' and b'' in I' such that $f(a'') = p$ and $f(b'') = q$, and let I'' be the closed interval with end points a'' and b''. Because $f[I'']$ is an interval containing the points p and q, it includes the interval $[p; q] = J$. By the minimality property of I' we get $I'' = I'$, hence $a'' = a'$ and $b'' = b'$, or just the other way round, $a'' = b'$ and $b'' = a'$. Conclusion: f assumes the values p and q on the interval I' only in the end points a' and b' of I'.

Now suppose that $f[I'] \supsetneq J$, i.e., assume that there is a point $x' \in I'$ such that either $f(x') > q$ or $f(x') < p$. In the first case, argue as follows (in the second case, the argument is similar). The continuous function f has the value $p < q$ in one of the end

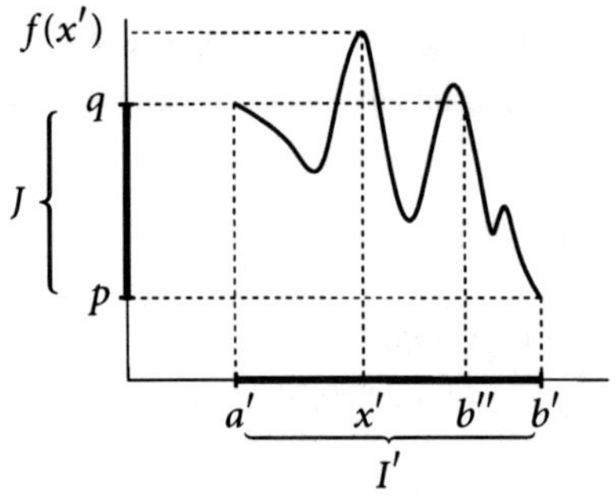

Fig. 2.6. Illustrating the proof of Claim (b); the interval I' is not minimal with respect to the condition $f[I'] \supseteq J$.

points of the interval I' and it has the value $f(x') > q$ in the interior point x' of I'. So by the intermediate value theorem, f assumes the value q somewhere between that end point and x', so in an interior point x'' of I'. See also Figure 2.6. This contradicts our earlier result that the value q can only be assumed in an end point of I'. Consequently, $f[I'] = I$. □

Remark. It follows from the proof of part 2 of the lemma that I' can be chosen such that the interior of I' is mapped onto the interior of J, and the two end points of I' onto the two end points of J.

Theorem 2.2.2 (Li and Yorke). *Suppose the dynamical system (X, f) has a periodic point with primitive period 3. Then for every $p \in \mathbb{N}$ there is a periodic point in X with primitive period p.*

Proof. Let $\{x_1, x_2, x_3\}$ be a periodic orbit with a primitive period of 3 and assume that $f: x_1 \mapsto x_3 \mapsto x_2 \mapsto x_1$. Essentially, there are only the following two possibilities for the ordering of the three points x_1, x_2 and x_3 in $\mathbb{R}$: either $x_1 < x_2 < x_3$ or $x_3 < x_2 < x_1$. Other possible orderings boil down to one of these two (up to the numbering of the points). See Figure 2.7. Let I_0 be the closed interval with end points x_1 and x_2, and let I_1 be the closed interval with end points x_2 and x_3. Then I_0 and I_1 are subintervals of X and, using that the continuous image of an interval is an interval, one easily shows that

$$f : I_0 \overset{\circ}{\rightsquigarrow} I_0 \cup I_1 , \tag{2.2-2}$$

$$f : I_1 \overset{\circ}{\rightsquigarrow} I_0 . \tag{2.2-3}$$

Formula (2.2-2) implies that $f : I_0 \overset{\circ}{\rightsquigarrow} I_0$. Hence it follows immediately from Lemma 2.2.1 (1) that there is an invariant point of f in I_0, hence in X.

Next, we show that there is a periodic point with primitive period 2. To this end, note that the formulas (2.2-2) and (2.2-3) imply that

$$I_0 \overset{\circ}{\rightsquigarrow} I_1 \overset{\circ}{\rightsquigarrow} I_0 . \tag{2.2-4}$$

So by Lemma 2.2.1 (2) there is a closed subinterval J_0 of I_0 such that

$$J_0 \twoheadrightarrow I_1 \overset{\circ}{\to} I_0 \supseteq J_0 \,. \tag{2.2-5}$$

It follows that $f^2 : J_0 \overset{\circ}{\to} J_0$, hence there is an invariant point z_0 of f^2 in J_0, hence in X. Then z_0 is a periodic point with period 2 under f. It remains to show that the point z_0 has *primitive* period 2, i.e., that it is not an invariant point under f. Assume the contrary. By (2.2-5), $f : J_0 \twoheadrightarrow I_1$ hence $z_0 = f(z_0) \in f[J_0] = I_1$. So z_0 is a common point of J_0 and I_1; consequently, it is a common point of I_0 and I_1. However, the only common point of the intervals I_0 and I_1 is x_2, which is not an invariant point. This contradiction show that the point z_0 is periodic with primitive period 2.

For $p = 3$ there is nothing to prove, so let $p \geq 4$. Consider the following diagram, which is based upon the formulas (2.2-2) and (2.2-3):

$$\underbrace{I_0 \overset{\circ}{\to} I_0 \overset{\circ}{\to} \dots \overset{\circ}{\to} I_0}_{p-2 \text{ arrows}} \overset{\circ}{\to} I_1 \overset{\circ}{\to} I_0 \,.$$

In view of Lemma 2.2.1 (3), for $k = 0, \dots, p-2$ there is a closed subinterval J_k of I_0 such that

$$J_{p-2} \twoheadrightarrow J_{p-3} \twoheadrightarrow \cdots \twoheadrightarrow J_0 \twoheadrightarrow I_1 \overset{\circ}{\to} I_0 \supseteq J_{p-2} \,. \tag{2.2-6}$$

Thus, we have $f^p : J_{p-2} \overset{\circ}{\to} J_{p-2}$, so by Lemma 2.2.1 (1) there is a periodic point $z_1 \in J_{p-2}$ with period p. It remains to prove that p is the primitive period of z_1. Assume the contrary: z_1 has primitive period q with $q < p$. Then Lemma 1.1.2 (c) implies that q is a divisor of p and because $p \geq 4$, this implies that $q \leq \frac{1}{2}p \leq p-2$. So in view of formula (2.2-6) the orbit $\{ z_1, f(z_1), \dots, f^{q-1}(z_1) \}$ of z_1 is included in $J_{p-2} \cup J_{p-3} \cup \dots \cup J_0 \subseteq I_0$. On the other hand, it follows from (2.2-6) that the orbit of z_1 has a point in the interval I_1. The only common point of I_0 and I_1 is x_2, hence $x_2 \in \mathcal{O}(z_1)$. But then also $x_3 = f^2(x_2) \in \mathcal{O}(z_1)$, contradicting the conclusion that the orbit of z_1 is completely included in the interval I_0. □

Remark. For a related results, see Exercise 2.7 and Proposition 7.4.1.

Theorem 2.2.2 was published in 1975 in a paper entitled *'Period three implies chaos'* (for the 'chaos'-part, see Chapter 7 ahead). Later it became clear that this theorem is a trivial consequence of a result that was published eleven years earlier by the Russian mathematician Šarkovskij. He defined an ordering on $\mathbb{N}$, nowadays called the

Fig. 2.7. The possible orderings of the points of a periodic orbit of period 3.

Šarkovskij ordering, as follows:

$$
\begin{aligned}
&3 \prec 5 \prec 7 \prec \cdots \prec 2k-1 \prec 2k+1 \prec \dots \\
&\qquad \prec 2\cdot 3 \prec 2\cdot 5 \prec \cdots \prec 2(2k-1) \prec 2(2k+1) \prec \dots \\
&\qquad \dots \quad \dots \quad \dots \quad \dots \quad \dots \quad \dots \quad \dots \quad \dots \quad \dots \\
&\qquad \prec 2^l\cdot 3 \prec 2^l\cdot 5 \prec \cdots \prec 2^l(2k-1) \prec 2^l(2k+1) \prec \dots \\
&\qquad \prec 2^{l+1}\cdot 3 \prec 2^{l+1}\cdot 5 \prec \cdots \prec 2^{l+1}(2k-1) \prec 2^{l+1}(2k+1) \prec \dots \\
&\qquad \dots \quad \dots \quad \dots \quad \dots \quad \dots \quad \dots \quad \dots \quad \dots \quad \dots \quad \dots \quad \dots \\
&\qquad \dots \quad \dots \quad \dots \quad \dots \quad \cdots \prec 2^{n+1} \prec 2^n \prec \cdots \prec 8 \prec 4 \prec 2 \prec 1 .
\end{aligned}
$$

This ordering starts with the number 3 so, indeed, Theorem 2.2.2 follows from the following theorem:

Theorem 2.2.3 (Šarkovskij). *Let (X, f) be a dynamical system on an interval X in $\mathbb{R}$. If f has a periodic point with primitive period n, than f has periodic points with primitive period m for all $m \in \mathbb{N}$ with $n \prec m$.*

Proof. See Section 2.5. □

In the next two sections we shall show by examples that everything that is allowed by Šarovskij's Theorem is possible. To this end, we reformulate this theorem in a more convenient form. For every $n \in \mathbb{N}$, let

$$\mathbf{S}(n) := \{n\} \cup \{ k \in \mathbb{N} \,\vdots\, n \prec k \} .$$

In addition, let

$$\mathbf{S}(2^\infty) := \{ 2^m \,\vdots\, m \in \mathbb{Z}^+ \} ,$$

where the symbol 2^∞ means something like 'an entity larger than all powers of 2'. The set $\mathbb{N} \cup \{2^\infty\}$ will be denoted by $\mathbb{N}_\infty$ and the sets of the form $\mathbf{S}(n)$ with $n \in \mathbb{N}_\infty$ will be called *Šarkovskij tails* of $\mathbb{N}$. Finally, for any dynamical system (X, f) we denote by $\mathbf{Per}(f)$ the set of all primitive periods of periodic orbits under f. Now Šarkovskij's Theorem can be formulated as follows:

Theorem 2.2.4 (Šarkovskij). *Let (X, f) be a dynamical system on an interval X in $\mathbb{R}$. If $\mathbf{Per}(f) \neq \emptyset$ then there exists $n \in \mathbb{N}_\infty$ such that $\mathbf{Per}(f) = \mathbf{S}(n)$.*

Proof. Observe that, if $\mathbf{Per}(f) \neq \emptyset$ then it has a first element in the Šarkovskij ordering. □

2.3 The truncated tent map

For every $\lambda \in [0; 1]$ we define *the tent map truncated by λ* as

$$T_\lambda(x) := \min\{ T(x), \lambda \} \quad \text{for } x \in [0; 1] .$$

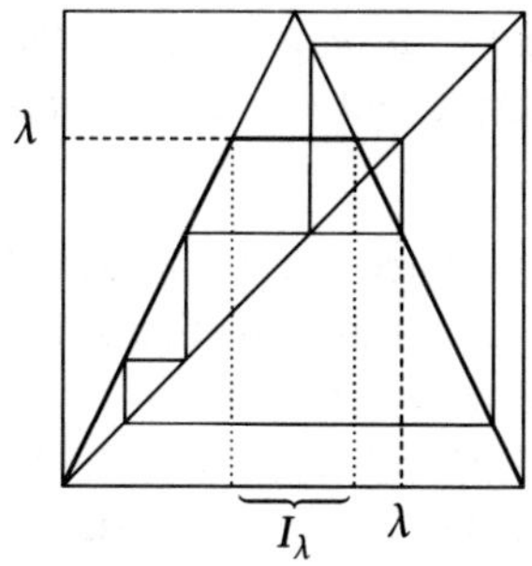

Fig. 2.8. The truncated tent map T_λ. In this example the point λ is both T_λ-periodic and T-periodic, with primitive periods 2 and 4, respectively.

Here T denotes the tent map defined in Example 0.4.2 ; see also 1.7.3. Below we shall repeatedly use the following trivial observation: if $x \in [0; 1]$ and x is not in the open interval $I_\lambda := (\frac{\lambda}{2}; 1 - \frac{\lambda}{2})$ where T is strictly greater than λ then $T(x) = T_\lambda(x)$. In fact,

$$T_\lambda(x) < T(x) \iff x \in I_\lambda \implies T_\lambda(x) = \lambda \,.$$

In general, points will not have the same orbits under T and T_λ. Obviously, a T-orbit is a T_λ-orbit, or a T_λ-orbit is a T-orbit, iff T and T_λ agree on that orbit, iff that orbit is included in the interval $[0; \lambda]$, iff that orbit has no points in I_λ. On the other hand, it is well possible that T and T_λ have periodic points in common which have different primitive periods, hence have different orbits. See Figure 2.8. Such a phenomenon can only occur if the periodic T-orbit includes a point x in I_λ, so that $T_\lambda(x) = \lambda$ and, in addition, the point λ is periodic under T_λ.

The truncated tent map T_1 equals the non-truncated tent map T. In the Introduction we have seen that for every $n \in \mathbb{N}$ there are periodic points with period n under T_1, but it is not yet clear that there are also points with *primitive* period n. However, there are $2^3 = 8$ periodic points with period 3, among which the two invariant points. So there are two periodic orbits with primitive period 3. Then by Li and Yorke's theorem, all possible *primitive* periods occur under T_1, that is: $\mathbf{Per}(T_1) = \mathbb{N} = \mathbf{S}(3)$.

The truncated tent map T_0 is identically equal to 0, hence it has the unique invariant point 0. Obviously, there are no other periodic points, so $\mathbf{Per}(T_0) = \{1\} = \mathbf{S}(1)$.

This shows that the largest and the smallest Šarkovskij tails can be represented as the set of primitive periods for some truncated tent map T_λ ($\lambda = 1$ or 0, respectively). We shall show now that every other Šarkovskij tail can occur as $\mathbf{Per}(T_\lambda)$ for a suitable choice of λ.

Values of λ such that $\mathbf{Per}(T_\lambda) = \mathbf{S}(n)$ for $n \in \mathbb{N}\backslash\{1, 3\}$ are obtained as follows: Recall that for every $n \in \mathbb{N}$ the non-truncated tent map has at least one and at most 2^n periodic orbits with primitive period n. From each of these orbits with primitive period n choose the largest element. Let λ_n be the smallest element so obtained:

$$\lambda_n := \min\{\max B \,\vdots\, B \text{ a periodic } T\text{-orbit with primitive period } n\} \,.$$

It is easily seen that λ_n is the smallest real number λ with the property that the interval $[0; \lambda]$ includes a periodic T-orbit with primitive period n.

Lemma 2.3.1. *Let $n \in \mathbb{N}$ and let $1 \geq \lambda \geq \lambda_n$. Then the point λ_n is periodic under the truncated tent map T_λ with primitive period n. Consequently, for all $n \in \mathbb{N}$ we have $\mathbf{S}(n) \subseteq \mathbf{Per}(T_\lambda)$, for every $\lambda \geq \lambda_n$.*

Proof. Let A be the periodic T-orbit with primitive period n of which λ_n is the maximal element, that is, $\lambda_n = \max\{A\}$. As all values of T on A are in A, they are at most λ_n, hence not larger than λ. Consequently, T and T_λ agree on A, so A is also a T_λ-orbit, i.e., it is a periodic orbit with primitive period n under T_λ. Hence $n \in \mathbf{Per}(T_\lambda)$ and Šarkovskij's Theorem implies that $\mathbf{S}(n) \subseteq \mathbf{Per}(T_\lambda)$. Moreover, since $\lambda_n \in A$, this shows in particular that the point λ_n is periodic under T_λ with primitive period n. □

Example. By brute computational force, using that the graphs of T^2, T^3 and T^4 consist of 2, 4 and 8 'tents', respectively, it is not difficult to find the invariant points of T^2, T^3 and T^4. In addition to the invariant points 0 and 2/3 for T, one finds periodic orbits with the following primitive periods:

Primitive period 2: $\frac{2}{5} \mapsto \frac{4}{5} \mapsto \frac{2}{5}$.
Primitive period 3: $\frac{2}{9} \mapsto \frac{4}{9} \mapsto \frac{8}{9} \mapsto \frac{2}{9}$ and $\frac{2}{7} \mapsto \frac{4}{7} \mapsto \frac{6}{7} \mapsto \frac{2}{7}$.
Primitive period 4: $\frac{2}{17} \mapsto \frac{4}{17} \mapsto \frac{8}{17} \mapsto \frac{16}{17} \mapsto \frac{2}{17}$, $\frac{6}{17} \mapsto \frac{12}{17} \mapsto \frac{10}{17} \mapsto \frac{14}{17} \mapsto \frac{6}{17}$ and $\frac{2}{15} \mapsto \frac{4}{15} \mapsto \frac{8}{15} \mapsto \frac{14}{15} \mapsto \frac{2}{15}$.

It follows that $\lambda_2 = \frac{4}{5}$, $\lambda_3 = \frac{6}{7}$ and that $\lambda_4 = \frac{14}{17}$. Lemma 2.3.4 below implies that the values of λ_n are between $\frac{4}{5}$ and $\frac{6}{7}$ for all $n \geq 2$.

The above lemma applies, in particular, to $\lambda = \lambda_n$. Consequently, the point λ_n is periodic under T_{λ_n}, with primitive period n. So $n \in \mathbf{Per}(T_{\lambda_n})$, hence $\mathbf{S}(n) \subseteq \mathbf{Per}(T_{\lambda_n})$. We shall prove in Theorem 2.3.5 below the converse of this inclusion.

Lemma 2.3.2. *Let $\lambda > 0$ and $k \in \mathbf{Per}(T_\lambda)$. If $\lambda \leq \lambda_k$ then λ is a periodic point under T_λ with primitive period equal to k.*

Proof. Let B be a periodic T_λ-orbit with primitive period k. It is sufficient to show that $\lambda \in B$. Assume the contrary. Then T cannot assume a value greater than or equal to λ in any point of B, for otherwise T_λ would have there the value λ. Consequently, T and T_λ agree on B. It follows that B is a periodic T-orbit with primitive period k. By the definition of λ_k, this implies that $\max\{B\} \geq \lambda_k \geq \lambda$. On the other hand, $B = T_\lambda[B] \subseteq [0;\lambda]$ and $\lambda \notin B$, so $\max\{B\} < \lambda$ (B is finite). Contradiction. □

Corollary 2.3.3. *If $k \in \mathbf{Per}(T_{\lambda_n})$ and $k \neq n$ then $\lambda_k < \lambda_n$.*

Proof. If $\lambda_n \leq \lambda_k$ then Lemma 2.3.2 would imply that λ_n is periodic under T_{λ_n} with primitive period k. But by Lemma 2.3.1 the point λ_n has primitive period n under T_{λ_n}. This would imply that $k = n$. □

Corollary 2.3.4. $\forall\, k, n \in \mathbb{N} : n \prec k \Rightarrow \lambda_k < \lambda_n$.

Proof. If $k, n \in \mathbb{N}$ and $n \prec k$ then by Lemma 2.3.1 and Šarkovskij's Theorem, $k \in \mathbf{S}(n) \subseteq \mathbf{Per}(T_{\lambda_n})$. Now apply Corollary 2.3.3. □

Theorem 2.3.5. $\forall\, n \in \mathbb{N} : \mathbf{Per}(T_{\lambda_n}) = \mathbf{S}(n)$.

Proof. We have seen already that the inclusion $\mathbf{S}(n) \subseteq \mathbf{Per}(T_{\lambda_n})$ holds. Assume that this inclusion is proper: there exists $k \in \mathbf{Per}(T_{\lambda_n}) \setminus \mathbf{S}(n)$. Since $n \in \mathbf{S}(n)$, it follows that $k \neq n$, so by Corollary 2.3.3 above we may conclude that $\lambda_k < \lambda_n$. On the other hand, the assumption that $k \notin \mathbf{S}(n)$ implies that $k \prec n$, hence $\lambda_n < \lambda_k$ by Lemma 2.3.4 (with k and n are interchanged). This contradiction completes the proof. □

In the proof of the above theorem we have nowhere assumed that n is not a power of 2, so the result also holds if n is a finite power of 2. We conclude this section by showing that there is a value λ_∞ such that $\mathbf{Per}(T_{\lambda_\infty}) = \mathbf{S}(2^\infty)$. It will turn out that the following number does the job:

$$\lambda_\infty := \sup\{\, \lambda_{2^n} \,\vdots\, n \in \mathbb{N} \,\}\,.$$

Lemma 2.3.6. *For all $n, m \in \mathbb{Z}^+$ and every odd integer $q \geq 3$ we have*

$$\lambda_{2^n} < \lambda_\infty < \lambda_{q2^m}\,.$$

Proof. In the Šarkovskij ordering we have $q \cdot 2^m \prec 3 \cdot 2^{m+1} \prec 2^{n+1} \prec 2^n$. So Lemma 2.3.4 above implies $\lambda_{2^n} < \lambda_{2^{n+1}} < \lambda_{3 \cdot 2^{m+1}} < \lambda_{q2^m}$. Using the definition of λ_∞ we get the desired inequalities. □

Theorem 2.3.7. $\mathbf{Per}(T_{\lambda_\infty}) = \mathbf{S}(2^\infty)$.

Proof. Consider arbitrary $n \in \mathbb{Z}^+$. Then by definition $\lambda_\infty \geq \lambda_{2^n}$, hence Lemma 2.3.1 (with n replaced by 2^n) implies that $2^n \in \mathbf{Per}(T_{\lambda_\infty})$. As this holds for every $n \in \mathbb{Z}^+$, this means that $\mathbf{S}(2^\infty) \subseteq \mathbf{Per}(T_{\lambda_\infty})$.

Assume that this inclusion is proper and let $k \in \mathbf{Per}(T_{\lambda_\infty}) \setminus \mathbf{S}(2^\infty)$. Then there are $m \in \mathbb{Z}^+$ and $q \in \mathbb{N}$, q odd and $q \neq 1$, such that $k = q \cdot 2^m$, hence Lemma 2.3.6 implies that $\lambda_\infty < \lambda_k$. So by Lemma 2.3.2 with $\lambda = \lambda_\infty$, k is the primitive period of the point λ_∞ under T_{λ_∞}. This implies that k is the unique element of $\mathbf{Per}(T_{\lambda_\infty}) \setminus \mathbf{S}(2^\infty)$. Therefore,

$$\mathbf{Per}(T_{\lambda_\infty}) = \mathbf{S}(2^\infty) \cup (*)\,,$$

where $(*)$ is a singleton subset of $\mathbb{N}$. But there is no possible choice for $(*)$ such that $\mathbf{S}(2^\infty) \cup (*)$ is a Šarkovskij tail, unless $(*)$ is the empty set. This contradiction shows that $\mathbf{Per}(T_{\lambda_\infty}) = \mathbf{S}(2^\infty)$. □

2.4 The double of a mapping

There is yet another method to obtain a mapping that has $\mathbf{S}(2^\infty)$ as its set of primitive periods. It is based on a construction that assigns to any continuous mapping

$f: [0; 1] \to [0; 1]$ a mapping $\tau(f): [0; 1] \to [0; 1]$ such that the restriction of the mapping $\tau(f)^2$ to the interval $[0; \frac{1}{3}]$ is conjugate to f. Therefore, $\tau(f)$ is sometimes called the *double* of f.

In the remainder of this section, f is an arbitrary continuous mapping of the unit interval $[0; 1]$ into itself. For every such a mapping f we define the continuous mapping $\tau(f) : [0; 1] \to [0; 1]$ by

$$\tau(f) := \begin{cases} \frac{1}{3}(f(3x) + 2) & \text{for } 0 \le x \le \frac{1}{3}\,, \\ (f(1) + 2)(-x + \frac{2}{3}) & \text{for } \frac{1}{3} \le x \le \frac{2}{3}\,, \\ x - \frac{2}{3} & \text{for } \frac{2}{3} \le x \le 1\,. \end{cases}$$

Lemma 2.4.1.

(1) *The mapping $\tau(f)$ has a unique invariant point and all other periodic points have even periods. In addition all non-invariant periodic orbits have a point in the interval $I := [0; \frac{1}{3}]$.*

(2) *The mapping $\varphi: x \mapsto \frac{1}{3}x: [0; 1] \to I$ is a conjugation from the dynamical system $([0; 1], f)$ to the system $(I, \tau(f)^2|_I)$.*

Proof. (1) Clearly, $\tau(f)$ maps the two intervals $I := [0; \frac{1}{3}]$ and $J := [\frac{2}{3}; 1]$ into each other: $\tau(f)[I] \subseteq J$ and $\tau(f)[J] = I$. This implies that a periodic point in one of these interval must have an even primitive period and that its orbit has a point in I. Moreover, the mapping $\tau(f)$ has an invariant point in the open interval $(\frac{1}{3}; \frac{2}{3})$ and it is easily seen to be the unique invariant point under $\tau(f)$. On this interval the derivative of $\tau(f)$ exists and is in absolute value greater than or equal to 2. Hence by Proposition 2.1.4, for any point x in this interval the distance of $\tau(f)^n(x)$ to the invariant point increases with n, unless x is equal to this invariant point or $\tau(f)^n(x)$ leaves this interval, i.e., ends up in I or J. In the latter case, it will never come back into $(\frac{1}{3}; \frac{2}{3})$ under any iterate of $\tau(f)$, so in that case the point x is not periodic. In particular, the orbit of any non-invariant periodic point is included in $I \cup J$, hence has a point in I and has even period.

(2) The proof consists of a straightforward computation, which we leave to the reader. However, the following geometric argument may be enlightening: on I, the mapping $\tau(f)$ followed by a translation over $-\frac{2}{3}$ is conjugate to f on $[0; 1]$, because it is just the mapping $\varphi \circ f \circ \varphi^{\leftarrow}$. However, the translation over $-\frac{2}{3}$ can be seen as another application of $\tau(f)$, because that is just the action of the mapping $\tau(f)$ on the interval $[\frac{2}{3}; 1]$. □

Lemma 2.4.2. *For every $n \in \mathbb{N}$, the mapping f has a periodic orbit with primitive period n iff the mapping $\tau(f)$ has a periodic orbit with primitive period $2n$.*

Proof. Let x be a periodic point under f with period n (not necessarily the primitive period). It follows from Lemma 2.4.1 (2) that the point $\frac{1}{3}x$ is periodic under $\tau(f)^2$ with period n, hence it is periodic under $\tau(f)$ with period $2n$. Conversely, by Lemma 2.4.1 (1) a periodic orbit under $\tau(f)$ has even period, say $2n$ with $n \in \mathbb{N}$, and it has a point in the interval I, say the point y. Then y is periodic under $\tau(f)^2$ with period n. Consequently, by Lemma 2.4.1 (2) the point $3y$ is periodic under f with period n.

By using these statements alternatingly, it is easy to see that if a point x has *primitive* period n under f, then the point $\frac{1}{3}x$ can have no periods smaller than $2n$ under $\tau(f)$, so $2n$ is the primitive period of $\frac{1}{3}x$ under $\tau(f)$. Similarly, if a point $y \in I$ has primitive period $2n$ under $\tau(f)$ then $3y$ has primitive period n under f, as it can have no smaller periods. □

Corollary 2.4.3. $\mathbf{Per}(\tau(f)) = \{\, 2n \,\vdots\, n \in \mathbf{Per}(f) \,\} \cup \{1\}$.

Proof. Clear from the Lemma's 2.4.1 (1) and 2.4.2. □

Lemma 2.4.4. *There exists a continuous mapping* $f_\infty : [0;1] \to [0;1]$ *such that* $\tau(f_\infty) = f_\infty$.

Proof. Let $\mathcal{C}$ be the set of all continuous mappings of the interval $[0;1]$ into itself, endowed with the metric ρ generated by the supremum norm:

$$\rho(f,g) := \sup_{x\in[0;1]} |f(x) - g(x)| = \max_{x\in[0;1]} |f(x) - g(x)|$$

for all $f, g \in \mathcal{C}$. With this metric, $\mathcal{C}$ is a complete metric space: see Example (2) after Theorem A.7.5 in Appendix A. It is easy to see that with respect to this metric the mapping $\tau : f \mapsto \tau(f) : \mathcal{C} \to \mathcal{C}$ is a contraction: for any two elements f and g of $\mathcal{C}$ we have $\rho(\tau(f), \tau(g)) = \frac{1}{3}\rho(f,g)$; see also Figure 2.9. Now Banach's Fixed Point Theorem – see Theorem A.7.9 in Appendix A – immediately implies the existence of $f_\infty \in \mathcal{C}$ such that $\tau(f_\infty) = f_\infty$. See Figure 2.10. □

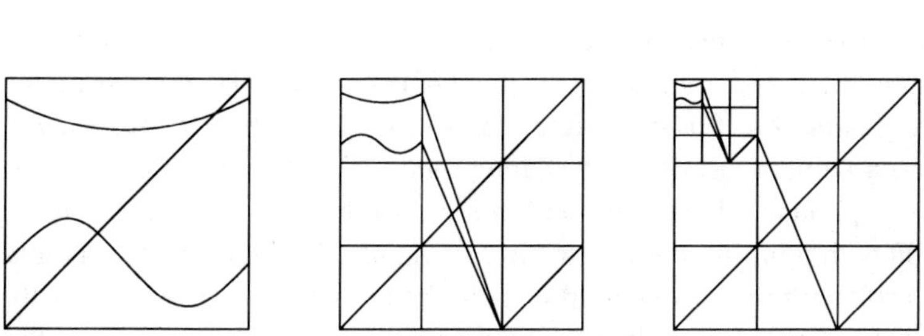

(a) (b) (c)

Fig. 2.9. The graphs of f, $\tau(f)$ and $\tau^2(f)$ for two different functions.

Proposition 2.4.5. $\mathbf{Per}(f_\infty) = \mathbf{S}(2^\infty)$.

Proof. Because $\tau(f_\infty) = f_\infty$, it follows from Corollary 2.4.3 that

$$\mathbf{Per}(f_\infty) = \{\, 2n \,\vdots\, n \in \mathbf{Per}(f_\infty) \,\} \cup \{1\}\,. \tag{2.4-1}$$

It is not difficult to show that $\mathbf{S}(2^\infty) = \{1, 2, 2^2, 2^3, \dots\}$ is the only set of integers satisfying (2.4-1), as follows: Since $1 \in \mathbf{Per}(f_\infty)$ it follows by induction from (2.4-1) that

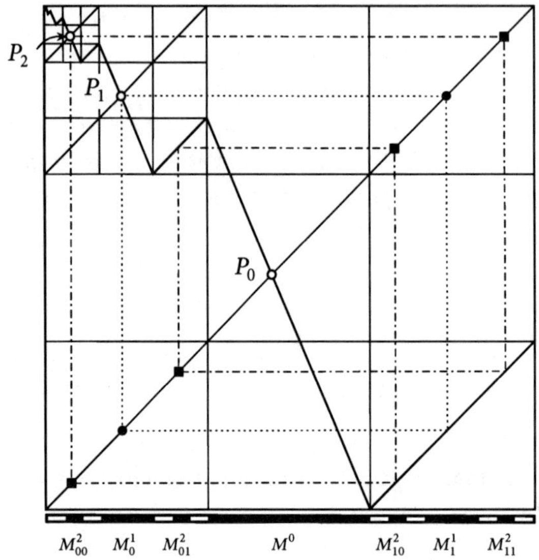

Fig. 2.10. The graph of f_∞ with the invariant point and the periodic orbits of periods 2 and 4.

$\mathbf{Per}(f_\infty) \supseteq \{1, 2, 2^2, 2^3, \dots\}$. Conversely, suppose that $\mathbf{Per}(f_\infty)$ contains a natural number greater than 1 which is not a power of 2 and let $q \cdot 2^m$ with $m \in \mathbb{N}$ and q odd, $q \geq 3$, be the smallest of such integers in $\mathbf{Per}(f_\infty)$. Then $q \cdot 2^{m-1} \notin \mathbf{Per}(f_\infty)$, so $q \cdot 2^m$ is not the double of a member of $\mathbf{Per}(f_\infty)$. This contradicts (2.4-1). □

2.4.6. In this example the Cantor set arises in a natural way. For notation and terminology, see Section B.1 in Appendix B.. By the proof of Lemma 2.4.1 (1), C_1 is invariant under f_∞. Using Lemma 2.4.1 (2) one readily shows by induction that for every $n > 1$ the set C_n is invariant under f_∞. Hence the Cantor set $C = \bigcap_{n=0}^{\infty} C_n$ is invariant as well.

The mapping f_∞ has an invariant point in the open interval M^0. The points of the periodic orbit with primitive period 2 are in M_0^1 and M_1^1. Similarly, the four points of the periodic orbit with primitive period 4 are in the intervals M_{ij}^2. See Figure 2.10. In general, each point of the periodic orbit with primitive period 2^n is in one of the intervals M_b^n with $b \in \{0, 1\}^n$. Conclusion:

- For every $n \in \mathbb{N}$, the distance of each of the points of the periodic orbit with primitive period 2^n to the Cantor set C is at most $3^{-n}/2$;
- Every neighbourhood of a point of C contains for almost all n an interval J_b^n for some $b \in \{0, 1\}^n$, hence is limit of a sequence of periodic points.
- The Cantor set C contains no f_∞-periodic points.

The proofs can easily be given by induction, Using Lemma 2.4.1 (2) and the observation that one gets C_{n+1} by multiplying C_n by $1/3$ and form the union of this scaled copy of C_n with a copy of itself that is translated over $2/3$.

NB. This example will be discussed again in Exercise 2.14, in 3.3.19, 4.2.10 and in 4.3.11 ahead.

2.5 The Markov graph of a periodic orbit in an interval

In this section we consider, again, a dynamical system (X, f) on an interval X in $\mathbb{R}$. Though X will not be assumed to be compact, all results in this section concern a bounded subset of X.

Consider a periodic point x of f with primitive period n. Denote the points of $\mathcal{O}(x)$ in increasing order by $x_1 < x_2 < \cdots < x_n$ and let $I_j := [x_j; x_{j+1}]$ for $j = 1, \ldots, n-1$. If $1 \le j \le n-1$ then the images $f(x_j)$ and $f(x_{j+1})$ of the end points of the interval I_j belong to the set $\{x_1, \ldots, x_n\}$. Consequently, the interval with end points $f(x_j)$ and $f(x_{j+1})$ is the union of consecutive intervals (at least one) of the form $I_{j'}$ with $1 \le j' \le n-1$. So the following convention makes sense: if $j, i \in \{1, \ldots, n-1\}$ then we shall write $I_j \longrightarrow I_i$ whenever $\langle f(x_j); f(x_{j+1})\rangle \supseteq I_i$. Here the expression $\langle a; b\rangle$ means $[a; b]$ if $a \le b$ and $[b; a]$ if $a \ge b$ for $a, b \in \mathbb{R}$. Note that $I_j \longrightarrow I_i$ implies[3] $I_j \overset{\circ}{\rightarrow} I_i$.

The *Markov graph* of the periodic orbit x is the directed graph whose vertices are the intervals I_j and where there is an edge from vertex I_j to vertex I_i whenever $I_j \longrightarrow I_i$ $(j, i \in \{1, \ldots, n-1\})$.

Example. In Figure 2.11 (a) and (b) the phase portraits of two periodic orbits with period 5 are sketched: one under the tent map and another one under the mapping f sketched in Figure 1.1 in Chapter 1. The Markov graphs of these orbits are presented in Figure 2.11 (c) and (d), respectively.

A *cycle* of length k ($k \in \mathbb{N}$) in the Markov graph of x is a path of the form

$$J_0 \longrightarrow J_1 \longrightarrow \ldots \longrightarrow J_{k-1} \longrightarrow J_0 \,. \tag{2.5-1}$$

where J_i for $i = 0, \ldots, k-1$ is a vertex in the Markov graph. That J_0 appears also at the end of the path is only to denote that a cycle is a path that ends in the vertex where it begins; it does not mean that J_0 is counted twice[4]; stated otherwise, the length of the cycle (2.5-1) is the number of arrows in (2.5-1). In general, the *length of a path* is the number of edges in the path (edges that appear more than once are counted accordingly). Note that for every $i = 0, \ldots, k-1$ the cycle (2.5-1) can also be represented as

$$J_i \longrightarrow \ldots \longrightarrow J_{k-1} \longrightarrow J_0 \longrightarrow J_1 \longrightarrow \ldots \longrightarrow J_i \,.$$

A cycle of the form (2.5-1) is said to be *primitive* whenever it is not a shorter cycle followed more than once.

Thus, a cycle of the form $J_0 \longrightarrow J_1 \longrightarrow J_2 \longrightarrow J_0 \longrightarrow J_1 \longrightarrow J_2 \longrightarrow J_0$ is not primitive: it is the cycle $J_0 \longrightarrow J_1 \longrightarrow J_2 \longrightarrow J_0$ followed twice. On the other hand, a cycle of the form $J_0 \longrightarrow J_1 \longrightarrow J_2 \longrightarrow J_0 \longrightarrow J_2 \longrightarrow J_3 \longrightarrow J_0$ is primitive, though it is the union of proper subcycles, namely, the cycles $J_0 \longrightarrow J_1 \longrightarrow J_2 \longrightarrow J_0$ and $J_0 \longrightarrow J_2 \longrightarrow J_3 \longrightarrow J_0$.

---

3 Of course, the converse need not be true.

4 Properly speaking, a cycle should be notated in a circular fashion (identifying the initial and final appearance of J_0). This would also better reflect the fact that each vertex J_i can be chosen as the beginning and end of the cycle (2.5-1).

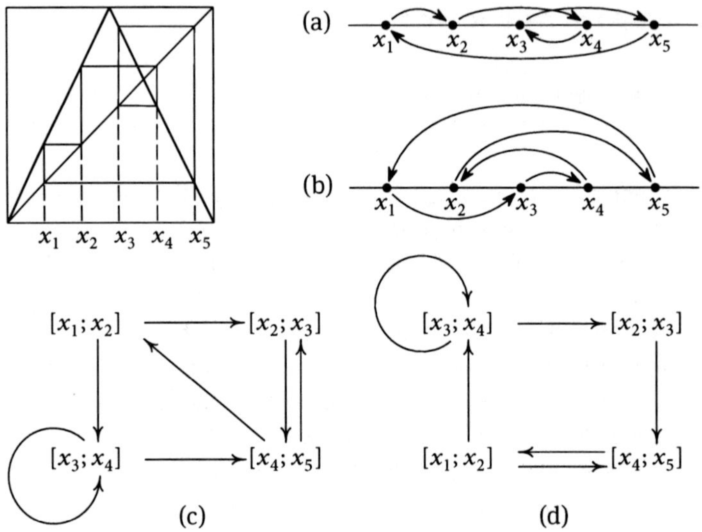

Fig. 2.11. Top left: The tent map with a periodic orbit of period 5.
(a) The phase portrait of the sketched period-5 orbit of under the tent map.
(b) Phase portrait of the period-5 orbit of the function sketched in Figure 1.1.
(c) The Markov graph of the periodic orbit in (a).
(d) The Markov graph of the periodic orbit in (b).

A cycle $J_0 \longrightarrow J_1 \longrightarrow \ldots \longrightarrow J_{n-1} \longrightarrow J_0$ is said the be *fundamental* whenever the interval J_0 has an end point c such that for all $k \in \{0, \ldots, n-1\}$ the point $f^k(c)$ is an end point of J_k. Note that, by definition, the length of a fundamental cycle equals the period n of x. Since $f^n(c) = c$ (recall that the end points of the intervals J_k are the points of a periodic orbit with primitive period n), this definition is independent of the choice of the vertex J_0 at the begin (= end) of the cycle. See Figure 2.13 below. We shall show now that the Markov graph of a periodic point always has a fundamental cycle.

Lemma 2.5.1.
(1) *In the Markov graph of a non-invariant periodic point there is a unique fundamental cycle. In this cycle each vertex of the Markov graph appears at most twice and at least one vertex appears exactly twice.*
(2) *The fundamental cycle can be decomposed into two primitive cycles.*

Proof. (1) We use the notation agreed upon in the definitions above. In particular, x is a periodic point with primitive period $n > 1$ and, in increasing order, the points of $\mathcal{O}(x)$ are $x_1 < x_2 < \cdots < x_n$. Moreover, $I_i := [x_i; x_{i+1}]$ for $i = 1, \ldots, n-1$. We shall describe an inductive procedure to find successively the vertices J_k of the fundamental cycle and the points $f^k(c)$ for $k = 0, \ldots, n-1$.
Put $J_0 := I_1 = [x_1; x_2]$ and $c := x_1$. For J_1 we must select a vertex of the Markov graph, i.e., one of the intervals I_i ($1 \le i \le n-1$), such that it has $f(c)$ as an end point and, in

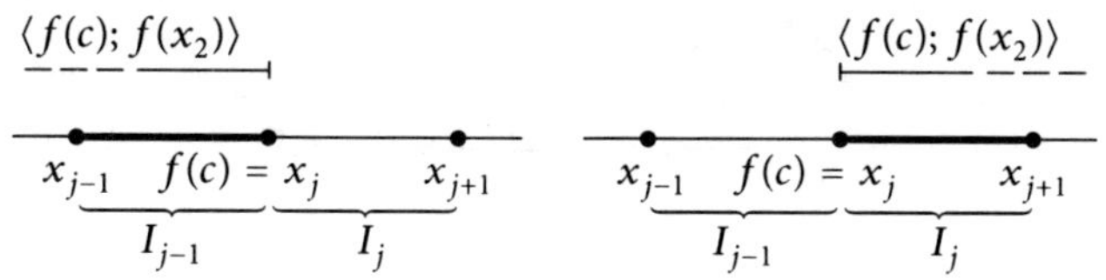

Fig. 2.12. Suppose $f(c) = x_j$ with $1 \le j \le n-1$. Left: if $f(x_2) < f(c)$ then $J_1 := I_{j-1}$. Right: if $f(x_2) > f(c)$ then $J_1 := I_j$.

addition, such that $I_i \subseteq \langle f(c); f(x_2)\rangle$. The first condition would allow for two choices of J_1, but in combination with the second condition it determines J_1 uniquely: only one of the intervals I_i that are covered by the interval $\langle f(c); f(x_2)\rangle$ borders on the end point $f(c)$ of the latter. See Figure 2.12.

By repeating this procedure we build a sequence $\{J_k : k = 0, \dots, n-1\}$ of intervals which are all vertices of the Markov graph of x, such that $f^k(c)$ is an end point of J_k and such that $J_k \longrightarrow J_{k+1}$ for $0 \le k \le n-1$. Thus, if $J_{k-1} =: [a; b]$ $(a, b \in \{x_1, \dots, x_n\})$ has already been found such that one of its end points a or b is $f^{k-1}(c)$, then J_k is the unique interval I_i included in the interval $\langle f(a); f(b)\rangle$ which has $f(f^{k-1}(c)) = f^k(c)$ as an end point.

Finally, we have J_{n-1} with end points $f^{n-1}(c)$ and some point x_l. The next step in the procedure produces J_n as the interval I_i in $\langle f^n(c); f(x_l)\rangle$ which has $f^n(c)$ as an end point. But $f^n(c) = c$ because c, as a point from the orbit of x, is periodic with period n. As c was chosen as the first (left-most) point in the orbit of x, the only possible choice for J_n is $I_1 = J_0$. Thus, $J_n = J_0$ and the construction of a fundamental cycle is completed.

Now let $K_0 \longrightarrow K_1 \longrightarrow \dots \longrightarrow K_{n-1} \longrightarrow K_0$ be another fundamental cycle and let the point c' be the end point of K_0 such that $f^k(c') \in K_k$ for $k = 0, \dots, n-1$. Since both c' and c are points of the orbit of x, we have $c = f^j(c')$ for some j with $0 \le j \le n-1$. Hence c is an end point of K_j. But I_1 is the only interval that has $c = x_1$ as an end point, hence $K_j = I_1 = J_0$. However, in the procedure described above we have seen that each vertex of the graph in a fundamental cycle, together with the selected end point, uniquely determines the remainder of the cycle. So, as an ordered n-tuple, the cycle $(K_j, \dots, K_{n-1}, K_0, \dots, K_{j-1}, K_j)$ is equal to the cycle $(J_0, \dots, J_{n-1-j}, J_{n-j}, \dots, J_{n-1}, J_0)$. This shows that the two fundamental cycles are equal to each other.

Next, consider any vertex I_k of the Markov graph $(1 \le k \le n-1)$. Then there are $i, j \in \{0, \dots, n-1\}$ such that $I_k = [f^i(c); f^j(c)]$ (recall that the points c and x have the same orbit). Consequently, the vertex I_k can occur in the fundamental cycle $(J_0, J_1 \dots, J_{n-1}, J_0)$ only as J_i or J_j. Stated otherwise, each vertex of the graph can occur at most twice in the fundamental cycle. On the other hand, the fundamental cycle has length n, but there are only $n-1$ vertices in the graph. Hence at least one vertex must appear twice in the cycle.

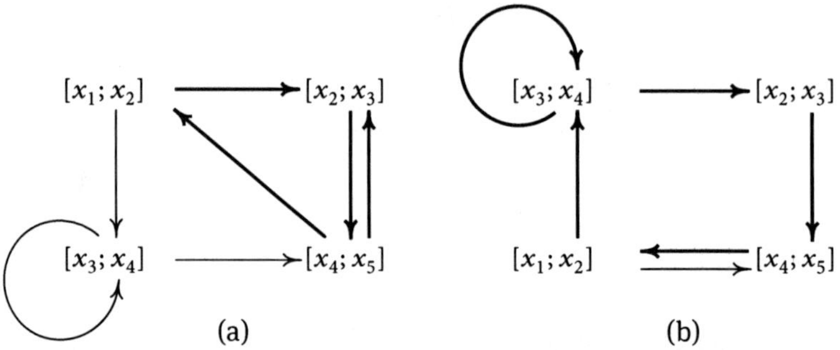

Fig. 2.13. Fundamental cycles in the graphs of the Figures 2.11 (c) and (d).

(2) Let J_i be the vertex that appears twice in the fundamental cycle: say that $J_i = J_j$ with $0 \le i < j \le n-1$. Then the fundamental cycle looks as follows:

$$\overbrace{J_0 \longrightarrow \ldots \longrightarrow J_i}^{C_{21}} \underbrace{\longrightarrow \ldots \longrightarrow J_j}_{C_1} \overbrace{\longrightarrow \ldots \longrightarrow J_0}^{C_{22}} .$$

Obviously, the fundamental cycle splits into two subcycles: the cycle C_1 from vertex J_i to its appearance as J_j, and the cycle C_2 that remains after deleting the vertices from J_{i+1} up to J_j from the fundamental cycle (this cycle is indicated above as the union of C_{21} and C_{22}). It is easy to see that the cycles C_1 and C_2 are primitive. For if C_1 were not primitive then J_i would appear between the appearance of J_i already mentioned and its appearance as J_j, which is impossible since J_i can appear not more than two times in the fundamental cycle. Similarly, C_2 is primitive for otherwise J_i would have, again, a third appearance in the fundamental cycle. □

Examples. Using the procedure outlined in the proof of the above lemma one finds in Figure 2.13 (a) the following fundamental cycle (the heavy arrows):

$$\begin{array}{ccccccccccc} [x_1;x_2] & \longrightarrow & [x_2;x_3] & \longrightarrow & [x_4;x_5] & \longrightarrow & [x_2;x_3] & \longrightarrow & [x_4;x_5] & \longrightarrow & [x_1;x_2] \\ x_1 & \mapsto & x_2 & \mapsto & x_4 & \mapsto & x_3 & \mapsto & x_5 & \mapsto & x_1\,. \end{array}$$

Note that two vertices appear twice in this fundamental cycle, and one vertex doesn't appear at all in it. The arrows $[x_1;x_2]\longrightarrow[x_3;x_4]\longrightarrow[x_3;x_4]\longrightarrow[x_4;x_5]\longrightarrow[x_1;x_2]$ do not form a fundamental cycle, though it is a cycle and one can choose the end points $x_2 \mapsto x_4 \mapsto x_3 \mapsto x_5 \mapsto x_1$ in the correct intervals. However, its length is not 5 as the definition requires, but 4.

Similarly, it is easily checked that the heavy arrows in Figure 2.13 (b) form a fundamental cycle:

$$\begin{array}{ccccccccccc} [x_1;x_2] & \longrightarrow & [x_3;x_4] & \longrightarrow & [x_3;x_4] & \longrightarrow & [x_2;x_3] & \longrightarrow & [x_4;x_5] & \longrightarrow & [x_1;x_2] \\ x_1 & \mapsto & x_3 & \mapsto & x_4 & \mapsto & x_2 & \mapsto & x_5 & \mapsto & x_1 \end{array}$$

or (starting at another vertex in the same cycle)

$$\begin{array}{ccccccccccc} [x_3; x_4] & \longrightarrow & [x_3; x_4] & \longrightarrow & [x_2; x_3] & \longrightarrow & [x_4; x_5] & \longrightarrow & [x_1; x_2] & \longrightarrow & [x_3; x_4] \\ x_3 & \mapsto & x_4 & \mapsto & x_2 & \mapsto & x_5 & \mapsto & x_1 & \mapsto & x_3 \end{array}.$$

The following lemma is fundamental for the proof of Šarkovskij's theorem.

Lemma 2.5.2 (Štefan). *Let (X, f) be a dynamical system on an interval and let $x \in X$ be a non-invariant periodic point. Let*[5] *$n \geq 2$ and suppose the Markov graph of x contains a primitive cycle $J_0 \longrightarrow J_1 \longrightarrow \ldots \longrightarrow J_{n-1} \longrightarrow J_0$ of length n. Then there exists a periodic point $y \in J_0$ with primitive period n such that $f^k(y) \in J_k$ for $0 \leq k \leq n-1$.*

Proof. Note that, for vertices I and K in the Markov graph of x, $I \longrightarrow K$ implies $I \overset{\circ}{\rightarrow} K$. Consequently, Lemma 2.2.1 (3) implies that for $0 \leq j \leq n-1$ there is a closed interval K_j of J_j such that

$$K_0 \twoheadrightarrow K_1 \twoheadrightarrow \cdots \twoheadrightarrow K_{n-1} \twoheadrightarrow J_0 \supseteq K_0\,.$$

Hence $f^n[K_0] \supseteq K_0$, so K_0 contains an invariant point y under f^n, that is, a periodic point under f with period n. Obviously, $f^k(y) \in f^k[K_0] = K_k \subseteq J_k$ for $0 \leq k \leq n-1$. Since the cycle is primitive, it follows that n is the primitive period of y, provided y is not an end point of J_0. For if n is not the primitive period of y then there is $p < n$ such that $f^p(y) = y \in J_0$. As also $f^p(y) \in J_p$, this would imply that $y \in J_0 \cap J_p$; in particular, it would follow that y is an end point of J_0.

Now suppose that y is an end point of J_0. Then y is in the orbit of x, hence the orbits of y and x coincide and the points y and x have the same primitive period p. Since y has also period n, this implies that n is a multiple of p. Now recall that the points of the orbit of x (i.e., those of the orbit of y) form the end points of the vertices of the Markov graph under consideration. It follows that, for $0 \leq k \leq n-1$, the point $f^k(y)$ is an end point of J_k. So the cycle $J_0 \longrightarrow J_1 \longrightarrow \ldots \longrightarrow J_{n-1} \longrightarrow J_0$ satisfies the definition of the fundamental cycle of x, except that its length is not p (the period of x) but n, a multiple of p. Now observe that the point $f^k(y)$ runs cyclically through the orbit of y, while $f^k(y) \in J_k$ for $0 \leq k \leq n-1$. It follows that the cycle under consideration is a multiple of the fundamental cycle of x. As this cycle is given to be primitive, this is impossible, unless $p = n$. □

Proposition 2.5.3. *Let (X, f) be a dynamical system on an interval which has a periodic point with an odd primitive period larger than 1. Let p be the smallest odd natural number, $p > 1$, for which there is a periodic point x with primitive period p and let c the*

5 In view of Lemma 2.2.1 (1) the result is trivially true for $n = 1$.

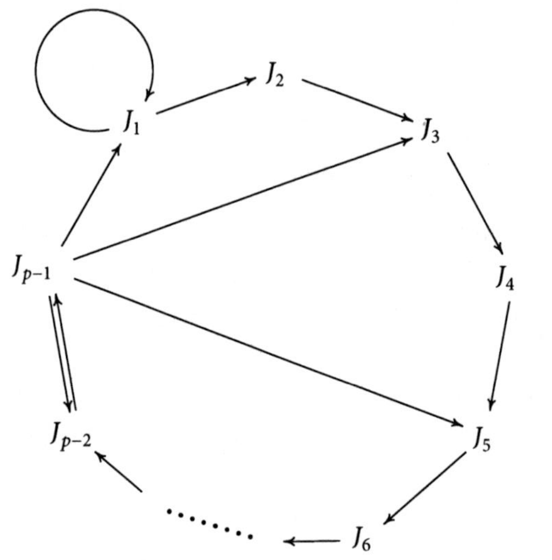

Fig. 2.14. The Markov graph of a periodic point with period the smallest odd number $p > 1$ for which there is a periodic point with primitive period p.

central point of the orbit of x. If $c < f(c)$ then the points of $\mathcal{O}(x) = \mathcal{O}(c)$ are ordered as follows:

$$f^{p-1}(c) < f^{p-3}(c) < \cdots < f^2(c) < c < f(c) < f^3(c) < \ldots < f^{p-2}(c).$$

Moreover, if

$$J_1 := \langle c; f(c)\rangle, \quad J_2 := \langle f^2(c); c\rangle,$$
$$J_3 := \langle f(c); f^3(c)\rangle, \quad J_4 := \langle f^4(c); f^2(c)\rangle,$$
$$\cdots\cdots\cdots\cdots$$
$$J_{p-2} := \langle f^{p-4}(c); f^{p-2}(c)\rangle, \quad J_{p-1} := \langle f^{p-1}(c); f^{p-3}(c)\rangle,$$

then the Markov graph of x is as in Figure 2.14. If $c > f(c)$ then the ordering of the points of the orbit of x is just the other way round and the Markov graph is the same.

Proof. We consider the Markov graph of the periodic point x with primitive period p. The fundamental cycle in that graph has odd length p. According to Lemma 2.5.1 (2) it splits into two primitive cycles, one of which must have odd length (the sum of their lengths is the odd number p). But then Lemma 2.5.2 would imply that there is a periodic point with odd period less than p. This contradicts the choice of p as the smallest odd number strictly larger than 1 that occurs as the period of a periodic point, unless that odd period, hence the length of that primitive cycle of odd length, is equal to 1. Consequently, the fundamental cycle can be written as

$$J_1 \longrightarrow J_1 \longrightarrow J_2 \longrightarrow \ldots \longrightarrow J_{p-1} \longrightarrow J_1$$

with $J_k \neq J_1$ for $k \geq 2$, because J_1 cannot occur more than twice[6]. Its two primitive subcycles are $J_1 \longrightarrow J_1$ and $J_1 \longrightarrow J_2 \longrightarrow \dots \longrightarrow J_{p-1} \longrightarrow J_1$.

Suppose another vertex of the graph appears twice in the fundamental cycle: there are i and j with $2 \leq i < j \leq p-1$ and $J_i = J_j$. Then the cycles

$$J_1 \longrightarrow J_2 \longrightarrow \dots \longrightarrow J_i = J_j \longrightarrow \dots \longrightarrow J_{p-1} \longrightarrow J_1$$

and

$$J_1 \longrightarrow J_1 \longrightarrow J_2 \longrightarrow \dots \longrightarrow J_i = J_j \longrightarrow \dots \longrightarrow J_{p-1} \longrightarrow J_1$$

have lengths $p-(j-i)-1 < p$ and $p-(j-i) < p$, respectively, and one of these lengths is odd. The first cycle is part of the primitive cycle in the fundamental cycle, hence is primitive itself. Since its length is at least 2, adding the edge $J_1 \longrightarrow J_1$ before this cycle in order to get the second cycle does not change this (it does not double the cycle). So both cycles above are primitive, and we can apply Lemma 2.5.2 to the cycle of odd length. As before we arrive at a contradiction with the choice of p as the smallest odd number strictly larger than 1 that occurs as the period of a periodic point. Conclusion: the vertices $J_2, \dots, J_{p-1}$ are mutually different, so $(J_1, J_2, \dots, J_{p-1})$ is a permutation of all $p-1$ vertices of the graph.

In a similar way one shows: if the Markov graph has an edge $J_i \longrightarrow J_k$ with $1 \leq i, k \leq p-1$ and $k \geq i+2$, or an edge $J_i \longrightarrow J_1$ with $2 \leq i \leq p-2$ then one can make a shortcut to get a primitive cycle of odd length (if necessary, add the loop $J_1 \longrightarrow J_1$ to the cycle) with a length strictly less than p which, again by Lemma 2.5.2, contradicts the choice of p.

Resuming, the fundamental cycle contains all vertices of the Markov graph just once, except J_1, which occurs twice. Moreover, the Markov graph of x has, except the edges that occur in the fundamental cycle, no edges towards J_1 and no edges that point 'forwards', from J_i to J_k with $k \geq i+2$.

In accordance with the definition of a Markov graph, enumerate the points of the orbit of x in increasing order as $x_1 < x_2 < \dots < x_p$, and let $I_j := [x_j; x_{j+1}]$ for $1 \leq j \leq p-1$. Let k be such that $I_k = J_1$, the vertex where the fundamental cycle has its loop. From the vertex J_1 there are edges to J_1 and to J_2 and, according to what has just been shown, the Markov graph of x has no other edges starting in J_1. As the interval $\langle f(x_k); f(x_{k+1})\rangle$ is a union of vertices of the Markov graph, this implies that $\langle f(x_k); f(x_{k+1})\rangle = J_1 \cup J_2$. In particular, the intervals J_1 and J_2 are adjacent to each other. So either J_2 is the interval I_{k-1} immediately to the left of I_k or J_2 is the interval I_{k+1} to the right of I_k. As $J_1 \cup J_2$ is the interval with end points $f(x_k)$ and $f(x_{k+1})$, there are only the following possibilities:

(1) $J_2 = I_{k-1}$, so $f(x_k) = x_{k+1}$ and $f(x_{k+1}) = x_{k-1}$

(2) $J_2 = I_{k+1}$, so $f(x_{k+1}) = x_k$ and $f(x_k) = x_{k+2}$

6 Recall that the initial and final appearance of J_1 in this linearly notated cycle are identified.

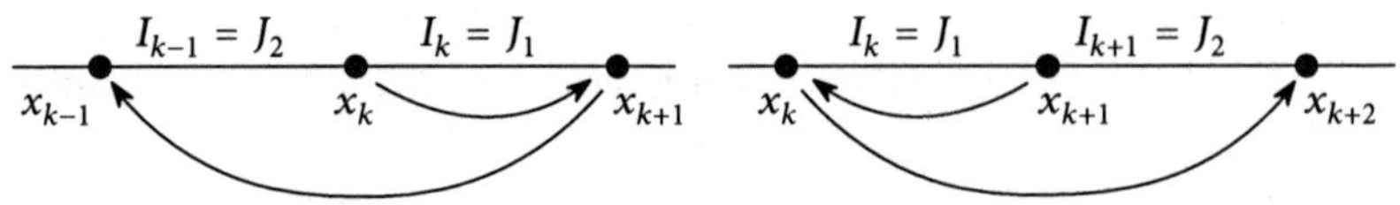

We shall work out the first case, the second being similar but giving rise to the reverse order. Write $c := x_k$. If $p = 3$ the proof of the statement concerning the ordering of the points of $\mathcal{O}(c)$ is complete (with $k = 2$ in this case). The Markov graph of $\mathcal{O}(c) = \mathcal{O}(x)$ is easily seen to look as follows:

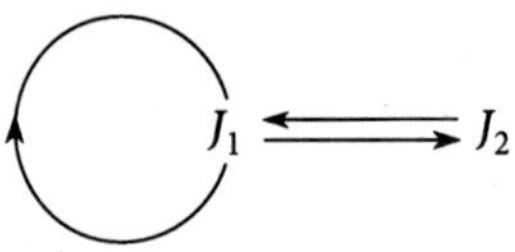

This is the graph of Figure 2.14 in the case that $p = 3$.

Now assume that $p > 3$ (so $p \geq 5$). In order to determine J_3 – the vertex to which the edge $J_2 \longrightarrow J_3$ in the fundamental cycle points – we have to look at the interval $\langle f(x_{k-1}); f(x_k) \rangle = \langle f^3(c); x_{k+1} \rangle$. Since the point c has a primitive period of at least 5 it is clear that $f^3(c) \notin \{x_{k-1}, x_k, x_{k+1}\}$ (recall that $f(c) = x_{k+1}$ and that $f^2(c) = x_{k-1}$), so either $f^3(c) < x_{k-1}$ or $f^3(c) > x_{k+1}$. The first possibility would imply that $[x_k; x_{k+1}]$ is included in $\langle f^3(c); x_{k+1} \rangle$, meaning that $J_2 \longrightarrow J_1$ would be an edge in the Markov graph of x. We have seen that this is not possible if $p > 3$. Conclusion: $f^3(c) > x_{k+1}$, that is, $f^3(c) = x_i$ with $i > k + 1$, and therefore $\langle f^3(c); x_{k+1} \rangle = [x_{k+1}; x_i]$. It follows that in the Markov graph of x there are no edges from J_2 to J_j for $j = 1, 2$ (which we knew already), and since there are no edges $J_2 \longrightarrow J_j$ for $j \geq 4$, the only edge in the Markov graph starting in J_2 is $J_2 \longrightarrow J_3$. Consequently, the only remaining possibility is that $f^3(c) = x_{k+2}$, implying that $J_3 = [x_{k+1}; x_{k+2}] = I_{k+1}$. Next, in order to determine J_4, the target vertex of the edge $J_3 \longrightarrow J_4$, we look at the interval $\langle f(x_{k+1}); f(x_{k+2}) \rangle = \langle x_{k-1}; f^4(c) \rangle$. Since $p \geq 5$ it is clear that $f^4(c) \neq f^j(c)$ for $j = 0, 1, 2$ or 3, hence $f^4(c) < x_{k-1}$ or $f^4(c) > x_{k+2}$. If $f^4(c) > x_{k+2}$ then $\langle f(x_{k-1}); f^4(c) \rangle \supseteq J_2 \cup J_1 \cup J_3$, which would imply the existence of an edge $J_3 \longrightarrow J_1$ and of an 1-edge cycle at J_3, neither of which exists. Consequently, $f^4(c) < x_{k-1}$, that is, $f^4(c) = x_i$ with $i < k - 1$, and therefore $\langle x_{k-1}; f^4(c) \rangle = [x_i; x_{k-1}]$. It follows that in the Markov graph of x there are no edges from J_3 to J_j for $j = 1, 2$ and 3 (for $j = 1$ and $j = 3$ we knew this already). Moreover, as there is no edge in the graph from J_3 to J_j for $j \geq 5$, $J_3 \longrightarrow J_4$ is the only edge starting in J_3. It follows that $f^4(c) = x_{k-2}$: if it were x_i with $i < k - 2$ then an additional edge would start in J_3. Consequently, $J_4 = I_{k-2} = [x_{k-2}; x_{k-1}] = [f^4(c); f^2(c)]$.

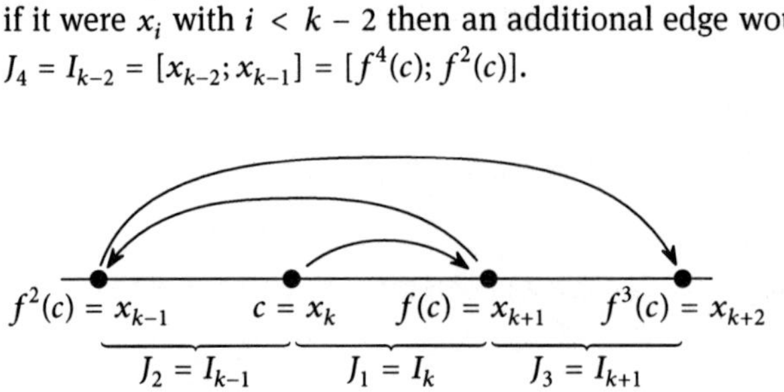

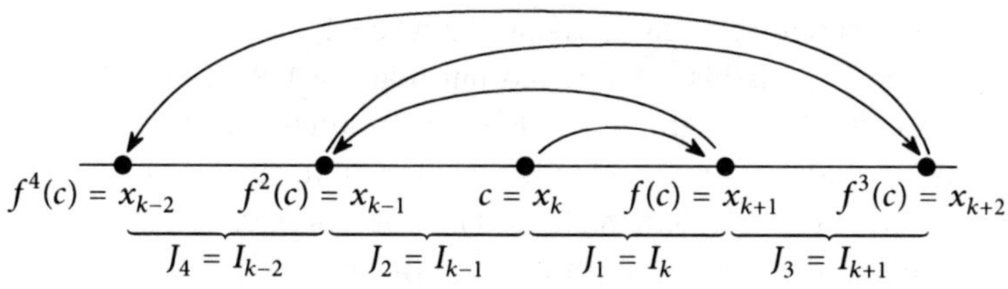

If $p = 5$ then the proof of the statement concerning the ordering of the points of $\mathcal{O}(c)$ is complete (with $k = 3$). Moreover, as $f(x_{k-2}) = f^5(c) = c$ it follows that $\langle f(x_{k-2}); f(x_{k-1})\rangle = J_1 \cup J_3$. Thus, in the Markov graph under consideration there are edges from J_4 to J_1 and J_3. This shows that for $p = 5$ the Markov graph of $\mathcal{O}(x) = \mathcal{O}(c)$ is as in Figure 2.14.

If $p > 5$ then we can proceed in a similar way to prove the proposition.

As a final note we remark that in all cases the point c turns out to be precisely the central point of the ordered finite sequence $x_1, \dots, x_n$. Moreover, recall that the case we have considered is the one with $c < f(c)$. As observed earlier, the only other possibility is that $c > f(c)$, in which case we get similar results with all inequalities reversed. □

The next corollary and lemma follow immediately from Šarkovskij's Theorem. Since they are needed in the proof of this theorem we have to give independent proofs.

Corollary 2.5.4. *Let (X, f) be a dynamical system on an interval which has a periodic point with an odd primitive period p larger than 1. Then for every $q \in \mathbb{N}$ such that either $q \geq p + 1$ or q is even there is a periodic point in X with primitive period q.*

Proof. Without limitation of generality we may assume that p is the smallest odd natural number greater than 1 for which there is a periodic point x with primitive period p. Then the Markov graph of x looks like the graph in Figure 2.14. In this graph there is for every even number $q \leq p-3$ a primitive cycle of even length (including the arrow from J_{p-1} to one of the vertices J_3, J_5, up to J_{p-2}). Obviously, there is also such a cycle for the value $q = p - 1$ (formed by the arrows $J_1 \longrightarrow \dots \longrightarrow J_{p-1} \longrightarrow J_1$). So by Lemma 2.5.2, for every even natural number $q \leq p - 1$ there is a periodic point with primitive period q.

Next, let $q \geq p + 1$ be an arbitrary natural number. By adding the loop at J_1 just $q - p$ times to the cycle $J_1 \longrightarrow \dots \longrightarrow J_{p-1} \longrightarrow J_1$ one gets a primitive cycle of length q. So yet another application of Lemma 2.5.2 completes the proof. □

By the above proposition, the existence of a non-invariant periodic point with odd primitive period has special consequences. The existence of a periodic point with even primitive period is not special at all:

Lemma 2.5.5. *If f has a point with primitive period greater than 1 then f has a point with primitive period 2 and f has an invariant point.*

Proof. Let p be the smallest natural number greater than 1 such that f has a periodic point with primitive period p; call this point x_0. Assume that $p \geq 3$. By Lemma 2.5.1 (2) the fundamental cycle in the Markov graph of x_0 has two primitive subcycles, at least one of which has length strictly greater than 1 (otherwise the fundamental cycle would have length 2). On the other hand, the length of that subcycle is strictly smaller than p, so by Lemma 2.5.2 there would be a periodic point with period larger than 1 and smaller than p, a contradiction with the choice of p.

So there is a periodic point x_0 for f with primitive period 2. Then f maps the interval with end points x_0 and $f(x_0)$ over itself. Consequently, by Lemma 2.2.1 (1) there is an invariant point between x_0 and $f(x_0)$. □

Proof of Šarkovskij's Theorem. Let $m, n \in \mathbb{N}$ such that $n \prec m$ and assume that there is a periodic point x under f with primitive period n. Claim: f admits a periodic point with primitive period m.

First, we settle the case that n is arbitrary and $m = 1$, that is, if f has a periodic point then there is an invariant point. This follows immediately from Lemma 2.5.5.

Next, we consider the case that n is odd. Then Corollary 2.5.4 implies that the above claim is true – take into account that if $n \prec m$ then either m is even or m is odd, in which case $m \geq n + 2 > m + 1$.

It remains to consider the case that n is even, say, $n = 2^r p$ with $r \geq 1$ and $p \in \mathbb{N}$, p odd. We break up the proof for this case into three subcases, according to the values of p and m.

Case (a). $p = 1$ and $m = 2^s$ with $1 \leq s < r$ (so both m and n belong to the set $\mathbf{S}(2^\infty) \setminus \{1\}$).

By Proposition 1.1.4 (1) the point x is periodic under $f^{2^{s-1}}$ with primitive period $2^r/2^{s-1} > 1$. So by Lemma 2.5.5, $f^{2^{s-1}}$ has a periodic point with primitive period 2. Proposition 1.1.4 (2) implies that this point is periodic under f with primitive period $(2 \cdot 2^{s-1})/b$ with $b = 1$ – see also Example (2) after Proposition 1.1.4. This completes the proof of the claim in this case.

Case (b). $p > 1$ and $m = 2^r q$ with q even (this accounts for all values of m of the form $2^s q'$ with $s \geq r + 1$ and q' odd, including $q' = 1$), or $m = 2^s$ for arbitrary $s \in \mathbb{N}$.

First, we consider the case that $m = 2^r q$ with q even. Again by Proposition 1.1.4 (1), the point x is periodic under f^{2^r} with primitive period equal to $2^r p/2^r = p$. As q is even, Corollary 2.5.4 implies that there is a periodic point for f^{2^r} with primitive period q. Now Proposition 1.1.4 (2) tells us that this point is periodic under f with primitive period $2^r q/b$ with $b = 1$ – see also Example (2) after Proposition 1.1.4. This completes the proof of the claim in the case that $m = 2^r q$ with q even.

Note that this implies the existence of periodic points with primitive periods $2^s \cdot 1$ for all $s \geq r+1$, i.e., for all for all sufficiently high powers of 2 in the set $\mathbf{S}(2^\infty) \setminus \{1\}$. Then apply Case (a) in order to get the result for the remaining powers of 2. This concludes the proof in the case that $m = 2^s$ for arbitrary $s \in \mathbb{N}$.

Case (c). $p > 1$ and $m = 2^r q$ with q odd.

As in case (b), Proposition 1.1.4 (1) implies that the point x is periodic under f^{2^r} with primitive period $2^r p/2^r = p$. As q is odd and, by assumption, $2^r p = n \prec m = 2^r q$,

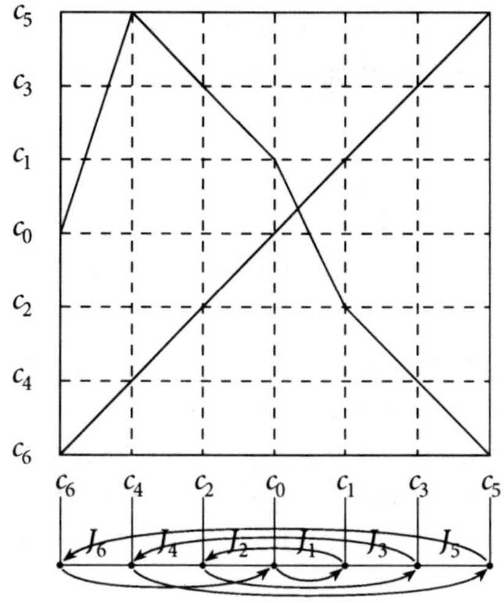

Fig. 2.15. Graph of the mapping $g_7 : [0;6] \to [0;6]$, together with the phase portrait of the periodic orbit of the point $c_0 := 3$.

we have $q \geq p + 1$. Consequently, Corollary 2.5.4 implies that there is a periodic point for f^{2^r} with primitive period q. In view of Proposition 1.1.4 (2), this point is periodic under f with primitive period $2^r q/b$ with b a divisor of 2^r. Thus, there is a periodic point for f with primitive period $n' = 2^i q$ with $i \leq r$. If $i = r$ then the proof is completed. If $i < r$ then write $m = 2^i \cdot 2^{r-i} q = 2^i q'$, where $q' := 2^{r-i} q$ is even. So now we are in the situation of Case (b) – with r replaced by i, p replaced by q and q replaced by q' – which shows that f has a periodic point with primitive period m. □

Example. Let $k \in \mathbb{N}$, k odd and $k \geq 3$. Use the ordering in Proposition 2.5.3 as a guide to construct a piecewise linear map g_k of the interval $[0; k-1]$ onto itself under which the point 0 is periodic with primitive period k. (For g_5, see Figure 1.1 and for g_7, see Figure 2.15.) By construction, the Markov graph of the point 0 looks like Figure 2.14. A careful consideration of this graph shows that there are no periodic points with odd primitive period smaller than k. Hence $\mathbf{Per}(g_k) = \mathbf{S}(k)$.

NB. More about this example is in Example (2) after Theorem 2.6.8 below and in Exercise 2.10 (2).

2.6 Transitivity of mappings of an interval

Also in this section (X, f) is a dynamical system on an interval X in $\mathbb{R}$ (open, closed or half-open, not necessarily bounded). In this case the following conditions are equivalent:

– f is transitive on X, i.e., there exists a point $x_0 \in X$ that is transitive under f or, equivalently, a point x_0 for which $\omega_f(x_0) = X$;
– f is topologically ergodic on X, i.e., if U, V are two non-empty open subsets of X then there exists $n \in \mathbb{Z}$ such that $f^n[U] \cap V \neq \emptyset$

(recall that X is a 2nd-countable Baire space). The first theorem in this section states that transitivity of f implies the existence of a dense set of periodic points. But first, a lemma.

Lemma 2.6.1. *Let J be a subinterval of X containing no f-periodic points. If $m, n \in \mathbb{N}$, $0 < m < n$, and the points z, $f^m(z)$ and $f^n(z)$ are in J, then either $z < f^m(z) < f^n(z)$ or $z > f^m(z) > f^n(z)$. Stated otherwise, if m is between 0 and n, then $f^m(z)$ is between $f^0(z)$ and $f^n(z)$.*

Proof. Assume that there exists a point $z \in J$ such that for certain values of $m, n \in \mathbb{N}$ with $0 < m < n$ both points $f^m(z)$ and $f^n(z)$ are in J, but that the point $f^m(z)$ is not situated between the points z and $f^n(z)$. So assume that $z < f^m(z)$ and that also $f^n(z) < f^m(z)$ (the other cases are similar); equality can be excluded, because there are no periodic points in J. For convenience, let $g := f^m$. So our assumption is, among others, that $z < g(z)$.

Claim: $z < g(z) \leq g^k(z)$ for all $k \geq 1$. For $k = 1$ this is true by assumption. Assume that this is true for some $k \in \mathbb{N}$ but not for $k + 1$, that is, assume that $g^k(z) > z$ but $g^k(g(z)) < g(z)$. Consequently, g^k would have an invariant point between the points z and $g(z)$, contradicting the assumption that f has no periodic points in J. This shows, by induction, that our claim is true for all $k \in \mathbb{N}$. In particular, it holds for $k := n - m$:

$$z < f^{(n-m)m}(z)\,. \tag{2.6-1}$$

Our other assumption was that $f^n(z) < f^m(z)$, i.e., $f^{n-m}(f^m(z)) < f^m(z)$. Let $h := f^{n-m}$ and $w := f^m(z)$. Then $h(w) < w$. Similar to the argument used above for g one shows that $h^k(w) \leq h(w) < w$ for all $k \in \mathbb{N}$. In particular, for $k := m$ we get

$$f^{(n-m)m}(w) < w\,. \tag{2.6-2}$$

The inequalities (2.6-1) and (2.6-2) imply that $f^{(n-m)m}$ has an invariant point between the points z en w in J. This contradicts the assumption that f has no periodic points in J. □

Theorem 2.6.2. *Let (X, f) be a transitive system on an interval X. Then the set of periodic points of f is dense in X.*

Proof. Let J be any open subinterval in X and consider three mutually disjoint subintervals J_{left}, J_{middle} and J_{right} of J. The set of transitive points is dense in X, so there is a transitive point z in J_{left}. As the point z is transitive, every subinterval of X contains points $f^i(z)$ for infinitely many values of $i \in \mathbb{N}$. It follows that there are $m, n \in \mathbb{N}$ with $0 < m < n$ such that $f^m(z) \in J_{\text{right}}$ and $f^n(z) \in J_{\text{middle}}$. Then the ordering of the three points z, $f^m(z)$ and $f^n(z)$ does not agree with the absence of periodic points in the interval J. □

Remark. Let (X, f) be a transitive system on an interval X. Then for every $n \in \mathbb{N}$ the set of periodic points with primitive period at least n is dense in X. See Exercise 2.11 (1).

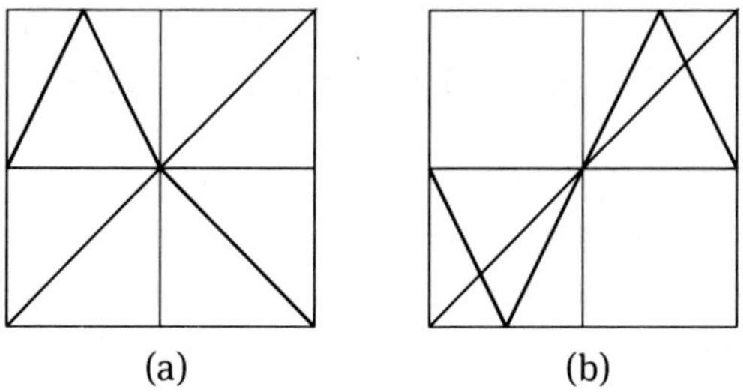

Fig. 2.16. (a) Graph of the piecewise linear map S of $[0; 1]$ into itself.
(b) The graph of S^2 with its invariant intervals $I_0 := [0; 1/2]$ and $I_1 := [1/2; 1]$.

2.6.3. As a motivation for our next result it is useful to consider two transitive maps of the unit interval: the tent map T and the piecewise linear map S whose graph is drawn in Figure 2.16 (a). That T is transitive has been shown Example (1) after Theorem 1.3.5. That S is transitive can most easily seen by considering the mapping S^2, whose graph is drawn in Figure 2.16 (b). The two intervals $I_0 := [0; 1/2]$ and $I_1 := [1/2; 1]$ are invariant under S^2, and it is not difficult to verify that both subsystems (I_0, S^2) and (I_1, S^2) of $([0; 1], S^2)$ are conjugate to the system $([0; 1], T)$. Consequently, these systems are transitive. Moreover, $S|_{I_1} : (I_1, S^2) \to (I_0, S^2)$ is a conjugation. So if $x \in I_1$ is a transitive point in the system (I_1, S^2), then $S(x)$ is transitive in (I_0, S^2). It follows easily that x is a transitive point in $[0; 1]$ under S.

The transitive mappings T and S are completely different in nature as to transitivity of higher iterates T^n and S^n for $n \geq 2$: in 1.7.3 it has been observed that T^n is transitive on $[0; 1]$ for every $n \geq 2$. On the other hand, it is clear that S^2 is not transitive on $[0; 1]$: the intervals I_0 and I_1 are invariant, hence no point can have a dense orbit in $[0; 1]$ under S^2. We shall show now that the examples discussed above represent all possible types of transitive maps on compact intervals.

Lemma 2.6.4. *Suppose that the system (X, f) is transitive. Then*
(1) *f has an invariant point c strictly between the end points of X.*
(2) *f is a semi-open mapping.*

Proof. (1) Assume the contrary: then by the intermediate value theorem, either $f(x) > x$ or $f(x) < x$ for all points x of X, except possibly in the end points. Hence by 2.1.1 and 2.1.2, every orbit in the interior of X strictly increases or decreases. This contradicts transitivity of the system (X, f).
(2) It is (necessary and) sufficient to show that for any non-degenerate subinterval J of X the interval $f[J]$ is non-degenerate as well. Assume the contrary: there is an open subinterval J of X such that $f[J] = \{y\}$ for some point $y \in X$. Let $x_0 \in J$ be a transitive point under f (recall that the transitive points are dense in X). Then $y = f(x_0)$, so the point y is transitive as well. Consequently, $f^m(y) \in J$ for some $m \in \mathbb{Z}^+$, hence $f^{m+1}(y) = y$. Thus, the transitive point y is periodic. It follows that X consists of the single periodic orbit of y. This is impossible (unless X is degenerate, in which case f is also semi-open!). □

Theorem 2.6.5. *Assume that X is a compact interval, say, $X := [a;b]$ with $a, b \in \mathbb{R}$, $a < b$. If the system (X, f) is transitive then there are the following, mutually exclusive, possibilities:*
(1) *f is totally transitive, that is, for all $n \in \mathbb{N}$ the system (X, f^n) is transitive;*
(2) *There exists a point $c \in (a;b)$ such that the (non-degenerate) intervals $I_0 := [a;c]$ and $I_1 := [c;b]$ are interchanged by f (that is, $f[I_0] = I_1$ and $f[I_1] = I_0$), hence are invariant under f^2, and the subsystems (I_0, f^2) and (I_1, f^2) of (X, f^2) are totally transitive.*

In the latter case, f^2 is not transitive on X and c is the unique invariant point in X under f.

Proof. We start with the final statement: assume that Case (2) holds. Since I_0 and I_1 are invariant under f^2 it is obvious that no point of X can have a dense orbit under f^2. Hence f^2 is not transitive on X. Moreover, as f interchanges I_0 and I_1, there is no f-invariant point in $[a;c) \cup (c;b]$. On the other hand, by Lemma 2.6.4 (1) there exists an invariant point in $(a;b)$, so c is the unique f-invariant point in X.

In order to prove the theorem, consider a transitive point x_0 in X. For every $n \in \mathbb{N}$ and for $0 \le i \le n-1$, put $F_i^n := \omega_{f^n}(f^i(x_0))$. As X is compact, these sets F_i^n are not empty. We shall show that for every $n \in \mathbb{N}$ there are points $a = a_0 < a_1 < \dots < a_k = b$ in X (depending on n, but this is not reflected in the notation) such that the collection $\mathcal{D}_n$ of closed intervals $[a_j; a_{j+1}]$ for $j = 0, \dots, k-1$ has the following properties:
(a) each of the sets F_i^n for $i = 0, \dots, n-1$ includes a member of $\mathcal{D}_n$;
(b) the members of $\mathcal{D}_n$ are cyclically permuted by f, as follows: there is a permutation $j_0, \dots, j_{k-1}$ of $0, \dots, k-1$ such that, if $I_i := [a_{j_i}; a_{j_{i+1}}]$ then $f[I_i] \subseteq I_{i+1 \pmod k}$ for $i = 0, \dots, k-1$.

This will be shown in the lemma's following this theorem.

Assuming that this has been proved, proceed as follows: by Lemma 2.6.4 (1) above there is an invariant point c of f strictly between the end points a and b of X. There are two possibilities:

Case 1. For every $n \in \mathbb{N}$ there exists $D \in \mathcal{D}_n$ such that $c \in D$ and c is not an end point of D.

Case 2. There exists $n \in \mathbb{N}$ such that c is one of the division points that define $\mathcal{D}_n$, say, $c = a_j$ with $1 \le j \le k-1$ (obviously, c cannot be one of the points a_0 or a_k).

We shall show now that these two cases correspond to the two possible situations that are mentioned in the theorem.

Case 1. In view of property (b) it is clear that $f[D] \subseteq D$ and that, consequently, the collection $\mathcal{D}_n$ has of only one member, namely, D. It follows that $D = X$. Now it follows from (a) that $F_0^n = \dots = F_{n-1}^n = X$. By the definition of F_0^n this means that $\omega_{f^n}(x_0) = X$, that is, the point x_0 is transitive under f^n. Since this holds for every $n \in \mathbb{N}$ it follows that f is totally transitive[7].

Case 2. As $a < c < b$ the collection $\mathcal{D}_n$ has in this case at least two members, namely, $D := [a_{j-1}; a_j]$ and $D' := [a_j; a_{j+1}]$. The fact that c is an invariant point implies, in view of property (b) above, that $f[D] \subseteq D$ or D' and that $f[D'] \subseteq D$ or D'. If $f[D] \subseteq D$, then D is a closed invariant set, implying that $D = X$, because D includes a transitive point. This is not compatible with the assumption that the collection $\mathcal{D}_n$ has at least two members. Similarly, the inclusion $f[D'] \subseteq D'$ is not possible. Consequently, $f[D] \subseteq D'$ and $f[D'] \subseteq D$. But then it follows from (b) that $\mathcal{D}_n = \{D, D'\}$, hence that $D = [a; c]$ and $D' = [c; b]$. If we rename D and D' to $I_0 := D$ and $I_1 := D'$ then $f[I_0] \subseteq I_1$ and $f[I_1] \subseteq I_0$. These inclusions are equalities: if $f[I_0]$ is a proper subset of I_1 then the points $I_1 \setminus f[I_0]$ are not in $f[X]$, contradicting the fact that f is transitive (by Remark (2) after Proposition 1.3.2, f is surjective). So $f[I_0] = I_1$ and, similarly, $f[I_1] = I_0$. In particular, the closed intervals I_0 and I_1 are invariant under f^2.

Assume that the transitive point x_0 of f is situated in I_0 (if $x_0 \in I_1$ then the arguments are similar). Then $f^k(x_0) \in I_0$ iff k is even and $f^k(x_0) \in I_1$ iff k is odd (note that for no value of k the point $f^k(x_0)$ is in both intervals, because for such a k we would have $f^k(x_0) = c$, in which case x_0 would be ultimately invariant, hence not transitive). Since the orbit of x_0 is dense in X, it follows that the points $f^k(x_0)$ with even k are dense in I_0. Stated otherwise, the orbit of x_0 under f^2 is dense in I_0. Similarly, the orbit of $f(x_0)$ under f^2 is dense in I_1. Consequently, the systems (I_0, f^2) and (I_1, f^2) are transitive.

Now apply the above analysis to each of the transitive systems (I_0, f^2) and (I_1, f^2). It should be clear that Case 2 does not occur for any of these systems: for $i = 1, 2$ the interval I_i cannot be split into two subintervals that are interchanged by f^2, because its end point c is invariant under f^2. So only Case 1 is possible, that is, the system (I_i, f^2) is totally transitive. □

Remark. In Case 2 of the theorem only the even iterates of f are not transitive on X. See Exercise 2.12.

It remains to show that for every $n \in \mathbb{N}$ there is a collection $\mathcal{D}_n$ of adjacent closed intervals satisfying the conditions (a) and (b) mentioned above. Recall that the point $x_0 \in X$ is transitive under f, so $\omega_f(x_0) = X$. Moreover, for every $n \in \mathbb{N}$ and for $i = 0, \dots, n-1$ we have defined $F_i^n := \omega_{f^n}(f^i(x_0))$. Since the interval X is compact, Theorem 1.4.5 implies that all sets F_i^n are non-empty. By Proposition 1.4.3 (2), the sets F_i^n are closed in X, hence compact, and invariant under f^n. Below we shall consider the interiors of the sets F_i^n. These will be the interiors in the relative topology on X. However, the difference of the interior of a set in X with its interior in $\mathbb{R}$ consists at most of one or two

7 Similarly, the points $f^i(x_0)$ for $i = 1, \dots, n-1$ are transitive under f^n.

of the end points a and b of X. This will hardly play any role in what follows and will, therefore, be neglected.

Lemma 2.6.6. *Let $n \in \mathbb{N}$. Then:*
(1) $X = F_0^n \cup \cdots \cup F_{n-1}^n$.
(2) *For all $i = 0, \ldots, n-1$ we have $f[F_i^n] = F_{i+1\ (\mathrm{mod}\ n)}^n$.*
(3) *For all $i = 0, \ldots, n-1$ the interior $(F_i^n)^\circ$ of F_i^n is not empty.*
(4) *For all $i, j \in \{0, \ldots, n-1\}$, if $(F_i^n)^\circ \cap (F_j^n)^\circ \neq \emptyset$ then $F_i^n = F_j^n$.*

Proof. (1) Let $x \in X = \omega_f(x_0)$. It follows from Lemma 1.4.1 (2) that x is the limit of a subsequence of the orbit of x_0 under f. Since $\mathbb{Z}^+$ is the union of the finitely many subsequences $i + n\mathbb{Z}^+$ with $i = 0, \ldots, n-1$, there is such an i so that the subsequence of the orbit of x_0 converging to x contains infinitely many terms of the form $f^{i+nk}(x_0) = (f^n)^k(f^i(x_0))$ with $k \in \mathbb{Z}^+$. The subsequence consisting of these terms also converges to x for $k \rightsquigarrow \infty$. This shows that a subsequence of the orbit of $f^i(x_0)$ under f^n converges to x, so by Lemma 1.4.1 (2) it follows that $x \in F_i^n$.
(2) Note that $f \circ f^n = f^n \circ f$ and that the mapping $f: X \to X$ is surjective by Remark 2 to Lemma 1.3.2. So $f: (X, f^n) \to (X, f^n)$ is a factor mapping. As X is compact, all f^n-orbits have compact closures, so by Remark 2 after Proposition 1.5.4, $f[\omega_{f^n}(f^i(x_0))] = \omega_{f^n}(f^{i+1}(x_0))$. For $0 \le i \le n-2$ this means, by definition, that $f[F_i^n] = F_{i+1}^n$. For $i = n-1$, Proposition 1.4.3 (1) implies that $f[F_{n-1}^n] = \omega_{f^n}(f^n(x_0)) = \omega_{f^n}(x_0) = F_0^n$.
(3) By 1 above and Baire's Theorem, at least one of the closed sets F_i^n has a non-empty interior. In view of 2 above and Lemma 2.6.4 (2), it follows that each of the sets F_i has a non-empty interior.
(4) Let $i, j \in \{0, \ldots, n-1\}$ and assume that $(F_i^n)^\circ \cap (F_j^n)^\circ \neq \emptyset$. Then this intersection is a neighbourhood of a point of the limit set of $f^i(x_0)$ under f^n, hence it – and, consequently, also the set F_j^n – includes a point of the orbit of $f^i(x_0)$ under f^n. Since the set F_j^n is closed and f^n-invariant and, by Proposition 1.4.3 (1), all points of an orbit have the same limit set, it follows that $\omega_{f^n}(f^i(x_0)) \subseteq F_j^n$, that is, $F_i^n \subseteq F_j^n$. Similarly, one shows that $F_j^n \subseteq F_i^n$. This completes the proof. □

Resuming, for every $n \in \mathbb{N}$ we have defined a collection of n non-empty closed sets F_i^n, not necessarily mutually different, such that two *different* members of this collection have disjoint interiors.

Let $\mathcal{E}_n$ denote the set whose elements are the connected components of the open subset $(F_0^n)^\circ \cup \cdots \cup (F_{n-1}^n)^\circ$ of X. By Example (2) in Section A.6 of Appendix A, $\mathcal{E}_n$ is a collection of mutually disjoint open subintervals of X. Moreover, as the sets $(F_i^n)^\circ$ for the different values of i either coincide or are disjoint, every member of $\mathcal{E}_n$ is completely included in precisely one of the sets $(F_i^n)^\circ$ – though its index i may be not unique[8].

8 Every member of $\mathcal{E}_n$ is an interval that is open in X. Consequently, the most left-hand member of $\mathcal{E}_n$ is possibly of the form $[a; r)$ and the most right-hand member may have the form $(s; b]$. In those cases we replace these intervals by $(a; r)$ and $(s; b)$, respectively.

Lemma 2.6.7. *Let $n \in \mathbb{N}$. Then:*

(1) *For every $C \in \mathcal{E}_n$ and $l \in \mathbb{Z}^+$ there exists a unique member C' of $\mathcal{E}_n$ such that $f^l[\,\overline{C}\,] \subseteq \overline{C'}$ and $f^l[\,\overline{C}\,] \cap D = \emptyset$ for all members D of $\mathcal{E}_n$ that are different from C'.*

(2) *For every pair $C, C'' \in \mathcal{E}_n$ there exists $l \in \mathbb{Z}^+$ such that $f^l[\,\overline{C}\,] \subseteq \overline{C''}$.*

(3) *The collection $\mathcal{E}_n$ is finite and the set $\bigcup \mathcal{E}_n$ is dense in X.*

(4) *There are points $a = a_0 < \cdots < a_k = b$ in X such that $\mathcal{E}_n$ equals the set of open intervals $(a_j; a_{j+1})$ for $j = 0, \ldots, k-1$. The set of closures of these intervals (denoted by $\mathcal{D}_n$) is permuted cyclically by f: there is a permutation $(j_0, \ldots, j_{k-1})$ of the numbers $0, \ldots, k-1$ such that, if $I_i := [a_{j_i}; a_{j_{i+1}}]$ then $f[I_i] \subseteq I_{i+1 \pmod k}$ for $i = 0, \ldots, k-1$.*

Proof. (1) For $l = 0$ the statement is trivial: take $C' = C$. Next, we prove the statement for $l = 1$. If $C \in \mathcal{E}_n$ then C is an interval, hence $f[C]$ is an interval, which is non-degenerate by Lemma 2.6.4 (2). Moreover, $C \subseteq (F_i^n)^\circ$ for some i, so Lemma 2.6.6 (2) implies that $f[C] \subseteq F^n_{i+1 \pmod n}$. If we omit the end points from the interval $f[C]$ (if there are any) then the remaining open interval is included in the interior of $F^n_{i+1 \pmod n}$, hence it is included in a component C' of this interior, which is a member of $\mathcal{E}_n$. Then the end points of $f[C]$ are in $\overline{C'}$, hence $f[C] \subseteq \overline{C'}$ and therefore, by continuity of f, also $f[\,\overline{C}\,] \subseteq \overline{C'}$. This concludes the proof of the existence of C'.
Now consider arbitrary $D \in \mathcal{E}_n$, $D \neq C'$. Then C' and D are disjoint open sets, so that $\overline{C'} \cap D = \emptyset$, hence $f[\,\overline{C}\,] \cap D = \emptyset$ as well, which was to be proved. It remains to prove unicity of C'. To this end, note that if $f[\overline{C}]$ were included in $\overline{D}$ for some $D \in \mathcal{E}_n$, $D \neq C'$, then by the above it would be included in $\overline{D} \setminus D$. Since $f[\overline{C}]$ includes the non-degenerate interval $f[C]$ and $\overline{D} \setminus D$ consists of at most two points this is impossible. This proves unicity of C'.
Finally, if $l > 1$ then the the desired results for $f^l[\,\overline{C}\,]$ follow easily by induction in l, as follows: Assume that for $l \in \mathbb{N}$ there is for every $C \in \mathcal{E}_n$ a member C' of $\mathcal{E}_n$ according to statement (1). By what has already been proved, there is $C'' \in \mathcal{E}_n$ such that $f[\,\overline{C'}\,] \subseteq \overline{C''}$ and $f[\,\overline{C'}\,] \cap D = \emptyset$ for all members D of $\mathcal{E}_n$ that are different from C''. It easily follows that C'' is a member of $\mathcal{E}_n$ such that $f^{l+1}[\,\overline{C}\,] \subseteq \overline{C''}$ and $f^{l+1}[\,\overline{C}\,] \cap D = \emptyset$ for all members D of $\mathcal{E}_n$ that are different from C''. Unicity of C'' is shown as above: $f^{l+1}[\,\overline{C}\,]$ includes a non-degenerate interval, which cannot be included in any set of the form $\overline{D} \setminus D$ (a set with at most two points) with $D \in \mathcal{E}_n$ different from C''.
(2) Consider $C, C'' \in \mathcal{E}_n$. Since C and C'' are two non-empty open subsets of X and the system (X, f) is topologically ergodic, there exists $l \in \mathbb{Z}^+$ such that $f^l[C] \cap C'' \neq \emptyset$, hence $f^l[\,\overline{C}\,] \cap C'' \neq \emptyset$. By (1) above, there is $C' \in \mathcal{E}_n$ such that $f^l[\,\overline{C}\,] \subseteq \overline{C'}$ and $f^l[\,\overline{C}\,] \cap D = \emptyset$ for all $D \in \mathcal{E}_n \setminus \{C'\}$. This obviously implies that $C' = C''$, so that $f^l[\,\overline{C}\,] \subseteq \overline{C''}$
(3) Assume that $\mathcal{E}_n$ has at least two members (otherwise $\mathcal{E}_n$ is certainly finite) and let C and C' be two different members of $\mathcal{E}_n$. Then by (2) there are $m, l \in \mathbb{N}$ such that $f^m[\,\overline{C}\,] \subseteq \overline{C'}$ and $f^l[\,\overline{C'}\,] \subseteq \overline{C}$, hence $f^{m+l}[\,\overline{C}\,] \subseteq \overline{C}$ with $m + l \geq 1$. Let k be the smallest integer ≥ 1 such that $f^k[\,\overline{C}\,] \subseteq \overline{C}$. Since, again by (2), all members of $\mathcal{E}_n$ are visited by a set of the form $f^m[\,\overline{C}\,]$ with $m \in \mathbb{Z}^+$, it should be clear that $\mathcal{E}_n$ has just k different members. This concludes the proof that $\mathcal{E}_n$ is finite.

Next, observe that $A := \overline{\bigcup \mathcal{E}_n} = \bigcup\{\overline{C} : C \in \mathcal{E}_n\}$ (this is because $\mathcal{E}_n$ is finite), so in view of 1 above the closed set A is invariant. Since its interior is (obviously) not empty, topological ergodicity of the system (X, f) implies that $A = X$: see Exercise 1.6 (3). Stated otherwise, $\bigcup \mathcal{E}_n$ is dense in X.

(4) This is just a reformulation of what we have proved in (1), (2) and (3). Indeed, if a finite collection of open subintervals of X is dense in X then this collection consists of a finite number of adjacent open intervals of the form described above. Moreover, by what was observed in the proof of 3, the members of $\mathcal{E}_n$ can be labelled as $C_0, \dots, C_{k-1}$ such that

$$C_0 \overset{f}{\rightsquigarrow} \dots \overset{f}{\rightsquigarrow} C_{k-1} \overset{f}{\rightsquigarrow} C_0 \,, \tag{2.6-3}$$

where $C_i \overset{f}{\rightsquigarrow} C_{i+1\,(\mathrm{mod}\,k)}$ means $f[\overline{C}_i] \subseteq \overline{C}_{i+1\,(\mathrm{mod}\,k)}$ for $i = 0, \dots, k-1$. Since each of the sets $\overline{C}_i$ for $i = 0, \dots, k-1$ has the form $[a_{j_i}; a_{j_{i+1}}]$ this completes the proof of 4. □

Remark. In (4) we actually have equalities. Suppose that for $C, C' \in \mathcal{E}_n$ with $f[\overline{C}] \subseteq \overline{C'}$ we have $f[\overline{C}] \neq \overline{C'}$. Then the set $\overline{C'} \setminus f[\overline{C}]$ includes a non-degenerate interval within $\overline{C'}$, which is not covered by any set of the form $f[\overline{D}]$ with $D \in \mathcal{E}_n$: by its definition it is not covered by $f[\overline{C}]$, and if $D \neq C$ then, by the periodicity in (2.6-3), $f[\overline{D}] \subseteq \overline{C''}$ with $C'' \neq C'$, so $f[\overline{D}]$ has at most an end point in common with $\overline{C'}$. It follows that the mapping f would not be surjective.

Our next result elaborates on the first possibility for a transitive interval mapping mentioned in Theorem 2.6.5.

Theorem 2.6.8. *Let (X, f) be a dynamical system on a compact interval. The following conditions are equivalent:*

(i) *f is transitive and there is a periodic point under f with odd primitive period greater than 1.*

(ii) *f^2 is transitive on X.*

(iii) *f is totally transitive (that is, for all $n \in \mathbb{N}$ the system (X, f^n) is transitive).*

Proof. “(i)⇒(iii)”: If f is transitive on X but not totally transitive then possibility (2) of Theorem 2.6.5 applies. In that case all non-invariant periodic points have an even period, so (i) cannot hold.

“(iii)⇒(ii)”: Obvious.

“(ii)⇒(iii)”: If f^2 is transitive then f is transitive as well. If (iii) does not hold then possibility (2) of Theorem 2.6.5 applies: f^2 leaves two proper subintervals of X invariant, contradicting that f^2 is transitive on X.

“(iii)⇒(i)”: If (iii) holds then, obviously, f is transitive, so it remains to prove that f admits a periodic point with odd primitive period distinct from 1 (i.e., that point is not invariant).

Recall from Proposition 1.1.1 that the set F of all f-invariant points is closed. Transitivity of f implies that F cannot be equal to X, hence $X \setminus F$ is a non-empty open subset of the interval X. Consequently, there is a non-degenerate closed subinterval J of X that includes no f-invariant points. We claim that $f^n[J] \supseteq J$ for almost all $n \in \mathbb{N}$.

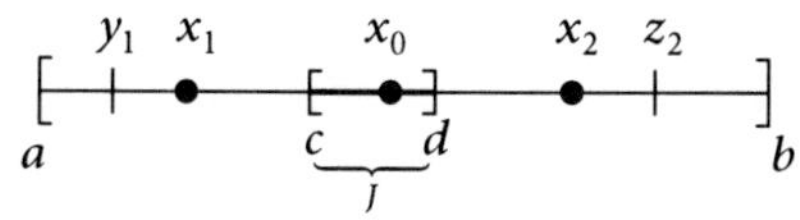

Fig. 2.17. Illustrating the proof of Theorem 2.6.8.

Assuming that this has been proved, proceed as follows: Consider an odd value of n for which the inclusion $f^n[J] \supseteq J$ holds. By Lemma 2.2.1 (1), the interval J contains an invariant point of f^n, i.e.„ a periodic point of f with period n. The primitive period of this point is a divisor of n, hence this primitive period is odd as well. Since J includes no invariant points of f, this shows that (i) holds.

We proceed with the proof of the above claim, namely, that $f^n[J] \supseteq J$ for almost all $n \in \mathbb{N}$. Let $X = [a; b]$ and let $J = [c; d]$. If necessary, we may replace J by a smaller closed non-degenerate interval, so that we may assume that $a < c < d < b$. As f is transitive, Theorem 2.6.2 implies that the f-periodic points are dense in X. Consequently, there are periodic points $x_0 \in J$, $x_1 \in (a; c)$ and $x_2 \in (d; b)$. If one or both of the points a or b happen to belong to $\mathcal{O}(x_1)$ then replace x_1 by a periodic point in the non-empty open set $(a; c) \setminus \mathcal{O}(x_1)$, which point will have an orbit disjoint from the 'old' orbit $\mathcal{O}(x_1)$. Thus, we may assume that $a, b \notin \mathcal{O}(x_1)$[9]. Similarly, we may assume that $a, b \notin \mathcal{O}(x_2)$. For $i = 1, 2$, let $y_i := \min \mathcal{O}(x_i)$ and let $z_i := \max \mathcal{O}(x_i)$. Then for $i = 1, 2$ one has $y_i, z_i \in \mathcal{O}(x_i)$, so these points have the same periods as the (periodic!) point x_i, and $\mathcal{O}(x_i) \subseteq [y_i; z_i]$. Moreover, since $a, b \notin \mathcal{O}(x_i)$ for $i = 1, 2$, it is clear that $a < y_1 \leq x_1 \leq z_1 < b$ and $a < y_2 \leq x_2 \leq z_2 < b$. Note also that, by the choice of the points x_1 and x_2, one has $J \subseteq [x_1; x_2] \subseteq [y_1; z_2]$. See Figure 2.16.

Let k be a common multiple of the primitive periods of the points x_0, x_1 and x_2 and let $g := f^k$. Then the point x_0, as well as the points x_1 and x_2 together with all points in their orbits, are invariant under g. In particular, the points y_1, z_1, y_2 and z_2 are invariant under g.

Now use the assumption that f is totally transitive: the mapping g is transitive on X, hence the system (X, g) is topologically ergodic. It follows that there exists $n_1 \in \mathbb{N}$ such that g^{n_1} maps a point from J into the open interval $(a; y_1)$. Then $g^{n_1}[J]$ is an interval that includes the g-invariant points y_1 and x_0. So these points are included in the interval $g^{n_1+j}[J]$ for all $j \in \mathbb{Z}^+$, hence $[y_1; x_0] \subseteq g^n[J]$ for almost all $n \in \mathbb{N}$. In a similar manner one shows that $[x_0; z_1] \subseteq g^n[J]$ for almost all $n \in \mathbb{N}$. Consequently,

$$\mathcal{O}_f(x_1) \subseteq g^n[J] \quad \text{for almost all } n \in \mathbb{N}\,.$$

Using similar arguments one shows

$$\mathcal{O}_f(x_2) \subseteq g^n[J] \quad \text{for almost all } n \in \mathbb{N}\,.$$

9 If both $a, b \in \mathcal{O}(x_1)$ then this one step is sufficient to get a periodic orbit with a point in $(a; c)$ that contains neither a nor b. If only $a \in \mathcal{O}(x_1)$ then by this one step we get a periodic orbit with a point in $(a; c)$ that does not contain a but, perhaps, contains the point b. If so, then we select in a similar way a third periodic orbit, different from the previous ones, which will contain neither a nor b.

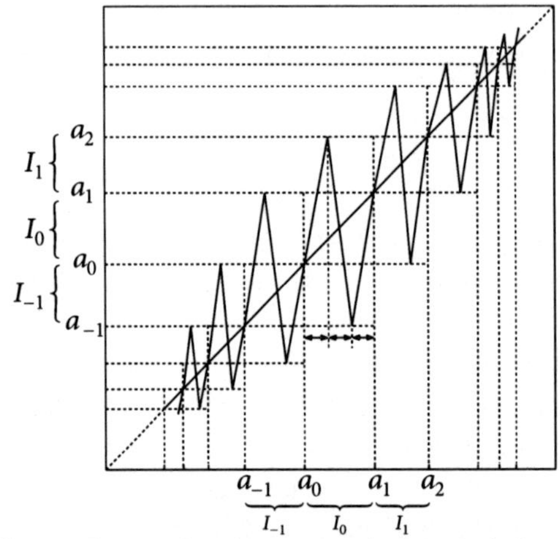

Fig. 2.18. The graph of the mapping defined in Example (2).

Hence there exists $N \in \mathbb{N}$ such that $\mathcal{O}_f(x_i) \subseteq g^N[J] = f^{kN}[J]$ for $i = 1, 2$. By applying iterates of f to both sides of this inclusion, taking into account that for $i = 1, 2$ the orbit $\mathcal{O}(x_i)$ is invariant under f, one easily sees that

$$y_i, z_i \in \mathcal{O}_f(x_i) \subseteq f^m[J] \quad \text{for all } m \geq kN\,.$$

Since $f^m[J]$ is an interval for every value of m this implies, in particular, that $J \subseteq [y_1; z_2] \subseteq f^m[J]$ for all $m \geq kN$. This completes the proof. □

Examples.

(1) The tent map T is totally transitive; see 1.7.3. And indeed, the unit interval is not divided in two non-degenerate subintervals that are interchanged by T. Moreover, condition (i) of Theorem 2.6.8 holds: the point $4/5$ has period three.

(2) The mapping g_k of $[0; k-1]$ onto itself ($k \in \mathbb{N}$, k odd and not 1) defined at the end of Section 2.5, turns out to be transitive; the proof will be postponed to Exercise 6.12. The unique invariant point of g_k does not divide the interval $[0; k-1]$ in two subintervals that are interchanged by g_k, so Theorem 2.6.5 implies that g_k is totally transitive. Of course, one might also use Theorem 2.6.8 (i)⇒(iii).

(3) Consider a bi-infinite monotonous sequence $(a_n)_{n\in\mathbb{Z}}$ in the unit interval $[0; 1]$, i.e., consider points $0 < \cdots < a_{-2} < a_{-1} < a_0 < a_1 < \cdots < 1$, such that $a_{-n} \rightsquigarrow 0$ and $a_n \rightsquigarrow 1$ for $n \rightsquigarrow \infty$. For every $i \in \mathbb{Z}$ let $I_i := [a_i; a_{i+1}]$. Let f be a piecewise linear continuous mapping of $[0; 1]$ onto itself which, for every $n \in \mathbb{Z}$, maps I_n onto the interval $I_{n-1} \cup I_n \cup I_{n+1} = [a_{n-1}; a_{n+2}]$, as indicated in Figure 2.18. The local extremes of f on the interior of I_n are supposed to be at $1/3$ and $2/3$ of the interval I_n, so the absolute value of the derivative of f is everywhere – where defined – greater than 3. Claim: f is transitive.
In order to prove this, it is sufficient to show that for every non-degenerate closed subinterval J of $[0; 1]$ and every $\varepsilon > 0$ there exists $n \in \mathbb{N}$ such that $f^n[J] \supseteq [\varepsilon; 1-\varepsilon]$. This is true if there exist $N \in \mathbb{N}$ and $i \in \mathbb{Z}$ such that $f^N[J] \supseteq I_i$. For this implies that $f^{N+k}[J] \supseteq f^k[I_i] = [a_{i-k}; a_{i+k+1}]$, and the latter interval includes $[\varepsilon; 1-\varepsilon]$ for almost

all k. Thus, we want to show that there exists $N \in \mathbb{N}$ such that $f^N[J]$ contains two points where f has a local extreme (let us call this 'critical points'). For if this is true then $f^N[J]$ certainly includes one of the intervals I_i.

Assume that a non-degenerate subinterval K of $[0;1]$ includes no critical point: then $|f[K]| \geq 3|K|$ (here $|\cdot|$ denotes the length of an interval). And if K includes one critical point then such a point divides K in two closed subintervals, one of which – call it K' – has length at least $\frac{1}{2}|K|$ (this is also true if the critical point coincides with an end point of K). Then f stretches K' by a factor at least 3, so

$$\big|f[K]\big| \geq \big|f[K']\big| \geq 3|K'| \geq \frac{3}{2}|K|\,.$$

Thus, if K includes at most one critical point then f stretches K by a factor of at least $3/2$. Consequently, if J is a non-degenerate subinterval of $[0;1]$ such that for every $n \in \mathbb{N}$ the interval $f^n[J]$ includes at most one critical point, then by induction one gets $\big|f^n[J]\big| \geq (3/2)^n|J|$. As $\big|f^n[J]\big| \leq 1$ for all n, this is impossible. Hence there exists $N \in \mathbb{N}$ such that $f^N[J]$ contains two critical points.

It follows from the proof above that for every non-degenerate closed subinterval J of the unit interval and for every $\varepsilon > 0$ the inclusion $f^n[J] \supseteq [\varepsilon; 1-\varepsilon]$ holds for almost all $n \in \mathbb{N}$. By Exercise 2.13 this implies that f is strongly mixing and totally transitive. The latter can also be seen directly: condition (2) of Theorem 2.6.5 is not fulfilled.

The following observation will be useful later on (in the Chapters 7 and 8).

Corollary 2.6.9. *Let (X, f) be a dynamical system on a compact interval. If f is transitive then there exists a periodic point whose period is not a power of 2.*

Proof. By Theorems 2.6.5 and 2.6.8, f has a periodic point either with period p or with period $2p$ for some odd integer greater than 1. □

Exercises

2.1. Let X be an interval in $\mathbb{R}$ and let $f: X \to X$ be continuous. Assume that f is monotonously increasing on a subinterval J of X and let x_0 be a point in J such that $\mathcal{O}(x_0) \subseteq J$. Then the orbit of x_0 is monotonous (increasing if $f(x_0) > x_0$, decreasing if $f(x_0) < x_0$ and eventually constant if $f^{k+1}(x_0) = f^k(x_0)$ for some value of $k \in \mathbb{Z}^+$), so it has a limit $z \in \mathrm{cl}_{\mathbb{R}}(J)$ (possibly $z = \infty$ or $-\infty$). If $z \in X$ then the point z is invariant under f.

2.2. Consider the quadratic mapping $f_\mu: \mathbb{R} \to \mathbb{R}$ with $\mu \geq 2$. Let α_μ be the unique solution of the equation $f_\mu(x) = 1/2$ on the interval $(0; 1/2)$:

$$\alpha_\mu = \tfrac{1}{2} - \tfrac{1}{2}\sqrt{1 - \tfrac{2}{\mu}}\,.$$

(a) Show that f_μ^2 is increasing on the interval $[0; \alpha_\mu]$ and decreasing on the intervals $[\alpha_\mu; 1/2]$ and $[\hat{p}_\mu; 1/2]$ (recall that $\hat{p}_\mu := 1/\mu$).
(b) Show that $f_\mu^2(1/2) \geq 1/2$ if $\mu \leq 3$, so that $f_\mu^2[\hat{p}_\mu; p_\mu] \subseteq [1/2; p_\mu]$ if $2 \leq \mu \leq 3$.
(c) Prove: if $2 \leq \mu \leq 3$ then $f_\mu^2(x) \geq x$ for all $x \in [0; p_\mu]$.
(d) For $3 < \mu < 1 + \sqrt{6}$ the quadratic mapping f_μ has an attracting periodic orbit of period 2 in $(0; 1)$.

2.3. Let $f: \mathbb{R} \to \mathbb{R}$ be a continuously differentiable mapping. Prove:
(a) Attracting invariant points of f are isolated from each other (that is, each has a neighbourhood that contains none of the others).
(b) Repelling invariant points of f are isolated from each other as well.
(c) If x_0 is a repelling invariant point of f then there is a neighbourhood of x_0 that is mapped by f homeomorphically onto a neighbourhood of x_0.
(d) The mapping

$$f: x \mapsto \begin{cases} x + x^3 \sin \frac{2\pi}{x} & \text{for } x \neq 0 \\ 0 & \text{for } x = 0 \end{cases}$$

is continuously differentiable, but the invariant point 0 of f is not an isolated invariant point (hence neither attracting nor repelling).

2.4. Show that for every closed subset F of $\mathbb{R}$ there is a continuous mapping $f: \mathbb{R} \to \mathbb{R}$ such that F is equal to the set of invariant points of f.

2.5. Let X be an interval in $\mathbb{R}$ (not necessarily bounded or open). Show that a homeomorphism $f: X \to X$ has no periodic points with primitive period 3 or more.

2.6. Show that in the definition of 'attracting periodic point' (under a continuously differentiable mapping) *any* period of that point can be used.

2.7. Let $f : [0; 3] \to [0; 3]$ be a continuous mapping. Assume that $f(0) = 1$, $f(1) = 3$, $f(2) = 0$ and $f(3) = 2$. Show that for every $p \in \mathbb{N}$ there is a periodic point for f with primitive period p.

2.8.
(1) Let (X, f) be a dynamical system on an interval in $\mathbb{R}$ and let $\{x_0, x_1\}$ be the orbit of a periodic point with period 2. Show that between x_0 and x_1 there is an invariant point.
(2) Prove the following counterpart of Lemma 2.2.1 (1): if I is a closed interval and f is a continuous mapping defined on I such that $f[I] \subseteq I$ then I contains an invariant point of f.

2.9. Show that for $0 \leq \lambda \leq \frac{2}{3}$ the point λ is invariant under the truncated tent map T_λ, and λ belongs to the orbit of every point of $[0; 1]$ (so every point is eventually invariant). Similarly, show that for $\frac{2}{3} < \lambda \leq \frac{4}{5}$ the point λ is periodic under T_λ with primitive period 2 and that λ belongs to the orbit of every point of $[0; 1]$ (so every point is eventually periodic).

2.10.
(1) Prove the following generalisation of Corollary 2.4.3:

$$\forall m \in \mathbb{N} : \mathbf{Per}(\tau^m(f)) = \{n \cdot 2^m \vdots n \in \mathbf{Per}(f)\} \cup \mathbf{S}(2^{m-1}).$$

(2) Let $k \in \mathbb{N}$, $k \geq 2$ and k odd, and let g_k be defined as in the final example in Section 2.5. Show that $\mathbf{Per}(\tau^m(g_k)) = \mathbf{S}(k \cdot 2^m)$ for all $m \in \mathbb{N}$ and $k \in \mathbb{N}$, k odd. Thus, every Šarkovskij tail can be realized as the set of periods of a function of the form $\tau^m(g_k)$ for $m \in \mathbb{Z}^+$ and k an odd natural number.

2.11.
(1) Let (X, f) be a transitive system on a non-degenerate interval X. Show that for every $n \in \mathbb{N}$ the set of periodic points with primitive period at least n is dense.
(2) Let (X, f) be a topologically ergodic system (on an arbitrary Hausdorff space) with a dense set of periodic points. Then either X consists of one single periodic orbit or for every $n \in \mathbb{N}$ the set of periodic points with primitive period at least n is dense in X.

2.12. Let (X, f) be a transitive system on a compact interval. Then for every odd p the mapping f^p is transitive on X.

2.13. Let X be the compact interval $[a; b]$. The following conditions are equivalent:
(i) f is totally transitive (see Theorem 2.6.8).
(ii) f is weakly mixing.
(iii) f is strongly mixing.
(iv) For every $\varepsilon > 0$ and for every non-degenerate (closed) subinterval J of X there exists $N \in \mathbb{N}$ such that $f^n[J] \supseteq [a + \varepsilon; b - \varepsilon]$ for all $n \geq N$.

NB. By the Examples (1) and (2) after Theorem 2.6.8, the tent map and the mappings g_k for odd $k \geq 3$ odd are totally transitive, hence strongly mixing.

2.14. Consider the mapping $f_\infty : [0; 1] \to [0; 1]$ defined in Section 2.4. Show that for every $n \in \mathbb{Z}^+$ the set $C_n = \bigcup\{J_b^n \vdots b \in \{0, 1\}\}$ is invariant by showing that the intervals J_b^n are mapped onto each other as indicated in the following picture, where the dotted arrows denote inclusions and the solid arrows show how the intervals are mapped onto each other.

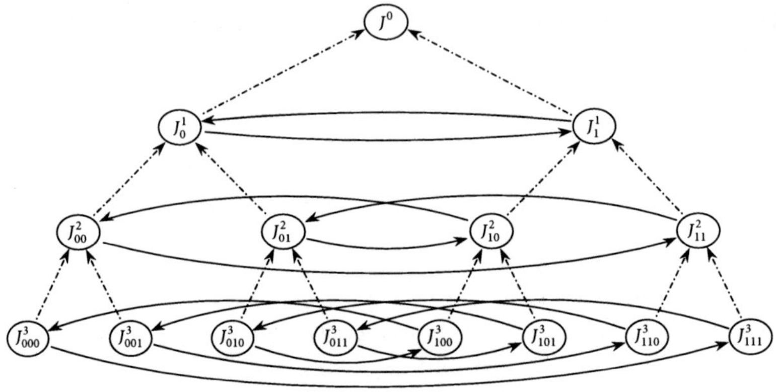

Fig. 2.19. The intervals J^n_b are mapped onto each other in a complicated way.

Notes

1 The method of graphical iteration is due to Picard, though it is also called the *method of Koenigs–Lemeray*. Historically, Picard used iteration to find solutions of fixed point problems, such as solving initial value problems of differential equations (see Note 1 at the end of the Introduction).
2 For certain interval maps the *Julia–Singer Theorem* gives an upper bound for the number of attracting periodic orbits. 'Attracting' is defined similar as in the case $\mu > 3$ in 2.1.5: the orbit of the periodic point x_0 with primitive period p is said to be *attracting* whenever $|(f^p)'(x_0)| < 1$; see 3.3.18 ahead. (Actually, the theorem gives an upper bound for the number of *asymptotically stable* periodic orbits – which might be larger, because every attracting periodic orbit is asymptotically stable; see 3.3.18 below.) It states:

Theorem (Julia–Singer). *Let f be an S-mapping with N critical points in X. Then the dynamical system (X, f) has at most $N + 2$ attracting periodic orbits.*

An S-mapping is a C^3 mapping $f: X \to \mathbb{R}$ for which the so-called *Schwarzian derivative*

$$Sf(x) := \frac{f'''(x)}{f'(x)} - \frac{3}{2}\left(\frac{f''(x)}{f'(x)}\right)^2$$

is negative at all points where it is defined, i.e., outside of the set of *critical points* (that are points x with $f'(x) = 0$). Every member of the quadratic family is an S-mapping, hence has at most three attracting periodic orbits. A close analysis of the proof shows that two of these are not present, so actually there is at most one attracting periodic orbit. This is in accordance what is stated in Note 4 below.
3 The Li–Yorke Theorem appeared in the paper T. Y. Li & J. A. Yorke [1975]. As to Lemma 2.2.1 (2), most texts use the following argument to prove the existence of a minimal interval I' such that $f[I'] \supseteq J$. First, let $a' := \sup\{\, y \in I : f[[y; b]] \supseteq J\,\}$ and, subsequently, let $b' := \inf\{\, z \in I : f[[a'; z]] \supseteq J\,\}$; then $I' := [a'; b']$ has the desired properties. However, the proofs that this inf and sup exist implicitly use the Axiom of Choice, so we decided to make this dependence more explicit by using Zorn's Lemma.
4 The Šarkovskij order is a linear order of $\mathbb{N}$ of order-type $\omega^\omega + {}^*\omega$, where ω is the order type of the ordered set $1, 2, 3, \dots$ and ${}^*\omega$ is the type of the ordered set $\dots, -3, -2, -1$. With the hypothetical element 2^∞ inserted at its proper place one would get a linearly ordered set of type $\omega^\omega + 1 + {}^*\omega$ (one of the best introductions to order types, ordinal numbers and ordinal arithmetic is, in my opinion, Chapter III of A. A. Fraenkel [1953]). Interestingly enough, when the 'tail' of powers of 2 is omitted from

the Šarkovskij order then one gets a *well-ordered set* whose ordinal number is ω^ω but whose cardinal number is $\aleph_0$ (and not $\aleph_0^{\aleph_0}$), showing that exponentiation of ordinal numbers does not agree with exponentiation of the corresponding cardinal numbers. This example is very much like the example on page 286 in Fraenkel's book mentioned above.

Recall that Šarkovskij's Theorem can be formulated as follows: for every interval map f there exists $n \in \mathbb{N}_\infty$ such $\mathbf{Per}(f) = \mathbf{S}(n)$. This unambiguously defined member of $\mathbb{N}_\infty$ is often called *the (Šarkovskij) type of* f.

Šarkovskij's Theorem is typically a result for dynamics on intervals in $\mathbb{R}$. There is even no counterpart for the circle. Indeed, for every $n \in \mathbb{N}$, all points of $\mathbb{S}$ are periodic with period n under the rotation $\varphi_{1/n}$ and there are no points with other periods. So the occurrence of a certain period has no consequence for the occurrence of other periods. There are variants for mappings of a square into itself, but their proofs are more complicated.

Several proofs of Šarkovskij's Theorem are known. The original proof was published in Russian. For simplifications, see B.-S. Du [2007]. Our choice for the proof based on Markov graphs, presented in Section 2.5 above, is motivated by the fact that we need Proposition 2.5.3 also in Chapter 8.

Šarkovskij's Theorem says nothing about the attracting properties of periodic orbits. For example, recall the period-doubling bifurcation of the quadratic family in the Introduction: there is a sequence $1 < \mu_1 < \mu_2 < \cdots < \mu_\infty = 3.57\ldots$ such that for values of μ with $\mu_n < \mu \leq \mu_{n+1}$ the quadratic mapping f_μ has an attracting periodic orbit with primitive period 2^{n-1} ($n \in \mathbb{N}$). If $n \geq 2$ then by Šarkovskij's Theorem there are also orbits with period 2^i for $i = 0, \ldots, n-2$, but they do not show up in numerical experiments because they are not attracting. It can be shown that for μ equal to the Feigenbaum point 3.57…the mapping f_μ has periodic points with period 2^n for every $n \in \mathbb{Z}^+$ and no periodic points with other periods. Moreover, for $\mu = 1+2\sqrt{2}$ the mapping f_μ has an attracting periodic orbit of period 3 (see Note 7 below), so f_μ has all periods for this value of μ. But only the period-3 orbit shows up in experiments.

5 The first proof in English that a transitive system on a closed interval has a dense set of periodic points is in M. Barge & J. Martin [1985], though the result was already published in Russian by Šarkovskij in 1964. The simple proof in Section 2.6 is taken from the paper M. Vellekoop & R. Berglund [1994].

6 Part of the conclusion of the proof of Theorem 2.6.5, or rather, the result of Exercise 2.12, holds also for the circle: if $f : \mathbb{S} \to \mathbb{S}$ is transitive and f has an invariant point then f^p is transitive on $\mathbb{S}$ for all odd $n \in \mathbb{N}$. See E. M. Coven & I. Mulvey [1986].

7 As explained in the Case $\mu > 3$ in 2.1.5, the value 3 which μ has to pass in order to see an orbit of period two appear can also be found by solving the two equations

$$f_\mu^2(x) = x \quad \text{and} \quad (f_\mu^2)'(x) = 1$$

(the condition that the graph of f_μ^2 touches the diagonal in an invariant point of f_μ^2). A similar method can be used to find a periodic orbit with period 3. We shall indicate how one may prove the following statement:

For μ slightly larger than $1 + 2\sqrt{2}$ the mapping f_μ has a periodic point with period 3, hence it has periodic orbits of all periods.

Proof (by intimidation). By analogy with what happens at the value $\mu = 3$ for f_μ^2, the value which μ must pass so that a periodic point x of period three occurs (a bifurcation point) is found from the equations

$$\begin{cases} f_\mu^3(x) = x\,, \\ (f_\mu^3)'(x) = 1\,. \end{cases}$$

Here

$$f_\mu^3(x) = \mu^3 x(1-x)(1-\mu x+\mu x^2)(1-\mu^2 x+\mu^2(\mu+1)x^2-2\mu^3x^3+\mu^3x^4)\,.$$

Since the second equation is the derivative of the first this means that we are looking for a solution of the first equation with multiplicity at least two. Hence one can eliminate x from this system by putting the discriminant of the polynomial $f_\mu^3(x)-x$ equal to 0. This gives the equation

$$\mu^{42}(\mu-1)^2(\mu^2+\mu+1)^2(\mu^2-2\mu-7)^3(\mu^2-5\mu+7)^4 = 0\,.$$

(If you want to do this by hand you will be busy for some time. Maybe you should start up a program like Maple, put

```
g:=m^3*x*(1-x)*(1-m*x+m*x^2)*(1-m^2*x+m^2*(m+1)*x^2-2m^3*x^3+m^3*x^4),
```

determine `d:=discrim(g,x)` and carry out `factor(d)`.) The third and fifth factor give no real solutions, so we have to take into account only the first, second and fourth factors. We find the real solutions

$$\mu_0 = 1+2\sqrt{2}\,,\quad \mu_1 = 1-2\sqrt{2}\,,\quad \mu_2 = 1\,,\quad \mu_3 = 0\,.$$

The second solution is negative, the third gives the invariant point 0 of f_1, and the fourth solution is not interesting at all. So μ_0 is the desired value of μ. □

8 In the proof of Corollary 2.6.9 it was observed that a transitive system (X, f) on a compact interval has a periodic point either with period p (in the case that f is totally transitive – or mixing: see Exercise 2.13) or with period $2p$ for some odd integer greater than 1 (in the case that f^2 is not transitive). It can be shown that in the latter case there is a periodic point with primitive period 6. See L. Block & E. M. Coven [1987]; see also Proposition 7.5.4 ahead.

3 Limit behaviour

Abstract. In this chapter we discuss methods to study the behaviour of dynamical systems in the long run. We distinguish two types of such limit behaviour. The first type occurs when an invariant set in the phase space 'attracts' all neighbouring points, or at least, when neighbouring points are not moving away from the invariant set (stability). The second type of limit behaviour applies to individual states and concerns variations of periodicity. For both types of limit behaviour the notion of limit set is of central importance. Therefore, we first revisit limit sets.

In the remainder of this chapter we deal only with the first type of limit behaviour mentioned above. Much attention will be paid to attraction by stable sets (asymptotic stability) in locally compact spaces. The second type of stability is discussed in Chapter 4.

3.1 Limit sets and attraction

As in Chapter 1, (X, f) will denote a dynamical system on an arbitrary Hausdorff phase space. The phase mapping is only required to be continuous. Recall from Section 1.4 the definition and the elementary properties of the *limit set* $\omega_f(x)$ of a point $x \in X$ – denoted as $\omega(x)$ if f is understood.

Examples.

(1) Let $[r, t]$ denote the point $(r \cos 2\pi t, r \sin 2\pi t)$ of $\mathbb{R}^2$ ($r \geq 0$ and $0 \leq t < 1$). Consider the mapping $f: [r, t] \mapsto [\sqrt{r}, \varphi_a([t])] : \mathbb{R}^2 \to \mathbb{R}^2$, where φ_a denotes the rigid rotation of the circle over $a \in \mathbb{R}$. The point $P := [0, 0]$ is invariant, hence $\omega(P) = \{P\}$. Next, consider a point $x = [r, t]$ with $r > 0$. If $a \in \mathbb{Q}$ dan is $\omega(x) = \mathcal{O}([1, t])$. If $a \in \mathbb{R} \setminus \mathbb{Q}$ then $\omega(x) = \mathbb{S}$ (for convenience, the unit circle in $\mathbb{R}^2$ is identified with $\mathbb{S}$).

(2) The following should be clear from Lemma 1.4.1 (2): if (X, f) is a dynamical system with a metric phase space, $x \in X$ and $\lim_{n \rightsquigarrow \infty} f^n(x) =: x_0$ exists then $\omega(x) = \{x_0\}$ (hence the point x_0 is invariant). Use this to determine the limit sets of the points of $\mathbb{R}$ under the quadratic mapping f_μ in the various cases considered in 2.1.5. For example, in the case that $0 < \mu < 1$ one has: $\omega_{f_\mu}(x) = \{0\}$ for $x \in (p_\mu; \hat{p}_\mu)$, $\omega_{f_\mu}(x) = \emptyset$ for $x \notin [p_\mu; \hat{p}_\mu]$, $\omega_{f_\mu}(0) = \{0\}$ and $\omega_{f_\mu}(p_\mu) = \omega_{f_\mu}(\hat{p}_\mu) = \{p_\mu\}$ for $x \in \{p_\mu, \hat{p}_\mu\}$.

Recall from Theorem 1.4.5 that for a point $x \in X$ with a compact orbit closure the limit set $\omega(x)$ is a non-empty compact invariant set; it is even completely invariant: see Exercise 3.2. If X is locally compact then the following converse of Theorem 1.4.5 holds:

Theorem 3.1.1. *Assume that X is locally compact. If $x \in X$ and $\omega(x)$ is compact and not empty then $\overline{\mathcal{O}(x)}$ is compact.*

Proof. By the statement after the Examples in Appendix A.2.1, $\omega(x)$ has an open neighbourhood U with compact closure $\overline{U}$. By Lemma 1.4.1 (1), $f^n(x) \in U \subseteq \overline{U}$ for infinitely many values of $n \in \mathbb{Z}^+$. It is sufficient to prove that $f^n(x) \in \overline{U}$ for almost all values of n. Assume the contrary; then there is a sequence of points $f^{n_i}(x) \in \overline{U}$ such that $f^{n_i+1}(x) \notin \overline{U}$ for all i. The first sequence has a accumulation point z – which must be in $\omega(x)$ – but then the second sequence has the accumulation point $f(z)$ outside of U, contradicting invariance of $\omega(x)$. □

Example. Let $\mathbb{Z}^* := \{-\infty\} \cup \mathbb{Z} \cup \{\infty\}$ with its usual ordering (i.e., $-\infty < n < \infty$ for $n \in \mathbb{Z}$). Let $X^* := (\mathbb{Z}^* \times \mathbb{N}) \cup \{(0, \infty)\}$ with the following topology: all points of $\mathbb{Z} \times \mathbb{N}$ are isolated, and in the points at infinity we have the following local bases:
- at $(-\infty, n)$ for $n \in \mathbb{N}$: all sets $[-\infty; m] \times \{n\}$ with $m \in \mathbb{Z}$,
- at (∞, n) for $n \in \mathbb{N}$: all sets $[m; \infty] \times \{n\}$ with $m \in \mathbb{Z}$,
- at $(0, \infty)$: all sets $(\mathbb{Z}^* \times [n; \infty)) \cup \{(0, \infty)\}$ with $n \in \mathbb{N}$.

Here $[-\infty; m]$, $[m; \infty]$ and $[n; \infty)$ denote the usual order-intervals in $\mathbb{Z}^*$ (for example, $[-\infty; m] := \{k \in \mathbb{Z}^* : k \leq m\}$, etc.).

It is straightforward to check that X^* is a compact Hausdorff space (compactness: outside of an open set containing the point $(0, \infty)$ there are only finitely many points of $\{-\infty, \infty\} \times \mathbb{N}$, and outside the union of open sets containing these points there are only finitely many isolated points).

Define $f^* : X^* \to X^*$ as indicated by the arrows in Figure 3.1; the point $(0, \infty)$ is invariant. Let $C := (\{-\infty\} \times \mathbb{N}) \cup (\{\infty\} \times \mathbb{N})$ and let $X := X^* \setminus C = (\mathbb{Z} \times \mathbb{N}) \cup \{(0, \infty)\}$. It is an easy to verify that $\omega_{f^*}(x) = C \cup \{(0, \infty)\}$ for every point $x \in X \setminus \{(0, \infty)\}$ and that $\omega_{f^*}(x) = \{(0, \infty)\}$ for every point $x \in C \cup \{(0, \infty)\}$.

Clearly, X is an invariant subset of X^*; let $f := f^*|_X$. It is easily seen that in the system (X, f) one has $\omega_f(x) = \{(0, \infty)\}$ for every point $x \in X$: the orbit of x enters infinitely often into any neighbourhood of the point $(0, \infty)$. So every point of X has a compact limit set but points of $X \setminus \{(0, \infty)\}$ have a non-compact orbit closure (namely, all of X). So local compactness of the phase space cannot be omitted from the hypothesis of Theorem 3.1.1.

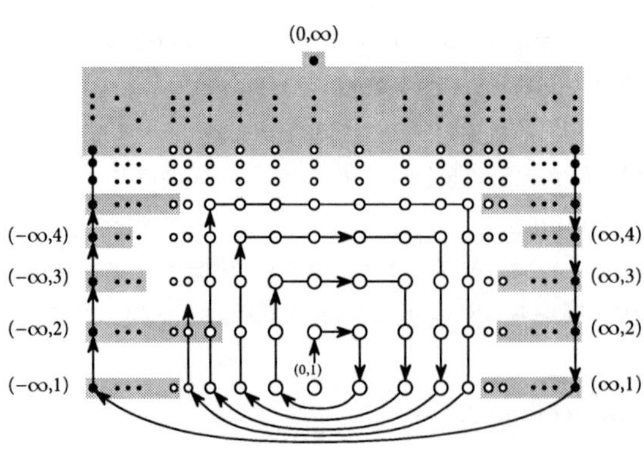

Fig. 3.1. The grey areas denote typical neighbourhoods of the point $(0, \infty)$ and of the points $(\pm\infty, n)$ (only indicated for $1 \leq n \leq 4$).

Next, we consider the behaviour of limit sets under morphisms:

Proposition 3.1.2. *Let $\varphi: (X, f) \to (Y, g)$ be a morphism of dynamical systems and let $x \in X$.*
(1) $\varphi[\omega_f(x)] \subseteq \omega_g(\varphi(x))$.
(2) *If φ is a conjugation or if $\overline{\mathcal{O}_f(x)}$ is compact then* $\varphi[\omega_f(x)] = \omega_g(\varphi(x))$.
(3) *If X is locally compact and $\omega_f(x)$ is compact and not empty then the equality $\varphi[\omega_f(x)] = \omega_g(\varphi(x))$ holds as well.*

Proof. Consider the following sequence of equalities and inclusions:

$$\begin{aligned}\varphi[\omega_f(x)] &\overset{(1)}{=} \varphi\Big[\bigcap_{n\geq 0}\overline{\mathcal{O}_f(f^n(x))}\Big] \overset{(2)}{\subseteq} \bigcap_{n\geq 0}\varphi\Big[\overline{\mathcal{O}_f(f^n(x))}\Big]\\ &\overset{(3)}{\subseteq} \bigcap_{n\geq 0}\overline{\varphi\big[\mathcal{O}_f(f^n(x))\big]} \overset{(4)}{=} \bigcap_{n\geq 0}\overline{\mathcal{O}_g(\varphi(f^n(x)))}\\ &\overset{(5)}{=} \bigcap_{n\geq 0}\overline{\mathcal{O}_g(g^n(\varphi(x)))} \overset{(6)}{=} \omega_g(\varphi(x))\,.\end{aligned}$$

The equalities (1) and (6) are justified by the definition of 'limit set', (4) is justified by Proposition 1.5.2 (1) and (5) by the definition of a morphism. The inclusions labelled (2) and (3) hold for every continuous mapping φ. However, if φ is a homeomorphism of X onto Y then these inclusions can be replaced by equalities. Moreover, if $\overline{\mathcal{O}_f(x)}$ is compact then all orbit closures above are compact, hence the inclusions (2) and (3) may also be replaced by equalities: (2) by Appendix A.3.5 and (3) in view of statement (4) in A.3.1. Finally, statement 3 in the proposition follows from statement 2 and Theorem 3.1.1 above. □

Examples.
(1) Consider the embedding mapping $\mathrm{id}_X : (X, f) \to (X^*, f^*)$, where X, X^*, f and f^* are as in the Example after Theorem 3.1.1 above. If $x \in X$ then $\omega_f(x) = \{(0, \infty)\}$, but $\omega_{f^*}(x)$ contains also points different from $(0, \infty)$. So the inclusion in Proposition 3.1.2 (1) can be proper.
(2) Let $X := ([0; 1] \times \{0, 1\}) \setminus \{(0, 1)\}$, $Y := [0; 1]$ and let the mappings $f : X \to X$ and $g : Y \to Y$ be defined by $f(s, i) := (s^2, i)$ for $(s, i) \in X$ and $g(s) := s^2$ for $s \in Y$. Let $\varphi: X \to Y$ be defined by $\varphi(s, i) := s$ for $(s, i) \in X$; see Example (2) after Proposition 1.5.4. Then φ is a factor mapping, but $\emptyset = \varphi[\omega_f(s, 1)] \subsetneqq \omega_g(\varphi(s, 1)) = \{0\}$ if $0 < s < 1$.
(3) Let (X^*, f^*) be as in the Example after Theorem 3.1.1 above, let $Z := X^* \setminus \{(0, \infty)\}$ and $h := f^*|_Z$. Obviously, if $x \in Z \setminus C$ then $\omega_h(x) = C = \omega_{f^*}(x) \setminus \{(0, \infty)\}$. Let $\varphi: Z \to X^*$ be the embedding mapping. Then $\varphi: (Z, h) \to (X^*, f^*)$ is a morphism that does not map $\omega_h(x)$ onto $\omega_{f^*}(x)$.

NB. In the first example, the limit set under consideration is non-empty and compact, but X is not locally compact (basic neighbourhoods of the point $(0, \infty)$ are closed and not compact). In the second, X is locally compact, but $\omega_f(x) = \emptyset$. In the third example, we have locally compact phase spaces, but $\omega_h(x)$ is not compact. So none of the conditions for equality in Proposition 3.1.2 (3) can be omitted.

Traditionally, attraction is defined in terms of limit sets, as follows: Let A be a non-empty closed invariant subset of X. We say that *A topologically attracts the point $x \in X$* or that *the point x is topologically attracted by the set A*, whenever $\emptyset \neq \omega(x) \subseteq A$. The set

$$\mathcal{B}_f(A) := \{x \in X \,\vdots\, \emptyset \neq \omega(x) \subseteq A\} \tag{3.1-1}$$

is called the *basin of attraction of A* under f, or just *the basin of A*. If f is understood we shall just write $\mathcal{B}(A)$ instead of $\mathcal{B}_f(A)$. If A is a singleton set, say, $A = \{x_0\}$, then we write $\mathcal{B}(x_0)$ instead of $\mathcal{B}(\{x_0\})$.

Examples.

(1) Use the results obtained in Example (2) before Theorem 3.1.1 to describe the basins of the invariant points of the quadratic mapping in the various cases considered in 2.1.5. For example, if $0 < \mu < 1$ then $\mathcal{B}_{f_\mu}(0) = (p_\mu; \hat{p}_\mu)$ and $\mathcal{B}_{f_\mu}(p_\mu) = \{p_\mu, \hat{p}_\mu\}$.

(2) Consider the mapping $f : x \mapsto \frac{\pi}{2} \sin x : \mathbb{R} \to \mathbb{R}$. The invariant point $\frac{\pi}{2}$ of f is attracting (according to the definition in Chapter 2: the derivative in that point has absolute value less than 1). The basin of this point is $\mathcal{B}\left(\frac{\pi}{2}\right) = \bigcup_{k \in \mathbb{Z}} (2k\pi; (2k+1)\pi)$.

(3) In the Example after Theorem 3.1.1, let $A := \{(0, \infty)\}$. Then $\mathcal{B}_{f^*}(A) = C \cup \{(0, \infty)\}$. Note that the points of $\mathbb{Z} \times \mathbb{N}$ are not included in $\mathcal{B}(A)$, because the limit sets of those point have points in C, that is, outside of A. On the other hand, and $\mathcal{B}_f(A) = X$ (now the points of C do not belong to X).

Lemma 3.1.3. *Let A be a non-empty closed invariant subset of X. Then:*

(1) *$\mathcal{B}(A)$ is invariant.*

(2) *$\mathcal{B}(A)$ is 'backwards' invariant, i.e., $(f^k)^{\leftarrow}[\mathcal{B}(A)] \subseteq \mathcal{B}(A)$ for all $k \in \mathbb{N}$.*

(3) *If B is another non-empty closed invariant subset of X and $A \cap B = \emptyset$ then $\mathcal{B}(A) \cap \mathcal{B}(B) = \emptyset$ as well.*

(4) *If B is a closed invariant set such that $B \cap A = \emptyset$ then $B \cap \mathcal{B}(A) = \emptyset$. In particular, if $x \in \mathcal{B}(A) \setminus A$ then the point x is not periodic.*

(5) *If A is compact then $A \subseteq \mathcal{B}(A)$.*

Proof. (1), (2) Use that $\omega(f^k(x)) = \omega(x)$ for all $x \in X$ and $k \in \mathbb{Z}^+$.

(3) If $x \in \mathcal{B}(A) \cap \mathcal{B}(B)$ then $\emptyset \neq \omega(x) \subseteq A \cap B$.

(4) If $x \in B \cap \mathcal{B}(A)$ then $\emptyset \neq \omega(x) \subseteq A$, and also $\omega(x) \subseteq \overline{\mathcal{O}(x)} \subseteq B$, whence $B \cap A \neq \emptyset$. In particular, if B is the orbit of a periodic point x in $\mathcal{B}(A)$ then $B = \omega(x) \subseteq \mathcal{B}(A)$, hence $A \cap B \neq \emptyset$ and, consequently, $x \in A$.

(5) Let $x \in A$. Then $\omega(x) \subseteq \overline{\mathcal{O}(x)} \subseteq A$ and, as A is compact, $\omega(x) \neq \emptyset$: see the Remark after Theorem 1.4.5. Hence $x \in \mathcal{B}(A)$. □

Example. Consider the subsystem on the invariant subset $X^* \setminus \{(0, \infty)\} = (\mathbb{Z} \times \mathbb{N}) \cup C$ of the system (X^*, f^*) defined in the example after Theorem 3.1.1. Then $\mathcal{B}(C) = \mathbb{Z} \times \mathbb{N}$, which does not include C (is even disjoint from C), because in this subsystem $\omega(x) = \emptyset$ for all $x \in C$.

Proposition 3.1.4. *Let $\varphi: (X, f) \to (Y, g)$ be a morphism of dynamical systems and let A be a non-empty closed invariant subset of X.*

(1) *If φ is a conjugation then $\varphi[\mathcal{B}_f(A)] = \mathcal{B}_g(\varphi[A])$.*
(2) *Assume that $\varphi[A]$ is a closed subset of Y and that all orbit closures of points of $\mathcal{B}_f(A)$ are compact. Then $\varphi[\mathcal{B}_f(A)] \subseteq \mathcal{B}_g(\varphi[A])$.*
(3) *If A is compact and X is locally compact then $\varphi[\mathcal{B}_f(A)] \subseteq \mathcal{B}_g(\varphi[A])$.*

Proof. The set $\varphi[A]$ is not empty, invariant under g and in each of the three cases it is closed in Y, so it makes sense to consider the basin $\mathcal{B}_g(\varphi[A])$.

Assume that all points of $\mathcal{B}_f(A)$ have a compact orbit closure or that φ is a conjugation. Then, by Proposition 3.1.2, $\omega_g(\varphi(x)) = \varphi[\omega_f(x)]$ is a non-empty subset of $\varphi[A]$ for all $x \in \mathcal{B}_f(A)$, hence $\varphi[\mathcal{B}_f(A)] \subseteq \mathcal{B}_g(\varphi[A])$. If φ is a conjugation then we can apply the same argument to φ^{-1} in order to get the equality $\varphi[\mathcal{B}_f(A)] = \mathcal{B}_g(\varphi[A])$. This proves 1 and 2.

As to 3, if A is compact then for every point $x \in \mathcal{B}_f(A)$ the set $\omega(x)$ is non-empty and compact. So by Theorem 3.1.1, this case reduces to 2. □

Remarks.

(1) For another situation where the inclusion $\varphi[\mathcal{B}_f(A)] \subseteq \mathcal{B}_g(\varphi[A])$ holds, see Corollary 3.2.4 ahead.
(2) In Example (3) before Lemma 3.1.3, let φ denote the inclusion mapping of X into X^*. The sets $\varphi[\mathcal{B}_f(A)]$ and $\mathcal{B}_{f^*}(\varphi[A])$ are non-empty and have only the point $(0, \infty)$ in common: none is included in the other. So in general there is no nice relationship between the sets $\varphi[\mathcal{B}_f(A)]$ and $\mathcal{B}_g(\varphi[A])$.

3.1.5. A non-empty closed invariant subset A of X is said to be *topologically attracting* if it has a neighbourhood W such that $\emptyset \neq \omega(x) \subseteq A$ for every point $x \in W$, that is, $W \subseteq \mathcal{B}(A)$. Stated otherwise, A is topologically attracting iff $\mathcal{B}(A)$ *is* a neighbourhood of A. An invariant point x_0 under f is said to be a *topologically attracting point* whenever the singleton set $\{x_0\}$ is topologically attracting.

If A is a topologically attracting set then, by definition, always $A \subseteq \mathcal{B}(A)$, even if A is not compact (compare this with Lemma 3.1.3 (5) above). However, in most applications below, A will be compact.

Examples.

(1) The following should be clear from Proposition 1.4.1 (2): if x_0 is an invariant point in a dynamical system (X, f) with a metric phase space and x_0 has a neighbourhood W such that $\lim_{n\rightsquigarrow\infty} f^n(x) = x_0$ for all $x \in W$ then the point x_0 is topologically attracting. Now use the results obtained in Example (2) before Theorem 3.1.1 to determine the topologically attracting invariant points of the the quadratic mapping $f_\mu: \mathbb{R} \to \mathbb{R}$ in the various cases considered in 2.1.5. For

example, if $0 < \mu < 1$ then 0 is the unique topologically attracting invariant point. Moreover, if $1 < \mu \leq 3$ then the point $p_\mu = 1 - {}^1\!/_\mu$ is the unique topologically attracting invariant point.

(2) If X is an interval in $\mathbb{R}$ and f is a C^1-function then by Proposition 2.1.3 and 1 above, every attracting invariant point of f is topologically attracting.

(3) The invariant point $[0]$ in $\mathbb{S}$ under the argument doubling map ψ is not topologically attracting. This follows easily from Lemma 3.1.3 (4) and the fact that $\mathbb{S}$ has a dense set of periodic points. Of course, one may also apply Proposition 3.1.7 below. Similarly, the point 0 in $[0; 1]$ it is not topologically attracting under the tent map.

(4) Let $f: \mathbb{S} \to \mathbb{S}$ be the mapping given by $f([t]) := [t^2]$ for $0 \leq t < 1$. The point $[0]$ is invariant under f and $\omega(x) = \{[0]\}$ for all $x \in \mathbb{S}$, so $\mathcal{B}([0]) = \mathbb{S}$ and $[0]$ is topologically attracting in the system $(\mathbb{S}, f)$. Note that in the example after Theorem 3.1.1 the point $(0, \infty)$ has a similar behaviour in the subsystem on the invariant subset $C \cup \{(0, \infty)\}$ of X^*.

Theorem 3.1.6. *Let A be a non-empty, closed invariant subset of X. The following conditions are equivalent:*
(i) *A is topologically attracting.*
(ii) *$\mathcal{B}(A)$ is an open subset of X and $A \subseteq \mathcal{B}(A)$.*
(iii) *$\mathcal{B}(A)$ is a neighbourhood of A.*

Proof. "(i)⇒(ii)": Assume that A is topologically attracting. We know already that in that case $A \subseteq \mathcal{B}(A)$ (and even that (iii) is true). Let W be an open neighbourhood of A such that $\emptyset \neq \omega(z) \subseteq A$ for all $z \in W$. We shall show that

$$\mathcal{B}(A) = \bigcup_{k=0}^{\infty} (f^k)^{\leftarrow}[W], \tag{3.1-2}$$

which clearly implies that $\mathcal{B}(A)$ is open.

"$\subseteq$": Let $x \in \mathcal{B}(A)$. Then, by definition, $\omega(x) \subseteq A$, so W is a neighbourhood of each point of $\omega(x)$. Then by Lemma 1.4.1 (1), W meets $\mathcal{O}(x)$ and therefore there exists $k \in \mathbb{Z}^+$ with $f^k(x) \in W$, that is, $x \in (f^k)^{\leftarrow}[W]$.

"$\supseteq$": by Lemma 3.1.3 (2), $(f^k)^{\leftarrow}[\mathcal{B}(A)] \subseteq \mathcal{B}(A)$ for every $k \in \mathbb{Z}^+$. In view of the choice of W we have $W \subseteq \mathcal{B}(A)$, hence $(f^k)^{\leftarrow}[W] \subseteq \mathcal{B}(A)$ for every $k \in \mathbb{Z}^+$.

"(ii)⇒(iii)": Obvious.

"(iii)⇒(i)": Clear from the definition. □

Proposition 3.1.7. *Let (X, f) be a transitive system. Then X has no topologically attracting proper subsets.*

Proof. Let A be a proper subset of X. By assumption, X has a dense subset of transitive points points, each with limit set all of X. Hence no neighbourhood of A consists exclusively of points x with $\omega(x) \subseteq A$. □

Topological attraction is a dynamical property:

Proposition 3.1.8. *Let $\varphi: (X, f) \simeq (Y, g)$ be a conjugation and let A be a non-empty closed invariant subset of X. Then A is topologically attracting under f iff $\varphi[A]$ is topologically attracting under g.*

Proof. By Proposition 3.1.4 (1), $\mathcal{B}_g(\varphi[A]) = \varphi[\mathcal{B}_f(A)]$. Since φ is a homeomorphism, this equality implies that $\mathcal{B}_g(\varphi[A])$ is an open neighbourhood of $\varphi[A]$ iff the set $\mathcal{B}_f(A)$ is an open neighbourhood of A. □

Example. Arbitrary factor maps do not preserve topologically attraction. For example, let $X := \mathbb{R} =: Y$, let $f(x) := x^3$ for $x \in \mathbb{R}$ and let the mappings $g: Y \to Y$ and $\varphi: X \to Y$ be given by

$$g(x) := \begin{cases} (x+1)^3 - 1 & \text{for } x \geq 0\,, \\ (x-1)^3 + 1 & \text{for } x \leq 0\,, \end{cases} \quad \text{and} \quad \varphi(x) := \begin{cases} 0 & \text{for } -1 \leq x \leq 1\,, \\ x - 1 & \text{for } x \geq 1\,, \\ x + 1 & \text{for } x \leq -1\,. \end{cases}$$

Then $\varphi: (X, f) \to (Y, g)$ is a factor mapping. The point 0 in X is topologically attracting under f, but the point $\varphi(0)$ is not topologically attracting under g.

Next, we shall discuss a stronger attraction property which, for lack of a better term, we call 'strong attraction'. A non-empty closed invariant subset A of X is said to *strongly attract a subset* B of X whenever

$$\forall U \in \mathcal{N}_A : \; f^n[B] \subseteq U \quad \text{for almost all } n \in \mathbb{N}\,. \tag{3.1-3}$$

If A strongly attracts B then every superset of A strongly attracts every subset of B. Moreover, every invariant set strongly attracts itself. We say that A *strongly attracts a point* x whenever it attracts the singleton set $\{x\}$. Clearly, this is the case iff every neighbourhood of A includes the 'tail' $\mathcal{O}(f^n(x))$ of $\mathcal{O}(x)$ for some $n \in \mathbb{N}$. The next two theorems show how strong attraction and limit sets are related.

Theorem 3.1.9. *Let $x \in X$ and assume that its orbit closure $\overline{\mathcal{O}(x)}$ is compact. Then the limit set $\omega(x)$ is a non-empty compact completely invariant set which strongly attracts the point x.*

Proof. That $\omega(x)$ is non-empty and invariant was shown in Section 1.4. For complete invariance, see Exercise 3.2. That it strongly attracts the point x follows easily from Lemma A.2.2 in Appendix A, applied to the descending sequence of compact sets $\overline{\mathcal{O}(f^n(x))}$ for $n \in \mathbb{Z}^+$. □

Example. In the example after Theorem 3.1.1, let $x \in X \setminus \{(0, \infty)\}$. Then $\omega_f(x) = \{(0, \infty)\}$, but the orbit the x leaves any neighbourhood of $\{(0, \infty)\}$ again and again, so the limit set $\omega(x)$ does not strongly attract the point x. This shows that the condition of a compact orbit closure cannot be omitted from the hypothesis of Theorem 3.1.9.

Actually, $\omega(x)$ is the smallest non-empty compact set that strongly attracts the point x:

Theorem 3.1.10. *Let $x \in X$ and let A be a non-empty compact subset of X. If A strongly attracts the point x then $\emptyset \neq \omega(x) \subseteq A$, that is, $x \in \mathcal{B}(A)$.*

Proof. By the assumptions on A and x, for every neighbourhood V of A there exists $k \in \mathbb{N}$ such that $\mathcal{O}(f^k(x)) \subseteq V$, hence

$$\omega(x) \subseteq \overline{\mathcal{O}(f^k(x))} \subseteq \overline{V}\,.$$

Since $A = \bigcap\{\overline{V} : V \in \mathcal{N}_A\}$ – see formula (A.2-1) in Appendix A – this implies that $\omega(x) \subseteq A$. It remains to show that $\omega(x) \neq \emptyset$. Assume the contrary: then Lemma 1.4.1 (1) implies that every point $a \in A$, being not a point of $\omega(x)$, has an open neighbourhood U_a such that there are only finitely many values of $n \in \mathbb{Z}^+$ with $f^n(x) \in U_a$. Cover A by finitely many of such sets U_a and let U be their union. Then there are only finitely many values of n such that $f^n(x) \in U$, contradicting the assumptions. □

For any non-empty closed invariant subset A of X, define the *basin of strong attraction* by

$$\mathcal{B}^*_f(A) := \{\, x \in X : A \text{ strongly attracts the point } x \,\}\,. \tag{3.1-4}$$

If f is understood then we write just $\mathcal{B}^*(A)$ for $\mathcal{B}^*_f(A)$. If x_0 is an invariant point in X then we write $\mathcal{B}^*(x_0)$ for $\mathcal{B}^*(\{x_0\})$.

The following examples show that, in general, there is no nice relationship between the sets $\mathcal{B}^*(A)$ and $\mathcal{B}(A)$.

Examples.
(1) Let $X := [0;\infty) \setminus \{1\}$ and $f : x \mapsto \sqrt{x} : X \to X$. The set $A := [0;1)$ is closed in X (but not compact), $\mathcal{B}(A) = \{0\}$ and $\mathcal{B}^*(A) = X$. So $\mathcal{B}(A) \subsetneq \mathcal{B}^*(A)$.
(2) Let (X, f) be the system defined in the Example after Theorem 3.1.1 and consider the closed set $A := \{(0,\infty)\}$. Then $\mathcal{B}^*_f(A) = A \subsetneq X = \mathcal{B}_f(A)$.

The properties of basins of strong attraction are quite similar to those of the ordinary basins of attraction, but sometimes they behave slightly better (see Proposition 3.1.13 below):

Lemma 3.1.11. *Let A be a non-empty closed invariant subset of X. Then:*
(1) *$\mathcal{B}^*(A)$ is invariant.*
(2) *$\mathcal{B}^*(A)$ is 'backwards' invariant, i.e., $(f^k)^{\leftarrow}[\mathcal{B}^*(A)] \subseteq \mathcal{B}^*(A)$ for all $k \in \mathbb{N}$.*
(3) *If B is a non-empty closed invariant subset of X such that A and B have disjoint neighbourhoods then $\mathcal{B}^*(A) \cap \mathcal{B}^*(B) = \emptyset$.*
(4) *If B is a non-empty closed invariant subset of X such that A and B are disjoint then $B \cap \mathcal{B}^*(A) = \emptyset$. In particular, if $x \in \mathcal{B}^*(A) \setminus A$ then the point x is not periodic.*
(5) *Always $A \subseteq \mathcal{B}^*(A)$.*

Proof. (1), (2) Use that $f^n(f(x)) = f(f^n(x)) = f^{n+1}(x)$ for all $x \in X$ and all $n \in \mathbb{Z}^+$.
(3) Obvious.
(4) If $y \in B$ then $f^n(y) \in B$ for all $n \in \mathbb{Z}^+$, so the neighbourhood $X \setminus B$ of A includes none of the points $f^n(y)$, hence $y \notin \mathcal{B}^*(A)$. In particular, if B is the orbit of a periodic point $x \in X \setminus A$ then $B \cap A = \emptyset$ (recall that A is invariant), hence $x \in B \subseteq X \setminus \mathcal{B}^*(A)$.
(5) Obvious, as A is invariant. □

Proposition 3.1.12. *Let A be a non-empty closed invariant subset of X.*
(1) *If A is compact then $\mathcal{B}^*(A) \subseteq \mathcal{B}(A)$.*
(2) *If all orbit closures of points of $\mathcal{B}(A)$ are compact then $\mathcal{B}(A) \subseteq \mathcal{B}^*(A)$.*
(3) *If A is compact and X is locally compact then $\mathcal{B}^*(A) = \mathcal{B}(A)$.*

Proof. Statement 1 is clear from Theorem 3.1.10. In the case of statement 2 it follows from Theorem 3.1.9 that every point x of $\mathcal{B}(A)$ is strongly attracted by $\omega(x)$, hence by its superset A. Now statement 3 follows from Theorem 3.1.1. □

Proposition 3.1.13. *Let $\varphi: (X, f) \to (Y, g)$ be a morphism of dynamical systems and let A be a non-empty closed invariant subset of X. Then $\varphi[\mathcal{B}_f^*(A)] \subseteq \mathcal{B}_g^*(\varphi[A])$, with equality if φ is a conjugation.*

Proof. Let $x \in \mathcal{B}_f^*(A)$ and let U be a neighbourhood of $\varphi[A]$. Then $\varphi^{\leftarrow}[U]$ is a neighbourhood of A, hence $f^n(x) \in \varphi^{\leftarrow}[U]$ for almost all n. It follows that $g^n(\varphi(x)) = \varphi(f^n(x)) \in U$ for almost all n, i.e., $\varphi(x) \in \mathcal{B}_g^*(\varphi[A])$. □

Remark. As an application we give an alternative proof of Proposition 3.1.4 (2) in the case that A is compact. Let A be f-invariant and compact and let the orbit closures of points in $\mathcal{B}(A)$ be compact. Then Proposition 3.1.12 (1),(2), Propositions 3.1.13 and 3.1.12 (1) imply: $\varphi[\mathcal{B}_f(A)] = \varphi[\mathcal{B}_f^*(A)] \subseteq \mathcal{B}_g^*(\varphi[A]) \subseteq \mathcal{B}_g(\varphi[A])$.

A non-empty closed invariant subset A of X is said to be *strongly attracting* whenever $\mathcal{B}^*(A)$ is a neighbourhood of A.

Examples.
(1) If x_0 is an invariant point in a dynamical system (X, f) with a metric phase space and x_0 has a neighbourhood W such that $\lim_{n \rightsquigarrow \infty} f^n(x) = x_0$ for all $x \in W$ then the point x_0 is strongly attracting. In particular, if X is an interval in $\mathbb{R}$ and f is a C^1-function then Proposition 2.1.3 implies that every attracting invariant point of f is strongly attracting.
(2) Let $f: \mathbb{S} \to \mathbb{S}$ be the mapping given by $f([t]) := [t^2]$ for $0 \le t < 1$. The point $[0]$ is strongly attracting in the system $(\mathbb{S}, f)$ with $\mathcal{B}^*([0]) = \mathbb{S}$. (Recall from Example (4) in 3.1.5 that the point $[0]$ is topologically attracting as well, with $\mathcal{B}([0]) = \mathbb{S}$.)

Theorem 3.1.14. *Let A be a non-empty, closed invariant subset of X. The following conditions are equivalent:*
(i) *A is strongly attracting.*
(ii) *$\mathcal{B}^*(A)$ is an open subset of X.*

Proof. "(i)⇒(ii)": Assume that A is strongly attracting. Then, by definition, there is an open subset W of X such that $A \subseteq W \subseteq \mathcal{B}^*(A)$. We shall show that $\mathcal{B}^*(A) = \bigcup_{k=0}^{\infty}(f^k)^{\leftarrow}[W]$ which implies that $\mathcal{B}(A)$ is open.

"⊇": by Lemma 3.1.11 (2) and in view of the choice of W it is clear that we have, for every $k \in \mathbb{Z}^+$, $(f^k)^{\leftarrow}[W] \subseteq (f^k)^{\leftarrow}[\mathcal{B}^*(A)] \subseteq \mathcal{B}^*(A)$.

"⊆": Let $x \in \mathcal{B}^*(A)$. Then A strongly attracts the point x, hence the neighbourhood W of A includes $f^n(x)$ for almost all n. In particular, there exists $k \in \mathbb{Z}^+$ with $x \in (f^k)^{\leftarrow}[W]$. □

Proposition 3.1.15. *Let A be a non-empty compact invariant subset of X. If A is strongly attracting then* $\mathcal{B}^*(A) = \mathcal{B}(A)$ *and A is topologically attracting.*

Proof. By Lemma 3.1.11 (2), $\bigcup_{k=0}^{\infty}(f^k)^{\leftarrow}[\mathcal{B}^*(A)] \subseteq \mathcal{B}^*(A)$. If $\mathcal{B}^*(A)$ is a neighbourhood of A then in the proof of "⊆" in Theorem 3.1.6 (i)⇒(ii), W may be replaced by $\mathcal{B}^*(A)$. It follows that $\mathcal{B}(A)$ is included in the left-hand side of the above inclusion. Hence $\mathcal{B}(A) \subseteq \mathcal{B}^*(A)$. Conversely, $\mathcal{B}(A)^* \subseteq \mathcal{B}(A)$, because A is compact. □

Example. The converse is not generally true: in Example (2) before Lemma 3.1.11 the (compact) set A is topologically attracting, but $\mathcal{B}^*(A) = A$ is not a neighbourhood of A.

3.2 Stability

In the study of 'physical' systems an important question is whether a given equilibrium point or periodic orbit is stable: if the state of the system is sufficiently close to that point or orbit, will it approach that point or orbit arbitrarily close (topological attraction), will it remain close to it (stability) or both (asymptotic stability)? Here follow the precise definitions:

A non-empty compact and *completely* invariant subset A of X is said to be *stable* whenever the following condition is fulfilled:

$$\forall U \in \mathcal{N}_A \;\; \exists V \in \mathcal{N}_A \; \vdots \; f^n[V] \subseteq U \quad \text{for all } n \in \mathbb{Z}^+ . \tag{3.2-1}$$

An invariant point x_0 under f is said to be *stable* or it is called a *stable equilibrium* whenever $\{x_0\}$ is a stable set. The idea of stability is best illustrated by the following examples:

Examples.

(1) Consider the mapping $f: x \mapsto -x: [-1; 1] \to [-1; 1]$. Then the invariant point 0 is stable: (3.2-1) holds, because every neighbourhood U of 0 includes a symmetric neighbourhood V, which is invariant.

(2) Let the mapping $f: \mathbb{S} \to \mathbb{S}$ be given by $f([t]) := [t^2]$ for $0 \le t < 1$. Then the invariant point $[0]$ is not stable in the system $(\mathbb{S}, f)$. In order to prove this, consider the neighbourhood $U := \mathbb{S} \setminus [1/2]$ of $[0]$. Then every neighbourhood V of $[0]$ includes

a point x such that $f^n(x) = 1/2 \notin U$ for some $n \in \mathbb{N}$. In fact, let $n \in \mathbb{N}$ be so large that $x := f^{-n}([1/2]) \in V$: such n exists because $f^{-n}([1/2]) \rightsquigarrow [1] = [0]$ for $n \rightsquigarrow \infty$.

In the stability condition (3.2-1) the inclusion $f^n[V] \subseteq U$ is required also for $n = 0$, which implies that $V \subseteq U$. On the other hand, if for some $U \in \mathcal{N}_A$ there exist $V \in \mathcal{N}_A$ and $k \in \mathbb{N}$ such that the required inclusion $f^n[V] \subseteq U$ only holds for $n \geq k$, then $V' := V \cap \bigcap_{i=0}^{k-1} (f^i)^{\leftarrow}[U]$ is a neighbourhood of A such that $f^n[V'] \subseteq U$ for all $n \geq 0$. So in that case A is stable as well.

For the proof that $A \subseteq V'$ in the final remark above one needs that if $A \subseteq U$ then $A \subseteq (f^i)^{\leftarrow}[U]$ for $0 \leq i < k$, which is true because A is invariant. *Complete* invariance of A is not needed, but if A collapses to a smaller set under f, it would not be quite reasonable to call A a stable set, even if A satisfies condition (3.2-1).

Lemma 3.2.1. *Let A be a stable set in X and let $x \in X$. If $\omega(x) \cap A \neq \emptyset$ then $x \in \mathcal{B}^*(A)$.*

Proof. Let U be any neighbourhood of A and select a neighbourhood V of A according to condition (3.2-1). Since A includes a point of $\omega(x)$, it is clear from Lemma 1.4.1 (1) that there exists $k \in \mathbb{N}$ with $f^k(x) \in V$. So for all $n \geq k$ we have, by the choice of V, $f^n(x) = f^{n-k}(f^k(x)) \in f^{n-k}[V] \subseteq U$. Hence $x \in \mathcal{B}^*(A)$. □

Corollary 3.2.2. *Let A be a stable subset of X.*
(1) $\mathcal{B}^*(A) = \mathcal{B}(A)$.
(2) *If $x \in X$ and $\omega(x) \cap A \neq \emptyset$ then $\emptyset \neq \omega(x) \subseteq A$, i.e., $x \in \mathcal{B}(A)$.*
(3) *Let B be transitive closed invariant set. If $B \cap A \neq \emptyset$ then $B \subseteq A$. Consequently, a transitive stable set has no proper stable subsets. In particular, a transitive system has no proper stable subsets.*

Proof. (1) Since A is compact we know already that $\mathcal{B}^*(A) \subseteq \mathcal{B}(A)$. Conversely, if $x \in \mathcal{B}(A)$ then, obviously, $\omega(x) \cap A \neq \emptyset$, hence $x \in \mathcal{B}^*(A)$.
(2) Clear from Lemma 3.2.1, since $\mathcal{B}^*(A) \subseteq \mathcal{B}(A)$ by compactness of A.
(3) If x is a transitive point of B then $\omega(x) = B$, hence $\omega(x) \cap A \neq \emptyset$. So by (2), $B = \omega(x) \subseteq A$. The final statements are easy consequences. □

Examples.
(1) In the Example (2) preceding Lemma 3.1.11 we have $\mathcal{B}^*(A) \subsetneq \mathcal{B}(A)$.
(2) The argument-doubling system and the system defined by the tent map on the unit interval are transitive; see 1.7.2 and 1.7.3. Hence no periodic orbit in these systems is stable.

Corollary 3.2.3.
(1) *Let x_0 be an invariant point in X. If either x_0 is stable or X is locally compact, then*[1]

$$\mathcal{B}(x_0) = \{ x \in X : \lim_{n \rightsquigarrow \infty} f^n(x) = x_0 \}. \tag{3.2-2}$$

1 By the definitions in Section A.4 in Appendix A, in non-metric topological spaces the expressions $\lim_{n \rightsquigarrow \infty} z_n = y$ and $z_n \rightsquigarrow y$ mean that every neighbourhood of y includes z_n for almost all n.

(2) *Let X be a metric space, say, with metric d and let A be a non-empty compact invariant subset of X. If A is stable or if X is locally compact, then*

$$\mathcal{B}(A) = \{ x \in X : \lim_{n \rightsquigarrow \infty} d(f^n(x), A) = 0 \} . \tag{3.2-3}$$

Proof. (1) The set in the right-hand side of (3.2-2) is equal to $\mathcal{B}^*(x_0)$ which, in turn, by Proposition 3.1.12 (3) or 3.2.2 (1) is equal to $\mathcal{B}(x_0)$.
(2) This, again follows from Proposition 3.1.12 (3) or Corollary 3.2.2 (1): the set in the right-hand side (3.2-3) equals $\mathcal{B}^*(A)$, because the sets $B_\varepsilon(A, d)$ with $\varepsilon > 0$ form a neighbourhood base of the compact set A: see Appendix A.7.3. □

Corollary 3.2.4. *Let $\varphi: (X, f) \to (Y, g)$ be a morphism of dynamical systems and let A be a non-empty compact and completely invariant subset of X. If either A or $\varphi[A]$ is stable then $\varphi[\mathcal{B}_f(A)] \subseteq \mathcal{B}_g(\varphi[A])$.*

Proof. If A is stable then Lemma 3.2.2 (1) and the Propositions 3.1.12 (1) and 3.1.13 imply $\varphi[\mathcal{B}(A)] = \varphi[\mathcal{B}^*(A)] \subseteq \mathcal{B}^*(\varphi[A]) \subseteq \mathcal{B}(\varphi[A])$.

If $\varphi[A]$ is stable then for every point $x \in \mathcal{B}_f(A)$ we have $\emptyset \neq \omega_f(x) \subseteq A$, hence $\emptyset \neq \varphi[\omega_f(x)] \subseteq \varphi[A]$. However, $\omega_g(\varphi(x)) \supseteq \varphi[\omega_f(x)]$ by Proposition 3.1.2 (1), so $\omega_g(\varphi(x)) \cap \varphi[A] \neq \emptyset$. Then Corollary 3.2.2 (2) implies that $\varphi(x) \in \mathcal{B}_g(\varphi[A])$. □

Proposition 3.2.5. *Let $\varphi: (X, f) \xrightarrow{\simeq} (Y, g)$ be a conjugation and let A be a non-empty compact and completely invariant subset of X. Then A is stable iff $\varphi[A]$ is stable.*

Proof. We need only prove the 'only if': applied to the conjugation φ^{-1} this gives the 'if'. So assume that A is stable. Then $\varphi[A]$ is compact and, by Proposition 1.5.2 (3), completely invariant. The remainder of the proof, namely, that condition (3.2-1) holds for $\varphi[A]$ if it holds for A, is left for the reader. □

Examples. Stability is not preserved by factor mappings. Let $X := [0; 1]$, $f(x) := x^2$ for $x \in X$, and let $Y := \mathbb{S}$, $g[t] := [t^2]$ for $0 \leq t < 1$. Then $\varphi: t \mapsto [t]: X \to Y$ defines a factor mapping from (X, f) onto (Y, g). The invariant point 0 in X is stable under f, but $\varphi[0]$ is not stable in Y under g: see Example (2) before Lemma 3.2.1.

3.2.6. A non-empty closed subset A of X is said to be *asymptotically stable* whenever it is stable and topologically attracting. An invariant point x_0 under f is said to be *asymptotically stable* whenever the set $\{x_0\}$ is asymptotically stable. Thus, a subset A of X is asymptotically stable iff

(a) A is not empty, compact, completely invariant and every neighbourhood U of A includes a neighbourhood V of A such that $f^n[V] \subseteq U$ for all $n \in \mathbb{Z}^+$, and
(b) A has a neighbourhood W such that $\emptyset \neq \omega(x) \subseteq A$ for all $x \in W$.

The Examples (1) and (2) below show that the two conditions (a) and (b) are independent of each other.

Examples.

(1) Consider the mapping $f: x \mapsto -x: \mathbb{R} \to \mathbb{R}$. Then the invariant point 0 is stable but not topologically attracting, hence not asymptotically stable.
(2) In Example (4) in 3.1.5, the invariant point 0 is topologically attracting but not stable: see Example (2) before Lemma 3.2.1.
(3) Let x_0 be an attracting invariant point under a C^1-mapping $f: X \to X$, X an interval in $\mathbb{R}$. By Proposition 2.1.3 and Corollary 3.2.3 (1), x_0 is an asymptotically stable invariant point. See also Example (2) in 3.1.5.
(4) Let $f: [0; 1] \to [0; 1]$ be defined by $f(x) := x + \frac{1}{6}x^3 \sin\frac{1}{x}$ if $x \neq 0$ and $f(0) := 0$. The invariant point 0 is stable but not asymptotically stable, In fact, f is monotonous, and if $a_n := \frac{1}{n\pi}$ for $n \in \mathbb{N}$ then the intervals $[a_{n+1}; a_n]$ for $n \in \mathbb{N}$ are invariant, and the invariant points a_n are alternatingly attracting and repelling, according as n is even or odd. In addition, each of the intervals $A_n := [0; a_n]$ for even n is asymptotically stable in $[0; 1]$, as is each interval $[a_{n+k}; a_n]$ for even n and k.

Theorem 3.2.7. *Let A be a stable subset of X. Then A is asymptotically stable iff $\mathcal{B}(A)$ is a neighbourhood of A (in which case $\mathcal{B}(A)$ is open).*

Proof. Clear from the definition and Theorem 3.1.6. □

Theorem 3.2.8. *A transitive asymptotically stable set is minimal with respect to inclusion in the class of all asymptotically stable sets. Hence distinct transitive asymptotically stable sets are mutually disjoint.*

Proof. Clear from Corollary 3.2.2 (3) □

Remarks.

(1) The converse of the first statement is not true: consider $X := \mathbb{R}$ and let $f(x) := \frac{1}{2}x$ for $x \le 0$, $f(x) := x$ for $0 \le x \le 1$ and $f(x) := \frac{1}{2}(x+1)$ for $x \ge 1$. Then $A := [0; 1]$ is asymptotically stable, A has no proper asymptotically stable subsets, but A is not transitive.
(2) Arbitrary asymptotically stable sets can intersect without being equal. For example, consider a system with three asymptotically stable invariant points x_1, x_2 and x_3, and let $A := \{x_1, x_2\}$ and $B := \{x_2, x_3\}$.

An invariant point in a dynamical system on an interval in $\mathbb{R}$ is already asymptotically stable as soon as it is topologically attracting:

Proposition 3.2.9. *Let (X, f) be a dynamical system on an interval X in $\mathbb{R}$ and let x_0 be a topologically attracting invariant point in X. Then x_0 is stable, hence asymptotically stable.*

Proof. Recall from the locally compact case of Corollary 3.2.3 (1) that the set $\mathcal{B}_f(x_0)$ can be characterized as the set of points $x \in X$ for which $\lim_{n \rightsquigarrow \infty} f^n(x) = x_0$. In particular, $\mathcal{B}_f(x_0)$ contains no invariant or periodic points different from x_0.

By Lemma 3.1.3 (1) and Theorem 3.1.6, $\mathcal{B}_f(x_0)$ is an invariant open (in X) neighbourhood of x_0. Hence $\mathcal{B}_f(x_0)$ includes a non-degenerate interval containing the point x_0. Let W be the union of all intervals included in $\mathcal{B}_f(x_0)$ that contains the point x_0; so W is the largest interval with this property. Obviously, $f[W]$ is again such an interval, hence $f[W] \subseteq W$: W is invariant. Claim:

$$\forall n \in \mathbb{N}\ \forall x \in W: \begin{cases} x > x_0 \Rightarrow f^n(x) < x\,, \\ x < x_0 \Rightarrow f^n(x) > x\,. \end{cases} \tag{3.2-4}$$

In the right-hand side we have strict inequalities: if we would have $f^n(x) = x$ for some $n \in \mathbb{N}$ then the point x would be periodic, which is impossible for a point $x \in W \setminus \{x_0\}$. Moreover, (3.2-4) implies that the (possibly only finitely many) points of the orbit $\mathcal{O}(x)$ that are situated right of x_0 form a (strictly) decreasing sequence in W. For if $m > k$ and $f^k(x) > x_0$, then (3.2-4) with $f^k(x)$ instead of x – recall that W is invariant – implies that $f^m(x) = f^{m-k}(f^k(x)) < f^k(x)$. Similarly, the points of $\mathcal{O}(x)$ that are situated left of x_0 form a (strictly) increasing sequence.

It is possible that there are points $x \in W \setminus \{x_0\}$ with $f^n(x) = x_0$ for some $n \in \mathbb{N}$. For other points the orbit may jump from the left side of x_0 to the right side and vice versa. So the sequence of points of an orbit at the right (or left) side of x_0 may have many gaps. Think about what might occur for $f: x \mapsto \frac{1}{2}x\sin(\frac{1}{x}): [-1;1] \to [-1;1]$ and $x_0 = 0$. Of course, this does not happen if x_0 is an end point of X.

First, we prove (3.2-4) for the case $n = 1$, that is, we show that $f(x) < x$ for all points $x \in W \cap (x_0;\infty)$ (the proof that $f(x) > x$ for all points x in W left of x_0 is similar). Suppose there is a point $y \in W \cap (x_0;\infty)$ such that $f(y) \geq y$. If we would have $f(x) \geq x$ for all $x \in W \cap (x_0;\infty)$ then the set $W \cap (x_0;\infty)$ would be invariant, and for any point x in this set it would be impossible for $f^n(x)$ to tend to x_0. Hence there exists $x_1 \in W \cap (x_0;\infty)$ such that $f(x_1) < x_1$. Then the function $x \mapsto f(x) - x$ changes sign on the interval W between the points x_1 and y. It follows that there is an invariant point between x_1 and y in W, which contradicts the fact that W contains no invariant points. This concludes the proof of (3.2-4) for the case $n = 1$.

For all other values of $n \in \mathbb{N}$, the above argument can be applied to the mapping f^n instead of f. To prove this, note that by the characterization of $\mathcal{B}_f(x_0)$ as the set of points whose orbit under f converges to x_0 – and, similarly, of $\mathcal{B}_{f^n}(x_0)$ as the set of points whose orbit under f^n converges to x_0 – it is easy to see that $\mathcal{B}_{f^n}(x_0) \supseteq \mathcal{B}_f(x_0) \supseteq W$ (see also Remark 1 after Lemma 3.3.11 (1) ahead). This implies that W is an interval included in $\mathcal{B}_{f^n}(x_0)$, containing the point x_0, and invariant under f^n. So what we have proved above for f also holds for f^n: if $x \in W$ then $f^n(x) < x$ if $x > x_0$ and $f^n(x) > x$ if $x < x_0$. This concludes the proof of (3.2-4).

After these preliminaries we come the proof x_0 is a stable invariant point; thus, we want to show that condition (3.2-1) holds for $A := \{x_0\}$, that is

$$\forall U \in \mathcal{N}_{x_0}\ \exists V \in \mathcal{N}_{x_0} \vdots\ f^n[V] \subseteq U \quad \text{for all } n \in \mathbb{Z}^+\,. \tag{3.2-4}$$

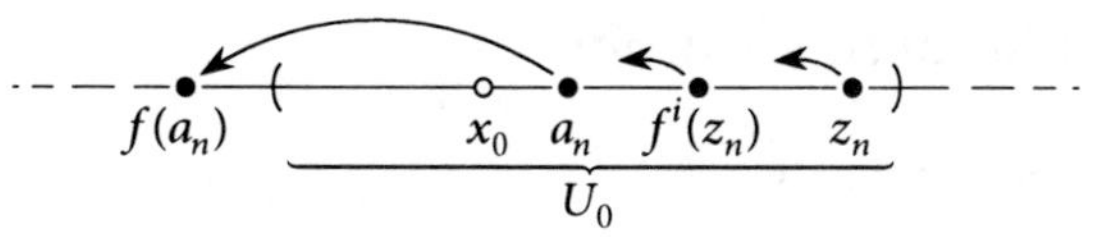

Fig. 3.2. Illustration of the final part of the proof of 3.2.9. The points $f(a_n)$ and a_n are at different sides of x_0, otherwise $f(a_n)$ would be situated between x_0 and z_n, hence in U_0.

If x_0 is an end point of X it is clear from (3.2-4) that condition (3.2-4) holds. So assume that x_0 is not an end point of X, and assume that (3.2-4) is false: there are a neighbourhood U_0 of x_0 and a sequence of points z_n converging to x_0 such that for every $n \in \mathbb{N}$ there exists $k_n \in \mathbb{Z}^+$ with the property that $f^{k_n}(z_n) \notin U_0$. Without limitation of generality we may assume that U_0 is an open interval around x_0, that $U_0 \subseteq W$, that $z_n \in U_0$ for every $n \in \mathbb{N}$ and that k_n is the first value of i for which $f^i(z_n) \notin U_0$, so that $f^i(z_n) \in U_0$ for $i = 0, \dots, k_n - 1$.

Claim: the points $f^i(z_n)$ for $i = 0, \dots, k_n - 1$ are all at the same side of the point x_0. For suppose the contrary and let j_n be the first member of the set $\{0, \dots, k_n - 1\}$ such that $f^{j_n}(z_n)$ is not at the same side of x_0 as z_n. Since $f^{j_n}(z_n) \in U_0$, it would follow from (3.2-4) applied to the points z_n and $f^{j_n}(z_n)$ that *all* points of the orbit of z_n would be situated between z_n and $f^{j_n}(z_n)$, hence in U_0. This is not the case, because $f^{k_n}(z_n) \notin U_0$. This contradiction proves the claim.

Consequently, for each $n \in \mathbb{N}$ the points z_n and $a_n := f^{k_n-1}(z_n)$ are at the same side of x_0 and therefore, by (3.2-4), a_n is between x_0 and z_n. It follows that $a_n \rightsquigarrow x_0$ if n tends to infinity. Since f is continuous, this implies that $f^{k_n}(z_n) = f(a_n) \rightsquigarrow f(x_0) = x_0$, which contradicts the fact that $f^{k_n}(z_n) \notin U_0$ for all n. □

Just for completeness, we note that asymptotic stability is a dynamical property:

Proposition 3.2.10. *Let $\varphi: (X, f) \rightrightarrows (Y, g)$ be a conjugation and let A be a non-empty compact invariant subset of X. Then A is asymptotically stable under f iff $\varphi[A]$ is asymptotically stable under g.*

Proof. Clear from the Propositions 3.1.8 and 3.2.5. □

In Section 4.5 ahead we will continue our study of asymptotic stability. Among others, in Theorem 4.5.9 we give a complete characterization of asymptotically stable sets in locally compact metric spaces that are minimal in the sense that they include no proper asymptotically stable subsets. But first we consider the preceding notions of stability, attraction and asymptotic stability for the special case of periodic orbits (however, the Lemma's (3.3.*i*) for $i \in \{1, 2, 3, 5, 6, 14, 15\}$ are of general interest).

3.3 Stability and attraction for periodic orbits

In this section we consider a periodic orbit with primitive period p. It will turn out that this is a stable or an asymptotically stable set iff any/every point of the orbit is a stable, respectively asymptotically stable, invariant point under f^p. For strong and topological attraction in the absence of stability the situation is slightly less satisfying: see the Example after Proposition 3.3.9 and Theorem 3.3.12 ahead.

The material in this section consists of four parts which are more or less similar to each other. In 3.3.1–3.3.4 we deal with stability. Next, strong attraction and topological attraction are treated in 3.3.5–3.3.9 and in 3.3.10–3.3.13, respectively. Finally, in 3.3.14–3.3.19 we collect the preceding material and apply it for asymtotic stability.

The following notation will be used: fix $p \in \mathbb{N}$, $p \geq 2$, and put $g := f^p$. The point x_0 is assumed to be periodic x_0 with primitive period p. Moreover, for $i = 0, \ldots, p-1$ let $x_i := f^i(x_0)$, so $\mathcal{O}_f(x_0) = \{x_0, \ldots, x_{p-1}\}$. Note that the points of $\mathcal{O}_f(x_0)$ are invariant under g.

In what follows, 'invariant', 'stable', etc., without additional condition will always mean 'under f', i.e., in the system (X, f). If confusion is likely to occur then we shall specify the system in which the notion is meant to be used (e.g., 'invariant under f', or 'g-invariant', etc.).

Lemma 3.3.1. *Let A be a non-empty compact completely invariant subset of X. Then A is stable iff every neighbourhood of A includes an invariant neighbourhood of A.*

Proof. "If": Obvious. "Only if": Let U and V be neighbourhoods of A as in (3.2-1). Then $V' := \bigcap_{k=0}^{\infty} (f^k)^{\leftarrow}[U] = \{x \in X : \mathcal{O}(x) \subseteq U\}$ is an invariant subset of U. As A is invariant it is easily seen that $A \subseteq V'$. Condition (3.2-1) implies that $V \subseteq V'$, so V' is a neighbourhood of A. □

Lemma 3.3.2. *Let A be a compact completely invariant subset of X and assume that A is the union of two mutually disjoint non-empty closed invariant subsets A_1 and A_2. Then A is stable iff both A_1 and A_2 are stable.*

Proof. "If": Straightforward.

"Only if": Assume that A is stable. First, note that the invariant sets A_1 and A_2 are completely invariant; the straightforward proof is left as an exercise for the reader.

Next, let U_1 and U_2 be arbitrary open neighbourhoods of A_1 and A_2, respectively. In view of Lemma 3.3.1, we want to show that for $i = 1, 2$ there is an invariant neighbourhood V_i of A_i such that $V_i \subseteq U_i$.

By Appendix A.2.1, we may assume that U_1 and U_2 are mutually disjoint, because A_1 and A_2 are compact. For $i = 1, 2$, let $W_i := U_i \cap f^{\leftarrow}[U_i]$. As the set A_i is invariant it follows that $A_i \subseteq W_i$, so W_i is a neighbourhood of A_i. Then $W_1 \cup W_2$ is a neighbourhood of A, hence Lemma 3.3.1 implies that A has an invariant neighbourhood V such that $V \subseteq W_1 \cup W_2$. Now for $i = 1, 2$, let $V_i := V \cap W_i$. Then it is clear that V_i is a neighbourhood of A_i and that $V_i \subseteq W_i \subseteq U_i$. It remains to show that V_i is an invariant set. To this end, note that $f[V_i] \subseteq f[V] \subseteq V$ and that $f[V_i] \subseteq f[W_i] \subseteq U_i$. It follows that $f[V_i] \subseteq V \cap U_i$. However, $V \cap U_i \subseteq V_i$, because the sets U_1 and U_2 are mutually disjoint and $V = V_1 \cup V_2$ with $V_i \subseteq U_i$. This completes the proof. □

Remarks. By induction one easily shows that a finite disjoint union of compact invariant sets is stable iff each of them is stable.

Lemma 3.3.3. *Let A be a non-empty completely invariant compact subset of X. Then A is stable under f iff A is stable under g.*

Proof. "Only if": Trivial.

"If": Let U and V be neighbourhoods of A such that $g^k[V] \subseteq U$ for all $k \in \mathbb{Z}^+$. Then the set $V' := \bigcap_{i=0}^{p-1} (f^i)^{\leftarrow}[V]$ is a neighbourhood of A, because A is invariant. Moreover, if $n \in \mathbb{Z}^+$ then $n = kp + i$ with $k \in \mathbb{Z}^+$ and $0 \leq i \leq p-1$, hence $f^n[V'] = f^{kp}[f^i[V']] \subseteq g^k[V] \subseteq U$. □

Theorem 3.3.4. *The following conditions are equivalent:*
(i) *$\mathcal{O}_f(x_0)$ is stable under f.*
(ii) *There exists $i \in \{0, \ldots, p-1\}$ such that x_i is a stable invariant point under g.*
(iii) *For each $i \in \{0, \ldots, p-1\}$ the point x_i is stable and invariant under g.*

Proof. "(i)⇒(iii)": Assume that $\mathcal{O}_f(x_0)$ is stable under f. Then $\mathcal{O}_f(x_0)$ is stable under g, by Lemma 3.3.3. As $\mathcal{O}_f(x_0)$ is the union of a finite number of g-invariant points, the Remark after Lemma 3.3.2 implies that each of these points is stable under g.

"(iii)⇒(i)": If (iii) holds then $\mathcal{O}_f(x)$, being a finite union of g-stable invariant points, is stable under g. So by Lemma 3.3.3, $\mathcal{O}_f(x)$ is f-stable.

"(iii)⇒(ii)": Obvious.

"(ii)⇒(iii)": Without limitation of generality, we may assume that the g-invariant point x_0 is g-stable. Consider any $j \in \{1, \ldots, p-1\}$ and let U_j be a neighbourhood of the point x_j. Since f^j maps x_0 onto x_j there is a neighbourhood U_0 of x_0 such that $f^j[U_0] \subseteq U_j$. As X is a Hausdorff space we may and shall assume that U_0 and U_j are mutually disjoint.

By our assumption the neighbourhood U_0 of x_0 includes a g-invariant neighbourhood N_0 of x_0. The mapping f^{p-j} sends x_j to x_0, so $(f^{p-j})^{\leftarrow}[N_0]$ is a neighbourhood of x_j. It follows that $N_j := U_j \cap (f^{p-j})^{\leftarrow}[N_0]$ is a neighbourhood of x_j. Obviously, $N_j \subseteq U_j$, so the proof is completed if we manage to show that N_j is g-invariant. In

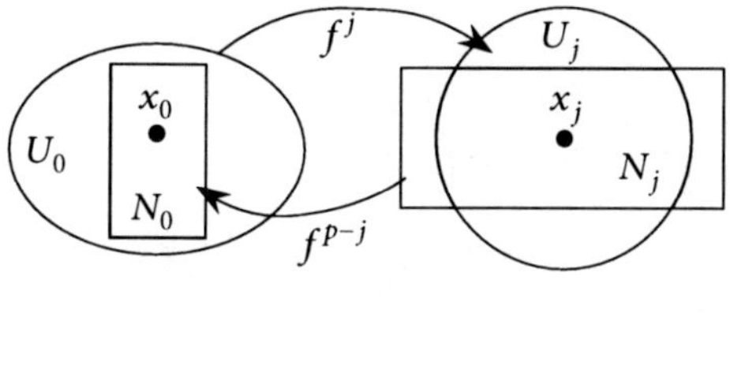

Fig. 3.3. Illustrating the proof of the implication (ii)⇒(iii) in Theorem 3.3.4.

order to do so, first note that

$$g[N_j] = f^j[f^{p-j}[N_j]] \subseteq f^j[N_0] \subseteq f^j[U_0] \subseteq U_j . \tag{3.3-1}$$

On the other hand, we have

$$f^{p-j}[g[N_j]] = g[f^{p-j}[N_j]] \subseteq g[N_0] \subseteq N_0 , \tag{3.3-2}$$

because N_0 is g-invariant. It follows that $g[N_j] \subseteq (f^{p-j})^{\leftarrow}[N_0]$. Together with (3.3-1) this implies that $g[N_j] \subseteq U_j \cap (f^{p-j})^{\leftarrow}[N_0] = N_j$. □

Remark. If f is a homeomorphism then the proof of the implication (ii)⇒(iii) is trivial: then $f^i : (X, g) \to (X, g)$ is a conjugation, hence by Proposition 3.2.5, if x_0 is g-stable then so is x_i $(0 \le i \le p-1)$.

Lemma 3.3.5. *Let A be a compact invariant subset of X and assume that A is the union of two mutually disjoint non-empty closed invariant subsets A_1 and A_2. Then $\mathcal{B}^*(A) = \mathcal{B}^*(A_1) \cup \mathcal{B}^*(A_2)$*

Proof. "⊇": Trivial.

"⊆": Let $x \in \mathcal{B}^*(A)$ and or i=1,2, let U_i be an open neighbourhood of A_i. We may assume that U_1 and U_2 are disjoint, because the sets A_1 and A_2 are compact. For $i = 1, 2$, let $W_i := U_i \cap f^{\leftarrow}[U_i]$. Since A_i is invariant it follows that $A_i \subseteq W_i$, hence W_i is a neighbourhood of A_i. Consequently, $W_1 \cup W_2$ is a neighbourhood of A. By the choice of x as a point of $\mathcal{B}^*(A)$ there exists $N \in \mathbb{Z}^+$ such that $f^n(x) \in W_1 \cup W_2$ for all $n \ge N$.

Assume that $f^N(x) \in W_1$. Then $f^{N+1}(x) = f(f^N(x)) \in f[W_1] \subseteq U_1$, but also $f^{N+1}(x) \in W_1 \cup W_2$. It follows that $f^{N+1}(x) \in (W_1 \cup W_2) \cap U_1 = W_1$. Proceeding by induction, it follows that $f^n(x) \in W_1$ for all $n \ge N$. In particular, this implies that $f^n(x) \in W_2$ for only finitely many values of n. Next, let W_1' be an arbitrary neighbourhood of A_1. Then $(W_1' \cap W_1) \cup W_2$ is a neighbourhood of A, hence $f^n(x) \in (W_1' \cap W_1) \cup W_2$ for almost all n. But $f^n(x) \in W_2$ for only finitely many values of n, so $f^n(x) \in W_1' \cap W_1 \subseteq W_1'$ for almost all n. It follows that $x \in \mathcal{B}^*(A_1)$. In a similar way one shows that if $f^N(x) \in W_2$ then $x \in \mathcal{B}^*(A_2)$. □

By the trivial inclusion '$\mathcal{B}^*(A) \supseteq \mathcal{B}^*(A_1) \cup \mathcal{B}^*(A_2)$', if A_1 and A_2 are strongly attracting then so is A. The implication the other way round is not true, as the following example shows.

Example. Define $f : \mathbb{S} \to \mathbb{S}$ by

$$f([t]) := \begin{cases} [2t^2] & \text{for } 0 \le t < \frac{1}{2} , \\ [2(t - \frac{1}{2})^2 + \frac{1}{2}] & \text{for } \frac{1}{2} \le t < 1 . \end{cases}$$

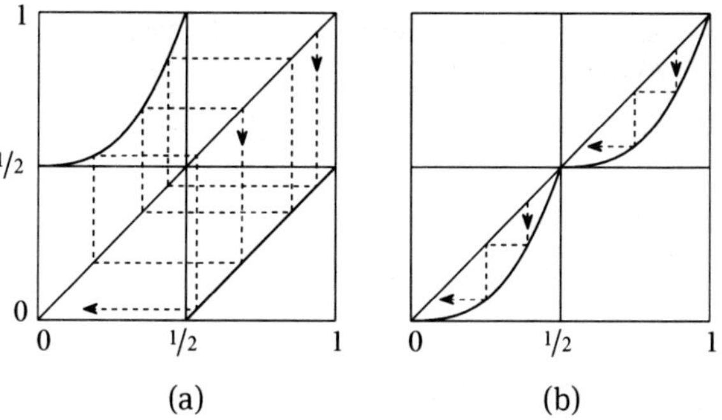

Fig. 3.4. The circle depicted as the unit interval (the points 0 and 1 should be identified.) (a) The graph of f. (b) The graph of f^2.

See Figure 3.4 (b) ahead. The system $(\mathbb{S}, f)$ has two invariant points, namely, $[0]$ and $[1/2]$. These points are, so to say, 'one-sided' strongly attracting: $\mathcal{B}^*[0] = [[0]; [1/2])$ and $\mathcal{B}^*[1/2] = [[1/2]; [1])$ (see 1.7.6 for the notation of arcs in $\mathbb{S}$). Thus, the two invariant sets $A_1 := \{[0]\}$ and $A_2 := \{[1/2]\}$ are not strongly attracting. But their union $A := \{ [0], [1/2] \}$ is strongly attracting, with $\mathcal{B}^*(A) = \mathbb{S}$.

Lemma 3.3.6. *Let A be a non-empty closed invariant subset of X. Then $\mathcal{B}^*_f(A) = \mathcal{B}^*_g(A)$. In particular, A is strongly attracting under f iff it is so under g.*

Proof. It is sufficient to prove that the stated equality holds: in that case, if one of $\mathcal{B}^*_f(A)$ and $\mathcal{B}^*_g(A)$ is a neighbourhood of A then so is the other.

"$\subseteq$": Easy.

"$\supseteq$": Let $x \in \mathcal{B}^*_g(A)$ and let U be any open neighbourhood of A. Then the set $\bigcap_{i=0}^{p-1} (f^i)^{\leftarrow}[U]$ is open and includes A (for A is invariant), so it is a neighbourhood of A as well, so $f^{np}(x) \in \bigcap_{i=0}^{p-1} (f^i)^{\leftarrow}[U]$ for almost all n, that is, $f^{np+i}(x) \in U$ for almost all n and for $i = 0, \dots, p-1$. As every $m \in \mathbb{Z}^+$ can be written as $m = np + i$ with $0 \le i \le p-1$, it follows easily that $f^m(x) \in U$ for almost all m. □

Corollary 3.3.7. $\mathcal{B}^*_f(\mathcal{O}_f(x_0)) = \bigcup_{i=0}^{p-1} \mathcal{B}^*_g(x_i)$.

Proof. By Lemma 3.3.5, the right-hand side of this equality is equal to the set $\mathcal{B}^*_g(\mathcal{O}_f(x_0))$ which, by Lemma (3.6), is equal to $\mathcal{B}^*_f(\mathcal{O}_f(x_0))$. □

Lemma 3.3.8. *For every $i \in \{0, \dots, p-1\}$ we have*

$$\mathcal{B}^*_g(x_0) = (f^i)^{\leftarrow}[\mathcal{B}^*_g(x_i)]\,. \tag{3.3-3}$$

and

$$\mathcal{B}^*_g(x_i) = (f^{p-i})^{\leftarrow}[\mathcal{B}^*_g(x_0)]\,. \tag{3.3-4}$$

Proof. Recall that the mappings f^i and f^{p-i} are endomorphisms of the system (X, g). So Proposition 3.1.13 implies that $f^i[\mathcal{B}^*_g(x_0)] \subseteq \mathcal{B}^*_g(x_i)$, so, $\mathcal{B}^*_g(x_0) \subseteq (f^i)^{\leftarrow}[\mathcal{B}^*_g(x_i)]$. A similar argument shows that the inclusion $\mathcal{B}^*_g(x_i) \subseteq (f^{p-i})^{\leftarrow}[\mathcal{B}^*_g(x_0)]$ holds. Hence

$$\begin{aligned}\mathcal{B}^*_g(x_0) &\subseteq (f^i)^{\leftarrow}[\mathcal{B}^*_g(x_i)] \subseteq (f^i)^{\leftarrow}[(f^{p-i})^{\leftarrow}[\mathcal{B}^*_g(x_0)]] \\ &= g^{\leftarrow}[\mathcal{B}^*_g(x_0)] \subseteq \mathcal{B}^*_g(x_0)\,,\end{aligned}$$

where we have used Lemma 3.1.11 (2) and the fact that $f^{p-i} \circ f^i = f^p = g$. Thus, the inclusions here are equalities. In particular, we get (3.3-3). The proof of (3.3-4) is similar. □

Proposition 3.3.9. *The following conditions are equivalent:*
(i) *There exists $i \in \{0, \dots, p-1\}$ such that x_i is a strongly attracting invariant point under g.*
(ii) *For each $i \in \{0, \dots, p-1\}$ the point x_i is a strongly attracting invariant point under g.*
and they imply that $\mathcal{O}_f(x_0)$ is strongly attracting under f.

Proof. The implication (ii)⇒(i) is obvious. The implication (i)⇒(ii) follows easily from Lemmas 3.3.8 and 3.1.14. Moreover, the final statement follows from Corollary 3.3.7. □

Examples. The following example shows that this proposition is the best possible result: let $X := \mathbb{S}$ and let $f : \mathbb{S} \to \mathbb{S}$ be given by

$$f([t]) := \begin{cases} [2t^2 + \frac{1}{2}] & \text{for } 0 \le t < \frac{1}{2}\,, \\ [t - \frac{1}{2}] & \text{for } \frac{1}{2} \le t < 1\,. \end{cases}$$

The points [0] and [1/2] form a periodic orbit with period 2; let us denote it by A. Every other point of $\mathbb{S}$ 'hops' from the upper half of $\mathbb{S}$ to the lower half and vice versa. In that process, the odd terms of the orbit approach one point of A and the even terms approach the other. So $\mathcal{B}^*_f(A) = \mathbb{S}$ and A is strongly attracting. Moreover, a straightforward computation shows that $g := f^2$ is equal to the mapping defined in the Example after Lemma 3.3.5. See Figure 3.4 (b). It follows that the points [0] and [1/2] are not strongly attracting under g.

For topological attraction the results are similar to those for strong attraction – or slightly worse, because we need additional conditions. Note that

$$\forall\, x \in X:\ \omega_g(x) \subseteq \omega_f(x)\,. \tag{3.3-5}$$

This is an easy consequence of Lemma 1.4.1 (1) and the obvious fact that an infinite subset of $\mathbb{N}p$ is an infinite subset of $\mathbb{N}$. See also Lemma 3.3.10 (1) below.

Lemma 3.3.10. *Let A be a non-empty completely invariant compact subset of X and let x be an arbitrary point of X.*

(1) $\omega_f(x) = \bigcup_{i=0}^{p-1} \omega_g(f^i(x))$.

(2) $\mathcal{B}_f(A) \subseteq \mathcal{B}_g(A)$. *Consequently, if A is topologically attracting under f then A is topologically attracting under g.*

Proof. (1) "$\supseteq$": Formula (3.3-5) and Proposition 1.4.3 (1) immediately imply that $\omega_g(f^i(x)) \subseteq \omega_f(f^i(x)) = \omega_f(x)$ for $i = 0, \dots, p-1$.
"$\subseteq$": Consider a point $y \notin \bigcup_{i=0}^{p-1} \omega_g(f^i(x))$ (note that if such a point does not exist then nothing remains to be proved). Lemma 1.4.1 (1) implies that for $i = 0, \dots, p-1$ the point y has a neighbourhood V_i such that there are only finitely many values of n for which $f^{np+i}(x) = g^n(f^i(x)) \in V_i$. Using the fact that every non-negative integer k can be written as $k = pn + i$ with $n \in \mathbb{Z}^+$ and $0 \leq i \leq p-1$ it follows that $f^k(x) \in W := \bigcap_{i=0}^{p-1} V_i$ for only finitely many values of k. As W is a neighbourhood of y, this implies that $y \notin \omega_f(x)$. This completes the proof.
(2) Consider a point $x \in \mathcal{B}_f(A)$, i.e., $\emptyset \neq \omega_f(x) \subseteq A$. Inclusion (3.3-5) implies that $\omega_g(x) \subseteq A$, so if we are able to show that $\omega_g(x) \neq \emptyset$ then, by definition, $x \in \mathcal{B}_g(A)$. Note that $\omega_f(x) \neq \emptyset$, so by 1 there is $i \in \{0, \dots, p-1\}$ such that $\omega_g(f^i(x)) \neq \emptyset$. As f^{p-i} is an endomorphism of the system (X, g), it follows from Proposition 3.1.2 (1) that

$$f^{p-i}[\omega_g(f^i(x))] \subseteq \omega_g(f^{p-i}(f^i(x))) = \omega_g(g(x)) = \omega_g(x),$$

where we have also used Proposition 1.4.3 (1) for the mapping g. This completes the proof that $\mathcal{B}_f(A) \subseteq \mathcal{B}_g(A)$. The remainder of statement 2 should be clear now from Theorem 3.1.6. □

Recall from Proposition 3.1.12 (3) and Corollary 3.2.2 (1) that if the phase space is locally compact or that the discussion is about a stable set, there is no difference between strong attraction and topological attraction. Take also into account that, by Theorem 3.3.4, the periodic orbit $\mathcal{O}_f(x_0)$ is stable under f iff all points x_i for $i = 0, \dots, p-1$ are stable under g. Hence the following results are clear from 3.3.6– 3.3.9:

Lemma 3.3.11. *Assume that X is locally compact or that $\mathcal{O}_f(x_0)$ is stable under f. Then:*

(1) $\mathcal{B}_f(\mathcal{O}_f(x_0)) = \mathcal{B}_g(\mathcal{O}_f(x_0))$, *so $\mathcal{O}_f(x_0)$ is topologically attracting under f iff $\mathcal{O}(x_0)$ is topologically attracting under g.*

(2) $\mathcal{B}_g(x_0) = (f^i)^{\leftarrow}[\mathcal{B}_g(x_i)]$ *and* $\mathcal{B}_g(x_i) = (f^{p-i})^{\leftarrow}[\mathcal{B}_g(x_0)]$ *for* $0 \leq i < p$.

(3) $\mathcal{B}_f(\mathcal{O}_f(x_0)) = \bigcup_{i=0}^{p-1} \mathcal{B}_g(x_i)$.

Remarks.

(1) In statement 1, $\mathcal{O}(x_0)$ may be replaced by any non-empty compact invariant set A (stable if X is not locally compact) and in $g := f^p$, p may be any natural number.

(2) By Example (2) in 3.3.13 the statements 1 and 2 are not true in general; by Example (1) in 3.3.13 statement 3 is not generally true.

Theorem 3.3.12. *Suppose that the phase space X is locally compact, or that the periodic orbit $\mathcal{O}_f(x_0)$ is stable*[2] *under f. Then the following conditions are equivalent:*
(i) *There exists $i \in \{0, \dots, p-1\}$ such that x_i is a topologically attracting invariant point under g.*
(ii) *For each $i \in \{0, \dots, p-1\}$ the point x_i is a topologically attracting invariant point under g.*

and they imply that $\mathcal{O}_f(x_0)$ is topologically attracting under f.

Proof. Clear from the above. □

Example. The example after Proposition 3.3.9 shows that the converse of the final statement may be not true, not even in the case that X is compact and f is a homeomorphism.

3.3.13. Statement 3 in Lemma 3.3.11 – or what amounts in the given situation to the same, Corollary 3.3.7 – has an interesting reformulation. Recall from Section A.4 from Appendix A that a sequence $(z_n)_{n\in\mathbb{N}}$ is said to converge in a topological space to a point z whenever every neighbourhood of z includes the point z_n for almost every n.

Let $x \in \mathcal{B}_g(x_i)$ for some $i \in \{0, \dots, p-1\}$. Corollary 3.2.3 (1) applied to g implies that in the given situation $f^{np}(x) = g^n(x) \rightsquigarrow x_i$ for $n \rightsquigarrow \infty$. By applying iterates of f we infer that for every $j \in \{0, \dots, p-1\}$ one has

$$f^{np+j}(x) \rightsquigarrow x_{(i+j) \pmod p} \qquad \text{for } k \rightsquigarrow \infty\,. \tag{3.3-6}$$

Conversely, if a point x satisfies these conditions for some $i \in \{0, \dots, p-1\}$ then every neighbourhood of $\mathcal{O}_f(x_0)$ contains the point $f^n(x)$ for almost all values of $n \in \mathbb{Z}^+$ and Proposition 3.1.12 (1) implies that $x \in \mathcal{B}_f(\mathcal{O}_f(x_0))$ (no additional conditions needed). This proves:

Assume that X is locally compact or that the periodic orbit $\mathcal{O}_f(x_0)$ is stable under f. Then a point $x \in X$ is in $\mathcal{B}_f(\mathcal{O}_f(x_0))$ iff $\mathcal{O}_f(x)$ splits into p subsequences as in (3.3-6), *each of which converges to a point of $\mathcal{O}_f(x_0)$.*

This behaviour occurs in Example (1) before Theorem 3.1.1 (if $a \in \mathbb{Q}$). The 'only if' in the above statement may be not true if X is not locally compact and the periodic orbit $\mathcal{O}_f(x_0)$ is not stable: in Example 3.3.13 (1) below it is not possible to select a subsequence of the orbit of the point with label 0 which converges to x_0 such that the remaining points of this orbit converge to x_1.

Examples. The following examples show that some of the preceding results may not hold if the phase space is not locally compact and the periodic orbit is not stable.
(1) Let $X := \{x_0\} \cup (\mathbb{Z}\times\mathbb{Z}) \cup \{x_1\}$ where x_0 and x_1 are two distinct points not belonging to $\mathbb{Z}\times\mathbb{Z}$. Give X the topology which is given in the following way by local bases: all points of $\mathbb{Z}\times\mathbb{Z}$ are isolated, the point x_0 has a local base consisting the sets

2 The case that $\mathcal{O}_f(x_0)$ is stable is treated more comprehensively in Theorem 3.3.16 below.

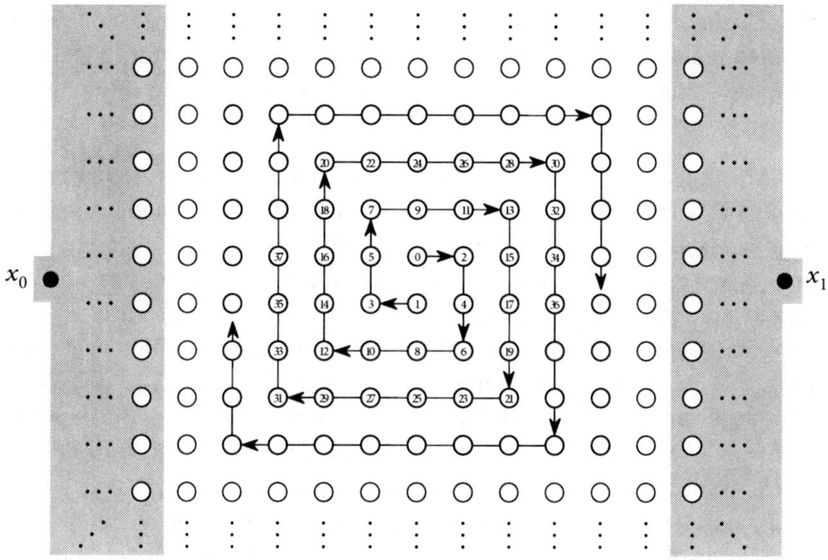

Fig. 3.5. Illustrating Example (1). The shaded areas indicate typical neighbourhoods of the points x_0 and x_1. The arows indicate the two orbits under f^2.

$U_k := \{x_0\} \cup ((-\infty; -k] \times \mathbb{Z})$, and the point x_1 has a local base consisting of all sets of the form $V_k := ([k; \infty) \times \mathbb{Z}) \cup \{x_1\}$ for $k \in \mathbb{N}$. In Figure 3.5 we have drawn two spiralling polygons in $\mathbb{Z} \times \mathbb{Z}$, one numbered by even non-negative integers, the other by odd positive integers. The mapping f is defined as follows: $f(i) := i + 1$ for every number i on one of the polygons. Note that if i is far left or far up then $f(i)$ is far right or far down, respectively; similarly with left and right or up and down interchanged. Moreover, put $f(x_0) := x_1$ and $f(x_1) := x_0$. It is not difficult to see that f is continuous at the points x_0 and x_1, so f is continuous on all of X. Obviously, $\{x_0, x_1\}$ is a periodic orbit with period 2. One easily sees that $\omega_f(x) = \{x_0, x_1\}$ for all $x \in X$. Consequently, $\mathcal{B}_f(x_0, x_1) = X$.
The mapping $g := f^2$ has two g-invariant points x_0 and x_1, and there are just two other orbits under g, the orbit of the point labelled 0 (all points on one of the polygons) and the orbit of the point labelled 1 (all points on the other polygon). The points of $\mathbb{Z}\times\mathbb{Z}$ spiral around under g – the example after Theorem 3.1.1 comes to mind – and we find that $\omega_g(x) = \{x_0, x_1\}$ for all $x \in \mathbb{Z}\times\mathbb{Z}$; moreover, $\omega_g(x) = \{x\}$ for $x \in \{x_0, x_1\}$. It follows that

$$\mathcal{B}_g(\{x_0, x_1\}) = X, \quad \mathcal{B}_g(x_0) = \{x_0\}, \quad \mathcal{B}_g(x_1) = \{x_1\}.$$

So in this case one has

$$\mathcal{B}_f(\{x_0, x_1\}) = \mathcal{B}_g(\{x_0, x_1\}) \supsetneqq \mathcal{B}_g(x_0) \cup \mathcal{B}_g(x_1).$$

NB. The mapping f can be made a homeomorphism by adding countably many isolated points that can serve as the past under f of the point with label 0. The various basins under f or g do not change if we do so. Hence in the modified system we have $\mathcal{B}_f(\{x_0, x_1\}) = \mathcal{B}_g(\{x_0, x_1\}) \supsetneqq \mathcal{B}_g(x_0) \cup \mathcal{B}_g(x_1)$ as well.

(2) Let $X := \{x_0\} \cup ((-\mathbb{N}) \times \mathbb{N}) \cup \{z_0\} \cup (\mathbb{N} \times \mathbb{N}) \cup \{x_1\}$. The topology of X looks much like the topology in the previous example: all points except x_0, x_1 and z_0 are isolated, local bases at x_0 and x_1 are formed by all sets of the form $\{x_0\} \cup ((-\infty; -k] \times \mathbb{N})$ and $([k; \infty) \times \mathbb{N}) \cup \{x_1\}$ ($k \in \mathbb{N}$), respectively, and the point z_0 has a local base consisting of all sets of the form $W_k := \{z_0\} \cup (\{1\} \times [k; \infty))$ with $k \in \mathbb{N}$. It is clear that X with this topology is a Hausdorff space. For the definition of the mapping $f: X \to X$ we refer to Figure 3.6. Label the points of $X \setminus \{x_0, x_1, z_0\}$ by the non-negative integers as shown in this picture and define f in these points by $f(i) := i + 1$ for $i \in \mathbb{Z}^+$. Let X_l and X_r be the sets of points with even and odd labels, respectively. In the remaining points, put $f(x_0) := x_1$, $f(x_1) := x_0$ and $f(z_0) := x_0$. It is easily checked that f is continuous. Moreover, $\omega_f(x) = \{x_0, x_1, z_0\}$ for all $x \in X \setminus \{x_0, x_1, z_0\}$ and $\omega_f(x) = \{x_0, x_1\}$ for $x \in \{x_0, x_1, z_0\}$. This implies that $\mathcal{B}_f(\{x_0, x_1\}) = \{x_0, x_1, z_0\}$.
In Figure 3.6 the mapping $g := f^2$ is also indicated, by the arrows. It is easy to see that $\omega_g(x) = \{x_0\}$ if $x \in X_l \cup \{x_0\}$, $\omega_g(x) = \{x_1, z_0\}$ if $x \in X_r$ and $\omega_g(x_1) = \omega_g(z_0) = \{x_1\}$. So

$$\mathcal{B}_g(\{x_0, x_1\}) = X_l \cup \{x_0, x_1, z_0\} \quad \text{and}$$

$$\mathcal{B}_g(x_0) = X_l \cup \{x_0\}, \quad \mathcal{B}_g(x_1) = \{x_1, z_0\}.$$

Consequently, in this example we have

$$\mathcal{B}_f(\{x_0, x_1\}) \subsetneqq \mathcal{B}_g(\{x_0, x_1\}) = \mathcal{B}_g(x_0) \cup \mathcal{B}_g(x_1).$$

and $\mathcal{B}_g(x_0) \neq f^{\leftarrow}[\mathcal{B}_g(x_1)]$ as well as $\mathcal{B}_g(x_1) \neq f^{\leftarrow}[\mathcal{B}_g(x_0)]$.

Finally, we deal with asymptotic stability. Actually, this concerns the stable case of Theorem 3.3.12. First we formulate and prove the counterparts of the Lemmas 3.3.2 and 3.3.3 for asymptotic stability.

Lemma 3.3.14. *Let A be a compact completely invariant subset of X and assume that A is the union of two mutually disjoint non-empty closed invariant subsets A_1 and A_2. Then A is asymptotically stable under f iff both A_1 and A_2 are asymptotically stable under f.*

Proof. "If": Clear from the 'if' in 3.3.2 and the obvious fact that the union of topologically attracting sets is topologically attracting.

"Only if": Assume that A is asymptotically stable. Then Lemma 3.3.2 implies that the sets A_1 and A_2 are stable. So for A, A_1 and A_2 the notions of strong attraction

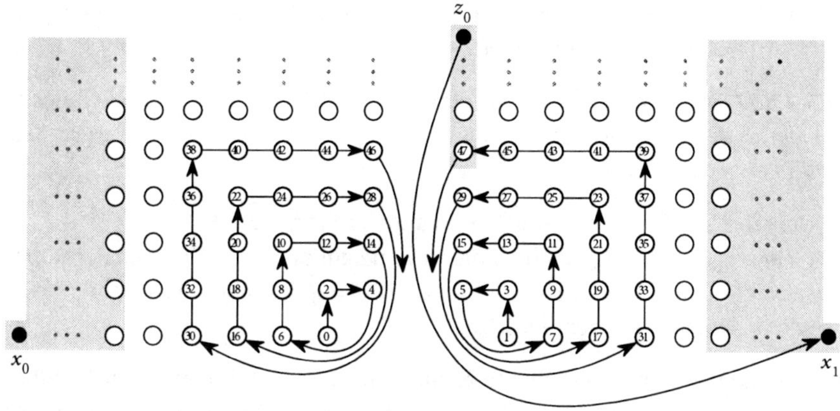

Fig. 3.6. Example (2). The shaded areas indicate typical neighbourhoods of the points x_0, x_1 and z_0. The arrows denote the orbits under f^2.

and topological attraction coincide. So our assumption implies that $\mathcal{B}^*(A)$ is a neighbourhood of A, hence of A_1 and of A_2. It follows from Lemma 3.3.1 that for $i = 1, 2$ the set A_i has an invariant neighbourhood U_i such that $U_i \subseteq \mathcal{B}^*_f(A)$. Without limitation of generality we may assume that U_1 and U_2 are disjoint. If $x \in U_1 \subseteq \mathcal{B}^*(A)$ then by Lemma 3.3.5, x is in $\mathcal{B}^*(A_1)$ or in $\mathcal{B}^*(A_2)$. As $f^n(x) \in U_1$ for all n it is impossible that $f^n(x) \in U_2$ for almost all n, and therefore $x \notin \mathcal{B}^*(A_2)$. This shows that $U_1 \subseteq \mathcal{B}^*(A_1)$. Similarly, $U_2 \subseteq \mathcal{B}^*(A_2)$. Consequently, the basins of strong attraction of the sets A_1 and A_2 are neighbourhoods of these sets, i.e., these sets are strongly attracting, hence topologically attracting. □

Lemma 3.3.15. *Let A be a non-empty completely invariant compact subset of X. Then A is asymptotically stable under f iff A is asymptotically stable under g.*

Proof. Clear from the Lemma's 3.3.3 and the stable case of Lemma 3.3.11 (1) – see also Remark 1 after Lemma 3.3.11 – taking into account Theorem 3.1.6 for f and for g. □

Theorem 3.3.16. *The following conditions are equivalent:*
(i) *$\mathcal{O}_f(x_0)$ is asymptotically stable under f.*
(ii) *There exists $i \in \{0, \dots, p-1\}$ such that x_i is an asymptotically stable invariant point under g.*
(iii) *For each $i \in \{0, \dots, p-1\}$ the point x_i is an asymptotically stable invariant point under g.*

Proof. "(i)⇒(iii)": Assume (i). Lemma 3.3.15 implies that $\mathcal{O}_f(x_0)$ is asymptotically stable under g, so Lemma 3.3.14 and a simple induction argument imply condition (iii).

"(iii)⇒(i)": Obviously, (iii) implies that $\mathcal{O}_f(x_0)$ is asymptotically stable under g. Now (i) follows from Lemma 3.3.15.

"(ii)⇔(iii)": Clear from the equivalence (ii)⇔(iii) in Theorem 3.3.4 and the equivalence (i)⇔(ii) in (the stable case of) Theorem 3.3.12. □

Corollary 3.3.17. *Assume that X is an interval in* $\mathbb{R}$. *Then* $\mathcal{O}_f(x_0)$ *is asymptotically stable under f iff some point of* $\mathcal{O}_f(x_0)$, *say* x_0, *has a neighbourhood U such that* $\lim_{n\rightsquigarrow\infty} g^n(x) = x_0$ *for all* $x \in U$.

Proof. "Only if": Use Theorem 3.3.16 (i)⇒(ii) and Corollary 3.2.3 (1).

"If": If $\lim_{n\rightsquigarrow\infty} g^n(x) = x_0$ for all points x in a neighbourhood of x_0 then x_0 is topologically attracting under g. Hence Proposition 3.2.9 implies that x_0 is asymptotically stable under g. Now apply Theorem 3.3.16. □

3.3.18. Assume that X is an interval in $\mathbb{R}$ and that $f: X \to X$ is continuously differentiable. Recall that an f-invariant point $z \in X$ is said to be attracting[3] whenever $|f'(z)| < 1$. In Example (3) in 3.2.6 it was observed that an attracting invariant point is asymptotically stable.

Now let $x_0 \in X$ be a periodic point with primitive period p. Then all points of $\mathcal{O}(x_0)$ are attracting under g iff some point of the orbit of x_0 is attracting under g: see Exercise 3.5. If these conditions are fulfilled then we say that *the periodic orbit* $\mathcal{O}_f(x_0)$ *is attracting*. In that case, the points of $\mathcal{O}_f(x_0)$ are asymptotically stable under g, hence by Theorem 3.3.16 the orbit $\mathcal{O}_f(x_0)$ is asymptotically stable under f.

Examples. The logistic map $f_\mu: \mathbb{R} \to \mathbb{R}$ has for $3 < \mu < 1 + \sqrt{6}$ an asymptotically stable periodic orbit of primitive period 2: see 2.1.5, the case $\mu > 3$.

Another example: let $f: \mathbb{R} \to \mathbb{R}$ be given by $f(x) = -\sqrt{x}$ for $x \geq 0$ and $f(x) = \sqrt{|x|}$ for $x < 0$. The system has a periodic orbit of period 2, namely, $\{-1, 1\}$. This orbit is asymptotically stable, because $f'(1) \cdot f'(-1) = 1/4$.

3.3.19 **(A more complicated example).** Consider the mapping f_∞ of the unit interval into itself, defined and discussed in Section 2.4. For notation, we refer also to Appendix B. and Figure 2.10. Recall that the Cantor set C is invariant under f_∞. In addition, recall that for every $n \in \mathbb{Z}^+$ the mapping f_∞ has just one periodic orbit with primitive period 2^n, which is included in the set M_n: the orbit of the point P_n. Denote the union of these periodic orbits by $P(f_\infty)$. Obviously, the set $P(f_\infty)$ is invariant. Note that $P(f_\infty)$ is countably infinite, and included in the set $\bigcup_{n=0}^{\infty} M_n$. Moreover, inspection of Figure 2.10 – see also the observations in 2.4.6 – shows that all cluster points of the set $P(f_\infty)$ are situated in C. Consequently, the set $C \cup P(f_\infty)$ is closed, hence compact.

Claim. *The set* $C \cup P(f_\infty)$ *is completely invariant and* $\mathcal{B}_{f_\infty}(C \cup P(f_\infty)) = [0; 1]$, *so the set* $C \cup P(f_\infty)$ *is topologically attracting under* f_∞.

3 Just 'attracting'; not strongly or topologically attracting.

It is obvious that $P(f_\infty)$ is completely invariant. We have seen above that C is invariant, but it is not obvious that C is completely invariant. It is, but the proof will be postponed to 4.2.10 below: it will be shown there that C is minimal under f_∞, so that $f_\infty[C] = C$. (This also implies that C includes no periodic points, so that $P(f_\infty)$ is the set of *all* periodic points in $[0; 1]$ under f_∞.) It follows immediately that $C \cup P(f_\infty)$ is completely invariant as well.

Consider a point $x \in M_0$ different from the invariant point P_0. On M_0 the function f_∞ is differentiable with derivative 7/3. Consequently, if the points $x, f_\infty(x), \dots, f_\infty^n(x)$ are all in M_0 then the mean value theorem implies that $|f_\infty^n(x) - P_0| \geq (7/3)^n|x - P_0|$. The set M_0 is bounded and $7/3 > 1$, hence there exists $i \in \mathbb{N}$ such that $f_\infty^i(x) \notin M_0$, that is, $f_\infty^i(x) \in C_1$. The set C_1 is invariant, so $f_\infty^n(x) \in C_1$ for all $n \geq i$. Similarly, by using Lemma 2.4.1 (2) and induction one can show: if $k > 1$ and x is in M_k but not in the orbit of the periodic point P_k then $f_\infty^n(x) \in C_{k+1}$ for almost all n.

Now consider an arbitrary point $x \in [0; 1]$. If $x \in C$ then $\emptyset \neq \omega(x) \subseteq C$. If $x \notin C$ then $x \in M_k$ for some $k \in \mathbb{Z}^+$. Then either x is not in the orbit of the periodic point P_k, or x is in the orbit of P_k. In the former case the previous paragraph implies that $f_\infty^{n_1}(x) \in C_{k+1}$ for some $n_1 \in \mathbb{Z}^+$. There are three mutually exclusive possibilities: (1) $f_\infty^{n_1}(x) \in C_{k+2}$, (2) $f_\infty^{n_1}(x) \in M_{k+1}$ but not in the orbit of the periodic point P_{k+1}, and (3) $f_\infty^{n_1}(x)$ is in the orbit of the periodic point P_{k+1}. By what we have proved in the above, in the cases (1) and (2) we have $f_\infty^n(x) \in C_{k+2}$ for almost all $n \in \mathbb{Z}^+$. Proceeding in this way, we find that either $x \in \bigcup_{j=0}^\infty (f^j)^\leftarrow[P(f_\infty)]$ or that there is a subsequence $n_1 < n_2 < \dots$ of $\mathbb{Z}^+$ such that for every $i \in \mathbb{N}$ we have $f_\infty^{n_i}(x) \in C_{k+i}$, hence $f_\infty^n(x) \in C_{k+i}$ for all $n \geq n_i$. In the latter case, for any $j \in \mathbb{N}$ there are only finitely many values of n such that $f_\infty^n(x) \in M_j$, and because $\omega(x) \neq \emptyset$ this clearly implies that $\omega(x) \subseteq C$. On the other hand, if $x \in \bigcup_{j=0}^\infty (f^j)^\leftarrow[P(f_\infty)]$ then there is $n \in \mathbb{Z}^+$ such that $f^n(x)$ is in the orbit of one of the periodic points P_i, so that $\omega(x) = \mathcal{O}(P_i) \subseteq P(f_\infty)$. Consequently, in all cases we have $\emptyset \neq \omega(x) \subseteq C \cup P(f_\infty)$. This completes the proof of the above claim.

However, the invariant point P_0 is not stable and taking into account the results of Lemma 2.4.1 (2), Proposition 3.2.5 and Theorem 3.3.4 one shows that none of the periodic orbits is stable. It follows easily from Lemma 3.3.2 that the set $C \cup P(f_\infty)$ is not stable either. *Consequently, the set $C \cup P(f_\infty)$ is topologically attracting but it is not asymptotically stable.*

Using similar methods, it is not too difficult to show that *the set C is stable under f_∞*. But C is not topologically attracting: the periodic orbits accumulate to C, so by Lemma 3.1.3 (4) no neighbourhood of C is included in the basin of C. See Exercise 3.6 (2) for a general description of this situation.

3.4 Asymptotic stability in locally compact spaces

Most results in this section are only useful for dynamical systems with a locally compact phase space; if that assumption is needed we shall say so explicitly. It will repeatedly be used that in a locally compact space *every non-empty compact subset has a neighbourhood base consisting of compact neighbourhoods*. See Appendix A.2.1

Theorem 3.4.1. *Let A be a non-empty completely invariant compact subset of X. If the set A strongly attracts a neighbourhood U_0 of itself then A is asymptotically stable and $U_0 \subseteq \mathcal{B}(A)$.*

Conversely, if A is stable then A strongly attracts every compact subset of $\mathcal{B}(A)$. In particular, if X is locally compact and A is asymptotically stable then A strongly attracts a neighbourhood of itself.

Proof. Assume that A strongly attracts a neighbourhood U_0 of itself. In order to show that A is stable, consider an arbitrary neighbourhood U of A. Without limitation of generality we may assume that $U \subseteq U_0$. By assumption, $f^n[U] \subseteq f^n[U_0] \subseteq U$ for almost all n, so by the discussion after formula (3.2-1), A is a stable set. Note that Corollary 3.2.2 (1) now implies that $\mathcal{B}(A) = \mathcal{B}^*(A)$. As it is obvious that $U_0 \subseteq \mathcal{B}^*(A)$, it follows that $\mathcal{B}(A)$ is a neighbourhood of A. So A is topologically attracting.

Conversely, assume that A is stable and let B be a compact subset $\mathcal{B}(A)$. Consider an arbitrary neighbourhood U of A. There is a neighbourhood V of A such that $f^i[V] \subseteq U$ for all $i \in \mathbb{Z}^+$. If $x \in B$ then $x \in \mathcal{B}(A) = \mathcal{B}^*(A)$, so there exists $n(x) \in \mathbb{N}$ such that $f^{n(x)}(x) \in V$. Continuity of f implies that there is an open neighbourhood V_x of x such that $f^{n(x)}[V_x] \subseteq V$. By the choice of V this implies that $f^i[V_x] \subseteq U$ for all $i \geq n(x)$. Cover the compact set B by finitely many of the open sets V_x and let N be the largest of the corresponding integers $n(x)$. If $z \in B$ then z is in one of the sets V_x of this finite subcover, hence $f^i(z) \in U$ for all $i \geq N$. This completes the proof that A strongly attracts B.

This part of the proof shows that if A is stable and A strongly attracts every point of a compact set B then A strongly attracts B. Informally, one might say that in this case 'pointwise strong attraction' implies 'uniform strong attraction'.

Finally, assume that X is locally compact and that A is asymptotically stable. Then $\mathcal{B}(A)$ is a neighbourhood of A, so by the initial remark in this section, A has a neighbourhood U such that $B := \overline{U} \subseteq \mathcal{B}(A)$ and B is compact. By what just has been proved, this set B is strongly attracted by A and, consequently, U is strongly attracted by A as well. □

Remark. If X is not locally compact then the final conclusion of the theorem may be false: see Example (1) after Theorem 3.4.3 below.

Theorem 3.4.2. *Let X be locally compact and let A be a non-empty compact subset of X. The following statements are equivalent:*

(i) *A is asymptotically stable.*

(ii) *A is completely invariant and A strongly attracts a neighbourhood of itself.*

(iii) *A has a compact invariant neighbourhood U_0 such that*

$$\bigcap_{n\geq 0} f^n[U_0] = A\,. \tag{3.4-1}$$

(iv) *A has arbitrarily small compact invariant neighbourhoods U such that*

$$\bigcap_{n\geq 0} f^n[U] = A\,. \tag{3.4–1*}$$

If one of these conditions is fulfilled then $U_0 \subseteq \mathcal{B}(A)$; moreover, A strongly attracts every compact subset of $\mathcal{B}(A)$.

Proof. "(i)⇔(ii)" and the final statement: Clear from Theorem 3.4.1.

"(ii)⇒(iv)": Let U_0 be any neighbourhood of A that is strongly attracted by A. Then it is easily seen that

$$\bigcap_{n\geq 0} f^n[U_0] \subseteq \bigcap\{W \,\vdots\, W \in \mathcal{N}_A\} = A\,,$$

where the equality follows from formula (A.2-1) in Appendix A. On the other hand, $\bigcap_{n\geq 0} f^n[U_0] \supseteq A$, because A is completely invariant. Thus, formula (3.4-2) holds for every neighbourhood U_0 of A that is strongly attracted by A. So it remains to prove that A has arbitrarily small compact invariant neighbourhoods that are strongly attracted by A.

As X is locally compact, it follows from the initial remark of this section that A has a neighbourhood basis consisting of compact sets. By Theorem 3.4.1, A is asymptotically stable, hence stable, so we may conclude from Lemma 3.3.1 that every neighbourhood of A includes an invariant neighbourhood, the closure of which is invariant as well. Conclusion: A has a neighbourhood base consisting of compact invariant sets. However, the members of this base that are included in U_0 form a neighbourhood base as well. Since A strongly attracts every subset of U_0 this concludes the proof.

"(iv)⇒(iii)": Obvious.

"(iii)⇒(ii)": Assume (iii). Invariance of U_0 implies that the sets $f^n[U_0]$ for $n \in \mathbb{Z}^+$ form a decreasing sequence of compact sets. Consequently, Lemma (A.2.2) implies that condition (3.1-3) holds for $B := U_0$, so U_0 is strongly attracted by A. Moreover, Lemma A.3.5 implies that

$$f[A] = f\Big[\bigcap_{n\geq 0} f^n[U_0]\Big] = \bigcap_{n\geq 0} f^{n+1}[U_0] = A\,.$$

This shows that A is completely invariant. □

Remarks.

(1) Local compactness of X is used explicitly only in the proofs of the implications (i)⇒(ii) and (ii)⇒(iv); this guarantees that A has arbitrarily small *compact* neighbourhoods at all. For the implication (iii)⇒(ii), compactness of U_0 replaces local compactness of the ambient space. Example (2) after Theorem 3.4.3 below shows that if compactness of U_0 in (iii) is omitted then (iii) may not imply (ii).

(2) In Theorem 3.4.3 below it will be shown that in (iii) the condition that U_0 is invariant can be omitted provided A is completely invariant.
(3) Proposition 2.1.3 shows that an attracting invariant point of a C^1-mapping on an interval satisfies Condition 3.4.2 (iii).
(4) In the conditions (iii) and (iv) A is not required to be completely invariant. In point of fact, if (3.4-1) holds for some neighbourhood U_0 of A (actually, for any superset U_0 of A) *then A is the largest completely invariant subset of U_0*. For assume that A' is a completely invariant set such that $A' \subseteq U$. Then one has $A' = \bigcap_{n=0}^{\infty} f^n[A'] \subseteq \bigcap_{n=0}^{\infty} f^n[U] = A$.

The proof of the following theorem requires some knowledge of convergence of nets as explained in Section A.4 of Appendix A. In particular, we use nets indexed by cofinal subsets of the directed set of neighbourhoods of a point x. Such a cofinal set is nothing but a local base at the point x.

Theorem 3.4.3. *Let X be a locally compact space and let A be a non-empty compact subset of X. The following conditions are equivalent:*
(i) *A is asymptotically stable.*
(ii) *A is completely invariant and there is a neighbourhood W_0 of A such that*

$$\bigcap_{n\geq 0} f^n[W_0] = A\,. \tag{3.4-2}$$

Proof. "(i)⇒(ii)": Obviously, condition (ii) is implied by Theorem 3.4.2 (iii).

"(ii)⇒(i)": Let W_0 be a neighbourhood of A such that $\bigcap_{n\geq 0} f^n[W_0] = A$. Because X is locally compact, there is an open neighbourhood V of A such that $\overline{V}$ is compact and $\overline{V} \subseteq W_0$. Let

$$U := \{\, x \in X : \mathcal{O}(x) \subseteq V \,\} = \bigcap_{n\geq 0} (f^n)^{\leftarrow}[V]$$

and let $U_0 := \overline{U}$. Then U is easily seen to be invariant, so U_0 is invariant as well. Moreover, $U \subseteq V$, hence $U_0 \subseteq \overline{V}$, and it follows that U_0 is compact. In addition, as $A \subseteq U \subseteq U_0$ and $U_0 \subseteq \overline{V} \subseteq W_0$, we get, taking into account that $f^n[A] = A$ for all $n \in \mathbb{Z}^+$,

$$A \subseteq \bigcap_{n\geq 0} f^n[U_0] \subseteq \bigcap_{n\geq 0} f^n[W_0] = A\,.$$

So $A = \bigcap_{n\geq 0} f^n[U_0]$. Consequently, U_0 satisfies the conditions of Theorem 3.4.2 (iii), except possibly that it is a neighbourhood of A. In order to complete the proof it is sufficient to show that the set U defined above is open or, equivalently, that its complement

$$F := X \setminus U = \{\, x \in X : \mathcal{O}(x) \cap (X \setminus V) \neq \emptyset \,\}$$

is closed. To this end, consider a point $x \in \overline{F}$. We want to show that $x \in F$. For every $U \in \mathcal{N}_x$ there exists a point $x_U \in U \cap F$ and, by the definition of F, there is a non-negative integer $k(U)$ such that $f^{k(U)}(x_U) \notin V$. We may and shall assume that $k(U)$

is the smallest non-negative integer for which this is the case. This means: if $x_U \notin V$ then $k(U) = 0$ and if $x_U \in V$ then $f^i(x_U) \in V$ for $i = 0, \dots, k(U) - 1$.

We distinguish two cases. The first case is that there is a local base $\mathcal{B}$ at x such that the subset $\{k(B) : B \in \mathcal{B}\}$ of $\mathbb{Z}^+$ is bounded (hence finite). In this case there is a local base $\mathcal{B}'$ at x such that the mapping $k: U \mapsto k(U)$ is constant on $\mathcal{B}'$: one of the subsets $\{U \in \mathcal{B} : k(U) = n\}$ for n in the finite set of values of k on $\mathcal{B}$ must be cofinal in $\mathcal{B}$ (if the union of finitely many subsets of $\mathcal{B}$ is all of $\mathcal{B}$ then one of those sets is cofinal[4]), hence is a local base at x as well. Let N be the common value of $k(U)$ for $U \in \mathcal{B}'$. Then

$$f^N(x) = \lim_{U \in \mathcal{B}'} f^N(x_U) = \lim_{U \in \mathcal{B}'} f^{k(U)}(x_U) \in \overline{X \setminus V} = X \setminus V\,,$$

which means that $x \in F$, as desired.

Next, we consider the case that for every local base $\mathcal{B}$ at the point x the set $\{k(U) : U \in \mathcal{B}\}$ is not bounded. In particular, it follows that if $\mathcal{B}$ is any local base $\mathcal{B}$ at x then there is $W \in \mathcal{B}$ such that $k(U) \geq 2$ for every $U \in \mathcal{B}$ with $U \subseteq W$: otherwise k would be bounded by 2 on a suitable local base that is a subset of $\mathcal{B}$. Consequently, we may and shall assume that $k(U) \geq 2$ for all $U \in \mathcal{B}$. Then for every $U \in \mathcal{B}$ one has

$$f^{k(U)}(x_U) = f(f^{k(U)-1}(x_U)) \overset{*}{\in} f[V] \subseteq f[\overline{V}]\,,$$

where $\overset{*}{\in}$ is justified by the special choice of the numbers $k(U)$ for $U \in \mathcal{B}$. The set $f[\overline{V}]$ is the continuous image of a compact set, hence is compact as well. So the net $\{f^{k(U)}(x_U) : U \in \mathcal{B}\}$ is situated in a compact set and has, consequently, a cluster point z. By definition, this means that for every neighbourhood W of z there is a cofinal subset $\mathcal{B}_W$ of $\mathcal{B}$ (i.e., a local base at x) such that $f^{k(U)}(x_U) \in W$ for every $U \in \mathcal{B}_W$. By our assumption the set $\{k(U) : U \in \mathcal{B}_W\}$ is not bounded, hence for every $j \in \mathbb{Z}^+$ there exists $U_j \in \mathcal{B}_W$ such that $k(U_j) \geq j$. Then

$$f^{k(U_j)}(x_{U_j}) = f^j(f^{k(U_j)-j}(x_{U_j})) \overset{*}{\in} f^j[V] \subseteq f^j[\overline{V}]\,,$$

which implies that $W \cap f^j[\overline{V}] \neq \emptyset$. This holds for every neighbourhood W of z, so $z \in f^j[\overline{V}]$ (recall that $f^j[\overline{V}]$ is compact, hence closed). As this holds for every $j \in \mathbb{Z}^+$ we get $z \in \bigcap_{j\geq 0} f^j[\overline{V}] \subseteq \bigcap_{j \geq 0} f^j[W_0] = A$. On the other hand, z is a cluster point of a net in the closed set $X \setminus V$, which implies that $z \in X \setminus V$. This is a contradiction, as $A \subseteq V$. So the second case cannot occur. □

Remark. In the theorem above we may replace W_0 by its interior, so we may assume that W_0 is open. By a similar argument, we may assume that W_0 is compact, or is open with compact closure.

4 This is also correct if the local base $\mathcal{B}$ is finite, in which case the point x is isolated and $\{x\}$ is the final element of $\mathcal{B}$.

We now provide the examples and counter examples promised above.

Examples.

(1) The final statement in Theorem 3.4.1 may not hold if X is not locally compact.
Let X be the *hedgehog space* with countably many spines: the disjoint union of countably many copies I_n of the unit interval ($n \in \mathbb{N}$) with all left end points 0 of these intervals identified with each other to one single point P (for a more exact description, see [Eng], 4.1.5). The intervals I_n for $n \in \mathbb{N}$ are called the *spines* of X. Define $\rho\colon X \times X \to \mathbb{R}^+$ by

$$\rho(x,y) := \begin{cases} |x-y| & \text{if } x \text{ and } y \text{ belong to the same spine,} \\ |x| + |y| & \text{if } x \text{ and } y \text{ belong to different spines.} \end{cases}$$

This definition is unambiguous, since the only point belonging to more than one spine is the point P (which belongs to all spines). It is easily seen that ρ is a metric on X. The distance of two points on the same spine is their Euclidean distance in the unit interval; the distance of two points on different spines is measured via the point P. With the topology defined by this metric the space X is easily seen to be not compact. Similarly, no closed ball around the point P – which is a hedgehog space with short spines – is compact. Thus, P has no compact neighbourhoods.
Define $f\colon X \to X$ by $f(x) := (1 - \frac{1}{n})x$ if $x \in I_n$ ($n \in \mathbb{N}$) (again, this definition is unambiguous). Then P is an invariant point. Since $f^k(x) \rightsquigarrow P$ for $k \rightsquigarrow \infty$ ($x \in X$) and since every open ball around P is invariant, it follows that the invariant point P is asymptotically stable. Now suppose that the point P strongly attracts a neighbourhood of itself. Then there exists $\varepsilon > 0$ such that for every $\delta > 0$ there exists $k_\delta \in \mathbb{N}$ such that $f^k[B_\varepsilon(P)] \subseteq B_\delta(P)$ for all $k \geq k_\delta$. Fix $k \geq k_\delta$; then for every $n \in \mathbb{N}$ and for every point x of I_n with $|x| < \varepsilon$ one has $|f^k(x)| < \delta$. This implies that $(1 - \frac{1}{n})^k \varepsilon \leq \delta$. As this is supposed to hold for every n, this implies that $\varepsilon \leq \delta$, contradicting that δ should be allowed to be arbitrarily small.

(2) The implication (iv)⇒(ii), and therefore the implication (iii)⇒(ii), in Theorem 3.4.2 may not hold if X is not locally compact: if (3.4-1) holds for arbitrarily small (but non-compact) invariant neighbourhoods U of A then A need not be asymptotically stable, in which case it does not strongly attract a neighbourhood of itself.
Let X be the space l_2 of all square summable sequences $x = (x_i)_{i\in\mathbb{N}}$ or $x_1x_2x_3\dots$ with its usual metric, defined by

$$d(x,y) := \sqrt{\sum_{i=1}^{\infty} |x_i - y_i|^2} \quad \text{for } x = (x_i)_{i\in\mathbb{N}} \text{ and } y = (y_i)_{i\in\mathbb{N}} \text{ in } X\,.$$

Define $f\colon X \to X$ by $f(x) := 0x_1x_2\dots$ for $x = (x_i)_{i\in\mathbb{N}} \in X$. It is easily seen that f is an isometry, hence continuous. For every $\varepsilon > 0$ the open ε-ball $U_\varepsilon := B_\varepsilon(\mathbf{0})$ around the point $\mathbf{0} = 000\dots$ is invariant under f. So the invariant point $\mathbf{0}$ is stable. However, if $x \neq \mathbf{0}$ then $\mathbf{0} \notin \omega(x)$, so the point $\mathbf{0}$ is not asymptotically stable. On the

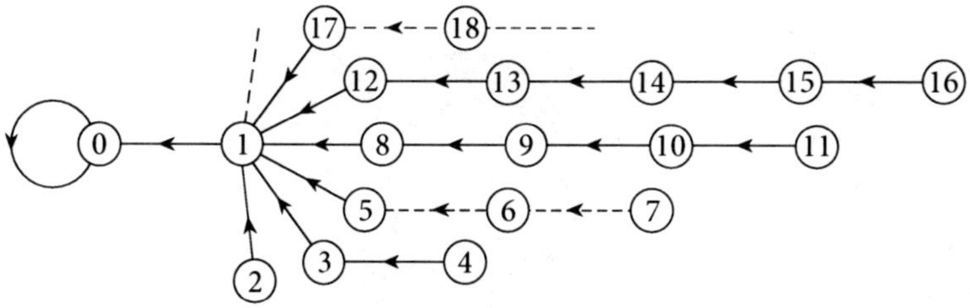

Fig. 3.7. For $A := \{0, 1\}$ one has $\bigcap_n f^n[X] = A$. Yet A is not completely invariant, hence not stable.

other hand, for every $\varepsilon > 0$ the neighbourhood U_ε of $\mathbf{0}$ satisfies condition (3.4–1*). (It follows that X is not locally compact. This is a well-known fact, usually proved as a consequence of the fact that l_2 as a vector space is infinite dimensional.)

NB. See also the final comment to Example (3) below.

(3) The following example shows that in Theorem 3.4.3 the condition expressed by formula (3.4-2) does not imply that A is completely invariant (not even if W_0 is invariant, but not compact, in which case A is invariant). Let $X := \mathbb{Z}^+$ with its usual (discrete) topology: a locally compact metric space. We shall define a function $f : X \to X$ such that the set $\bigcap_n f^n[X] = \{0, 1\}$ is not completely invariant. Consider the sequence $(n_k)_{k \in \mathbb{N}}$, given by $n_1 := 2$ and $n_{k+1} := n_k + k$ for $k \in \mathbb{N}$, and let f be defined by

$$f(j) := \begin{cases} j - 1 & \text{if } j \notin \{0, 1\} \cup \{n_k : k \in \mathbb{N}\}, \\ 0 & \text{if } j \in \{0, 1\}, \\ 1 & \text{if } j \in \{n_k : k \in \mathbb{N}\}. \end{cases}$$

See Figure 3.7. Thus, the set $\bigcap_n f^n[X]$ is not completely invariant (but it is topologically attracting and it has itself as an invariant neighbourhood).

NB. This example (with $U_0 := X$) also shows that compactness of U_0 is essential to get complete invariance of A in the implication (iii)⇒(ii) in Theorem 3.4.2.

Corollary 3.4.4. *Let X be an arbitrary Hausdorff space and let W be a compact subset of X with non-empty interior W°. If $f[W] \subseteq W^\circ$ then the set $A := \bigcap_{n \geq 0} f^n[W]$ is non-empty, compact and completely invariant; it strongly attracts the neighbourhood W of itself, hence it is asymptotically stable.*

Proof. Note that W is invariant. So by the implication (iii)⇒(ii) of Theorem 3.4.2 (where local compactness of the phase space is not used explicitly: see Remark 1 after that theorem) it is sufficient to show that W is a neighbourhood of A. This is obvious, because $A \subseteq f[W] \subseteq W^\circ$. □

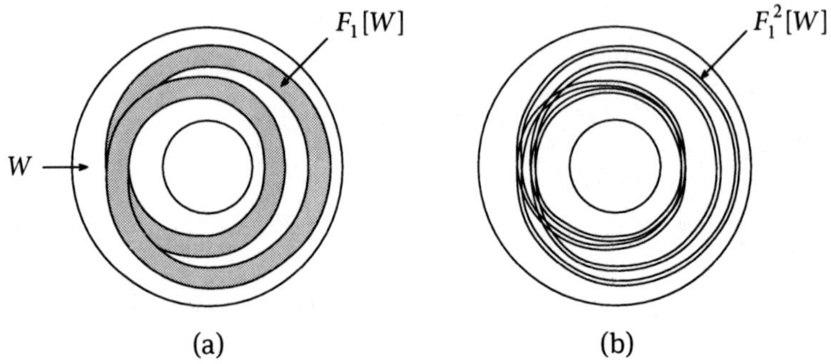

Fig. 3.8. The sets W, $F_1[W]$ and $F_1^2[W]$.

Remarks.

(1) Recall from Remark 4 after Theorem 3.4.2 that in this corollary the set A is the largest completely invariant subset of W.

(2) Compactness of W is essential in the above corollary: consider, e.g., the system $(\mathbb{R}, f)$ with $f(x) = x + 1$ for $x \in \mathbb{R}$, and let $W = [0; \infty)$.

3.4.5 (Application). Let for $0 \le \lambda \le 1$ the continuous mapping $F_\lambda : \mathbb{R} \to \mathbb{R}$ be given by $F_\lambda(r, \theta) = (\frac{3}{2} + \frac{1}{4}r + \frac{1}{2}\lambda\cos\theta, 2\theta)$ in polar coordinates, and W is the (compact) annulus $\{(r,\theta) \vdots 1 \le r \le 3\}$ in $\mathbb{R}^2$. The mapping F_0 contracts the set W radially onto the annulus $\{(r,\theta) \vdots 1\frac{3}{4} \le r \le 2\frac{1}{4}, 0 \le \theta \le 2\pi\}$, but it stretches W so that $F_0[W]$ winds itself twice around the origin. A similar behaviour occurs for $\lambda = 1$, except that $F_1[W]$ looks slightly different: see Figure 3.8 (a). If λ grows from 0 to 1 then $F_0[W]$ is deformed gradually into $F_1[W]$. It is clear that $F_\lambda[W] \subset W^\circ$ for every λ under consideration, hence W includes the asymptotically stable subset $A_\lambda = \bigcap_{n\ge0} F_\lambda^n[W]$. The set $F_0^n[W]$ is an annulus of width $2/4^n$ (winding itself 2^n times around the origin) and, consequently, A_0 is equal to the circle $\{(1,\theta) \vdots 0 \le \theta \le 2\pi\}$. For $\lambda = 1$ the situation is much more complicated. Figure 3.8 (b) gives a sketch of the set $F_1^2[W]$; we leave it for the reader to get an idea of the set $F_1^3[W]$. By extrapolating this idea one finds that, geometrically, A_1 must be a quite complicated 'curve' that winds itself infinitely often around the origin. Is is also dynamically complicated: the subsystem in this asymptotically stable set turns out to be sensitive to initial conditions (see Chapter 7 for the meaning of this statement; the proof of the statement is outside the scope of this book): it is a 'strange attractor'.

All asymptotically stable sets in locally compact systems are obtained as described in the preceding corollary:

Proposition 3.4.6. *Assume that X is a locally compact Hausdorff space. Then every asymptotically stable set A has arbitrarily small*[5] *compact neighbourhoods W such that $f[W] \subseteq W^\circ$ and $A = \bigcap_{n\ge0} f^n[W]$.*

5 So we may assume that $W \subseteq \mathcal{B}(A)$. See also the final statement in Theorem 3.4.2.

Proof. Let A be an asymptotically stable set and let U be an open neighbourhood of A. We want to show that U includes a compact neighbourhood of A satisfying the conditions of the theorem. We may assume that $\overline{U}$ is compact, that $\overline{U} \subseteq \mathcal{B}(A)$ and that condition (3.4–1*) holds.

Select an additional compact neighbourhood N of A such that $N \subset U$ (since A is stable, N can be chosen to be invariant, but we do not need that). By Appendix A.3.4, applied to the compact subspace $\overline{U}$ of X, there is a continuous function $h: \overline{U} \to [0; 1]$ such that[6] $h[N] = \{0\}$ and $h[\overline{U} \setminus U] = \{1\}$ (recall that U is open). We can extend h to all of X by defining $h(x) = 1$ for $x \in X \setminus \overline{U}$. It is easily seen that h is continuous at every point of X. Now define a mapping $G: \mathcal{B}(A) \to \mathbb{R}$ by

$$G(x) := \sum_{n=0}^{\infty} 2^n\, h(f^n(x)) \quad \text{for } x \in \mathcal{B}(A)\,.$$

Proposition 3.1.12 (1) implies that for every $x \in \mathcal{B}(A)$ one has $f^n(x) \in N$ for almost all n, so the series for $G(x)$ has only finitely many non-zero terms. It follows that G is well-defined on $\mathcal{B}(A)$. In order to prove that G is continuous we extend this argument in the following way: if $x \in \mathcal{B}(A)$ then there is a neighbourhood O_x of x with compact closure included in $\mathcal{B}(A)$. By the second part of Theorem 3.4.1 there exists $n(x) \in \mathbb{N}$ such that $f^n[O_x] \subseteq N$ for all $n > n(x)$. Consequently, $G(y) = \sum_{n=0}^{n(x)} 2^n h(f^n(y))$ for all $y \in O_x$: a finite sum with the same number of terms for all $y \in O_x$. As h is continuous, it follows that G is continuous on O_x. In particular, G is continuous at the point x. Therefore, G is continuous on $\mathcal{B}(A)$.

Observe that if $x \in A$ then $f^n(x) \in A$ for all $n \in \mathbb{N}$, because A is invariant. Hence $G(x) = 0$ for all $x \in A$. Moreover, $G(x) \geq h(x) = 1$ for every $x \in \mathcal{B}(A) \setminus U$. Consequently, $W := \{\, x \in \mathcal{B}(A) \,\vdots\, G(x) \leq 1/2 \,\}$ is a closed neighbourhood of A which is included in U. In particular, W is included in $\overline{U}$, hence W is compact. Finally, if $x \in \mathcal{B}(A)$ then

$$G(f(x)) = \sum_{n=0}^{\infty} 2^n h(f^{n+1}(x)) = \frac{1}{2} \sum_{n=1}^{\infty} 2^n h(f^n(x)) \leq \frac{1}{2}\, G(x)\,.$$

Using the fact that G is continuous it follows easily that $f[W] \subseteq W^\circ$.

Finally, A is completely invariant and $A \subseteq W$, hence $A = f^n[A] \subseteq f^n[W]$ for every $n \in \mathbb{Z}^+$, so $A \subseteq \bigcap_{n\geq 0} f^n[W]$. Conversely, U was chosen such that formula (3.4–1*) holds. Hence $\bigcap_{n\geq 0} f^n[W] \subseteq \bigcap_{n\geq 0} f^n[U] = A$. □

Remark. Example (1) after Theorem 3.4.3 shows that this proposition is not necessarily true if X is not locally compact, even if one omits the condition that W be compact: for every sufficiently small neighbourhood W of the point P one has $A = \bigcap_{n\geq 0} f^n[W]$ but never $f[W] \subseteq W^\circ$.

6 If U is clopen we may take $h(x) = 0$ for all $x \in U$ and $h(x) = 1$ for all $x \in X \setminus U$.

Corollary 3.4.7. *Let X be a locally compact metric space and let A be an asymptotically stable subset of X. Then for every neighbourhood U of A there exists a uniform neighbourhood $\mathcal{W}$ of f in the space of all continuous self-mappings of X such that for every $g \in \mathcal{W}$ the dynamical system (X, g) has an asymptotically stable subset included in U.*

Proof. By Proposition 3.4.6, A has a compact neighbourhood W such that $W \subseteq U$ and $f[W] \subseteq W^\circ$, i.e., $f[W] \cap (X \setminus W^\circ) = \emptyset$. Let ε be the distance of the compact set $f[W]$ to the closed set $X \setminus W^\circ$; then $\varepsilon > 0$. If $g\colon X \to X$ is a continuous mapping such that $d(f(x), g(x)) < \varepsilon$ for all $x \in X$ then it is easy to see that $g[W] \subseteq W^\circ$, so that, by Corollary 3.4.4, there is an asymptotically stable set for g included in W, hence in U. □

Remarks.

(1) In the above proof it is sufficient that $d(f(x), g(x)) < \varepsilon$ for $x \in W$ only. Thus, the approximation of f by g is only required to be good on a suitable compact neighbourhood of the asymptotically stable set.

(2) The metric in X is only needed for the definition of a 'uniform neighbourhood' of f. A reader familiar with uniform spaces will be able to interpret and prove the corollary for any uniform space whose underlying space has a locally compact Hausdorff topology.

3.4.8. Corollary 3.4.7 expresses a *robustness* property: a small perturbation of f does not cause a property of the system to disappear. In the present case one finds an asymptotically stable set in perturbed system 'close to' the asymptotically stable of the original system.

Robust properties are important for applications: if the mathematical model f is a sufficiently good approximation of the 'real' (but unknown) phase mapping g and the system (X, f) has a robust property, then the 'real' system has the property as well. In particular, if one finds an asymptotically stable set A for f then it is reasonable to expect that in a neighbourhood of A there is an asymptotically stable set for the 'real' system[7]. The corollary only states that the asymptotically stable set for g is not much larger than that for f (there is no 'explosion'). But it can be much smaller (i.e., an 'implosion' is possible).

Example. Let $f\colon x \mapsto x - \frac{1}{2}x(x-1)^2(x-2)\colon \mathbb{R} \to \mathbb{R}$. One easily shows that f maps the interval $W := [\frac{1}{2}; \frac{5}{2}]$ into its interior and that the interval $A_0 := [1; 2]$ is the asymptotically stable set whose existence is claimed by Corollary 3.4.4; this can also easily be verified by the methods explained in Section 2.1. For every $\lambda \in \mathbb{R}$, let $g_\lambda\colon x \mapsto f(x) + \lambda\colon \mathbb{R} \to \mathbb{R}$. If λ is close to 0 then the methods of Section 2.1 show that g_λ has an asymptotically stable set A_λ. If $\lambda < 0$ then A_λ is an interval with end points close to the end points of A_0 (no implosion). On the other hand, if $\lambda > 0$ then the set A_λ consists of one invariant point only, close to the point 2.

7 As g is unknown one cannot really know whether the approximation is good enough.

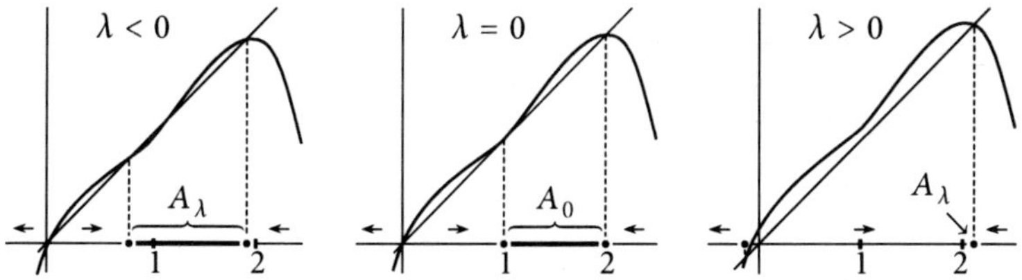

Fig. 3.9. For $\lambda > 0$ the asymptotically stable set A_λ shrinks to a point.

3.5 The structure of (asymptotically) stable sets

Even in cases where an asymptotically stable set occupies only a 'negligible' part of its basin – either topologically (e.g., a nowhere dense set) or in any other relevant sense (e.g., it is a null-set with respect to some invariant measure) – the system's behaviour in that basin is heavily influenced by the behaviour within the asymptotically stable set. In particular, if the latter is complicated then so is the former. To have a better understanding of the system it is, consequently, necessary to study the structure of the system inside its asymptotically stable sets.

One possibility for the internal structure of an asymptotically stable set A is that it contains a proper subset B that is asymptotically stable in the subsystem on A. In that case, if the phase space is locally compact, B will shown to be asymptotically stable in the full system. Moreover, we study the components of (asymptotically) stable sets in locally connected spaces (that are locally compact). The study of asymptotically stable sets will be continued in Section 4.5 ahead.

Lemma 3.5.1. *Let X be a locally compact space, let A be a stable subset of X and let B be an asymptotically stable set in the subsystem on A. Then the set B is stable in the full system (X, f).*

Proof. As the set A is stable under f, it follows from Lemma 3.3.1 that every neighbourhood of A includes an invariant neighbourhood of A. This means that the set $\mathcal{W}_A$ of all invariant neighbourhoods is cofinal in the set $\mathcal{N}_A$ of all neighbourhoods of A (recall that $\mathcal{N}_A$ is a directed set with respect to the usual ordering by inclusion; see the first example in Section A.4 of Appendix A for details). In particular, it follows from Appendix A.2.1 that $\bigcap\{V : V \in \mathcal{W}_A\} = A$.

By Theorem 3.4.1[8], the set B strongly attracts a neighbourhood of itself in A. This neighbourhood is of the form $A \cap W_0$ for some neighbourhood W_0 of B in X. Obviously, B strongly attracts every subset of $A \cap W_0$. In particular, if W is a neighbourhood of B in X such that $W \subseteq W_0$ then B strongly attracts $A \cap W$. Hence the neighbourhood $A \cap W^\circ$ of B in A includes the set $f^n[A \cap W]$ for almost all n, i.e., there exists $p_W \in \mathbb{N}$

8 This is the only place where we use that B is *asymptotically* stable.

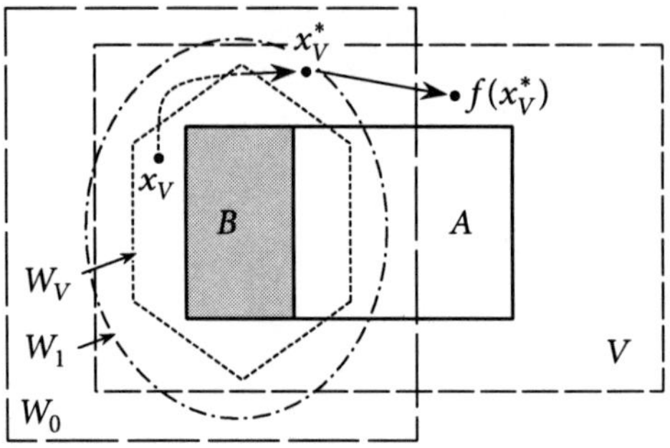

Fig. 3.10. Illustrating the proof of Lemma 3.5.1.

such that

$$f^n[A \cap W] \subseteq A \cap W^\circ \quad \text{for all } n \geq p_W \tag{3.5-1}$$

(here W° is the interior of W in X, so the set $A \cap W^\circ$ is a neighbourhood of B in A).

Assume that the set B is not stable in X under f. Then there exists a neighbourhood W_1 of B in X with the property that every neighbourhood of B in X contains a point x such that $\mathcal{O}_f(x) \nsubseteq W_1$. Since every neighbourhood of B included in W_1 has this property as well, we may assume from the outset that W_1 is compact, hence closed, and that $W_1 \subseteq W_0$. Note that this implies that (3.5-1) holds with $W = W_1$. Let $p := p_{W_1}$; then (3.5-1) implies

$$f^p[A \cap W_1] \subseteq A \cap W_1^\circ . \tag{3.5-1*}$$

For every $V \in \mathcal{W}_A$, put $W_V := V \cap \bigcap_{n=0}^{p} (f^n)^{\leftarrow}[W_1]$. Then W_V is a neighbourhood of B in X, so the criterion used for the choice of W_1 implies that there is a point $x_V \in W_V$ such that $\mathcal{O}_f(x_V) \nsubseteq W_1$. Let i be the first element of $\mathbb{Z}^+$ such that $f^i(x) \notin W_1$. Since $f^j(x_V) \in W_1$ for $0 \leq j \leq i-1$, it follows from the definition of W_V that $i > p$. So we can define $x_V^* := f^{i-p}(x_V)$; then $x_V^* \in W_1$ and $f^p(x_V^*) \notin W_1$. Moreover, $x_V \in V$ and V is invariant under f, so both x_V^* and $f^p(x_V^*)$ belong to V.

The net $\{x_V^*\}_{V \in \mathcal{W}_A}$ is situated in the compact set W_1, hence it has a cluster point $z \in W_1$. Because $\bigcap\{V : V \in \mathcal{W}_A\} = A$ and $x_V^* \in V$ for every $V \in \mathcal{W}_A$, it should be clear that $z \in A$, hence $z \in A \cap W_1$. Consequently, (3.5-1) implies that $f^p(z) \in A \cap W_1^\circ$. However, continuity of f implies that $f^p(z)$ is a cluster point of the net $\{f^p(x_V^*)\}_{V \in \mathcal{W}_A}$, and because $\{f^p(x_V^*)\}_{V \in \mathcal{W}_A}$ is a net in $X \setminus W_1$, it follows that $f^p(z) \in \overline{X \setminus W_1} = X \setminus W_1^\circ$. As $f^p(z) \in A \cap W_1^\circ$, this is impossible. So B is stable under f. □

Theorem 3.5.2. *Let X be a locally compact space, let A be an asymptotically stable subset of X and let B be an asymptotically stable set in the subsystem on A. Then B is asymptotically stable in the full system (X, f).*

Proof. By the previous lemma, B is stable in X under f, so it remains to show that B is topologically attracting in the full system on X.

As B is topologically attracting in the subsystem on A, $\mathcal{B}_{f|_A}(B)$ is an open neighbourhood of B in A, so there is a neighbourhood U of B in X such that $A \cap U \subseteq \mathcal{B}_{f|_A}(B)$.

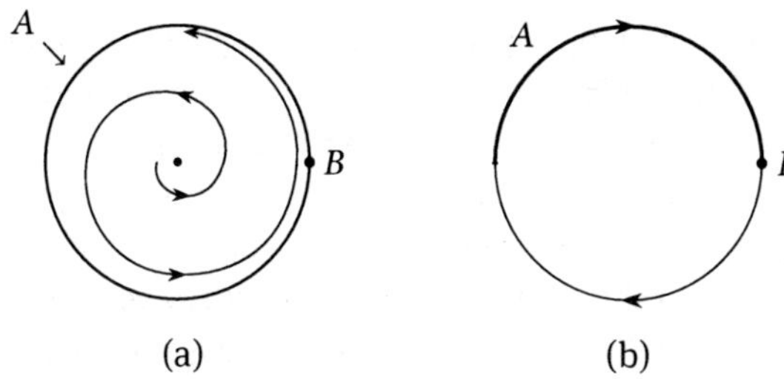

Fig. 3.11. (a) A is stable in X and B is stable in A, but not in X. (b) A is topologically attracting in X and B is so in A, but not in X.

Moreover, we may assume that $U \subseteq \mathcal{B}_f(A)$, for $\mathcal{B}_f(A)$ is open in X and includes B. Without limitation of generality we may assume that U is closed. By the definition of stability in Section 3.5, B has a neighbourhood V in X such that $\mathcal{O}_f(x) \subseteq U$ for every point $x \in V$; in particular, $V \subseteq U$. We shall show that $V \subseteq \mathcal{B}_f(B)$, which will imply that B is topologically attracting in X under f.

Consider any point $x \in V$. Then $x \in \mathcal{B}_f(A)$, so $\emptyset \neq \omega_f(x) \subseteq A$. On the other hand, one has $\omega_f(x) \subseteq \overline{\mathcal{O}_f(x)} \subseteq U$. Consequently, $\omega_f(x) \subseteq A \cap U \subseteq \mathcal{B}_{f|_A}(B)$. As $\omega_f(x)$ is an $f|_A$-invariant subset of A this implies, in view of Lemma 3.1.3 (4), that $\omega_f(x) \cap B \neq \emptyset$. Since B is stable, Corollary 3.2.2 (2) now implies that $\omega_f(x) \subseteq B$, which means that $x \in \mathcal{B}_f(B)$. This concludes the proof that $V \subseteq \mathcal{B}_f(B)$. □

The counterparts of the above theorem for stability without topological attraction, or for topological attraction without stability, are not true, not even in compact spaces with a phase mapping that is a homeomorphism:

Examples.

(1) Let X be the closed unit disk D_2 in the complex plane and define the mapping f by $f([r,t]) := [\sqrt{r}, t + (1-r)]$ (recall that $[r,t] = r\cos 2\pi t + i \sin 2\pi t$). Then $A := \mathbb{S}$ is an asymptotically stable set in X and $B := \{[1,0]\}$ is stable in the subsystem on A because all points of A are invariant. But B is not stable in X. See Figure 3.11 (a).

(2) Consider the system $(\mathbb{S}, f^2)$ defined in the example after Theorem 3.3.9. Let A be the closed upper half of $\mathbb{S}$, $A := \{[t] \,\vdots\, 0 \leq t \leq 1/2\}$, and let $B := \{[0]\}$. Then A is topologically attracting in $\mathbb{S}$ and B is topologically attracting in A, but B is not topologically attracting in $\mathbb{S}$. See Figure 3.11 (b).

As explained in Note 5 at the end of this chapter, asymptotically stable sets that have no proper asymptotically stable subsets are the candidates for what might be called 'attractors'. We have shown in a Theorem 3.2.8 that *transitive* asymptotically stable sets have this property. Therefore, in the remainder of this chapter we consider transitive (asymptotically) stable sets and, in particular, we shall study the components of such sets. (Note that transitive stable and asymptotically stable sets need not be connected; a quick example: an asymptotically stable periodic orbit with primitive period at least 2.) Recall the notation agreed on in 1.5.8: if A is an invariant subset of X then R will denote the equivalence relation on A defined by the partition of A into its components, $q: A \to A/R$ will denote the quotient mapping, and $\tilde{f}: A/R \to A/R$ will denote the continuous mapping induced by $f|_A$, for which the relation $q \circ f|_A = \tilde{f} \circ q$ holds.

By Theorem 1.5.9 there are two possibilities for the components of a compact *transitive* subset A: (1) A has only finitely many components, and these are cyclically permuted by f, and (2) the space A/R is an infinite 0-dimensional compact Hausdorff space which is transitive under $\tilde{f}$. Our next theorem states that in a locally compact, locally connected space only the first possibility applies to a transitive asymptotically stable set.

Lemma 3.5.3. *Let X be a locally connected space, let A be a compact subset of X and let U be a neighbourhood of A. Then U includes the union of finitely many mutually disjoint sets $V_1, \ldots, V_n$ such that for $1 \leq i \leq n$ one has $A_i := A \cap V_i \neq \emptyset$ and V_i is a connected neighbourhood of A_i.*

Proof. By local connectedness of X, every point of A has a connected neighbourhood that is included in the interior of U. The compact set A can be covered by finitely many of such neighbourhoods. Thus, we get a finite collection $\mathcal{C}$ of connected sets with nonempty interiors, each of which meets A; moreover, $V := \bigcup \mathcal{C}$ is a neighbourhood of A included in U.

Select any member W of $\mathcal{C}$ and let $\mathcal{C}_1$ be the collection of all members W' of $\mathcal{C}$ for which there is a finite chain $W = W_1, W_2, \ldots, W_k = W'$ in $\mathcal{C}$ such that $W_i \cap W_{i+1} \neq \emptyset$ for $i = 1, \ldots k - 1$. Then $V_1 := \bigcup \mathcal{C}_1$ is connected and V_1 is disjoint from all members of $\mathcal{C} \setminus \mathcal{C}_1$ (if any). Repeat this procedure with the collection $\mathcal{C} \setminus \mathcal{C}_1$ instead of $\mathcal{C}$ (assuming it is not empty, otherwise we are finished): we get a subfamily $\mathcal{C}_2$ of $\mathcal{C}$ such that the set $V_2 := \bigcup \mathcal{C}_2$ is connected, disjoint from V_1 and disjoint from all members of $\mathcal{C} \setminus (\mathcal{C}_1 \cup \mathcal{C}_2)$. If the family $\mathcal{C} \setminus (\mathcal{C}_1 \cup \mathcal{C}_2)$ is not empty we can repeat the above procedure with this family, etc. After finitely many steps $\mathcal{C}$ is exhausted and we end up with a finite number of mutually disjoint connected sets $V_1, \ldots, V_n$ such that $A \subseteq V = V_1 \cup \ldots \cup V_n \subseteq U$.

To complete the proof, observe that each of the sets V_i ($i = 1, \ldots, n$) meets A, because every member of $\mathcal{C}$ is supposed to meet A. Moreover, V_i is a neighbourhood of the set $A_i := A \cap V_i$. The reason is that every point of A is an interior point of one of the members of the full collection $\mathcal{C}$, so that every element of A_i is an interior point of one of the members of the subfamily $\mathcal{C}_i$ (and not of a member of $\mathcal{C}_j$ with $j \neq i$, because those are disjoint from A_i), hence an interior point of V_i. □

Theorem 3.5.4. *Let X be a locally compact, locally connected space and let A be an asymptotically stable set in X. Then A has only finitely many connected components.*

Proof. Let U_0 be a neighbourhood of A with a compact closure included in $\mathcal{B}(A)$. By Lemma 3.5.3, A has a neighbourhood V such that $V \subseteq U_0$ and $V = V_1 \cup \cdots \cup V_n$ for mutually disjoint connected sets V_i. Since V is a neighbourhood of A with compact closure included in the closure of U_0, hence included in $\mathcal{B}(A)$, we can apply Theorem 3.4.1 to the effect that V is strongly attracted by A.

We shall show now that A has at most n components. Assume the contrary: A/R has at least $n+1$ points. Since A/R is not connected (by Appendix A.6.3 it is even 0-dimensional), it splits up into two disjoint non-empty closed sets. These sets, in turn, are not connected (except when a singleton), hence at least one of them splits up into two disjoint non-empty closed sets. This procedure can at least be repeated until we have $n+1$ mutually disjoint non-empty closed subsets of A/R whose union is equal to A/R. The complete pre-images of these sets under the quotient map $q\colon A \to A/R$ form a collection of $n+1$ mutually disjoint non-empty closed subsets $F_1, \dots, F_{n+1}$ of A such that $A = F_1 \cup \dots \cup F_{n+1}$. These sets form a disjoint collection of compact subsets of the Hausdorff space X, hence they have mutually disjoint open neighbourhoods in X. Thus, there are mutually disjoint open sets U_j in X such that $F_j \subseteq U_j$ ($j = 1, \dots, n+1$).

The union $U := U_1 \cup \dots \cup U_{n+1}$ is a neighbourhood of A. Since V is strongly attracted by A, there exists $k \in \mathbb{N}$ such that $f^k[V] \subseteq U$. In particular, for each $i \in \{1, \dots, n\}$ we have $f^k[V_i] \subseteq U = U_1 \cup \dots \cup U_{n+1}$. The set $f^k[V_i]$, being the continuous image of a connected set, is connected. As the sets U_j for $j = 1, \dots, n+1$ are open and mutually disjoint, it follows that every set $f^k[V_i]$ has to be completely included in one of the sets U_j: there exists a unique $j(i) \in \{1, \dots, n+1\}$ such that $f^k[V_i] \subseteq U_{j(i)}$. In addition, each of the sets U_j turns out to include (at least) one of the sets $f^k[V_i]$. This is because f^k maps A onto itself – A is completely invariant – so that for any $x \in U_j \cap A$ there exists $y \in A$ with $x = f^k(y)$. If i is such that $y \in V_i$ then $f^k[V_i] \cap U_j \neq \emptyset$, hence $f^k[V_i] \subseteq U_j$, that is, $j = j(i)$.

Consequently, we have obtained a surjection $i \mapsto j(i)$ from the n-element set $\{1, \dots, n\}$ onto the $(n+1)$-element set $\{1, \dots, n+1\}$. This is impossible, which shows that A has at most n components. □

Corollary 3.5.5. *Let X be a locally compact, locally connected space and let A be an asymptotically stable set in X. If A is transitive then A has finitely many connected components which are cyclically permuted by f.*

Proof. Clear from Theorem 1.5.9 – see also the Remark after Theorem 1.5.9 – and Theorem 3.5.4 above. □

In the above situation, A/R consists of a single periodic orbit. If A is not asymptotically stable, but just stable, then A/R turns out to be at least minimal under $\tilde{f}$. Local compactness of X is not needed for this result.

Theorem 3.5.6. *Let X be locally connected and let A be a stable subset of X. If A is transitive then the system $(A/R, \tilde{f})$ is minimal.*

Proof. If A/R is finite then Theorem 1.5.9 implies that it is a single periodic orbit (hence minimal) under $\tilde{f}$. So it remains to consider the case that A/R is infinite (but the following proof also applies if A/R is finite).

Let F be a non-empty closed proper subset of A/R and let z be a point of A/R not in F. By Theorem 1.5.9, the space A/R is 0-dimensional, hence the point z has a clopen

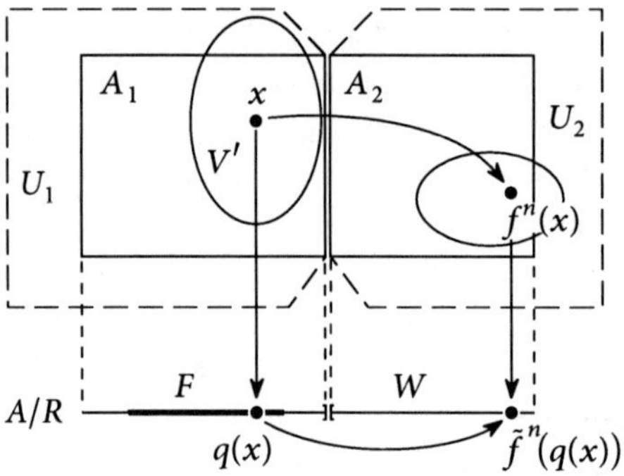

Fig. 3.12. Illustrating the proof of Theorem 3.5.6. There exists $n \in \mathbb{N}$ such that f^n maps the set V', hence the point x, into U_2.

neighbourhood W in A/R which is disjoint from F. Then $A_1 := q^{\leftarrow}[(A/R) \setminus W]$ and $A_2 := q^{\leftarrow}[W]$ are mutually disjoint clopen subsets of A. In particular, A_1 and A_2 are mutually disjoint compact subsets of X, so they have disjoint open neighbourhoods U_1 and U_2, respectively. Then $U := U_1 \cup U_2$ is a neighbourhood of A in X, so by stability of A there exists a neighbourhood V of A such that $f^n[V] \subseteq U$ for every $n \geq 0$.

Consider an arbitrary point x in $q^{\leftarrow}[F]$ and let V' be a connected neighbourhood of x in X such that $V' \subseteq V$. Then $f^n[V'] \subseteq U = U_1 \cup U_2$ for every $n \in \mathbb{Z}^+$. Since the set $f^n[V']$ is connected and the sets U_1 and U_2 are mutually disjoint and open, it follows that for every $n \geq 0$ we have either $f^n[V'] \subseteq U_1$ or $f^n[V'] \subseteq U_2$. In particular, the case for $n = 0$ implies that $V' \subseteq U_1$, because $x \in q^{\leftarrow}[F] \subseteq A_1 \subseteq U_1$.

Recall from Theorem 1.3.5 that the transitive system $(A, f|_A)$ is topologically ergodic. So there exists $n \in \mathbb{Z}^+$ such that $f^n[V' \cap A] \cap (U_2 \cap A) \neq \emptyset$, that is, $f^n[V'] \cap U_2 \neq \emptyset$. By what was observed in the preceding paragraph, this implies that $f^n[V'] \subseteq U_2$ and, consequently, $f^n(x) \in U_2$. Since $f^n(x) \in A$ as well, this implies that $f^n(x) \in A_2$. Now apply q to this result in order to get $\tilde{f}^n(q(x)) \in q[A_2] = W$. Since W is disjoint from F and $q(x) \in F$ by the choice of x, it follows that F is not invariant under $\tilde{f}$. This shows that the space A/R has no closed $\tilde{f}$-invariant proper subsets. □

Corollary 3.5.7. *If a transitive compact set in a locally connected space has infinitely many components and it includes a periodic orbit then it is not stable.*

Proof. Assume the contrary: then $(A/R, \tilde{f})$ is minimal and includes a periodic orbit (the image under q of the periodic orbit in A). So A/R would be finite. □

Corollary 3.5.8. *Let (X, f) be a dynamical system on an interval in $\mathbb{R}$ and let A be a compact transitive stable subset of X. Then there are the following, mutually exclusive, possibilities:*

(1a) *A is a finite periodic orbit.*

(1b) *A is the union of finitely many non-degenerate closed intervals that are cyclically permuted by f.*

(2) *A is a Cantor space, minimal under f.*

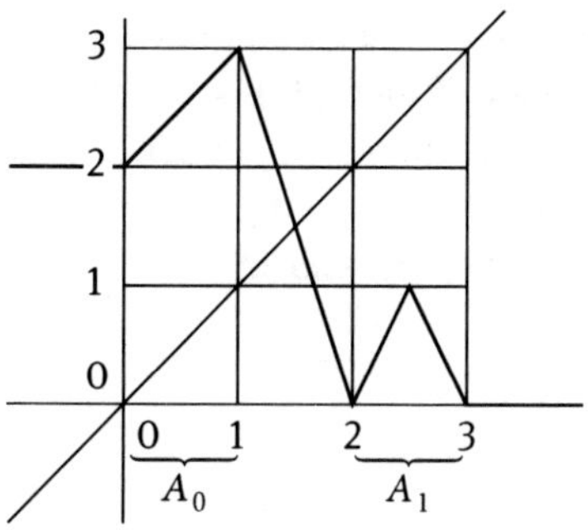

Fig. 3.13. Graph of the function f considered in Example (3).

Proof. This is Corollary 1.5.12 with in Case (2) minimality of A under f added – recall from the proof of Corollary 1.5.12 that in this case $(A, f|_A)$ is conjugate to $(A/R, \tilde{f})$, which is a minimal system by Theorem 3.5.6. □

Corollary 3.5.9. *Let X be locally connected and let A be a transitive stable subset of X. Then for every point x in $\mathcal{B}(A)$ the limit set $\omega(x)$ meets every component of A.*

Proof. If $x \in \mathcal{B}(A)$ then $\omega(x)$ is a non-empty compact invariant subset of A, so minimality of A/R under $\tilde{f}$ – which follows from Theorem 3.5.6 – implies that $q[\omega(x)] = A/R$. This is equivalent to the statement that $\omega(x)$ meets every component of A. □

Examples.

(1) Every (asymptotically) stable periodic orbit in a dynamical system on an interval provides an example of possibility (1a) of Corollary 3.5.8.

(2) in 3.3.19 we have a dynamical system $([0; 1], f_\infty)$ in which the Cantor set C occurs as a stable and transitive invariant subset. However, the proof that C is transitive under the mapping f_∞ will be postponed to 4.2.10 ahead, where it will be shown that C is minimal under f_∞ – in accordance with Case (2) of Corollary 3.5.8.

(3) Let $f : \mathbb{R} \to \mathbb{R}$ be defined by the following equations:

$$f(x) := \begin{cases} 2 & \text{for } x \leq 0 \\ x + 2 & \text{for } 0 \leq x \leq 1 \\ -3x + 6 & \text{for } 1 \leq x \leq 2 \\ T(x-2) & \text{for } 2 \leq x \leq 3 \\ 0 & \text{for } x \geq 3 \end{cases} \tag{3.1}$$

where T is the tent map. See Figure 3.13.
Let $A_0 := [0; 1]$, $A_1 := [2; 3]$ and $A := A_0 \cup A_1$. Then A is easily seen to be asymptotically stable, either by straightforward computation of limit sets of points of $\mathbb{R}$ or by application of Theorem 3.4.2, taking into account that f maps a neighbourhood of the form $[-\varepsilon; \frac{4}{3}] \cup [\frac{5}{3}; 3 + \varepsilon]$ of A onto A. In order to show that f is transitive on A, first observe that $f^2|_{A_0} = T$, the tent map, so A_0 is transitive under f^2. Moreover,

it is easy to check that $f|_{A_0}: x \mapsto x + 2: A_0 \to A_1$ is a conjugation from (A_0, f^2) onto (A_1, f^2). So if x is a transitive point in A_0 under f^2 then $f(x)$ is a transitive point of A_1 under f^2. It follows easily that x is transitive in $A_0 \cup A_1$ under f. So A is a transitive asymptotically stable set. Its components A_0 and A_1 are cyclically permuted by f, which is in accordance with Corollaries 3.5.5 and 3.5.8 (1b).

Exercises

3.1.
(1) Let Y be an invariant subset of X and let $y \in Y$. Then $\omega_{f|_Y}(y) = Y \cap \omega_f(y)$. If, in addition, Y is closed then $\omega_f(y) \subseteq Y$, hence $\omega_{f|_Y}(y) = \omega_f(y)$.
(2) Let $n \in \mathbb{N}$ and $x \in X$. Assume that $\overline{\mathcal{O}(x)}$ is compact or that f is a homeomorphism. If $\omega_f(x)$ is infinite then so is $\omega_{f^n}(f^i(x))$ for $i = 0, \dots, n-1$.
(3) Let X be locally compact. If $x \in X$ and $\omega(x)$ is finite then $\omega(x)$ consists of one single periodic orbit.

3.2. Let $x \in X$. If $\overline{\mathcal{O}(x)}$ is compact or if f is a homeomorphism, then $\omega(x)$ is completely invariant. In particular, this is the case if X is locally compact and $\omega(x)$ is non-empty and compact. Show (by an example) that this conclusion may not hold if $\overline{\mathcal{O}(x)}$ is not compact and f is not a homeomorphism.

3.3. Assume that X be a metric space with metric d.
(1) If $x \in X$ and $\overline{\mathcal{O}(x)}$ is compact then $\omega(x)$ is the smallest non-empty closed subset K of X with the property that $\lim_{n \rightsquigarrow \infty} d(f^n(x), K) = 0$.
(2) Let X be locally compact and let x_0 be a periodic point in X with primitive period p. If $x \in \mathcal{B}(\mathcal{O}(x_0))$ then there exists a point $x_i \in \mathcal{O}(x_0)$ such that $\lim_{n \rightsquigarrow \infty} d(f^n(x), f^n(x_i)) = 0$. (Points x and x_i with this property are said to be *asymptotic to each other*; see Section 7.3.)

3.4. Show that Lemma 3.3.11 (1),(2) and Theorem 3.3.12 also hold if f is a homeomorphism. Actually, in that case (notation is as in Section 3.3):
(a) If A is a non-empty completely invariant compact set then $\mathcal{B}_f(A) = \mathcal{B}_g(A)$.
(b) $\mathcal{B}_g(x_0) = (f^i)^{\leftarrow}[\mathcal{B}_g(x_i)]$ and $\mathcal{B}_g(x_i) = (f^{p-i})^{\leftarrow}[\mathcal{B}_g(x_0)]$.

3.5. Let X be an interval in $\mathbb{R}$ and let $f: X \to X$ be continuously differentiable. Recall that an f-invariant point $z \in X$ is said to be attracting whenever $|f'(z)| < 1$. Let $x_0 \in X$ be a periodic point with primitive period p. Show that the following conditions are equivalent:
(i) All points of $\mathcal{O}(x_0)$ are attracting under f^p.
(ii) Some point of the orbit of x_0 is attracting under f^p.

3.6.

(1) Prove that a non-empty A satisfying condition (3.2-1) is necessarily invariant.

(2) Let X be locally compact and let A be a stable subset of X which is not asymptotically stable. Show that for every neighbourhood U of A there is $x \in U$ such that $\emptyset \neq \omega(x) \subseteq U \setminus A$. Thus, limit sets disjoint from A accumulate to A.

3.7.

(1) Let Y be an invariant subset of X and let $A \subseteq Y$. Assume that A is compact and completely invariant. Prove: if A is stable or topologically attracting in (X, f) then A is stable or topologically attracting, respectively, in $(Y, f|_Y)$.
The converse is not generally true (give an example), but it is true if Y has non-empty interior Y° in X and $A \subseteq Y^\circ$. See also Theorem 3.5.2.

(2) Prove the following converse of the final statement of Theorem 3.4.2: if U is a compact invariant nbd of a non-empty set A such that (3.4-1) holds then $U \subseteq \mathcal{B}(A)$.

3.8. The *limit set* of a non-empty subset Y of X is the subset

$$\omega_f(Y) := \bigcap_{n=0}^{\infty} \overline{\bigcup_{m \geq n} f^m[Y]}$$

of X. If f is understood we write $\omega(Y)$ instead of $\omega_f(Y)$. For every $x \in X$ we have $\omega(\{x\}) = \omega(\mathcal{O}(x)) = \omega(y)$ as defined in Section 1.4. Prove the following statements:

(1) $\omega(Y)$ is closed in X and invariant under f.

(2) If Y is invariant then $\omega(Y) = \bigcap_{n=0}^{\infty} \overline{f^n[Y]}$; if Y is closed and completely invariant then $\omega(Y) = Y$. If Y is compact and invariant then $\omega(Y) = \bigcap_{n=0}^{\infty} f^n[Y] \neq \emptyset$.

(3) $\omega(\bigcup_{m \geq 0} f^m[Y]) = \omega(Y)$.

(4) $\omega(Y) = \omega(\overline{Y})$

(5) Let $\widetilde{Y} := \overline{\bigcup_{m \geq 0} f^m[Y]} = \overline{\bigcup_{y \in Y} \mathcal{O}(y)}$. Then $\omega(\widetilde{Y}) = \omega(Y)$.

(6) If $\widetilde{Y}$ is compact then $\omega(Y) \neq \emptyset$ and $\omega(Y)$ is the largest completely invariant subset of $\widetilde{Y}$. In particular, if Y is invariant and compact then $\omega(Y)$ is the largest completely invariant non-empty subset of Y.

NB. If X is a compact metric space and Y is a closed subset of X then $\omega(Y)$ is the limit of the sequence $\left(\overline{\bigcup_{m \geq n} f^m[Y]}\right)_{n \in \mathbb{Z}^+}$ in the hyperspace of X. See (A.5) and (A.6)7 in [deV].

3.9. Let X be a locally compact space and let A be a non-empty compact subset of X. Then the following conditions are equivalent:

(i) A is asymptotically stable.

(ii) $A = \omega(W)$ for an open neighbourhood W of A such that $\widetilde{W}$ is compact.

3.10.

(1) Let A_1 and A_2 be compact invariant subsets of X and assume that $A_1 \cap A_2 \neq \emptyset$. Then $\mathcal{B}(A_1 \cap A_2) = \mathcal{B}(A_1) \cap \mathcal{B}(A_2)$. Hence, if both A_1 and A_2 are topologically attracting then so is $A_1 \cap A_2$. If both A_1 and A_2 are stable or asymptotically stable then the set $A_1 \cap A_2$ has the same property iff it is completely invariant.

(2) For $i = 1, 2$ define $\alpha_i : [0; 1] \to \mathbb{R}^2$ by

$$\alpha_i(t) := \begin{cases} (\frac{1}{2}, t) & \text{for } 0 \le t \le \frac{1}{2}, \\ ((2-i) + (2i-3)t, t) & \text{for } \frac{1}{2} \le t \le 1, \end{cases}$$

let $A_i := \alpha_i[[0; 1]]$, $Y := A_1 \cup A_2$, and define $f: Y \to Y$ by $f(\alpha_i(t)) := \alpha_i(t^2)$ for $0 \le t \le 1$ and $i = 1, 2$. Then A_1 and A_2 are asymptotically stable sets in the dynamical system (Y, f), but $A_1 \cap A_2$ is not completely invariant.

(3) Let $(A_n)_{n \in \mathbb{N}}$ be a descending sequence of non-empty compact completely invariant sets. Then $A := \bigcap_{n \in \mathbb{N}} A_n$ is completely invariant. If each A_n is topologically attracting then A need not be topologically attracting: give an example.

3.11. Let X be a compact Hausdorff space and let $A := \bigcap_{i=0}^{\infty} f^i[X]$. Show that A is asymptotically stable: it is the largest asymptotically stable subset of X.

3.12.
(1) Let A be a subset of X such that $f[A] \supseteq A$. If A strongly attracts a neighbourhood of itself (or any superset of A) then A is invariant (hence completely invariant).
(2) Let X be a locally compact Hausdorff space and let A be a non-empty compact subset of X that strongly attracts a neighbourhood of itself. Show that A includes an asymptotically stable set.
(3) If X is a 2nd countable locally compact space then the collection $\mathcal{A}$ of all asymptotically stable subsets of X is countable.

3.13.
(1) Let X be locally compact and locally connected and let A be an asymptotically stable subset of X. Show that if A is minimal then A is totally minimal iff A is connected.
(2) Prove: Let X be locally connected and let A be a transitive stable subset of X. If A contains a periodic point with primitive period p then A has finitely many connected components, and the number of components of A divides p.

Notes

1 If $\overline{\mathcal{O}(x)}$ is compact then the point x is said to be (positively) *stable in the sense of Lagrange*, or (positively) *Lagrange stable*. This really is a stability-concept: the limit set $\omega(x)$ of a Lagrange stable point x is a non-empty compact set, so the points $f^n(x)$ 'tend to concentrate in some limited area' for $n \rightsquigarrow \infty$; see Theorem 3.1.9 or Exercise 3.3 (1).

The reason that that this is called 'stability' is, that in a complete metric space for every $\varepsilon > 0$ the complete orbit of a Lagrange stable point is determined, up to accuracy ε, by a finite initial segment of that orbit:

Theorem. *Let (X, f) be a dynamical system, let $x_0 \in X$ and assume that X is a complete metric space. Then $\overline{\mathcal{O}(x_0)}$ is compact iff the following condition is fulfilled:*

$$\forall \varepsilon > 0\ \exists N_\varepsilon \in \mathbb{N} : \mathcal{O}(x_0) \subseteq B_\varepsilon(\{ f^n(x_0) : 0 \le n \le N_\varepsilon\}). \qquad (*)$$

Proof. "If": Assume that $(*)$ holds. By Theorem A.7.5, we have to show that $\overline{\mathcal{O}(x_0)}$ is totally bounded. For every $N \in \mathbb{N}$, let $\mathcal{O}(x_0, N) := \{ f^n(x_0) : 0 \le n \le N\}$. If $x \in \overline{\mathcal{O}(x_0)}$ and $\varepsilon > 0$ then $B_\varepsilon(x) \cap \mathcal{O}(x_0) \ne \emptyset$, hence by condition $(*)$, $B_{2\varepsilon}(x) \cap \mathcal{O}(x_0, N_\varepsilon) \ne \emptyset$. Consequently, $x \in \bigcup\{B_{2\varepsilon}(y) : y \in \mathcal{O}(x_0, N_\varepsilon)\}$. This shows that $\overline{\mathcal{O}(x_0)}$ is covered by finitely many 2ε-balls. Since ε is arbitrary, it follows that $\overline{\mathcal{O}(x_0)}$ is totally bounded.

"Only if": Let $\varepsilon > 0$. For every $x \in \overline{\mathcal{O}(x_0)}$ there exists a point $y_x \in B_\varepsilon(x) \cap \mathcal{O}(x_0)$, so the sets $B_\varepsilon(y_x)$ for $x \in \overline{\mathcal{O}(x_0)}$ form an open cover of $\overline{\mathcal{O}(x_0)}$. It has a finite subcover, which has the form $\{B_\varepsilon(y_x) : x \in A_\varepsilon\}$ for some finite subset A_ε of $\mathcal{O}(x_0)$. This implies that $(*)$ holds. □

2 The idea behind the definition of a basin is to describe for a given invariant set the set of all points that are 'attracted' by it. In models of 'physical' systems the fact that a state x is in the basin of a set A usually means that for sufficiently large n, due to inaccuracy of measurements, rounding errors, etc., the state $f^n(x)$ is 'in' A.

In the definitions of topological and strong attraction we have included the condition that the set A under consideration is invariant. This does not follow automatically from the condition that $\emptyset \ne \omega(x) \subseteq A$ for all x in some neighbourhood of A, or from condition Lemma (3.1-3) with $B = A$ or B a neighbourhood of A. Consider, for example, $A = [0; 1]$ under the mapping $f : \mathbb{R} \to \mathbb{R}$ that is defined as follows: $f(x) := -|x - 1/2| + 3/2$ for $x \le 1$ and $f(x) = \frac{3}{4}x + \frac{1}{4}$ for $x \ge 1$. Then A is not an invariant set, but $\omega(x) = \{1\} \subseteq A$ for all $x \in \mathbb{R}$, so $\mathcal{B}(A) = \mathbb{R}$. Moreover, condition (3.1-3) is fulfilled for $B := (-1; 2)$. However, see Exercise 3.12 (1).

3 In some texts, topologically attracting sets are called *limit stable* sets. However, Example (4) in 3.1.5 shows that this notion lacks an essential part of what intuitively should be a property of a set called 'stable': staying close to the set when returning to it after a small perturbation (this fails for points $[t]$ with t between 0 and 1 but close to 1).

4 In this book we consider notions of stability only for compact completely invariant sets. Stability can also be defined for non-invariant sets; see Section 7.1 ahead. See also Note 5 at the end of Chapter 1. That we restrict our attention to *compact* stable sets is mainly for technical reasons (e.g., for non-compact sets the conclusion of 3.1.3 (5) does not hold). A standard text for stability – for systems with continuous time – is N. P. Bhatia & G. P. Szegö [1970] (also for non-compact sets). For systems with discrete time, see J. Buescu [1997], from which much of Section 3.4 was borrowed. For Theorem 3.4.3, see J. Milnor [1985].

5 In some publications a compact non-empty set is called an *attracting set* whenever it satisfies the condition of Theorem 3.4.2 (iii), or just that of Theorem 3.4.3 (ii). This does not conflict with the definition of an attracting invariant point or periodic orbit in a C^1-system on an interval in $\mathbb{R}$ as given in 3.3.18, because such a point or orbit is asymptotically stable, hence satisfies Condition 3.4.3 (ii). A set A as considered in Corollary 3.4.4 is sometimes called a *trapped attracting set* and W is called a *trapping neighbourhood.*

In the literature there is a variety of definitions of the notion of an 'attractor'. Clearly, 'asymptotically stable' would not be a good definition of an 'attractor': in any dynamical system the phase space is asymptotically stable. Moreover, the union of two disjoint asymptotically stable sets is asymptotically stable, a property that is undesirable for 'attractors'. Another undesirable property is that an asymptotically stable set can properly include another such set. All definitions in the literature agree with each other in the following respects:

(a) An attractor has an attraction-property (like asymptotic stability).
(b) An attractor is 'irreducible': an invariant subset of an attractor is not an attractor.

Such an irreducibility-property could be transitivity – see Theorem 3.2.8 – or minimality. There are other irreducibility-properties in circulation, some of which have a measure-theoretic character. Apart from the above conditions, one would like that the attractors in a given dynamical system also have the following properties:

(c) The union of the basins of the attractors is dense in the phase space.
(d) Each attractors is 'robust' under perturbations of the system.
(e) The collection of all systems not satisfying the conditions (c) and (d) is negligible (e.g., a first category set in the suitably topologized set of all systems).

This means that in 'almost all' systems, 'almost all' points are attracted by an attractor that does not disappear after a small perturbation of the system. Various examples show that in the category of all dynamical systems on compact metric spaces the last three conditions torpedo all proposals for the definition of an attractor. See for example M. Hurley [1982]. For this reason, one often studies this type of problems only for differentiable mappings on differentiable manifolds.

6 Most results in this section are well-known or at least folklore, though some of the proofs may be new. In particular, Theorem 1.5.9, Lemma 3.5.3 and Theorem 3.5.4 are taken from Buescu's book mentioned above. The proofs given there did not convince me, so I have given alternative proofs. Also Theorem 3.5.6 and its corollaries are borrowed from this source.

Assume that in the situation of Theorem 3.5.6 the space X is also locally compact and metrizable, and that A/R is infinite, hence a Cantor space. Then the system $(A/R, \tilde{f}\,)$ can be shown to be conjugate to a so-called *adding machine*. (The most simple case of an adding machine is treated in 4.2.8 ahead.) See J. Buescu [1997], Theorem 2.3.1.

The examples after Theorem 3.1.1 and after 3.3.13 are less pathological than they may seem at first sight. The example after 3.1.1 is, essentially, a version of Example (1).5.6 of Buescu's book. However, rigorous proofs of the relevant properties are quite elaborate in Buescu's model, while they are obvious in our model.

7 Exercise 3.9 comes from C. Conley [1978].

8 Invertible vs. non-invertible systems. In the study of bilateral systems (X, f), where f is a homeomorphism, our notion of limit set is called the *positive limit set* or *ω-limit set*. In that case it makes sense to also consider the *negative* or *α-limit set* of a point under f, which is the positive limit set of that point under the mapping f^{-1}. The results in Section 3.1 hold without much modification for invertible systems. But some results can be improved in the case of invertible systems. In particular, if f is a homeomorphism then for every point x the positive limit set $\omega(x)$ is two-sided invariant: see Exercise 3.2.

For most of the Sections 3.2 through 3.5 it makes no difference whether the setting is an invertible or a non-invertible system. However, bear in mind that in an invertible system 'invariant' means 'completely invariant'. And, of course, in invertible systems some of the proofs are much easier, notably those of Theorem 3.3.4 and Lemma 3.3.11 (2),(3). One of the reasons is that under a homeomorphism the image of a neighbourhood of a point is a neighbourhood of the image of the point. In particular, Theorem 3.5.2 has a very simple proof:

Proof of Theorem 3.5.2 for the case that f is a homeomorphism and $f[A] = A$. By Theorem 3.4.2 there is a neighbourhood U of A such that $A = \bigcap_{n\geq 0} f^n[U]$. Similarly, there is a neighbourhood V of B such that $B = \bigcap_{n\geq 0} f^n[A \cap V]$. The additional condition on f implies that $f^n[A \cap V] = A \cap f^n[V]$, so

$$B = \bigcap_{n\geq 0} f^n[V] \cap A = \bigcap_{n\geq 0} (f^n[V] \cap f^n[U]) = \bigcap_{n\geq 0} f^n[V \cap U] \supseteq \bigcap_{n\geq 0} f^n[B] = B\,.$$

This implies that $B = \bigcap_{n\geq 0} f^n[V \cap U]$. Now apply Theorem 3.4.3. □

4 Recurrent behaviour

Abstract. In this chapter we deal with modifications of periodic behaviour: without being periodic a state may approach itself arbitrarily close and infinitely often (recurrent point), possibly within uniformly bounded intervals of time (almost periodic point), or neighbouring states may approach the original state (non-wandering point). Of special interest is the set of all non-wandering points of a dynamical system: it includes all limit sets of the system. For this reason it is usually regarded as the most interesting part of a dynamical system.

In order to account for inaccuracy of measurements, rounding errors in simulations of systems, etc., we also consider the concepts of chain-recurrence and chain-transitivity, defined in terms of so-called pseudo-orbits (roughly: orbits as might appear from measurements). These notions turn out to shed additional light on asymptotic stability.

4.1 Recurrent points

A point x in a dynamical system (X, f) is said to be *recurrent* whenever

$$\forall U \in \mathcal{N}_x\ \exists n \in \mathbb{N} : f^n(x) \in U\,. \tag{4.1-1}$$

Note that $x = f^0(x) \in U$ for every $U \in \mathcal{N}_x$, which is why in this definition we require that $n \in \mathbb{N}$, i.e., $n \geq 1$. Obviously, the point $x \in X$ is recurrent iff it belongs to the orbit closure of the point $f(x)$. The set of recurrent points in the dynamical system (X, f) will be denoted by $R(X, f)$. It is well possible that $R(X, f) = \emptyset$ (e.g., let $X := \mathbb{Z}$ and $f(x) := x + 1$ for $x \in X$).

Every periodic point is recurrent, but not every recurrent point is periodic: see the examples below. A non-periodic eventually periodic point is easily seen to be not recurrent.

A transitive point x is recurrent: for every neighbourhood U of x there are infinitely many values of n with $f^n(x) \in U$. But not every recurrent point is transitive: see the examples below and also Exercise 4.1-2.

Recall that for a point $x \in X$ and a subset A of X the dwelling set of x in A is defined as the set $D_f(x, A) := \{n \in \mathbb{Z}^+ : f^n(x) \in A\}$; if f understood we simply write $D(x, A)$ instead of $D_f(x, A)$. Thus, a point $x \in X$ is recurrent iff $D(x, U) \supsetneq \{0\}$ for every neighbourhood U of x. In Proposition 4.1.1 (ii) below it will be shown that in this case the set $D(x, U)$ is unbounded. See also Exercise 4.2.

If x is a periodic point then the set $D(x, U)$ does not depend on U for sufficiently small U (in fact, so small that $U \cap \mathcal{O}(x) = \{x\}$). The converse is also true: see Exercise 4.1 (1). In many cases, however, the set $D(x, U)$ will get sparser in $\mathbb{Z}^+$ if U gets smaller. See Exercise 5.12 (1) ahead.

Examples.

(1) Consider the rigid rotation $(\mathbb{S}, \varphi_a)$ with $a \in \mathbb{R}\setminus\mathbb{Q}$. Every point of $\mathbb{S}$ has a dense orbit under φ_a, hence is recurrent. But no point is periodic.

(2) Every infinite transitive system (like the tent map on the unit interval or the argument-doubling transformation on the circle) has a dense sets of transitive, hence recurrent, points none of which is periodic. On the other hand, a periodic point in an infinite system is an example of a non-transitive recurrent point.

(3) Consider the mapping $\tau_a := \varphi_a \times \mathrm{id}_{\mathbb{S}} : \mathbb{S}^2 \to \mathbb{S}^2$, where $a \notin \mathbb{Q}$. Each 'horizontal' circle $S_t := \mathbb{S} \times [t]$ $(0 \leq t < 1)$ is invariant and minimal under τ_a, because the restricted system (S_t, τ_a) is conjugate to the system $(\mathbb{S}, \varphi_a)$. Consequently, no orbit is transitive in $\mathbb{S}^2$ and no point of $\mathbb{S}^2$ is periodic under τ_a, but every point of $\mathbb{S}^2$ is recurrent under τ_a. Another example of a recurrent point which is neither periodic nor transitive will be given in 5.6.1 (3) ahead.

If A is an invariant subset of X and $x \in A$ then the point x is recurrent in the subsystem on A iff x is recurrent in the full system (X, f). To prove this, note that the neighbourhoods of the point x in A are the sets $U \cap A$ with U a neighbourhood of x in X and that $D(x, U) = D(x, U \cap A)$, because the set A is invariant. We may summarize this by the equality $R(X, f) \cap A = R(A, f)$. In particular, *a point x is recurrent in the full system iff it is recurrent in the subsystem on its orbit.* So recurrence is a property of the orbit alone[1].

Proposition 4.1.1. *Let x be a point in X. The following conditions are equivalent:*

(i) *The point x is recurrent.*

(ii) *For every neighbourhood U of x the set $D(x, U)$ is infinite.*

(iii) $x \in \omega(x)$.

(iv) $\overline{\mathcal{O}(x)} = \omega(x)$.

Proof. "(ii)⇔(iii)": Clear from Lemma 1.4.1 (1).

"(iii)⇔(iv)": Clearly, (iii) holds iff $\overline{\mathcal{O}(x)} \subseteq \omega(x)$ (the limit set $\omega(x)$ is closed and invariant). Recall also that always $\omega(x) \subseteq \overline{\mathcal{O}(x)}$.

"(i)⇔(ii)": The implication (ii)⇒(i) is clear from the definition, so we need only prove (i)⇒(ii). Assume that the point x is recurrent. If x is periodic then it is obvious that (ii) holds. So we may assume that the point x is not periodic. Then for every neighbourhood U of x and every $k \in \mathbb{N}$ there is a neighbourhood V of x such that $V \subseteq U$ and $f^i(x) \notin V$ for $i = 1, \ldots, k$. As the point x is recurrent there exists $n \in \mathbb{N}$ such that $f^n(x) \in V$. Because of the choice of V it is clear that $n > k$. This shows that for every $k \in \mathbb{N}$ there exists $n > k$ with $f^n(x) \in U$. □

1 Such properties are called *orbital properties*. For example, the property of being eventually periodic is orbital. The property for a point to be transitive is not orbital: by Proposition 4.1.1 (iv) a recurrent point is transitive in its orbit closure, hence in its orbit, but not necessarily in the full system.

Proposition 4.1.2. *The set $R(X, f)$ of all recurrent points is invariant.*

Proof. Use that $D(f(x), U) = D(x, f^{\leftarrow}[U])$ for all $x \in X$ and $U \subseteq X$. (Suggestion for an alternative proof: use Proposition 4.1.1 (iii) and Theorem 1.4.3 (1).) □

Remark. The set $R(X, f)$ is not necessarily closed. For example, under the tent map on $[0; 1]$ or the argument-doubling transformation on $\mathbb{S}$ the set of all periodic (hence of all recurrent) points is dense in the phase space. However, there exist also non-recurrent points (e.g., all non-invariant eventually invariant points). See Also 5.6.1 (2) ahead.

Theorem 4.1.3. *Every point in a minimal system is recurrent. Consequently, every non-empty compact invariant subset in any dynamical system contains a recurrent point. In particular, if a point has a compact orbit closure then its limit set contains a recurrent point.*

Proof. The first statement is clear from the fact that all points in a minimal system are transitive. To prove the remaining statements, take into account the Theorems 1.4.5 and 1.2.7. □

Proposition 4.1.4. *Let $x \in X$. The following conditions are equivalent:*
(i) *The point x is recurrent but not periodic.*
(ii) *$\mathcal{O}(x)$ is an infinite subspace of X without isolated points.*

Proof. "(i)⇒(ii)": By the (easy) implication (ii)⇒(i) of Theorem 1.1.5, if the orbit of x were finite then x would be eventually periodic. Since a non-periodic eventually periodic point cannot be recurrent, this shows that if (i) holds then $\mathcal{O}(x)$ is infinite. In order to show that no point of $\mathcal{O}(x)$ is isolated, consider an arbitrary point $y \in \mathcal{O}(x)$ and an arbitrary neighbourhood U of y. By Proposition 4.1.2 above, if x is recurrent then y is recurrent, in which case there exists an integer $n \geq 1$ such that $f^n(y) \in U$. Obviously, $f^n(y) \neq y$, for otherwise the point y would be periodic, contradicting what was observed above, namely, that the point x is not eventually periodic. As $f^n(y) \in \mathcal{O}(x)$, it follows that U includes a point of $\mathcal{O}(x)$ different from y. This completes the proof that the point y is not isolated in $\mathcal{O}(x)$.

"(ii)⇒(i)": Assume (ii). Since $\mathcal{O}(x)$ is infinite, the point x is not periodic. In addition, the point x is not isolated in $\mathcal{O}(x)$, so every neighbourhood of x contains a point of this orbit different from x, i.e., contains a point $f^n(x)$ with $n \neq 0$. □

Remark. If X is metrizable then $\mathcal{O}(x)$ is a countable metric space. In that case, a theorem of Sierpiński's[2] states that condition (ii) is equivalent with the condition that $\mathcal{O}(x)$ is homeomorphic with $\mathbb{Q}$, the space of rational numbers. See also Exercise 4.3.

Corollary 4.1.5. *Let X be a Čech-complete space and let $x \in X$. Then the following conditions are equivalent:*
(i) *The point x is periodic.*
(ii) $\omega(x) = \mathcal{O}(x)$.

---

2 See [Eng], Exercise 6.2.A(d).

Proof. "(i)⇒(ii)": This is Proposition 1.4.2 (1).

"(ii)⇒(i)": If (ii) holds then $x \in \omega(x)$, so the point x is recurrent. Since now $\mathcal{O}(x)$ is a closed subspace of X, it is a Baire space (with its relative topology in X) and the argument used in the proof of Theorem 1.1.5 shows that some point of the orbit of x is isolated in $\mathcal{O}(x)$ (see also Exercise 1.1). Hence by Proposition 4.1.4, the point x is periodic. □

Remark. If X is just a Hausdorff space then the implication (ii)⇒(i) may not hold: see the remark after Proposition 1.4.2.

Proposition 4.1.6. *Let $\varphi: (X, f) \to (Y, g)$ be a morphism of dynamical systems. Then $\varphi[R(X, f)] \subseteq R(Y, g)$, i.e., the image under φ of a recurrent point is recurrent. If φ is a conjugation then $\varphi[R(X, f)] = R(Y, g)$.*

Proof. Use the equality $D_g(\varphi(x), U) = D_f(x, \varphi^{\leftarrow}[U])$ for $x \in X$ and U a neighbourhood of $\varphi(x)$ in Y. □

Remark. The inclusion $\varphi[R(X, f)] \subseteq R(Y, g)$ may be strict, even if φ is a factor mapping, i.e., recurrence is not lifted by factor maps. Consider, for example, the trivial morphism $\varphi: (X, f) \to (Y, g)$ of a system without recurrent points onto the trivial system on a singleton set. However:

Theorem 4.1.7. *Let $\varphi: (X, f) \to (Y, g)$ be a factor mapping and assume that X is compact. Then $\varphi[R(X, f)] = R(Y, g)$. In particular, if $x_0 \in X$ and $\varphi^{\leftarrow}[\varphi(x_0)] = \{x_0\}$, and $\varphi(x_0)$ is recurrent under g then x_0 is recurrent under f.*

Proof. In view of Proposition 4.1.6 it remains to show that if $y_0 \in R(Y, g)$ then the fibre $\varphi^{\leftarrow}[y_0]$ contains a point of $R(X, f)$. So consider a point $y_0 \in R(Y, g)$ and let $Y_0 := \overline{\mathcal{O}_g(y_0)}$. Since y_0 has a dense orbit in the subsystem (Y_0, g) of (Y, g) and the point y_0 is recurrent in this subsystem, it is a transitive point in this subsystem; we leave the easy proof as an exercise for the reader (see Exercise 4.1-2). Now apply Proposition 1.5.5 (1): the fibre $\varphi^{\leftarrow}[y_0]$ includes a point that is transitive in some subsystem of (X, f). In particular, this point is recurrent in this subsystem, hence it is recurrent in (X, f). □

Example. Let $f: ([s], [t]) \mapsto ([s + a], [t + 2s + a]): \mathbb{S}^2 \to \mathbb{S}^2$, with $a \in \mathbb{R}$. The canonical projection of $\mathbb{S}^2$ into its first coordinate is a factor map of $(\mathbb{S}^2, f)$ onto $(\mathbb{S}, \varphi_a)$ and every point of $\mathbb{S}$ is recurrent under φ_a – it is either periodic or transitive – so for every $s \in \mathbb{R}$ the subset $S_s := [s] \times \mathbb{S}$ of $\mathbb{S}^2$ contains a recurrent point. For every pair $b_1, b_2 \in \mathbb{R}$ the mapping $\rho: ([s], [t]) \mapsto ([s + b_1], [t + b_2]): \mathbb{S}^2 \to \mathbb{S}^2$ is an automorphism of the system $(\mathbb{S}^2, f)$, mapping the recurrent point onto any other point of $\mathbb{S}^2$ (for a suitable choice of b_1 and b_2). It follows that every point of $\mathbb{S}^2$ is recurrent under f.

4.2 Almost periodic points and minimal orbit closures

A point $x \in X$ is said to be *almost periodic* whenever for every neighbourhood U of x the set $D(x,U) := \{n \in \mathbb{Z}^+ : f^n(x) \in U\}$ has *bounded gaps*, that is, by definition, there exists $l \in \mathbb{N}$ such that the difference between any two consecutive elements of $D(x,U)$ is at most l. Equivalently: the point x is almost periodic iff

$$\forall U \in \mathcal{N}_x \ \ \exists l \in \mathbb{Z}^+ : [n; n+l] \cap D(x,U) \neq \emptyset \quad \text{for all } n \in \mathbb{Z}^+, \tag{4.2-1}$$

that is, every set of $(l+1)$ consecutive elements in $\mathbb{Z}^+$ contains an element of $D(x,U)$. Note: if $l = 0$ for every $U \in \mathcal{N}_x$ then the point x is invariant.

Obviously, every periodic point is almost periodic, and every almost periodic point is recurrent. The examples following Corollary 4.2.3 below show that the converse of these statement is not true, so 'almost periodic' is really between 'periodic' and 'recurrent'.

If A is an invariant subset of X then a point in A is almost periodic in the subsystem (A,f) iff it is almost periodic in (X,f). The proof is similar to the proof of corresponding statement for recurrent points and is left to the reader. *In particular, whether a point is almost periodic or not depends only on the orbit of that point*[3] .

Lemma 4.2.1. *Let $x \in X$. The following statements are equivalent:*
(i) *The point x is almost periodic.*
(ii)

$$\forall U \in \mathcal{N}_x \ \exists l \in \mathbb{Z}^+ : \mathcal{O}(x) \subseteq \bigcup_{j=0}^{l} (f^j)^{\leftarrow}[U]\,. \tag{4.2-2}$$

If these conditions are fulfilled then the following is fulfilled as well:
(iii)

$$\forall U \in \mathcal{N}_x \ \exists l \in \mathbb{Z}^+ : \mathcal{O}(x) \subseteq \bigcup_{j=0}^{l} f^j[U]\,. \tag{4.2-3}$$

If f is injective then also (iii)⇒(i).

Proof. "(i)⇔(ii)": For any neighbourhood U of x, and any $n, l \in \mathbb{Z}^+$ and $j \in \{0,\dots,l\}$ one has $n + j \in D(x,U)$ iff $f^n(x) \in (f^j)^{\leftarrow}[U]$. Using this it is easy to show that (4.2-1) and (4.2-2) are equivalent.

"(i)⇒(iii)": Assume (i) and let U be an arbitrary neighbourhood of x. Choose $l \in \mathbb{Z}^+$ according to (4.2-1) and let $n \in \mathbb{Z}^+$. If $0 \le n \le l$ then we obviously have $f^n(x) \in f^n[U] \subseteq \bigcup_{j=0}^{l} f^j[U]$. And if $n \ge l$ then apply formula (4.2-1) with n replaced by $n - l$: one gets $[n-l; n] \cap D(x,U) \neq \emptyset$, i.e., there exists $j \in \{0,1,\dots,l\}$ such that $n - j \in D(x,U)$. This means that $f^n(x) = f^j(f^{n-j}(x)) \in f^j[U]$. This shows that (4.2-3) holds.

Conversely, if (iii) holds, U is a neighbourhood of x and l is as in (4.2-3) then for every integer n with $n \ge l$ there exists an integer $j \in [0; l]$ such that $f^n(x) \in f^j[U]$,

3 So 'almost periodic' is an orbital property.

hence $f^{n-j}(x) \in (f^j)^{\leftarrow}[f^j[U]]$. If f is injective, then the right-hand side is equal to U, so in that case we have $n - j \in D(x, U)$ and, consequently, $[n - l; n] \cap D(x, U) \neq \emptyset$. This shows that the gaps in $D(x, U)$ are bounded (by l). □

Remarks. In general, (iii) does not imply (i): let X have at least two points and let f map X onto a single (invariant) point $x_0 \in X$. Then every point $x \in X \setminus \{x_0\}$ satisfies (iii) but not (i).

There is a close connection between minimality and almost periodicity:

Theorem 4.2.2 (Birkhoff). *Let $x \in X$.*

(1) *Assume that X is a regular Hausdorff space. If the point x is almost periodic then its orbit closure $\overline{\mathcal{O}(x)}$ is minimal. If, in addition, X is locally compact then $\overline{\mathcal{O}(x)}$ is compact.*

(2) *If the orbit closure $\overline{\mathcal{O}(x)}$ is compact and minimal then x is an almost periodic point.*

Proof. For convenience, let $A := \overline{\mathcal{O}(x)}$.

(1) Assume that X is a regular Hausdorff space and that x is an almost periodic point. In view of Proposition 1.2.6, we want to show that every point of A has a dense orbit in A, that is, if $y \in A$ then $A = \overline{\mathcal{O}(y)}$. It is sufficient to show that $x \in \overline{\mathcal{O}(y)}$, for then $A = \overline{\mathcal{O}(x)} \subseteq \overline{\mathcal{O}(y)} \subseteq A$.

So let V be any neighbourhood of x in X. As X is a regular space, we may assume that V is closed in X. By (4.3-2), there exists $l \in \mathbb{Z}^+$ such that $\mathcal{O}(x) \subseteq \bigcup_{n=0}^{l} (f^n)^{\leftarrow}[V]$. Because for every $n \in \mathbb{Z}^+$ the set $(f^n)^{\leftarrow}[V]$ is closed (it is the preimage of a closed set under a continuous mapping), the right hand side of this inclusion is closed as well, hence it includes $A = \overline{\mathcal{O}(x)}$. In particular, $y \in (f^n)^{\leftarrow}[V]$ for some $n \in \mathbb{Z}^+$, so that $\mathcal{O}(y) \cap V \neq \emptyset$. This holds for every neighbourhood V of y, hence $x \in \overline{\mathcal{O}(y)}$.

Now assume that X is locally compact and that V be a compact neighbourhood of x. By (4.2-3), there exists $l \in \mathbb{Z}^+$ such that $\mathcal{O}(x) \subseteq \bigcup_{n=0}^{l} f^n[V]$. Here the right-hand side, a finite union of continuous images of a compact set, is compact and closed. It follows that $\overline{\mathcal{O}(x)}$ is a closed subset of a compact set, hence is compact.

(2) Let $x \in X$ and assume that A is compact and minimal. Consider an arbitrary open neighbourhood U of x in X. Because A is a minimal set, Proposition 1.2.6 implies that every point y in A has a dense orbit in A, that is, $f^n(y) \in U$ for some $n \in \mathbb{Z}^+$. This shows that $A \subseteq \bigcup_{n=0}^{\infty} (f^n)^{\leftarrow}[U]$. Since A is compact and every set $(f^n)^{\leftarrow}[U]$ for $n \in \mathbb{Z}^+$ is open, it follows that there is $l \in \mathbb{Z}^+$ such that $\mathcal{O}(x) \subseteq A \subseteq \bigcup_{n=0}^{l} (f^n)^{\leftarrow}[U]$. So by Lemma 4.2.1 above, the point x is almost periodic. □

Remarks. If X is locally compact then compactness of $\overline{\mathcal{O}(x)}$ in statement 1 also follows from Exercise 1.5 (6). Moreover, in statement 2 compactness of $\overline{\mathcal{O}(x)}$ cannot be omitted: see Example (3) below.

Corollary 4.2.3. *Every point in a compact minimal set is almost periodic. In particular, every dynamical system with a compact phase space contains an almost periodic point.*

Proof. The first statement is clear from Theorem 4.2.2. The second statement now follows from Theorem 1.2.7. □

Examples.

(1) An infinite compact minimal system contains no periodic points, so all its points are almost periodic and non-periodic. Examples: the rigid rotation, see Example (2) after Proposition 1.2.6; other examples are in 1.7.6 and in Proposition 4.2.9, Proposition 5.6.4, Proposition 5.6.7, Theorem 5.6.14 and 6.3.7 below.

(2) Transitive points (which are always recurrent) are not necessarily almost periodic. In a non-minimal transitive system on a regular Hausdorff space no transitive point is almost periodic, otherwise by Theorem 4.2.2 (1) its orbit closure (the phase space) would be minimal. This applies to the tent map on $[0; 1]$ and the argument-doubling transformation on $\mathbb{S}$. See also the initial remarks in 5.6.2 ahead.

(3) Let Z be the – invariant – set of transitive points in a non-minimal system on a compact Hausdorff space (see 2 above). Then every point of Z has a dense orbit in Z, so the subsystem on Z is minimal. Yet no point in Z is almost periodic in Z, because it is not almost periodic in the full system.

Proposition 4.2.4. *The set of almost periodic points is invariant.*

Proof. Let x be an almost periodic point in X. If U is a neighbourhood of $f(x)$, then $D(x, f^{\leftarrow}[U]) \subseteq D(f(x), U)$. The former set has bounded gaps, so is the latter has bounded gaps as well. □

Remarks. The set of all almost periodic points need not be closed: the tent map on $[0; 1]$ has a dense sets of (almost) periodic points, but the set of all almost periodic points is not the full phase space (see Example (2) above), hence it is not closed.

Proposition 4.2.5. *Let X be a regular Hausdorff space. If $x \in X$ is an almost periodic point then all points of $\overline{\mathcal{O}(x)}$ are almost periodic as well.*

Proof. Let $y \in \overline{\mathcal{O}(x)}$ and assume that y is not almost periodic: there is a neighbourhood U of y such that

$$\forall l \in \mathbb{Z}^+ \ \exists n_l \in \mathbb{Z}^+ : [n_l; n_l + l] \cap D(y, U) = \emptyset , \tag{4.2-4}$$

that is, $f^{n_l+k}(y) \notin U$ for $k = 0, \dots, l$. We shall show that this implies that there are a neighbourhood V' of x and an element $k_0 \in \mathbb{Z}^+$ such that

$$\forall l \geq k_0 \ \exists n'_l \in \mathbb{Z}^+ : [n'_l; n'_l + l] \cap D(x, V') = \emptyset , \tag{4.2-5}$$

which contradicts almost periodicity of the point x, i.e., that the set $D(x, V')$ should have bounded gaps.

Because the space X is regular there is an open neighbourhood V of y such that $\overline{V} \subseteq U$. Consider any $l \in \mathbb{Z}^+$ and let $n_l \in \mathbb{Z}^+$ be as in (4.2-4). The set of continuous functions $\{ f^{n_l+k} : 0 \leq k \leq l\}$ is finite and each of these functions maps the point y outside of U, hence inside of the open set $X \setminus \overline{V}$. Consequently, there is a neighbourhood W_l of y such that $f^{n_l+k}[W_l] \cap \overline{V} = \emptyset$ for $k = 0, \dots, l$. As y is in the orbit closure

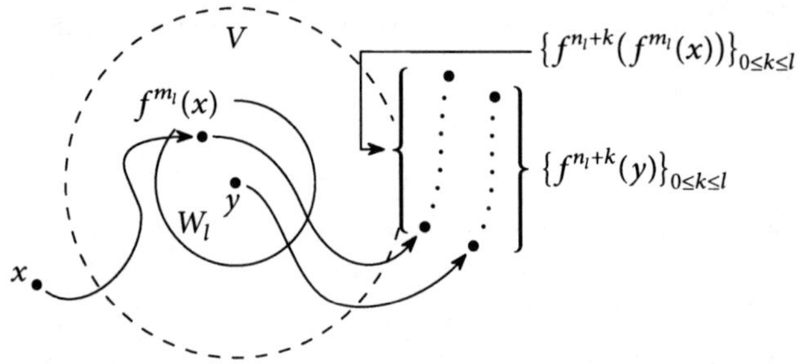

Fig. 4.1. Illustrating the proof of Proposition 4.2.5.

of x, the neighbourhood W_l of y meets the orbit of x, so there exists $m_l \in \mathbb{Z}^+$ such that $f^{m_l}(x) \in W_l$. By the choice of W_l this implies that $f^{n_l+k+m_l}(x) \notin V$ for $k = 0, \ldots, l$.

Now let $k_0 \in \mathbb{Z}^+$ be the first element of $\mathbb{Z}^+$ with the property that $f^{k_0}(x) \in V$ and let $V' := (f^{k_0})^{\leftarrow}[V]$. Since without limitation of generality we may assume that $W_l \subseteq V$, it is clear that $k_0 \le m_l$, hence $n_l' := n_l + m_l - k_0 \ge 0$, for every $l \in \mathbb{Z}^+$. Obviously, V' is a neighbourhood of x and it is easily checked that (4.2-5) holds. □

Remarks.

(1) Quick proof for the case that X is locally compact: if x is almost periodic then by Theorem 4.2.2 (1), $\overline{\mathcal{O}(x)}$ is a compact minimal set, so each of its points is almost periodic by Theorem 4.2.2 (2).

(2) A similar result does not hold for recurrent points. For example, a transitive point is recurrent and has a dense orbit, but if not all points in the phase space are recurrent (as is the case for the tent map) then not all points in the orbit closure of the transitive point are recurrent.

Next, we deal with preservation and lifting of almost periodicity. The reader should compare Theorem 4.2.7 below with Theorem 4.1.7 above.

Proposition 4.2.6. *Let $\varphi: (X, f) \to (Y, g)$ be a morphism of dynamical systems. If the point $x \in X$ is almost periodic under f then the point $\varphi(x)$ is almost periodic under g. If φ is a conjugation then x is almost periodic under f iff $\varphi(x)$ is almost periodic under g.*

Proof. Similar to the proof of Proposition 4.1.6. □

Remark. In particular, it follows that almost periodicity is a dynamical property.

Theorem 4.2.7. *Let $\varphi: (X, f) \to (Y, g)$ be a factor mapping and assume that X and Y are compact. If the point $y \in Y$ is almost periodic under g then there exists an almost periodic point $x \in \varphi^{\leftarrow}[y]$.*

In particular, if $x_0 \in X$ and $\varphi^{\leftarrow}[\varphi(x_0)] = \{x_0\}$ then almost periodicity of $\varphi(x_0)$ under g implies almost periodicity of x_0 under f.

Proof. Clear from Proposition 1.5.5 (2) (see also Corollary 1.5.6) and Theorem 4.2.2. □

Remark. By a similar argument, Exercise 1.8-2 implies: if the factor mapping $\varphi\colon (X, f) \to (Y, g)$ of compact systems is irreducible and $y \in Y$ is almost periodic under g then every point of $\varphi^{\leftarrow}[y]$ is almost periodic.

4.2.8 (The adding machine). Let $G := \{0, 1\}^{\mathbb{Z}^+}$, the product of countably many copies of the set $\{0, 1\}$ endowed with the product topology (where each copy of $\{0, 1\}$ has the discrete topology). Thus, the elements of G are the sequences $x = x_0\, x_1\, x_2 \dots$ with $x_n = 0$ or 1 for all $n \in \mathbb{Z}^+$. The topology of G is characterized as follows: a base for the neighbourhood system of a point x in G is formed by the collection of all sets of the form $\widetilde{B}_k(x) := \{\, y \in G \,\vdots\, y_n = x_n \text{ for } n = 0, \dots, k-1 \,\}$. For details, see Appendix A.5.2. By Tychonov's Theorem A.5.4, G is a compact Hausdorff space. (More details about G and its topology can be found in Section 5.1 ahead, where this space will be called Ω_2.)

Define a mapping $f : G \to G$ in the following way: if $x \in G$ then $f(x)$ is the sequence obtained by adding the sequences $x_0x_1x_2x_3\dots$ and $1000\dots$ coordinate-wise modulo 2, with unrestricted carry-over to the right:

$$f(x) := \begin{cases} 1\, x_1\, x_2\, x_3 \cdots\cdots\cdots & \\ 0 \cdots 0\, 1\, x_{n+1}\, x_{n+2} \cdots & \text{if } x_0 = \cdots = x_{n-1} = 1 \text{ and } x_n = 0\,, \\ 0\,0\,0 \cdots\cdots\cdots\cdots & \text{if } x_n = 1 \text{ for all } n \in \mathbb{Z}^+\,. \end{cases}$$

Claim: the mapping $f\colon G \to G$ is continuous, so that (G, f) is a dynamical system, called *the adding machine*. In order to prove this claim, note that if two points x and y in G have an initial block of coordinates of length k in common ($k \geq 1$), then the points $f(x)$ and $f(y)$ of G have an initial block of coordinates of length k in common as well. Stated otherwise, if $x, y \in G$ and $y \in \widetilde{B}_k(x)$ then $f(y) \in \widetilde{B}_k(f(x))$. This obviously implies that f is continuous. See also Exercise 5.2-2.

Clearly, f is a bijection of G onto itself; its inverse is the mapping that subtracts from each $x \in G$ the sequence $1000\dots$, which amounts to the same as coordinate-wise addition modulo 2 of the sequence $111\dots$ with carry-over to the right. Because G is a compact Hausdorff space, bijectivity of f implies that f is a homeomorphism. It is easily seen that f^{-1} looks as follows:

$$f^{-1}: \begin{cases} 0\,0\,0 \cdots\cdots & \mapsto\quad 1\,1\,1 \cdots\cdots \\ 1\, x_1 x_2 \cdots\cdots & \mapsto\quad 0\, x_1 x_2 \cdots\cdots \\ 0^k 1\, x_{k+1} \cdots & \mapsto\quad 1^k 0\, x_{k+1} \cdots \end{cases} \tag{4.2-6}$$

Let $\iota\colon G \to G$ be the mapping that interchanges the digits 0 and 1 in the sequences of coordinates of points of G. Then ι is a homeomorphism and it is easily verified that $\iota \circ f^{-1} = f \circ \iota$. Consequently, $\iota\colon (G, f) \mathrel{\underset{\sim}{\to}} (G, f^{-1}$ is a conjugation.

Proposition 4.2.9. *The system (G, f) is minimal.*

Proof. It is sufficient to show that the point $\mathbf{0} := 000\dots$ is almost periodic and that it has a dense orbit. In order to prove this, first notice that for every $n \in \mathbb{Z}^+$ the sequence of coordinates of the point $f^n(\mathbf{0})$ starts with the binary representation of n, followed by 0's (think of the action of f as a binary odometer). In particular, if $k \in \mathbb{N}$ and n is an integer multiple of 2^k then $f^n(\mathbf{0})$ starts with k 0's, i.e., the point $f^n(\mathbf{0})$ belongs to the basic neighbourhood $\widetilde{B}_k(\mathbf{0})$ of $\mathbf{0}$. This shows that the gaps in $D(\mathbf{0}, \widetilde{B}_k(\mathbf{0}))$ are bounded by 2^k. This completes the proof that the point $\mathbf{0}$ is almost periodic.

Next, consider a point $y \in G$ and let $n \in \mathbb{Z}^+$ be the integer with binary representation $y_0 \dots y_{k-1}$. Then the sequence of coordinates of $f^n(\mathbf{0})$ starts with the block $y_0 \dots y_{k-1}$, which means that $f^n(\mathbf{0})$ belongs to the basic neighbourhood $\widetilde{B}_k(y)$ of y. This completes the proof that the point $\mathbf{0}$ has a dense orbit. □

4.2.10 (The system $([0;1], f_\infty)$ revisited). Consider the mapping $f_\infty : [0;1] \to [0;1]$ defined in Section 2.4 and further investigated in 3.3.19. In 3.3.19 the proof that C is completely invariant under f_∞ is missing. We shall now give the missing proof by showing that the system $(C, f_\infty|_C)$ is conjugate to the adding machine. As $f[G] = G$, this implies that $f_\infty[C] = C$, so C is completely invariant under f_∞.

Recall that each point x of C has a ternary expansion $x = \sum_{n=0}^{\infty} x_n 3^{-n-1}$ with $x_n \in \{0, 2\}$ for every $n \in \mathbb{Z}^+$. We shall express this by saying that x is represented by the sequence $x_0x_1x_2\dots$ and we shall write $x \equiv x_0x_1x_2\dots$. If $x \in C$ and $x \equiv x_0x_1x_2\dots$ then $\frac{1}{3}x \equiv 0x_0x_1x_2\dots$; moreover, if $x \in C$ and $0 \le x \le \frac{1}{3}$ then x is represented by a sequence of the form $0x_1x_2\dots$, and $3x \equiv x_1x_2\dots$. In order to see how f_∞ acts on these representations of points of C we distinguish three cases:

(a) $x \equiv 000\dots$, that is, $x = 0$.
It should be clear that $f_\infty(0) = 1$; see also Figure 2.10. This means:

$$f_\infty : 000\dots \mapsto 222\dots$$

(b) $x \equiv 2x_1x_2\dots$, that is, $\frac{2}{3} \le x \le 1$.
Since $f_\infty = \tau(f_\infty)$, the definition of $\tau(f_\infty)$ implies that $f_\infty(x) = x - \frac{2}{3}$. As the ternary expansion of the real number $\frac{2}{3}$ is $2000\dots$, this means:

$$f_\infty : 2\,x_1x_2\dots \mapsto 0\,x_1x_2\dots\,.$$

(c) $x \equiv 0^k 2x_{k+1}\dots$ with $k \in \mathbb{N}$, that is, $0 < x \le \frac{1}{3}$.
We shall show by induction that for all $k \in \mathbb{Z}^+$ (so including $k = 0$)

$$f_\infty : 0^k 2\,x_{k+1}\dots \mapsto 2^k 0\,x_{k+1}\dots\,.$$

Note that this is correct if $k = 0$: this is case (b) above. Assume that this is correct for a certain $k \ge 0$, and let $x \equiv 0^{k+1}2\,x_{k+2}\dots$. Since $0 \le x \le \frac{1}{3}$, the definitions and the induction hypothesis imply that

$$\begin{aligned} f_\infty(x) &= \tau(f_\infty)(x) = \tfrac{1}{3} f_\infty(3x) + \tfrac{2}{3} \\ &= \tfrac{1}{3} f_\infty(0^k 2\,x_{k+2}\dots) + \tfrac{2}{3} \equiv \tfrac{1}{3}\cdot(2^k 0\,x_{k+2}\dots) + 200\dots \\ &= 0\,2^k 0\,x_{k+2}\dots + 200\dots = 2^{k+1} 0\,x_{k+2}\dots\,. \end{aligned}$$

This completes the proof of case (c).

It is straightforward to check that the mapping

$$h: x_0x_1x_2 \cdots \mapsto \frac{x_0}{2}\frac{x_1}{2}\frac{x_2}{2}\ldots : C \to G$$

is a homeomorphism; see also Exercise 4.8. The above analysis of how f_∞ is represented in terms of the ternary representations of the points of C implies that the mapping $g := h \circ f_\infty \circ h^{-1} : G \to G$ looks as follows:

$$g: \begin{cases} 0\,0\,0\,\ldots\cdots & \mapsto \quad 1\,1\,1\,\ldots\cdots \\ 1\,x_1x_2\cdots\cdots & \mapsto \quad 0\,x_1x_2\,\cdots\cdots \\ 0^k1\,x_{k+1}\ldots & \mapsto \quad 1^k0\,x_{k+1}\,\ldots \end{cases}$$

Comparison with (4.2-6) shows that this is the inverse of the phase mapping f of the adding machine. Since, by the final observation in 4.2.8, the system (G, g) is conjugate to the system (G, f), we get:

Conclusion. *The system (C, f_∞) is conjugate to the adding machine (G, f) and is, consequently, minimal.*

4.3 Non-wandering points

Recurrence can be seen as a generalization of periodicity: a recurrent point does not return exactly to itself, but it returns to any (arbitrarily small) neighbourhood of itself. The following notion involves an additional weakening of this definition: now also the initial state is 'blurred' to a neighbourhood. A point $x \in X$ is said to be *non-wandering* (under f) whenever

$$\forall U \in \mathcal{N}_x \exists n \in \mathbb{N} : f^n[U] \cap U \neq \emptyset. \tag{4.3-1}$$

Hence a point $x \in X$ is non-wandering iff for every neighbourhood U of x one has $U \cap (f^n)^{\leftarrow}[U] \neq \emptyset$ for some $n \geq 1$. Of course, the possibility $n = 0$ has to be ruled out, because otherwise (4.3-1) would be fulfilled for every point $x \in X$. The set of all non-wandering points in (X, f) will be denoted by $\Omega(X, f)$. It is called the *non-wandering set* of X. Note that it is well possible that $\Omega(X, f) = \emptyset$ (e.g., let $X := \mathbb{R}$ and $f(x) := x + 1$ for $x \in X$). If $\Omega(X, f) = X$ then we say that the *system (X, f) is non-wandering*.

A point that is not non-wandering is called a *wandering* point. Thus, a point is wandering iff it has a neighbourhood U_0 such that $f^n[U_0] \cap U_0 = \emptyset$ for all $n \in \mathbb{N}$ – equivalently, $U_0 \cap (f^n)^{\leftarrow}[U_0] = \emptyset$ for all $n \in \mathbb{N}$. Of course, without restriction of generality we may assume that U_0 is open.

If a point is non-wandering in a subsystem (A, f) of (X, f) then it is non-wandering in the full system (X, f): take into account that, for any (open) set U of X and any $n \in \mathbb{N}$ we have $f^n[U \cap A] \cap (U \cap A) \subseteq f^n[U] \cap U$. In general, the converse is not true, as a point of U that returns into U need not belong to A, not even if U is a neighbourhood of a point of A. So in general one has $\Omega(A, f) \subsetneq \Omega(X, f) \cap A$.

Example. Let (X^*, f^*) be the system described in the example after Theorem 3.1.1. Claim: every point of the set $C \cup \{(0,\infty)\}$ is non-wandering. We prove this only for the point $(-\infty, 1)$; for all other points the proof is similar. Consider an arbitrary basic neighbourhood $V_m := [-\infty; -m]\times\{1\}$ of the point $(-\infty, 1)$ in X^* ($m \in \mathbb{N}$). Then the point $(-m, 1)$ belongs to V_m and after $4(m+1)$ steps under f^* it is back in V_m. This proves the claim. As all other points of X^* are isolated and not periodic they are, obviously, wandering. Conclusion: $\Omega(X^*, f^*) = C \cup \{(0,\infty)\} =: A$.

Note that A is closed invariant under f^*. The invariant point $(0,\infty)$ is non-wandering in the subsystem on A. The points of C are isolated in A and as they are not periodic they are, consequently, wandering in the subsystem on A. It follows that $\Omega(A, f^*) = \{(0,\infty)\} \neq A = \Omega(X^*, f^*) \cap A$.

Proposition 4.3.1.
(1) *Every recurrent point is non-wandering.*
(2) *If X has a dense subset of recurrent points then $\Omega(X, f) = X$. In particular, this is the case if X has a dense set of periodic points.*
(3) *If (X, f) is transitive or topologically ergodic then $\Omega(X, f) = X$.*

Proof. (1) Clear from the definitions.
(2) Consider $x \in X$ and let U be an open neighbourhood of x. Then U contains a recurrent point y. Since U is a neighbourhood of the point y there exists an $n \geq 1$ such that $f^n(y) \in U$. So the point x is non-wandering.
(3) The transitive case: X has a dense set of transitive points, which are all recurrent; now use 2. For the ergodic case: see Exercise 4.4 (1). □

Examples.
(1) Every rigid rotation of the circle is non-wandering: either the system consists of only periodic points, or the system is minimal. In both cases, all points are recurrent.
(2) The systems of the the tent map on $[0; 1]$ and of the argument-doubling transformation on $\mathbb{S}$ are non-wandering, because they have dense sets of periodic points. But not all points are recurrent: see the Remark after Proposition 4.1.2. Similarly, the points of the set C in the example preceding Proposition 4.3.1 are non-wandering, but they are not recurrent.
(3) Shift systems, to be defined in Chapter 5, form an important class of non-wandering systems. See Corollary 5.2.6 (2) ahead (and the remark following it).

Proposition 4.3.2. *Let $x \in \Omega(X, f)$. Then for every neighbourhood U of x the set $D(U, U) := \{n \in \mathbb{N} : f^n[U] \cap U \neq \emptyset\}$ is infinite.*

Proof. Let U be an arbitrary neighbourhood of x. If x is a periodic point then every period of x is included in $D(U, U)$, so in that case $D(U, U)$ is infinite. So we may assume that the point x is not periodic. In that case we shall show that for every $k \in \mathbb{N}$ there exists $n > k$ such that $f^n[U] \cap U \neq \emptyset$.

If $k \in \mathbb{N}$ then for each $i \in \{1, \ldots, k\}$ the points x and $f^i(x)$ are different, hence they have disjoint neighbourhoods V_i and W_i, respectively. Then $V_i' := (f^i)^{\leftarrow}[W_i] \cap V_i$ is a neighbourhood of x such that V_i' and $f^i[V_i']$ are disjoint. So $U_k := \bigcap_{i=1}^k V_i'$ is a neighbourhood of x and $U_k \cap f^i[U_k] = \emptyset$ for $i = 1, \ldots, k$. Now let $U_k' := U \cap U_k$. Then U_k' is a neighbourhood of x, and since the point x is non-wandering, there exists $n \in \mathbb{N}$ such that $f^n[U_k'] \cap U_k' \neq \emptyset$. By the choice of U_k' it is clear that $n > k$. Moreover, since $U_k' \subseteq U$ it follows that $f^n[U] \cap U \neq \emptyset$. □

Proposition 4.3.3. *$\Omega(X, f)$ is closed and invariant in X.*

Proof. Without limitation of generality we may assume that $\Omega(X, f) \neq \emptyset$.

"$\Omega(X, f)$ is closed": Let $x \in X$ be a wandering point. Then x has an open neighbourhood U_0 such that $f^n[U_0] \cap U_0 = \emptyset$ for all $n \in \mathbb{N}$. As U_0 is a neighbourhood of all of its points, this implies that all points of U_0 are wandering. Thus, the set of wandering points is open and, consequently, its complement $\Omega(X, f)$ is closed.

"$\Omega(X, f)$ is invariant": Let $x \in \Omega(X, f)$ and consider an arbitrary neighbourhood U of $f(x)$. It is easily checked that $D(f^{\leftarrow}[U], f^{\leftarrow}[U]) \subseteq D(U, U)$ Since $f^{\leftarrow}[U]$ is a neighbourhood of the point x the left-hand side of this inclusion, hence the right-hand side as well, includes a positive natural number. Consequently, $f(x) \in \Omega(X, f)$. □

Corollary 4.3.4. $\overline{R(X, f)} \subseteq \Omega(X, f)$.

Proof. Clear from Propositions 4.3.1 (1) and 4.3.3. □

Proposition 4.3.5. *For every point $x \in X$ we have $\omega(x) \subseteq \Omega(X, f)$.*

Proof. If $\omega(x) = \emptyset$ there is nothing to prove, so assume that $\omega(x) \neq \emptyset$. Let $y \in \omega(x)$ and let U be a neighbourhood of y. By Lemma 1.4.1 (1) there are at least two different values $k, l \in \mathbb{Z}^+$, say with $l \geq k + 1$, such that the points $y_k := f^k(y)$ and $y_l := f^l(y)$ belong to U. Since $f^{l-k}(y_k) = y_l \in U$, it follows that $f^{l-k}[U] \cap U \neq \emptyset$. As $l - k \geq 1$ it follows that the point y is non-wandering. □

Remark. If $x \in R(X, f)$ then $x \in \omega(x) \subseteq \Omega(X, f)$. Similarly, if X has a transitive point x then $\omega(x) = X$, hence $\Omega(X, f) = X$. This provides alternative proofs of Proposition 4.3.1 (1), (3). Note that Proposition 4.3.1 (2) would follow immediately from Corollary 4.3.4.

Corollary 4.3.6. *Assume that X is compact. Then $\Omega(X, f)$ is a non-empty closed invariant set such that $\mathcal{B}^*(\Omega(X, f)) = \mathcal{B}(\Omega(X, f)) = X$. So the set $\Omega(X, f)$ is a strongly and topologically attracting.*

Proof. Use Propositions 4.3.5 and 3.1.12 (3). □

Example. In general, not all points of $\Omega(X, f)$ belong to the limit set of a point in X. Modify the mapping f^* in the system described in the Example after Theorem 3.1.1 as follows: $f^*(m, 1) := (-m + 1, 1)$ for $m \geq 1$; at all other points of X^* the mapping f^* remains as before, so all points in $X^* \setminus (C \cup \{(0, \infty)\})$ are now periodic. Then all points of X^* are non-wandering but the points of C are not situated in any limit set.

Recall that $\Omega(A, f) \subseteq A \cap \Omega(X, f)$ for any non-empty invariant subset A of X. So if $A \subseteq \Omega(X, f)$ then $\Omega(A, f) \subseteq A$, but even in this case we do not always have an equality: see the example preceding Proposition 4.3.1. We shall now define a closed invariant set A of $\Omega(X, f)$ such that $\Omega(A, f) = A$.

The *the centre* of the dynamical system (X, f) is defined as the closure of the set of recurrent points in (X, f). It is denoted by $Z(X, f)$. Thus[4],

$$Z(X, f) := \overline{R(X, f)} \subseteq \Omega(X, f)\,. \tag{4.3-2}$$

Proposition 4.3.7. *The centre $Z(X, f)$ is a closed and invariant subset of X. Moreover, $\Omega(Z(X, f), f) = Z(X, f)$. Consequently, all points of $Z(X, f)$ are non-wandering in the subsystem on $Z(X, f)$.*

Proof. If the set $Z(X, f)$ is empty there is nothing to prove, so we may assume that $Z(X, f) \neq \emptyset$ or, equivalently, that $R(X, f) \neq \emptyset$. Then the Propositions 4.1.2 and 1.2.3 imply that $Z(X, f)$ is closed and invariant.

In order to prove the final statement, recall that a point in X is recurrent in the system (X, f) iff it is recurrent in every subsystem to which it belongs. Since $R(X, f) \subseteq Z(X, f)$, it follows that $R(X, f)$ coincides with the set of all recurrent points in the subsystem on $Z(X, f)$. Consequently, this subsystem has a dense set of recurrent points. By Proposition 4.3.1 (2) or Corollary 4.3.4, applied to this subsystem, it follows that $\Omega(Z(X, f), f) = Z(X, f)$. □

Remark. In the proof it was shown that $R(Z(X, f), f) = R(X, f)$. It follows that $Z(Z(X, f), f) = Z(X, f)$.

The following result implies that in a metric Baire space the converse of Proposition 4.3.1 (2) holds.

Theorem 4.3.8. *Let X be a metric Baire space. If $\Omega(X, f) = X$ then the set $R(X, f)$ includes a dense G_δ-set.*

Proof. Let d denote the metric of X. For every $m \in \mathbb{N}$, let

$$E(m) := \left\{ x \in X : d(x, f^k(x)) < \frac{1}{m} \text{ for some } k \in \mathbb{N} \right\}.$$

For every $m \in \mathbb{N}$ the set $E(m)$ is open in X: in order to see this, take into account that the mapping $x \mapsto d(x, f^k(x)) : X \to \mathbb{R}^+$ is continuous. We claim that this set is dense in X. Consider any point $x \in X$ and any open neighbourhood U of x such that $U \subseteq B_{1/2m}(x)$. The point x is non-wandering, hence there is a point $y \in U$ such that $f^k(y) \in U$ for some $k \geq 1$, that is, $d(y, f^k(y)) < \frac{1}{m}$, hence $y \in E(m)$. This shows that $U \cap E(m) \neq \emptyset$. This holds for every sufficiently small neighbourhood of x, hence x is in the closure of $E(m)$. This completes the proof of our claim.

4 For an alternative definition, see Note 7 at the end of this chapter.

Thus, for every $m \in \mathbb{N}$ the set $E(m)$ is open and dense in X. As X is a Baire space, it follows that the G_δ-set $\bigcap_{m\in\mathbb{N}} E(m)$ is dense in X. It is easily seen that every point from this intersection is recurrent. This completes the proof. □

Remark. By Baire's Theorem, if X is a locally compact metric space then X is a Baire space, so if in that case $\Omega(X, f) = X$ then $\overline{R(X, f)} = X$. It can be shown that this is true in *all* locally compact spaces. Similarly, in the next corollary the metrizability condition on the phase space can be omitted.

Corollary 4.3.9. *Let X be a metric Čech-complete space*[5]*. Then $Z(X, f)$ is the largest non-empty subset A of X with the following property:*
(P) A is a closed invariant subset of X such that $\Omega(A, f) = A$.
Thus, $Z(X, f)$ has property (P) and if a non-empty subset A of X has property (P) then $A \subseteq Z(X, f)$.

NB. If $Z(X, f) = \emptyset$ then there are no non-empty sets with property *(P)*.

Proof. In Proposition 4.3.7 it was shown that $Z(X, f)$ has property *(P)*. Let A be a subset of X with property *(P)*. Since X is a metric Čech complete space and A is a closed subset of X, it follows that A is a metric Baire space. Theorem 4.3.8 implies that the system (A, f) has a dense set of recurrent points (recurrent in A, hence recurrent in X). Stated otherwise, there is a subset B of $R(X, f)$ such that $A \subseteq \overline{B}$. This implies immediately that $A \subseteq \overline{R(X, f)} = Z(X, f)$. □

Example. We shall construct a dynamical system (X, f) such that all points of X are non-wandering and no point is recurrent under f. This will show that the conclusion of Theorem 4.3.8 does not hold if X is just a Hausdorff space.

As a first try consider the system depicted in Figure 4.2. The phase space is the set $X_0 := \mathbb{N} \times \{0, 1\}$ endowed with the topology in which all points of $\mathbb{N} \times \{1\}$ are isolated and the points of $\mathbb{N} \times \{0\}$ have neighbourhoods as indicated in the picture. In order to describe these neighbourhoods more precisely we need some definitions.

For $k \in \mathbb{Z}^+$, let $a_k := \frac{1}{2}k(k+1)$, so that $a_k - a_{k-1} = k$ for every $k \geq 1$. The natural numbers $1 + a_k$ for $k \geq 0$ are the points in the first column of the set $\mathbb{N} \times \{1\}$. More generally, if $n \in \mathbb{N}$ then the natural numbers $n + a_k$ for $k \geq n - 1$ are the points in the n-th column of the set $\mathbb{N} \times \{1\}$. Next, for all $i, n \in \mathbb{N}$ let

$$S_i(n) := \{\, n + a_k \,\vdots\, k \geq i + n - 2 \,\}, \tag{4.3-3}$$

which is the n-th column of $\mathbb{N} \times \{1\}$ from which the first $i-1$ elements are deleted. Then for every $n \in \mathbb{N}$ a local base at the point $(n, 0)$ is formed by the collection of all sets $U_i(n) := \{(n, 0)\} \cup (S_i(n) \times \{1\})$ with $i \in \mathbb{N}$. We leave it to the reader to show that in this way a Hausdorff topology on X_0 is obtained.

5 Equivalently, X is completely metrizable; cf. [Eng], Theorem 4.3.26.

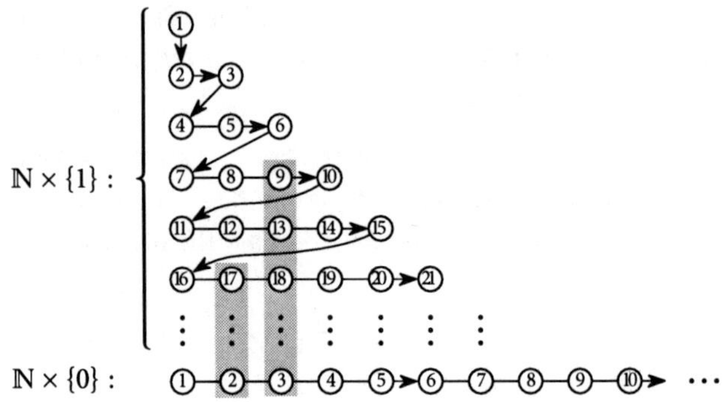

Fig. 4.2. The space X_0. The grey areas indicate the neighbourhoods $U_5(2)$ and $U_2(3)$ of the points $(2, 0)$ and $(3, 0)$.

Define $f: X_0 \to X_0$ by $f(n, j) := (n+1, j)$ for $n \in \mathbb{N}$ and $j = 0, 1$; see the arrows in Figure 4.2. Taking into account that $f[U_i(n)] = U_{i-1}(n+1)$ for $i \geq 2$ it is easy to check that f is continuous, and it is clear from Figure 4.2 that every point of $\mathbb{N} \times \{0\}$ is non-wandering. Alternatively, one might apply Proposition 4.3.5 above, as every point of $\mathbb{N} \times \{0\}$ is in $\omega(1, 1)$. On the other hand, for $j = 0, 1$ the subset $\mathbb{N} \times \{j\}$ has the discrete topology (all points are isolated) and since none of its points is periodic under f, all its points are non-recurrent (in the subspace, hence in the full space). Thus, no point of X_0 is recurrent under f.

Of course, this example is not what we are looking for: all points of the subset $\mathbb{N} \times \{1\}$ of X_0 are wandering. We might remedy this by glueing an additional copy of $\mathbb{N}$ to the subset $\mathbb{N} \times \{1\}$ of X_0 in the same way as $\mathbb{N} \times \{1\}$ is glued to $\mathbb{N} \times \{0\}$. Thus, we add the set $\mathbb{N} \times \{2\}$ to X_0 and define a local base at any point $(m, 1)$ as the collection of all sets $\{(m, 1)\} \cup (S_i(m) \times \{2\})$ with $i \in \mathbb{N}$. However, if we do not modify the local bases at the points of $\mathbb{N} \times \{0\}$ then we do not get a genuine topology: no set of the form $U_i(n)$ is a neighbourhood of any of its points of the form $(m, 1)$. Therefore, at any point $(n, 0)$ with $n \in \mathbb{N}$ we define as a local base the collection of all sets of the form $U_i(n) \cup \bigcup\{(S_{i_k}(k) \times \{2\}) : (k, 1) \in U_i(n)\}$ where $i_k \in \mathbb{N}$ for every k with $(k, 1) \in U_i(n)$. With f defined by $f(n, j) := (n+1, j)$ for $n \in \mathbb{N}$ and $j = 0, 1, 2$ all points of the extended space are non-recurrent, and all points of $\mathbb{N} \times \{0, 1\}$ are non-wandering. But now the points of $\mathbb{N} \times \{2\}$ are wandering .... It should be clear that this procedure of extending the space should be repeated infinitely often in order to push the wandering points completely out of the space.

So let $X := \mathbb{N} \times \mathbb{Z}^+$. We shall call a set of the form $\{(n, l)\} \cup (S \times \{l+1\})$ with $S \subseteq \mathbb{N}$ a *plume* at the point $(n, l) \in \mathbb{N} \times \mathbb{Z}^+$ with *trail* $S \times \{l+1\}$. An *admissible plume* at the point $(n, l) \in \mathbb{N} \times \mathbb{Z}^+$ is a plume of the special form $\{(n, l)\} \cup (S_i(n) \times \{l+1\})$ with $i \in \mathbb{N}$. The set $S_i(n) \times \{l+1\}$ is called an *admissible trail* for the point (n, l). A subset U of X is called *special* whenever it has the form $U := \bigcup\{S_l \times \{l\} : l \geq l_0\}$ where, for every $l \geq l_0$, the set $S_{l+1} \times \{l+1\}$ is a union of admissible trails, one for every point of $S_l \times \{l\}$. Obviously,

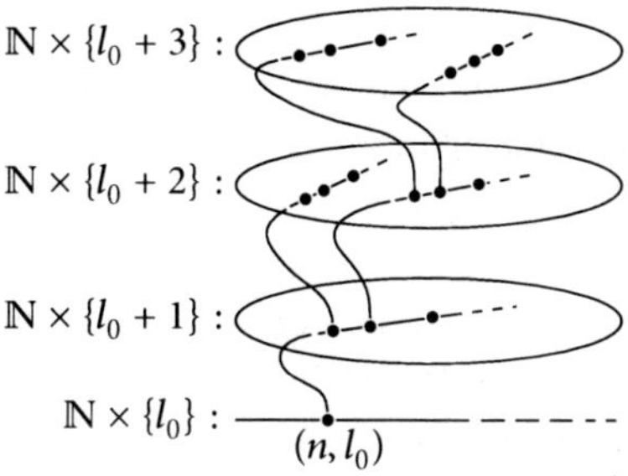

Fig. 4.3. Schematic representation of a special set (a basic neighbourhood of the point $(n, l_0) \in X$). At each level $l > l_0$ we show only trails of two points of one of the trails on the previous level.

if U is a special subset of X then for every point $(n, l) \in U$ some admissible plume at (n, l) is included in U.

After these preparations we define the topology for X by specifying a local base at each point: a local base at the point $(n, l_0) \in X$ is formed by the collection of all special subsets of the form $\bigcup\{ S_l \times \{l\} : l \geq l_0 \}$ with $S_{l_0} = \{n\}$. The reader should check that such a set is a neighbourhood of each of its points (a necessary condition for the definition of a topology by means of local bases). It is straightforward to show that this is a Hausdorff topology, as follows: If (n, l) and (m, l) are two points of X at the same 'level' l and $m \neq n$ then, taking into account that any admissible trail for the point (n, l) is disjoint from each admissible trail for the point (m, l), one easily shows by induction that every basic neighbourhood of the point (n, l) is disjoint from every basic neighbourhood of the point (m, l): at each level all trails in the one neighbourhood are disjoint from all trails in the other. If we have two points at different levels, say, the points (n, l) and (n', l') with $l' > l$, then a basic neighbourhood of the point (n, l) in which all trails at level l' are of the form $S_i(m)$ with i sufficiently large is disjoint from every basic neighbourhood of the point (n', l') (again, use induction on the levels greater than l').

Define $f : X \to X$ by $f(n, l) := (n + 1, l)$ for all $(n, l) \in \mathbb{N} \times \mathbb{Z}^+$. The proof that f is continuous with respect to the topology defined above is left to the reader. Moreover, similar as for (X_0, f) one shows that for every $l \in \mathbb{Z}^+$ all points of $\mathbb{N} \times \{l\}$ are non-wandering in (X, f). It follows that every point of X is non-wandering under f. Similarly, every orbit is included in a 'level set' $\mathbb{N} \times \{l\}$, which is, with its relative topology in X, a discrete space. Since no point is periodic, it follows that no point of X is recurrent under f.

Proposition 4.3.10. *Let $\varphi : (X, f) \to (Y, g)$ be a morphism of dynamical systems. Then $\varphi[\Omega(X, f)] \subseteq \Omega(Y, g)$, i.e., the image under φ of a non-wandering point in X is non-wandering in Y. Moreover, if φ is a conjugation then $\varphi[\Omega(X, f)] = \Omega(Y, g)$.*

Proof. The straightforward proof is left to the reader. □

Examples.

(1) Consider the tent map $T\colon \mathbb{R} \to \mathbb{R}$. The unit interval has a dense set of periodic points under T – see 1.7.3 – which are recurrent under T in the full system on $\mathbb{R}$. Consequently, $Z(\mathbb{R}, T) \supseteq [0; 1]$, hence $\Omega(\mathbb{R}, T) \supseteq [0; 1]$. On the other hand, all points outside of $[0; 1]$ are wandering. It follows that $\Omega(\mathbb{R}, T) = Z(\mathbb{R}, T) = [0; 1]$. Now Proposition 4.3.7 would imply that all points of $[0; 1]$ are non-wandering in the subsystem $([0; 1], T)$, a result that can also easily be derived from the observation that this subsystem has a dense set of periodic points.

(2) The quadratic map f_4 and the tent map are conjugate on $[0; 1]$. Hence it follows from the above that all points of $[0; 1]$ are non-wandering (in the subsystem on $[0; 1]$, hence in the full system). As points outside the unit interval are wandering under f_4, it follows that $\Omega(\mathbb{R}, f_4) = [0; 1]$.

(3) For $\mu \geq 2 + \sqrt{5}$ the non-wandering set in the system $(\mathbb{R}, f_\mu)$ is included in the set Λ, defined in 1.7.5: all points outside of Λ are easily seen to be wandering. We shall see later that all points in the subsystem (Λ, f_μ) are non-wandering, hence they are non-wandering in the full system $(\mathbb{R}, f_\mu)$. This will follow from 6.3.6 (2) ahead, taking into account Corollary 5.2.6 (2). So $\Omega(\mathbb{R}, f_\mu) = \Lambda$.

(4) The trivial morphism $\varphi\colon (X, f) \to (Y, g)$, where (X, f) is a dynamical system without non-wandering points and Y is a singleton set (one non-wandering point) shows that, in general, the inclusion $\varphi[\Omega(X, f)] \subseteq \Omega(Y, g)$ is strict, even if φ is surjective.

4.3.11 (The non-wandering set of the system $([0; 1], f_\infty)$). Consider the mapping f_∞ defined in Lemma 2.4.4, discussed in Proposition 2.4.5, in 3.3.19 and in 4.2.10. We have seen in 3.3.19 that the set $C \cup P(f_\infty)$ includes all limit sets of points of $[0; 1]$. Here C is the Cantor set and $P(f_\infty)$ is the (countable) set of all periodic points in $[0; 1]$ under f_∞. Actually, the arguments used to prove this show that, for every $n \in \mathbb{Z}^+$, a point of the (open) set M_n which is not in the orbit of a periodic point will leave M_n and never return in it at a later moment. So by Proposition 4.3.2, such a point is wandering. Consequently, $\Omega([0; 1], f_\infty) \subseteq C \cup P(f_\infty)$.

In fact, we have an equality here: every point of $C \cup P(f_\infty)$ is non-wandering. For the periodic points of $P(f_\infty)$ this is obvious. As to the points of C, in 4.2.10 it was shown that C is minimal under f_∞ so that, in particular, all points of C are recurrent, hence non-wandering, under f_∞.

4.4 Chain-recurrence

In the first paragraph of Section 4.3 it was suggested that the properties of a point of being recurrent or non-wandering might be seen as periodicity for that point under not-too-precise observation. In the present section we give another approach to account for inaccuracy of measurements, rounding errors, etc.. In this approach one considers pseudo-orbits – approximations of the 'real' orbits – rather than the orbits

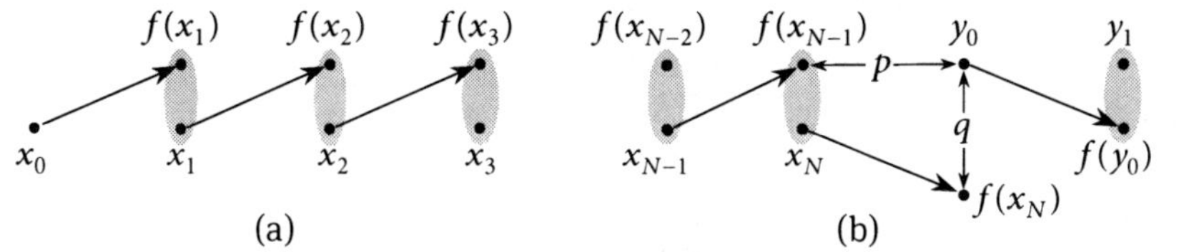

Fig. 4.4. (a) Schematic representation of the ε-chain (x_0, x_1, x_2, x_3). The grey ellipses connect point with distance at most ε. (b) Two ways to concatenate two ε-chains that have no common point.

themselves. Unless stated otherwise we consider a dynamical system (X, f) on a metric space X with metric d.

4.4.1. Let ε be an arbitrary positive real number and let $x, y \in X$ (not necessarily different). An *ε-chain*, also called an *ε-pseudo-orbit*, *from x to y* is an ordered sequence $(x_0, \dots, x_N)$ with $x_0 = x$, $x_N = y$ and $N \geq 1$ such that

$$d\left(f(x_n), x_{n+1}\right) \leq \varepsilon \quad \text{for } n = 0, \dots, N-1\,.$$

See Figure 4.4 (a). We say that this ε-chain *starts* at the point x and that it *ends* at the point y, or that it *connects* x and y (in this order). More generally, if A and B are non-empty subsets of X then an ε-chain from a point of A to a point of B is said *to begin in A* and *to end in B*, or that it *connects A and B*. The following properties of pseudo-orbits are immediate consequences of the definition and will be used without further reference.

(1) ε-Chains can be concatenated: if $(x_0, \dots, x_{N-1}, x_N)$ and $(y_0, \dots, y_M)$ are two ε-chains and $x_N = y_0$ then $(x_0, \dots, x_{N-1}, y_0, \dots, y_M)$ is also an ε-chain.
(2) If $(x_0, \dots, x_N)$ is an ε-chain and $0 \leq i < j \leq N$ then $(x_i, \dots, x_j)$ is an ε-chain.
(3) If $\varepsilon' < \varepsilon$ then every ε'-chain is an ε-chain.

If the condition $x_N = y_0$ is omitted from statement 1, then we have a $\max\{\varepsilon, p\}$-chain $(x_0, \dots, x_{N-1}, y_0, \dots, y_M)$, with $p := d(f(x_{N-1}), y_0)$, and a $\max\{\varepsilon, q\}$-chain $(x_0, \dots, x_{N-1}, x_N, y_0, \dots, y_M)$, with $q := d(f(x_N), y_0)$. See Figure 4.4 (b). Note that $p \leq \varepsilon$ if $x_N = y_0$, in which case we get 1 above.

Examples.

(1) If $x \in X$ and $\varepsilon > 0$ then for $N \geq 1$ the initial segment $(x, f(x), \dots, f^N(x))$ of the orbit of x is an ε-chain. It follows that for every $\varepsilon > 0$ there is an ε-chain from x to any point of $\mathcal{O}(x) \setminus \{x\}$. More generally, if $y \in \overline{\mathcal{O}(x)} \setminus \{x\}$ then there exists $n \in \mathbb{N}$ such that $d(f^n(x), y) < \varepsilon$, so $(x, \dots, f^{n-1}(x), y)$ is an ε-chain from x to y. A similar reasoning shows: if the point x is recurrent then for every $\varepsilon > 0$ there is an ε-chain from the point x to itself.
(2) Let X be a connected space, let $f := \mathrm{id}_X$ and let $\varepsilon > 0$. Then every finite sequence $(x_0, x_1, \dots, x_N)$ in X such that $d(x_n, x_{n+1}) < \varepsilon$ for $n = 0, \dots, N-1$ is an ε-chain. It is straightforward to show that for every point $x_0 \in X$ the (non-empty) set

$$K_\varepsilon(\{x_0\}) := \{\, y \in X : \text{there is an } \varepsilon\text{-chain from } x_0 \text{ to } y \,\}$$

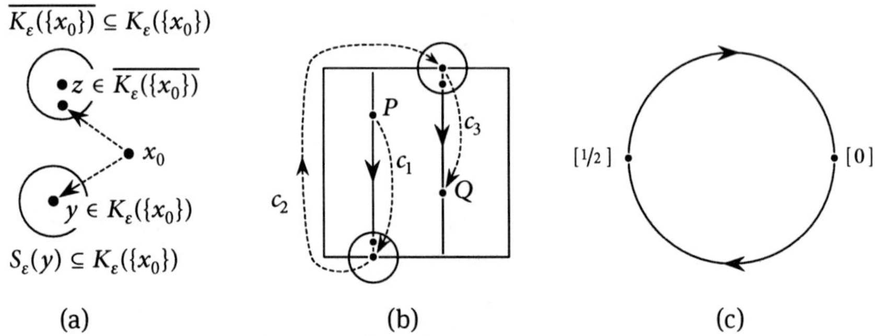

Fig. 4.5. The dotted lines indicate ε-pseudo-orbits. (a) Example (2): suggestion of proof that $K_\varepsilon(\{x_0\})$ is both closed and open. (b) Example (3): an ε-chain from P to Q. (c) Example (4).

is clopen; see Figure 4.5 (a). As X is connected it follows that $K_\varepsilon(\{x_0\}) = X$. This holds for every point $x_0 \in X$, hence every two points in X can be connected by an ε-chain. Note that (x_0, x_0) is an ε-chain, connecting the point x_0 with itself.

(3) Let $X := [0;1] \times [0;1]$ and let $f\colon (x, y) \mapsto (x, y - 4x(1-x)y(1-y))\colon X \to X$. By combining the ideas of the Examples (1) and (2) it is easily seen that any two points of X can be connected by an ε-chain. See Figure 4.5 (b), where c_1 represents an ε-chain as at the end of Example (1) above, c_2 is an ε-chain consisting of invariant points as in Example (2), and c_3 consists of an 'ε-jump' into a point in the past of y followed by a segment of the orbit of that point. (Such a *detour* may also be necessary if Q is on the same vertical line segment as P but not within distance ε from the orbit of P.)

(4) Consider the system $(\mathbb{S}, f)$ defined in the Example after Lemma 3.3.5. Every pair of points of $\mathbb{S}$ can be connected by an ε-chain: follow orbits and make ε-jumps at the invariant points [0] and [1/2], taking care that (if necessary) one jumps into the past of the target.

These examples show that, even for arbitrarily small ε, the ε-pseudo-orbits can deviate considerably from real orbits. This is because the errors add up: the point x_n does not approximate $f^n(x_0)$, but $f(x_{n-1})$, where x_{n-1} itself is already an approximation. However, one can approximate any initial segment of an orbit as close as one wants by a δ-chain for sufficiently small δ:

Proposition 4.4.2. *Let $x_0 \in X$ and let $N \geq 1$. Then for every $\varepsilon > 0$ there exists $\delta > 0$ such that*

$$d(x_n, f^n(x_0)) < \varepsilon \quad \text{for } n = 0, 1, \ldots, N$$

for every δ-chain $(x_0, x_1, \ldots, x_N)$.

Proof. The proof is by induction in N. For $N = 1$ the result is obviously true for every $\varepsilon > 0$, with $\delta = \varepsilon$. Now suppose the statement is true for some $N \in \mathbb{N}$ and every $\varepsilon > 0$.

As f is continuous at the point $f^N(x_0)$ there exists $\delta_1 > 0$ such that

$$d(f(z), f^{N+1}(x_0)) < \tfrac{1}{2}\varepsilon \quad \text{for all } z \in B_{\delta_1}(f^N(x_0)). \tag{4.4-1}$$

Without restriction of generality we may assume that $\delta_1 \leq \varepsilon$. By the induction hypothesis (with δ_1 instead of ε) there exists $\delta > 0$ such that for every δ-chain $(x_0, \ldots, x_N)$ one has $d(x_n, f^n(x_0)) < \delta_1$ for $n = 0, 1, \ldots, N$. We may assume that $\delta \leq \frac{1}{2}\varepsilon$.

Now consider an arbitrary δ-chain $(x_0, \ldots, x_N, x_{N+1})$. By the induction hypothesis we have $d(x_n, f^n(x_0)) < \delta_1 \leq \varepsilon$ for $n = 0, 1, \ldots, N$. For the final value $n = N + 1$ we infer from (4.4-1) and the triangle equality that

$$\begin{aligned} d(x_{N+1}, f^{N+1}(x_0)) &\leq d(x_{N+1}, f(x_N)) + d(f(x_N), f^{N+1}(x_0)) \\ &< \delta + \tfrac{1}{2}\varepsilon \leq \varepsilon . \end{aligned}$$

as (x_N, x_{N+1}) is part of a δ-chain and $x_N \in B_{\delta_1}(f^N(x_0))$. So the statement is true for $N + 1$ instead of N. □

A point $x \in X$ is said to be *chain-recurrent* whenever for every $\varepsilon > 0$ there is an ε-chain from x to x. The set of all chain-recurrent points under f is called the *chain-recurrent set*; it is denoted by $CR(X, f)$.

We shall see below that, despite the nomenclature, chain-recurrent points look more like non-wandering points than like recurrent points.

Proposition 4.4.3. $\Omega(X, f) \subseteq CR(X, f)$.

Proof. Let $x_0 \in \Omega(X, f)$ and let $\varepsilon > 0$. By continuity of f there is a neighbourhood U of x_0 such that $d(f(y), f(x_0)) \leq \varepsilon$ for all $y \in U$. We may assume that $U \subseteq S_\varepsilon(x_0)$. Since the point x_0 is non-wandering there is a point $y_0 \in U$ such that $f^n(y_0) \in U$ for some $n \in \mathbb{N}$. Now it is easily checked that $(x_0, f(y_0), \ldots, f^{n-1}(y_0), x_0)$ is an ε-chain. □

Corollary 4.4.4. *For all $x \in X$ one has $\omega(x) \subseteq CR(X, f)$. In particular, if X is compact then $CR(X, f) \neq \emptyset$.*

Proof. The first statement follows immediately from Proposition 4.3.5. This implies the second statement, as in a compact space $\omega(x) \neq \emptyset$ for every point x. □

Examples.

(1) In general, $P(X, f) \subseteq AP(X, f) \subseteq R(X, f) \subseteq \Omega(X, f) \subseteq CR(X, f)$, where $P(X, f)$ and $AP(X, f)$ denote the sets of periodic and almost periodic points, respectively, of the dynamical system (X, f). All inclusions can be strict: see the Examples (1) and (2) after Corollary 4.2.3, Example (2) after Proposition 4.3.1 and, finally, Example (3) below.

(2) In a transitive system all points are chain-recurrent. More generally, in a topologically ergodic system every point is chain-recurrent: use Exercise 4.4 (1).

(3) In general, $\Omega(X, f) \neq CR(X, f)$: see the Examples (3) and (4) in 4.4.1, where all points are chain-recurrent but many points are wandering.

This follows also from the following example: remove the two invariant points [0] and [1/2] from the phase space in Example (4) of 4.4.1. Then we have a system on a locally compact (= Baire) metric space where still all points are chain-recurrent, but which has no non-wandering points. As there are no recurrent points, this example also shows that Theorem 4.3.8 does not hold with $\Omega(X, f)$ replaced by $CR(X, f)$.

(4) Let $f: [0; \infty) \to [0; \infty)$ be a monotonously increasing function such that $f(0) = 0$ and $f(1) = 1$. Moreover, assume that $f(x) > x$ for $0 < x < 1$ and that $f(x) < x$ for $x > 1$ (for example, $f(x) = \sqrt{x}$ for $x \geq 0$). By Example (1) above, the invariant points 0 and 1 are chain-recurrent. The following argument shows that there are no other chain-recurrent points: let $0 < x_0 < 1$ and let $\varepsilon := 1/2(f(x_0) - x_0)$. By induction one easily shows that $x_n > x_0$ for $n = 0, \ldots, N$ for every ε-chain $(x_0, \ldots, x_N)$: if $x_{n-1} > x_0$ then $x_n \geq f(x_{n-1}) - \varepsilon \geq f(x_0) - \varepsilon = x_0 + \varepsilon$. In particular, $x_N \neq x_0$. So the point x_0 is not chain-recurrent. A similar argument works for $x_0 > 1$.

NB 1. With slightly more effort one can show that the monotonicity condition can be omitted from this example, provided the intervals $[0; 1]$ and $[1; \infty)$ are invariant.

NB 2. Similar arguments show that if $f(0) = 0$ and either $f(x) > x$ or $f(x) < x$ for every $x > 0$, and f is monotonously increasing, then 0 is the only chain-recurrent point in $[0; \infty)$.

(5) Consider the mapping $f: \mathbb{R}^2 \to \mathbb{R}^2$ defined by

$$f(x) := \left(\operatorname{sign}(x_1)\sqrt{|x_1|}, \tfrac{1}{2}x_2\right) \quad \text{for } x = (x_1, x_2) \in \mathbb{R}^2 ,$$

where $\operatorname{sign}(t) := 1$ or -1 if $t \geq 0$ or $t < 0$, respectively. See Figure 4.6. One easily sees that $CR(\mathbb{R}^2, f) = \{(-1, 0), (0, 0), (1, 0)\}$ (three invariant points). The inclusion "$\supseteq$" is obvious and "$\subseteq$" follows from Example (4) above: for every non-invariant point there is an $\varepsilon > 0$ such that for every ε-chain starting in that point either the first coordinate or the second coordinate (or both) of the end-point is different from the corresponding coordinate of the starting point.

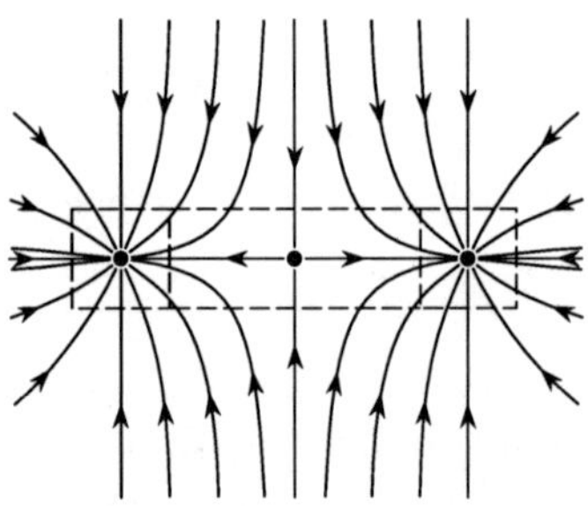

Fig. 4.6. The curves with the arrows indicate invariant sets along which the points of $\mathbb{R}^2$ hop under application of f. (For the dotted rectangles, see Example (3) at the end of this section.)

Proposition 4.4.3 states that non-wandering points are chain-recurrent under any compatible metric. In general, however, chain-recurrence depends on the metric being used. So it is not a dynamical property.

Example. Let (X, f) be the system of Example (4) in 4.4.1 from which the points $[0]$ and $[1/2]$ are omitted. All points of this system are chain-recurrent. Define a system (Y, g) as follows: Let $Y := \mathbb{R} \times \{-1, 1\}$ and let $\varphi\colon X \to Y$ be the projection of X (viewed as a subset of the unit circle in $\mathbb{R}^2$) onto Y, with the origin as the centre of projection. So $\varphi([t]) := ((\tan 2\pi t)^{-1}, 1)$ for $0 < t < 1/2$ and $\varphi([t]) := (-(\tan 2\pi t)^{-1}, -1)$ for $1/2 < t < 1$. It is straightforward to show that $\varphi\colon X \to Y$ is a homeomorphism. Next, define $g\colon Y \to Y$ as $g := \varphi \circ f \circ \varphi^{-1}$. Then $\varphi\colon (X, f) \to (Y, g)$ is an isomorphism of dynamical systems. But it is clear from geometric considerations that no point in the system (Y, g) is chain-recurrent with respect to the metric in Y inherited from the Euclidean metric of $\mathbb{R}^2$.

However, chain recurrence is a dynamical property in the class of dynamical systems with compact metric phase spaces:

Theorem 4.4.5. *Let $\varphi\colon (X, f) \to (Y, g)$ be a morphism of dynamical systems and assume that X is compact. Then $\varphi[CR(X, f)] \subseteq CR(Y, g)$. In particular, if φ is a conjugation then a point $x \in X$ is chain-recurrent under f iff the point $\varphi(x)$ is chain-recurrent under g.*

Proof. Note that φ is uniformly continuous with respect to the metrics in X and Y. Hence for every $\varepsilon > 0$ there exists $\delta > 0$ such that the image under φ of a δ-chain in X is an ε-chain in Y. This implies the result. □

Theorem 4.4.6. *The chain-recurrent set is closed and invariant.*

Proof. First, we show that the set $CR(X, f)$ is closed.

Consider a point $x_0 \in \overline{CR(X, f)}$; the proof that $x_0 \in CR(X, f)$ looks much like the proof of Proposition 4.4.3. Let $\varepsilon > 0$ be arbitrary. We want to show that there is an ε-chain from x_0 to x_0. Continuity of f at the point x_0 and the fact that x_0 is in the closure of the set $CR(X, f)$ imply that there is within distance $\varepsilon/2$ of x_0 a chain-recurrent point y_0 such that $f(y_0)$ has distance at most $\varepsilon/2$ to $f(x_0)$. Let $(y_0, y_1, \dots, y_{n-1}, y_n = y_0)$ be an $\varepsilon/2$-chain. See Figure 4.7 (a).

Since the distances of $f(y_0)$ to y_1 and to $f(x_0)$ both are at most $\varepsilon/2$, it follows that the distance of y_1 to $f(x_0)$ is at most ε. Moreover, the distances of $y_n = y_0$ to $f(y_{n-1})$ and to x_0 are at most $\varepsilon/2$, hence the distance of x_0 to $f(y_{n-1})$ is at most ε. Consequently, $(x_0, y_1, \dots, y_{n-1}, x_0)$ is an ε-chain. This completes the proof that $CR(X, f)$ is a closed set.

Next, we show that the set $CR(X, f)$ is invariant. Let $x_0 \in CR(X, f)$; we have to show that $f(x_0) \in CR(X, f)$. Let $\varepsilon > 0$ be arbitrary. Continuity of f in the point $f(x_0)$ implies that there exists a $\delta > 0$ such that $f(x)$ is within distance $\varepsilon/2$ from $f^2(x_0)$ for all $x \in B_\delta(f(x_0))$. We may assume that $\delta < \varepsilon/2$. As x_0 is a chain-recurrent point, there is a δ-chain $(x_0, x_1, \dots, x_{n-1}, x_0)$ from x_0 to itself. Then, obviously, $(x_0, x_1, \dots, x_{n-1}, x_0, f(x_0))$

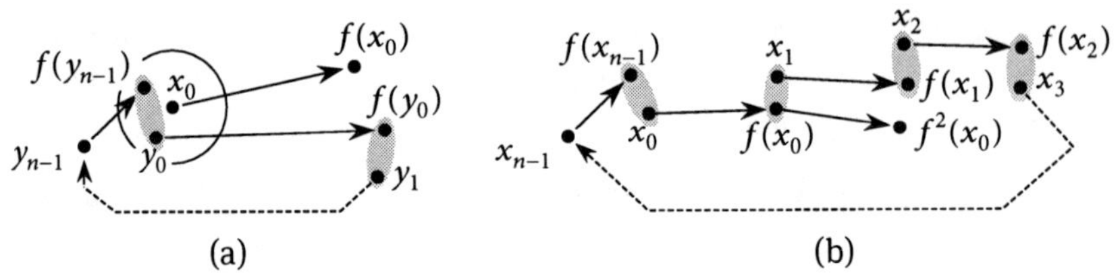

Fig. 4.7. The dotted lines represent parts of pseudo-orbits and the grey ellipses connect pairs of points that are close because of the definition of a pseudo-orbit.

is a δ-chain as well, and therefore $(x_1, \dots, x_{n-1}, x_0, f(x_0))$ is also a δ-chain. Note that x_1 has distance at most δ to $f(x_0)$. Hence the distance of $f^2(x_0)$ to $f(x_1)$ is at most $\varepsilon/2$. As the distance of $f(x_1)$ to x_2 is at most δ it follows that the distance of $f^2(x_0)$ to x_2 is at most $\varepsilon/2 + \delta \leq \varepsilon$. Consequently, $(f(x_0), x_2, \dots, x_{n-1}, x_0, f(x_0))$ is an ε-chain. See Figure 4.7 (b). This completes the proof. □

Remark. The reader may be worried about the fact that we silently have assumed that the δ-chain $(x_0, x_1, \dots, x_{n-1}, x_0)$ is long enough so as to contain a point x_2. But what to do if the δ-chain under consideration has the form (x_0, x_1) with $x_1 = x_0$? In that case, put $x_2 := x_0$ and $x_3 := x_1$. Then (x_0, x_1, x_2, x_3) is also a δ-chain from x_0 to itself, and the above proof can be given without any changes. See also Exercise 4.9.

Just like for non-wandering points, a point that is chain-recurrent in a subsystem is chain-recurrent in the full system. And just like for non-wandering points, the converse is not true in general.

Examples.

(1) In Example (3) in 4.4.1 every point is chain-recurrent in the full system, but in the subsystem on any of the invariant sets $\{t\} \times [0; 1]$ only the two invariant end points are chain-recurrent. In the subsystem on $\{t\} \times (0; 1)$ no point is chain-recurrent.

(2) In Example (4) in 4.4.1 the subsystem on any one of the two invariant closed half circles has only two chain-recurrent points (the invariant points). The subsystem on any of the two invariant open half circles has no chain-recurrent points at all.

(3) Consider the system (X^*, f^*) of the Example preceding Proposition 4.3.1 – which is the example described right after Lemma 3.1.1 – and let $X := X^* \setminus (\{\infty\} \times \mathbb{N})$ and $f := f^*|_X$. In addition, let $C' := \{-\infty\} \times \mathbb{N}$. The space X^* is compact and second countable, hence metrizable; see Appendix A.7.7. So X is metrizable as well (but not compact) and it makes sense to discuss chain-recurrence in this example (even though we refrain from an explicit description of a metric). As in the example preceding Proposition 4.3.1 one shows that all points of the set $C' \cup \{(0, \infty)\}$ are non-wandering in (X, f), hence chain-recurrent (no knowledge of the metric is needed for this conclusion). This implies that $C' \cup \{(0, \infty)\} \subseteq CR(X, f)$. We shall show that this inclusion is an equality.

Let x_0 be a point in the orbit of the point $(0, 1)$. Then x_0 is isolated, so if ε is sufficiently small then $\{x_0\}$ is a closed ε-ball around the point x_0, in which case an ε-chain $(x_0, \ldots, x_{N-1}, x_N = x_0)$ from x_0 to itself satisfies the condition $f(x_{N-1}) = x_0$. Thus, the point x_0 has a predecessor in the orbit of $(0, 1)$ and the ε-chain includes this predecessor. As this is true for every sufficiently small ε, it follows that this predecessor is chain-recurrent as well. By induction, it follows that if the orbit of the point $(0, 1)$ contains a chain-recurrent point then the point $(0, 1)$ itself is chain-recurrent. Thus, we arrive at a contradiction, as this point has no predecessor. This completes the proof that $CR(X, f) = C' \cup \{(0, \infty)\}$.
Finally, we claim that the points of C' are not chain-recurrent in the subsystem on the set $CR(X, f)$. For these points are isolated in $C' \cup \{(0, \infty)\}$ and a reasoning similar to the above – with the point $(0, 1)$ replaced by $(\infty, 1)$ – proves the claim.

Example (3) above shows that, in general, $CR(CR(X, f), f)$ can be a proper subset of $CR(X, f)$. In compact spaces the situation is better:

Theorem 4.4.7. *Let X be a compact metric space. Then every chain-recurrent point of the system (X, f) is chain-recurrent in the subsystem on $CR(X, f)$. Consequently,*

$$CR(CR(X, f), f) = CR(X, f).$$

Proof. Let $x_0 \in CR(X, f)$. We have to prove that for every $\varepsilon > 0$ there exists an ε-chain in $CR(X, f)$ from x_0 to x_0. To this end we shall construct a subset C of X (depending on the choice of x_0) such that $x_0 \in C$ and such that for each point $y \in C$ there is, for every $\varepsilon > 0$, an ε-chain from y to itself consisting of points of C. Then, obviously, $C \subseteq CR(X, f)$, hence the points of the ε-chains for the points y of C mentioned above are all in $CR(X, f)$. In particular, this holds for the point x_0 (recall that $x_0 \in C$), which was to be proved.

We start with the construction of the set C. For every $n \in \mathbb{N}$ there is a $1/n$-chain $(x_0 = x_0^{(n)}, \ldots, x_{N(n)}^{(n)} = x_0)$ in X. Extend this chain to a one-sided infinite periodic sequence: put $x_i^{(n)} := x_{i \,(\mathrm{mod}\, N(n))}^{(n)}$ for $i \in \mathbb{Z}^+$. Obviously, *every finite segment of this infinite sequence is a $1/n$-chain.* Finally, let $C_n := \{\, x_0^{(n)}, \ldots, x_{N(n)-1}^{(n)} \,\} = \{\, x_i^{(n)} : i \in \mathbb{Z}^+ \,\}$ and define

$$C := \bigcap_{k \in \mathbb{N}} \overline{\bigcup_{n \geq k} C_n} .$$

Note that $x_0 \in C$, so $C \neq \emptyset$.

We claim that for every point in C and for every $\varepsilon > 0$ there is an ε-chain from that point to itself consisting of points in C. So let $y \in C$ and let $\varepsilon > 0$. As f is uniformly continuous on X there exists $\delta > 0$ such that

$$d(f(x'), f(y')) < \tfrac{1}{3}\varepsilon \quad \text{for all } x', y' \in X \quad \text{with } d(x', y') < \delta . \tag{4.4-2}$$

We may assume that $\delta < \varepsilon/3$. The definition of C implies that every neighbourhood of the point y meets the set C_n for infinitely many values of $n \in \mathbb{N}$. Moreover, by

Appendix A.2.2, every neighbourhood of C includes the set $\overline{\bigcup_{n\geq k} C_n}$ for almost all k, hence it includes almost all sets C_n. Hence there is an $n \in \mathbb{N}$ such that

(1) $1/n < \varepsilon/3$;

(2) $B_\delta(y)$ meets the set C_n: there exists $i \in \mathbb{Z}^+$ such that $d(y, x_i^{(n)}) < \delta$;

(3) $C_n \subseteq B_\delta(C)$: if $x' \in C_n$ then $d(x', y') < \delta$ for some $y' \in C$.

Fix $n \in \mathbb{N}$ and $i \in \mathbb{Z}^+$ according to 1, 2 and 3 and select for $j = 0, \dots, N(n)$ a point $y_j \in C$ such that $d(y_j, x_{i+j}^{(n)}) < \delta$ and such that $y_{N(n)} = y_0 = y$, as follows: For a start, put $y_0 := y$. Clearly, $y_0 \in C$ by the choice of the point y, and condition 2 just states that $d(y_0, x_{i+0}^{(n)}) < \delta$. Next, for $1 \leq j \leq N(n) - 1$ the point $x_{i+j}^{(n)}$ belongs to C_n, hence condition 3 above guarantees that there exists a point $y_j \in C$ such that[6] $d(x_{i+j}^{(n)}, y_j) < \delta$. Finally, put $y_{N(n)} := y_0$ and recall that $x_{i+N(n)}^{(n)} = x_i^{(n)}$, so that for the point $y_{N(n)}$ the required inequality is fulfilled, because it is fulfilled for y_0.

The proof will be finished if we can show that $(y_0, y_1, \dots, y_{N(n)} = y_0)$ is an ε-chain, for then this will be an ε-chain in C from the point y to itself. This is straightforward: if $0 \leq j \leq N(n) - 1$ then the triangle equality implies that that the distance $d(f(y_j), y_{j+1})$ is at most equal to the sum of the three numbers $d(f(y_j), f(x_{i+j}^{(n)}))$, $d(f(x_{i+j}^{(n)}), x_{i+j+1}^{(n)})$ and $d(x_{i+j+1}^{(n)}, y_{j+1})$. By the choice of the point y_j, formula (4.4-2) and condition 1 above, this is less than or equal to $\frac{1}{3}\varepsilon + \frac{1}{n} + \delta < \varepsilon$. Here we have used that $(x_i^{(n)}, \dots, x_{i+N(n)}^{(n)})$ is a $1/n$-chain. See also Figure 4.8. □

Example. Consider the system (X^*, f^*) of the example after Theorem 3.1.1. It is easily seen that $CR(X^*, f^*) = C \cup \{(0, \infty)\}$: use arguments similar to those used in Example (3) preceding Theorem 4.4.7. But in contradistinction with the latter example we now have $CR(CR(X^*, f^*), f^*) = CR(X^*, f^*)$ as a consequence of Theorem 4.4.7. This can also be shown directly, using that for any compatible metric on X^*, every ε-ball

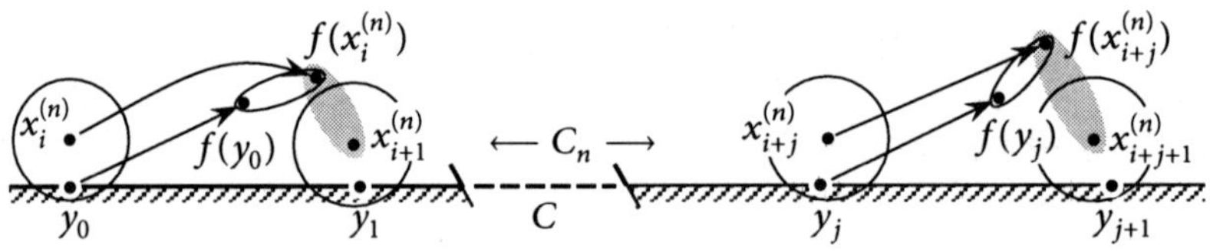

Fig. 4.8. Very schematic illustration of the proof of Theorem 4.4.7. The circles denote δ-neighbourhoods, the grey ellipses connect points with distance at most $1/n$ and the white ellipses connect points with distance at most $\varepsilon/3$.

6 In the proof of Theorem 4.4.11 below a set C will be constructed in the same way as in the present proof. There we shall need the following: the sequence $(x_k^{(n)})_{k\in\mathbb{Z}^+}$ is periodic with period $N(n)$, hence $x_{i+j}^{(n)} = x_0^{(n)} = x_0 \in C$ for some $j \in \{0, \dots, N(n) - 1\}$ (in point of fact, $j = N(n) - i$). For that value of j one may choose $y_j = x_0$.

around the point $(0, \infty)$ contains almost all points of $\{-\infty\} \times \mathbb{N}$ and of $\{\infty\} \times \mathbb{N}$: one can construct ε-chains in C that 'jump' over the point $(0, \infty)$.

NB. Note that $\Omega(\Omega(X^*, f^*), f) \subsetneqq \Omega(X^*, f^*)$. See the Example preceding Proposition 4.3.1.

A non-empty subset K of X is said to be *chain-transitive* whenever for any two points x and y in K and every $\varepsilon > 0$ there is an ε-chain in X from x to y. The ε-chain in the definition need not consist of points of K (except for its begin and its end). In particular, K is not required to be invariant. In addition, it follows that a non-empty subset of a chain-transitive set is chain-transitive as well.

> If for every $\varepsilon > 0$ an ε-chain exists consisting of points of K then K is said to be *internally chain-transitive*. It is not too difficult to show that a compact internally chain-transitive set is invariant; the proof resembles the 'invariance' part of the proof of Theorem 4.4.6.

The points x and y occur symmetrically in this definition: in addition to the existence of ε-chains from x to y for every $\varepsilon > 0$ it requires also the existence of ε-chains from y to x. This observation leads to:

Lemma 4.4.8. *Let K be a non-empty subset of X. The following conditions are equivalent:*
(i) *K is chain-transitive.*
(ii) $\forall\, x, y \in K\ \forall \varepsilon > 0 : \exists \varepsilon$*-chain from x to x containing y.*
(iii) $\exists\, x_0 \in K \,\vdots\, \forall x \in X\ \forall \varepsilon > 0 : \exists \varepsilon$*-chain from x_0 to x_0 containing x.*
(iv) $\exists\, x_0 \in K \,\vdots\, \forall x \in X\ \forall \varepsilon > 0 : \exists \varepsilon$*-chain from x to x containing x_0.*

Proof. "(i)⇒(ii)": Concatenate a chain from x to y with a chain from y to x.

"(ii)⇒(iii)⇔(iv)": Obvious.

"(iv)⇒(i)": If (iv) holds and $x, y \in K$ then for every $\varepsilon > 0$ there are ε-chains from x to x_0 and vice versa, and from x_0 to y and vice versa. Concatenate these chains. □

Examples.
(1) In the Examples (2), (3) and (4) in 4.4.1 the whole phase space is chain-transitive. In Example (3), every set $\{t\} \times [0; 1]$ with $0 \leq t \leq 1$ is chain-transitive.
(2) In Example (3) after Theorem 4.4.6, the set $C' \cup \{(0, \infty)\} = CR(X, f)$ is chain-transitive. The proof is similar to the proof that all points of this set are chain-recurrent. Note that $C' \cup \{(0, \infty)\} = \omega(1, 0)$, so this example will also follow from Proposition 4.4.9 below.

Proposition 4.4.9. *If $x \in X$ and $\omega(x) \neq \emptyset$ then the set $\omega(x)$ is chain-transitive.*

Proof. This proof combines the ideas of the proofs of Propositions 4.3.5 and 4.4.3. If $y, z \in \omega(x)$ then $f(y) \in \omega(x)$ as well, and Lemma 1.4.1 (1) implies that for every $\varepsilon > 0$ there are $m, n \in \mathbb{N}$ with $m < n$, such that $f^m(x) \in S_\varepsilon(f(y))$ and $f^n(x) \in S_\varepsilon(z)$. Then $(y, f^m(x), \dots, f^{n-1}(x), z)$ is an ε-chain. □

Recall that if a set is chain-transitive then every non-empty subset is chain-transitive as well. In particular, every singleton-subset of a chain-transitive set is chain-transitive, which obviously means that its unique point is chain-recurrent (note that in the definition we do not require that the points x and y are distinct); see also the implication (i)⇒(ii) in Lemma 4.4.8. *Thus, every chain-transitive set is a subset of $CR(X, f)$.* Conversely, if a point x of X is chain-recurrent then the set $\{x\}$ is chain-transitive. So the singletons $\{x\}$ for $x \in CR(X, f)$ are the smallest chain-transitive sets. In the other direction it may be interesting to look for *maximal* chain-transitive sets. A non-empty subset of X is called a *chain-component of X* or a *basic set* whenever it is a maximal chain-transitive set. So a non-empty subset K of X is a basic set iff K is chain-transitive and if B is a chain-transitive subset of X such that $B \supseteq K$ then $B = K$.

Examples.

(1) In the systems in Example (1) after Lemma 4.4.8 the whole phase space is a basic set.

(2) Consider the system $([0; 1], f)$ with $f(x) := \sqrt{x}$ for $0 \le x \le 1$. In Example (4) after Corollary 4.4.4 it was shown that $CR([0; 1], f) = \{0, 1\}$. Similar arguments show that for sufficiently small ε there can be no ε-chain from the point 1 to the point 0. Consequently, $\{0\}$ and $\{1\}$ are maximal chain-transitive sets.

(3) Recall that in Example (5) after Corollary 4.4.4, $CR(\mathbb{R}^2, f) = \{(-1, 0), (0, 0), (1, 0)\}$. By arguments similar to those used in the proof in Example (2) above it is easily seen that the only chain-transitive sets are $K_{-1} := \{(-1, 0)\}$, $K_0 := \{(0, 0)\}$ and $K_1 := \{(1, 0)\}$. So it is clear that these three sets are the basic sets.

Proposition 4.4.10.

(1) *The closure of a chain-transitive set is chain-transitive. In particular, every basic set is closed.*

(2) *Every basic set in invariant.*

(3) *If X is compact then every basic set is completely invariant.*

Proof. (1) Let K be a chain-transitive set. In order to prove that $\overline{K}$ is chain-transitive, select a point $x_0 \in K$ and consider an arbitrary point $x \in \overline{K}$. By Lemma 4.4.8, it is sufficient to prove the following claim: for every $\varepsilon > 0$ there is a 2ε-chain from x to x_0 and a 2ε-chain from x_0 to x.

So let ε be an arbitrary non-negative real number. The open neighbourhood $B_\varepsilon(x)$ of x contains a point y of K, so there is an ε-chain $(x_0, \dots, x_{N-1}, y)$ from x_0 to y. By the triangle inequality, $d(f(x_{N-1}), x) \le d(f(x_{N-1}), y) + d(y, x) \le 2\varepsilon$, which implies that $(x_0, \dots, x_{N-1}, x)$ is a 2ε-chain from x_0 to x.

In order to prove the existence of a 2ε-chain in the other direction, note that there is neighbourhood U of x such that $d(f(x), f(x')) < \varepsilon$ if $x' \in U$. Then U includes a point z of K, and there is an ε-chain $(z, z_1, \dots, z_{N-1}, x_0)$ from z to x_0. Obviously, $d(f(x), z_1) \le d(f(x), f(z)) + d(f(z), z_1) < 2\varepsilon$, so that $(x, z_1, \dots, z_{N-1}, x_0)$ is a 2ε-chain from x to x_0.

This completes the proof that $\overline{K}$ is chain-transitive. In particular, if K is a basic set then $\overline{K}$ is chain-transitive, so (by maximality of K) $\overline{K} = K$, i.e., K is closed.

(2) Let K be a basic set and consider a point $x_0 \in K$. We shall show that the set $K \cup \{f(x_0)\}$ is chain-transitive. As K is a maximal chain-transitive set it follows that $K \cup \{f(x_0)\} = K$, that is, $f(x_0) \in K$. This is true for every $x_0 \in K$, so K is invariant.

Since K is chain-transitive, there is for every $\delta > 0$ a δ-chain from x_0 to itself. Given $\varepsilon > 0$, the proof of the invariance of the chain-recurrent set in Theorem 4.4.6 shows how to find $\delta > 0$ such that a δ-chain as just mentioned can be modified into an ε-chain from $f(x_0)$ to itself which contains the point x_0. As this can be done for every $\varepsilon > 0$, it follows from Lemma 4.4.8 that the set $K \cup \{f(x_0)\}$ is chain-transitive.

(3) Let K be a basic set and assume that X is compact. If K consists of only one point then the fact that K is invariant obviously implies that K is completely invariant, and we are done. So we may assume that K includes at least two points.

Let $x \in K$. By the assumption that K has at least two points, there is a point $z \in K$ different from x. Since K is chain-transitive there exists for every $n \in \mathbb{N}$ a $1/n$-chain from z to itself that contains the point x, say,

$$C_n := (z = y_0^{(n)}, \ldots, y_{k(n)-1}^{(n)}, y_{k(n)}^{(n)} = x, \ldots, y_{N(n)}^{(n)} = z)$$

with $1 \le k(n) < N(n)$. If $k(n) = 1$ for some value of n then for this value of n one has $d(f(z), x) = d(f(y_0^{(n)}), y_1^{(n)}) \le 1/n$. If this is the case for infinitely many values of n then $x = f(z)$ and the proof is complete. Consequently, we may assume that $k(n) \ge 2$ for almost all n.

Since X is compact, we may assume (by passing to a suitable subsequence) that the sequence $(y_{k(n)-1}^{(n)})_{n\in\mathbb{N}}$ converges in X to a point $y \in X$ (recall that these points are not required to be situated in K). Continuity of f implies that the sequence $(f(y_{k(n)-1}^{(n)}))_{n\in\mathbb{N}}$ converges to $f(y)$. For each n we have $d(f(y_{k(n)-1}^{(n)}), x) \le 1/n$, hence $f(y) = x$. It remains to show that $y \in K$. To this end we shall prove that $K \cup \{y\}$ is a chain-transitive set. As K is a maximal chain-transitive set, this will imply that $y \in K$.

In order to prove that the set $K \cup \{y\}$ is chain-transitive it is sufficient to show that for every $\varepsilon > 0$ there is a 2ε-chain from z to itself which contains the point y. If $\varepsilon > 0$ then select $n \in \mathbb{N}$ such that $k(n) \ge 2$, that $1/n < \varepsilon$ and that the points $f(y_{k(n)-1}^{(n)})$ and y have distance at most ε. Then it is straightforward to show that if we replace in C_n the point $y_{k(n)-1}^{(n)}$ by the point y we get a 2ε-chain: the distance of the points $f(y_{k(n)-2}^{(n)})$ and y is at most $1/n + \varepsilon \le 2\varepsilon$, and the distance of the points $f(y)$ and x is zero. This completes the proof. □

It should be clear that a chain-transitive subset in a subsystem of (X, f) is chain-transitive in the full system. The converse is not generally true: see the Examples (1) and (2) after Theorem 4.4.6. Moreover, Theorem 4.4.13 (1) below implies that every basic set of a subsystem, being chain-transitive in the full system, is included in a basic set of the full system. The examples just mentioned show that this inclusion may be strict. However, our next theorem implies that in a compact metric space a basic subset is chain-transitive in the subsystem on that basic set. The proof is a modification of the proof of Theorem 4.4.7. (In point of fact, Theorem 4.4.7 would be an easy consequence of this result; we leave it for the reader to prove this.)

Theorem 4.4.11. *Let X be a compact metric space. Then every basic set is chain-transitive (hence a basic set) in the subsystem on that basic set*[7] *.*

Proof. Let K be a basic set in X under f and let $x_0, z_0 \in K$. We have to show that for every $\varepsilon > 0$ there is an ε-chain in K from z_0 to itself containing the point x_0. In order to do so, we proceed as in the proof of Theorem 4.4.7 with only a minor modification.

Construct a set C as in the proof of Theorem 4.4.7, using $1/n$-chains in X from the point x_0 to itself that also contain the point z_0 (as K is chain-transitive, this is possible). Then we get not only $x_0 \in C$, but also $z_0 \in C$. Moreover, note that in the part of the proof showing that every point y of C belongs to the set $CR(X, f)$, we may assume that the ε-chain $(y = y_0, y_1, \ldots, y_{N(n)} = y)$ in C constructed there includes the point x_0. This shows in the first place that y can be added to any chain-transitive set containing the point x_0 without destroying chain-transitivity; in particular, y belongs to the maximal chain-transitive set that contains the point x_0, that is $y \in K$. This shows that $C \subseteq K$. Consequently, for every point $y \in C$ and for every $\varepsilon > 0$ the ε-chain in C constructed above, from y to itself that contains the point x_0, is included in K. This holds, in particular, for the point $z_0 \in C$. This completes the proof. □

Example. Example (3) just before Theorem 4.4.7 – where $CR(X, f) = C' \cup \{(0, \infty)\}$ consists of just one basic set – shows that compactness of the phase space cannot be omitted from the assumptions of the theorem.

We come now to one of the main results of this section: X can be decomposed into the (mutually disjoint) basins of the basic sets and the set of point that 'go nowhere'.

Lemma 4.4.12. *Let $\{K_\alpha\}_{\alpha \in A}$ be a family of chain-transitive sets such that every two members of this family have a non-empty intersection. Then the set $\bigcup_{\alpha \in A} K_\alpha$ is chain-transitive as well.*

Proof. Let $x, y \in \bigcup_{\alpha \in A} K_\alpha$, say, $x \in K_\alpha$ and $y \in K_\beta$ for $\alpha, \beta \in A$. There exists $z \in K_\alpha \cap K_\beta$, hence for every $\varepsilon > 0$ there are ε-chains between x and z and between y and z. Concatenate these chains. □

Theorem 4.4.13.

(1) *Every chain-transitive set is included in a maximal chain-transitive set, i.e., in a basic set.*
(2) *The basic sets form a partition of $CR(X, f)$ in closed invariant sets.*

7 Stated otherwise: every basic set is internally chain-transitive.

Proof. (1) Let K be a chain-transitive set and let $\mathcal{K}$ be the collection of all chain-transitive sets in which K is included. Then $K \in \mathcal{K}$, so $\mathcal{K} \neq \emptyset$. Consequently, $K_0 := \bigcup \mathcal{K} \neq \emptyset$, and it follows easily from Lemma 4.4.12 that the set K_0 is chain-transitive. Then $K \subseteq K_0$ and it is straightforward to show that K_0 is a maximal chain-transitive set.
(2) We have already observed that all chain-transitive sets – hence all basic sets – are included in $CR(X, f)$. Conversely, if $x \in CR(X, f)$ then the set $\{x\}$ is a chain-transitive, hence part 1 of this theorem implies that x is included in a basic set. It follows that the set $CR(X, f)$ is the union of all basic sets. Moreover, two distinct basic sets are mutually disjoint, otherwise their union is chain-transitive, contradicting their maximality as basic sets. Together with Proposition 4.4.10 (1),2 this completes the proof. □

Corollary 4.4.14. *If $x \in X$ and $\omega(x) \neq \emptyset$ then $\omega(x)$ is included in a basic set of X.*

Proof. Clear from Theorem 4.4.13 (1) and Proposition 4.4.9. □

Example. Not every point of a basic set is necessarily included in a limit set: see the Examples (3) and (4) in 4.4.1, where the phase space is a basic set.

Theorem 4.4.15. *Let $\mathcal{K}$ denote the collection of all basic sets of X. Then*

$$X = \bigcup \{ \mathcal{B}(K) : K \in \mathcal{K} \} \cup \{ x \in X : \omega(x) = \emptyset \}. \tag{4.4-3}$$

This union actually is a partition: the sets $\mathcal{B}(K)$ for $K \in \mathcal{K}$ are mutually disjoint and disjoint from the set of points with empty limit set.

Proof. In order to prove equality (4.4-3), consider any point in X for which $\omega(x) \neq \emptyset$. Then Corollary 4.4.14 implies that there is $K \in \mathcal{K}$ such that $x \in \mathcal{B}(K)$. This completes the proof of the equality.

Next, we show that (4.4-3) defines a decomposition of X into mutually disjoint sets. Let $K, L \in \mathcal{K}$, $K \neq L$. Then $K \cap L = \emptyset$ by Theorem 4.4.13 (2), hence $\mathcal{B}(K) \cap \mathcal{B}(L) = \emptyset$ by Lemma 3.1.3 (3). Moreover, if $x \in \mathcal{B}(K)$ then, by definition, $\omega(x)$ is not empty. It follows that the sets $\bigcup\{\mathcal{B}(K) : K \in \mathcal{K}\}$ and $\{x \in X : \omega(x) = \emptyset\}$ are disjoint. □

Discussion.
(1) Usually, Theorem 4.4.15 is formulated under the assumption that X is compact (in which case the part $\{x \in X : \omega(x) = \emptyset\}$ is missing, because it is empty) or, at least, that all basic sets are compact. We shall point out now the consequence of the assumption that all members of $\mathcal{K}$ are compact, or rather, the consequence of not assuming this. If every member of $\mathcal{K}$ is compact then, by Lemma 3.1.3 (5), $K \subseteq \mathcal{B}(K)$ for every $K \in \mathcal{K}$, hence $CR(X, f) = \bigcup\{K : K \in \mathcal{K}\} \subseteq \bigcup\{\mathcal{B}(K) : K \in \mathcal{K}\}$. Consequently, the sets $CR(X, f)$ and $\{x \in X : \omega(x) = \emptyset\}$ are disjoint. This is not necessarily true if not all members of $\mathcal{K}$ are assumed to be compact. To see what this means, split every member K of $\mathcal{K}$ in two mutually disjoint sets K' and K'', as follows: let $K' := \{x \in K : \emptyset \neq \omega(x) \subseteq K\} = K \cap \mathcal{B}(K)$ and let $K'' := K \setminus K' = K \setminus \mathcal{B}(K)$. If for some point $x \in K''$ it would be the case that $\omega(x) \neq \emptyset$ then $\omega(x) \subseteq K$ (for $x \in K$ and K is closed and invariant), hence $x \in \mathcal{B}(K)$, contradicting the choice of $x \in K''$. It follows that $K'' = \{x \in K : \omega(x) = \emptyset\}$.

For a given basic set K, either of K' or K'' can be empty (but not both, of course). For example, if K is compact then $K \subseteq \mathcal{B}(K)$, hence $K' = K$ and $K'' = \emptyset$. It is also possible that $\mathcal{B}(K) = \emptyset$ for some $K \in \mathcal{K}$, in which case $K' = \emptyset$ and $K'' = K$; for example, this is the case for the basic set C in Example (2) below.

It follows from the discussion above that there are two kinds of points with empty limit set: those that are chain-recurrent – in the above notation, the points of $\bigcup_{K\in\mathcal{K}} K''$ – and those that are not. To get a more uniform representation, omit from both sides of equality (4.4-3) the points of $CR(X, f)$, i.e., omit all points that belong to one of the basic sets. Thus, omit the points of K' from $\mathcal{B}(K)$ for each $K \in \mathcal{K}$ and omit $\bigcup_{K\in\mathcal{K}} K''$ from the set $\{x \in X : \omega(x) = \emptyset\}$. The result is

$$X \setminus CR(X, f) = \bigcup_{K\in\mathcal{K}} (\mathcal{B}(K) \setminus K) \cup \{x \in X \setminus CR(X, f) : \omega(x) = \emptyset\}. \tag{4.4-4}$$

So X is decomposed in the set of all chain-recurrent points, sets of non-chain-recurrent points (which are outside of any basic set) moving towards a basic set and the set of non-chain-recurrent points with empty limit set.

(2) Nothing has been said about basic sets being topologically attracting, so the basins mentioned above need not be open sets. In point of fact, if X is compact (or if otherwise the set $\{x \in X : \omega(x) = \emptyset\}$ is empty), and X is connected and there is more than one basic set then not all basic sets can be topologically attracting: otherwise (4.4-3) would give a decomposition of X in (at least two) mutually disjoint non-empty open sets, contradicting connectedness. See also Example (3) below.

Examples.

(1) Let (X, f) be the system of the example preceding Proposition 4.3.1 from which the point $(0, \infty)$ is omitted. There is one basic set, viz., C; see the arguments in the example after Theorem 4.4.7 and Example (2) after Lemma 4.4.8. Clearly, $\mathcal{B}(C) = X \setminus C$, so $C' = C \cap \mathcal{B}(C) = \emptyset$ and $C'' = C$.

(2) Let (X, f) be the system of the example preceding Proposition 4.3.1, modified in the following way: the point $(0, \infty)$ is omitted, $f := (f^*)^{-1}$ on $X^* \setminus \{(0, 1)\}$ and the point (0,1) is invariant. Let $K := \{(0, 1)\}$. Using arguments similar to those used in the example after Theorem 4.4.7 and in Example (2) after Lemma 4.4.8 one shows that $CR(X, f) = C \cup K$ and that C and K are the only basic sets. Note that $\mathcal{B}(C) = \emptyset$, in accordance with the fact that $C = \{x \in X : \omega(x) = \emptyset\}$. Hence $C' = \emptyset$ and $C'' = C$. Moreover, $\mathcal{B}(K) = X \setminus C$, so $K \subseteq \mathcal{B}(K)$ and, consequently, $K' = K$ and $K'' = \emptyset$.

(3) Consider Example (5) after Corollary 4.4.4. Recall that $CR(\mathbb{R}^2, f) = \{(-1, 0), (0, 0), (1, 0)\}$ and that $K_{-1} := \{(-1, 0)\}$, $K_0 := \{(0, 0)\}$ and $K_1 := \{(1, 0)\}$ are the basic sets. Since there are no points with empty limit set, decomposition (4.4-3) reads in this case as $\mathbb{R}^2 = \mathcal{B}(K_{-1}) \cup \mathcal{B}(K_0) \cup \mathcal{B}(K_1)$. Straightforward computations (or Corollary 3.4.4: use the dotted rectangles in Figure 4.6) show that the sets $A := [-1; 1] \times \{0\}$, K_{-1} and K_1 are asymptotically stable, but that K_0 is not. In point of fact, $\mathcal{B}(K_{-1})$ is the open left half plane, $\mathcal{B}(K_1)$ is the open right half plane and $\mathcal{B}(K_0)$ is the x_2-axis (which is not open).

4.5 Asymptotic stability and basic sets

In this section we consider a dynamical system on a locally compact metric space with metric d. We shall combine the results from the previous section with those of the Sections 3.4 and 3.5.

We start with the introduction of yet another type of stable sets. A non-empty compact completely invariant[8] subset A of X is said to be *chain-stable* whenever for every neighbourhood U of A there is an $\varepsilon > 0$ such that every ε-chain starting in A is included in U.

Example. Consider the mapping $f: x \mapsto x^2 : [0;1] \to [0;1]$. Then 0 is a chain-stable invariant point. In order to prove this, show by induction that for every ε-chain $(0 = x_0, x_1, \dots, x_N)$ with $\varepsilon < 1/4$ one has $x_n \leq 2\varepsilon$ for $n = 0, \dots N$ (use that $x_n \leq f(x_{n-1}) + \varepsilon$ and that $x^2 \leq x/2$ for $0 \leq x \leq 1/2$).

The following proposition provides us with a large class of chain-stable sets (of which the above example is a special case):

Proposition 4.5.1. *Every asymptotically stable subset of X is chain-stable.*

Proof. Let U be a neighbourhood of A. By Theorem 3.4.6, A has a compact neighbourhood W such that $W \subseteq U$ and $f[W] \subseteq W^\circ$ (so W is invariant). Since the compact set $f[W]$ and the closed set $X \setminus W^\circ$ are disjoint they have a positive distance d_W. Let $0 < \varepsilon < d_W$. Claim: every ε-chain $(x_0, \dots, x_N)$ with $x_0 \in W$ is included in W. This immediately implies that every ε-chain that starts in A remains in U.

The proof of the claim is by induction. For $N = 1$ the claim is certainly true: if $x_0 \in W$ then the point $f(x_0)$ is situated in $f[W]$, hence it has distance at least ε to the set $X \setminus W^\circ$, so the point x_1 – which has distance at most ε to $f(x_0)$ – cannot be in $X \setminus W^\circ$. It follows that $x_1 \in W^\circ \subseteq W$. Next, assume that the claim is true for every ε-chain $(x_0, \dots, x_N)$ with $x_0 \in W$. Then for any ε-chain $(x_0, \dots, x_N, x_{N+1})$ with $x_0 \in W$ one has $x_i \in W$ for $i = 0, \dots, N$ (by hypothesis), and a reasoning similar to the one above shows that $x_{N+1} \in W$. □

Example. The converse of this proposition is not true. Consider Example (4) in 3.2.6: a system on the unit interval $[0;1]$ in which the invariant point 0 is not asymptotically stable. However, the point 0 is easily seen to be chain-stable: it has arbitrary small asymptotically stable, hence chain-stable, neighbourhoods.

Notwithstanding this example there is a result in the other direction: chain-stable sets are 'limits' of sequences of asymptotically stable sets. For the proof it will be convenient to make the following definition: if A is a non-empty subset of X and ε is a positive real number then

$$\mathcal{P}_\varepsilon(A) := \{ x \in X : \text{there is an } \varepsilon\text{-chain in } X \text{ from } A \text{ to } x \}$$

8 This definition is slightly redundant: see Exercise 4.10-2.

is called the *ε-attainable set* of A. If $0 < \delta < \varepsilon$ then, obviously, every δ-chain is an ε-chain, hence $\mathcal{P}_\delta(A) \subseteq \mathcal{P}_\varepsilon(A)$. As every initial segment of an orbit is an ε-chain, it is clear that $\bigcup_{n\geq 1} f^n[A] \subseteq \mathcal{P}_\varepsilon(A)$. Consequently, if $A \subseteq f[A]$ then $A \subseteq \mathcal{P}_\varepsilon(A)$.

Some definitions could have been formulated in terms of ε-attainable sets. For example, a non-empty subset K of X is chain-transitive iff $K \subseteq \mathcal{P}_\varepsilon(\{x\})$ for every $x \in K$ and every $\varepsilon > 0$. Similarly, a non-empty compact completely invariant subset A of X is chain-stable iff for every neighbourhood U of A there exists $\varepsilon > 0$ such that $\mathcal{P}_\varepsilon(A) \subseteq U$.

Lemma 4.5.2. *Let*[9] *A be a non-empty subset of X and let* $\varepsilon > 0$.
1. *If* $\kappa > \varepsilon$ *then* $\overline{\mathcal{P}_\varepsilon(A)} \subseteq \mathcal{P}_\kappa(A)$.
2. *The set* $\mathcal{P}_\varepsilon(A)$ *has a non-empty interior and if A is completely invariant then* $\mathcal{P}_\varepsilon(A)$ *is a neighbourhood of A.*
3. *Always*

$$f[\,\overline{\mathcal{P}_\varepsilon(A)}\,] \subseteq \mathcal{P}_\varepsilon(A)^\circ \subseteq (\,\overline{\mathcal{P}_\varepsilon(A)}\,)^\circ . \tag{4.5-1}$$

and if $\mathcal{P}_\varepsilon(A) \neq X$ *then even*

$$\mathrm{dist}(f[\,\overline{\mathcal{P}_\varepsilon(A)}\,], X \setminus \mathcal{P}_\varepsilon(A)^\circ) \geq \varepsilon . \tag{4.5-2}$$

Proof. (1) The proof of 1 is quite standard: if $x \in \overline{\mathcal{P}_\varepsilon(A)}$ then the $(\kappa - \varepsilon)$-ball around x contains the end point y of an ε-chain starting in A. Replace y by x and get in this manner a $((\kappa - \varepsilon) + \varepsilon)$-chain from A to x.

(2) The statements to be proved will turn out to be straightforward consequences of the following observation:

$$B_\varepsilon(f[A]) \subseteq \mathcal{P}_\varepsilon(A) . \tag{4.5-3}$$

In order to prove this inclusion, consider a point $y \in B_\varepsilon(f[A])$. Then there is a point $x \in A$ such that $d(f(x), y) < \varepsilon$. This means that the 2-element sequence (x, y) is an ε-chain. It starts in A and, consequently, $y \in \mathcal{P}_\varepsilon(A)$. This completes the proof of (4.5-3). Obviously, (4.5-3) implies that all points of the set $f[A]$ are interior points of $\mathcal{P}_\varepsilon(A)$. As $f[A]$ is not empty, it follows that $\mathcal{P}_\varepsilon(A)$ has a non-empty interior. Moreover, if A is completely invariant then $A = f[A]$. So in that case (4.5-3) implies that $\mathcal{P}_\varepsilon(A)$ is a neighbourhood of A.

(3) The statements in 3 turn out to be consequences of the following variation of the inclusion (4.5-3) above:

$$B_\varepsilon(f[\mathcal{P}_\varepsilon(A)]) \subseteq \mathcal{P}_\varepsilon(A) . \tag{4.5-4}$$

In order to prove this inclusion, consider a point $y \in B_\varepsilon(f[\mathcal{P}_\varepsilon(A)])$ and let $x \in \mathcal{P}_\varepsilon(A)$ be such that $d(f(x), y) < \varepsilon$. Then (x, y) is an ε-chain; concatenate it with an ε-chain from a point of A to x – which exists because $x \in \mathcal{P}_\varepsilon(A)$ – and conclude that $y \in \mathcal{P}_\varepsilon(A)$. This proves (4.5-4).

9 In this lemma local compactness of X is not needed.

We proceed by proving (4.5-1) and (4.5-2). If $\mathcal{P}_\varepsilon(A) = \emptyset$ then (4.5-1) is obviously true, so we may and shall assume that $\mathcal{P}_\varepsilon(A) \neq \emptyset$. In that case, the inclusion (4.5-4) implies

$$\operatorname{dist}(f[\mathcal{P}_\varepsilon(A)], X \setminus \mathcal{P}_\varepsilon(A)) \geq \varepsilon . \tag{4.5-5}$$

Thus, on the subset $f[\mathcal{P}_\varepsilon(A)] \times (X \setminus \mathcal{P}_\varepsilon(A))$ of $X \times X$ the metric d is greater than or equal to ε . Continuity of d on $X \times X$ then implies that d is also at least ε on the closure of this set in $X \times X$, i.e., on $\overline{f[\mathcal{P}_\varepsilon(A)]} \times \overline{X \setminus \mathcal{P}_\varepsilon(A)}$. Using that $f[\,\overline{\mathcal{P}_\varepsilon(A)}\,] \subseteq \overline{f[\mathcal{P}_\varepsilon(A)]}$ and $\overline{X \setminus \mathcal{P}_\varepsilon(A)} = X \setminus \mathcal{P}_\varepsilon(A)^\circ$, we get

$$\operatorname{dist}(f[\,\overline{\mathcal{P}_\varepsilon(A)}\,], X \setminus \mathcal{P}_\varepsilon(A)^\circ\,) \geq \varepsilon .$$

This proves (4.5-2). It also follows that $f[\,\overline{\mathcal{P}_\varepsilon(A)}\,] \subseteq \mathcal{P}_\varepsilon(A)^\circ$, which trivially implies the final inclusion in (4.5-1). □

Theorem 4.5.3. *Let A be a non-empty compact and completely invariant subset of X. The following conditions are equivalent:*
(i) *A is chain-stable.*
(ii) *A is the intersection of a descending sequence of asymptotically stable sets.*

Proof. "(i)⇒(ii)": Assume that A is chain-stable. Define for every $k \in \mathbb{N}$

$$P_k(A) := \{\, x \in X : \text{ there is a } \frac{1}{k}\text{-chain in } X \text{ from } A \text{ to } x \,\}$$

(so $P_k(A) := \mathcal{P}_{1/k}(A)$ for $k \in \mathbb{N}$). Note that for $k, m \in \mathbb{N}$ with $k > m$ one has $P_k(A) \subseteq P_m(A)$. It follows that the sets $P_k(A)$ for $k \in \mathbb{N}$ – hence also their closures – form a descending sequence.

The assumption that A is chain-stable implies that for every neighbourhood V of A there exists $\varepsilon > 0$ such that $\mathcal{P}_\varepsilon(A) \subseteq V$. Since for all $k \in \mathbb{N}$ with $1/k < \varepsilon$ we have $P_k(A) \subseteq \mathcal{P}_\varepsilon(A) \subseteq V$, it follows that $\overline{P_k(A)} \subseteq \overline{V}$. By formula (A.2–1) in Appendix A and the observation that $A \subseteq P_k(A)$ (which follows from Lemma 4.5.2 (2), as A is completely invariant) this implies that

$$\bigcap_{k \in \mathbb{N}} \overline{P_k(A)} = A . \tag{4.5-6}$$

Since X is locally compact the sets $\overline{V}$ may be assumed to be compact, hence almost all sets $\overline{P_k(A)}$ are compact.

For every $k \in \mathbb{N}$, let

$$A_k := \bigcap_{n \geq 0} f^n[\,\overline{P_k(A)}\,] .$$

For almost every k – say, for $k \geq k_0$ – the set $W_k := \overline{P_k(A)}$ is compact and has a non-empty interior W_k° such that, by (4.5-1), $f[W_k] \subseteq W_k^\circ$. In view of Corollary 3.4.4 this implies that for $k \geq k_0$ the set A_k is asymptotically stable. Finally, for every k one has $A \subseteq A_k$ because $A \subseteq P_k(A)$ and A is completely invariant, and one also has $A_k \subseteq \overline{P_k(A)}$. Consequently, it follows from (4.5-6) that $A = \bigcap_{k \in \mathbb{N}} A_k = \bigcap_{k \geq k_0} A_k$, the intersection of a descending sequence of asymptotically stable sets.

"(ii)⇒(i)": Let $A = \bigcap_{n\in\mathbb{N}} A_n$, where $A_{n+1} \subseteq A_n$ and A_n is asymptotically stable for every $n \in \mathbb{N}$. Let U be a neighbourhood of A. By Lemma (A.2.2), there exists $n \in \mathbb{N}$ such that $A_n \subseteq U$. Since by Proposition 4.5.1 the set A_n is chain-stable, there exists $\varepsilon > 0$ such that every ε-chain starting in A_n – hence those starting in A as well – are included in U. This shows that A is chain-stable. □

Corollary 4.5.4. *Let A be a chain-stable subset of X and define for every $\varepsilon > 0$ a set W_ε by $W_\varepsilon := \overline{\mathcal{P}_\varepsilon(A)}$. Then $f[W_\varepsilon] \subseteq W_\varepsilon^\circ$ for every $\varepsilon > 0$, and W_ε is compact for all sufficiently small ε. In addition, the sets W_ε for $\varepsilon > 0$ form a neighbourhood base of A.*

Proof. See the proof of Theorem 4.5.3, taking into account Lemma 4.5.2 (2). □

In view of the decomposition (4.4-4) it makes sense to characterize points that are situated in a set of the form $\mathcal{B}(K) \setminus K$ for a basic set K. We first consider asymptotically stable sets instead of basic sets.

Theorem 4.5.5. *Let $x_0 \in X$. The following conditions are equivalent:*
(i) *There is an asymptotically stable set A such that $x_0 \in \mathcal{B}(A) \setminus A$.*
(ii) *$x_0 \notin CR(X, f)$ and the set $\overline{\mathcal{P}_{\varepsilon_0}(\{x_0\})}$ is compact for some $\varepsilon_0 > 0$.*

Proof. "(i)⇒(ii)": Let A be an asymptotically stable subset of X and assume that $x_0 \in \mathcal{B}(A) \setminus A$. We show first that the point x_0 is not chain-recurrent.

By Proposition 4.5.1, there exists $\varepsilon_1 > 0$ such that every ε_1-chain starting in A is included in the neighbourhood $X \setminus \{x_0\}$ of A, which means that $x_0 \notin \mathcal{P}_{\varepsilon_1}(A)$. Lemma 4.5.2 (2) implies that $\mathcal{P}_{\varepsilon_1}(A)$ is a neighbourhood of A, hence of $\omega(x_0)$, so Lemma 1.4.1 (1) implies that $f^N(x_0) \in \mathcal{P}_{\varepsilon_1}(A)^\circ$ for some $N \in \mathbb{N}$. In particular, there exists $\varepsilon > 0$ such that

$$B_{2\varepsilon}(f^N(x_0)) \subseteq \mathcal{P}_{\varepsilon_1}(A)\,. \tag{4.5-7}$$

See Figure 4.9. By Lemma 3.1.3 (4), the point x_0 is not periodic, so the points $f^i(x_0)$ for $i = 0, 1, \ldots, N$ are all different from x_0. It follows that ε can be chosen such that also

$$2\varepsilon < d(f^i(x_0), x_0) \quad \text{for } i = 1, \ldots, N\,. \tag{4.5-8}$$

In view of Proposition 4.4.2 there exists $\delta > 0$ such that every δ-chain starting in x_0 approximates the first $N + 1$ points of the orbit of x_0 up to ε, that is, for every δ-chain $(x_0, \ldots, x_N)$ one has

$$d(x_i, f^i(x_0)) \leq \varepsilon \quad \text{for } i = 0, \ldots, N\,. \tag{4.5-9}$$

We may and shall assume that $\delta < \varepsilon_1$.

Consider an arbitrary δ-chain $(x_0, \ldots, x_M)$ that starts in the point x_0. If $M \leq N$ then (4.5-8) and (4.5-9) imply that $x_M \neq x_0$. If, on the other hand, $M > N$ then $x_N \in \mathcal{P}_{\varepsilon_1}(A)$ by (4.5-7) and (4.5-9), hence there is an ε_1-chain from a point of A to x_N. As $\delta < \varepsilon_1$, the δ-chain $(x_N, \ldots, x_M)$ is also an ε_1-chain. Concatenation of these chains shows that $x_M \in \mathcal{P}_{\varepsilon_1}(A)$, which implies that $x_M \neq x_0$. Conclusion: there is no δ-chain from x_0 to itself. Thus, the point x_0 is not chain-recurrent.

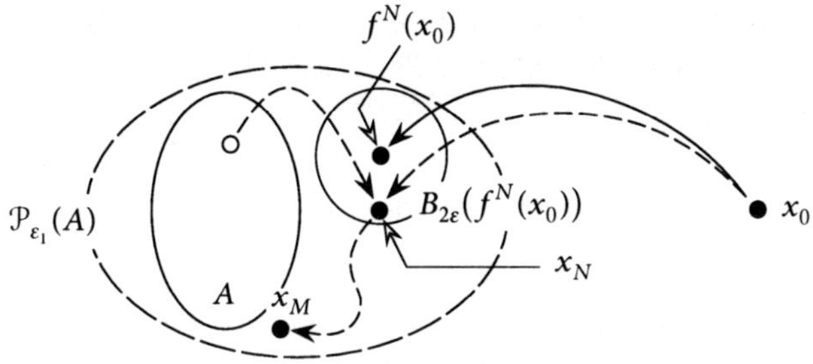

Fig. 4.9. Schematic representation of the proof of the implication (i)⇒(ii) in Theorem 4.5.5. The solid arrow represents an initial segment of the orbit of x_0, the dotted arrows represent pseudo-orbits.

In the above proof we have seen that $x_M \in \bigcup_{i=0}^N S_\varepsilon(f^i(x_0))$ if $M \leq N$, and that $x_M \in \mathcal{P}_{\varepsilon_1}(A)$ if $M > N$. As this holds for every δ-chain starting in x_0 this shows that

$$\mathcal{P}_\delta(\{x_0\}) \subseteq \bigcup_{i=0}^{N} S_\varepsilon(f^i(x_0)) \cup \mathcal{P}_{\varepsilon_1}(A)\,.$$

As X is locally compact we may assume that ε was chosen so small that the sets $S_\varepsilon(f^i(x_0))$ for $i = 0, \ldots, N$ are compact. In view of Corollary 4.5.4 we may also assume that $\mathcal{P}_{\varepsilon_1}(A)$ has a compact closure. Consequently, $\mathcal{P}_\delta(\{x_0\})$ has a compact closure as well.

"(ii)⇒(i)": Assume (ii). As the point x_0 is not chain-recurrent there exists $\varepsilon > 0$ such that $x_0 \notin \mathcal{P}_{2\varepsilon}(\{x_0\})$. We may replace ε by any smaller value, so we may assume that $\varepsilon < \varepsilon_0$. Then $\mathcal{P}_\varepsilon(\{x_0\}) \subseteq \mathcal{P}_{\varepsilon_0}(\{x_0\})$, hence the set $W := \overline{\mathcal{P}_\varepsilon(\{x_0\})}$ is compact by the choice of ε_0 according to (ii). Note that $x_0 \notin W$, because Lemma 4.5.2 (1) implies that $W_\varepsilon \subseteq \mathcal{P}_{2\varepsilon}(\{x_0\})$. Moreover, formula (4.5-1) implies that $f[W] \subseteq W^\circ$. So by Corollary 3.4.4, the set W includes the asymptotically stable set $A := \bigcap_{n\in\mathbb{N}} f^n[W]$. Obviously $x_0 \notin A$ (recall that $x_0 \notin W$), and we claim that that $x_0 \in \mathcal{B}(A)$.

The proof of this claim is standard. If U is a neighbourhood of A then, by Lemma A.2.2 in Appendix A, $f^n[W] \subseteq U$ for almost all n, hence Theorem 3.1.10 implies that $W \subseteq \mathcal{B}(A)$. Moreover, formula (4.5-3) implies that $f(x_0) \in \mathcal{P}_\varepsilon(\{x_0\}) \subseteq W$. Hence $f(x_0) \in \mathcal{B}(A)$ and, as $\mathcal{B}(A)$ is backwards invariant – see Lemma 3.1.3 (2) – also $x_0 \in \mathcal{B}(A)$. □

Remark. In the above theorem the asymptotically stable set A is included in the set $\overline{\mathcal{P}_{\varepsilon_0}(\{x_0\})}$. It is interesting to compare the equivalence in the above theorem with the following mutually equivalent statements[10] for a point $x_0 \in X$:

(i) There is a non-empty completely invariant compact set A such that $x_0 \in \mathcal{B}(A) \setminus A$.

(ii) x_0 is not recurrent and $\overline{\mathcal{O}(x_0)}$ is compact.

---

10 Note that in statement (i) A need not be asymptotically stable. As in the theorem above, local compactness of X is not needed for the implication (ii)⇒(i).

[Hints for the proofs: "(i)⇒(ii)": use Theorem 3.1.1 and Proposition 4.1.1. "(ii)⇒(i)": Let $A := \omega(x_0)$ and use Theorem 1.4.5, Proposition 4.1.1 and Exercise 3.2.]

So one might say that in Theorem 4.5.5 the set $\overline{\mathcal{P}_{\varepsilon_0}(\{x_0\})}$ takes over the role of the orbit closure of x_0 in the above statement. Actually, compactness of the set $\overline{\mathcal{P}_{\varepsilon_0}(\{x_0\})}$ implies compactness of the orbit closure of the point x_0 because, by Example (1) in 4.4.1, $\overline{\mathcal{O}(x_0)} \subseteq \mathcal{P}_\varepsilon(\{x_0\}) \cup \{x_0\}$ for all $\varepsilon > 0$.

Corollary 4.5.6. *Assume that X is compact or that for some other reason for every point x in $X \setminus CR(X, f)$ there exists $\varepsilon > 0$ such that the set $\mathcal{P}_\varepsilon(x)$ has a compact closure. Let $\mathcal{A}$ denote the set of all asymptotically stable subsets of X. Then*

$$X \setminus CR(X, f) = \bigcup \{\, \mathcal{B}(A) \setminus A : A \in \mathcal{A} \,\} . \tag{4.5-10}$$

Equivalently, $CR(X, f) = \bigcap_{A\in\mathcal{A}}(A \cup A^*)$, where for every $A \in \mathcal{A}$ the set A^* is defined as $A^* := \{x \in X : \omega(x) \cap A = \emptyset\}$.

Proof. Equality (4.5-10) is clear from the implication (ii)⇒(i) of the previous theorem. The equivalent equality is obtained by taking complements in equality (4.5-10), taking into account that for every $A \in \mathcal{A}$ the complement of the set $\mathcal{B}(A) \setminus A = (X \setminus A) \cap \mathcal{B}(A)$ is equal to the set $A \cup A^*$: recall from Corollary 3.2.2 (2) that for every $x \in X$, $\omega(x) \subseteq A$ iff $\omega(x) \cap A \neq \emptyset$. □

Examples. To get an idea of how Corollary 4.5.6 works, see Figure 4.10. The reader is also invited to illustrate the above theorem and its corollary by proving the following statements about the examples following Theorem 4.4.15.

(1) There are no asymptotically stable sets, but the set $X \setminus CR(X, f)$ is not empty, so formula (4.5-10) does not hold: for every point x of the orbit of (0,1) and for every $\varepsilon > 0$ the closure of the set $\mathcal{P}_\varepsilon(x)$ is not compact. Note that equality (4.5-10) holds in the system (X^*, f^*), with $\mathcal{A} = \{C \cup \{(0, \infty)\}\}$.
(2) There are two basic sets, C and K, one of which, namely, K, is asymptotically stable. The set $\mathcal{B}(K) \setminus K$ consists of the point (0,2) and its complete past. If x is a point in this set then for sufficiently small ε the set $\mathcal{P}_\varepsilon(\{x\})$ is equal to $\mathcal{O}(x)$, which is a finite, hence compact, set. Accordingly, these points are not chain-recurrent.
(3) Now $\mathcal{A} = \{K_{-1}, K_1, A\}$. All points of $\mathcal{B}(A) \setminus A = \mathbb{R}^2 \setminus A$ satisfy condition (i) of Theorem 4.5.5. So for every point x of $\mathbb{R}^2 \setminus A$ there exists $\varepsilon > 0$ such that $\mathcal{P}_\varepsilon(\{x\})$ has a compact closure. In addition, these points are not chain-recurrent. By using similar arguments with K_{-1} or K_1 instead of A one shows that the above conclusions hold for all points of $\mathbb{R}^2$ which are not in the set $K_{-1} \cup K_0 \cup K_1$, which is equal to the set $CR(\mathbb{R}^2, f)$. In particular, equality (4.5-10) holds. Note that we need all members of $\mathcal{A}$ in order to get this equality, even though K_{-1} and K_1 are subsets of A.

Let us make two observations, both of which are illustrated by Example (3) above. The first observation is that the sets $\mathcal{B}(A) \setminus A$ in formula (4.5-10) are not necessarily mutually disjoint. Moreover, it may be not possible to select a subset $\mathcal{A}'$ of $\mathcal{A}$ such that

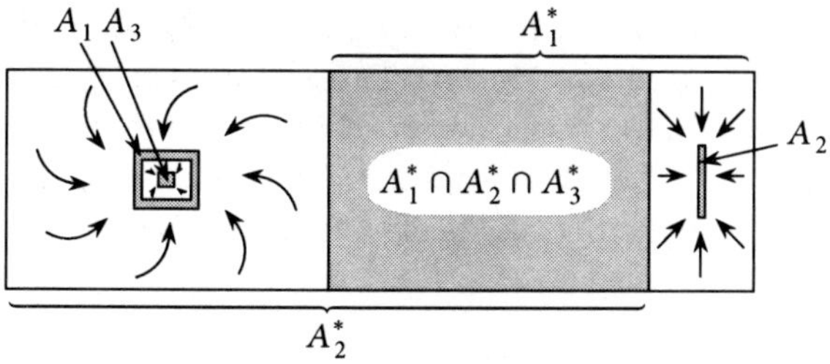

Fig. 4.10. Schematic representation of a compact system with three asymptotically stable subsets A_1, A_2 and A_3 such that $A_1 \cap A_2 = \emptyset$ and $\mathcal{B}(A_3) \subsetneq A_1$. The set A_3^* is supposed to consist of the union of the ring-shaped grey area within A_1 and the complement of A_1.

the sets $\mathcal{B}(A) \setminus A$ for $A \in \mathcal{A}'$ are mutually disjoint and still cover (i.e., form a partition of) the set $X \setminus CR(X, f)$.

Our second observation is the following. Suppose that X satisfies the conditions of Corollary 4.5.6. If $x \in X \setminus CR(X, f)$ then there is an asymptotically stable set A such that $x \in \mathcal{B}(A) \setminus A$. However, formula (4.4-4) implies that there is a basic set K such that $x \in \mathcal{B}(K) \setminus K$ (the possibility that $\omega(x) = \emptyset$ does not apply, because $\emptyset \neq \omega(x) \subseteq A$). It would be nice if $K = A$: then the asymptotically stable set A would be a basic set. The example above for points on the x_2-axis shows that this is not to be expected. But what we can say in general is that $\emptyset \neq \omega(x) \subseteq A \cap K$. By the next proposition this implies that $K \subseteq A$. Of course, if $K = A$ then A is a basic set. In the other case, if K is a proper subset of A then A is cannot be a basic set, as distinct basic sets are disjoint.

Proposition 4.5.7. *Let K be a basic set. If A is an asymptotically stable set and $A \cap K \neq \emptyset$ then $K \subseteq A$. Consequently, $K \subseteq A$ or $K \subseteq A^*$.*

Proof. Let $x_0 \in A \cap K$. Because $x_0 \in K$, the definition of a basic set implies that for every $\varepsilon > 0$ one has $K \subseteq \mathcal{P}_\varepsilon(\{x_0\})$. As $x_0 \in A$, it follows that $\mathcal{P}_\varepsilon(\{x_0\}) \subseteq \mathcal{P}_\varepsilon(A)$. Consequently, $K \subseteq \mathcal{P}_\varepsilon(A)$. On the other hand, by Proposition 4.5.1, the set A is chain-stable, so for every neighbourhood U of A there is an $\varepsilon > 0$ such that $\mathcal{P}_\varepsilon(A) \subseteq U$, hence $K \subseteq U$. Now it follows from formula (A.1-3) in Appendix A that $K \subseteq A$.

The final statement of the proposition is an easy consequence of what has been proved already and the fact that all points of K are chain-recurrent, so that, by in Corollary 4.5.6, $K \subseteq A \cup A^*$. □

Remark. See Exercise 4.11 (1) for a more general statement.

So the general picture is that for every asymptotically stable set A there are one or more basic sets K included in A such that $\mathcal{B}(A) \setminus A$ is the (disjoint) union of the sets $\mathcal{B}(K) \setminus K$ for those basic sets K. In this context it is interesting to have a characterization of the asymptotically stable sets that are basic sets. To this end we first prove a lemma:

Lemma 4.5.8. *The intersection of two asymptotically stable sets, if not empty, includes another asymptotically stable set.*

Proof. Let A_1 and A_2 be asymptotically stable sets such that $A_1 \cap A_2 \neq \emptyset$. By the implication (i)⇒(iii) in Theorem 3.4.2, A_i has a compact invariant neighbourhood U_i such that $A_i = \bigcap_{n\geq 0} f^n[U_i]$ $(i = 1, 2)$. Obviously, $U_1 \cap U_2$ is a compact invariant neighbourhood of $A_1 \cap A_2$. Let $A := \bigcap_{n=0}^{\infty} f^n[U_1 \cap U_2]$. Then A is compact and non-empty, because the sets $f^n[U_1 \cap U_2]$ for $n \geq 0$ form a descending sequence of non-empty compact sets. Moreover,

$$A \subseteq \bigcap_{n=0}^{\infty} \left(f^n[U_1] \cap f^n[U_2] \right) = \bigcap_{n=0}^{\infty} f^n[U_1] \cap \bigcap_{n=0}^{\infty} f^n[U_2] = A_1 \cap A_2 ,$$

so the neighbourhood $U_1 \cap U_2$ of $A_1 \cap A_2$ is also a neighbourhood of A. Hence by the implication (iii)⇒(i) in Theorem 3.4.2, A is asymptotically stable. □

Remark. In general, $A_1 \cap A_2$ itself is not completely invariant and, consequently, not an asymptotically stable set. But if $A_1 \cap A_2$ is completely invariant then it is an asymptotically stable. See Exercise 3.10-(1),(2).

The preceding lemma will be used in the following way: assume that A is an asymptotically stable subset of X that has no proper subsets that are asymptotically stable in X. If B is another asymptotically stable subset of X such that $A \cap B \neq \emptyset$ then the lemma and the assumption on A imply that a subset of $A \cap B$ equals A, hence $A \subseteq B$.

Theorem 4.5.9. *Let A be an asymptotically stable set. The following conditions are equivalent:*
(i) *A has no proper asymptotically stable subsets.*
(ii) *A is a basic set.*
(iii) *A is chain-transitive.*

Proof. "(i)⇒(iii)": Assume that A has no proper asymptotically stable subsets. We shall show first that all points of A are chain-recurrent. Suppose the contrary and let $z \in A \setminus CR(X, f)$. Proposition 4.5.1 and Corollary 4.5.4 imply that there exists $\varepsilon_0 > 0$ such that the set $\mathcal{P}_{\varepsilon_0}(A)$ has a compact closure. Consequently, its subset $\mathcal{P}_{\varepsilon_0}(\{z\})$ has a compact closure as well. By assumption, the point z is not chain-recurrent, so by Theorem 4.5.5 there exists an asymptotically stable subset A' of X such that $z \in \mathcal{B}(A') \setminus A'$. As $\emptyset \neq \omega(z) \subseteq A'$ and $\omega(z) \subseteq A$ (recall that $z \in A$ and that A is closed and invariant), it follows that $A \cap A' \neq \emptyset$, hence $A \subseteq A'$ by the observation preceding this theorem. This contradicts the choice if z as a point of A and the choice of A' such that $z \notin A'$. This completes the proof that $A \subseteq CR(X, f)$.

Now Theorem 4.4.13 (2) implies that every point of A belongs to a basic set which, by Proposition 4.5.7, is included in A. Thus, A is the disjoint union of basic sets. Let K be such a basic set. We want to show that $A = K$.

Claim. $A \subseteq \mathcal{P}_{2\varepsilon}(K)$ for every $\varepsilon > 0$.

Assume that this claim is true. As K is chain-transitive, concatenation of chains implies that for every $\varepsilon > 0$ and for every choice of points $x \in K$ and $y \in A$ there is a 2ε-chain from x to y. Now suppose that there is a second basic set K' included in A. Then K' has the same property as K. In particular, for every $\varepsilon > 0$ points of K and K' can be connected by 2ε-chains, in both directions. This would imply that the set $K \cup K'$ is chain-transitive, contradicting that K and K' are maximal chain-transitive sets. Consequently, A consists of just one basic set and is, consequently, chain-transitive (besides, this proves (ii)).

It remains to prove that the above claim holds true. To this end, define for every $\varepsilon > 0$ a set $W_\varepsilon := \overline{\mathcal{P}_\varepsilon(K)}$. Since $K \subseteq A$ it is clear that $W_\varepsilon \subseteq \overline{\mathcal{P}_\varepsilon(A)}$, hence W_ε is compact for sufficiently small ε (recall from Corollary 4.5.4 that $\overline{\mathcal{P}_\varepsilon(A)}$ is compact for sufficiently small ε). Moreover, formula (4.5-1) implies that $f[W_\varepsilon] \subseteq W_\varepsilon^\circ$. So by Corollary 3.4.4, the set $A_\varepsilon := \bigcap_{n\geq 0} f^n[W_\varepsilon]$ is asymptotically stable, provided ε is sufficiently small. We shall show below that $A \subseteq A_\varepsilon$ for sufficiently small ε, hence for all $\varepsilon > 0$ (note that W_ε, and therefore A_ε as well, grows if ε shrinks). Since it is clear that $A_\varepsilon \subseteq W_\varepsilon$, whereas $W_\varepsilon \subseteq \mathcal{P}_{2\varepsilon}(K)$ by Lemma 4.5.2 (1), this proves our claim above.

In order to show that $A \subseteq A_\varepsilon$ it is sufficient to show that $K \subseteq A_\varepsilon$: this will imply that $A \cap A_\varepsilon \neq \emptyset$ (recall that $K \subseteq A$) which gives the desired result by the observation preceding this theorem. First, note that $K \subseteq W_\varepsilon$, because K is chain-transitive. This obviously implies that $\bigcap_{n\geq 0} f^n[K] \subseteq A_\varepsilon$. Also, $\bigcap_{n\geq 0} f^n[K] \subseteq K$. Here the left-hand side is not empty, because it is the intersection of a descending chain of non-empty compact sets: K is compact and invariant. It follows that $K \cap A_\varepsilon \neq \emptyset$. So Proposition 4.5.7 implies that $K \subseteq A_\varepsilon$. This concludes the proof.

"(iii)⇒(ii)": If A is chain-transitive then by Theorem 4.4.13 (1) it is included in a basic set K. Then Proposition 4.5.7 implies that $A = K$.

"(ii)⇒(i)": Let B be an asymptotically stable subset of A. Then Proposition 4.5.7 implies that the basic set A is included in the asymptotically stable set B, hence $A = B$. □

Remarks.

(1) In the case that X is a locally compact metric space, Theorem 3.2.8 follows from the above result, as every transitive set is chain-transitive.

(2) By the observation preceding Proposition 4.5.7, if X satisfies the conditions of Corollary 4.5.6 then every asymptotically stable set A includes a basic set. If A itself is not a basic set then A also properly includes at least one other asymptotically stable set. It is possible that A includes infinitely many asymptotically stable sets, nested or mutually disjoint. See Example (4) in 3.2.6, where there are various families of asymptotically stable sets (e.g., the nested sets A_n for even n, the mutually disjoint basic sets $\{a_n\}$ for even n, or the intervals $[a_{n+2}; a_{n-2}]$ for a sufficiently sparse set of even values of n). In this system one can also easily indicate asymptotically stable sets that are not included in $CR(X, f)$, or that are included in $CR(X, f)$, but that are unions of more than one basic set (e.g., two attracting points).

Let us reconsider Corollary 4.5.6 above. If A is an asymptotically stable set then the set $A^* := \{x \in X : \omega(x) \cap A = \emptyset\}$ is called the *repeller associated with*[11] A, and (A, A^*) is called an *attractor-repeller pair*. As $A^* = X \setminus \mathcal{B}(A)$, $f^{\leftarrow}[\mathcal{B}(A)] \subseteq \mathcal{B}(A)$ and $\mathcal{B}(A)$ is open – use Corollary 3.2.2 (2), Lemma 3.1.3 (2) and Theorem 3.2.7, respectively – it follows that A^* is a (possibly empty) closed invariant set. Moreover, as $A \subseteq \mathcal{B}(A)$ it is clear that $A \cap A^* = \emptyset$.

Examples.

(1) Let $X := [-1; 1]$ and $f(x) := x^2$ for $x \in X$. Then $A := \{0\}$ is an asymptotically stable set and the associated repeller is the set $A^* = \{-1, 1\}$. Note that for any neighbourhood $V := (-\varepsilon; \varepsilon)$ of 0 with $0 < \varepsilon < 1$ the set $F := X \setminus V$ is a compact invariant neighbourhood of A^* such that $A^* = \bigcap_{n\geq 0}(f^n)^{\leftarrow}[F]$.

(2) Let $X := [0; 1]$ and $f(x) := \frac{1}{2}x$ for $x \in X$. Then $A := \{0\}$ is an asymptotically stable set and the associated repeller A^* is empty. Accordingly, if F is defined as in 1 above then $(f^n)^{\leftarrow}[F] = \emptyset$ for large n.

(3) Consider Example (4) in 4.4.1. Then there is no proper asymptotically stable subset of X at all. This is in accordance with Theorem 4.5.9: X is an asymptotically stable set in (X, f) and X is chain-transitive.

Lemma 4.5.10. *Let A be an asymptotically stable set in the dynamical system (X, f) and let A^* be the associated repeller.*

(1) *A^* is closed and invariant, and $A \cap A^* = \emptyset$.*

(2) *If $A \neq X$ then there a closed subset F of X with non-empty interior F° such that $A^* \subseteq F^\circ$, $f^{\leftarrow}[F] \subseteq F^\circ$ and $A^* = \bigcap_{n\geq 0}(f^n)^{\leftarrow}[F]$.*

(3) *If X is compact, $f[X] = X$ and $A \neq X$ then $A^* \neq \emptyset$.*

Proof. (1) This was observed already after the definitions above.

(2) By Proposition 3.4.6[12], A has an open neighbourhood V with a compact closure included in $\mathcal{B}(A)$ such that $f[\overline{V}] \subseteq V$ and $A = \bigcap_{n\geq 0} f^n[V]$: let $V := W^\circ$ with W as in Proposition 3.4.6. Since $A \neq X$ we may assume that $\overline{V} \neq X$. Put $F := X \setminus V$. Then F is a closed subset of X with non-empty interior $F^\circ = X \setminus \overline{V}$. Moreover, $F^\circ = X \setminus \overline{V} \supseteq X \setminus \mathcal{B}(A) = A^*$. In addition, $f^{\leftarrow}[F] = X \setminus f^{\leftarrow}[V] \subseteq X \setminus \overline{V} = F^\circ$. Finally, formula (3.1-2) implies that $A^* = X \setminus \mathcal{B}(A) = \bigcap_{n\geq 0} (f^n)^{\leftarrow}[F]$.

(3) Let F be as in statement 2 of the lemma. Because f is surjective it is clear that $(f^n)^{\leftarrow}[F] \neq \emptyset$ for every $n \in \mathbb{Z}^+$. So by 2, A^* is the intersection of a descending sequence of non-empty closed subsets of a compact space, hence not empty. □

Example. The condition that $f[X] = X$ cannot be omitted from statement 3: under the mapping $f : x \mapsto x^2 : [0; 1/2] \to [0; 1/2]$ the repeller associated with the asymptotically stable set $A\{0\}$ is empty.

11 Also the terms *dual* repeller and *complementary* repeller are in use.

12 Recall our standing hypothesis that X is a locally compact metric space.

If X is compact and $f[X] = X$ then X is an asymptotically stable set. By Theorem 4.5.9, either X is chain-transitive (i.e., X is a basic set) and X has no proper asymptotically stable subsets, or X is not chain-transitive, in which case X has a proper asymptotically stable subset A. In the latter case, statement 3 of the lemma states that the associated repeller A^* of A is not empty. A similar reasoning can be applied to the subsystem on any asymptotically stable subset of X.

By Corollary 4.5.6, the study of the chain recurrent means studying the attractor-repeller pairs of the system. For each such a pair (A, A^*) the dynamics within A or A^* can be very complicated, but the dynamics of the points 'between' A and A^* is very simple: they move away from A^* towards A, and this movement is governed by a kind of 'potential function':

Proposition 4.5.11. *Assume that X is compact and let (A, A^*) be an attractor-repeller pair. Then there is a continuous function $g\colon X \to [0; 1]$ such that $g^{\leftarrow}[0] = A$, $g^{\leftarrow}[1] = A^*$ and $g(f(x)) < g(x)$ for all $x \in X \setminus (A \cup A^*)$.*

Proof. Define a mapping $\varphi\colon X \to [0; 1]$ by

$$\varphi(x) := \frac{d(x, A)}{d(x, A) + d(x, A^*)} \quad \text{for } x \in X\,.$$

Then φ is continuous, $\varphi^{\leftarrow}[0] = A$ and $\varphi^{\leftarrow}[1] = A^*$ (recall that that for any closed subset F of X, $d(x, F) = 0$ iff $x \in F$ and that A and A^* are disjoint). Next, let for every $x \in X$

$$\tilde{\varphi}(x) := \sup\{\,\varphi[\mathcal{O}(x)]\,\} = \sup\{\,\varphi(f^n(x)) : n \in \mathbb{Z}^+\,\}\,.$$

Obviously, $\tilde{\varphi}$ maps X into $[0; 1]$ and $0 \le \varphi(x) \le \tilde{\varphi}(x)$ for all $x \in X$. Moreover, as A is invariant and φ is 0 on A, we get $\tilde{\varphi}(x) = 0$ for $x \in A$. Conversely, if $x \in X$ and $\tilde{\varphi}(x) = 0$ then $\varphi(x) = 0$, which means that $x \in A$. Thus, $\tilde{\varphi}^{\leftarrow}[0] = A$.

Next, we show that $\tilde{\varphi}^{\leftarrow}[1] = A^*$. If $x \in A^*$ then $1 = \varphi(x) \le \tilde{\varphi}(x) \le 1$. So $\tilde{\varphi}^{\leftarrow}[1] \supseteq A^*$. Conversely, consider a point $x \in X$ with $\tilde{\varphi}(x) = 1$ and assume that $x \notin A^*$. By the definition of $\tilde{\varphi}$, there is a subsequence $(n_i)_{i\in\mathbb{N}}$ of $\mathbb{Z}^+$ such that $\varphi(f^{n_i}(x)) \rightsquigarrow 1$. By passing to a suitable subsequence, we may assume that there is a point $z \in X$ such that $f^{n_i}(x) \rightsquigarrow z$; note that $z \in \omega(x)$ by Lemma 1.4.1 (2). Then $\varphi(f^{n_i}(x) \rightsquigarrow \varphi(z)$, which implies that $\varphi(z) = 1$ and, consequently, that $z \in A^*$. So $A^* \cap \omega(x) \neq \emptyset$. On the other hand, we have assumed that $x \notin A^* = X \setminus \mathcal{B}(A)$, hence $\omega(x) \subseteq A$: a contradiction (recall that $A \cap A^* = \emptyset$). Consequently, $\tilde{\varphi}^{\leftarrow}[1] \subseteq A^*$.

We continue the proof of the proposition by showing that $\tilde{\varphi}$ is continuous on X. Continuity at every point of A^* is easy: if $x \in A^*$ and $\varepsilon > 0$ then, by continuity of φ, there is a neighbourhood U of x in X such that $\varphi(x') > 1 - \varepsilon$, hence $\tilde{\varphi}(x) = 1 \ge \tilde{\varphi}(x') \ge \varphi(x') > 1 - \varepsilon = \tilde{\varphi}(x) - \varepsilon$, for all $x' \in U$.

Next, consider a point $x \in A$. Let $\varepsilon > 0$; continuity of φ implies that there is a neighbourhood V of A such that $0 \le \varphi(x') < \varepsilon$ for all $x' \in V$. By Theorem 3.4.1 we may assume that A strongly attracts V: in particular, there exists $N \in \mathbb{N}$ such that $f^n[V] \subseteq V$ for all $n \ge N$, i.e., $\varphi(f^n(x')) < \varepsilon$ for all $x' \in V$ and $n \ge N$. Continuity of the (finitely many) functions $\varphi \circ f^n$ for $n = 0, \dots, N-1$ implies that there is a neighbourhood W of A in X such that $\varphi(f^n(x')) < \varepsilon$ for all $x' \in W$ and $n = 0, \dots, N-1$. Consequently, if $x' \in V \cap W$ then $\varphi(f^n(x')) < \varepsilon$ for all $n \in \mathbb{Z}^+$, hence $\tilde{\varphi}(0) = 0 \le \tilde{\varphi}(x') \le \varepsilon$. This establishes continuity of $\tilde{\varphi}$ at every point of A.

Finally, consider a point x in $X \setminus (A \cup A^*) = \mathcal{B}(A) \setminus A$. As $\mathcal{B}(A) \setminus A$ is open, there is a compact neighbourhood W of the point x such that $W \subseteq \mathcal{B}(A) \setminus A$. Let $r := \min\{\varphi(x') : x' \in W\}$; because φ assumes its minimum on the (compact) set W and W is disjoint from the set $\varphi^{\leftarrow}[0] = A$, it is clear that $r > 0$. By Theorem 3.4.1, W is strongly attracted by A, hence there exists a natural number N such that for every $n \ge N$ the set $f^n[W]$ is included in the neighbourhood $\{x' \in X : 0 \le \varphi(x') < \frac{1}{2}r\}$ of A. This means: if $x' \in W$ then for all $n \ge N$ we have $\varphi(f^n(x')) < \frac{1}{2}r$. On the other hand, if $x' \in W$ then $\tilde{\varphi}(x') \ge \varphi(x') \ge r$ and it necessarily follows that for the supremum of the values $\varphi(f^n(x'))$ for $n \in \mathbb{N}$ we only need $n \in \{0, \dots, N-1\}$. So if $x' \in W$ then $\tilde{\varphi}(x') = \sup\{\varphi(f^n(x')) : n = 0, \dots, N-1\}$. Stated otherwise, on W the function $\tilde{\varphi}$ is the supremum of finitely many continuous functions of the form $\varphi \circ f^n$. Consequently, $\tilde{\varphi}$ is continuous on W. This completes the proof that $\tilde{\varphi}$ is continuous at every point x of $\mathcal{B}(A) \setminus A$. Thus, we have a continuous function $\tilde{\varphi}: X \to [0; 1]$ such that $\tilde{\varphi}^{\leftarrow}[0] = A$ and $\tilde{\varphi}^{\leftarrow}[1] = A^*$. In addition, it follows easily from the definition of $\tilde{\varphi}$ that

$$\forall x \in X : 0 \le \tilde{\varphi}(f(x)) \le \tilde{\varphi}(x) . \tag{4.5-11}$$

It remains to modify $\tilde{\varphi}$ into a mapping g such that this inequality becomes strict for $x \notin A \cup A^*$. To this end, put

$$g(x) := \sum_{n=0}^{\infty} \frac{\tilde{\varphi}(f^n(x))}{2^{n+1}} \quad \text{for } x \in X .$$

Since this series converges uniformly on X (it is majorated by the convergent series with constant terms $\sum_n 2^{-n-1}$), the function g is continuous. Moreover, it is easily checked that g maps X into the interval $[0; 1]$. In addition, it is obvious that $g(x) = 0$ or 1 iff all terms of the series for g are 0 or 1, respectively, which is the case iff $x \in A$ or $x \in A^*$, respectively.

Finally, inequality (4.5-11) implies that $g(f(x)) \le g(x)$ for every $x \in X$. In addition, if $x \in X$ then

$$g(f(x)) - g(x) = \sum_{n=0}^{\infty} \frac{\tilde{\varphi}(f^{n+1}(x)) - \tilde{\varphi}(f^n(x))}{2^{n+1}} .$$

Consequently, $g(f(x)) = g(x)$ iff $\tilde{\varphi}(f^{n+1}(x)) = \tilde{\varphi}(f^n(x))$ for every $n \in \mathbb{Z}^+$, iff $\tilde{\varphi}(f^n(x)) = \tilde{\varphi}(x)$ for all $n \in \mathbb{Z}^+$. However, if $x \notin A^*$ then $\emptyset \ne \omega(x) \subseteq A$, so a subsequence of $(f^n(x))_{n \in \mathbb{Z}^+}$ converges to a point of A, where $\tilde{\varphi}$ is equal to 0. Hence $\tilde{\varphi}(x) = 0$, i.e., $x \in A$. So if $x \notin A \cup A^*$ then $g(f(x)) \ne g(x)$. This completes the proof that $g(f(x)) < g(x)$ if $x \notin A \cup A^*$. □

Example. Let $X := [0; 3]$ and let $f: X \to X$ be defined by $f(x) := x^2$ for $0 \le x \le 1$, $f(x) := 1 + \sqrt{x-1}$ for $1 \le x \le 2$ and $f(x) := 2 + (x-2)^2$ for $2 \le x \le 3$. This system

has the following asymptotically stable subsets: $A_1 := \{0\}$, $A_2 := \{2\}$, $A_3 := [0;2]$, $A_4 := \{0,2\}$ and $A_5 := [2;3]$ (we consider only proper subsets). The corresponding repellers are $A_1^* = [1;3]$, $A_2^* = [0;1] \cup \{3\}$, $A_3^* = \{3\}$, $A_4^* = \{1,3\}$ and $A_5^* = [0;1]$. The construction in the proof of the proposition above gives:
- $\varphi_1(x) = x$ for $0 \le x \le 1$ and 1 elsewhere,
- $\varphi_2(x) = |2 - x|$ for $1 \le x \le 3$ and 1 elsewhere,
- $\varphi_3(x) = x - 2$ for $2 \le x \le 3$ and 0 elsewhere,
- $\varphi_4(x) = x$ for $0 \le x \le 1$ and $|2 - x|$ elsewhere, and
- $\varphi_5(x) = 1$ for $0 \le x \le 1$, $1 - x$ for $1 \le x \le 2$ and 0 elsewhere.

It is easily checked that $\tilde{\varphi}_j = \varphi_j$ for $j = 1, 2, 3, 4, 5$. Moreover, there is no need to consider the functions g_j, because we already have $\varphi_j(f(x)) < \varphi_j(x)$ (with a strict inequality) for $x \in \mathcal{B}(A_j) \setminus A_j$: this is because $a^2 < a$ for $0 < a < 1$.

A function g as described in the proposition, i.e., a function that is strictly decreasing along the orbits in $\mathcal{B}(A) \setminus A$ and constant on A and on A^* for an asymptotically stable set A is called a *Lyapunov function* for A. We shall show now that the Lyapunov functions for all attractor-repeller pairs in a compact metric space can be glued together to a global Lyapunov-like function. First, a lemma:

Lemma 4.5.12. *Let X be a compact metric space. If $x, y \in CR(X, f)$ then x and y belong to the same basic set iff for every attractor-repeller pair (A, A^*) in X either $x, y \in A$ or $x, y \in A^*$.*

Proof. “Only if”: Suppose there is a basic set K such that both x and y are in K. Let A be an asymptotically stable set. Then Proposition 4.5.7 implies that $x, y \in K \subseteq A$ or that $x, y \in K \subseteq A^*$.

“If”: Assume that for every asymptotically stable subset A of X we have either $x, y \in A$ or $x, y \in A^*$. By Theorem 4.4.13 (2) there is a basic set K such that $x \in K$. If we can show that, for every $\varepsilon > 0$ we have $y \in \mathcal{P}_{2\varepsilon}(\{x\})$ and $x \in \mathcal{P}_{2\varepsilon}(\{y\})$ then the set $K \cup \{y\}$ is chain-transitive. As K is a maximal chain-transitive set it follows that $y \in K$ as well.

So let $\varepsilon > 0$ be arbitrary. By Lemma 4.5.2 (3), $f[\overline{\mathcal{P}_\varepsilon(\{x\})}] \subseteq \mathcal{P}_\varepsilon(\{x\})^\circ$, hence Corollary 3.4.4 implies that the set $A := \bigcap_{n \ge 0} f^n[\overline{\mathcal{P}_\varepsilon(\{x\})}]$ is asymptotically stable and that $\overline{\mathcal{P}_\varepsilon(\{x\})}$ is a neighbourhood of A that is strongly attracted by A, hence, by Theorem 3.4.1, is included in $\mathcal{B}(A)$. Because the point x is chain-recurrent, we have $x \in \mathcal{P}_\varepsilon(\{x\}) \subseteq \overline{\mathcal{P}_\varepsilon(\{x\})}$ and, consequently, $x \in \mathcal{B}(A)$. In particular, $x \notin A^*$, so by assumption $x, y \in A$. This implies that $y \in \overline{\mathcal{P}_\varepsilon(\{x\})} \subseteq \mathcal{P}_{2\varepsilon}(\{x\})$, where the final inclusion comes from Lemma 4.5.2 (1). In a similar way one shows that $x \in \mathcal{P}_{2\varepsilon}(\{y\})$. □

Theorem 4.5.13. *Let X be a compact metric space and assume that f maps X onto X. There exists a continuous function $\Phi: X \to [0;1]$ such that*

(a) $\forall x \in X \setminus CR(X, f) : \Phi(f(x)) < \Phi(x)$.

(b) $\forall x, y \in CR(X, f) : \Phi(x) = \Phi(y)$ *iff x and y belong to the same basic set.*

(c) $\Phi[CR(X, f)]$ *is compact and nowhere dense in* $\mathbb{R}$.

Proof. As X is a compact metric space it has a countable base. Consequently, there is only a finite or a countably infinite number of asymptotically stable subsets; see Exercise 3.12 (3). Enumerate these as[13] $(A_n)_{n\in\mathbb{N}}$. In view of Lemma 4.5.10 (3), for every $n \in \mathbb{N}$ the repeller A_n^* associated with the asymptotically stable set A_n is not empty, so by Proposition 4.5.11 there is a continuous function $g_n : X \to [0; 1]$ such that $g_n^{\leftarrow}[0] = A_n$, $g_n^{\leftarrow}[1] = A_n^*$ and $g_n(f(x)) < g_n(x)$ for all $x \in X \setminus (A_n \cup A_n^*)$. As g_n is constant on the invariant sets A_n and A_n^* it follows that $g_n(f(x)) \leq g_n(x)$ for all $x \in X$. Put

$$\Phi(x) := 2 \sum_{n=1}^{\infty} \frac{g_n(x)}{3^n} \quad \text{for } x \in X .$$

This series is uniformly convergent (it is absolutely majorated by the convergent series $\sum_n 3^{-n}$). As every term of this series is a continuous function, it follows that Φ is continuous on X.

For every every $x \in X$ and $n \in \mathbb{N}$ we have $g_n(f(x)) \leq g_n(x)$ hence, obviously, $\Phi(f(x)) \leq \Phi(x)$. If $x \notin CR(X, f)$ then by Corollary 4.5.6 there exists $n \in \mathbb{N}$ such that $x \notin A_n \cup A_n^*$, hence $g_n(f(x)) < g_n(x)$. Consequently, in that case $\Phi(f(x)) \neq \Phi(x)$, and therefore $\Phi(f(x)) < \Phi(x)$. This shows that Φ has the property mentioned in (a).

We continue with the proof that Φ has property (c). If $x \in CR(X, f)$ then, again by Corollary 4.5.6, we have $x \in A_n \cup A_n^*$ and, consequently, $g_n(x) = 0$ or 1 for each $n \in \mathbb{N}$. Thus, $\Phi(x) = \sum_n a_n/3^n$ with $a_n = 0$ or 2 for every $n \in \mathbb{N}$. It follows that $\Phi[CR(X, f)] \subseteq C$, the Cantor set; see Appendix B.1.1. It follows that $\Phi[CR(X, f)]$ is nowhere dense in $\mathbb{R}$; see Appendix B.1.2 (c). Since $CR(X, f)$ is closed in X, hence compact, it follows that its continuous image under the continuous mapping Φ is compact as well.

Finally, we prove that Φ satisfies condition (b). Let $x, y \in CR(X, f)$ and assume that $\Phi(x) = \Phi(y)$. As above, put $a_n := 2g_n(x)$ and $b_n := 2g_n(y)$ for $n \in \mathbb{N}$. Then $a_n = 0$ or 2 according to $x \in A_n$ or $x \in A_n^*$ and, similarly, $b_n = 0$ or 2 if $y \in A_n$ or $y \in A_n^*$, respectively, and

$$\sum_{n=1}^{\infty} \frac{a_n}{3^n} = \Phi(x) = \Phi(y) = \sum_{n=1}^{\infty} \frac{b_n}{3^n} . \tag{4.5-12}$$

As an ambiguity in the ternary development of any real number r would involve the appearance of the digit 1 (followed by 0's) in one of the developments of r it follows that equality (4.5-12) can only hold if $a_n = b_n$ for every $n \in \mathbb{N}$. Consequently, for every $n \in \mathbb{N}$, x and y are both in A_n or they are both in A_n^*. So Lemma 4.5.12 implies that x and y belong to the same basic set. Conversely, if x and y belong to the same basic set then, in the above notation, $a_n = b_n$ for every $n \in \mathbb{N}$, hence $\Phi(x) = \Phi(y)$. □

Example. Let (X, f) be as in Example (1) after Proposition 4.5.11. We have seen there that as the Lyapunov function g_n for the asymptotically stable set A_n we may take the

13 Notationally, it is most convenient to assume in the remainder of the proof that there are infinitely many asymptotically stable sets. It is a trivial task to adapt the proof if there are only finitely many of them.

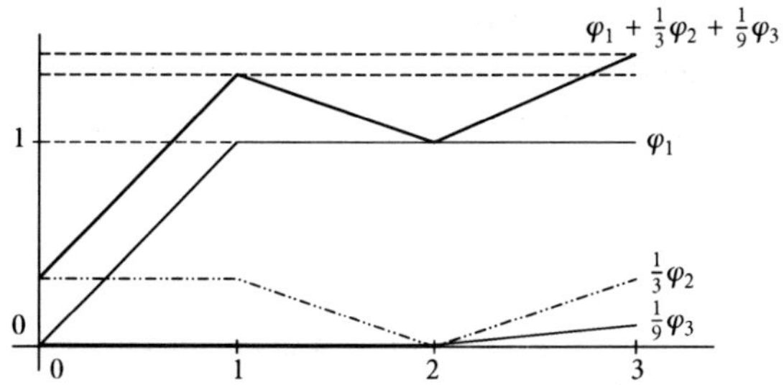

Fig. 4.11. A global Lyapunov function for the system (X, f) in the Example following Proposition 4.5.11.

function φ_n $(n = 1, 2, 3, 4, 5)$. Then up to a scaling factor the global Lyapunov function of the system is $\Phi = \varphi_1 + \frac{1}{3}\varphi_2 + \frac{1}{9}\varphi_3 + \frac{1}{27}\varphi_4 + \frac{1}{91}\varphi_5$. See Figure 4.11 (we did not draw the contributions of φ_4 and φ_5 as this would not change the graph of Φ essentially).

Exercises

4.1.

(1) A recurrent point x is periodic iff there is a neighbourhood U_0 of x such that the set $D(x, U)$ does not depend on the neighbourhood U of x provided $U \subseteq U_0$, iff there is a neighbourhood U_0 of x such that $\bigcap_{U \subseteq U_0} D(x, U) \setminus \{0\} \neq \emptyset$.

(2) Show that a point is transitive iff it is recurrent and has a dense orbit.

(3) Show that the sets $R(X, f)$ and $AP(X, f)$ are completely invariant if X is compact and f is surjective (for $AP(X, f)$, see Example (1) after Corollary 4.4.4).

(4) If x is a recurrent point then $x \in \overline{f[X]}$. Hence, if X has a dense set of recurrent points then $f[X]$ is dense in X.

4.2. (See H. Furstenberg [1981], Thm 2.17.) A subset P of $\mathbb{N}$ is said to be an *IP-set* whenever there is a sequence $p_1, p_2, p_3, \dots$ in P such that all finite sums $p_{i_1} + p_{i_2} + \cdots + p_{i_k}$ with $i_1 < i_2 < \cdots < i_k$ belong to P (the elements p_i need not be distinct, but in the expression $p_{i_1} + p_{i_2} + \cdots + p_{i_k}$ the indices are mutually different). Show that, if x_0 is a recurrent point in a dynamical system (X, f), then for every neighbourhood U_0 of x the set $D(x_0, U_0)$ is an IP-set.

4.3. Call a point $x \in X$ *eventually recurrent* whenever its orbit contains a recurrent point. Show that x is eventually recurrent iff $\mathcal{O}(x) \cap \omega(x) \neq \emptyset$.

(1) The following conditions are equivalent for a point $x \in X$:
   (i) x is not eventually recurrent.
   (ii) $\mathcal{O}(x)$ is an infinite discrete space.
   (iii) $\mathcal{O}(x)$ is locally compact but not compact.

(2) The following conditions are equivalent for a point $x \in X$:
   (i) x is eventually recurrent and not eventually periodic.
   (ii) $\exists\, m \in \mathbb{Z}^+ : \mathcal{O}(f^m(x))$ is a 1st-category space.
   (iii) $\exists\, m \in \mathbb{Z}^+ : \mathcal{O}(f^n(x))$ is a 1st-category space for all $n \geq m$.

4.4.
(1) If (X, f) is topologically ergodic then $\Omega(X, f) = X$.
(2) If (X, f) is topologically ergodic then for every pair of non-empty open subsets U and V of X the set $D(U, V)$ is infinite.

4.5. Assume that $f^n[X]$ is closed for every $n \in \mathbb{N}$ (for example, let X be compact; it would also be the case if $f[X] = X$, but then this exercise would be extremely trivial). Then $\Omega(X, f) \subseteq \bigcap_{n=0}^{\infty} f^n[X]$.
NB. See Exercise 3.11 for more about the set $\bigcap_{n=0}^{\infty} f^n[X]$.

4.6. Let (X, f) be a dynamical system on an interval in $\mathbb{R}$. Prove:
(1) Let $P(X, f)$ denote the set of all periodic points in X. Then $\overline{P(X, f)} = \overline{R(X, f)}$.
NB. Consequently, any initial segment of the orbit of a recurrent point can be approximated arbitrarily close by a segment of the orbit of some periodic point. Moreover, if $P(X, f)$ is closed then $R(X, f) = P(X, f)$.

(2) $\Omega(\Omega(X, f), f) = Z(X, f)$.
NB. As explained in Note 7, this means that the *depth* of the centre is 2.

4.7.
(1) Show that a point $x \in X$ is almost-periodic iff

$$\forall U \in \mathcal{N}_x \exists K \subset \mathbb{Z}^+ : \; K \text{ is finite and } K + D(x, U) = \mathbb{Z}^+ .$$

(2) Let X be a compact metric space and let $x \in X$. Show that the point x is almost periodic iff

$$\forall \varepsilon > 0 \; \exists l \in \mathbb{N} : \; \mathcal{O}(x) \subseteq \bigcap_{k \in \mathbb{Z}^+} B_\varepsilon(\{f^k(x), \ldots, f^{k+l}(x)\}) . \qquad (*)$$

So an almost periodic point in a compact metric space is characterized by the condition that for every $\varepsilon > 0$ its orbit can be approximated up to ε by every finite segment of the orbit of a certain length (depending on ε).
(3) Assume that X is a complete metric space and that the point $x \in X$ satisfies condition $(*)$ in 2 above. Show that $\overline{\mathcal{O}(x)}$ is a compact minimal set.
(4) Prove the remaining implications in the diagram at the next page.

4.8. Let G be as in Section 4.4. Prove that the mapping

$$x_0\, x_1\, x_2 \cdots \mapsto \sum_{i=0}^{\infty} 2x_i 3^{-i-1} : G \to [0; 1]$$

defines a homeomorphism of G onto the Cantor set C.

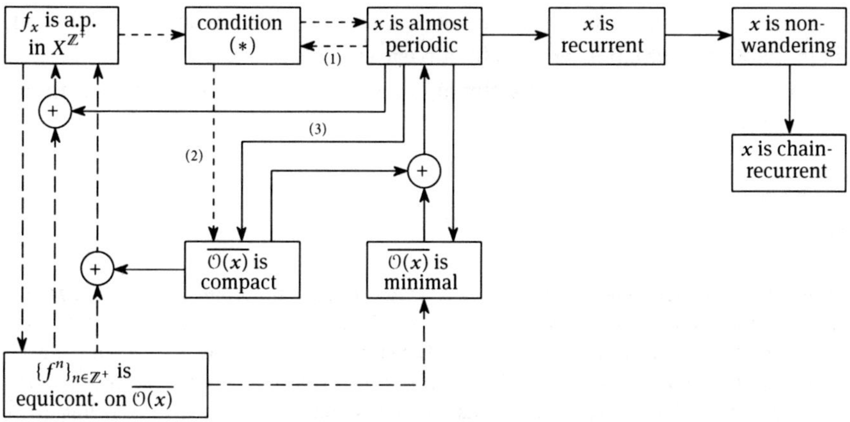

Diagram: Implications between the various notions of almost periodicity and recurrence in a system (X, f). The dotted arrows make only sense in metric spaces (but can be interpreted in uniform spaces). The definition of $f_x \in X^{\mathbb{Z}^+}$ is in Note 3 below. Additional conditions: (1) X is compact, (2) X is complete, (3) X is locally compact.

4.9. Assume that X is a metric space. Let $x_0 \in X$ and suppose there exists N such that for every ε there is an ε-chain from x_0 to x_0 of length at most N. Then the point x_0 is periodic.

4.10. Let X be a metric space.

(1) Define a relation $\sim$ on $CR(X, f)$ by

$$x \sim y \iff \forall \varepsilon > 0\ \exists \varepsilon\text{-chain from } x \text{ to } x \text{ containing } y$$

(see Lemma 4.4.8 for suggestions of equivalent alternative formulations). Show that $\sim$ is an equivalence relation on $CR(X, f)$ and that the equivalence classes under $\sim$ are the basic sets defined in Section 4.4.

(2) Let A be a closed subset of X such that for every neighbourhood U of A there exists $\varepsilon > 0$ such that every ε-chain starting in A remains in U. If X is a T_3-space, or if X is a Hausdorff space and A is compact, then A is invariant.

4.11. Let X be a locally compact metric space.

(1) Show that the first part of Proposition 4.5.7 holds for a chain-stable set A and a chain-transitive set K. Similarly, in Lemma 4.5.8 'asymptotically stable' can be replaced by 'chain-recurrent'

(2) Prove that the following conditions are equivalent for a chain-stable set A (cf. Theorem 4.5.9):

(i) A has no proper chain-stable subsets.

(ii) A is a basic set.

(iii) A is chain-transitive.

(3) Prove that every chain-stable set includes a minimal (with respect to inclusion) chain-stable set, which is by 2(i)⇒(ii) a basic set.
Give an example showing that a similar statement does not hold for asymptotically stable sets.

Notes

1 A recurrent point is often called *positively Poisson stable* (Poisson contributed substantially to the study of the motions of the planets around the sun, which are – due to mutual perturbation – not purely periodic). Also the name of Poincaré is closely connected to the notion of recurrence. This is due to the following method to prove the existence of recurrent points in certain dynamical systems, which is especially popular among physicists: many of the dynamical systems (X, f) modelling physical systems have an *invariant measure*: a Borel measure μ on X such that $\mu(f^{\leftarrow}[A]) = \mu(A)$ for every Borel set A of X. If μ is a finite invariant measure and A is a Borel set such that $\mu(A) > 0$ then with certainty points of A will return in A: there exists $n \geq 1$ such that $\mu(A \cap (f^n)^{\leftarrow}[A]) > 0$ (Poincaré's Recurrence Theorem). If, in addition, X is a separable metric space then it is rather easy to derive from this that μ-almost all points of X are recurrent.

Every dynamical system (X, f) on a compact Hausdorff space turns out to have an invariant measure; so in that case, if in addition, X is metrizable, almost all points in the support of this measure are recurrent. The existence result in Theorem 4.1.3 does not require metrizability of the phase space.

An important generalization of the existence theorem of recurrent points in compact systems is the following result by Furstenberg: consider a finite number of continuous mappings $f_i : X \to X$ $(0 \leq i \leq n)$. A point $x \in X$ is said to be *multiply recurrent* under these mappings whenever for every neighbourhood U of x there exists $k \in \mathbb{N}$ such that $f_i^k(x) \in U$ for $i = 1, \ldots, n$. Thus, a multiply recurrent point x is not only a common recurrent point of the mappings f_i, but it is approached by iterates $f_i^k(x)$ with the same k for all i.

Theorem (Furstenberg). *If X is a compact metric space then every finite commuting set of continuous mappings of X into itself admits a multiply recurrent point.*

Proof. See Theorem 2.6 in H. Furstenberg [1981]. □

2 In the Example after Theorem 4.1.7, the point $([0], [0])$ is recurrent under f, hence there is a subsequence $(n_i)_i$ of $\mathbb{N}$ such that $f^{n_i}([0], [0]) \rightsquigarrow ([0], [0])$. Since the orbit of $([0], [0])$ under f is $([0], [0]) \mapsto ([a], [a]) \mapsto ([2a], [4a]) \mapsto \cdots \mapsto ([na], [n^2 a]) \mapsto \ldots$, one gets: for every $a \in \mathbb{R}$ and every $\varepsilon > 0$ there are $n \in \mathbb{N}$ and $k \in \mathbb{Z}$ such that $|n^2 a - k| < \varepsilon$, hence

$$\left| a - \frac{k}{n^2} \right| < \frac{\varepsilon}{n^2}$$

(if $a \notin \mathbb{Q}$ then this is a 'good' approximation of a by rationals). See Section I.2 of H. Furstenberg [1981] for this and other results in this direction.

3 A subset of $\mathbb{R}^+$ with bounded gaps is often called *a relatively dense set in* $\mathbb{R}^+$. For example, sets like $10^6\,\mathbb{N}$ and $\{ x \in \mathbb{R}^+ : \sin x + \sin x\sqrt{2} = 0 \}$ are relatively dense. So a subset of $\mathbb{R}^+$ or $\mathbb{Z}^+$ is syndetic in $\mathbb{R}^+$ or $\mathbb{Z}^+$ iff it has bounded gaps. In a dynamical system on a metric space X the elements of $D_f(x, B_\varepsilon(x))$ $(x \in X, \varepsilon > 0)$ are often called *the ε-almost periods of x*. Consequently, in that situation the point x

is almost periodic iff for every $\varepsilon > 0$ the set of its ε-almost periods is relatively dense. In [GH][14], a subset D of a topological (semi)group S is said to be *syndetic in* S whenever S has a compact subset K such that $KD = S$.

Birkhoff's original formulation of Theorem 4.2.2 looks different from ours, because his terminology is different (his results were about continuous flows, but that is not very essential here). He used the term 'recurrent' where we use 'almost periodic', defined differently, but in a compact metric space equivalent to our notion of almost periodicity. In fact, condition $(*)$ of Exercise 4.7-2 is the discrete version of Birkhoff's notion of recurrence. Our notion of recurrence was called *positive Poisson stability* by Birkhoff. In the literature that uses this terminology – often based on the influential book V. V. Nemytskiĭ & V. V. Stepanov [1960] ([NS] for short) – our notion of almost periodicity is often called *almost recurrence.* In [NS] there is also a definition of 'almost periodicity' (of a point in a continuous-time system); for systems with discrete-time this definition would be: for every $\varepsilon > 0$ the set $\{n \in \mathbb{Z}^+ : d(f^{n+k}(x), f^k(x)) < \varepsilon \text{ for all } k \in \mathbb{Z}^+\}$ is relatively dense in $\mathbb{Z}^+$ (the phasespace is assumed to be a metric space). This means: consider the system $(X^{\mathbb{Z}^+}, \tau)$, where the space $X^{\mathbb{Z}^+}$ of all mappings from $\mathbb{Z}^+$ to X is endowed with the metric $\rho: (\varphi, \psi) \mapsto \sup_{k\in\mathbb{Z}^+} d(\varphi(k), \psi(k))$ and where $\tau: X^{\mathbb{Z}^+} \to X^{\mathbb{Z}^+}$ is defined by $\tau(\varphi)(k) := \varphi(k+1)$; then in this system the element $f_x : k \mapsto f^k(x) : \mathbb{Z}^+ \to X$ is an almost periodic point (in our sense of the definition). But also other terminology is in use. For instance, in H. Furstenberg [1981] our notion of almost periodicity is called *uniform recurrence.* In the following table we summarize the different terminology.

Our terminology	**[GH]**	**[NS]**
f_x is a.p. in $X^{\mathbb{Z}^+}$	–	almost periodic
condition $(*)$	weakly almost periodic on $\mathcal{O}(x)$	recurrent
almost periodic	almost periodic	almost recurrent
recurrent	recurrent	Poisson stable
non-wandering	regionally recurrent	non-wandering

4 The space G defined in 4.2.8 is a topological group with coordinate-wise addition modulo 2 with carry-over to the right as group operation. Consequently, the adding machine is a special case of the construction mentioned at the end of Note 7 in Chapter 1.

A compact Hausdorff topological group G is said to be *monothetic* whenever there is an element $a \in G$ such that the subgroup $H_a := \{a^n : n \in \mathbb{Z}\}$ is dense in G. It is well-known that for a compact Hausdorff group G the following conditions are equivalent:

(i) There exists $a \in G$ such that H_a is dense, that is, G is monothetic.
(ii) There exists $a \in G$ such that the semigroup $S_a := \{a^n : n \in \mathbb{Z}^+\}$ is dense.
(iii) There exists $a \in G$ such that the dynamical system (G, λ_a) is minimal.

Proofs. (i)⇒(ii): Consider the limit set $\omega(e)$ of the unit element e of G under the left translation λ_a. By compactness of G, every point is almost periodic, hence recurrent, under λ_a. Hence $a^m \in \omega(a^m) = \omega(\lambda_a^m(e)) = \omega(e)$, where we have used Proposition 1.4.3 (1). Similarly, $a^{-m} \in \omega(a^{-m}) = \omega(\lambda_a^m(a^{-m})) = \omega(e)$. This implies that $H_a \subseteq \omega(e)$, hence $\omega(e) = G$. In particular, the orbit S_a of e under λ_a is dense. (ii)⇒(i): Trivial. (ii)⇔(iii): Trivial; see also Note 4 in Chapter 1. (Essentially, our proof of (i)⇒(ii) is in E. Hewitt [1956].)

NB. All equicontinuous minimal systems are of this type; see Corollary 1.6.10. □

---

14 Warning: interpretation of results from [GH] in our context requires an interchange of left and right in all notation, due to the fact that [GH] writes xf for $f(x)$.

5 The adding machine occurs in the literature under many different names: dyadic group, odometer, infinite register shift, ... It can be shown that a stable limit set in a 1-dimensional dynamical system is either a periodic orbit, a cycle of intervals, or an adding machine: see Chapter 2 in J. Buescu [1997]. This is in agreement with Corollary 3.5.8.

6 In the logistic system $([0;1], f_\mu)$ with $\mu = \mu_\infty$ the situation is similar to that of the system $([0;1], f_\infty)$ described in 4.3.11 and 4.2.10 – here μ_∞ is the Feigenbaum point described in Example 0.4.1 in the Introduction (see also Note 4 in Chapter 2: there is a countable set of periodic orbits (repelling, with periods 2^n for $n \in \mathbb{N}$), which accumulate to a set C' which can be shown to be homeomorphic to the Cantor set C. The union of these periodic orbits with C' is the non-wandering set of f_μ, the so-called *Feigenbaum attractor* (it is topologically attracting – compare this with 3.3.19 — but not asymptotically stable).

7 Birkhoff's original definition of the centre of a system was as follows: form the subsystem on the non-wandering set of the system; then form the subsystem on the non-wandering set of this subsystem, etc.: repeat this procedure until its 'stops' (i.e., gives no smaller sets) after a possibly transfinite number of steps. The final result is called the centre of the system. In a compact system the centre so obtained is not empty and it turns out to be equal to the closure of the set of recurrent points. The (possibly transfinite) number of steps in Birkhoff's procedure to arrive at the centre is called the *depth of the centre*. The results of Exercise 4.6 are originally due to A. N. Šarkovskij (1964) and were later rediscovered by E. M. Coven & G. A. Hedlund [1980] and Z. Nitecki [1980], respectively.

8 The theory of pseudo-orbits, chain-recurrence, chain-transitivity and their applications to the theory of attractors was originated in C. Conley [1978] (for invertible systems with continuous time). Here he proved his Fundamental Theorem – see Corollary 4.5.6 and Theorem 4.5.13 – and developed his homotopy index theory (we pay no attention at all to this aspect of his theory). Later, the theory was extended to systems with discrete time, and to non-compact systems; see e.g., J. Franks [1988] and M. Hurley [1998].

Though computer simulations produce at best pseudo-orbits, if the precision δ is sufficiently good then by Proposition 4.4.2 a finite δ-chain can be approximated up to ε by a finite segment of a real orbit. Question: are infinite δ-chains in this manner traceable by real orbits? This motivates the following definition: a dynamical system on a metric space is said to have the *shadowing property*, also called the *pseudo-orbit tracing property*, whenever for every $\varepsilon > 0$ there exists $\delta > 0$ such that every infinite δ-chain $(x_n)_{n\in\mathbb{Z}^+}$ in X is ε-shadowed by the orbit of some point y_0, that is, $d(x_i, f^i(y_0)) \leq \varepsilon$ for all $i \in \mathbb{Z}^+$. Well-known examples of mappings with this property are so-called Axiom A homeomorphisms on their non-wandering sets (see Proposition 3.6 in R. Bowen [1975]) and subshifts of finite type (to be defined in Chapter 5; see Exercise 5.16 (3) ahead, which result is due to P. Walters [1979]). Another example is the tent map on the unit interval (and hence the logistic function f_4); see E. M. Coven, I. Kan & J. A. Yorke [1988]. In a system with the shadowing property one has $\Omega(X, f) = CR(X, f)$.

The shadowing property has applications in, among others, the theory of stability of mappings. Roughly, a continuous mapping (homeomorphism) $f: X \to X$ is said to be *stable in the class of continuous self-maps of X (homeomorphisms, respectively)* whenever f is a factor of every sufficiently small perturbation of f under a morphism that is close to the identity map. See e.g., K. Sakai [1987].

9 Invertible vs. non-invertible systems. In the literature on invertible dynamical systems the notion we have defined by formula (4.1-1) is usually called *positive recurrence*. So by Proposition 4.1.1, a point x is positively recurrent iff for every neighbourhood U of x the set $D(x, U) := \{n \in \mathbb{Z} : f^n(x) \in U\}$ is not bounded from above, iff $x \in \omega(x)$. Similarly, a point x is said to be *negatively recurrent* under a homeomorphism f whenever for every neighbourhood U of x the set $D(x, U)$ is not bounded from below, iff $x \in \alpha(x)$. A point x is said to be *recurrent* whenever it is both positively and negatively recurrent. Accordingly, one distinguishes the sets $R^+(X, f)$, $R^-(X, f)$ and $R(X, f) := R^+(X, f) \cap R^+(X, f)$. It is easy to see that these sets are *two-sided* invariant.

In this language, Theorem 4.1.3 states that every point in a positively minimal set is positively recurrent. It is easy to modify the proof so as to show that all points of a two-sided minimal set under a homeomorphism are both positively and negatively recurrent.

The usual definition of a non-wandering point under a homeomorphism is as follows: a point x is *non-wandering* (under the homeomorphism f) whenever for every neighbourhood U of x the set $D(U,U) \setminus \{0\} = \{n \in \mathbb{Z} \setminus \{0\} : f^n[U] \cap U \neq \emptyset\}$ is not empty. As the set $D(U,U)$ is symmetric, that is, $D(U,U) = -D(U,U)$, this definition is equivalent to the one we have given in Section 4.3. Moreover, Proposition 4.3.2 states: if x is a non-wandering point then for every neighbourhood U of x the set $D(U,U)$ is not bounded from above. By symmetry, this implies that $D(U,U)$ is not bounded from below. Thus, the point x is non-wandering iff for every neighbourhood U of x the set $D(U,U)$ is not bounded from above and not from below.

In this context, Proposition 4.3.1 (1),2 is about 'positively recurrent'; obviously, it is also true for 'negatively recurrent'. Similarly, Proposition 4.3.1 (3) is about 'positively transitive'. But one also finds the following statement in the literature: if (X,f) is ergodic (according to the definition for invertible systems, as explained in Note 12 to Chapter 1) and there are no isolated points in X (hence also: if there is a point with dense full orbit and there are no isolated points) then all points of X are non-wandering. In an invertible system that is both ergodic and non-wandering, for any two open sets U and V the set $D(U,V)$ is not bounded from above or below, hence it contains positive elements (see [deV], II(4.8); compare this with Exercise 4.4-2). Hence such a system is ergodic according to our definition in Section 1.3. This explains why in the invertible theory one often finds the combination 'ergodic with no isolated points' or 'ergodic and non-wandering' where in the present book we only need 'ergodic' (or 'transitive'). For example, in the invertible theory Proposition 5.3.7 below would read: a shift space is irreducible iff it is ergodic and non-wandering.

The example described in Note 12 at the end of Chapter 1 shows that an invertible ergodic system with isolated points need not be non-wandering, this in contrast to non-invertible systems (see Exercise 4.4 (1)).

In the invertible theory almost periodicity is defined similar to our definition in Section 4.2, with the obvious modification that $D(x,U)$ is relatively dense in $\mathbb{Z}$ (has uniformly bounded gaps) for every neighbourhood U of x. There is an abundant literature on almost periodic points; see [GH] and [deV] and Note 3 above. The results of Section 4.2 are also valid in the invertible theory, with obvious modifications.

5 Shift systems

Abstract. In this chapter we discuss an important class of dynamical systems: the shift systems. They form an almost inexhaustible source of examples and counter examples, but they are also important in their own right and they have many applications. In this book we can only lift a tip of the veil. In fact, after the presentation of the basic definitions and facts in the present chapter we only discuss, in Chapter 6, the basics of 'symbolic dynamics': how can shift systems be used to represent other systems?

Everything in this chapter has a rather combinatorial flavour. Yet it is topology: the space in which everything takes place is the well-known Cantor space. Actually, it is the possibility to reduce topological questions to combinatorial ones that makes shift systems so useful.

5.1 Notation and terminology

5.1.1. In this chapter, S will be a finite set with at least two elements. Its elements will be called *symbols* and S will be called the *symbol set* or *alphabet*. The number of elements of S will be denoted by s. For convenience we identify S with an initial segment of $\mathbb{Z}^+$: $\mathrm{S} = \{0, 1, \dots, s-1\}$.

The use of integers here is purely symbolic: they are just names for the symbols. Properly speaking, we should denote the symbols more abstractly by $\alpha_0, \dots, \alpha_{s-1}$. But to avoid clumsy lower indices we use i instead of α_i. In examples we sometimes use a different notation, like $\mathrm{S} = \{a, b, c\}$. In the special case that $s = 2$ we sometimes interpret the symbols 0 and 1 as the integers 0 and 1 with addition modulo 2.

The *sequence space* Ω_S is the space of all infinite sequences $x = (x_n)_{n\in\mathbb{Z}^+}$ with $x_n \in \mathrm{S}$ for all $n \in \mathbb{Z}^+$. Thus,

$$\Omega_\mathrm{S} := \mathrm{S}^{\mathbb{Z}^+} .$$

In the special case that $\mathrm{S} = \{0, 1\}$ we write Ω_2 instead of Ω_S. The elements of Ω_S will be denoted without commas and parentheses, as follows:

$$x = x_0\, x_1\, x_2\, \dots\, x_n\, \dots$$

with $x_n \in \mathrm{S}$ for all $n \in \mathbb{Z}^+$. Properly speaking, an element $x \in \Omega_\mathrm{S}$ is a function $x\colon \mathbb{Z}^+ \to \mathrm{S}$; its value at an element $n \in \mathbb{Z}^+$ is denoted by x_n rather than by the more customary $x(n)$. We call x_n the n-th *coordinate* of x and we also say that the symbol x_n

has position n in x or that it *occurs at position*[1] n. The mapping $\pi_n : \Omega_S \to S$ that assigns to every point of Ω_S its n-th coordinate will be called the *n-th projection* ($n \in \mathbb{Z}^+$). Then

$$x = \pi_0(x)\,\pi_1(x)\,\pi_2(x)\;\dots\;\pi_n(x)\dots$$

for every $x \in \Omega_S$. So if $\alpha \in S$ and $n \in \mathbb{Z}^+$ then $x \in \overleftarrow{\pi_n}[\alpha]$ iff α occurs at position n in x, iff $x_n = \alpha$.

5.1.2. A *word* or *block over* S is an element of S^k for some $k \in \mathbb{N}$, that is, a finite ordered sequence of elements of S, as follows:

$$b := b_0 \dots b_{k-1} \in S^k = \underbrace{S \times \dots \times S}_{k \text{ times}} .$$

In this case, the natural number k is called the *length* of the block b. The length of a block b will be denoted by $|b|$. A block with length k will also be called a *k-block*. There is a unique block with length 0, the unique member of the set S^0. It will be called the *empty block* or *the empty word* and is denoted by $\varnothing$. The set of all blocks over S will be denoted by S^*. Stated otherwise, $S^* := \bigcup_{k \in \mathbb{Z}^+} S^k$.

For blocks we use the notions of 'coordinate' and 'position' just as for sequences. Thus, if $b = b_0 \dots b_{|b|-1} \in S^*$ then for $n = 0, \dots, |b| - 1$ the symbol b_n will be called the n-th *coordinate* of b and we shall say that it is the symbol at *position n* in b.

If $x \in \Omega_S$ and $k, l \in \mathbb{Z}^+$, $0 \le k < l$, then the block $x_k \dots x_l$ will be denoted by $x_{[k\,;\,l]}$ and also by $x_{[k\,;\,l+1)}$. The latter notation is particularly convenient for a block with its last position at $l - 1$ for some $l > k$: $x_{[k\,;\,l)} = x_{[k\,;\,l-1]}$.

New words can be formed by *concatenation* of two or more words: if b and c are finite words and b or c is empty then $bc = c$ or b, respectively; if neither b nor c is empty then $bc := b_0 \dots b_{|b|-1} c_0 \dots c_{|c|-1}$, i.e., it is the word with length $|b| + |c|$ whose n-th coordinate is

$$(bc)_n := \begin{cases} b_n & \textit{if } \; 0 \le n \le |b| - 1\,, \\ c_{n-|b|} & \textit{if } \; |b| \le n \le |b| + |c| - 1\,. \end{cases}$$

Concatenation is easily seen to be associative: if b, c and d are words then $(bc)d = b(cd)$; we shall denote this block by bcd. Consequently, concatenation of an arbitrary finite number of words can be defined and can unambiguously be written without parentheses.

Such a phrase can be found in any book on group theory and the proof (with induction on the number of factors in a product) is usually left to the reader, as we do here. The recipe is: just write down all coordinates in the right order.

Let $b, c \in S^*$. We say that b *occurs in* c or, equivalently, that c *contains* b, or that b is a *subblock* of c, whenever there are $p, q \in S^*$ (possibly empty) such that $c = pbq$. In that

---

1 Note the following discrepancy: the numbering of positions starts with 0, so the symbol in position n is at what in informal usage would be called the $(n + 1)$st place.

case we also say that b occurs in c *at position* $|p|$. If $p = \varnothing$ then we say that c *begins with* the block b, or that b the *initial* $|b|$-*block* of c, or that b is in *initial position* in c. In the case that $q = \varnothing$ we say that c *ends with* the block b, or that b is the *final* $|b|$-*block* of c (at position $|c| - |b|$ in c).

Example. The block 01 has three occurrences in the word 01101001: one at the begin (position 0), one at the end (position 6) and one at position 3. Formally, the empty block $\varnothing$ occurs in every block, but it makes no sense to define its position.

The definitions and the terminology on concatenation of finite blocks given above will also be used for infinite sequences (as far as meaningful). Thus, if $y \in \Omega_S$ and p and b are blocks with lengths k and l, respectively, then the sequence $x := pby \in \Omega_S$ is defined by

$$pby = \underbrace{p_0 \dots p_{k-1}}_{p} \underbrace{b_0 \dots b_{l-1}}_{b} y_0 \, y_1 \, y_2 \dots ,$$

and we say that x *begins with* the block p, or that p is an *initial block* of x, and that the block b occurs in x at position k. In this example, $b = x_{[k\,;\,k+l)}$. Thus, a finite word b occurs in x at position k iff $b = x_{[k\,;\,k+|b|)}$.

Obviously, concatenation of two infinite sequences is not defined. On the other hand, concatenation of an infinite number of finite blocks can be easily defined: if $(b^{(k)})_{k \in \mathbb{Z}^+}$ is a sequence of non-empty finite blocks then

$$x := b^{(0)} \, b^{(1)} \, b^{(2)} \dots b^{(k)} \dots$$

is the element of Ω_S, defined in the obvious way by successively writing down the coordinates of these blocks (in the right order).

Formally, let $\nu_0 := 0$ and for every $j \in \mathbb{N}$, let $\nu_j := \sum_{k=0}^{j-1} |b^{(k)}|$. For every $n \in \mathbb{Z}^+$ there is a unique value of j in $\mathbb{Z}^+$ such that $\nu_j \leq n \leq \nu_{j+1} - 1$. Then put $x_n := b^{(j)}_{n-\nu_j}$. Stated otherwise, for every $j \in \mathbb{Z}^+$ the block $b^{(j)}$ occurs in x at position ν_j.

The following conventions will be convenient: if $b \in \mathbf{S}^*$ and $\alpha \in \mathbf{S}$ then

- for every $k \in \mathbb{Z}^+$, $b^k \in \mathbf{S}^*$ is the concatenation of of k copies of b in particular, $\alpha^k = \alpha \dots \alpha$ (k times);
- $b^\infty \in \Omega_S$ is the concatenation of countably many copies of b in particular, $\alpha^\infty = \alpha\alpha\alpha\dots$, the point all of whose coordinates are α.

Let b be a non-empty block. The set of all points of Ω_S that contain the block b in initial position will be called the *cylinder on* b and it will be denoted by $C_0[b]$. Thus,

$$\begin{aligned} C_0[b] &:= \{ x \in \Omega_S \,\vdots\, x_{[0\,;\,|b|)} = b \} \\ &= \{ x \in \Omega_S \,\vdots\, x_n = b_n \text{ for } n = 0, \dots, |b| - 1 \} \\ &= \bigcap \{ \overleftarrow{\pi_n} [b_n] \,\vdots\, n = 0, \dots, |b| - 1 \} . \end{aligned} \qquad (5.1\text{-}1)$$

We reserve a special notation for cylinders that are based on an initial block of a point $x \in \Omega_S$, i.e., cylinders of the form $C_0[x_{[0;k)}]$ with $k \in \mathbb{N}$:

$$\widetilde{B}_k(x) := C_0[x_{[0;k)}] = \{ y \in \Omega_S : y_n = x_n \text{ for } 0 \le n \le k-1 \}. \tag{5.1-2}$$

5.1.3. Consider the finite set **S** as a topological space with the discrete topology: each point (hence every subset) is clopen. With this topology, **S** is a compact Hausdorff space. We endow $\Omega_S = \mathbf{S}^{\mathbb{Z}^+}$ with the product topology: the weakest topology for which all projections $\pi_n : \Omega_S \to \mathbf{S}$ for $n \in \mathbb{Z}^+$ are continuous.

Observation. *Every cylinder set in Ω_S is clopen in this topology.*

Proof. By (5.1-1), every cylinder set is the intersection of a finite number of sets of the form $\pi_n^{\leftarrow}[\alpha]$ with $n \in \mathbb{Z}^+$ and α an element of **S**. The inverse images of such clopen singleton sets under the continuous projections are clopen, hence each cylinder is the intersection of finitely many clopen sets, hence clopen as well. □

As explained in Appendix A.5.2, this topology on Ω_S is characterized by the fact that each point x in Ω_S has a local base consisting of sets of the form $\prod_{n\in\mathbb{Z}^+} V_n$ where, for every $n \in \mathbb{Z}^+$, V_n is a (basic) neighbourhood of the point $x_n = \pi_n(x)$ in **S**, and $V_n = \mathbf{S}$ for almost all $n \in \mathbb{Z}^+$.

Without limitation of generality we may replace here the phrase ‘for almost all n’ by ‘for all $n \ge k$ for some $k \in \mathbb{N}$’. Moreover, as a basic neighbourhood of a point in **S** we can always take the singleton set containing only that point. Consequently, the topology on Ω_S is characterized as follows: every point $x \in \Omega_S$ has a local base consisting of all sets V for which there is $k \in \mathbb{Z}^+$ such that $V = \prod_{n\in\mathbb{Z}^+} V_n$ with $V_n = \{x_n\}$ for $0 \le n \le k-1$ and $V_n = \mathbf{S}$ for all $n \ge k$. Such a set V can be written as

$$V = \{ y \in \Omega_S : y_n = x_n \text{ for } 0 \le n \le k-1 \} = \widetilde{B}_k(x). \tag{5.1-3}$$

Conclusion. *For every point x of Ω_S the cylinders based on initial blocks of x form a local base at x.*

In the sequel, we shall consider Ω_S as a topological space endowed with the product topology discussed above. We shall show that Ω_S is a Cantor space.

Theorem 5.1.4. *Ω_S is a compact 0-dimensional space without isolated points.*

Proof. As noted earlier, Ω_S is the product of compact spaces, hence compact (Tychonov’s Theorem; but see also Exercise 5.1 (4)). Moreover, each point has a local base consisting of cylinder sets. Since these are all clopen, it follows that Ω_S is 0-dimensional. In order to show that Ω_S has no isolated points, it is sufficient to observe that every cylinder set contains at least two points. This is obvious: a point $y \in \Omega_S$ belongs to a given cylinder set iff in finitely many positions the coordinates of y have a prescribed value. All coordinates of y in the other positions are free and for each of them there are at least two choices. So the cylinder contains infinitely many points. □

5.1.5. Being a product of countably many metric spaces, Ω_S is metrizable as well. However, we shall not base ourselves on this general result. Instead, we shall define a metric on Ω_S and show that it is compatible with the product topology defined above. For two points $x, y \in \Omega_S$, let

$$d(x, y) := \begin{cases} \left(1 + \min\{\, j \in \mathbb{Z}^+ : x_j \neq y_j \,\}\right)^{-1} & \text{if } x \neq y\,, \\ 0 & \text{if } x = y\,. \end{cases}$$

So if $k \in \mathbb{Z}^+$ is the first position where the coordinates of x and y differ:

$$\left.\begin{array}{lcl} x & = & a_0 \dots a_{k-1}\, x_k\, x_{k+1}\, \dots \\ y & = & a_0 \dots a_{k-1}\, y_k\, y_{k+1}\, \dots \end{array}\right\} \quad \text{with } x_k \neq y_k\,,$$

then $d(x, y) = 1/(1 + k)$. Equivalently, if a is the largest initial block that x and y have in common, then

$$d(x, y) = \frac{1}{1 + |a|}\,.$$

Note that in this formula $a = \emptyset$ iff $x_0 \neq y_0$, iff $d(x, y) = 1$.

Claim 1. *d is a metric on Ω_S.*

Proof. Consider three points x, y and z in Ω_S. We have to show that:
(i) $d(x, y) \geq 0$ and $d(x, y) = 0$ iff $x = y$.
(ii) $d(x, y) = d(y, x)$.
(iii) $d(x, z) \leq d(x, y) + d(y, z)$, even: $d(x, z) \leq \max\{d(x, y), d(y, z)\}$.

The proofs of (i) and (ii) are trivial. For the proof of (iii), first note that these inequalities are obvious if $x = y$ or $y = z$. So assume that both $x \neq y$ and $y \neq z$. Suppose that the common initial block of x and y has length k and that the common initial block of y and z has length l. In addition, suppose that $k \leq l$ (if $k > l$, the proof is similar). Then the common initial block of x and z has a length of at least k. Since $k = \min\{k, l\}$, it follows that $d(x, z) \leq 1/(1 + k) = \max\{1/(1 + k), 1/(1 + l)\} = \max\{d(x, y), d(y, z)\}$. This concludes the proof that d is a metric on Ω_S. □

Claim 2. *The topology generated by the metric d coincides with the product topology.*

Proof. Let x and y be two points of Ω_S, let $n \in \mathbb{N}$ and consider the following four statements, which are easily seen to be mutually equivalent:

(1) $d(x, y) < 1/n$.
(2) $d(x, y) \leq 1/(1 + n)$.
(3) The common initial block of x and y has length at least n.
(4) $y \in C_0[x_0 \dots x_{n-1}] = \widetilde{B}_n(x)$.

In accordance with Appendix A.7.1, let $B_r(x, d)$ denote the open ball (with respect to the metric d) centred at the point x and with radius $r > 0$. Then the equivalence of the statements (1) through (4) can be summarised as follows:

$$B_{1/n}(x,d) = \left\{ y \in \Omega_S : d(x,y) \leq \frac{1}{1+n} \right\} = \widetilde{B}_n(x) . \tag{5.1-4}$$

Thus, for every $n \in \mathbb{N}$ the open ball $B_{1/n}(x,d)$ – a basic neighbourhood of the point x in the metric topology – coincides with the basic neighbourhood $\widetilde{B}_n(x)$ of x in the product topology. This holds for every point x of Ω_S and every $n \in \mathbb{N}$. Therefore, the two topologies under consideration coincide. □

If $x, y \in \Omega_S$, $r > 0$ and $y \in B_r(x)$ then $z \in B_r(x)$ iff $z \in B_r(y)$. In particular, it follows that $B_r(y) = B_r(x)$. Hence each point y in the ball of radius r centred at x is the centre of that ball. This property characterizes d as a *non-Archimedean metric*.

Theorem 5.1.6. *Ω_S is a Cantor space.*

Proof. Combine the results of Theorem 5.1.4 and 5.1.5. □

We shall now define a dynamical system with Ω_S as phase space. There are several mappings of Ω_S into itself that can be considered as a phase mapping (see, for example, 4.2.8 – the adding machine), but there is one that stands out: the shift mapping, to be defined in the next section.

5.2 The shift mapping

5.2.1. The *shift mapping* $\sigma : \Omega_S \to \Omega_S$, often also called just[2] the *shift*, is the mapping which is defined as follows: if $x \in \Omega_S$ then $\sigma(x)$ is the point whose n-th coordinate equals the $(n+1)$-st coordinate of x ($n \in \mathbb{N}$):

$$(\sigma x)_n := x_{n+1} \quad \text{for all } n \in \mathbb{Z}^+ .$$

Then for every $x \in \Omega_S$ and $k \in \mathbb{Z}^+$:

$$(\sigma^k x)_n = x_{n+k} \quad \text{for all } n \in \mathbb{Z}^+ .$$

Thus, under σ, all coordinates shift one position to the left and the initial coordinate is dropped. For example, if $x = 101000\dots$ then $\sigma x = 01000\dots$, $\sigma^2 x = 1000\dots$ and $\sigma^3 x = 000\dots$.

Proposition 5.2.2. *The shift is a continuous and open mapping of Ω_S onto itself: a surjective s-to-1 mapping, hence not injective.*

Proof. It is a straightforward exercise to show that $\sigma[\widetilde{B}_k(x)] = \widetilde{B}_{k-1}(x)$ for every $k \in \mathbb{N}$, $k \geq 2$, and $x \in \Omega_S$. Since for every point $x \in \Omega_S$ the sets $\widetilde{B}_n(x)$ for $n \in \mathbb{N}$ form a local

2 The *system* (Ω_S, σ) is also often called 'the (full) shift'.

base at x, this not only shows that the mapping σ is continuous, but also that it is open.

Alternative proof of continuity: by the definition of σ it is clear that $\pi_n \circ \sigma = \pi_{n+1}$ for every $n \in \mathbb{Z}^+$. It follows that $\pi_n \circ \sigma$ is continuous for every $n \in \mathbb{Z}^+$, because π_{n+1} is continuous. So by the defining property of a product topology, σ is continuous.

If $x \in \Omega_S$ then a point $y \in \Omega_S$ is mapped onto x by σ iff there exists $\alpha \in S$ such that $y := \alpha x_0 x_1 x_2 \dots$. This shows that σ is surjective and also that σ is an s-to-1 mapping: there are s possible choices for α. □

Remark. Obviously, for every $\alpha \in S$ the restriction of σ to the cylinder set $C_0[\alpha]$ is a bijection, hence a homeomorphism, of that cylinder set onto Ω_S. Thus, σ maps (or rather, expands[3]) each member of the clopen partition $\{C_0[\alpha] \,\vdots\, \alpha \in S\}$ of Ω_S – recall from 5.1.3 that every cylinder set is clopen – homeomorphically onto all of Ω_S.

In the remainder of this chapter we shall study the dynamical properties of the system (Ω_S, σ). We shall often refer to this system as the *(full) shift*. Often the study of this dynamical system is called 'Symbolic Dynamics', because of the fact that the elements of Ω_S are sequences of symbols. But we shall use that term in a more restricted sense for the study of dynamical systems by means of of symbolic sequences (see Chapter 6).

Proposition 5.2.3. *Let $x \in \Omega_S$.*
1. *The point x is invariant under σ iff $x_n = x_0$ for all $n \in \mathbb{N}$ (that is, the sequence x is constant).*
2. *The point x is periodic under σ with period p (not necessarily primitive) iff $x_{n+kp} = x_n$ for all $n, k \in \mathbb{Z}^+$, iff $x_n = x_{n \,(\mathrm{mod}\, p)}$ for all $n \in \mathbb{Z}^+$ (that is, x is a periodic sequence with period p).*

Proof. Statement 1 is obvious and for the proof of statement 2, use Corollary 1.1.3, taking into account that $(\sigma^n x)_0 = x_n$ for all $x \in \Omega_S$ and all $n \in \mathbb{Z}^+$. □

Thus, a point $x \in \Omega_S$ is periodic iff it has the form $x = bbb\cdots = b^\infty$ for some finite block b. In that case we call b a *period block* of x, and $|b|$ is a period of x. Clearly, $|b|$ is the primitive period of x iff the block b cannot be written as a concatenation of more than one copy of an initial subblock of b. For example, the point $(10^{p-1})^\infty$ has primitive period p ($p \in \mathbb{N}$) and the point $(0101)^\infty$ has primitive period 2. In particular, for every $p \in \mathbb{N}$ there is a periodic point in Ω_S with primitive period p.

Corollary 5.2.4.
1. *For every $p \in \mathbb{N}$ there are exactly s^p periodic points with period p in Ω_S (i.e., points whose primitive period is a divisor of p). In particular, there are s invariant points, namely, the points α^∞ with $\alpha \in S$.*
2. *The periodic points form a countably infinite dense subset of Ω_S.*

3 This property is formally different from expansiveness as defined in Section 6.2.

Proof. (1) A periodic point with period p is completely determined by its initial block of length p. For such a block there are exactly s^p possibilities.
(2) For every $n \in \mathbb{N}$, denote the set of periodic points with primitive period n by $\mathbf{Per}_n(\sigma)$. By part 1 of the corollary, each set $\mathbf{Per}_n(\sigma)$ is finite, so the set $\bigcup_n \mathbf{Per}_n(\sigma)$ of all periodic points is countable. Moreover, these sets are mutually disjoint and each of them is non-empty, hence the set of all periodic points is infinite.
In order to show that the set of all periodic points is dense in Ω_S it is sufficient to observe that for every point x in Ω_S and every natural number k the basic neighbourhood $\widetilde{B}_k(x)$ of x contains a periodic point. In fact, if $y := b^\infty$ with $b := x_{[0;k)}$ then $y \in \widetilde{B}_k(x)$ and, obviously, y is periodic. □

Proposition 5.2.5. *Let $x, z \in \Omega_S$. Then $z \in \overline{\mathcal{O}_\sigma(x)}$ iff every subblock of z occurs in x, iff every initial subblock of z occurs in x.*

Proof. "If": Assume that every initial subblock of z occurs in x and consider any basic neighbourhood $\widetilde{B}_n(z)$ of z ($n \in \mathbb{N}$). By assumption, the block $z_{[0;n)}$ occurs in x, say, at position m. This means that $\sigma^m x$ starts with the block $z_{[0;n)}$, so $\sigma^m(x) \in \widetilde{B}_n(z)$, hence $\widetilde{B}_n(z) \cap \mathcal{O}_\sigma(x) \neq \emptyset$. This holds for every basic neighbourhood of z, so $z \in \overline{\mathcal{O}_\sigma(x)}$.

"Only if": If $z \in \overline{\mathcal{O}_\sigma(x)}$ then every neighbourhood of z meets $\mathcal{O}_\sigma(x)$. In particular, for every $n \in \mathbb{Z}^+$ there exists $m \in \mathbb{Z}^+$ such that $\sigma^m x \in \widetilde{B}_n(z)$. This means that the initial n-block of z occurs in x (at position m). □

Corollary 5.2.6.
(1) *The shift system (Ω_S, σ) is transitive.*
(2) *All points in Ω_S are non-wandering under σ.*

Proof. (1) Since σ is surjective it is, by Proposition 1.3.2, sufficient to show that there exists a point z in Ω_S with a dense orbit under σ, that is, such that $\overline{\mathcal{O}_\sigma(z)} = \Omega_S$ (we might refer to Corollary 1.3.3 as well). By Proposition 5.2.5, every subblock of any point of Ω_S, hence every possible finite block, must occur in z. We shall construct such a point, as follows:
For every $k \in \mathbb{N}$ the set $\mathbf{S}^k$ is finite, hence the set $\mathbf{S}' := \mathbf{S}^* \setminus \{\square\}$ of all non-empty blocks is countable. Let $b^{(1)}, b^{(2)}, b^{(3)}, \dots$ be any enumeration of $\mathbf{S}'$ and let z be the point obtained by concatenation of these blocks, that is, let $z := b^{(1)} b^{(2)} b^{(3)} \dots$. Now it is clear that every possible finite block occurs in z, hence $\overline{\mathcal{O}_\sigma(z)} = \Omega_S$.
(2) Clear from 1 above and Propositions 4.3.1 (3) one might also apply Proposition 4.3.1 (2) and Corollary 5.2.4 (2). □

Remark. Not every point of Ω_S is transitive (e.g., a periodic point is not), so a non-wandering point in a dynamical system is not necessarily transitive. Not every point of Ω_S is recurrent – see 5.6.1 (2) below – so a non-wandering point is not necessarily recurrent.

Points with special properties (recurrence, almost periodicity, or with a minimal weakly mixing orbit closure) will be discussed at the end of this chapter, in Section 5.6.

Limit sets can be characterized as easily as orbit closures. We leave it to the reader to prove the following: if $x, z \in \Omega_S$ then $z \in \omega_\sigma(x)$ iff every initial block of z occurs infinitely often in x, iff every block of z occurs infinitely often in x (note the difference with Proposition 5.2.5 above). For an example, see Exercise 5.3 (1).

Recall from Corollary 3.2.2 (3) and Proposition 3.1.7 that a transitive system has no proper stable subsets and no proper topologically attracting subsets. In particular the system (Ω_S, σ) does not include such sets. In point of fact, the shift system is everywhere locally unstable (the technical term is *expansive* – see Section 6.2 and Chapter 7) and for this reason (asymptotic) stability is not an issue in shift systems.

5.3 Shift spaces

A *shift space (over the symbol set* S*)* is a non-empty, closed invariant subset of the full shift system (Ω_S, σ). If X is a shift space then the restriction $\sigma|_X$ of σ to X will be denoted by σ_X. However, if no confusion is likely to arise we shall simply write σ for σ_X. The dynamical system (X, σ) is called the *subshift on* X. Often the shift space X itself will also be called a subshift.

Periodic orbits and orbit closures in (Ω_S, σ) are shift spaces. Also, if T is a finite symbol set with t elements, $2 \le t \le s$, then Ω_T occurs as a subshift in Ω_S. For T can be identified with the subset $\{0, \dots, t-1\}$ of S, hence Ω_T can be identified with the non-empty subset $\{x \in \Omega_S : x_n \in T \text{ for all } n\}$ of Ω_S, which is easily seen to be closed and invariant. Moreover, the relative topology of this set in Ω_S coincides with the natural topology on Ω_T (e.g., because the metrics defined according to 5.1.5 agree with each other).

Proposition 5.3.1. *A shift space is a Cantor space iff it has no isolated points.*

Proof. Every shift space is a compact, metrizable and 0-dimensional space because it is a closed subspace of such a space. So the only property that is possibly missing for it to be a Cantor space is that there are no isolated points. □

In an infinite shift system the shift mapping is never a homeomorphism: (so we really need the 'non-invertible' theory; see Note 8 at the end of this chapter).

Proposition 5.3.2. *Let* (X, σ_X) *be a subshift of* Ω_S*. If* σ_X *is injective then the set* X *is finite.*

Proof. That σ_X is injective means: if $x, y \in X$ and $x_i = y_i$ for all $i \ge 1$ then $x = y$, hence $x_0 = y_0$. We prove a stronger version of this statement:

$$\exists n \in \mathbb{N} : \forall x, y \in X : \; x_{[1;n)} = y_{[1;n)} \;\Rightarrow\; x_0 = y_0 \,. \tag{5.3-1}$$

Assume that this is false: for every $n \in \mathbb{N}$ there are points $x^{(n)}, y^{(n)} \in X$ such that

$$x_i^{(n)} = y_i^{(n)} \quad \text{for } i = 1, \ldots, n-1, \text{ but } x_0^{(n)} \neq y_0^{(n)}. \tag{5.3-2}$$

The sequences $(x^{(n)})_{n\in\mathbb{N}}$ and $(y^{(n)})_{n\in\mathbb{N}}$ have convergent subsequences, say with limits x and y, respectively. Passing to these limits in (5.3-2), taking into account that all projections $\pi_i|_X : X \to \mathbf{S}$ are continuous, we get $x_i = y_i$ for every $i \in \mathbb{N}$. Also $x_0 \neq y_0$, because otherwise $x_0^{(n)} = x_0 = y_0 = y_0^{(n)}$ for almost all n. Hence $x \neq y$ and $\sigma_X(x) = \sigma_X(y)$, contradicting the injectivity of σ_X on X. This contradiction proves[4] (5.3-1).

Let $k \in \mathbb{N}$. By applying (5.3-1) successively to the points $\sigma_X^i x$ and $\sigma_X^i y$ with (in this order) $i = k, \ldots, 1$ we find for all points $x, y \in X$:

$$x_{[k\,;\,k+n)} = y_{[k\,;\,k+n)} \Rightarrow x_i = y_i \quad \text{for } i = 0, \ldots, k-1 \tag{5.3-3}$$

with the same value of n as in (5.3-1).

Cover X by finitely many, say N, basic open sets $\widetilde{B}_n(z)$ with $z \in X$. We shall show that X has at most N points. Assume the contrary and select a finite subset F of X containing $N+1$ different points. So the number of points of F is strictly lager than the number of sets $\widetilde{B}_n(z)$ by which F is covered. Hence for every value of $k \in \mathbb{N}$ there are (at least) two points in F whose images under σ_X^k belong to the same set $\widetilde{B}_n(z)$ (this would even be true if σ_X were not injective). There are only finitely many different pairs of mutually distinct points in F, and there are only finitely many different sets $\widetilde{B}_n(z)$ to consider, hence there are two different points $u, v \in F$ such that for infinitely many values of k the points $\sigma_X^k u$ and $\sigma_X^k v$ belong to the same set $\widetilde{B}_n(z)$, i.e., $u_{[k;k+n)} = v_{[k;k+n)}$ for infinitely many values of k. Obviously, (5.3-3) now implies that $u_i = v_i$ for all $i \in \mathbb{Z}^+$, contradicting that $u \neq v$. This contradiction shows that X has at most N points, hence is finite. □

5.3.3. Unless the contrary is stated, every shift space under consideration will be over the symbol set $\mathbf{S}$: a closed, non-empty invariant subset of $\Omega_\mathbf{S}$. We shall characterize shift spaces as the non-empty subsets of $\Omega_\mathbf{S}$ for which there exists a (possibly infinite) list of 'excluded blocks': a point of $\Omega_\mathbf{S}$ belongs to the shift space iff none of its subblocks is in that list.

For the precise formulation and the proof of this statement we need the following notation and terminology: if $X \subseteq \Omega_\mathbf{S}$ then the set of all blocks that do *not* occur in any point of X will be denoted by $\mathcal{A}(X)$:

$$\mathcal{A}(X) := \{\, b \in \mathbf{S}^* \,\vdots\, b \text{ does not occur in any point of } X \,\}. \tag{5.3-4}$$

The members of $\mathcal{A}(X)$ will be called *blocks that are absent from X* or the *X-absent blocks*. If $x \in \Omega_\mathbf{S}$ then $\mathcal{A}(\{x\})$ is the set of all blocks that do not occur in x. Hence (5.3-4) can be formulated as follows: $\mathcal{A}(X) = \bigcap_{x\in X} \mathcal{A}(\{x\})$.

4 For another proof of formula (5.3-1), see Exercise 5.6 (4).

It is obvious that $\mathcal{A}(X) = \mathbf{S}^*$ iff $X = \emptyset$. Recall that the empty block $\not{\square}$ occurs in every point of Ω_S; so if $X \neq \emptyset$ then $\not{\square} \notin \mathcal{A}(X)$. Also, if $X = \Omega_S$ then $\mathcal{A}(X) = \emptyset$. The converse need not be true (example: $X := \Omega_S \setminus \{0^\infty\}$). However, it follows easily from the next proposition that this converse is true if X is a shift space. Moreover, if the set $\mathcal{A}(X)$ is not empty then it is infinite: if b is absent from X then for all finite blocks c and d the block cbd is absent from X as well.

The following definition is, in a sense, dual to the previous one: let $\mathcal{B}$ be a set of non-empty blocks. Define the subset $\mathfrak{X}(\mathcal{B})$ of Ω_S by[5]

$$\mathfrak{X}(\mathcal{B}) := \{\, x \in \Omega_S : \text{no member of } \mathcal{B} \text{ occurs in } x \,\}\,. \tag{5.3-5}$$

We call $\mathcal{B}$ a (not: the) *set of forbidden blocks for* $\mathfrak{X}(\mathcal{B})$. It is easy to see that $\mathfrak{X}(\mathcal{B}) = \Omega_S$ iff $\mathcal{B} = \emptyset$: "if" is obvious and "only if" follows from the observation that e.g. $b^\infty \notin \mathfrak{X}(\mathcal{B})$ for any $b \in \mathcal{B}$. Moreover, if $\mathcal{B} = \mathbf{S}^* \setminus \{\not{\square}\}$ then clearly $\mathfrak{X}(\mathcal{B}) = \emptyset$. However, $\mathfrak{X}(\mathcal{B})$ can also be empty for a finite set $\mathcal{B}$ of forbidden blocks, for example, if $\mathcal{B} = \mathbf{S}^k$ for some $k \geq 1$.

Remarks.

(1) Suppose that $\mathcal{B}$ and $\mathcal{B}'$ are sets of non-empty blocks such that $\mathcal{B} \subseteq \mathcal{B}'$. Then $\mathfrak{X}(\mathcal{B}) \supseteq \mathfrak{X}(\mathcal{B}')$. However, if $\mathcal{B}'$ contains, apart from the blocks of $\mathcal{B}$, only blocks that are already absent from $\mathfrak{X}(\mathcal{B})$, that is, if $\mathcal{B}' \subseteq \mathcal{A}(\mathfrak{X}(\mathcal{B}))$, then clearly $\mathfrak{X}(\mathcal{B}) = \mathfrak{X}(\mathcal{B}')$: if $x \in \mathfrak{X}(\mathcal{B})$ then, by the assumption on $\mathcal{B}'$, no member of $\mathcal{B}'$ occurs in x, hence $x \in \mathfrak{X}(\mathcal{B}')$.

(2) If $\mathcal{B}$ is a set of non-empty blocks then, obviously, $\mathcal{B} \subseteq \mathcal{A}(\mathfrak{X}(\mathcal{B}))$, that is, all forbidden blocks for $\mathfrak{X}(\mathcal{B})$ are absent from $\mathfrak{X}(\mathcal{B})$. In general, we have no equality here. In fact, if $\mathcal{B} \neq \emptyset$ then, by the above inclusion, also $\mathcal{A}(\mathfrak{X}(\mathcal{B})) \neq \emptyset$, hence $\mathcal{A}(\mathfrak{X}(\mathcal{B}))$ is infinite, but $\mathcal{B}$ can be finite. For conditions on $\mathcal{B}$ to be equal to $\mathcal{A}(\mathfrak{X}(\mathcal{B}))$ we refer to Exercise 5.4.

Proposition 5.3.4. *Let X be a subset of Ω_S. The following statements are equivalent:*

(i) *X is a shift space in Ω_S.*

(ii) *$X \neq \emptyset$ and $X = \mathfrak{X}(\mathcal{B})$ for some set $\mathcal{B}$ of non-empty blocks.*

(iii) *$X \neq \emptyset$ and $X = \mathfrak{X}(\mathcal{A}(X))$.*

Proof. (i)⇒(iii): First note that, by the definition of a shift space, $X \neq \emptyset$. Moreover, it is clear from the definition of $\mathcal{A}(X)$ that $X \subseteq \mathfrak{X}(\mathcal{A}(X))$. To prove the converse, consider an arbitrary point $z \in \mathfrak{X}(\mathcal{A}(X))$. We shall show that every basic neighbourhood $\widetilde{B}_n(z)$ ($n \in \mathbb{N}$) of z in Ω_S has a non-empty intersection with X. Since X is a closed subset of Ω_S, this will imply that $z \in X$. Let $n \in \mathbb{N}$. Because $z \in \mathfrak{X}(\mathcal{A}(X))$, the block $z_{[0;n)}$ is not an element of $\mathcal{A}(X)$, hence it occurs in some point $x \in X$, say in position k. Then $\sigma^k x$ begins with the block $z_{[0;n)}$, which means that $\sigma^k x \in \widetilde{B}_n(z)$. But X is invariant under σ, so $\sigma^k x \in X$ as well. So $\widetilde{B}_n(z) \cap X \neq \emptyset$, as claimed. (Compare this with the proof of Proposition 5.2.5 above.)

5 If we would allow for $\not{\square} \in \mathcal{B}$ then $\mathfrak{X}(\mathcal{B})$ would always be empty: the empty block occurs in every point of Ω_S, so every point would be excluded from $\mathfrak{X}(\mathcal{B})$.

(iii)⇒(ii): Obvious.

(ii)⇒(i): Let $X = \mathfrak{X}(\mathcal{B})$ for a subset $\mathcal{B}$ of $\mathbf{S}^* \setminus \{\square\}$. Then X is invariant under σ: if $x \in \mathfrak{X}(\mathcal{B})$ then every block that occurs in σx also occurs in x, hence is not in $\mathcal{B}$, hence $\sigma x \in \mathfrak{X}(\mathcal{B})$. So if we assume that $X \neq \emptyset$ then it remains to show that X is closed in $\Omega_{\mathbf{S}}$. Thus, we want to show: if $z \in \Omega_{\mathbf{S}} \setminus X$ then some neighbourhood of z does not meet X. To this end, observe that the fact that $z \notin X = \mathfrak{X}(\mathcal{B})$ implies that there are $k \in \mathbb{Z}^+$ and $l \in \mathbb{N}$ such that $z_{[k;k+l)} \in \mathcal{B}$. Then the neighbourhood $\widetilde{B}_{k+l}(z)$ of z is disjoint from the set $\mathfrak{X}(\mathcal{B})$, because the block $z_{[0;k+l)}$, hence also the forbidden block $z_{[k;k+l)}$, occurs in every point of $\widetilde{B}_{k+l}(z)$. □

Examples.

(1) Consider two elements of $\mathbf{S}$; say, 0 and 1. For every $n \in \mathbb{N}$, let $x^{(n)} := 1^n 0^\infty$; moreover, let $x^{(0)} := 0^\infty$ and let $x^{(\infty)} := 1^\infty$. Then the sequence $(x^{(n)})_{n\in\mathbb{Z}^+}$ converges in $\Omega_{\mathbf{S}}$ to the point $x^{(\infty)}$. The set consisting of this sequence together with its limit is closed. This set is also invariant under σ, for σ acts in the following way on these points: $x^{(n)} \mapsto x^{(n-1)} \mapsto \cdots \mapsto x^{(0)}$, and the points $x^{(0)}$ and $x^{(1)}$ are invariant. Consequently, this set is a shift space. In fact, it is the shift space $\mathfrak{X}(\mathcal{B})$ for the set $\mathcal{B} = (\mathbf{S} \setminus \{0, 1\}) \cup \{01\}$.

(2) Let $\mathbf{S} = \{0, 1\}$ and $\mathcal{B} = \{00\}$. Then $\mathfrak{X}(\mathcal{B})$ is not empty, hence a shift space. Note that $x \in \mathfrak{X}(\mathcal{B})$ iff the occurrences of the symbol 0 are isolated in x. This shift space is called the *golden mean shift*. It has no isolated points: see Exercise 5.5 (1). Therefore, it is a Cantor space, hence homeomorphic to Ω_2. But the golden mean shift is not conjugate to the full shift on Ω_2: in the golden mean system there is only one invariant point and in the full shift there are two of them.

(3) Let $\mathbf{S} = \{0, 1\}$ and $\mathcal{B} = \{0000, 10\}$. Thus, the symbol 1 in a member of $\mathfrak{X}(\mathcal{B})$ can never be followed by a 0 and such a sequence can have at most three initial 0's. So the shift space $\mathfrak{X}(\mathcal{B})$ has only four elements: $\mathfrak{X}(\mathcal{B}) = \{0001^\infty, 001^\infty, 01^\infty, 1^\infty\}$.

(4) Let $\mathbf{S} = \{0, 1\}$ and let $\mathcal{B}$ be the set of all blocks 10^k1 for *odd* $k \in \mathbb{N}$. Then $\mathfrak{X}(\mathcal{B})$ is the set of all points x in Ω_2 such that between any two subsequent occurrences of the symbol 1 there is an even number (including zero) of occurrences of the symbol 0. Clearly, $\mathfrak{X}(\mathcal{B}) \neq \emptyset$ so it is a shift space. The subshift on $\mathfrak{X}(\mathcal{B})$ is called the *even shift*. Note that there is no restriction on the parity of the length of an initial block of the form 0^k ($k \in \mathbb{N}$) of a point of $\mathfrak{X}(\mathcal{B})$ (such a restriction would violate invariance of the set $\mathfrak{X}(\mathcal{B})$ under σ). The even shift space has no isolated points, hence is a Cantor space. See Exercise 5.5 (1).

(5) Let $\mathbf{S} = \{0, 1\}$ and let $\mathcal{B}$ be the set of all blocks 10^k1 for k *not* a prime number. Then $\mathfrak{X}(\mathcal{B})$ is the set of all points x in Ω_2 such that between any two subsequent occurrences of the symbol 1 the number of occurrences of the symbol 0 is zero or a prime number. This shift space is called the *prime gap shift*. It is a Cantor space: see Exercise 5.5 (1).

(6) Let X be the set of all elements of Ω_2, in which the 1's appear infinitely often, and such that the number of 0's between two successive occurrences of a 1 is 1, 2, or 3;

before the first occurrence of a 1 it is 0, 1, 2 or 3. So $X = \mathfrak{X}(\mathcal{B})$ with $\mathcal{B} = \{0000, 11\}$. It is clear that $X \neq \emptyset$, so X is a shift space, called the (1,3) *run-length limited shift*. This shift space is also a Cantor space; see Exercise 5.5 (1).

(7) Let $S = \{a, b, c\}$ and let X be the set of points of Ω_S in which a block of the form ab^kc^ma with $k, m \in \mathbb{Z}^+$ may only occur if $m = k$. Then $X \neq \emptyset$ and $X = \mathfrak{X}(\mathcal{B})$ for the (infinite) set of blocks ab^kc^ma with $k \neq m$. So X is a shift space. It is called the *context free shift*. This shift space is easily seen to be a Cantor space.

(8) The set of all periodic points in Ω_2 is not empty and it is invariant under the shift, but it is not a shift space: it is dense in Ω_2 but not equal to Ω_2, hence not closed.

5.3.5. If a shift space X is defined as $X = \mathfrak{X}(\mathcal{B})$ for some set $\mathcal{B}$ of forbidden blocks, then the elements of the complement $\mathcal{B}^c$ of $\mathcal{B}$ in the set $\mathbf{S}^*$ of all finite blocks will be called the *allowed blocks* (or *allowed words*) for X. Obviously, the definition of $\mathfrak{X}(\mathcal{B})$ can be given in terms of allowed blocks:

$$\mathfrak{X}(\mathcal{B}) = \{x \in \Omega_S : \text{every subblock of } x \text{ is allowed}\}. \tag{5.3-6}$$

For every shift space X, the elements of the complement $\mathcal{A}(X)^c$ of the set $\mathcal{A}(X)$ in the set $\mathbf{S}^*$ of all finite blocks are called the blocks or words *present in* X, or the X-*present* blocks or words. It follows from the definition of $\mathcal{A}(X)$ that a word is X-present iff there exists a point in X in which that word occurs. For this reason, the set of all X-present words is also called the *language* of X; we shall denote it by $\mathcal{L}(X)$. So, by definition, $\mathcal{L}(X) := \mathcal{A}(X)^c$.

The set $\mathcal{L}(X)$ is always infinite because there is no upper bound for the lengths of the X-present blocks. Even in the simple case that $X = \{\alpha^\infty\}$ for some $\alpha \in \mathbf{S}$ we have $\mathcal{L}(X) = \{\alpha^k : k \in \mathbb{Z}^+\}$, an infinite set.

Examples.

(1) The language of the full shift space Ω_2 is the set $\mathbf{S}^*$ (including the empty block).

(2) The language of the golden mean shift is the set of all finite blocks over $\{0, 1\}$ that do not contain the block 00: $\{\square, 0, 1, 01, 10, 11, 010, 011, 101, 110, 111, 0101, \ldots\}$. An elegant way of representing of the language of this shift will be given in Section 5.5.

(3) Let $x \in \Omega_S$ and let X be the orbit closure of x under the shift, that is, $X := \overline{\mathcal{O}_\sigma(x)}$. Then X is a shift space and Proposition 5.2.5 implies that $\mathcal{L}(X)$ is the set of all finite blocks that occur in x.

We can reformulate Proposition 5.3.4 (iii) in terms of the language $\mathcal{L}(X)$ of a shift space X, as follows:

Proposition 5.3.6. *Let X be a shift space and let $x \in \Omega_S$. Then $x \in X$ iff all blocks that occur in x are in the language $\mathcal{L}(X)$ of X.*

Proof. Proposition 5.3.4 (iii) states that $x \in X$ iff none of its blocks does not occur anywhere in X, that is, iff all its blocks occur somewhere in X. □

The reader may recognise this as a *compactness theorem*: a statement saying that an object has a certain property whenever all its finite subobjects are of a certain type.

Note that the terms 'X-absent block' and 'X-present' block are 'absolute': they depend only on the shift space X itself. On the other hand, the terms 'forbidden block' and 'allowed block' refer to the way X happens to be defined by a set of blocks – which is not unique. As observed in Remark 2 in 5.3.3, in general we have $\mathcal{B} \subseteq \mathcal{A}(\mathfrak{X}(\mathcal{B}))$, hence $\mathcal{L}(\mathfrak{X}(\mathcal{B})) \subseteq \mathcal{B}^c$, that is, *forbidden blocks are absent from* $\mathfrak{X}(\mathcal{B})$, *and words in the language of* $\mathfrak{X}(\mathcal{B})$ *are allowed*. These inclusions may be proper: it is possible that not every $\mathcal{B}$-allowed block (i.e., member of $\mathcal{B}^c$) actually occurs in an element of $\mathfrak{X}(\mathcal{B})$.

Example (1). Let $\mathbf{S} = \{0, 1\}$ and $\mathcal{B} := \{10, 11\}$. Then $X := \mathfrak{X}(\mathcal{B}) = \{0^\infty\}$, because the symbol 1 cannot occur without creating a forbidden block, so the block 01 is absent in X, but it is not an element of $\mathcal{B}$. Stated otherwise, the block 01 is allowed, but it is not in the language of X.

In a shift space the set of periodic points may be not dense, and there may be no transitive points. For example, in the shift space $X = \{01^\infty, 1^\infty\}$ there is only one periodic point and no transitive point. In order to mimic the proof of the existence of a transitive point as given in Proposition 5.2.6 (1) we would like to write down a sequence of symbols containing all X-present blocks – but the problem is that we do not know how to put them together in such a way that the concatenations are allowed. A similar obstruction seems to prevent the construction of periodic points in a shift space. To remedy this we introduce the following notion.

A shift space X is said to be *irreducible* whenever for any ordered pair of non-empty X-present words u and v there exists a word w such that uwv is X-present. Here the word w is, of course, X-present. Also, we may always assume that w is not the empty word. For if uv is already X-present then, by the definition of irreducibility, there is a word w' such that $(uv)w'(uv)$ is X-present, so we can take $w := vw'u$. Then $w \neq \not{\square}$, even if $w' = \not{\square}$.

Example (2). Obviously, the full shift is irreducible. Also the golden mean shift, the even shift and the prime gap shift are irreducible, as is the (1,3) run-length limited shift (in all cases, take for the block w the symbol 1, preceded and followed by a suitable number of 0's). Similarly, the context free shift is irreducible (take for w the 1-block a preceded and followed by a suitable number of c's and b's, respectively).

Proposition 5.3.7. *A shift space X is irreducible iff it is transitive.*

Proof. Suppose X is irreducible. Let $u^{(1)}, u^{(2)}, \dots, u^{(n)}, \dots$ be an enumeration of all X-present blocks. As X is irreducible, there is a block $w^{(1)}$ such that the block $u^{(1)}w^{(1)}u^{(2)}$ is X-present. Now proceed by induction: let $k \geq 2$ and suppose that for $i = 1, \dots, k-1$

we have already found blocks $w^{(i)}$ such that the block $b^{(k)} := u^{(1)}w^{(1)} \dots u^{(k-1)}w^{(k-1)}u^{(k)}$ is X-present. Because X is irreducible, there exists a block $w^{(k)}$ such that the block $b^{(k)}w^{(k)}u^{(k+1)}$ is X-present. In this way we find the infinite concatenation

$$z := u^{(1)}w^{(1)}u^{(2)}w^{(2)} \dots u^{(k)}w^{(k)}\, u^{(k+1)} \dots ,$$

denoting a point in Ω_S in which all finite subblocks belong to $\mathcal{L}(X)$ (because they all occur in a suitable X-present initial segment of z) and in which all X-present blocks occur. Then Proposition 5.3.6 implies that $z \in X$, and similar to in the proof of Proposition 5.2.6 (1) one shows that the orbit of z is dense in X. In order to prove that the point z is transitive, we would have to show that every X-present block occurs infinitely often in z. This is not difficult, but instead we shall extend the inductive procedure used above. Let the blocks $b^{(k)}$ for $k \in \mathbb{N}$ be as above. By induction, we find blocks $v^{(k)}$ such that all blocks $b^{(1)}v^{(1)}b^{(2)} \dots b^{(k)}v^{(k)}b^{(k+1)}$ are X-present. Then

$$z' := b^{(1)}v^{(1)}b^{(2)}v^{(2)} \dots b^{(k)}v^{(k)} \dots$$

is a point in X (because all subblocks are X-present) in which all X-present blocks $u^{(i)}$ ($i \in \mathbb{N}$) occur infinitely often.

Conversely, assume that X is transitive, say, with transitive point x, and consider two X-present blocks u and v. Then there are points y and z in X in which these blocks occur. Because X is invariant under σ we may assume that these blocks occur in initial position, u in y and v in z. The transitive point x visits the cylinders based on the initial blocks u of y and v of z infinitely often, hence it contains the blocks u and v at arbitrarily large positions. Consequently, there are natural numbers k and l with $l > k + |u|$ such that the blocks u and v occur in x at positions k and l, respectively. Obviously, these occurrences of u and v do not overlap, and the block w can be defined as the block of coordinates between these occurrences of u and v. Then the block uwv occurs in x, so it is X-present. □

Example. By the example just before Proposition 5.3.7, the golden mean shift, the even shift, the prime gap shift, the (1,3) run-length limited shift and the context free shift are irreducible, hence transitive. So it follows from Proposition 4.3.1 (3) that in each of these subshifts every point is non-wandering. As transitive points are recurrent, each of these systems has a dense set of recurrent points. This is in accordance with Theorem 4.3.8. These facts can also be obtained as trivial consequences of the fact that all of these shift spaces have a dense set of periodic points: see Exercise 5.9 (5).

5.3.8. A particular class of shift spaces is formed by those that can be given by a *finite* list of forbidden blocks: a shift space X and the corresponding subshift are said to be of *finite type* whenever there is a finite set $\mathcal{B}$ of non-empty blocks such that $X = \mathfrak{X}(\mathcal{B})$. We shall use the abbreviation *SFT* for the phrase '(sub)shift of finite type'.

If $\mathcal{B}$ is a finite set of blocks then the set of lengths of its elements has a maximum, say N. Conversely, if there exists $N \in \mathbb{N}$ such that every element of $\mathcal{B}$ has length at most N then $\mathcal{B}$ has at most s^N elements, hence $\mathcal{B}$ is finite. So if X is a subshift then it is of finite type iff there exists a natural number N such that $X = \mathfrak{X}(\mathcal{B})$ for a set $\mathcal{B}$ of non-empty words, all with length at most N. In this case we say that X is an SFT of *order* N.

The order N of an SFT refers to the set of forbidden blocks that happens to be used to define that shift space. We may assume that these blocks all have length precisely N. In fact, the following general lemma states that the blocks in a (possibly infinite) set $\mathcal{B}$ can be extended without affecting the shift space $\mathfrak{X}(\mathcal{B})$.

Lemma 5.3.9. *Let $\mathcal{B}$ be a set of blocks and let $\kappa\colon \mathcal{B} \to \mathbb{Z}^+$ be an arbitrary mapping. For every $b \in \mathcal{B}$, let $\mathcal{D}_b$ be the set of all possible extensions of the block b to a block of length $|b| + \kappa(b)$, that is, $\mathcal{D}_b := \{bc : c \in \mathbf{S}^{\kappa(b)}\}$, and let $\mathcal{D} := \bigcup\{\mathcal{D}_b : b \in \mathcal{B}\}$. Then $\mathfrak{X}(\mathcal{B}) = \mathfrak{X}(\mathcal{D})$.*

Proof. Every member of $\mathcal{D}$ contains a member of $\mathcal{B}$, so $\mathfrak{X}(\mathcal{B}) \subseteq \mathfrak{X}(\mathcal{D})$. Conversely, if $x \in \Omega_S$ and $x \notin \mathfrak{X}(\mathcal{B})$ then some $b \in \mathcal{B}$ occurs in x, say at position n. But then the block $x_{[n;n+|b|+\kappa(b))}$ is in $\mathcal{D}$, hence $x \notin \mathfrak{X}(\mathcal{D})$. □

Corollary 5.3.10. *Let X be an SFT of order N. Then $X = \mathfrak{X}(\mathcal{B})$ for a finite set $\mathcal{B}$ of forbidden blocks, all of whose members have a length precisely equal to N. In addition, X is an SFT of order M for every $M \geq N$.*

Proof. If $X = \mathfrak{X}(\mathcal{B})$ for a finite set $\mathcal{B}$, and if all members of $\mathcal{B}$ have length at most N, then we can replace every block in $\mathcal{B}$ that is too short by the set of all its extensions to length N. In this way we get a set $\mathcal{D}$ of blocks that all have length N and such that $X = \mathfrak{X}(\mathcal{D})$. The same argument shows that X has order M for every $M \geq N$. □

In particular, it follows that the order of an SFT is not a property of the shift space itself, but of the set of forbidden blocks by which it is defined. To obtain a property of the shift space one could consider the minimum of its orders. But Corollary 5.4.7 below shows that this is not a dynamical property: the SFT $(X^{(k)}, \sigma_{X^{(k)}})$ of order 2 mentioned there is conjugate to the original SFT (X, σ_X).

Examples.

(1) Every orbit of a periodic point in Ω_S with primitive period $p \in \mathbb{N}$ is an SFT of order p: the orbit is $\mathfrak{X}(\mathcal{B})$, where $\mathcal{B}$ is the set of all p-blocks that are not a cyclic permutation of the primitive period block.

(2) The golden mean shift is, by definition, an SFT of order 2. In accordance with Corollary 5.3.10, it can also be considered as an SFT of order 3, with as set of forbidden blocks the collection $\mathcal{D} := \{000, 001\}$ (apply Lemma 5.3.9 to $\mathcal{B} = \{00\}$ – of course, one can also see this directly; note that $\mathcal{D} := \{000, 001, 100\}$ would also do but, apparently, we can dispense with the block 100).

(3) Similarly, the (1,3) run-length limited shift is an SFT of order 4.

(4) The even shift is not of finite type. For suppose the contrary and assume that the even shift is $\mathfrak{X}(\mathcal{B})$ for a finite set $\mathcal{B}$ of blocks. Let N be the length of the largest block in $\mathcal{B}$. Then no subblock of the point $x := 10^{2N+1}10^{\infty}$ is in $\mathcal{B}$. In order to verify this, we need to consider only blocks of length at most N. These occur all

in one of the points 10^∞ and $0^N 10^\infty$, which belong to the even shift, so that none of their subblocks is in $\mathcal{B}$. Hence $x \in \mathfrak{X}(\mathcal{B})$, which is obviously false.

(5) Similar arguments shows that the prime gap shift and the context free shift are not of finite type. See also the Examples (3) and (4) in 5.5.3 ahead.

By Proposition 5.3.6, a shift space is determined by its language, which is always an infinite set. An SFT turns out to be determined by a *finite* list of allowed blocks, which (just like a finite list of forbidden blocks) enables an effective verification of whether of a point belongs to that subshift or not. To facilitate the formulation of the next proposition, we introduce the following notation: if X is a shift space and $k \in \mathbb{N}$ then

$$\mathcal{A}_k(X) := \mathbf{S}^k \cap \mathcal{A}(X) \quad \text{and} \quad \mathcal{L}_k(X) := \mathbf{S}^k \cap \mathcal{L}(X)\,.$$

Thus, $\mathcal{A}_k(X)$ is the set of all X-absent k-blocks, and $\mathcal{L}_k(X)$ is the set of all X-present k-blocks.

Proposition 5.3.11. *Let X be a shift space and let $N \in \mathbb{N}$. The following conditions are equivalent:*

(i) *X is an SFT of order N.*

(ii) *There exists a subset $\mathcal{B}$ of $\mathbf{S}^N$ such that $X = \mathfrak{X}(\mathcal{B})$.*

(iii) *There is a subset $\mathcal{L}$ of $\mathbf{S}^N$ such that*

$$X = \mathfrak{F}_N(\mathcal{L}) := \{\, x \in \Omega_{\mathbf{S}} \,\vdots\, x_{[n;n+N)} \in \mathcal{L} \text{ for all } n \in \mathbb{Z}^+ \,\},$$

If these conditions are fulfilled then it may be assumed that $\mathcal{L} = \mathbf{S}^N \setminus \mathcal{B}$ and that $\mathcal{B}$ is the set of all blocks of length N that do not occur in any point of X, and that every block from the set $\mathcal{L}$ actually occurs in a point of X, i.e., that $\mathcal{B} = \mathcal{A}_N(X)$ and $\mathcal{L} = \mathcal{L}_N(X)$.

Proof. (i)⇒(ii): Clear from Corollary 5.3.10.

(ii)⇒(iii): Suppose (ii) holds, and let $\mathcal{L} := \mathbf{S}^N \setminus \mathcal{B}$. Then for every point x of $\Omega_{\mathbf{S}}$, $x \in X$ iff no subblock of x of length N is in $\mathcal{B}$, iff all blocks in x of length N are in $\mathcal{L}$. So condition (iii) holds.

(iii)⇒(ii): If (iii) holds then (ii) holds with $\mathcal{B} := \mathbf{S}^N \setminus \mathcal{L}$.

(ii)⇒(i): Clear from the definition of 'shift of finite type'.

Final remark: In the proofs of the equivalence of (ii) and (iii) we have seen already that we may assume that $\mathcal{L} = \mathbf{S}^N \setminus \mathcal{B}$. Moreover, suppose we have $\mathcal{B}$ as in (ii). Then $\mathcal{B} \subseteq \mathcal{A}_N(X) \subseteq \mathcal{A}(X)$, hence by Remark 1 in 5.3.3 we get $\mathfrak{X}(\mathcal{A}_N(X)) = \mathfrak{X}(\mathcal{B}) = X$. Thus, the set $\mathcal{B}$ can be replaced by $\mathcal{A}_N(X)$. If we do so, then in (iii), $\mathcal{L} = \mathbf{S}^N \setminus \mathcal{B}$ is replaced by $\mathbf{S}^N \setminus \mathcal{A}_N(X) = \mathcal{L}_N(X)$. □

Remarks.

(1) In (iii) above we have defined an operator set $\mathfrak{F}_N$ that assigns to any set $\mathcal{L}$ of N-blocks the set of points in $\Omega_{\mathbf{S}}$ – a subshift if not empty – for which every N-subblock is in $\mathcal{L}$. The subscript N indicates that we need to check only N-blocks.

(2) In general, for $\mathcal{L}$ as in (iii) above one has $\mathcal{L}_N(X) \subseteq \mathcal{L}$. This inclusion can be proper: see Example (1) after Proposition 5.3.6, where $\mathcal{L} := \{0,1\}^2 \setminus \mathcal{B} = \{00, 01\}$, but where $\mathcal{L}_2(X) = \{00\}$.

Proposition 5.3.12. *A shift space X is an SFT iff $\sigma_X : X \to X$ is an open mapping.*

Proof. "Only if": Let N be the order of X as an SFT. It is sufficient to show that

$$\forall x \in X \; \forall n \geq N : \sigma_X[X \cap \widetilde{B}_n(x)] = X \cap \widetilde{B}_{n-1}(\sigma_X x) . \tag{5.3-7}$$

So consider an arbitrary point $x \in X$. The inclusion "$\subseteq$" in (5.3-7) is trivial. In order to prove "$\supseteq$", consider any point $x' \in X \cap \widetilde{B}_{n-1}(\sigma_X x)$. Then x' has the form $x' = x_1 \dots x_{n-1} x'_n x'_{n+1} \dots$ with $x'_i \in \mathbf{S}$ for $i \geq n$. Let $x'' := x_0 x' = x_0 x_1 \dots x_{n-1} x'_n x'_{n+1} \dots$. It is clear that $x'' \in \widetilde{B}_n(x)$. In addition, since $N \leq n$, every N-block of x'' either occurs in x (namely, the initial N-block of x'') or in x' (all other N-blocks of x''), hence is X-present. Now Proposition 5.3.11 (iii) implies that $x'' \in X$. As $\sigma_X x'' = x'$ this completes the proof.

"If": We introduce the following *ad hoc* notation: for every finite block b let $C_X[b] := C_0[b] \cap X$. By hypothesis, for every $\alpha \in \mathbf{S}$ the set $\sigma_X[C_X[\alpha]]$ is open on X (possibly empty), hence a union of intersections of cylinder sets with X. The set $\sigma_X[C_Z[\alpha]]$ is compact, so it is a union of finitely many of such sets. As the symbol set $\mathbf{S}$ is finite, one sees after some reflection that there exists $p \in \mathbb{N}$ such that all of those cylinder sets, for all $\alpha \in \mathbf{S}$, can be based on blocks of length p. Thus, for every $\alpha \in \mathbf{S}$ there exists a set $B_\alpha \subseteq \mathbf{S}^p$ such that

$$\sigma_X[C_X[\alpha]] = \bigcup \{ C_X[b] : b \in B_\alpha \} .$$

Without limitation of generality we may assume that if $\sigma_X[C_X[\alpha]] \neq \emptyset$ then $C_X[b] \neq \emptyset$ for every $b \in B_\alpha$. An easy onsequence of the above equality is: if $\alpha \in \mathbf{S}$ and b is any p-block such that $\sigma_X[C_X[\alpha]] \cap C_X[b] \neq \emptyset$ then, because cylinders based on different p-blocks are mutually disjoint, $b \in B_\alpha$ and $C_X[b] \subseteq \sigma_X[C_X[\alpha]]$.

We are now ready to show that X is an SFT of order $p + 1$. In view of Proposition 5.3.11 we have to prove: if x is a point of $\Omega_\mathbf{S}$ such that every subblock of x of length $p+1$ is in $\mathcal{L}(X)$ then $x \in X$. By Proposition 5.6.3 it is sufficient to show that every subblock of such a point x belongs to $\mathcal{L}(X)$. The proof is by induction on the length m of such blocks.

For $m = p+1$ the result is clear. Let $m > p+1$ and assume that for every point $x \in \Omega_\mathbf{S}$ with the property that each subblock of length $p + 1$ is in $\mathcal{L}(X)$ also every subblock of length m is in $\mathcal{L}(X)$. Consider a point x of $\Omega_\mathbf{S}$ such that each subblock of length $p + 1$ is in $\mathcal{L}(X)$. We shall show that its initial $(m + 1)$-block $x_{[0\,;\,m+1)}$ is in $\mathcal{L}(X)$; by applying this conclusion to $\sigma_X^k x$ for $k \geq 1$ one may conclude that every subblock of x of length $m + 1$ is in $\mathcal{L}(X)$, and the proof by induction is completed.

By hypothesis, the m-block $x_{[1\,;\,m+1)}$ is in $\mathcal{L}(X)$, so there exists $y \in X$ such that $y_{[0\,;\,m)} = x_{[1\,;\,m+1)}$, hence $y \in C_X[x_{[1\,;\,m+1)}]$. On the other hand, by the choice of x we have[6] $x_{[0\,;\,p+2)} \in \mathcal{L}(X)$, so there is a point $z \in X$ such that $z_{[0\,;\,p+2)} = x_{[0\,;\,p+2)}$ This implies that $\sigma_X z \in C_X[x_{[1\,;\,p+2)}]$ and, of course, also that $z \in C_X[x_0]$, so $\sigma_X[C_X[x_0]] \cap C_X[x_{[1\,;\,p+2)}] \neq \emptyset$. As observed above, this means that $C_X[x_{[1\,;\,p+2)}] \subseteq \sigma_X[C_X[x_0]]$. Since $m+1 \geq p+2$ it is clear that $C_X[x_{[1\,;\,m+1)}] \subseteq C_X[x_{[1\,;\,p+2)}]$, so the point $y \in C_X[x_{[1\,;\,m+1)}]$ selected above is in $\sigma_X[C_X[x_0]]$. This means that the point $x_0 y$ is in X. It follows that its initial $(m+1)$-block $x_0 y_0 \dots y_{m-1} = x_0 x_1 \dots x_m$ is in $\mathcal{L}(X)$. □

Remarks.
(1) So the topological property that $\sigma_X : X \to X$ is an open mapping characterizes SFT's. For another characterization, see Exercise 5.16 (3).
(2) Another important fact is that every SFT includes a periodic point, and that in a transitive (i.e., irreducible) SFT the set of periodic points is dense. See the Exercises 5.7 (1) and 7.9 (4).
(3) It is clear from Remark (2) above that *an infinite minimal subshift cannot be an SFT*. There is a more topologically oriented proof of this fact. Let (X, σ_X) be an infinite SFT which is minimal. By Proposition 5.3.12 above, $\sigma_X : X \to X$ is an open mapping. The discussion after Theorem 1.2.8 implies that σ_X is a homeomorphism, so by Proposition 5.3.2 above, X is finite.

5.4 Factor maps

In this section we consider, in addition to **S**, a second finite symbol set **T** (which may be equal to **S**). Let X be a shift space over **S** and, as before, let $\mathcal{L}_k(X)$ be the set of all X-present k-blocks ($k \in \mathbb{N}$). For every mapping $\Phi : \mathcal{L}_k(X) \to \mathbf{T}$ the following rule defines a mapping $\varphi : X \to \Omega_{\mathbf{T}}$: if $x \in X$ then the coordinates of $\varphi(x) \in \Omega_{\mathbf{T}}$ are given by

$$\forall n \in \mathbb{Z}^+ : \; \varphi(x)_n := \Phi(x_{[n\,;\,n+k)}) \,. \tag{5.4-1}$$

We call φ the *sliding block code with anticipation k*, or the *k-block code defined by* (or: *generated by*) Φ.

We shall denote this mapping φ by $\Phi_\infty^{(k)}$ or, if k is understood, simply by Φ_∞. The following picture illustrates the action of a sliding block code: for any point $x \in X$, put a window of width k over the sequence of coordinates of x and use the coordinates of x that appear in this window to compute a coordinate of the point $\Phi_\infty^{(k)}(x)$. To compute the next coordinate the window is slid one position to the right.

6 We use only that the initial $(p+1)$-block of x belongs to $\mathcal{L}(X)$ in order to show that its initial $(m+1)$-block is in $\mathcal{L}(X)$. However, in order to prove that every $(m+1)$-subblock of x belongs to $\mathcal{L}(X)$ we have to assume that the initial $(p+1)$-block of $\sigma^k x$ is in $\mathcal{L}(X)$ for every $k \geq 1$, i.e, that every $(p+1)$-subblock of x is in $\mathcal{L}(X)$.

$$\cdots x_{n-1}\ \boxed{x_n\ x_{n+1}\ \cdots\ x_{n+k-1}}\ x_{n+k}\ \cdots$$
$$\downarrow \Phi$$
$$\cdots y_{n-1}\ \boxed{y_n}\ y_{n+1}\ \cdots$$

If X and Y are shift spaces, $X \subseteq \Omega_{\mathbf{S}}$ and $Y \subseteq \Omega_{\mathbf{T}}$, and $\varphi: X \to \Omega_{\mathbf{T}}$ is a sliding block code such that $\varphi[X] \subseteq Y$ then we also say that we have a sliding block code (or a k-block code) from X to Y. If a mapping $\varphi: X \to Y$ is a k-block code then φ is also an l-block code for every $l > k$. For if φ is generated by $\Phi: \mathcal{L}_k(X) \to \mathbf{T}$ and if $l > k$ then Φ can be extended to a mapping from $\mathcal{L}_l(X)$ to $\mathbf{T}$ by assigning to each X-present l-block the value of Φ on its initial k-block.

Examples.

(1) Let $X = \Omega_{\mathbf{S}}$; then the shift mapping σ is the 2-block map generated by the mapping $\Phi: a_0a_1 \mapsto a_1 : \mathbf{S}^2 \to \mathbf{S}$, that is, $\sigma = \Phi_\infty^{(2)}$.

(2) Let $\mathbf{S} = \{0, 1\}$, let $\mathbf{T} = \{a, b\}$, and define $\Phi: \mathbf{S} \to \mathbf{T}$ by $\Phi(0) := a$, $\Phi(1) := b$. Then the 1-block code $\Phi_\infty^{(1)} : \Omega_{\mathbf{S}} \to \Omega_{\mathbf{T}}$ is a bijection. We shall see in Corollary 5.4.2 below that in this case $\Phi_\infty^{(1)}$ is a conjugation (not very surprising: Φ is just a renaming of symbols).

(3) Let $\mathbf{S} = \{0, 1\}$, let X be the golden mean shift, and define $\Phi: \mathbf{S} \to \mathbf{S}$ by $\Phi(0) := 1$ and $\Phi(1) := 0$. Then $\Phi_\infty^{(1)} : X \to \Omega_2$ is injective, mapping X onto the set Y of all points of Ω_2 in which the block 11 does not occur. Clearly, this is a shift space (the proof is exactly the same as for the golden mean shift). Actually, it follows from Corollary 5.4.2 below that $\Phi_\infty^{(1)}$ is a conjugation from (X, σ_X) with (Y, σ_Y). In point of fact, this is a special case of Example (2) above (with $a = 1$ and $b = 0$).

(4) Let $\mathbf{S} = \{0, 1\}$ and let $\Phi: a_0a_1 \mapsto a_0 + a_1 \pmod 2): \mathbf{S}^2 \to \mathbf{S}$. Then $\varphi := \Phi_\infty^{(2)}$ is a surjection of Ω_2 onto itself, but φ it is not injective: e.g., $\varphi(0^\infty) = 0^\infty = \varphi(1^\infty)$.

(5) Let $\mathbf{S} = \mathbf{T} = \{0, 1\}$, let X be the golden mean shift and let Y be the even shift. Define $\Phi: \mathcal{L}_2(X) \to \{0, 1\}$ by $\Phi(01) := \Phi(10) := 0$, $\Phi(11) := 1$. Claim: $\Phi_\infty^{(2)}[X] = Y$.
In order to prove this claim, observe the following. If $x \in X$ and for some $k \in \mathbb{Z}^+$ the block 10^k1 occurs in $\Phi_\infty^{(2)}(x)$, then it comes from the block $1(10)^r11$ in x with $k = 2r$. So k is even. This shows that $\Phi_\infty^{(2)}[X] \subseteq Y$. Using this observation it is also clear that every $y \in Y$ is the image of a point $x \in X$.

Proposition 5.4.1. *Let X and Y be shift spaces (not necessarily over the same symbol sets) and let $\varphi : X \to Y$ be a sliding block code. Then φ is continuous and $\varphi \circ \sigma_X = \sigma_Y \circ \varphi$, i.e., the following diagram commutes:*

$$\begin{array}{ccc} X & \xrightarrow{\ \sigma_X\ } & X \\ \varphi\downarrow & & \downarrow\varphi \\ Y & \xrightarrow[\ \sigma_Y\ ]{} & Y \end{array}$$

Consequently, $\varphi: (X, \sigma_X) \to (Y, \sigma_Y)$ is a morphism of dynamical systems.

Proof. Let $X \subseteq \Omega_S$, $Y \subseteq \Omega_T$, and suppose that φ is generated by the mapping $\Phi : \mathcal{L}_k(X) \to \mathbf{T}$ with $k \in \mathbb{N}$. For every $x \in X$ and $n \in \mathbb{Z}^+$ we have:

$$\big((\sigma_Y \circ \varphi)(x)\big)_n = \varphi(x)_{n+1} = \Phi(x_{[n+1\,;\,n+k+1)})$$

and

$$\big((\varphi \circ \sigma_X)(x)\big)_n = \big(\varphi(\sigma_X x)\big)_n = \Phi((\sigma_X x)_{[n\,;\,n+k)}) = \Phi(x_{[n+1\,;\,n+k+1)})\,.$$

This shows that $\varphi \circ \sigma_X = \sigma_Y \circ \varphi$. In order to show that φ is continuous it is sufficient to show that, for every $n \in \mathbb{Z}^+$, the mapping $\pi_n \circ \varphi: X \to \mathbf{T}$ is continuous, where π_n is the canonical projection of Ω_T onto $\mathbf{T}$. Consider arbitrary $n \in \mathbb{Z}^+$ and denote the restriction to X of the 'generalised' projection $x \mapsto x_{[n\,;\,n+k)} : \Omega_S \to \mathbf{S}^k$ by π_n^k. It is clear that π_n^k is continuous (its compositions with the canonical projections of $\mathbf{S}^k$ onto $\mathbf{S}$ are ordinary projections, hence continuous). By the definition of φ we have $\pi_n \circ \varphi = \Phi \circ \pi_n^k$, where the right-hand side is continuous (Φ is continuous because $\mathbf{S}^k$ and $\mathbf{T}$ are discrete). So the left-hand side is continuous as well. □

Corollary 5.4.2. *Let X and Y be shift spaces and let $\varphi : X \to Y$ be a sliding block code. Then $\varphi[X]$ is a shift space. If φ is injective then the subshifts (X, σ_X) and $(\varphi[X], \sigma_{\varphi[X]})$ are conjugate.*

Proof. By Proposition 1.5.4 (3), $\varphi[X]$ is an invariant subset of Y, hence invariant in the full shift space Ω_T of which Y is a subshift. It is not empty because X is not empty, and it is compact, hence closed in Ω_T, because it is the continuous image of the compact space X. This shows that $\varphi[X]$ is a shift space. Moreover, if φ is injective then by Theorem A.3.2 in Appendix A it is a homeomorphism. □

Remarks.

(1) Renaming symbols is an example of an injective sliding block code, so it produces shifts that are conjugate to the original shifts. See also the Examples (2) and 3 above.

(2) By Example (5) above, the even shift is a factor of the golden mean shift. Consequently, a factor of an SFT may not be of finite type.

Theorem 5.4.3 (Curtis, Lyndon and Hedlund). *Let X and Y be shift spaces and let $\varphi : (X, \sigma_X) \to (Y, \sigma_Y)$ be a morphism of dynamical systems. Then φ is a sliding block code.*

Proof. Let $X \subseteq \Omega_S$ and $Y \subseteq \Omega_T$. All cylinder sets in this proof will be in the space Ω_T, so there will be no ambiguity in the notation if we denote them just by $C_0[b]$ for finite non-empty blocks b over $\mathbf{T}$.

For every symbol $\alpha \in \mathbf{T}$ the set $C_0[\alpha] \cap Y$ is the intersection of a clopen subset of Ω_T with Y, hence clopen in Y. Then $A_\alpha := \varphi^{\leftarrow}[\,C_0[\alpha] \cap Y]$ is clopen in X, hence certainly compact. The sets $C_0[\alpha] \cap Y$ with $\alpha \in \mathbf{T}$ are mutually disjoint, so the sets A_α in X are mutually disjoint as well. By the statement just after formula (A.7-3) in Appendix A, these subsets have positive mutual distances. Because there are only finitely many of such sets, there is a $\delta > 0$ with the following property: if two points x and z are in different sets A_α with $\alpha \in \mathrm{T}$ then $d(x, z) \geq \delta$. Because the sets A_α cover X this implies: if $x, z \in X$ and $d(x, z) < \delta$ then there is a (unique) $\alpha \in \mathrm{T}$ such that both $x, z \in A_\alpha$. Using the definition of the metric in Ω_S (which is inherited by X) this means: there is a block length k such that

$$\forall\, x, z \in X:\ x_{[0;k)} = z_{[0;k)} \Rightarrow \varphi(x)_0 = \varphi(z)_0\,. \tag{5.4-2}$$

We want to define a mapping $\Phi\colon \mathcal{L}_k(X) \to \mathrm{T}$ so that $\varphi = \Phi_\infty^{(k)}$. To do so, first observe that $\mathcal{L}_k(X)$ is the set of all *initial* k-blocks of the points of X. In fact, each initial k-block of a point of X is, by definition, in $\mathcal{L}_k(X)$ and, conversely, each block in $\mathcal{L}_k(X)$ occurs in some point x' of X, say at position n, hence it occurs in the point $\sigma_X^n(x')$ in initial position, and this point belongs to X as well. Using this observation, we see that the following rule defines Φ on all of $\mathcal{L}_k(X)$: $\Phi(x_{[0;k)}) := \varphi(x)_0$ for all $x \in X$. By (5.4-2), this definition is unambiguous.

It remains to show that $\varphi = \Phi_\infty^{(k)}$. To this end, observe that for $x \in X$ and $n \in \mathbb{Z}^+$:

$$(\varphi(x))_n = ((\sigma_Y^n \circ \varphi)(x))_0 = (\varphi(\sigma_X^n x))_0 = \Phi((\sigma_X^n x)_{[0;k)}) = \Phi(x_{[n;n+k)})\,.$$

This proves that $\varphi = \Phi_\infty^{(k)}$, so φ is a sliding block code. □

We proceed with two applications of Theorem 5.4.3.

Theorem 5.4.4. *Let X and Y be shift spaces and assume that X is of finite type. Let $\varphi\colon (X, \sigma_X) \to (Y, \sigma_Y)$ be a factor map and let y be a periodic point in Y. Then $\varphi^\leftarrow[y]$ contains a periodic point of X.*

Proof. By Theorem 5.4.3, φ is a sliding block code, say, with anticipation k. Thus, there is a mapping $\Phi\colon \mathcal{L}_k(X) \to \mathrm{T}$ such that $\varphi = \Phi_\infty^{(k)}$. Let N be the order of the SFT X. By Corollary 5.3.10 we may assume that $N > k$.

Let x be an arbitrary point in $\varphi^\leftarrow[y]$ and let p be a period of y (the primitive period will do). Our aim is to find blocks w and v over $\mathbf{S}$ with the following properties:

(a) The block wvw occurs in x and is, consequently, X-present.
(b) w has length N.
(c) The length of wv is an integer multiple of p.

If we have achieved this, then we claim that the periodic point $x' := (wv)^\infty$ is in X and that $\varphi(x')$ is in the orbit of y. Before proving this claim we show how this implies the conclusion of the theorem. So let x' be the periodic point just defined which is claimed to be in X, and suppose that $\varphi(x') = \sigma_Y^i(y)$ for some $i \in \mathbb{Z}^+$. As y has period p, we may assume that $0 \leq i \leq p-1$. Then $x := \sigma_X^{p-i} x'$ is also a periodic point in X, and $\varphi(x) = \sigma_Y^p y = y$, which completes the proof of the theorem.

First we shall prove the claim. So let w and v be blocks over $\mathbf{S}$ satisfying the conditions (a), (b) and (c), and let $x' := (wv)^\infty$. Obviously, x' is a periodic point in Ω_S.

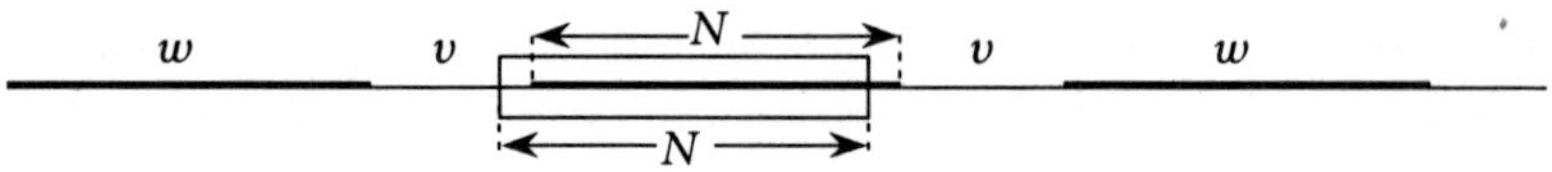

Fig. 5.1. A window with width $N = |w|$ never catches both the tail of a copy of v and the beginning of the next copy of v in $(wv)^\infty$.

Moreover, (b) implies that every block of length N in the sequence $(wv)^\infty$ always occurs within a copy of the block wvw (see Figure 5.1), hence is X-present by (a). Consequently, $x' \in X$. Next, we show that $\varphi(x')$ is in the orbit of y. To this end, observe that when we apply the sliding block code Φ to the coordinates of the point x and the first position of the sliding window is successively over the coordinates of wv in the subblock wvw of x, then a block u of coordinates of y is produced with length $|wv|$. In this process only coordinates of x are used that are within the block wvw: this is because the length N of the final block w in wvw is larger than the width k of the window used in the sliding block code. In the point $x' = wv\,wv\,wv\,wv\,\dots$ the block wvw is repeated again and again after $|wv| = |u|$ positions. Consequently, if we apply the sliding block code to x' we get the point $y' := uuuu\dots$. By condition (c) above, the length of u is an integer multiple of the period of y. Since it occurs in y, it follows easily from 5.2.3 (2) that either $y = u^\infty$ or $y = y_{[0;i)}u^\infty$ for some $i \in \{1 \dots, |u| - 1\}$. In both cases, y' is in the orbit of y under σ_Y. This completes the proof of the claim.

It remains to show that there exist subblocks w and v of x which satisfy the conditions (a), (b) and (c) above. To prove this, observe that there are infinitely many pairs of non-overlapping N-blocks in x whose positions differs an integer multiple of p, i.e., pairs of blocks of the form $x_{[n;n+N)}$ and $x_{[n+mp;n+mp+N)}$ with $n \in \mathbb{Z}^+$ and $m \in \mathbb{N}$ and such that $mp > N$. Because there are only finitely many different pairs of N-blocks, at least one of such pairs consists of mutually equal blocks. Denote this N-block by w and let v be the subblock of x that connects these two occurrences of w. Then $|w| = N$, $|wv|$ is a multiple of p and the block wvw occurs in x. This completes the proof that the conditions (a), (b) and (c) are fulfilled. □

Theorem 5.4.5. *A shift space that is conjugate to an SFT is an SFT.*

Proof. Let $X \subseteq \Omega_S$ be a shift space of finite type of order N, let $Y \subseteq \Omega_T$ be a shift space and let $\varphi\colon (X, \sigma_X) \to (Y, \sigma_Y)$, $\psi\colon (Y, \sigma_Y) \to (X, \sigma_X)$ be morphisms of dynamical systems such that $\varphi \circ \psi = \mathrm{id}_Y$. By Theorem 5.4.3, φ and ψ are sliding block codes, defined by mappings $\Phi\colon \mathcal{L}_m(X) \to \mathbf{T}$ and $\Psi\colon \mathcal{L}_n(Y) \to \mathbf{S}$, respectively. We shall show that Y is an SFT of order $(N+n)$. To this end, we have to show: if $y \in \Omega_T$ and every $(N+n)$-subblock of y is in $\mathcal{L}_{N+n}(Y)$, then $y \in Y$.

As a preliminary remark, note the following. If b is any Y-present block of length $(N+n)$ then there is $y' \in Y$ in which b occurs. Apply the sliding block code Ψ to the coordinates of y': then the first N coordinates of the block b produce a block of length N of coordinates of the point $\psi(y')$ in X. Denote this block by $\Psi^*(b)$; its position in $\psi(y')$

is the same as the position of b in y. Since the sliding window at the positions under consideration remains completely within this occurrence of the block b, it is clear that the block $\Psi^*(b)$ is independent of the choice of the point $y' \in Y$ that contains b and of the position of b in y'. Note also that the block $\Psi^*(b)$ is X-present, because $\psi(y') \in X$. Moreover, if we apply the sliding block code Φ to the coordinates of $\psi(y')$, this block produces a block of length N in the sequence of coordinates of $\varphi(\psi(y')) \in Y$. This block will be denoted by $\Phi^*(\Psi^*(b))$. This new block depends also on the $m-1$ coordinates following upon the block $\Psi^*(b)$ in $\psi(y')$, hence it might depend on the choice of y'. But $\varphi(\psi(y')) = y'$ and a moment's reflection shows that the block $\Phi^*(\Psi^*(b))$ starts in the same position as the block b. Therefore, $\Phi^*(\Psi^*(b))$ is the initial N-block of b.

Now consider a point $y \in \Omega_{\mathbf{T}}$ and assume that every $(N+n)$-subblock of y is Y-present. We cannot apply φ to the point y, for we do not know yet whether $y \in Y$, but we can apply the sliding block code Ψ to the sequence of coordinates of y because every n-subblock of y can be extended to an $(N+n)$-subblock of y, which, by assumption, is Y-present. This application of Ψ produces a point $x \in \Omega_{\mathbf{S}}$ all of whose N-blocks have the form $\Psi^*(b)$ for some Y-present $(N+n)$-block b. By what was observed above, these N-blocks are X-present, and because X is an SFT of order N this implies that $x \in X$. Consequently, $\varphi(x) \in Y$. But under the sliding block code Φ, any N-block $\Psi^*(b)$ in x with b an $(n+N)$-block in y is transformed into the initial N-block of b, which implies that $y = \varphi(x)$.

Conclusion. $y \in Y$. □

Remarks.

(1) In the proof above we have used only that $\varphi \circ \psi = \mathrm{id}_Y$. Thus, if Y is a subshift of the SFT X and Y is a factor of X under a factor map $\varphi\colon X \to Y$ such that $\varphi|_Y = \mathrm{id}_Y$ – this is sometimes expressed by saying that *Y is an equivariant retract of X* – then Y is an SFT as well. Note that being a subshift of an SFT and simultaneously being a factor of another SFT is not sufficient: the even shift is a subshift of Ω_2 and a factor of golden mean shift (both are SFT's), but is itself not an SFT.

(2) There are dynamical systems that do not arise as shift spaces but which are nevertheless conjugate to an SFT. See, for example, 6.3.5 ahead. Obviously, the theorem does not apply to such systems.

We close this section by a construction that will be needed later. Let X be a subshift over the symbol set $\mathbf{S}$, let $k \in \mathbb{N}$ and consider the set of X-present k-blocks as a new symbol set $\mathbf{T}$. So let $\mathbf{T} := \mathcal{L}_k(X)$. Then the identity mapping $\mathrm{I}\colon \mathcal{L}_k(X) \to \mathbf{T}$ generates a k-block code $\mathrm{I}_\infty^{(k)}\colon X \to \Omega_{\mathbf{T}}$, which is easily seen to be injective. So if $X^{(k)} := \mathrm{I}_\infty^{(k)}[X]$ then Corollary 5.4.2 implies that *$X^{(k)}$ is a shift space and that the system $(X^{(k)}, \sigma_{X^{(k)}})$ is conjugate to the system (X, σ_X).* It is called the k-th *higher block representation* of X.

Note that, for every $x \in X$,

$$\begin{aligned} \mathrm{I}_\infty^{(k)}(x) &= \mathrm{I}(x_{[0;k)})\,\mathrm{I}(x_{[1;k+1)})\,\mathrm{I}(x_{[2;k+2)})\dots\mathrm{I}(x_{[n;k+n)})\dots \\ &= \begin{bmatrix} x_0 \\ \vdots \\ x_{k-1} \end{bmatrix} \begin{bmatrix} x_1 \\ \vdots \\ x_k \end{bmatrix} \begin{bmatrix} x_2 \\ \vdots \\ x_{k+1} \end{bmatrix} \dots \begin{bmatrix} x_n \\ \vdots \\ x_{k+n-1} \end{bmatrix} \dots \end{aligned} \tag{5.4-3}$$

For convenience, a member c of $\mathbf{T}$ is denoted here as the column of coordinates of the k-block $\mathrm{I}^{-1}(c)$. Using this notation, it is easy to recover the coordinates of x from those of $\mathrm{I}_\infty^k(x)$. Thus, if $y = y_0y_1y_2\dots y_n\dots \in X^{(k)}$ then[7] one has $(\mathrm{I}_\infty^{(k)})^{-1}(y) = (y_0)_0\,(y_1)_0\,(y_2)_0\dots(y_n)_0\dots$, the sequence of 0-th coordinates of the k-blocks y_n. Consequently, the mapping $(\mathrm{I}_\infty^{(k)})^{-1}$ is the 1-block code generated by the projection $b \mapsto b_0 : \mathbf{T} \to \mathbf{S}$, which assigns to each X-present k-block its initial coordinate.

Lemma 5.4.6. *If (X,σ) is an SFT of order $N+1$ then $(X^{(N)}, \sigma_{X^{(N)}})$ is an SFT of order 2.*

Proof. As in the above, let $\mathbf{T} := \mathcal{L}_N(X)$ and let $\mathcal{L}_2^*$ be the set of all 2-blocks ab over $\mathbf{T}$ such that a and b (as N-blocks over $\mathbf{S}$) overlap in the same way as two successive coordinates in a point of $X^{(N)}$, i.e., $a_{[1;N)} = b_{[0;N-1)}$ and such that, in addition, the $(N+1)$-block ab_{N-1} – which is equal to the block a_0b – is X-present. Claim:

$$X^{(N)} = \{\, y \in \Omega_{\mathbf{T}} : y_iy_{i+1} \in \mathcal{L}_2^* \quad \text{for all } i \in \mathbb{Z}^+ \,\},$$

which shows that Y is an SFT of order 2.

"$\subseteq$": Clear from the construction.

"$\supseteq$": Let $y \in \Omega_{\mathbf{T}}$ and assume that $y_iy_{i+1} \in \mathcal{L}_2^*$ for all $i \in \mathbb{Z}^+$. Then for all $i \in \mathbb{Z}^+$ and all $j \in \{0,\dots,N-1\}$ we have $(y_i)_{j+1} = (y_{i+1})_j$, so that the $(N+1)$-blocks $(y_i)_0\dots(y_{i+N-1})_0(y_{i+N})_0$ and $(y_i)_0\dots(y_i)_{N-1}(y_{i+1})_{N-1}$ are equal to each other. The latter block is X-present by the definition of $\mathcal{L}_2^*$. Consequently, all $(N+1)$-blocks in the point $x := (y_0)_0(y_1)_0(y_2)_0\dots$ of $\Omega_{\mathbf{S}}$ are X-present. As X is an SFT of order $N+1$, the point x belongs to X. Moreover, $y = \mathrm{I}_\infty^{(N)}(x) \in X^{(N)}$ because, for all $n \in \mathbb{Z}^+$,

$$\begin{aligned} (\mathrm{I}_\infty^{(N)}(x))_n &= \mathrm{I}((y_n)_0\dots(y_{n+N-1})_0) \\ &= \mathrm{I}((y_n)_0\dots(y_n)_{N-1}) = \mathrm{I}(y_n) = y_n\,. \end{aligned}$$ □

Corollary 5.4.7. *If X is an SFT of order N then for all $k \geq N-1$ the k-th higher block representation $X^{(k)}$ of X is an SFT of order 2.*

7 The set $\mathbf{T}$ can be considered as the set of names of the members of $\mathcal{L}_k(X)$; then $\mathrm{I}(b)$ is the name of the block $b \in \mathcal{L}_k(X)$. Conversely, an element $c \in \mathbf{T}$ is the name of the k-block $\mathrm{I}^{-1}(c)$ over $\mathbf{S}$. In what follows the mapping I^{-1} will be suppressed: we don't make a clear distinction between a word and its name. In particular, if $c \in \mathbf{T}$ then we shall write c_i – the i-th coordinate of c when viewed as a k-block – instead of $(\mathrm{I}^{-1}(c))_i$.

Proof. An SFT X of order N also has order $k+1$ for all $k \geq N-1$. □

Proposition 5.4.8. *Let X be a shift space and let $\varphi\colon X \to Y$ be a factor map which is a k-block code. Then $\varphi \circ (\mathrm{I}_\infty^{(k)})^{-1}\colon X^{(k)} \to Y$ is a 1-block code.*

Proof. Let Y be a subshift over the symbol set $\mathbf{T}'$ and suppose that φ is the sliding block code generated by the mapping $\Phi\colon \mathcal{L}_k(X) \to \mathbf{T}'$. Recall that $X^{(k)}$ is a shift space over the symbol set $\mathbf{T} := \mathcal{L}_k(X)$. For convenience of notation, let $\psi := (\mathrm{I}_\infty^{(k)})^{-1}$, and recall that this is the 1-block code generated by the projection $\Psi\colon b \mapsto b_0\colon \mathbf{T} \to \mathbf{S}$. See also the following diagram:

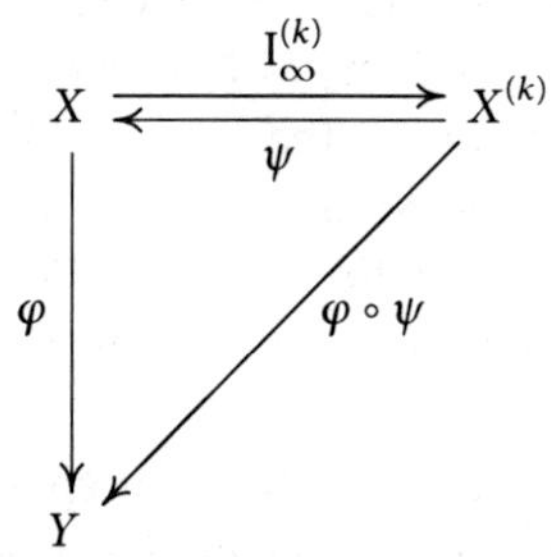

Using the way φ and ψ are generated by Φ and Ψ, respectively, we find for every point $z \in X^{(k)}$ and $i \in \mathbb{Z}^+$:

$$\begin{aligned}
\varphi(\psi(z))_i &= \Phi(\psi(z)_{[i\,;\,i+k-1)}) \\
&= \Phi((z_i)_0(z_{i+1})_0 \dots (z_{i+k-1})_0) \\
&\overset{(*)}{=} \Phi((z_i)_0(z_i)_1 \dots (z_i)_{k-1}) = \Phi(z_i)\,.
\end{aligned}$$

Here the equality $\overset{(*)}{=}$ follows from the fact that the subsequent coordinates of the point z are partially overlapping k-blocks as described in the definition of $X^{(k)}$. Note that in the final expression $\Phi(z_i)$ the coordinate z_i of z has to be seen as a single symbol from $\mathbf{T}$ – even though it is a k-block[8]. So from the equality obtained above it is clear that $\varphi \circ \psi$ is a 1-block code. □

Remark. Recall that an m-block code is a k-block code for every $k \geq m$. If we combine this observation With 5.4.7 then it follows that for all dynamical purposes a factor map $\varphi\colon X \to Y$ with X an SFT may be assumed to be a 1-block code, with X an SFT of order 2.

---

8 If we would not have suppressed the mapping I^{-1} in this discussion then we would have got the equality $\varphi(\psi(z))_i = (\Phi \circ \mathrm{I}^{-1})(z_i)$. So $\varphi \circ \psi$ is generated by the mapping $\Phi \circ \mathrm{I}^{-1}\colon \mathbf{T} \to \mathbf{T}'$.

5.5 Subshifts and graphs

Recall from the definitions and Proposition 5.3.11 that there is a close correspondence between the family $\mathfrak{P}(\mathbf{S}^2)$ of all sets of 2-blocks and SFT's of order 2. First of all, if $\mathcal{B} \in \mathfrak{P}(\mathbf{S}^2)$ – that is, $\mathcal{B}$ is a set of 2-blocks – and $\mathfrak{X}(\mathcal{B})$ is not empty then, by definition, it is an SFT of order 2. In Proposition 5.3.11 (iii) we introduced a different notation for this SFT: if $\mathcal{L} \in \mathfrak{P}(\mathbf{S}^2)$ then

$$\mathfrak{F}_2(\mathcal{L}) := \{ x \in \Omega_{\mathbf{S}} \vdots \forall i \in \mathbb{Z}^+ : x_i x_{i+1} \in \mathcal{L} \} = \mathfrak{X}(\mathcal{L}^c) \tag{5.5-1}$$

is either empty or it is an SFT of order 2 for which all elements of $\mathcal{L}$ are allowed (here $\mathcal{L}^c$ is the complement of $\mathcal{L}$ in $\mathbf{S}^2$). Every SFT of order 2 can be obtained in this way. In point of fact, if X is an SFT of order 2 then the set $\mathcal{L}_2(X)$ of all X-present 2-blocks, is an element of $\mathfrak{P}(\mathbf{S}^2)$ and, by Proposition 5.3.11, $X = \mathfrak{F}_2(\mathcal{L}_2(X))$.

So we have a mapping $\mathcal{L}_2$ from the set of all SFT's of order 2 into $\mathfrak{P}(\mathbf{S}^2)$ such that $\mathfrak{F}_2 \circ \mathcal{L}_2$ is the identity mapping on the set of all SFT's of order 2; consequently, $\mathcal{L}_2$ is injective.

There is an elegant geometrical method to represent sets of 2-blocks by means of directed graphs. Recall that a *directed graph* consists of a finite set of *vertices* (in illustrations often depicted as fat dots), together with, for every *ordered* pair of vertices (V_1, V_2), a finite (possibly empty) set[9] of *edges from* V_1 *to* V_2, which are also said to *start* in V_1 and to *end* in V_2 (often depicted as arrows).

We call a graph G *faithfully vertex-labelled* by the symbol set $\mathbf{S}$ whenever there is an injective mapping of the set of vertices of G into $\mathbf{S}$: every vertex gets a label and different vertices get different labels (but possibly not every symbol is used as a label). If G is labelled in this way then $\mathcal{E}_2(G)$ is defined as the following set of 2-blocks: the 2-block $\alpha\beta \in \mathbf{S} \times \mathbf{S}$ is in $\mathcal{E}_2(G)$ iff there is an edge in G from the vertex with label α to the vertex with label β.

For this definition the only thing that counts is whether for two given vertices there is an edge from the first vertex to the second one; if so, then the actual number of such edges is irrelevant. Therefore, *we shall assume that in a vertex-labelled graph there is, for every ordered pair of vertices, at most one edge from the first to the second vertex.*

If the reader is instructed to faithfully vertex-label a graph G then it is assumed (or must be verified by the reader) that this condition is fulfilled.

Conversely, if $\mathcal{L}$ is a set of 2-blocks over $\mathbf{S}$ then $\mathbf{G}(\mathcal{L})$ will denote the directed graph whose vertices are faithfully vertex-labelled with the symbols that occur in the members of $\mathcal{L}$, where there is a directed edge (just one) from the vertex with label α to the vertex with label β iff $\alpha\beta \in \mathcal{L}$. Obviously, the operators $\mathcal{E}_2$ on graphs and $\mathbf{G}$ on sets of

9 Actually, we define here a *multigraph*. We just omit the prefix 'multi'. Often we shall omit the adjective 'directed' as well. So by a graph we mean a directed multigraph.

2-blocks are inverse to each other. For easy reference, we write this down in formulas: if $\mathcal{L}$ is a set of 2-blocks and G is a faithfully vertex-labelled directed graph then

$$\mathcal{E}_2(\mathbf{G}(\mathcal{L})) = \mathcal{L} \quad \text{and} \quad \mathbf{G}(\mathcal{E}_2(G)) = G\,. \tag{5.5-2}$$

Note that for the second equality to be true it is essential that for every ordered pair of vertices in G there is at most one edge from the first to the second vertex. It is also essential that the labelling of G is faithful, i.e., that each label corresponds to a unique vertex.

Often we shall refer to the 1,1-correspondence expressed by (5.5-2) rather loosely by calling $\mathbf{G}(\mathcal{L})$ 'the graph corresponding to $\mathcal{L}$', by calling $\mathcal{E}_2(G)$ 'the set of 2-blocks defined by G', or by using phrases like 'the graph and the corresponding set of 2-blocks'.

The above-mentioned 1,1-correspondence between sets of 2-blocks and faithfully vertex-labelled directed graphs enables us now to define an SFT of order 2 by means of a faithfully vertex-labelled directed graph instead of by a set of allowed blocks: the set of allowed blocks will be given by means of the corresponding graph.

Examples.

(1) With the graph alongside – call it G – corresponds the set $\mathcal{E}_2(G) = \{00, 01\}$ of 2-blocks. With this as a set of allowed blocks one gets the shift space $\mathfrak{F}_2(\{00, 01\}) = \{0^\infty\}$.

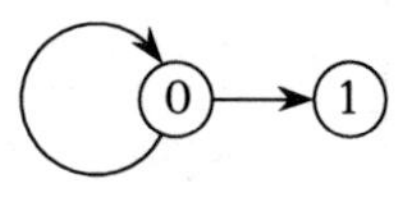

(2) In the table on the next page essentially all other graphs on two symbols are included. The graph on the three symbols 0, 1 and 2 will be considered later.

In the table on the next page we have used the following additional notation:

$$\mathfrak{M}_v := \mathfrak{F}_2 \circ \mathcal{E}_2 \quad \text{and} \quad \mathbf{G}_s := \mathbf{G} \circ \mathcal{L}_2\,.$$

Thus, if G is a faithfully vertex-labelled directed graph then $\mathfrak{M}_v(G)$ is – if not empty – the SFT of order 2 defined according to (5.5-1) by the set $\mathcal{E}_2(G)$ as a set of allowed 2-blocks. Conversely, if X is an SFT of order 2 then the graph corresponding to the set $\mathcal{L}_2(X)$ is denoted by $\mathbf{G}_s(X)$. Observe that by the first equality in (5.5-2) – with $\mathcal{L} = \mathcal{L}_2(X)$ – this definition implies the equality

$$\mathcal{E}_2(\mathbf{G}_s(X)) = \mathcal{L}_2(X)\,. \tag{5.5-3}$$

It should be clear from the above remarks about the operators $\mathfrak{F}_2$ and $\mathcal{L}_2$ that $\mathfrak{M}_v \circ \mathbf{G}_s$ is the identity operator on the set of SFT's of order 2: if X is an SFT of order 2 then

$$\mathfrak{M}_v(\mathbf{G}_s(X)) = X\,. \tag{5.5-4}$$

We could prove this formally, but the following argument (which repeats the above discussion) gives more insight: $\mathbf{G}_s(X)$ is the graph corresponding to the set of all X-present 2-blocks, so the left-hand side of (5.5-4) is the SFT of order 2 defined by this set as the set of allowed 2-blocks, which is, by Proposition 5.3.11, just X.

G	$\mathcal{E}_2(G)$	$\mathfrak{M}_v(G)$
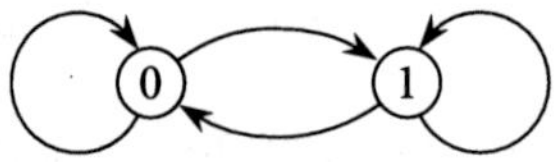	$\{00, 01, 10, 11\}$	Ω_2
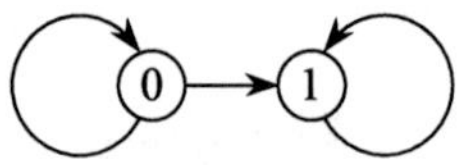	$\{00, 01, 11\}$	$\{0^k 1^\infty : k \in \mathbb{Z}^+\} \cup \{0^\infty\}$
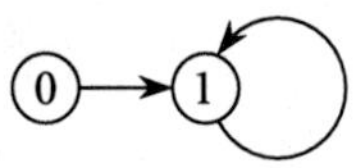	$\{01, 11\}$	$\{01^\infty, 1^\infty\}$
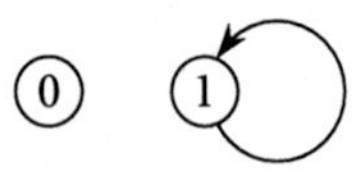	$\{11\}$	$\{1^\infty\}$
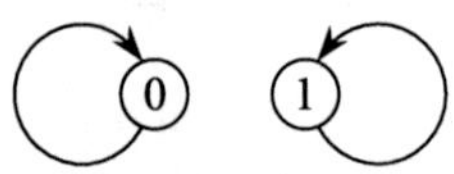	$\{00, 11\}$	$\{0^\infty, 1^\infty\}$
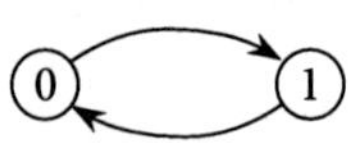	$\{01, 10\}$	$\{(01)^\infty, (10)^\infty\}$
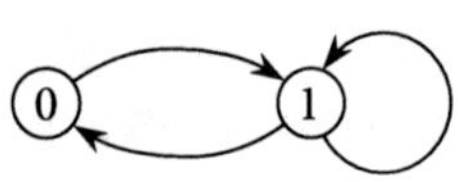	$\{01, 10, 11\}$	the golden mean shift
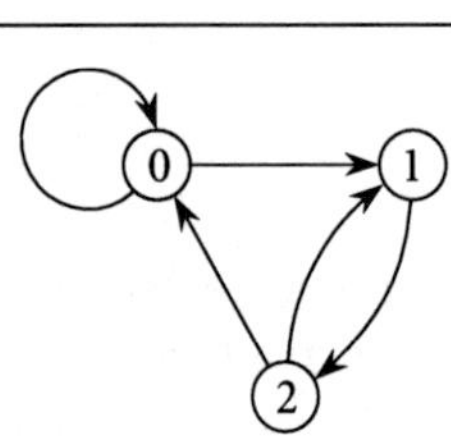	$\{00, 01, 12, 21, 20\}$	See Figure 5.4 (b)

On the other hand, $\mathbf{G}_s \circ \mathfrak{M}_v$ is not quite the identity operator on the family of faithfully vertex-labelled graphs: if G is such a graph, then by definition, $\mathbf{G}_s(\mathfrak{M}_v(G))$ corresponds to the set of all $\mathfrak{M}_v(G)$-present 2-blocks, whereas the original graph G corresponds to the (possibly larger) set of allowed blocks for $\mathfrak{M}_v(G)$. Thus, $\mathbf{G}_s(\mathfrak{M}_v(G))$ can be identified with a subgraph of G. In the Remark following Lemma 5.5.1 below shall characterize graphs G such that $\mathbf{G}_s(\mathfrak{M}_v(G)) = G$.

In the diagram below we recapitulate the above definitions. The following terminology is used: a set $\mathcal{L}$ of 2-blocks is 'allowed' whenever $\mathfrak{F}_2(\mathcal{L}) \neq \emptyset$ (hence is a shift space), and it is called a 2-language whenever there exists an SFT X of order 2 such that $\mathcal{L} = \mathcal{L}_2(X)$, i.e., whenever $\mathcal{L} = \mathcal{L}_2(\mathfrak{F}_2(\mathcal{L}))$. In Lemma 5.5.1 below the corresponding sets of graphs are characterized.

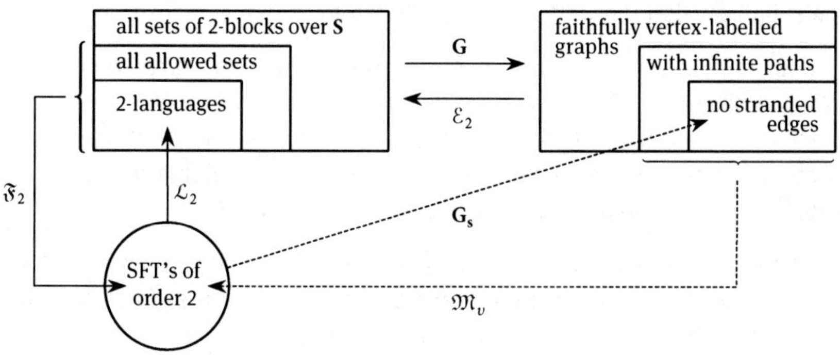

Associated with any directed graph G is its *adjacency matrix*. Let the vertices of G be numbered from 1 up to s and for $1 \le i, j \le s$, let a_{ij} be the number of edges from vertex i to vertex j. Then the matrix $A := [a_{ij}]$ is called the *adjacency matrix* of G. Obviously, every square matrix $A := [a_{ij}]$ with entries in $\mathbb{Z}^+$ can be interpreted as the adjacency matrix of a directed graph. If G is faithfully vertex-labelled by $\mathbf{S}$ then the SFT $\mathfrak{M}_v(G)$ can be described directly in terms of the adjacency matrix A of G. To this end, establish an explicit 1,1-correspondence between $\mathbf{S}$ and the numbers 1 through s, so that the rows/columns of A are labelled by $\mathbf{S}$. Then $\mathfrak{M}_v(G)$ is given by

$$\forall x \in \Omega_S : x \in \mathfrak{M}_v(G) \iff \forall i \in \mathbb{Z}^+ : a_{x_i x_{i+1}} = 1 .$$

In the case that $\mathbf{S} = \{0, \dots, s-1\}$ we interpret the symbols as numbers and let the symbol i correspond to the $(i+1)$-st row/column of A $(i = 0, \dots, s-1)$. Then the matrices for the shift spaces in the table on page 246 are, from top to bottom:

$$\begin{bmatrix}1&1\\1&1\end{bmatrix},\ \begin{bmatrix}1&1\\0&1\end{bmatrix},\ \begin{bmatrix}0&1\\0&1\end{bmatrix},\ \begin{bmatrix}0&0\\0&1\end{bmatrix},\ \begin{bmatrix}1&0\\0&1\end{bmatrix},\ \begin{bmatrix}0&1\\1&0\end{bmatrix},\ \begin{bmatrix}0&1\\1&1\end{bmatrix},\ \begin{bmatrix}1&1&0\\0&0&1\\1&1&0\end{bmatrix}.$$

One of the eigenvalues of the adjacency matrix of the golden mean shift is the golden mean ratio $\frac{1}{2}(1+\sqrt{5})$ – which gives this shift its name. See also Exercise 5.8.

Until now we have used (faithfully vertex-labelled) graphs only as a convenient means of representing admissible 2-blocks. Though this enables a couple of very useful computational techniques – see the Exercises 5.8 (2), (3) and 5.9 (2) for some simple examples – we shall not continue in this direction. Instead, we shall introduce a small shift in our attention which will give us the opportunity to introduce a new type of shift spaces.

A *finite walk* (or a *finite path*) on a directed graph G is a finite ordered set $(E_1, \dots, E_n)$ $(n \ge 1)$ of edges – the E_i are names for the edges – such that for every $i \in \{2, \dots, n\}$ the edge E_i starts in the vertex where the edge E_{i-1} ends. In this case, if the edge E_i starts at the vertex V_i and ends at the vertex V_{i+1} $(1 \le i \le n)$ then the ordered set $(V_1, \dots, V_{n+1})$ is also called a finite walk. Infinite walks on G are defined similarly. Obviously, if the vertices of the graph are labelled by symbols from $\mathbf{S}$ (not necessarily faithfully) then the finite walks on G define finite blocks over $\mathbf{S}$ by writing down the labels of the visited vertices in the right order. Similarly. infinite walks define elements of $\Omega_\mathbf{S}$. In that case we shall say that the walks *represent* those blocks or elements from $\Omega_\mathbf{S}$.

Example. Consider the graph for the golden mean shift (the 7th graph from above on page 246). It has three edges, one from the vertex with label 1 to itself (call this edge E_1), one from the vertex with label 1 to the vertex with label 0 (call this edge E_2) and one from the vertex with label 0 to the vertex with label 1 (call this edge E_3). The block $101^2 01^3 01$ is represented by the path $E_2E_3E_1E_2E_3E_1E_1E_2E_3E_1$.

Conversely, however, a finite block over $\mathbf{S}$ may not be represented by a finite path in the graph, even if all 2-subblocks are represented by edges (i.e., pairs of vertices connected by an edge in the right direction): for example, if G is the vertex-labelled graph ① ⟶ ⓪ ⟶ ① then the 3-block 010 is not represented by a finite path in G, though 01 and 10 are represented so. The reason is, of course, that the vertices with label 1 in 01 and in 10 are different from each other. If the graph is *faithfully* vertex-labelled by $\mathbf{S}$ and b is a finite block over $\mathbf{S}$ such that the 2-block b_ib_{i+1} is represented by an edge for every $i \in \{0, \dots |b| - 1\}$, then the vertex where the edge b_ib_{i+1} ends (the unique vertex with label b_{i+1}) is the same as the vertex where the edge $b_{i+1}b_{i+2}$ begins $(0 \le i \le |b| - 2)$, so these edges can be connected 'head-to-tail' to a finite path. Similarly, an infinite walk on G defines an element x of Ω_2 which is in $\mathfrak{M}_v(G)$, because every 2-block of x is in $\mathcal{E}_2(G)$. Conversely, the above discussion shows that each element of $\mathfrak{M}_v(G)$ is represented by a (unique) infinite walk on G, provided the vertex-labelling is faithful.

Lemma 5.5.1. *Let G be a directed graph, faithfully vertex-labelled by symbols from $\mathbf{S}$. Then $\mathfrak{M}_v(G)$ is the set of points of $\Omega_\mathbf{S}$ that are represented by an infinite walk on G. This set is either empty or an SFT of order 2, and every SFT X of order 2 is obtained in this way from the graph $\mathbf{G}_s(X)$.*

Proof. The first statement is clear from the above discussion. For the final statements, use the definitions and the equality in (5.5-4). □

Remark. Thus, the set of all $\mathfrak{M}_v(G)$-present 2-blocks is represented by the edges – 2-walks – in G that are part of (can be extended to) an infinite walk in G. So an edge

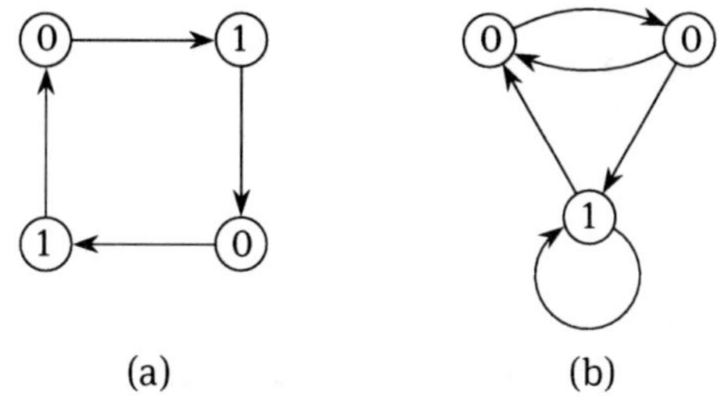

Fig. 5.2. Examples of non-faithfully vertex-labelled graphs G. (a) $\mathfrak{W}_v(G)$ is a periodic orbit with period 2 (an SFT of order 2; it can also be represented as in the 6th example on page 246). (b) $\mathfrak{W}_v(G)$ is the even shift (not an SFT).

represents an allowed but non-$\mathfrak{M}_v(G)$-present 2-block iff all possible extensions of this edge strand in a vertex from which no edge departs. Call an edge *stranded* whenever it terminates in a vertex from which no edge departs. *So the set of all $\mathfrak{M}_v(G)$-present 2-blocks is equal to the set $\mathcal{E}_2(G)$ of all allowed 2-blocks iff G has no stranded edges.* Recall that $\mathbf{G}_s(\mathfrak{M}_v(G))$ is the subgraph of G representing the $\mathfrak{M}_v(G)$-present 2-blocks. It follows that $\mathbf{G}_s(\mathfrak{M}_v(G)) = G$ iff there are no stranded edges in G.

So now we have an alternative method to obtain an SFT of order 2 from a faithfully vertex-labelled directed graph: in the first instance the graph is only used as a convenient means to represent a set of 2-blocks, but now the graph is used as an area to make infinite walks. This suggests that we can define a shift space directly as *the set of infinite walks on the vertex-labelled directed graph*. This definition can also be given if the labelling of the graph is not faithful, that is, if different vertices may have the same label. But in that case it is not meaningful to view the graph as a representation of allowed 2-blocks: this is only meaningful for the definition of an SFT of order 2, and we shall see below that infinite walks on such a graph may define subshifts that are not of finite type.

If G is a vertex-labelled graph then $\mathfrak{W}_v(G)$ – vertex walks – will denote the set of points of Ω_S that are represented by an infinite walk on G; we shall see in a moment that this set is either empty or a shift space. See Figure 5.2. Note that Lemma 5.5.1 states that $\mathfrak{W}_v(G) = \mathfrak{M}_v(G)$ if the graph G is faithfully vertex-labelled. If the labelling is not faithful then $\mathfrak{M}_v(G)$ is not defined.

Lemma 5.5.2. *Let G be a directed graph, vertex-labelled by the symbol set $\mathbf{S}$. Then the set $\mathfrak{W}_v(G)$ of all points of Ω_S that are represented by an infinite walk on G is either empty or a shift space.*

Proof. Suppose $\mathfrak{W}_v(G)$ is not empty, i.e., G admits infinite walks. Let $\mathbf{T}$ be the set of vertices of G, let $\Lambda : \mathbf{T} \to \mathbf{S}$ be the mapping that assigns to each vertex its label from $\mathbf{S}$ and let $\Lambda_\infty : \Omega_T \to \Omega_S$ be the sliding block code defined by Λ. Since G is faithfully vertex-labelled by $\mathbf{T}$, Lemma 5.5.1 implies that the infinite walks on G define a subshift Z of Ω_T. We claim that $\mathfrak{W}_v(G) = \Lambda_\infty[Z]$. Then by Corollary 5.4.2, $\mathfrak{W}_v(G)$ is a shift space.

The proof of the claim is straightforward: every point of Z or of $\mathfrak{W}_v(G)$ is an infinite walk on G, coded by $\mathbf{T}$ or $\mathbf{S}$, respectively, and the second coding is obtained from the

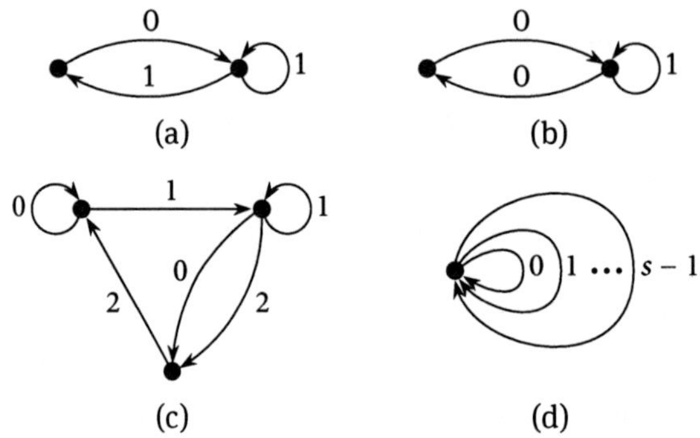

Fig. 5.3. Edge-labelled graphs. (a) The golden mean shift. (b) The even shift. (c) A graph with a 'double' edge. (d) The full shift over S.

first by replacing each coordinate (a vertex of G, i.e., a symbol from T) by its value under Λ (the label from S of that vertex). □

In general, the shift space so obtained is not of finite type: though in Figure 5.2 (b) the 2-blocks 10 and 01 are allowed, the block 101 is not (the block 01 cannot occur after the symbol 1). Actually, between two subsequent occurrences of a 1 there must be an even number (including zero) of occurrences of 0. Thus, the shift space defined here is the even shift, which is not of finite type: see Example (4) after Corollary 5.3.10.

5.5.3. It is also possible to label the *edges* of a directed graph by the symbols of S. Usually one does not require such an edge-labelling to be faithful. Figure 5.3 shows some examples. In such *edge-labelled* graphs we allow *multiple edges*: for two vertices there may be more than one edge from the first to the second vertex (provided such 'parallel' edges are labelled differently). See Figure 5.3 (c) and (d). Each infinite walk on such a graph G defines a point of Ω_S: write down the edge-labels in the right order. The subset of Ω_S so obtained will be denoted by $\mathfrak{W}_e(G)$ (the edge walks). In the next proposition we shall show that if $\mathfrak{W}_e(G)$ is not empty then it is a shift space. The shift spaces so obtained are called *sofic shifts*.

Examples.

(1) In Corollary 5.5.6 below it will be shown that every SFT is a sofic shift.

(2) It is clear from Figure 5.3 (b) that the even shift is a sofic shift. This would also follow from Theorem 5.5.5 below and the fact that the even shift is a factor of the golden mean shift; see Remark 2 after Corollary 5.4.2. Consequently, not every sofic shift is an SFT: recall that the even shift is not of finite type.

(3) The prime gap shift is not sofic. For suppose that this shift is $\mathfrak{W}_e(G)$ for some edge-labelled graph G. In order to get a block of the form 10^p1 for some prime number p there must be a path in G consisting of p edges with label 0 between two edges with label 1. Such a path can contain no loops. (Recall that a *loop* or *cycle* is a path that begins and ends in the same vertex.) In fact, a loop here would allow blocks with, between two consecutive occurrences of a 1, blocks of the form $0^k(0^l)^m0^n$ with arbitrary $m \in \mathbb{Z}^+$ ($k, n \geq 0$, $l \geq 1$: first k edges, then a loop of l edges

which is traversed m times and then n edges). However, for fixed k, l and n there are infinitely many values of m for which $ml + (k + n)$ is not prime: e.g., take for m a multiple of $k + n$. Since there are arbitrarily large primes – known already to Euclides – G must admit arbitrary long walks without loops. This is impossible for a finite graph.

(4) The same technique shows that the context free shift is not sofic. For suppose that this shift is $\mathfrak{W}_e(G)$ for some edge-labelled graph G. Let G have r edges and consider an infinite path representing the word $ab^{r+1}c^{r+1}a$. Then the subpath representing the word b^{r+1} must contain a loop, implying that there is an infinite path in G containing a subpath that represents a word of the form $ab^kc^{r+1}a$ with $k > r + 1$.

Proposition 5.5.4. *Let G be a directed graph, edge-labelled by the symbol set $\mathbf{S}$. Then the set $\mathfrak{W}_e(G)$ is either empty or a shift space. If the labelling is faithful and $\mathfrak{W}_e(G) \neq \emptyset$ then $\mathfrak{W}_e(G)$ is an SFT of order 2.*

Proof. Let $\mathfrak{W}_e(G) \neq \emptyset$. First, assume that the labelling of G is faithful. In this case $\mathfrak{W}_e(G)$ is the SFT of order 2 defined by the following collection of allowed 2-blocks: a 2-block b_0b_1 over $\mathbf{S}$ is allowed iff the vertex where the edge with label b_0 ends coincides with the vertex where the edge with label b_1 begins (briefly: b_0 is followed by b_1). The proof is similar to the proof of Lemma 5.5.1, interchanging 'vertices' and 'edges' (e.g., instead of 'there is an edge from the vertex with label x_i to the vertex with label x_{i+1}', read 'the edge with label x_i is followed by the edge with label x_{i+1}').

Next, assume that the labelling is not faithful. Let $\mathbf{T}$ be the set of all edges of G. Similar to the proof of Lemma 5.5.2 it follows that $\mathfrak{W}_e(G)$ is a factor of an SFT of order 2 over the symbol set $\mathbf{T}$. Now use Corollary 5.4.2 in order to complete the proof. □

Remark. The above proof shows: if G is faithfully edge-labelled then the set of allowed 2-blocks which defines $\mathfrak{W}_e(G)$ is the collection of all blocks b_0b_1 such that the edge b_0 is followed by the edge b_1.

Theorem 5.5.5. *A shift space X is sofic iff it is a factor of an SFT.*

Proof. "Only if": In the proof of Proposition 5.5.4 it was shown that a sofic shift (namely, any shift space of the form $\mathfrak{W}_e(G)$ for some edge-labelled graph G) is a factor of an SFT of order 2.

"If": Let X be a shift space over the symbol set $\mathbf{S}$ and suppose that there is a factor map $\varphi\colon Z \to X$, where Z an SFT, say of order k_1. By the Curtis–Lyndon–Hedlund Theorem, φ is a sliding block code, say with anticipation k_2. Let $k := \max\{k_1, k_2\}$. Since $k \geq k_1$, Corollary 5.3.10 implies that Z is an SFT of order k. Similarly, φ may be assumed to have anticipation k. By Proposition 5.4.8, X is a factor of $Z^{(k)}$ under a 1-block map ψ. If we denote the symbol set for $Z^{(k)}$ by $\mathbf{T}$ then this means that there is a mapping $\Psi\colon \mathbf{T} \to \mathbf{S}$ such that $\psi = \Psi_\infty^{(1)}$ (at this point it is irrelevant that $\mathbf{T}$ consists of certain k-blocks over the symbol set of Z).

Recall from Corollary 5.4.7 that $Z^{(k)}$ is an SFT of order 2, so that it can be obtained as the shift space $\mathfrak{M}_v(G)$ for a suitable faithfully vertex-labelled directed graph G. Now label the edges of G as follows: every edge starting in a vertex with label $\alpha \in \mathbf{T}$ gets label $\Psi(\alpha) \in \mathbf{S}$. This turns G into an edge-labelled graph. After some reflection it should be clear now that for every infinite walk in G the edge labels represent the point $\psi(z) \in X$, where $z \in Z^{(k)}$ is the point that is vertex-represented by this walk. This shows that $\mathfrak{W}_e(G) = \psi[\mathfrak{M}_v(G)] = \psi[Z^{(k)}] = X$. □

Corollary 5.5.6. *Every SFT is sofic, i.e., if X is an SFT then there is an edge-labelled graph G such that $X = \mathfrak{W}_e(G)$.*

Proof. Obviously, an SFT is a factor of itself. □

Remarks.

(1) Not every SFT of order 2 can be obtained from a *faithfully* edge-labelled graph. For example, it is not possible to obtain the golden mean shift from a graph with two edges that are faithfully labelled by the symbols 0 and 1. There are only three of such graphs, from which only the shift spaces Ω_2, $\{0^\infty, 1^\infty\}$ and $\{(01)^\infty, (10)^\infty\}$ can be obtained. None of these is conjugate to the golden mean shift (the number of invariant points does not agree).

(2) Obviously, a (finite) graph that admits infinite walks must contain a cycle. Then an infinite walk that consists of running through a cycle again and again represents a periodic point. Consequently, sofic shifts have periodic points; of course, this would also follow from Theorem 5.5.5 and Exercise 5.7. By a similar argument, in Theorem 5.4.4 'of finite type' can be replaced by 'a sofic shift'; we leave the proof for the reader.

The proof of Theorem 5.5.5 with $X = Z$ and $\psi = \mathrm{id}_X$ shows that the graph G in Corollary 5.5.6 is obtained in the following way: let G be the vertex-labelled graph of a higher block representation $X^{(k)}$ of X that is an SFT of order 2 (by Corollary 5.4.7 it is sufficient that $k \geq N - 1$, where N is the order of X). Recall that the vertices of G are faithfully labelled by the elements of $\mathcal{L}_k(X)$, the X-present k-blocks over the symbol set of X. For the edges in G we refer to the definition of the shift space $X^{(k)}$ preceding Lemma 5.4.6 – the 'overlapping blocks' idea. Label every edge of G with the 0-th coordinate of the k-block by which the vertex is labelled where that edge starts: in this particular case of the proof of Theorem 5.5.5, the mapping Ψ is equal to the inverse of the mapping $\mathrm{I}_\infty^{(k)}$ given in (5.4-3). Then $X = \mathfrak{W}_e(G)$. In Figure 5.4 this construction is illustrated for four cases.

Corollary 5.5.7. *A shift space that is a factor of a sofic shift is sofic. In particular, a shift space conjugate to a sofic shift is sofic.*

Proof. Use the fact that the composition of two factor mapping is a factor mapping. □

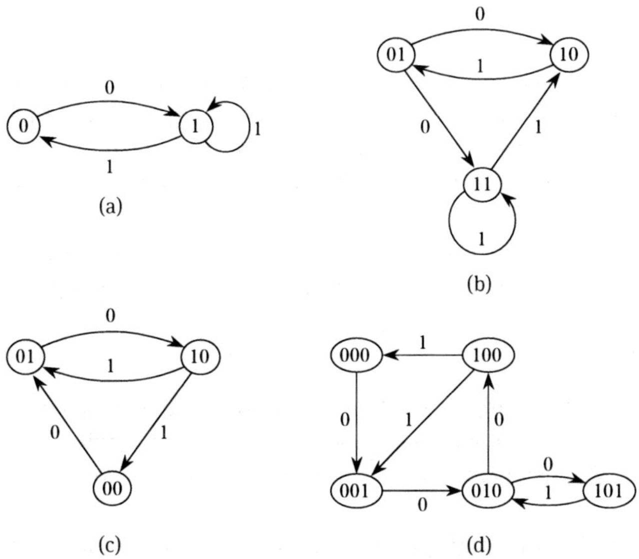

Fig. 5.4. SFT's represented as sofic shifts: (a) The golden mean shift as an SFT of order 2 and (b) as an SFT of order 3. (c) The (1,2) run-length limited shift $\mathfrak{X}(\{000, 11\})$ (order 3). (d) The (1,3) run-length limited shift (order 4).

5.6 Recurrence, almost periodicity and mixing

In the preceding sections the topology is often hidden behind combinatorial properties of blocks and sequences. Most attention is devoted to (combinatorial) definitions of certain subshifts and the only dynamical properties of subshifts we have considered are transitivity and periodicity. Note, however, that a result like Theorem 5.4.4 – despite its combinatorial proof – essentially uses topology: its proof uses the Curtis–Lyndon–Hedlund Theorem, which depends on compactness of the domain of the morphism under consideration.

In this section we construct points in the full shift space Ω_S with some special dynamical properties. In all cases the construction is based on the following principle: let a sequence $\{\, b(n) \,\vdots\, n \in \mathbb{Z}^+ \,\}$ of finite blocks over **S** be given such that

- $|b(n)| \rightsquigarrow \infty$ for $n \rightsquigarrow \infty$;
- $\forall\, n \in \mathbb{Z}^+$: the block $b(n+1)$ starts with the block $b(n)$.

Then there is a unique point $x \in \Omega_S$ such that, for every $n \in \mathbb{Z}^+$, $b(n)$ is the initial $|b(n)|$-block of x. The proof is easy. In order to determine the k-th coordinate x_k of x, select m so large that $|b(m)| > k$; this is possible by the first condition on the blocks $b(n)$. Then put $x_k := b(m)_k$; in view of the second condition on the blocks $b(n)$, this definition is unambiguous, and it is obvious that for every $n \in \mathbb{Z}^+$ the point x so defined starts with the block $b(n)$.

5.6.1 (Recurrence).

(1) Taking into account that basic neighbourhoods of a point in a shift system are cylinders based on initial blocks of that point, it is easy to show that *a point x in a shift system is recurrent iff every initial block of x occurs in x at a position greater than or equal to 1, iff every initial block of x occurs infinitely often in x* (use Proposition 4.1.1 (ii) for the final statement).

It follows that a point $x \in \Omega_S$ is recurrent iff there are $\alpha \in S$ and a sequence $\{w^{(n)} : n \in \mathbb{N}\}$ of (not necessarily non-empty) finite blocks such that x is given by

$$x = \Big\{\Big[(\alpha\, w^{(0)}\alpha)\, w^{(1)} (\alpha\, w^{(0)}\alpha)\Big]\; w^{(2)}\; \Big[(\alpha\, w^{(0)}\alpha)\, w^{(1)} (\alpha\, w^{(0)}\alpha)\Big]\Big\}\; w^{(3)} \Big\{\ldots$$

meaning that x is defined in the following way: define inductively a sequence of blocks $b(n)$ for $n \in \mathbb{Z}^+$ by $b(0) := \alpha$, $b(n+1) := b(n)w^{(n)}b(n)$, and let x be the unique point of Ω_S that for every $n \in \mathbb{Z}^+$ starts with the block $b(n)$ (according to the initial remark in this section).

It is clear that a point defined in this way is recurrent: every initial block (which is included in $b(n)$ for a suitable n) occurs in x at positions greater than 1. Conversely, if z is recurrent, let $\alpha := x_0$; then α occurs in x at a position greater than 1, so there is a (possibly empty) block $w^{(0)}$ such that x starts with the block $b(1) := \alpha w^{(0)}\alpha$. Since the block $b(1)$ occurs infinitely often in x it occurs in a position that has no overlap with the initial block $b(1)$. The coordinates of x between these non-overlapping occurrences of the block $b(1)$ form a (possibly empty) block $w^{(1)}$. Thus, there is a block $w^{(1)}$ in x such that x starts with the block $b(2) := b(1)w^{(1)}b(1)$. It is possible (but not compulsory) to select $w^{(1)}$ as the minimal block such that x starts with the block $b(2) = b(1)w^{(1)}b(1)$, i.e., the second copy of $b(1)$ in $b(2)$ is the first occurrence of a copy of $b(1)$ in x after the initial block $b(1)$ of x that has no overlap with this initial block. Subsequently, there there are infinitely many copies of the block $b(2)$ in x, so there is a (minimal) block $w^{(2)}$ such that x starts with the block $b(3) := b(2)w^{(2)}b(2)$. Etc.

(2) In a full shift system the set $R(\Omega_S, \sigma)$ of recurrent points is not closed. In fact, the set $R(\Omega_S, \sigma)$ is dense (it includes the dense set of all periodic points), but not all points of Ω_S are recurrent: e.g., the eventually periodic point 1110^∞ and the point $1\,0\,1\,0^2\,10^3\ldots1\,0^n1\ldots$ are not.

(3) Not all recurrent points in full shift systems are periodic or transitive.

For example, consider the point in Ω_2 that is defined as the recurrent point in 1 above with $\alpha := 0$ and $w^{(n)} := 1^{|b(n)|}$ for all $n \in \mathbb{Z}^+$. Thus, define the blocks $b^{(n)}$ inductively by

$$\begin{aligned} b^{(0)} &:= 0\,, \\ b^{(n+1)} &:= b^{(n)}\, \underbrace{1\ldots\ldots1}_{|b^{(n)}|\text{times}}\, b^{(n)} \quad (n \geq 0)\,, \end{aligned}$$

and let ρ be the unique point in Ω_2 that for every $n \in \mathbb{Z}^+$ starts with the block $b^{(n)}$:

$$\rho := \underbrace{\underbrace{\underbrace{0\,1\,0}_{b^{(1)}}\;1\,1\,1\;\underbrace{0\,1\,0}_{b^{(1)}}}_{b^{(2)}}\;1\,1\,1\,1\,1\,1\,1\,1\,1\;\underbrace{0\,1\,0\,1\,1\,1\,0\,1\,0}_{b^{(2)}}}_{b^{(3)}}\;1\,1\,1\,1\;\ldots\ldots$$

Then in view of the observations in 1 above the point ρ is recurrent.
Next, we show that the point ρ is not periodic (it is not even almost periodic: see Exercise 5.12 (1) , or the first paragraph in 5.6.2 below). Suppose it were periodic with primitive period p. Then each subblock of ρ of length greater than $3p$ would contain a full period block. As ρ has subblocks of consecutive 1's of arbitrary length, it would follow that 1^p is a period block of ρ, hence that ρ would be the invariant point 1^∞ – which is not the case.
Finally, the point ρ is not transitive: using Proposition 5.2.5 it is a straightforward exercise to show that the point 0^∞ is not in the orbit closure of ρ. More generally, it is not too difficult to show that 1^∞ is the unique periodic point in the orbit closure of ρ. See Exercise 5.12 (2). As the periodic points are dense in Ω_2, it follows that the orbit closure of ρ is nowhere dense.

5.6.2 (Almost periodicity). Taking into account that the basic neighbourhoods of a point in a shift system are cylinders based on initial blocks of that point, it is clear that *a point x in a shift system is almost periodic iff every initial block of x occurs infinitely often in x with bounded gaps.* It follows that a recurrent point as described in 5.6.1 (1) above is almost periodic iff the blocks $w^{(n)}$ for $n \in \mathbb{Z}^+$ can be chosen so that their lengths form a bounded set. In particular, the recurrent point ρ discussed in 5.6.1 (3) above is not almost periodic. Note also that no transitive point of Ω_S can be almost periodic, because otherwise Theorem 4.2.2 (1) would imply that Ω_S is minimal under σ, which would contradict Corollary 5.2.4.

We shall present now a famous example of an almost periodic point in the shift system (Ω_2, σ): the Morse–Thue sequence. Its orbit closure – a minimal system – will be called the Morse–Thue minimal system.

The *Morse–Thue sequence* is the element of Ω_2 which is defined in the following way. First a convention: if b is an arbitrary block over the symbol set $\{0, 1\}$ then $\overline{b}$ is the block that we get by interchanging 0's and 1's. For example, $\overline{1011} = 0100$. Note that for every block b we have $\overline{\overline{b}} = b$. Using this convention, define by induction the sequence of blocks $q(n)$ by

$$\begin{aligned} q(0) &:= 0\,, \\ q(n+1) &:= q(n)\,\overline{q(n)} \qquad (n \in \mathbb{Z}^+). \end{aligned}$$

The first five elements of this sequence of blocks are:

$$q(0) = 0\,, \quad q(1) = 01\,, \quad q(2) = 01\,10\,,$$
$$q(3) = \underbrace{0110}_{q(2)}\,\underbrace{1001}_{\overline{q(2)}}\,, \quad q(4) = \underbrace{0110\,1001}_{q(3)}\;\underbrace{1001\,0110}_{\overline{q(3)}}\,.$$

The following statements can be proved easily by induction in n: in most cases the statement for $n = 0$ is trivial, while the step from n to $n+1$ proceeds as follows: if the statement is true for $q(n)$ then it is easy to see that it is true for $q(n)\,\overline{q(n)}$, that is, for $q(n+1)$. In addition, some hints are given between square brackets.

(1) $\forall n \in \mathbb{Z}^+ : |q(n)| = 2^n$.

(2) $\forall k, n \in \mathbb{Z}^+ : n \geq k+1 \Rightarrow q(n)$ starts with the block $q(k)\,\overline{q(k)}$.

(3) $\forall k, n \in \mathbb{Z}^+ : n \geq k \Rightarrow q(n)$ is a concatenation of 2^{n-k} blocks of length 2^k, each of which is a copy of $q(k)$ or $\overline{q(k)}$. In particular, each of those blocks occurs in position $i \cdot 2^k$ with $i = 0, \dots, 2^{n-k} - 1$.

[Take into account that, if the block $q(k)$ or $\overline{q(k)}$ occurs in $q(n)$, then $\overline{q(k)}$ or $q(k)$, respectively, occurs in $\overline{q(n)}$ in the same position.]

(4) $\forall k, n \in \mathbb{Z}^+ : n \geq k+1 \Rightarrow$ the block $q(n)$ does not begin with the block $q(k)q(k)$ and it does not end with the block $\overline{q(k)}\,\overline{q(k)}$.

['Begin': Use 2 above. 'End': Similar, using the hint in 3.]

(5) $\forall\, k, n \in \mathbb{Z}^+ : n \geq k+2 \Rightarrow q(k)q(k)q(k)$ and $\overline{q(k)}\,\overline{q(k)}\,\overline{q(k)}$ do not occur in $q(n)$.

[Let the statement be true for $n-1$. Assume that for some $k \leq n-2$ the block $q(k)q(k)q(k)$ or the block $\overline{q(k)}\,\overline{q(k)}\,\overline{q(k)}$ occurs in $q(n)$. Since by assumption these blocks occur neither in $q(n-1)$, nor in $\overline{q(n-1)}$, it follows that they are distributed over the end of the subblock $q(n-1)$ and the beginning of the subblock $\overline{q(n-1)}$ of $q(n)$. Hence $q(n-1)$ ends with $q(k)q(k)$ or $\overline{q(k)}\,\overline{q(k)}$, or $\overline{q(n-1)}$ starts with $q(k)q(k)$ or $\overline{q(k)}\,\overline{q(k)}$. Both possibilities contradict statement (4) above.]

By statement (2) above, for every $n \in \mathbb{Z}^+$ the block $q(n+1)$ starts with $q(n)$, and by statement (1) above, $|q(n)| \rightsquigarrow \infty$ for $n \rightsquigarrow \infty$. Hence there is an element $\mu \in \Omega_2$ that, for every $n \in \mathbb{Z}^+$, starts with the block $q(n)$:

$$\mu = \underbrace{\underbrace{\underbrace{0110}_{q(2)}\ 1001}_{q(3)}\ 1001\,0110}_{q(4)}\ 1001\,0110\ 0110\,1001\ 1001\ \dots$$

The sequence μ is called the *Morse–Thue sequence*.

The statements (2), (3) and (5) above have the following consequences for μ:

(6) $\forall\, k \in \mathbb{Z}^+ : \mu$ begins with the block $q(k+1) = q(k)\overline{q(k)}$.

(7) $\forall\, k \in \mathbb{Z}^+ : \mu$ is a concatenation of blocks of length 2^k, each of which is a copy of $q(k)$ or $\overline{q(k)}$. In particular, every block of length 2^k occurring in μ in position $i{\cdot}2^k$ $(i \in \mathbb{Z}^+)$

is either $q(k)$ or $\overline{q(k)}$. We shall call these positions of the blocks $q(k)$ and $\overline{q(k)}$ the *natural positions*[10].

[By (6), the block $\mu_{[i2^k;(i+1)2^k)}$ occurs in $q(n)$ for sufficiently large n. Now use statement (3) above.]

The following can be shown: in the sequence μ, replace every occurrence of 0 by a copy of $q(k)$ and every occurrence of 1 by a copy of $\overline{q(k)}$. The sequence so obtained is, again, μ. Stated otherwise, the blocks $q(k)$ and $\overline{q(k)}$ occur in their natural positions in the same order as the 0's and 1's in μ.

(8) $\forall k \in \mathbb{Z}^+$: there are no three subsequent occurrences of the block $q(k)$ or of the block $\overline{q(k)}$ in μ.

[Such an occurrence would be part of $q(n)$ for sufficiently large n; now apply statement (5) above.]

The orbit closure $M := \overline{\mathcal{O}_\sigma(\mu)}$ of the point μ under the shift is a non-empty, closed and invariant subset of Ω_2. The subsystem (M, σ) of (Ω_2, σ) is called the *Morse–Thue system*.

Proposition 5.6.3. *The point μ is almost periodic. Consequently, M is minimal under σ.*

Proof. By Theorem 4.2.2 (1), the final statement is a consequence of the first. So we need only prove almost periodicity of the point μ in (Ω_2, σ).

Let b be an initial block of μ. In view of the initial remark in 5.6.2 we have to show that for every initial block b of μ the gaps in the set

$$D(\mu, C_0[b]) := \{n \in \mathbb{Z}^+ : b \text{ occurs in } \mu \text{ at position } n\}$$

are bounded. Select $k \in \mathbb{N}$ so that b is included in the initial block $q(k)$ of μ. Then it is sufficient to prove that the set $D(\mu, C_0[q(k)])$ has bounded gaps. But this follows immediately from the statements (7) and (8) in 5.6.2: those gaps cannot be larger than $3 \cdot 2^k$, the worst possible situation being that an occurrence of $q(k)$ is followed by two occurrences of $\overline{q(k)}$. □

Remark. Direct proof that M is minimal: it is sufficient to show that $\mu \in \overline{\mathcal{O}_\sigma(x)}$ for every point $x \in M$, i.e., that every initial block of μ occurs in x. This is obvious: if $k \in \mathbb{N}$, and c is any subblock of x of length $4 \cdot 2^k$ then c occurs in μ, because $x \in \overline{\mathcal{O}_\sigma(\mu)}$; see also Proposition 5.2.5. Hence by 5.6.2 (7),8 the block $q(k)$ occurs in c, hence in x. Consequently, every initial block of μ occurs in x.

Proposition 5.6.4. *The point μ is not ultimately periodic, so (M, σ) is an infinite minimal system.*

Proof. If the point μ were ultimately periodic then M would be a finite discrete space, in which the point μ would be isolated. We show that this is not the case. To this end,

10 These blocks can also occur in other positions. E.g., $q(2)$ occurs in $q(4)$ at position 6.

define for every $n \in \mathbb{N}$ a point $\mu^{(n)} := q(n)\,\mu \in \Omega_2$. Note that $\mu^{(n)} \neq \mu$ for all $n \in \mathbb{N}$, because $\mu^{(n)}$ starts with the block $q(n)q(n)$ while μ starts with the block $q(n)\overline{q(n)}$. On the other hand, since both $\mu^{(n)}$ and μ begin with the block $q(n)$ it is clear that

$$d(\mu^{(n)}, \mu) \leq (1 + |q(n)|)^{-1} = (1 + 2^n)^{-1}\,.$$

Hence $\mu^{(n)} \rightsquigarrow \mu$ if n tends to infinity. So in order to show that μ is not isolated in M it is sufficient to show that $\mu^{(n)} \in M$ for all $n \in \mathbb{N}$.

Consider an arbitrary integer $n \in \mathbb{N}$. By Proposition 5.2.5, it is sufficient to show that every initial block of $\mu^{(n)}$ occurs in μ. Any initial block of $\mu^{(n)}$ occurs in a block of the form $q(n)q(m)$ for some $m \geq 0$. In order to prove that this particular block occurs in μ we need the following observations, which are easily proved by induction on k, using the hints given below:

(a) $\forall\, k \in \mathbb{N}$: the block $q(n + 2k)$ ends with $q(n)$.
(b) $\forall\, k \in \mathbb{N}$: the block $q(k)q(k)$ occurs in μ.

For (a), use that $q(n+2(k+1)) = q(n+2k)\,\overline{q(n+2k)}\;\overline{q(n+2k)}\,q(n+2k)$. To prove (b), take into account that $\overline{q(k+2)} = \overline{q(k)}\,q(k)\,q(k)\,\overline{q(k)}$, where by Statement 5.6.2 (6), $\overline{q(k+2)}$ occurs in μ.

In order to prove that the block $q(n)q(m)$ occurs in μ, select $k \in \mathbb{N}$ so large that $n + 2k \geq m$. Then $q(m)$ occurs as initial block in $q(n + 2k)$ and, by (a), $q(n)$ occurs as final block in $q(n + 2k)$. Consequently, $q(n)q(m)$ occurs in $q(n + 2k)q(n + 2k)$ – see Figure 5.5 – which, by (b) above, occurs in μ. It follows that the block $q(n)q(m)$ occurs in μ as well. □

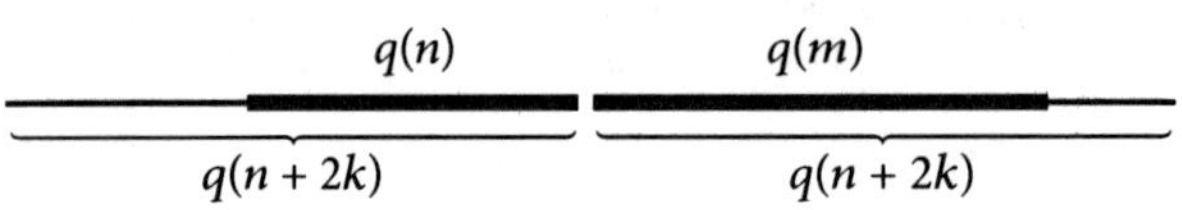

Fig. 5.5. Illustrating the proof of 5.6.4.

Remarks.

(1) By the Remark after Proposition 1.3.4, the infinite minimal system (M, σ) has no isolated points. So M is a Cantor space.
(2) The restriction of σ to M is not injective.

Proof. Similar to the proof that $\mu^{(n)} \in M$ one shows that $\nu^{(n)} := \overline{q(n)}\mu \in M$. Then $\mu^{(n)} \neq \nu^{(n)}$ and $\sigma^{2^n}(\mu^{(n)}) = \mu = \sigma^{2^n}(\nu^{(n)})$. So σ cannot be injective on M. This is in accordance with Proposition 5.3.2. □

5.6.5 (Strong and weak mixing).

(1) *The full shift* (Ω_S, σ) *is strongly (hence weakly) mixing.*

Taking into account that basic open sets in a shift system are cylinders based on finite blocks, it is sufficient to prove that for any two finite blocks a and b over $\mathbf{S}$ we have $\sigma^n C_0[a] \cap C_0[b] \neq \emptyset$ for almost all n. The proof is straightforward: if $n \geq |a|$ then the point $x := a\,0^{n-|a|}b\,0^\infty$ belongs to $C_0[a]$ and $\sigma^n x$ belongs to $C_0[b]$, hence $\sigma^n C_0[b] \cap C_0[a] \neq \emptyset$. (Of course, here the block $0^{n-|a|}$ may be replaced by any block of length $n - |a|$ and the 'tail' 0^∞ may be replaced by any infinite sequence of symbols.)

(2) Let X be a non-empty closed invariant subset of (Ω_S, σ) (a subshift). By formulating the definition of 'strong mixing' in terms of cylinder sets one easily sees that the subsystem (X, σ) of (Ω_S, σ) is strongly mixing iff for any two blocks $a, b \in \mathcal{L}(X)$ – the language of X – there exists $k \in \mathbb{N}$ such that for every $m \geq k$ there is a point in X in which the blocks a and b occur at positions that differ by m from each other. Indeed, for every $m \in \mathbb{N}$ one has $(\sigma^m C_0[a] \cap X) \cap (C_0[b] \cap X) \neq \emptyset$ iff there exists a point in X in which the blocks a and b occur at positions 0 and m, respectively, i.e., in positions that differ by m.

The subshift (X, σ) *is strongly mixing iff for every pair of blocks* $a, b \in \mathcal{L}(X)$ *and for almost every* $n \in \mathbb{Z}^+$ *there exists* $w \in \mathcal{L}(X)$ *with length* $|w| = n$ *such that* $awb \in \mathcal{L}(X)$. In order to prove the 'only if', note that if there is for almost all m a point in X in which the blocks a and b occur at positions that differ by m then for almost all (sufficiently large) m they occur in non-overlapping positions, in which case there is a word w of non-negative length $m - |a|$ between a and b. For the proof of the 'if', use that if a word awb is in $\mathcal{L}(X)$ then by the definition of $\mathcal{L}(X)$ there is $x \in X$ such that awb occurs in x. For examples of strongly mixing subshifts we refer to Exercise 5.13 (1).

(3) The proof of the following statement is left as an exercise for the reader; it consists of a reformulation of the definition of weak mixing in terms of cylinder sets, similar to the method used in 2 above. *A subshift* (X, σ) *is weakly mixing iff for every choice of four blocks* $a, b, c, d \in \mathcal{L}(X)$ *there are blocks* $v, w \in \mathcal{L}(X)$ *such that the blocks* av *and* cw *have the same length and* $avb, cwd \in \mathcal{L}(X)$. For the proof of 'only if', use Exercise 1.6 (6) in order to get non-overlapping positions of the relevant blocks.

For an example of a weakly mixing subshift that is not strongly mixing, see Exercise 5.13 (2). The Morse–Thue system is not weakly mixing (hence not strongly mixing); see Exercise 5.14.

5.6.6 (A topological version of Chacón's system). The shift system to be defined here is the topological version of a measure-preserving system due to Chacón. It is a subshift of Ω_2[11]. Define inductively a sequence of blocks $b(n)$ for $n \in \mathbb{Z}^+$ by

$$b(0) := 0 \quad \text{and} \quad b(n+1) := b(n)\,b(n)\,1\,b(n)\,.$$

11 By the second and third of the properties of the blocks $b(n)$ below, in every point of B all occurrences of 1 are isolated and the blocks of consecutive 0's have length 1, 2 or 3. So B is included in the (1,3) run-length limited shift, defined in Example (6) in Proposition 5.3.4.

For example, $b(1) = 0010$, $b(2) = 0010\,0010\,1\,0010$, etc. The following properties of the blocks $b(n)$ for $n \geq 1$ are easily proved by induction; they will be used without explicit reference. Let $n \in \mathbb{N}$ and $k \in \mathbb{Z}^+$; then:

- $b(n)$ begins with 001 and it ends with 10;
- the blocks of consecutive 0's in $b(n)$ have length 1, 2 or 3;
- all 1's (and the final occurrence of 0) are isolated in $b(n)$;
- if $n \geq k$ then $b(n)$ begins with $b(k)\,b(k)\,1$ and it ends with $1\,b(k)$;
- if $n \geq k$ then nowhere in $b(n)$ there are more than three consecutive occurrences of the block $b(k)$.

Since for every $n \in \mathbb{Z}^+$ the block $b(n+1)$ starts with $b(n)$, there is an infinite sequence of 0's and 1's, i.e., an element $\beta \in \Omega_2$ which, for every $n \in \mathbb{Z}^+$, starts with the block $b(n)$. Let B be the orbit closure of the point β in Ω_2 under σ. By Proposition 5.2.5, B is the set of all points x in Ω_2 with the property that every subblock of x occurs in β. Since each subblock of β occurs in a sufficiently long initial block of β, it follows that *a point x of Ω_2 belongs to B iff every subblock of x occurs in $b(n)$ for some $n \in \mathbb{N}$.*

Next, we identify two sequences of points in B that converge to β. In particular, it will follow that the point β is not isolated in B. To this end, consider for any $n \in \mathbb{Z}^+$ the following points of Ω_2:

$$\beta^{(n)} := b(n)\,\beta \quad \text{and} \quad \gamma^{(n)} := b(n)\,1\beta\,.$$

Claim. $\forall n \in \mathbb{Z}^+ : \beta^{(n)}, \gamma^{(n)} \in B$ and $\beta^{(n)}, \gamma^{(n)} \rightsquigarrow \infty$ for $n \rightsquigarrow \infty$.

Proof. We shall prove this claim only for the points $\gamma^{(n)}$; the proof for the points $\beta^{(n)}$ is similar. Note that every subblock of $\gamma^{(n)}$ occurs in an initial block of $\gamma^{(n)}$ of the form $b(n)\,1\,b(m)$ for some $m \in \mathbb{N}$, and we may assume that $m \geq n$. As $b(n)$ occurs as a final block in $b(m)$, the block under consideration occurs in $b(m)\,1\,b(m)$, hence in $b(m+1)$. This completes the proof that $\gamma^{(n)} \in B$.

Next, note that for every $n \in \mathbb{N}$ the points β and $\gamma^{(n)}$ both start with the block $b(n)$, which has length at least 3^n. Hence the distance of the points β and $\gamma^{(n)}$ is less than 3^{-n}. Consequently, $\gamma^{(n)} \rightsquigarrow \beta$ for $n \rightsquigarrow \infty$. □

The points $\beta^{(0)} = 0\beta$ and $\gamma^{(0)} = 01\beta$ are in B; hence $\beta' := \sigma\gamma^{(0)} = 1\beta$ is in B as well. Obviously, $\beta' \neq \beta^{(0)}$ and $\sigma\beta' = \beta = \sigma\beta^{(0)}$. So σ is not injective on B. (This would also follow from Proposition 5.3.2 and the fact that B is infinite – but the easiest way to prove that B is infinite is to consider the points $\beta^{(n)}$ and/or $\gamma^{(n)}$ for $n \in \mathbb{Z}^+$, so we had to introduce these points anyway.) It follows that B is not a single periodic orbit.

Proposition 5.6.7. *(B, σ) is an infinite minimal system. So B cannot have isolated points, hence is a Cantor space.*

Proof. That B is infinite was observed above. In order to prove that an arbitrary point x of B has a dense orbit it suffices to show that every subblock of β occurs in x. As every subblock of β occurs in a sufficiently long initial block of β it is sufficient to show that every block $b(n)$ for $n \in \mathbb{Z}^+$ occurs in x. To this end, consider a sufficiently large subblock of x, say, of length $3|b(n)|$. This block occurs in β, and every subblock of β with this length includes a copy of $b(n)$. This completes the proof. (Alternative proof: show that the point β is almost periodic.) □

If $n \in \mathbb{Z}^+$ then by a *spaced concatenation of blocks* $b(n)$ we mean a finite or (countably) infinite concatenation of copies of $b(n)$ and *isolated* 1's between (some of) these copies of $b(n)$. The occurrences of $b(n)$ used to form the spaced concatenation are called the *natural occurrences* of $b(n)$ or the occurrences in *natural position*. The isolated 1's between these natural occurrences of $b(n)$ are called *n-spacers*.

Example. The blocks $b(n)\,b(n)\,b(n)\,b(n)$ and $b(n)1b(n)1b(n)1b(n)$ are spaced concatenations of $b(n)$, but $1b(n)\,b(n)$ and $b(n)\;11\,b(n)$ are not. The block

$$b(n+1)\;1\;b(n+1) = b(n)\,b(n)\;1\;b(n)\;\;1\;\;b(n)\,b(n)\;1\;b(n)$$

is a spaced concatenation of blocks $b(n+1)$ with one $(n+1)$-spacer and two natural occurrences of $b(n+1)$; it is also a spaced concatenation of blocks $b(n)$ with three n-spacers (one of which is also an $(n+1)$-spacer) and six natural occurrences of $b(n)$.

For every $n \in \mathbb{Z}^+$ and every $k \geq n$ the block $b(k)$ is a spaced concatenation of blocks $b(n)$; in particular, β is a spaced concatenation of blocks $b(n)$. The straightforward proof (by induction) is left for the reader. The first two examples above show that, conversely, a spaced concatenation of blocks $b(n)$ does not necessarily occur in a block $b(k)$ for any $k \geq n$.

The following is essential for the sequel:

Lemma 5.6.8. *Let $n \in \mathbb{Z}^+$. In any spaced concatenation of blocks $b(n)$ all copies of $b(n)$ occur only at natural positions.*

Proof. We have to show that configurations like

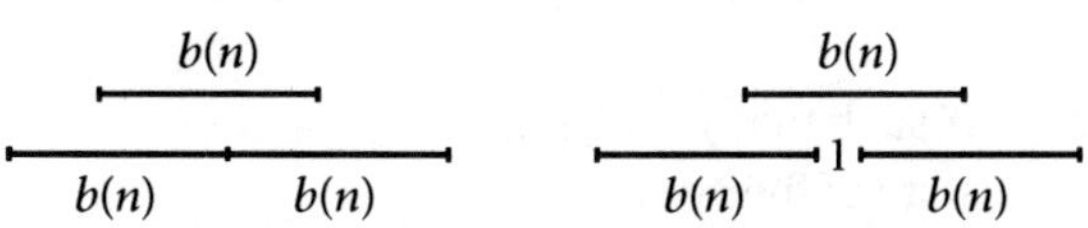

are impossible. (In these pictures, the upper block $b(n)$ indicates an occurrence of $b(n)$ in the lower concatenation not coinciding with one of the occurrences of $b(n)$ there.) The proof is by induction in n.

It is obvious that for $n = 0$ these configurations cannot occur; moreover, it is instructive to check this also for $n = 1$. Now assume that these configurations are impossible for some $n \in \mathbb{N}$ and consider similar configurations for $n + 1$ (draw the pictures yourself). By the induction hypothesis, the occurrences of the blocks $b(n)$ in the upper block $b(n+1) = b(n)\,b(n)\;1\;b(n)$ must coincide with the occurrences of $b(n)$ in the lower block. In the first configuration this would imply that the n-spacer in the upper block must coincide with one of the n-spacers in the lower concatenation which, in turn, implies that the upper block $b(n+1)$ coincides with one of the lower occurrences of $b(n+1)$. In the second configuration a similar argument shows that the upper copy of the block $b(n+1)$ coincides with one of the lower copies, or else that we have the following configuration:

$$\begin{array}{lll} & b(n)\,b(n) & 1\;b(n) \\ b(n)\,b(n)\,1\,b(n) & b(n)\,b(n)\,1\,b(n) & 1\;b(n)\,b(n)\,1\,b(n) \end{array}$$

which is impossible because the block $b(n)$ – the first occurrence in the upper block – does not end with a 1. (Moreover, this would contradict the induction hypothesis concerning the blocks $b(n)$.) □

Remark. Consequently, in a spaced concatenation of blocks $b(n)$ an occurrence of $b(n)$ is followed by an occurrence of $b(n)$ or $1b(n)$ (provided it is not the final occurrence) and it is preceded by an occurrence of $b(n)$ or $b(n)1$ (provided it is not the initial occurrence).

5.6.9. In order to prove that the system (B, σ) is weakly mixing we need some additional terminology and observations. First, recall that a complete past of a point $x \in \Omega_2$ is a sequence $(x^{(n)})_{n\in\mathbb{Z}^-}$ of points in Ω_2 with the property that $\sigma x^{(n-1)} = x^{(n)}$ for all $n \in \mathbb{Z}^-$ and $x^{(0)} = x$. In contrast with the orbit – the future – of a point, a complete past is not unique: every point of Ω_2 has uncountably many complete pasts.

The union of a complete past and the orbit of a point is called a *complete history* of that point. Formally, a complete history of a point $x \in \Omega_2$ is a mapping $H_x : \mathbb{Z} \to \Omega_2$ such that $H_x(0) = x$ and $\sigma(H_x(n-1)) = H_x(n)$ for all $n \in \mathbb{Z}$, hence $\sigma^k H_x(m) = H_x(m+k)$ for all $m \in \mathbb{Z}$ and $k \in \mathbb{Z}^+$. However, we shall also call the range $H_x[\mathbb{Z}]$ of this mapping a complete history[12].

There is a close relationship between complete histories of points that are all in the complete history of a given point. In fact, let $x, y \in \Omega_2$ and let H_x be a complete history of x. Then the following conditions are easily seen to be equivalent:

(i) $x \in \mathcal{O}(y)$ or $y \in \mathcal{O}(x)$.
(ii) $y \in H_x[\mathbb{Z}]$, that is, there exists $k \in \mathbb{Z}$ such that $y = H_x(k)$.
(iii) $K_y : n \mapsto H_x(n+k)$ is a complete history of the point y.
(iv) There is a complete history K_y of y such that $K_y[\mathbb{Z}] = H_x[\mathbb{Z}]$.

These equivalences may be paraphrased by saying that *a complete history is a complete history of each of its points.*

Next, we define the notion of a ‘bi-sequence’. A *bi-sequence* is a two-sided infinite sequence of the symbols 0’s and 1’s, i.e., an element of $\{0,1\}^{\mathbb{Z}}$. We shall also have occasion to employ the notion of a *left-sequence*: an element of $\{0,1\}^{(-\infty\,;0]}$. In this context, elements of $\Omega_2 = \{0,1\}^{[0\,;\infty)}$ will sometimes be called *right-sequences*. All no-

12 This is similar to the usage for curves, which are defined as mappings from an interval to a space, but where the range of such a mapping as also often called a curve.

tation and terminology used for members of Ω_2 (as far as meaningful) will also be used for left- and bi-sequences. In a bi-sequence the 0-coordinate will be indicated by a dot above the coordinate in question. Thus, in $\dots x_{-1}\dot{x}_0 x_1 \dots$ the 0-coordinate is x_0, in $\dots x_{k-1}\dot{x}_k x_{k+1} \dots$ it is x_k.

A bi-sequence $\dots x_{-2}x_{-1}\dot{x}_0 x_1 x_2 \dots$ *may be viewed as a representation of a complete history of the point* $x := x_0 x_1 x_2 \dots$ *of* Ω_2 *under the shift*: if we put $H_x(n) := x_{[n\,;\infty)}$ for $n \in \mathbb{Z}$ then H_x obviously is a complete history of the point x. This complete history consists of all 'tails' $x_{[n\,;\infty)}$ $(n \in \mathbb{Z})$ of the given bi-sequence.

Recall our convention about coordinates: all finite blocks and all (right-) sequences representing elements of Ω_2 start with the coordinate 0. So if $n \in \mathbb{Z}$ then the coordinates of $H_x(n)$ are given by $(H_x(n))_i = x_{i+n}$ for $i \geq 0$. In particular, the bi-sequence can be recovered from the history by the equality $x_n = (H_x(n))_0$.

Conversely, any complete history $H_x : \mathbb{Z} \to \Omega_2$ (x in Ω_2) is generated by a unique bi-sequence, namely, the bi-sequence $\dots x_{-2}x_{-1}\dot{x}_0 x_1 x_2 \dots$ defined by $x_n := (H_x(n)))_0$ for $n \in \mathbb{Z}$. It is easily checked that $x_{[n\,;\infty)} = H_x(n)$ for every $n \in \mathbb{Z}$ so, indeed, this bi-sequence represents the given complete history. Moreover, this is the unique bi-sequence for which this is the case (see the small print above). So there is a 1-to-1-correspondence between the set of all 'dotted' bi-sequences in Ω_2 and the set of all complete histories of points of Ω_2 under σ.

Now consider two points $x, y \in \Omega_2$ that satisfy the equivalences (i) through (iv) above, i.e., $y = H_x(k)$ for some $k \in \mathbb{Z}$, and let $\dots x_{-1}\dot{x}_0 x_1 \dots$ and $\dots y_{-1}\dot{y}_0 y_1 \dots$ be the bi-sequences representing the complete histories H_x of x and K_y of y. Then by the above,

$$y_n = (K_y(n))_0 = (H_x(n+k))_0 = x_{n+k}$$

for every $n \in \mathbb{Z}$. Thus, the bi-sequence representing H_y is equal to the bi-sequence representing H_x shifted over k positions: the bi-sequence representing the complete history of y is $\dots x_{k-1}\dot{x}_k x_{k+1} \dots$ (note the dot above x_k). We may paraphrase this as follows: if the points share a complete history then the corresponding bi-sequences are equal up to the place of the dot. So by 'forgetting the dot' we retain the 1,1-correspondence between complete histories (as subsets of Ω_2) and bi-sequences (without a dot, or rather, potentially a dot above any coordinate).

Finally, if x is a point of B then the future (i.e., the orbit) of x is included in B, because B is invariant) but it is not to be expected that every complete past of x is included in B. However, $\sigma[B] = B$ because the subsystem (B, σ) of the shift system (Ω_2, σ) is minimal, so every point of B actually has a complete past, hence a complete history, in B. It is easy to see that a bi-sequence represents a complete history in B of a point of B iff every subblock of the bi-sequence occurs in β, iff every subblock of the bi-sequence occurs in a block $b(n)$ for some $n \in \mathbb{Z}^+$.

Example. Define a left-sequence $\tilde{\beta}$ as follows: for every $n \in \mathbb{Z}^+$ the block $b(n+1)$ ends with the block $b(n)$, hence there is a left-sequence $\tilde{\beta}$ which ends, for every $n \in \mathbb{N}$, with the block $b(n)$ (hence with the block $1\,b(n)$, so $\tilde{\beta}$ is *not* the mirror image of β). Then the bi-sequences $\tilde{\beta}\beta$ and $\tilde{\beta}\,1\beta$ represent two different complete histories of the point β. All right sequences in these bi-sequences represent points of B, because they represent points in the orbits of $\beta^{(n)}$ or $\gamma^{(n)}$ for sufficiently large n.

5.6.10. There is an elegant way to construct complete histories of points of B. In the construction of the blocks $b(n)$, the block $b(n+1)$ is obtained from $b(n)$ by appending the block $b(n)1b(n)$ after $b(n)$. But we may also obtain a nested sequence of copies of the blocks $b(0), b(1), b(2), \dots$ where for each n the block $b(n)$ is not necessarily the first block of $b(n+1)$, but maybe the second or third. To make this more precise, recall that a block b can be seen as – actually, *is* – a function from the interval $\{0, \dots, |b|-1\}$ in $\mathbb{Z}^+$ into the symbol set $\{0,1\}$. When seen in this way, the function $b(n+1)$ is an extension of the function $b(n)$. However, we can also consider 'generalized finite blocks': functions from arbitrary finite intervals in $\mathbb{Z}^+$ into $\{0,1\}$. Stated otherwise, the numbering of coordinates of such a generalized finite block may start at any (possibly negative) integer. We shall now define a sequence of such generalized blocks, closely related to the points of B.

Consider a 'triadic' sequence $\xi \in \{1,2,3\}^{\mathbb{N}}$ and define inductively generalized blocks $d(n)$ for $n \in \mathbb{Z}^+$ by the following rules:

$$d_\xi(0) := b(0) = 0\,,$$

where the single symbol 0 in this block has coordinate number 0, i.e., $d_\xi(0)$ is the function defined on the interval $\{0\}$ of $\mathbb{Z}^+$ that maps the single point of this interval onto the symbol 0. In addition, for every $n \in \mathbb{Z}^+$:

$$d_\xi(n+1) := \begin{cases} \mathbf{d}_\xi(\mathbf{n})\, b(n)\, 1\, b(n) & \text{if } \xi(n+1) = 1\,, \\ b(n)\, \mathbf{d}_\xi(\mathbf{n})\, 1\, b(n) & \text{if } \xi(n+1) = 2\,, \\ b(n)\, b(n)\, 1\, \mathbf{d}_\xi(\mathbf{n}) & \text{if } \xi(n+1) = 3\,, \end{cases}$$

and the coordinates of $d_\xi(n+1)$ are numbered in such a way that the generalized block $d_\xi(n)$ keeps its coordinate numbering when considered as a subblock of $d_\xi(n+1)$. When seen as functions, $d_\xi(n+1)$ is an extension of the function $d_\xi(n)$ to a larger interval; which interval that is depends on the values of $\xi(1), \dots, \xi(n)$ (which determine the domain of $d_\xi(n)$ as a function) and the value of $\xi(n+1)$, in the way suggested by the above definition. It follows from these definitions that the block $d_\xi(0)$ occurs in every block $d_\xi(n)$ at position 0, i.e., every generalized block $d_\xi(n)$ has the symbol 0 at position 0. This position will be indicated by a dot.

It is straightforward to show by induction that for every $n \in \mathbb{N}$, as a finite sequence of 0's and 1's, *the blocks $d_\xi(n)$ and $b(n)$ are equal to each other*. Stated otherwise, if '$d_\xi(n)$' is the ordinary block obtained from the generalized block $d_\xi(n)$ by re-numbering the coordinates (starting with 0), then '$d_\xi(n)$'$= b(n)$. *Consequently, for every pair of $k, n \in \mathbb{Z}^+$ with $k \geq n$ the generalized block $d_\xi(k)$ is a spaced concatenation of blocks $b(n)$.*

Examples.

(1) If $\xi = 111\dots$ then the blocks $d_\xi(n)$ expand to the right for increasing n, as follows:

$$d_\xi(3) = \underbrace{\underbrace{\dot{0}010}_{d_\xi(1)}\ \underbrace{0010}_{b(1)}\ 1\ \underbrace{0010}_{b(1)}}_{d_\xi(2)}\quad \underbrace{0010\ 0010\ 1\ 0010}_{b(2)}\quad 1\quad \underbrace{0010\ 0010\ 1\ 0010}_{b(2)}$$

(2) If $\xi = 333\dots$ then the blocks $d_\xi(n)$ expand to the left for increasing n, as follows:

$$d_\xi(3) = \underbrace{0010\ 0010\ 1\ 0010}_{b(2)}\quad \underbrace{0010\ 0010\ 1\ 0010}_{b(2)}\quad 1\quad \underbrace{\underbrace{0010}_{b(1)}\ \underbrace{0010}_{b(1)}\ 1\ \underbrace{001\dot{0}}_{d_\xi(1)}}_{d_\xi(2)}$$

(3) If $\xi = 222\dots$ then the blocks $d_\xi(n)$ expand to both sides for increasing n:

$$d_\xi(3) = \underbrace{0010\ 0010\ 1\ 0010}_{b(2)}\quad \underbrace{0010\ \underbrace{0\dot{0}10}_{d_\xi(1)}\ 1\ 0010}_{d_\xi(2)}\quad 1\quad \underbrace{0010\ 0010\ 1\ 0010}_{b(2)}$$

For every sequence $\xi \in \{1,2,3\}^{\mathbb{N}}$ the blocks $d_\xi(n)$ for $n \in \mathbb{Z}^+$ form an expanding nested sequence, defining a right-sequence (see Example (1)), a left-sequence (see Example (2)) or a bi-sequence (see Example (3)) ξ^* of 0's and 1's.

Above, we have observed that if $n \in \mathbb{Z}^+$ then for every $k \geq n$ the generalized block $d_\xi(k)$ is a spaced concatenation of blocks $b(n)$. *It follows that for every $\xi \in \{1,2,3\}^{\mathbb{N}}$ and for every $n \in \mathbb{Z}^+$ the (left-, right or bi-) sequence ξ^* is also a spaced concatenation of blocks $b(n)$.*

In order to avoid clumsy formulations we say that a *bi-sequence* is a *universally spaced concatenation* whenever it is for every $n \in \mathbb{Z}^+$ a spaced concatenation of blocks $b(n)$. Thus, *if $\xi \in \{1,2,3\}^{\mathbb{N}}$ and ξ^* is a bi-sequence then ξ^* is a universally spaced concatenation.*

5.6.11. The following observations lead to a proof that the system (B, σ) is weakly mixing. The first observation is obvious:

(1) *If $\xi \in \{1,2,3\}^{\mathbb{N}}$ then ξ^* is a bi-sequence iff the generalized blocks $d_\xi(n)$ eventually expand to both sides as n increases, iff $\xi(n)$ is not eventually 1 or 3.* □

Let us call a ternary sequence $\xi \in \{1,2,3\}^{\mathbb{N}}$ an *ordinary sequence* whenever it is not eventually 1 or 3. If ξ is an ordinary sequence then ξ^* is a bi-sequence, representing a complete history of a point in Ω_2, namely, that of the point $(\xi^*)_{[0\,;\infty)}$. This point is in B, because every subblock of the bi-sequence ξ^* occurs in a generalized block of the form $d_\xi(n)$ for some $n \in \mathbb{N}$, i.e., it occurs in a copy of $b(n)$ for some $n \in \mathbb{N}$. A point of B obtained in this way as $(\xi^*)_{[0\,;\infty)}$ for some ordinary sequence ξ will be called a *ordinary point*, and we shall say that it is *generated* by the ordinary sequence ξ.

(2) *If ξ and η are ordinary sequences and $\xi \neq \eta$ then $(\xi^*)_{[0\,;\infty)} \neq (\eta^*)_{[0\,;\infty)}$.*

Proof. Let n be the smallest value in $\mathbb{Z}^+$ such that $\xi(n+1) \neq \eta(n+1)$. If $n = 0$, i.e., if $\xi(1) \neq \eta(1)$, then $d_\xi(1)$ and $d_\eta(1)$ are obtained as mutually different choices from the three generalized blocks

$$\dot{0}\,0\,1\,0\,,$$
$$0\,\dot{0}\,1\,0\,,$$
$$0\,0\,1\,\dot{0}\,,$$

It is easy to see that different choices lead to different bi-sequences. For example, if $\xi(1) = 2$ and $\eta(1) = 1$ then $d_\xi(1) = 0\dot{0}10$, hence $\xi_1^* = 1$, and $d_\eta(1) = \dot{0}010$, hence $\eta_1^* = 0$. Consequently, in this case we have $\xi^* \neq \eta^*$. Similarly, if $\xi(1) = 2$ and $\eta(1) = 3$ then $\xi_{-1}^* = 0$ and $\eta_{-1}^* = 1$. The case that $\xi(1) = 1$ and $\eta(1) = 3$ is slightly more involved: then in η^* – a spaced concatenation of blocks $b(1)$ in natural positions – the generalized block $d_\eta(1)$ is either followed by 0010 or by a 1-spacer; in both cases there is a mismatch of 1's and 0's in ξ^* and η^*. All other possibilities follow from these cases by interchanging the roles of ξ and η.
Next, consider the case that $n > 0$. As $\xi(i) = \eta(i)$ for $i = 1, \ldots, n$, the blocks $d_\xi(n)$ and $d_\eta(n)$ occur at the same positions in ξ^* and η^*, respectively. The argument showing that ξ^* and η^* are different is quite similar to the above arguments: just replace the 0's by copies of $b(n)$ and detect positions where a spacer in the one bi-sequence appears in the same position as the initial or the final coordinate of a copy of $b(n)$ in the other bi-sequence. □

(3) *There are uncountably many ordinary points.*

Proof. In order to prove this it is, in view of 2 above, sufficient to show that there are uncountably many ordinary sequences. This can be shown in the following way: for each infinite subset S of $\mathbb{N}$, let $\xi_S := 1 + \chi_S$, where χ_S is the characteristic function of S. Then $\xi_S \in \{1, 2\}^{\mathbb{N}}$ and ξ_S is not eventually 1, so it is an ordinary sequence. Now observe that there are uncountably many infinite subsets of $\mathbb{N}$ (there are only countably many finite subsets). □

It follows from statement (2) above that *for every ordinary point the ordinary sequence by which it is generated is unique*. Consequently, we can speak of *the* generator of an ordinary point. Thus, an ordinary point has a 'natural' complete history, namely, the one that is represented by the bi-sequence ξ^* for its generator ξ. By the final observation in 5.6.10 above, this bi-sequence is a universally spaced concatenation.

Example. By Example (3) in 5.6.10, $010\,1\,b(1)\,1\,b(2)\,1\,b(3)\,1\,b(4)\,1\,\ldots$ is an ordinary point, generated by the sequence $\xi = 222\ldots$. The right-sequence of coordinates of this point is not a spaced concatenation of blocks $b(k)$ for any $k \in \mathbb{N}$, but for every k a suitable 'tail' of this right sequence is a spaced concatenation of blocks $b(k)$.

It is conceivable that an ordinary point has other complete histories as well, represented by other bi-sequences. However, it will follow from 4 below that every ordinary point has only one complete history within the set of all complete histories that are represented by a universally spaced concatenation. This unique complete history is, of course, its natural history.

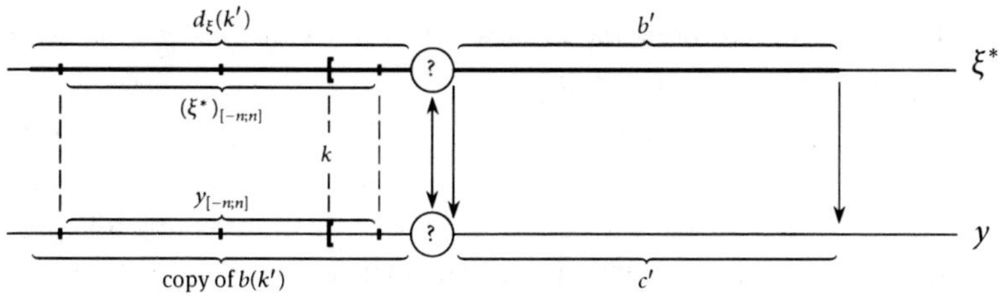

Fig. 5.6. Illustrating the proof of Statement 4. The question mark indicates either the empty block or an isolated occurrence of 1.

Ordinary points and all points from their natural histories will be called *non-exceptional points*[13]. Thus, a point x is non-exceptional iff there is an ordinary sequence $\xi \in \{1, 2, 3\}^{\mathbb{N}}$ such that $x = (\xi^*)_{[m\,;\infty)}$ for some $m \in \mathbb{Z}$. In that case, the bi-sequence ξ – with the dot above $(\xi^*)_m$ – represents a complete history of x. *So every non-exceptional point has a complete history which is represented by a bi-sequence that is a universally spaced concatenation.* (By Exercise 5.15 (1) this is true for *every* point of B, but we do not need this here.) Obviously, the set of non-exceptional points is non-empty and completely invariant in B. It is dense in B, because the system (B, σ) is minimal. It is not equal to B: according to Exercise 5.15 (4) the point β is not a non-exceptional point.

(4) *Let $\xi \in \{1, 2, 3\}^{\mathbb{N}}$ be an ordinary sequence and let y be any universally spaced concatenation. If there exists $k \in \mathbb{Z}^+$ such that $(\xi^*)_{[k\,;\infty)} = y_{[k\,;\infty)}$ then $y = \xi^*$.*

Proof. Recall that, for every $k' \in \mathbb{Z}^+$, the bi-sequence ξ^* is a spaced concatenations of blocks $b(k')$, and that in ξ^* and y the blocks $b(k')$ occur only at natural positions. In particular, the Remark after Lemma 5.6.8 will be used, both for ξ^* and for y.
Let $n \in \mathbb{N}$, $n > k$. By assumption, ξ is not eventually 1 or 3, so it follows from Statement 1 above that there exists $k' \in \mathbb{N}$ such that $d_\xi(k')$ covers the block $(\xi^*)_{[-n\,;n]}$. Let b' denote the copy of $b(k')$ in ξ^* following after $d_\xi(k')$. Then b' occurs in $(\xi^*)_{[k\,;\infty)}$, hence in $y_{[k\,;\infty)}$ and that in the same position. Stated otherwise, a copy c' of $b(k')$ occurs in y in the same position as b' occurs in ξ^*. It follows that the copy of $b(k')$ preceding c' in y is in the same position as $d_\xi(k')$: a possible k'-spacer between $d_\xi(k')$ and b' is reproduced in $y_{[k\,;\infty)}$ as a k'-spacer left of c' and, conversely, a possible k'-spacer in y immediately to the left of c' is reproduced in $(\xi^*)_{[k\,;\infty)}$ between $d_\xi(k')$ and b'. In particular, it follows that $y_{[-n\,;n]} = (\xi^*)_{[-n\,;n]}$. This conclusion holds for every $n \in \mathbb{N}$, hence $y = \xi^*$. □

(5) *Let $\xi \in \{1, 2, 3\}^{\mathbb{N}}$ be an ordinary sequence. Then the point $(\xi^*)_{[0\,;\infty)}$ has a unique complete past in the set of all non-exceptional points, namely the set of all points*

---

13 An ordinary point always has 0-coordinate 0, but the natural history of an ordinary point includes many points with 0-coordinate 1. So the ordinary points form a proper subset of the set of all non-exceptional points. There are also 'exceptional' points in B, such as the points β, $\beta^{(n)}$ and $\gamma^{(n)}$ for $n \in \mathbb{Z}^+$: see Exercise 5.15 (4).

$(\xi^*)_{[-n\,;\infty)}$ *for* $n \in \mathbb{N}$. *It follows that there are at most countably many non-exceptional points that have the point* $(\xi^*)_{[0\,;\infty)}$ *in their orbit.*

Proof. Consider an non-exceptional point $z \in B$, say, $z = (\zeta^*)_{[m\,;\infty)}$ with ζ an ordinary sequence and $m \in \mathbb{Z}$ and assume that there exists $k \geq 1$ such that $\sigma^k z = x$, i.e., such that $(\zeta^*)_{[m+k\,;\infty)} = (\xi^*)_{[0\,;\infty)}$. Renumber the coordinates of ζ^* by diminishing all coordinate-numbers by $m + k$. Thus, we consider the bi-sequence y with $y_i := (\zeta^*)_{i+m+k}$ for $i \in \mathbb{Z}$. Then the bi-sequence y is, just like ζ^*, a universally spaced concatenation and, obviously, $y_{[0\,;\infty)} = (\zeta^*)_{[m+k\,;\infty)} = (\xi^*)_{[0\,;\infty)}$. So 4 above implies that $y = \xi^*$. Consequently, $(\zeta^*)_i = y_{i-m-k} = (\xi^*)_{i-m-k}$ for every $i \in \mathbb{Z}$. In particular, it follows that $z = (\zeta^*)_{[0\,;\infty)} = (\xi^*)_{[-m-k\,;\infty)}$.
The concluding statement is a trivial consequence of the fact that the point $(\xi^*)_{[0\,;\infty)}$ has a unique complete past within the set of all non-exceptional points and that this complete past is countable. □

(6) *There exists a pair of ordinary points x and y in B such that y does not belong to the orbit of x and x does not belong to the orbit of y.*

Proof. Select any ordinary point x. As the orbit of x is countable, there remains an uncountable set of ordinary points from which y can be chosen in such a way that y is not in the orbit of x. Moreover, it follows from 5 above that there are only countably many ordinary points that have the point x in their orbit. So y can also be chosen outside of any past of x, which means that x is not in the orbit of y. □

(7) *If x and y are ordinary points in B such that y does not belong to the orbit of x and x does not belong to the orbit of y then the point (x, y) has a dense orbit in $B \times B$ under $\sigma \times \sigma$.*

Proof. First, we reformulate what we want to prove as a combinatorial condition. Taking into account that basic open sets in $B \times B$ are products of cylinder sets based on arbitrary words in $\mathcal{L}(B)$, it is easily seen that we have to show: if $c^{(1)}, c^{(2)} \in \mathcal{L}(B)$ then $c^{(1)}$ occurs in x and $c^{(2)}$ occurs in y in the same position. So consider two arbitrary elements $c^{(1)}$ and $c^{(2)}$ of $\mathcal{L}(B)$. Then there exists $k \in \mathbb{N}$ so large that both $c^{(1)}$ and $c^{(2)}$ occur in the block $b(k)$, say at the positions p_1 and p_2, respectively. Assume that $p_1 \leq p_2$ (the case that $p_1 > p_2$ is treated similarly, by reversing the roles of x and y; this is possible by the symmetry in the given data). Then we want to show that $b(k)$ occurs in x at a position l_1 and in y at a position l_2 such that $l_1 + p_1 = l_2 + p_2$, that is, $l_2 = l_1 - (p_2 - p_1)$, where $0 \leq p_2 - p_1 \leq |b(k)| - 1$. So it is sufficient to prove that for every $k \in \mathbb{Z}^+$ and for every $i \in \{0, \dots, |b(k)| - 1\}$ there exists $l \geq 0$ such that there are occurrences of a copy of $b(k)$ in x and in y at positions l and $l - i$, respectively. In such a case we shall call i the *delay* of the occurrences of $b(k)$ in x and y. When a copy of $b(k)$ occurs in x at a position where y has a k-spacer then we say that there is a delay of $|b(k)|$: there is are occurrences of $b(k)$ in x at position l, say, and in y preceding this k-spacer, at position $l - |b(k)|$.

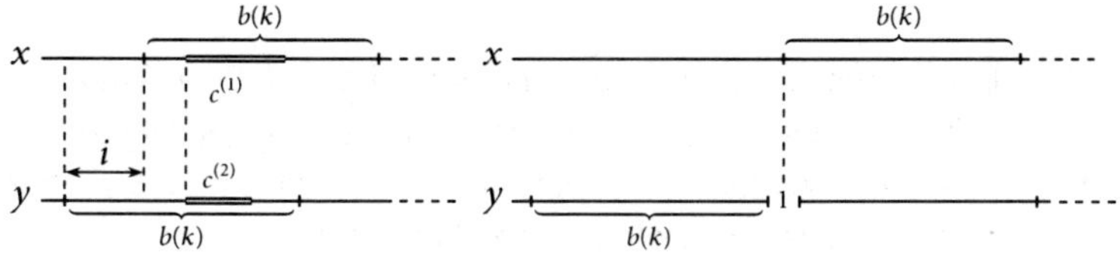

Delay i with $0 \leq i \leq |b(k)| - 1$. Delay $|b(k)|$.

For every $k \in \mathbb{Z}^+$, let D_k be the set of all delays i with $0 \leq i \leq |b(k)|$ of occurrences of copies of $b(k)$ in x and y. It follows from the discussion above that it is sufficient to show that $\{0, \ldots, |b(k)|-1\} \subseteq D_k$ for every $k \in \mathbb{Z}^+$. In any case it is clear that $D_k \neq \emptyset$: the right-sequences of coordinates of x and y are spaced concatenations of blocks $b(k)$ – see the final statement in 5.6.10 – so for every occurrence of a copy of $b(k)$ in y there is a copy of $b(k)$ in x with a delay of at most $|b(k)|$.

So let $k \in \mathbb{Z}^+$ be arbitrary, and select for every $i \in D_k$ a position $n(i)$ at which a copy of the block $b(k)$ occurs in x with delay i; this implies that a copy of $b(k)$ occurs in y at position $n(i) - i \geq 0$. Now recall that x and y are ordinary points, so there are ordinary sequences $\xi, \eta \in \{1, 2, 3\}^{\mathbb{N}}$ such that $x = (\xi^*)_{[0\,;\infty)}$ and $y = (\eta^*)_{[0\,;\infty)}$. By Statement 1 above, it is possible to select $k' \in \mathbb{N}$ so large that for every $i \in D_k$ the generalized block $d_\xi(k')$ includes the occurrence $x_{[n(i)\,;n(i)+|b(k)|)}$ of $b(k)$ in x and such that, in addition, $d_\eta(k')$ includes the occurrence $y_{[n(i)-i\,;n(i)+|b(k)|-i)}$ of $b(k)$ in y. We may also select $k' > k$, so that the block $b(k')$ is a spaced concatenation of copies of $b(k)$.

Thus, it is sufficient to require that the coordinates $x_{n(i)}$ and $x_{n(i)+|b(k)|-1}$ are both in $d_\xi(k')$, and that $y_{n(i)-|b(k)|}$ and $y_{n(i)}$ are in $d_\eta(k')$. The reader should be not worried about the fact that $d_\xi(k')$ and $d_\eta(k')$ may be not included in $(\xi^*)_{[0\,;\infty)} = x$ or $(\eta^*)_{[0\,;\infty)} = y$, respectively, because the copies of $b(k)$ we are interested in do occur in x and y and, moreover, later on we will only consider copies of $b(k')$ further to the right (i.e., at higher positions) which are, consequently, fully included in x and/or y.

Stated otherwise, there is an occurrence u of a copy of $b(k')$ in ξ^* and an occurrence v of a copy of $b(k')$ in η^* which satisfy the following conditions:

(∗) $\forall i \in D_k$: $x_{n(i)} \in \acute{u}$ and $y_{n(i)} \in \grave{v}$; here $\acute{u}$ is the block u from which the final $|b(k)| - 1$ coordinates are removed and $\grave{v}$ is the block v from which the initial $|b(k)|$ coordinates are removed.

Claim. The occurrences u and v of $b(k')$ in ξ^* and η^*, respectively, can be chosen in such a way that condition (∗) holds and, in addition, one of u and v is followed by a k'-spacer and the other is not.

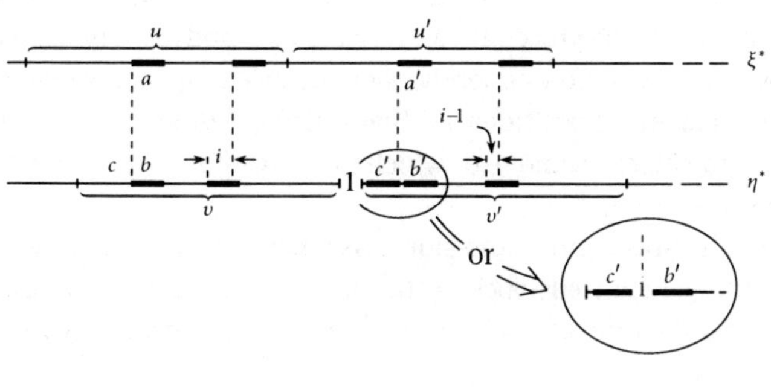

Suppose that this claim is true, and assume that u is immediately followed by a copy u' of $b(k')$ and that v is followed by $1v'$, Then u' is just u, shifted over $|b(k')|$ positions to the right, and v' is v shifted over $|b(k')| + 1$ positions to the right. It follows that the copies of $b(k)$ in v are shifted over one position more than the copies of $b(k)$ in u. Hence all delays are diminished by 1.

To be precise, if $i \in D_k$, $i \geq 1$, then there are a copy of $b(k)$ in u at position $n(i)$ and a copy of $b(k)$ in v at position $n(i) - i$ (positions in x and y, respectively. Then the shifted copies occur at positions $n(i) + |b(k')|$ and $(n(i) - i) + |b(k')| + 1 = (n(i) + |b(k')|) - (i - 1)$, respectively. So now there is a delay of $i - 1$. This shows: if $i \in D_k$ and $i \geq 1$, then $i - 1 \in D_k$. Similarly, if $i - 1 \geq 1$ then also $i - 2 \in D_k$, etc. In particular, as $D_k \neq \emptyset$, it follows that $0 \in D_k$.

For the copies of $b(k)$ that appear in u and v with delay 0 at position $n(0)$, call them a and b, compare the shifted copy a' of a in u' with the copy c' of $b(k)$ that precedes the shifted copy b' of b in v'. This is the shifted copy of the occurrence c of $b(k)$ immediately preceding the block b in η^* (note that c is entirely included in v). Then the initial coordinate of a' has the same position as the final coordinate of c', or as the k-spacer between c' and b', if there is one. So now we have a delay of $|b(k)| - 1$ or of $|b(k)|$. Stated otherwise, $|b(k)| - 1 \in D_k$ or $|b(k)| \in D_k$. In the latter case we also get $|b(k)| - 1 \in D_k$ by what we have proved already. By again applying the fact that for every element of D_k also its predecessor is in D_k, it follows that $\{0, \dots, |b(k)| - 1\} \subseteq D_k$.

The case that u but not v is followed by a k'-spacer is treated in a similar way: now for every $i \in D_k$ with $i \leq |b(k)| - 1$ also $i + 1 \in D_k$, so that $|b(k)| \in D_k$, and this, in turn, implies that $0 \in D_k$. Consequently, also in this case we have $\{0, \dots, |b(k)| - 1\} \subseteq D_k$.

This completes the proof that the system $(B \times B, \sigma \times \sigma)$ includes a point with dense orbit – but it remains to prove the claim that the occurrence u of $b(k')$ in ξ^* and the occurrence v of $b(k')$ in η^* can be chosen in such a way that condition $(*)$ holds and that one of u and v is followed by a k'-spacer and the other is not. In order to prove this, it is sufficient to show that the following procedure stops after finitely many steps:

- put $u := d_\xi(k')$ and $v := d_\eta(k')$;
- if one of u and v is followed by a k'-spacer and the other is not then stop;
- if not, replace u and v by the copies of $b(k')$ immediately to their right and repeat the previous step.

If the procedure stops then we have copies u and v of $b(k')$ in ξ^* and η^* with exactly the same *relative* positions with respect to each other as the original generalized blocks $d_\xi(k')$ and $d_\eta(k')$: at each step in the procedure we go, both in ξ^* and in η^*, either $|b(k')|$ or $|b(k')| + 1$ positions to the right. Consequently, every delay $i \in D_k$ occurs for copies of $b(k)$ within u and v, occurring at positions $n(i)$ that satisfy condition $(*)$. Moreover, it is obvious that one of the blocks u and v for which the procedure stops, is followed by a k'-spacer and the other is not.

It remains to show that the above procedure stops. Assume that this is not the case: then following upon the generalized blocks $d_\xi(k')$ and $d_\eta(k')$ the bi-sequences ξ^* and η^* have the identical patterns of copies of $b(k')$ and k'-spacers. Consequently, if p and q denote the first positions that are not in $d_\xi(k')$ and $d_\eta(k')$, respectively, then $(\xi^*)_{[p\,;\infty)} = (\eta^*)_{[q\,;\infty)}$. Since both p and q are positive, this implies that $x_{[p\,;\infty)} = y_{[q\,;\infty)}$ or, $\sigma^p x = \sigma^q y$. Assuming that $q \geq p$ it follows that for $z := \sigma^{q-p} y$ we have

$$\sigma^p z = \sigma^p(\sigma^{q-p} y) = \sigma^q y = \sigma^p x\,,$$

hence $z_{[p\,;\infty)} = x_{[p\,;\infty)}$. However, z is an non-exceptional point, with complete history represented by a shifted copy[14] h of the bi-sequence η^* and, consequently, the equality $z_{[p\,;\infty)} = x_{[p\,;\infty)}$ means that $h_{[p\,;\infty)} = (\xi^*)_{[p\,;\infty)}$. As h (like η^*) is a universally spaced concatenation we infer from Statement 4 above that $h = \xi^*$. In particular, it follows that $z = h_{[0\,;\infty)} = (\xi^*)_{[0\,;\infty)} = x$, i.e., $\sigma^{q-p} y = x$. This contradicts the assumption that x is not in the orbit of y. Similarly, the assumption that $p \geq q$ leads to a contradiction with the assumption that y is not in the orbit of x. These contradictions show that the above procedure stops after a finite number of steps. □

Theorem 5.6.12. *The system* (B, σ) *is minimal and weakly mixing.*

Proof. That the system (B, σ) is minimal was already observed in Proposition 5.6.7. It follows that σ maps B onto B. Consequently, $\sigma \times \sigma$ is a surjection of $B \times B$ onto itself. Hence the point with dense orbit in $B \times B$ under $\sigma \times \sigma$ is transitive: see Proposition 1.3.2. But then Theorem 1.3.5 implies that the system $(B \times B, \sigma \times \sigma)$ is topologically ergodic, that is, (B, σ) is weakly mixing. □

5.6.13 (A Toeplitz system). If in 5.6.1 (1) all blocks $w^{(n)}$ have equal length then the positions at which any initial block of x occurs in x form an arithmetic progression. In that case the sequence of coordinates of x is called a *Toeplitz sequence* and its orbit closure in (Ω_S, σ) – which is minimal, because the point x is almost periodic – is called a *Toeplitz system*. If x is a Toeplitz sequence then for every neighbourhood U of x the set $D(x, U)$ includes a set of the form $m\mathbb{Z}^+$ for some $m \in \mathbb{N}$; a point (in any system) with this property is called a *regularly almost periodic point* (also: a quasi-periodic point).

Example in Ω_2: let $\alpha = 0$, $w^{(n)} = 1$ if n is even and $w^{(n)} = 0$ if n is odd (mutually different in order to get a non-periodic sequence) we obtain a point τ with the following sequence of coordinates:

$$\begin{array}{cccccccccc}
0 & 1\;0\;0 & 0\,1\,0 & 1 & 010\;0\;010 & 0 & 010\;0\;010 & 1 & 010\;0\;010 & 1\;\ldots\ldots\\
\uparrow & \uparrow & & \uparrow & & \uparrow & & & & \uparrow\\
w^{(0)} & w^{(1)} & & w^{(2)} & & w^{(3)} & & & & w^{(4)}
\end{array}$$

Let T be the orbit closure of the point τ in the shift system (Ω_2, σ). The system (T, τ) is an example of *a Toeplitz system*. By the above, the point τ is (regularly) almost periodic; consequently, the system (T, σ) is minimal.

14 This argument is similar to the one used in the proof of Statement (5).

Theorem 5.6.14. *The Toeplitz system (T, σ_T) has an equicontinuous factor, namely, the adding machine (G, f). Consequently, the Toeplitz system (T, σ_T) is not weakly mixing.*

Proof (outline). The metric used for T is the restriction to T of the metric of Ω_2, and the metric used for G is the metric of Ω_2 as well; both will be denoted by d. Denote the point 0^∞ of G by $\underline{0}$.

Define φ on the orbit of the point τ by $\varphi(\sigma^k \tau) := f^k(\underline{0})$ for every $k \in \mathbb{Z}^+$. The point τ is not ultimately periodic (see below), so φ is well defined. Moreover, φ turns out to be uniformly continuous on $\mathcal{O}_\sigma(\tau)$. As G is a complete metric space, it follows that φ has a continuous extension to all of the closure T of $\mathcal{O}_\sigma(\tau)$, which we shall also denote by φ. Obviously, φ maps T onto G (for $\varphi[T]$ is compact, hence closed, and $\varphi[T]$ includes the dense orbit of $\underline{0}$) and since the identity $\varphi \circ \sigma = f \circ \varphi$ holds on the dense subset $\mathcal{O}_\sigma(\tau)$ of T, it holds on all of T. Thus, $\varphi\colon (T, \sigma) \to (G, f)$ is a factor map. In what follows we provide some hints for intermediate steps in the above proof.

(a) Define inductively a sequence of blocks $(t(n))_{n\in\mathbb{Z}^+}$ by means of the following rules:

$$t(0) := 0 \quad \text{and} \quad t(n+1) := \begin{cases} t(n)\,0\,t(n) & \text{if } n \text{ is odd} \\ t(n)\,1\,t(n) & \text{if } n \text{ is even} \end{cases} \qquad (n \in \mathbb{Z}^+)\,.$$

Note that $|t(n+1)| = 2|t(n)| + 1$, hence $|t(n)| = 2^{n+1} - 1$ $(n \in \mathbb{Z}^+)$.
Since for every $k, n \in \mathbb{Z}^+$ with $k > n$, the block $t(k)$ starts with the block $t(n)$, there exists a (unique) point in Ω_2 that starts, for every $n \in \mathbb{Z}^+$, with the block $t(n)$. It is not too difficult to verify that this point is equal to the point τ defined in 5.6.13 above.

(b) If $k, n \in \mathbb{Z}^+$ and $k \le n$ then $t(n) = t(k)\, s_1\, t(k)\, s_2 \dots s_{a(k,n)}\, t(k)$, where the blocks $t(k)$ occur at the positions $i \cdot 2^{k+1}$ for $i = 0, 1, \dots, a(k, n)$ and $s_i \in \{0, 1\}$ for $i = 1, \dots, a(k, n)$; here $a(k, n) := 2^{n-k} - 1$. (These positions of the blocks $t(k)$ will be called the *natural positions* of these blocks; we shall also call these occurrences of $t(k)$ the *natural occurrences* of these blocks. The coordinates s_i will be called *separators*.) Prove this by induction in n for fixed k.
In particular, the block $t(n)$ starts and ends with the block $t(k)$. Moreover, if k is odd then $s_1 = 0$ and $s_2 = 1$; if k is even then $s_1 = 1$ and $s_2 = 0$. Consequently, if $n \ge k + 2$ then the blocks $t(k)\,0\,t(k)$ and $t(k)\,1\,t(k)$ both occur in $t(n)$.

NB. There may be other positions at which the block $t(k)$ occurs in $t(n)$. For example, the block $t(1) = 010$ occurs in $t(3) = (010\ 0\ 010)\,1\,(010\ 0\ 010)$ not only at the natural positions 0, 4, 8 and 12, but also at position 6.

(c) As was remarked earlier, T is a minimal subshift of Ω_2. The point τ is not isolated in T: for every $k \in \mathbb{Z}^+$ the point $\tau^{(k)} := t(k)\,0\,\tau$ is in T and $\tau^{(k)} \rightsquigarrow \tau$ for $k \rightsquigarrow \infty$. In particular, the point τ is not ultimately periodic. In order to prove this, first note that $\tau^{(k)} \ne \tau$ if k is even. Next, observe that every initial block of $\tau^{(k)}$ occurs in τ, so $\tau^{(k)} \in T$ for all k.

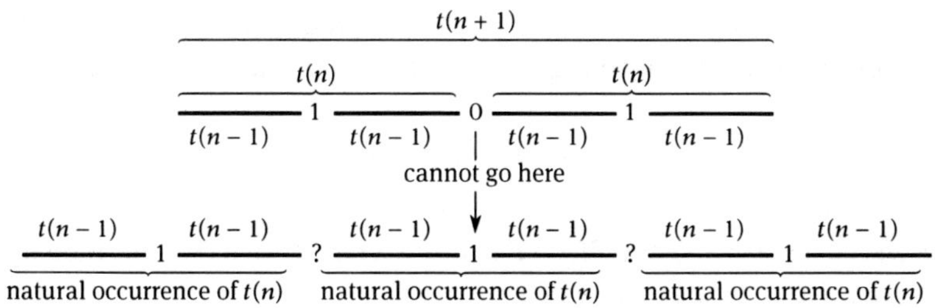

Fig. 5.7. All blocks $t(n-1)$ are in natural position, so the central separator 0 in $t(n+1)$ can go only to a question mark. Hence each of the blocks $t(n)$ induced by the occurrence of $t(n+1)$ coincides with a natural occurrence of $t(n)$.

(d) For every $k \in \mathbb{Z}^+$ the sequence of coordinates of τ consists of copies of the block $t(k)$ at their natural positions $i \cdot 2^{k+1}$ for $i \in \mathbb{Z}^+$, separated by 0's or 1's at positions $i \cdot 2^{k+1} - 1$ for $i \in \mathbb{N}$.
(e) Let $n \in \mathbb{N}$. Each occurrence of the block $t(n)$ in τ, say, at position p, induces two occurrences of $t(n-1)$ in τ, namely, at the positions p and $p + 2^n$. Then for every $n \in \mathbb{N}$ these occurrences of the block $t(n-1)$ are in their natural positions (even though $t(n)$ may be not in a natural position). The proof is by induction. For $n = 1$: all occurrences of 1 in τ have an odd position, hence the two 0's in every occurrence of $t(1)$ occur at even positions in τ. Assume that the statement is true for some $n \in \mathbb{N}$ and let n be odd (for even n the proof is similar). Any occurrences of the block $t(n+1)$ induces four occurrences of the block $t(n-1)$ which are, by the induction hypothesis, in natural position. Now see Figure 5.7.
(f) If $k < l$ and $\sigma^k\tau$ and $\sigma^l\tau$ have an initial block of length $3|t(n)|$ in common for some $n \in \mathbb{Z}^+$ then $l - k \in 2^n\mathbb{Z}^+$. For the points $\sigma^k\tau$ and $\sigma^l\tau$ have a copy of $t(n)$ in the same position, so by (e), the natural occurrences of $t(n-1)$ in $\sigma^k\tau$ have the same positions as the natural occurrences of $t(n)$ in $\sigma^l\tau$. Hence $l - k$ is a multiple of $|t(n-1)| + 1$.
(g) By the result of (c), the mapping $\varphi\colon \mathcal{O}_\sigma(\tau) \to G$ is unambiguously defined by $\varphi(\sigma^n\tau) := f^n(\underline{0})$ for $n \in \mathbb{Z}^+$. Then φ is uniformly continuous on $\mathcal{O}_\sigma(\tau)$. For by (f), if $d(\sigma^k\tau, \sigma^l\tau) < (1/3) \cdot (1/2^{n+1})$ then the coordinate sequences of $f^k(\underline{0})$ and $f^l(\underline{0})$ differ by $0^n w^{(n)} 0^\infty$ for some sequence $w^{(n)}$ of 0's and 1's, hence they have an initial block of length n in common, that is, $d(f^k(\underline{0}), f^k(\underline{0})) < 1/(n+1)$. □

Exercises

5.1.

(1) Prove that $\Omega_S = \bigcup\{C_0[b] : b \in S^m\}$ for all $m \in \mathbb{N}$. Consequently: if $\mathcal{T}$ is any topology on Ω_S and all cylinders are open with respect to $\mathcal{T}$ then they are also closed with respect to $\mathcal{T}$. Conversely, if all cylinders are closed then they are open as well.

(2) According to the well-know result that a countable product of metric spaces is metrizable, the product topology on Ω_S is generated by the metric ρ, defined by

$$\rho(x, y) = \sum_{n=0}^{\infty} \frac{1}{2^n} \delta(x_n, y_n) \quad \text{for all } x, y \in \Omega_S ,$$

where $\delta(\alpha, \beta) = 0$ if $\alpha \neq \beta$ and $\delta(\alpha, \beta) = 1$ if $\alpha = \beta$ (Kronecker-delta). Show directly that the metrics d and ρ are equivalent.

(3) Let $\{ x^{(j)} : j \in \mathbb{N} \}$ be a a sequence in Ω_S. Show that it converges in Ω_S with limit x iff for every $k \in \mathbb{N}$ there is $N_k > 0$ such that $x^{(j)}_{[0;k)} = x_{[0;k)}$ for all $j \geq N_k$.

(4) Prove that Ω_S is compact with the metric d defined in 5.1.5 by showing that every sequence in Ω_S has a convergent subsequence.

5.2.

(1) Prove the continuity of σ by showing that $d(\sigma x, \sigma y) \leq 2d(x, y)$ for all $x, y \in \Omega_S$.

(2) Recall the definition of the adding machine in 4.2.8 and note that the space G defined there is nothing but Ω_2. Let $f : G \to G$ be the mapping defined in 4.2.8. Show that $d(f(x), f(y)) = d(x, y)$ for all $x, y \in G$, i.e., that f is an isometry (so the set of mappings $\{f^n : n \in \mathbb{Z}^+\}$ is uniformly equicontinuous on G).

(3) Show that for every $k \in \mathbb{N}$ the family $\{ C_0[b] : b \in S^k \}$ is a clopen partition of Ω_S and that σ^k maps each of its members homeomorphically onto Ω_S.

(4) Prove: for every $k \in \mathbb{N}$ there exists a point in Ω_S that has a dense orbit under σ^k.

5.3.

(1) Consider the point $x := b^{(1)}b^{(2)}b^{(3)} \cdots \in \Omega_2$, where $b^{(k)} := 0^k 1^k$ $(k \in \mathbb{N})$. Show that $\omega_\sigma(x) = \{0^k 1^\infty : k \in \mathbb{Z}^+\} \cup \{1^k 0^\infty : k \in \mathbb{Z}^+\}$. Define an invariant subset X of Ω_2 by $X := \Omega_2 \setminus \bigcup_{k \in \mathbb{Z}^+} (\sigma^k)^{\leftarrow}[0^\infty]$ and consider the dynamical system (X, σ_X) (not a shift system, as X is not closed). Determine the limit set $\omega_{\sigma_X}(x)$ in the system (X, σ_X) and show that it is not true that $\lim_{n \rightsquigarrow \infty} d(\sigma^n(x), \omega(x)) = 0$; compare this with Theorem 3.1.9.

(2) Let $P(\sigma)$ be the set of all points in Ω_S that are periodic under σ. Obviously, the subsystem $(P(\sigma), \sigma)$ of (Ω_S, σ) is not transitive; show that it is topologically ergodic (compare this with Theorem 1.3.5).

(3) Let X be a shift space. A point $x \in X$ has a dense orbit in X iff every element of $\mathcal{L}(X)$ occurs in x, iff $\mathcal{L}(X)$ is the set of all blocks occurring in x. A point $x \in X$ is transitive in X iff every element of $\mathcal{L}(X)$ occurs infinitely often in x.

5.4. Let $\mathcal{B}$ be a set of blocks over S, let $\mathcal{L}$ be the language of the shift space $\mathfrak{X}(\mathcal{B})$ and let $\mathcal{B}^c := S^* \setminus \mathcal{B}$, where $S^* := \bigcup_{n=0}^{\infty} S^n$. The following statements are equivalent[15] :

(i) $\mathcal{B}$ is equal to the set of all absent blocks of the shift space $\mathfrak{X}(\mathcal{B})$.

(ii) $\mathcal{L} = \mathcal{B}^c$.

---

15 In (iii)(b) en (iv)(b) "symbol $a \in S$" can be replaced by "block $c \in S^*$".

(iii) $\mathcal{B}$ satisfies the following two conditions:
   (a) If $b \in \mathcal{B}$ and b occurs in a block c then $c \in \mathcal{B}$.
   (b) If $b \in \mathbf{S}^*$ and $b\alpha \in \mathcal{B}$ for every symbol $\alpha \in \mathbf{S}$ then $b \in \mathcal{B}$.

(iv) $\mathcal{B}^c$ satisfies the following two conditions:
   (a) If $b \in \mathcal{B}^c$ and a block c occurs in b then $c \in \mathcal{B}^c$.
   (b) If $b \in \mathcal{B}^c$ then there is a symbol $\alpha \in \mathbf{S}$ such that $b\alpha \in \mathcal{B}^c$.

If these conditions are fulfilled and it is not true that $\mathbf{S} \subseteq \mathcal{B}$ – that is, not all blocks consisting of one symbol are forbidden – then $\mathfrak{X}(\mathcal{B}) \neq \emptyset$.

5.5.
(1) Show that the phase spaces of the golden mean shift, the even shift, the prime gap shift, the (1,3) run-length limited shift and the context free shift have no isolated points, hence are Cantor spaces. Prove the same for the Toeplitz system.
(2) If X and Y are SFT's over the same symbol set and $X \cap Y \neq \emptyset$ then $X \cap Y$ is an SFT as well. Give an example showing that $X \cup Y$ need not be an SFT.

5.6.
(1) Prove that the full shift systems $(\Omega_\mathbf{S}, \sigma)$ and $(\Omega_\mathbf{T}, \sigma)$ are conjugate iff the symbol sets $\mathbf{S}$ and $\mathbf{T}$ have the same number of elements.
(2) Let X and Y be shift spaces such that $X \subseteq Y \subseteq \Omega_\mathbf{S}$. Show that every morphism of dynamical systems $\varphi\colon (X, \sigma_X) \to (\Omega_\mathbf{T}, \sigma)$ can be extended to a morphism $\psi\colon (Y, \sigma_Y) \to (\Omega_\mathbf{T}, \sigma)$.
(3) The identity mapping from the even shift onto itself cannot be extended to a morphism of the full shift Ω_2 to the even shift (so in 2 above, the target space $\Omega_\mathbf{T}$ cannot be replaced by an arbitrary shift space).
(4) Let X be a shift space and assume that σ_X is injective. Prove formula (5.3-1) using the Curtis–Lyndon–Hedlund Theorem 5.4.3.

5.7. Show that every SFT includes a periodic point.

5.8. Let G be a directed graph, let A be its adjacency matrix, and consider any $m \in \mathbb{N}$.
(1) The entry $(A^m)_{ij}$ of the matrix A^m is equal to the number of paths of length m (the number of edges in the path) from vertex i to vertex j.
(2) The trace $\mathrm{Tr}(A^m)$ of A^m is equal to the number of cycles[16] in G. If G is a faithfully vertex-labelled directed graph then $\mathrm{Tr}(A^m)$ is equal to the number of periodic points with period m in the shift space $\mathfrak{M}_v(G)$.
(3) The number of periodic points with period m in the golden mean shift is equal to $a_m + a_{m+2}$, where $(a_m)_{m\in\mathbb{N}}$ is the Fibonacci sequence.

   NB. For another application of 1 above see Exercise 5.9 (2).

16 Recall that cycle or loop is a path that begins end ends in the same vertex. If such a path consists of m edges then there are m vertices that can be taken as begin/end point and therefore this path counts as m cycles.

5.9.

(1) Prove that a directed graph G has a unique subgraph H without stranded edges such that $\mathfrak{M}_v(G) = \mathfrak{M}_v(H)$. Actually, H is the maximal subgraph of G without stranded edges and $H = \mathbf{G}_s(\mathfrak{M}_v(G))$. As an example, see the following picture, where the maximal subgraph without stranded edges is indicated by the heavy arrows.

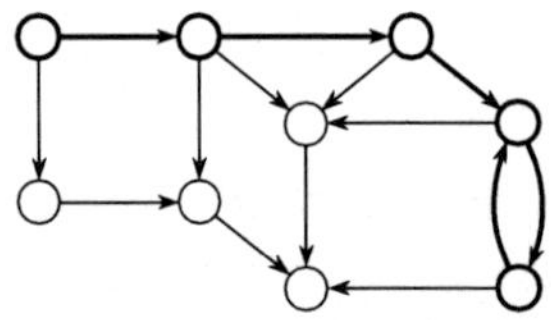

Fig. 5.8. A graph with stranded edges.

(2) Let G be a faithfully vertex-labelled graph without stranded edges or isolated vertices. The following conditions are equivalent:
  (i) The shift space $\mathfrak{M}_v(G)$ is irreducible.
  (ii) G is strongly connected, i.e., for every ordered pair of vertices there is a path from the first vertex to the second one.
  (iii) The adjacency matrix A of G is irreducible, that is, if $1 \leq i, j \leq s$ then there is $n \in \mathbb{N}$ such that $(A^n)_{ij} > 0$.

(3) Irreducibility of a subshift X and irreducibility of the mapping $\sigma_X : X \to X$ are unrelated.

(4) An irreducible SFT has a dense set of periodic points.

(5) The following shift spaces, though not of finite type, have dense sets of periodic points: the even shift, the prime gap shift and the context free shift.

5.10. Let $\varphi : X \to Y$ a factor map of the shift spaces X and Y over the symbol sets $\mathbf{S}$ and $\mathbf{T}$, respectively.

(1) Suppose $X = \mathfrak{M}_v(G)$ for some faithfully vertex-labelled graph G and that φ is a 2-block code, generated by $\Phi : \mathbf{S}^2 \to \mathbf{T}$, i.e., $\varphi = \Phi^2_\infty$. Let H be the edge-labelled graph that is obtained in the following way: as an unlabelled directed graph, $H = G$; if an edge in G goes from a vertex with label α to a vertex with label β then as an edge of H it gets the label $\Phi(\alpha\beta)$. Show that $Y = \mathfrak{W}_e(H)$.
Example: Compare the factor map in Example (5) just before Proposition 5.4.1 with the edge-labelled graph for the even shift in Figure 5.3 (b).

(2) Suppose $X = \mathfrak{W}_e(G)$ for some edge-labelled graph G and that φ is a 1-block code, generated by $\Phi : \mathbf{S} \to \mathbf{T}$, i.e., $\varphi = \Phi^{(1)}_\infty$. Let H be the edge-labelled graph that is obtained in the following way: as an unlabelled directed graph, $H = G$; if an edge in G has label $\alpha \in \mathbf{S}$ then as an edge of H it gets the label $\Phi(\alpha)$. Show that $Y = \mathfrak{W}_e(H)$.

5.11.

(1) Let G be the edge-labelled graph denoted in Figure 5.9 (a). Then $\mathfrak{W}_e(G)$ is not an SFT (so the restricted shift mapping $\sigma_{\mathfrak{W}_e(G)} : \mathfrak{W}_e(G) \to \mathfrak{W}_e(G)$ is not open).

(2) Prove: if X is the (sofic) even shift, or the (non-sofic) prime gap shift then the mapping $\sigma_X : X \to X$ is is semi-open and it is open at every point of $X \setminus \{10^\infty\}$.

(3) Let G be the edge-labelled graph denoted in Figure 5.9 (b). Then $\mathfrak{W}_e(G)$ is a sofic shift space, but on it the shift mapping is not semi-open. (With 2 above this shows that 'semi-open' does not characterize sofic shifts.)

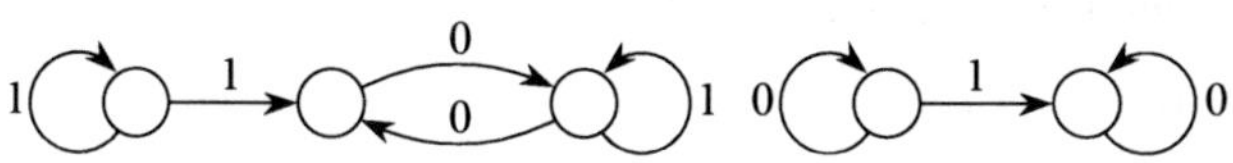

(a) (b)

Fig. 5.9. Two sofic shifts none of which is an SFT.

5.12. Let $\rho \in \Omega_S$ be the point defined and discussed in 5.6.1 (3).

(1) Show that $\bigcap\{D(\rho, U) : U \in \mathcal{N}_\rho\} = \{0\}$ and that the gaps in $D(\rho, U)$ are not bounded (i.e., the point ρ is not almost periodic).

(2) Show that 1^∞ is the only periodic point in $\overline{\mathcal{O}_\sigma(\rho)}$.

5.13.

(1) The golden mean shift, the even shift, the prime gap shift, the (1,3) run-length limited shift are strongly mixing. Similarly, the context free shift is strongly mixing.

NB. A sufficient (but certainly not necessary) condition for an irreducible SFT of order 2 to be strongly mixing is that its graph contains a loop (an edge that starts and ends in the same vertex).

(2) Let P be a subset of $\mathbb{Z}^+$ with $0 \in P$ and let

$$\mathcal{B}_P := \{ b \in \{0,1\}^* : b_i = b_j = 1 \text{ implies } |i - j| \in P \}$$

(words in which the occurrences of 1 have prescribed distances).
Let

$$X_P := \mathfrak{X}(\{0,1\}^* \setminus \mathcal{B}_P) = \{x \in \Omega_2 : \text{every subblock of } x \text{ is in } \mathcal{B}_P\}.$$

Since $X_P \neq \emptyset$ (for $0^\infty \in X_P$), X_P is a shift space.

(a) If the set $\mathbb{Z}^+ \setminus P$ is infinite then the system (X_P, σ) is not strongly mixing.

(b) Assume that P has the property that for every finite subset F of $\mathbb{Z}^+$ there exists $k \geq 0$ such that $k + F \subseteq P$. Then the system (X_P, σ) is weakly mixing.

Let P be the union of mutually disjoint intervals in $\mathbb{Z}^+$ of length 1, 2, 3, ..., taking care that between successive intervals there is at least one element of $\mathbb{Z}^+$. Then P is replete and P has an infinite complement (by choosing the intervals with gaps of lengths 1,2, 3, ... the complement of P will be replete as well). *For this choice of P, the system (X_P, σ) is weakly mixing but not strongly mixing.*

(3) Show that semi-Sturmian systems are not weakly mixing.

5.14. The following results show that the Morse–Thue system is not weakly mixing (hence not strongly mixing).

(1) Show that the block 0110 = $q(2)$ occurs in the Morse–Thue sequence μ only at even positions and conclude that for every point in M either all occurrences of $q(2)$ are at even positions or all occurrences of $q(2)$ are at odd positions.
(2) Let $M_1 := \{x \in M : q(2)$ occurs in x only at even positions$\}$. Then show that $M_1 \neq \emptyset$, that $M_1 \cap \sigma[M_1] = \emptyset$ and that $M = M_1 \cup \sigma[M_1]$.
(3) Show that M_1 is a clopen subset of M.
(4) The Morse–Thue system is not weakly mixing.

5.15. Let notation be as in 5.6.6– 5.6.12.

(1) Every point of B has a complete history represented by a universally spaced concatenation.
(2) Every ordinary point of B has a unique complete past in B.
(3) Every non-exceptional point of B has a unique complete past in B.
(4) The points β has no *unique* complete past, hence it is exceptional (i.e., not an non-exceptional point).

5.16.

(1) Let X be any shift space. The following are equivalent:
   (i) X is chain-transitive.
   (ii) For every $n \in \mathbb{N}$ and every two X-present n-blocks u and v of length n there exist $N \geq 1$ and an $(n + N)$-block $a_0 \dots a_{n+N-1}$ every n-subblock of which is X-present and such that $u = a_{[0\,;\,n)}$ and $v = a_{[N;n+N-1]}$.

   In particular, if these conditions are fulfilled then for every X-present block v there is a symbol α such that the block αv is X-present (in this connection, cf. Note 8).
(2) Let X be an SFT. Then X is chain-transitive iff X is transitive.
(3) A subshift is an SFT iff it has the pseudo-orbit tracing property (the definition is in Note 8 at the end of Chapter 4).

Notes

Preliminary remark on notation and terminology. Properly speaking, an element x of Ω_S is a function $x : \mathbb{Z}^+ \to S$ with value x_n at $n \in \mathbb{Z}^+$. Similarly, for $k \in \mathbb{Z}^+$ a block of length k is a function $b : \{0, \dots, k-1\} \to S$ whose value at $i \in \{0, \dots, k-1\}$ is denoted by b_i. Thus, a subblock of $x \in \Omega_S$ is the restriction of x to a subintervalof $\mathbb{Z}^+$. In particular, $x_{[k\,;\,l)}$ is the restriction of x to the subset $[k;l) \cap \mathbb{Z}^+$ of the domain of x. The chapter on 'Symbolic dynamics' of [GH] (Chapter C.) employs this function-and-restriction terminology. Our usage is in the spirit of D. Lind & B. Marcus [1995].

1 We shall not attempt to give a historical introduction to shift spaces, nor to give an overview over existing literature. Much information, including many references to the literature, can be found in G. A. Hedlund [1969]. A wealth of information can be found in D. Lind & B. Marcus [1995] and in B. Kitchens [1998].

2 The Morse–Thue sequence used to be called the Morse-sequence until it dawned on the mathematical community that this sequence had been studied by the Norwegian mathematician Thue in 1906, long before Morse defined it in 1921. Actually, the French mathematician Prouhet used the sequence already in 1851, but he did not write it down explicitly. This sequence was also discovered by non-mathematicians, e.g., by the Dutch chess grandmaster Max Euwe, who used the cube-free property (see below) for a problem in chess. The Danish composer Per Nørgård used the sequence in some of his compositions (e.g., in his third symphony and in 'Daktylos', a work for percussion). See also J.-P. Allouche & J. Shallit [1999].

Much more is known about the Morse–Thue sequence than we reproduced in our text. Of the many remarkable properties of the sequence we mention its self-similarity (which is also often referred to as its fractal structure):

(a) Delete all terms in μ at odd positions: then one gets μ again:

$$\begin{array}{ccccccccc} 0110 & 1001 & 1001 & 0110 & 1001 & 0110 & 0110 & 1001 & \dots \\ 0\cdot1\cdot & 1\cdot0\cdot & 1\cdot0\cdot & 0\cdot1\cdot & 1\cdot0\cdot & 0\cdot1\cdot & 0\cdot1\cdot & 1\cdot0\cdot & \dots \end{array}$$

(b) Replace every occurrence of 0 by a copy of $q(k)$ and every occurrence of 1 by a copy of $\overline{q(k)}$. The sequence so obtained is, again, μ; see [GH], 12.30(9). So the blocks $q(k)$ and $\overline{q(k)}$ occur in their natural positions in the same order as the 0's and 1's in μ.

The observation that a succession of three blocks $q(n)$ or $\overline{q(n)}$ does not occur in μ is a particular case of the fact that μ is *cube-free*: there is no subblock b of μ such that bbb occurs in μ (which obviously implies that μ is not eventually periodic). See M. Morse & G. A. Hedlund [1944]. The proof in Proposition 5.6.4 that μ is not (eventually) periodic under σ is 'dynamical' in that it uses the topology and some results from the theory of dynamical systems. It is inspired by a proof on p. 18 in the book J. Auslander [1988]. There is also a purely combinatorial proof, reproduced e.g. in [deV], III(2.29).

3 The proof of Proposition 5.3.2 is adapted from E. M. Coven & M. Keane [2006].

4 The characterization of SFTs in Proposition 5.3.12 comes from W. Parry [1966]. The characterization in Exercise 5.16 (3) is due to P. Walters [1997].

There is an extensive 'algebraic' theory of SFT's of order 2, where properties of these shift spaces are related to properties of the adjacency matrices of the corresponding graphs. For a simple example, see Exercise 5.8. In particular, the eigenvalues of such matrices play an important role, e.g., in the classification of SFT's of finite type (how to recognise conjugacy). For an overview of an important classification problem of SFT's in terms of their transition matrices, see J. B. Wagoner [2004].

5 Readers familiar with Markov chains will recognize in a vertex shift a Markov chain where all probabilities involved are either 0 or 1. This is why vertex shifts are also called *topological Markov chains*. This is also the reason why we denote the operator that maps faithfully vertex labelled directed graphs to SFT's of order 2 by $\mathfrak{M}_v$.

6 The techniques used in the proofs of Proposition 5.5.4 and Theorem 5.5.5 are less *ad hoc* than they might seem to be at first sight: they rely on some general constructions on (directed) graphs that are often used in the literature. For completeness we describe them briefly.

(a) Let G be a vertex-labelled directed graph. Assign to every edge of G the label of the vertex where the edge starts. In this way one gets an edge-labelled graph G'. It is obvious that $\mathfrak{W}_e(G') = \mathfrak{M}_v(G)$. If we apply this idea to the 'standard' vertex-labelled graph for the golden mean shift (see page 246) then we get the graph of Figure 5.5 (a). This construction is implicit in the proof of Theorem 5.5.5.

(b) Let G be a directed graph. Define a new graph $G^{\hat{}}$ whose set of vertices is the set of the edges of G. In $G^{\hat{}}$ there is an edge from vertex i to vertex j iff in G the edge i ends where the edge j begins. By this construction, a walk in G (finite or infinite) carries over to a walk in $G^{\hat{}}$, and vice versa. Now suppose that G is edge-labelled by a set $\mathbf{S}$. This labelling of G is carried over to a vertex-labelling of $G^{\hat{}}$, faithful iff the original labelling is faithful. Obviously, an element of Ω_S is edge-represented by a walk in G iff the corresponding walk in $G^{\hat{}}$ vertex-represents the same element of Ω_S, hence $\mathfrak{M}_v(G^{\hat{}}) = \mathfrak{W}_e(G)$. If G is the edge-labelled graph in Figure 5.5 (a) – for the golden mean shift – then $G^{\hat{}}$ is the graph at the bottom of page 246, with the label 2 replaced by a 1. This construction is implicit in the proof of Proposition 5.5.4.

The above remarks imply that every shift space obtained from a vertex-labelled graph can be obtained from an edge-labelled graph, and vice versa. But the constructions also suggest that vertex-labelled graphs tend to be less economic. Therefore in D. Lind & B. Marcus [1995] the stress is on edge-labelled graphs. The reason that we nevertheless spend some attention to the representation of SFT's of order 2 by vertex-labelled graphs is that there is much literature using this as a starting point. Moreover, in many articles the representation techniques are taken for granted so that for an inexperienced reader it is not always clear which definition is used. In fact, the main reason for including Section 5.5 in the present book is to clarify the relationship and to explain the differences between the various methods.
7 The main ingredient of the proof of Theorem 5.6.14 is a special case of a more general result: a minimal system has a regularly almost periodic point iff it is an almost 1,1-extension of an adding machine. See N. Markley & G. M. E. Paul [1979].

A popular way to construct the Toeplitz sequence considered in 5.6.13 is as follows: write alternatingly 0's and 'blanks'; then fill in the blanks alternatingly with 1's and blanks; then fill in the remaining blanks alternatingly with 0's and blanks, and so on, and so forth. See the following scheme, where (in the notation of 5.6.1 (1) the first row represents the occurrences of the α's, the second row the blocks $w^{(0)}$, the third row the blocks $w^{(1)}$, etc.

$$
\begin{array}{l}
0 - 0 - 0 - 0 - 0 - 0 - 0 - 0 - 0 - 0 - 0 - 0 - 0 - 0 - \ldots\ldots \\
\;\;1 \;\;\;-\;\;\; 1 \;\;\;-\;\;\; 1 \;\;\;-\;\;\; 1 \;\;\;-\;\;\; 1 \;\;\;-\;\;\; 1 \;\;\;-\;\;\; 1 \;\;\;-\ \ldots\ldots \\
\;\;\;\;\;\;0 \;\;\;\;\;\;\;\;-\;\;\;\;\;\;\;\; 0 \;\;\;\;\;\;\;\;-\;\;\;\;\;\;\;\; 0 \;\;\;\;\;\;\;\;-\;\;\;\;\;\;\;\; 0\ \ldots\ldots
\end{array}
$$

After some reflection one sees that

$$
x_j = \begin{cases} 0 & \text{for } j = 4^k(2n+1) - 1 \\ 1 & \text{for } j = 4^k(4n+2) - 1 \end{cases} \quad (k, n \in \mathbb{Z}^+)\,.
$$

8 Invertible vs. non-invertible systems. In this chapter we have adapted the theory of the bilateral (invertible two-sided) shifts to the non-invertible one-sided shifts. Basically, we did so by replacing $\mathbb{Z}$ by $\mathbb{Z}^+$ in the definitions. To get the bilateral theory back one has to proceed the other way round. For full details of the theory of two-sided shifts we refer to Chapter 12 of [GH], to D. Lind & B. Marcus [1995] and the to Sections 2 and 5 in Chapter III of [deV]. In this theory one considers the invertible system (Σ_S, σ), where S is a finite set (the symbol set). Its phase space is $\Sigma_S := S^{\mathbb{Z}}$, the elements of which are the two-sided infinite sequences $x = \ldots x_{-2}\, x_{-1}\, x_0\, x_1 \ldots$ with $x_i \in S$ for all $i \in \mathbb{Z}$. This is a Cantor space, and a suitablemetric on Σ_S is given by the formulas $d(x, y) := 0$ if $x = y$ and $d(x, y) := (1+k)^{-1}$ if $x \neq y$ and k is the largest non-negative integer such that $x_{(-k\,;\,k)} = y_{(-k\,;\,k)}$. The phase mapping σ of the system is given by $\sigma(x)_n := x_{n+1}$ for $n \in \mathbb{Z}$ and $x \in \Sigma_S$. We shall call the system (Σ_S, σ) the *two-sided shift system*. (Two-sided) shift spaces are the non-empty closed invariant sets in this system.

Most modifications of our definitions for the one-sided shift to definitions for two-sided shifts are obvious. For example, the two-sided Morse–Thue sequence is defined as the (right-infinite) sequence defined in 5.6.2, preceded by its (left-infinite) mirror image; see Chapter 12 in [GH].

The two-sided topological version of Chacón's system is discussed in A. del Junco [1982], parts of which are included in [deV] III(2.44)–III(2.52). There it is shown that this bilateral system, in which the shift is a homeomorphism is a minimal weakly mixing system. By Note 12 to Chapter 1, this system is then positively minimal and positively weakly mixing. Stated otherwise, if we consider this as a dynamical system as studied in this book (as a semi-dynamical system, in the terminology of Note

12 to Chapter 1) then we have a system that is minimal and weakly mixing in the sense as defined in Chapter 1. Then it is straightforward to show that this system remains minimal and weakly mixing if we modify its phase space by considering only the right-sequences $x_{[0\,;\infty)}$ for x in the two-sided system. We have avoided this detour by the introduction of bi-sequences and complete histories.

In two-sided infinite shifts, sliding block codes can also have, apart from anticipation k, positive *memory* m in: $\varphi(x)_n := \Phi(x_{[n-m\,;\,n+k)})$ for all $n \in \mathbb{Z}$ (X a shiftspace, $x \in X$, $m, n \in \mathbb{Z}^+$, and Φ a mapping from $\mathcal{L}_{m+k}(X)$ to $\mathbf{S}$); in that case φ is denoted by $\Phi_\infty^{m,k}$. Finally, we mention that in Section 5.5 one should consider two-sided infinite paths in the graphs under consideration.

Most of the results in this chapter carry over without much modification in formulation or proof to the two-sided case. The most important difference is that $\sigma : \Sigma_\mathbf{S} \to \Sigma_\mathbf{S}$ is a homeomorphism instead of a continuous s-to-1 mapping (where s is the number of elements of $\mathbf{S}$). Moreover, the two-sided shift system is bilaterally transitive, hence ergodic (even strongly mixing) and non-wandering. We shall not go into details about the similarity between the two-sided and the one-sided shift systems, but we mention two minor differences. First, Proposition 5.3.2 is a typically one-sided result (see also the remark in Note 9 in Chapter 6 concerning positively expansive mappings). Moreover, Proposition 5.3.7 reads in the two-sided version: a subshift is irreducible iff it is ergodic and non-wandering (i.e., ergodic without isolated points: see also Note 9 in Chapter 4). Of course, the 'only if' in Proposition 5.3.12 is correct but trivial and the 'if' is obviously false for invertible SFT's.

9 We must warn the reader here that our definition of a one-sided shift space in Section 5.3 differs slightly from the definition in D. Lind & B. Marcus [1995]: there a one-sided shift space is defined as the set $X^+ := \{\, x_{[0\,;\infty)} \vdots x \in X \,\}$, where X is a two-sided shift space. If one employs graphs to define such a one-sided shift space then one considers only infinite walks that are part of (can be extended to) a two-sided infinite walk; in particular, vertices from which edges depart but where no edge ends do not occur in such a walk. Most of the theory as we have presented it is not affected by this discrepancy. Some examples might be different: e.g., the third graph on page 246 would define the shift space $\{1^\infty\}$ instead of $\{01^\infty, 1^\infty\}$. The most notable difference is in Exercise 5.4: condition 5.4(iv)(b) has to be changed (and 5.4(iii)(b) has to be changed accordingly) into

(b′) If $b \in \mathcal{B}^c$ then there are symbols $\alpha, \beta \in \mathbf{S}$ such that $\beta b \alpha \in \mathcal{B}^c$.

Equivalently, a shift space X (according to our definition) is a one-sided shift according to this more restrictive definition iff $\sigma|_X$ maps X *onto* X.

6 Symbolic representations

Abstract. In this chapter we discuss a method of modelling certain dynamical systems by means of shift spaces. The idea is to use the geometry/topology of a the phase space dynamical system (X, f) to code orbits by sequences of symbols and to find in this way a shift system of which the original system is a factor. Such a shift system will be called a 'symbolic model' or a 'symbolic representation' of the system (X, f). One hopes that the symbolic model can be proved to have some useful properties which are carried over to (X, f) – and which would be difficult to prove otherwise. Useful geometric or topological features for the construction of a symbolic representation of (X, f) are the property that the mapping f is semi-open and 'expanding' in some sense, or that the space X is 0-dimensional. Applications are only given for one-dimensional systems.

6.1 Topological partitions

In this section (X, f) will always denote an arbitrary dynamical system, unless stated otherwise. From 6.1.7 on X will be assumed to be compact.

6.1.1. Let $\mathcal{P} := \{P_0, \dots, P_{s-1}\}$ be a partition of X into s pieces ($s \in \mathbb{N}$). We can get an idea of the orbit of a point of X by considering the sequence of pieces that are successively visited. If $x \in X$ then for every $n \in \mathbb{Z}^+$ the point $f^n(x)$ is situated in a unique member P_{z_n} of $\mathcal{P}$. In this way we get a sequence $z = z_0z_1z_2z_3\dots$ of elements from $\mathbf{S} := \{0, \dots, s-1\}$, i.e., an element of $\Omega_\mathbf{S}$. It is called the *itinerary* of the point x and denoted by $\iota(x)$. So by definition,

$$\forall n \in \mathbb{Z}^+ : \ f^n(x) \in P_{\iota(x)_n} . \tag{6.1-1}$$

Also in the case that the sets P_i do not cover X but are still disjoint it is possible that certain elements of X have an itinerary according to (6.1-1): a point $x \in X$ has an itinerary iff $f^n(x) \in P_0 \cup \dots \cup P_{s-1}$ for every $n \in \mathbb{Z}^+$, iff x belongs to the set

$$X^*(\mathcal{P}, f) := \bigcap_{n \in \mathbb{Z}^+} (f^n)^{\leftarrow}[P_0 \cup \dots \cup P_{s-1}] . \tag{6.1-2}$$

If this set is empty then no point of X has an itinerary. If $x \in X^*(\mathcal{P}, f)$ then $f^n(f(x)) = f^{n+1}(x) \in P_{\iota(x)_{n+1}}$ for every $n \in \mathbb{Z}^+$, which clearly implies that $f(x) \in X^*(\mathcal{P}, f)$ and that the itinerary of $f(x)$ is obtained from the itinerary of x by applying the shift operator to it. So the set $X^*(\mathcal{P}, f)$ is invariant under f and on this set the equality $\sigma \circ \iota = \iota \circ f$ holds. In particular, it follows that the set of all itineraries of points of X – a subset of $\Omega_\mathbf{S}$ – is invariant under σ.

Though itineraries look promising, this approach has serious drawbacks. First of all, ι is not defined on all of X, unless $\mathcal{P}$ covers X, i.e., $\mathcal{P}$ is a partition. But in that case, $\iota[X]$ may be not a shift space, not even if X is compact. See Exercise 6.1 (2) for an example. Moreover, in the case that ι is defined on all of X the mapping $\iota\colon X \to \Omega_{\mathbf{S}}$ is not continuous, unless $\mathcal{P}$ is a clopen partition. See Exercise 6.1 (3). However, the condition that $\mathcal{P}$ is a clopen partition puts a heavy restriction on the class of systems for which itineraries might turn out to be useful. Even if $\iota\colon X \to \Omega_{\mathbf{S}}$ is well-defined and continuous then this often provides no additional information. For experience – see Note 2 – learns that mapping or embedding a system into a 'nice' system almost never reveals hitherto unknown dynamical properties of the original system. On the other hand, it often turns out to be very useful to know that there is a factor mapping of a 'nice' system onto the system under consideration. Consequently, our attention will be devoted to the construction of a factor mapping from a suitable shift space onto a given dynamical system (but itineraries will be playing a role). The above also suggest that genuine partitions are not really what we need, so we shall consider so-called 'topological partitions'.

6.1.2. A *topological partition* of X is a finite family $\mathcal{P} = \{ P_0, \dots, P_{s-1} \}$ of mutually disjoint non-empty open subsets of X whose closures cover X:

$$P_i \cap P_j = \emptyset \quad \text{for } i \neq j \ (i, j = 0, \dots, s-1) \quad \text{and}$$
$$X = \overline{P_0} \cup \dots \cup \overline{P_{s-1}} = \overline{P_0 \cup \dots \cup P_{s-1}}\,.$$

So a topological partition has in common with a genuine partition that it consists of mutually disjoint sets. However, the union of these sets may not be equal to X, but it has to be dense in X.

Let $\mathcal{P} = \{ P_0, \dots, P_{s-1} \}$ be a topological partition of X and, in accordance With 6.1.1 above, let $\mathbf{S} := \{ 0, \dots, s-1 \}$. In addition, let

$$U_{\mathcal{P}} := \bigcup \mathcal{P} = P_0 \cup \dots \cup P_{s-1}\,.$$

Then $U_{\mathcal{P}}$ is a dense open subset of X.

An itinerary as defined in 6.1.1 with respect to $\mathcal{P}$ will be called a *full itinerary*. Recapitulating, a point $x \in X$ has a full itinerary $\iota(x) \in \Omega_{\mathbf{S}}$ whenever $f^n(x) \in U_{\mathcal{P}}$ for every $n \in \mathbb{Z}^+$, in which case $\iota(x)$ is characterized by condition (6.1-1). As in 6.1.1, the set of points having a full itinerary is denoted by $X^*(\mathcal{P}, f)$, so $X^*(\mathcal{P}, f) = \bigcap_{n=0}^{\infty} (f^n)^{\leftarrow}[U_{\mathcal{P}}]$; see also formula (6.1-2). Note that $X^*(\mathcal{P}, f)$ is the intersection of countably many open sets: a G_δ-set. When $\mathcal{P}$ and f are understood the set $X^*(\mathcal{P}, f)$ will be denoted simply by X^*. If $x \in X^*$ then its full itinerary $\iota(x)$ is the element of $\Omega_{\mathbf{S}}$ characterized by formula (6.1-1), which can be rewritten as

$$x \in \bigcap_{n=0}^{\infty} (f^n)^{\leftarrow}[P_{\iota(x)_n}]\,. \tag{6.1-3}$$

Apart from full itineraries we shall also consider *partial* itineraries. A finite block $b = b_0 \dots b_{k-1}$ over $\mathbf{S}$ is said to be a *partial itinerary* of a point $x \in X$ whenever $f^n(x) \in P_{b_n}$

for $0 \le n \le k-1$. Thus, if for every $k \in \mathbb{N}$ and every block b of length k we define[1]

$$D_k(b) := \bigcap_{n=0}^{k-1} (f^n)^{\leftarrow}[P_{b_n}] \tag{6.1-4}$$

then the block b is a partial itinerary of the point $x \in X$ iff $x \in D_k(b)$. *In that case, the set $D_k(b)$ is an open neighbourhood of the point x.*

One more notational convention: if $z \in \Omega_S$ and $k \in \mathbb{N}$ then the clumsy expression $D_k(z_{[0\,;\,k)})$ will be simplified to $D_k(z)$; so

$$D_k(z) := D_k(z_{[0\,;\,k)}) = \bigcap_{n=0}^{k-1} (f^n)^{\leftarrow}[P_{z_n}] \tag{6.1-5}$$

Obviously, if $x \in X^*$ then every initial block of the full itinerary $\iota(x)$ is a partial itinerary of x. In fact, formula (6.1-3) implies that for all $k \in \mathbb{N}$

$$x \in \bigcap_{n=0}^{\infty} (f^n)^{\leftarrow}[P_{\iota(x)_n}] \subseteq \bigcap_{n=0}^{k-1} (f^n)^{\leftarrow}[P_{\iota(x)_n}] = D_k(\iota(x)), \tag{6.1-6}$$

Lemma 6.1.3. *Let notation be as above and assume that $X^* \neq \emptyset$.*
(1) *If $x \in X^*$ and $k \in \mathbb{N}$ then $D_k(\iota(x))$ is an open neighbourhood of x in X.*
(2) *X^* is an f-invariant G_δ-set in X.*
(3) *The mapping $\iota\colon X^* \to \Omega_S$ is continuous.*

Proof. (1) Clear from (6.1-6) and the observation following (6.1-4).
(2) That X^* is invariant was observed in 6.1.1 and that it is a G_δ-set was observed in 6.1.2.
(3) Consider a point $x \in X^*$, let $z := \iota(x)$ and consider the basic neighbourhood $\widetilde{B}_k(z)$ of the point z in Ω_S ($k \in \mathbb{N}$). By 1 above, $D_k(z)$ is a neighbourhood of the point x in X, so $D_k(z) \cap X^*$ is a neighbourhood of the point x in X^*. If $y \in D_k(z) \cap X^*$ then $\iota(y)$ is defined and for $0 \le n < k$ we have $f^n(y) \in P_{z_n}$, so $\iota(y)_n = z_n$. This means that $\iota[D_k(z) \cap X^*] \subseteq \widetilde{B}_k(z)$. This completes the proof that ι is continuous on X^*. □

6.1.4. For any topological partition $\mathcal{P} = \{P_0, \dots, P_{s-1}\}$ of X we define in the following way a shift space over the symbol set **S**. Call a block b over **S** of length $k \ge 1$ *forbidden* with respect to the pair $(\mathcal{P}, f)$, or just $(\mathcal{P}, f)$-*forbidden*, whenever $D_k(b) = \emptyset$. The set of $(\mathcal{P}, f)$-forbidden blocks will be denoted by $\mathcal{B}_{\mathcal{P},f}$. In accordance with 5.3.5 the elements of the set $\mathbf{S}^* \setminus \mathcal{B}_{\mathcal{P},f}$ are called the $(\mathcal{P}, f)$-*allowed* words or blocks. Thus, a finite block b is $(\mathcal{P}, f)$-allowed iff $D_k(b) \neq \emptyset$, iff there is a point $x \in X$ such that $f^n(x) \in P_{b_n}$ for $0 \le n \le k-1$, iff the block b is a partial itinerary of some point x of X.

Recall from Section 5.3 that the set $\mathcal{B}_{\mathcal{P},f}$ defines a subset $\mathfrak{X}(\mathcal{B}_{\mathcal{P},f})$ of Ω_S, which is a shift space if it is not empty. If this is the case we say that the topological partition $\mathcal{P}$

1 We should write $D(k, \mathcal{P}, f)(b)$ instead of $D_k(b)$, but $\mathcal{P}$ and f are always understood.

is *f-adapted* and we call this shift space the *symbolic model of* (X, f), *generated by* $\mathcal{P}$. It will be denoted by $Z(\mathcal{P}, f)$.

It follows from the definitions – see also formula (5.3-6) – that a point z of Ω_S belongs to $Z(\mathcal{P}, f)$ iff every block occurring in z is $(\mathcal{P}, f)$-allowed, iff every block occurring in z is a partial itinerary, iff every initial block of z is a partial itinerary, iff $D_k(z) \neq \emptyset$ for every $k \in \mathbb{N}$.

Remark. The discussion following Proposition 5.3.6 suggests that the set of all $(\mathcal{P}, f)$-allowed words can be strictly larger than the language of $Z(\mathcal{P}, f)$. See Exercise 6.2 for situations where one has equality. We give an example which shows that inequality can also be realized in the present setting. See Figure 6.1, where X is a closed interval, f is given by its graph and $\mathcal{P}$ consists of three adjacent open intervals P_0, P_1 and P_2.

By looking for partial itineraries we see that in example (a) all 2-blocks starting with the symbol 1 are forbidden and that 021 is the only allowed word of length 3. Consequently, every word of length four is forbidden. So $Z(\mathcal{P}, f) = \emptyset$ and the language of $Z(\mathcal{P}, f)$ is empty. Consequently, no allowed word is included in this language.

A similar reasoning shows that in example (b) the symbol 1 cannot occur in an element of $Z(\mathcal{P}, f)$. Hence the allowed block 01 is not in the language of $Z(\mathcal{P}, f)$. In this case, however, $Z(\mathcal{P}, f) \neq \emptyset$. In fact, by starting close enough near the point 0 one sees that arbitrarily long blocks of 0's are allowed. Hence $0^\infty \in Z(\mathcal{P}, f)$ Moreover, using the notation of Section 2.2, we see that $\overline{P}_0 \overset{\circ}{\rightarrow} \overline{P}_0 \cup \overline{P}_2$ and that $\overline{P}_2 \overset{\circ}{\rightarrow} \overline{P}_0$. So by the proof of Theorem 2.2.2, the set $\overline{P}_0 \cup \overline{P}_2$ contains a non-invariant periodic point of period 3 (actually, if $P_i := [i; i+1]$ for $i = 0, 1, 2$ and f has its summit at $1/2$ then the point $2/23$ has period 3); hence the system has periodic points of every (primitive) period. Obviously, these periodic points have a full itinerary and Proposition 6.1.5 (1) below implies that $Z(\mathcal{P}, f) \neq \emptyset$, as claimed. In point of fact, $Z(\mathcal{P}, f)$ is the golden mean shift, represented as the subset of $\Omega_{\{0,1,2\}}$ in which the symbol 1 does not occur and the symbol 2 occurs isolated: see Exercise 6.2 (2).

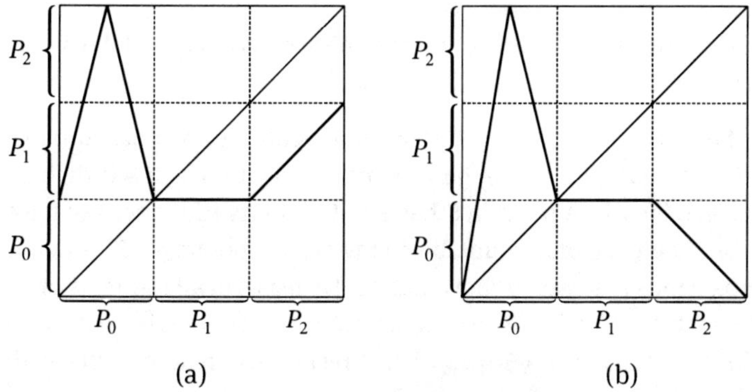

Fig. 6.1. In (a), $\mathfrak{X}(\mathcal{B}_{\mathcal{P},f}) = \emptyset$ so the shift space $Z(\mathcal{P}, f)$ does not exist. In (b) the shift space $Z(\mathcal{P}, f)$ exists, but not every $(\mathcal{P}, f)$-allowed word is in the language of $Z(\mathcal{P}, f)$.

Proposition 6.1.5. *Let notation be as above.*
(1) $\iota[X^*] \subseteq Z(\mathcal{P}, f)$; *in particular, if* $X^* \neq \emptyset$ *then* $\mathcal{P}$ *is* f*-adapted.*
(2) *If* X^* *is dense in* X *then* $\iota[X^*]$ *is dense in* $Z(\mathcal{P}, f)$.

Proof. (1) Consider a point $x \in X^*$ and let $z := \iota(x)$. Then (6.1-6) implies that for all $k \in \mathbb{N}$ one has $x \in D_k(z)$, hence $D_k(z) \neq \emptyset$. This implies that every initial block of z is $(\mathcal{P}, f)$-allowed, so $z \in Z(\mathcal{P}, f)$.
(2) Assume that X^* is dense in X and consider a point $z \in Z(\mathcal{P}, f)$. We want to show that for every $k \in \mathbb{Z}^+$ there exists a point x in X^* such that its itinerary $\iota(x)$ starts with the initial block $z_{[0;k)}$ of z. To this end, note that $z_{[0;k)}$ is a $(\mathcal{P}, f)$-allowed block, so the open set $D_k(z)$ is not empty. Consequently, it has a non-empty intersection with the dense set X^*. For any point $x \in X^* \cap D_k(z)$ one has $f^n(x) \in P_{z_n}$, which implies that $\iota(x)_n = z_n$, for $0 \leq n < k$. So the initial k-block of $\iota(x)$ is equal to $z_{[0;k)}$. □

Remark. If X^* is dense in X then by part 2 of the proposition, $Z(\mathcal{P}, f)$ is the closure of the set $\iota[X^*]$ in Ω_S.

The proof of 2 can easily be adapted so as to show: if every partial itinerary can be extended to a full itinerary then $\iota[X^*]$ is dense in the shift space $Z(\mathcal{P}, f)$. This condition implies that the language of $Z(\mathcal{P}, f)$ coincides with the set of all $(\mathcal{P}, f)$-allowed words. See also Exercise 6.5.

Theorem 6.1.6. *Assume that* X *is a Baire space and that* f *is a semi-open mapping. Then* X^* *is a dense* G_δ*-set in* X *and* $\iota[X^*]$ *is dense in* $Z(\mathcal{P}, f)$. *In particular, every topological partition* $\mathcal{P}$ *is* f*-adapted.*

Proof. In view of the preceding proposition it only remains to prove that X^* is a dense subset of X. Recall that $X^* = \bigcap_{n=0}^{\infty} (f^n)^{\leftarrow}[U_{\mathcal{P}}]$, where the set $U_{\mathcal{P}}$ is dense in X. Since for every $n \in \mathbb{Z}^+$ the mapping f^n is semi-open, it follows that $(f^n)^{\leftarrow}[U_{\mathcal{P}}]$ is dense in X as well: see the Lemmas A.3.6 and A.3.7. Of course, for every $n \in \mathbb{Z}^+$ the set $(f^n)^{\leftarrow}[U_{\mathcal{P}}]$ is also open in X, so X^* is the intersection of countably many dense open sets. Because X is a Baire space this implies that X^* is dense in X. In particular, it follows that $X^* \neq \emptyset$. □

Remark. The conclusion that $\mathcal{P}$ is f-adapted if f is semi-open is also valid if X is not a Baire space: see Exercise 6.2 (3).

Example. Consider the argument-doubling system $(\mathbb{S}, \psi)$ with the topological partition $\mathcal{P} := \{P_0, P_1\}$ of $\mathbb{S}$, where P_0 and P_1 are the two open arcs into which $\mathbb{S}$ is divided by the points $[0]$ and $[^1\!/_2]$. Clearly, a point in $\mathbb{S}$ has a full itinerary iff it does not have the point $[0]$ in its orbit. There are only countably many of such points, so $\mathbb{S}^*$ is dense in $\mathbb{S}$. Using binary expansions it is easy to show that $\iota[\mathbb{S}^*]$ consists of all elements of Ω_2 that do not terminate with 1^∞ or 0^∞ (see also Exercise 6.1 (1)). This set is easily seen to be dense in Ω_2. Consequently, $Z(\mathcal{P}, \psi) = \Omega_2$. These results are in accordance with the above theorem, because the mapping ψ is semi-open.

6.1.7. From now on we assume that (X, f) is a dynamical system with a *compact Hausdorff phase space* X. If $\mathcal{P} = \{P_0, P_1, \dots, P_{s-1}\}$ is an f-adapted topological partition of X then the shift space $Z(\mathcal{P}, f)$ exists. Recall that, by definition, for every point $z \in Z(\mathcal{P}, f)$ and for every $k \in \mathbb{N}$ the initial block $z_{[0;k)}$ of z is $(\mathcal{P}, f)$-allowed, which means that the set $D_k(z)$ is non-empty. Consequently, for every point z of $Z(\mathcal{P}, f)$ the sets $\overline{D_k(z)}$ for $k = 1, 2, 3, \dots$ form a decreasing sequence of non-empty closed sets in the compact space X, hence they have a non-empty intersection.

The topological partition $\mathcal{P}$ of X is called a *pseudo-Markov partition* whenever it is f-adapted and for every point $z \in Z(\mathcal{P}, f)$ the set $\bigcap_{k=1}^{\infty} \overline{D_k(z)}$ consists of just one point. If the shift space $Z(\mathcal{P}, f)$ is, in addition, of finite type then the topological partition $\mathcal{P}$ is called a *Markov partition.*

If $\mathcal{P}$ is a pseudo-Markov partition then for every $z \in Z(\mathcal{P}, f)$ the unique point of the intersection $\bigcap_{k=1}^{\infty} \overline{D_k(z)}$ will be denoted by $\psi_{\mathcal{P},f}(z)$. In this way we obtain a mapping $\psi_{\mathcal{P},f} : Z(\mathcal{P}, f) \to X$. So by definition, we have

$$\forall\, z \in Z(\mathcal{P}, f): \quad \{\psi_{\mathcal{P},f}(z)\} = \bigcap_{k=1}^{\infty} \overline{D_k(z)}\,.$$

In the sequel, we shall denote the shift space $Z(\mathcal{P}, f)$ just by Z and the mapping $\psi_{\mathcal{P},f}$ by ψ. In Proposition 6.1.8 below we shall show that ψ is a morphism of dynamical systems, so the subset $\psi[Z]$ of X is closed (Z is compact) and invariant, defining a subsystem of (X, f). If ψ is a surjection we call the morphism $\psi : (Z, \sigma_Z) \to (X, f)$ a *symbolic representation* of the dynamical system (X, f). Also in the case that ψ is not surjective, or that we do not yet know it to be surjective, we (sloppily) call the system (Z, σ_Z) a symbolic representation of (X, f).

Example. We revisit the argument-doubling system $(\mathbb{S}, f)$ – in order to avoid confusion with the representation mapping defined above we denote in this example the phase mapping by f – with the topological partition $\mathcal{P} := \{P_0, P_1\}$, where P_0 and P_1 are the open arcs into which $\mathbb{S}$ is divided by the points $[0]$ and $[1/2]$. We have seen already that $Z(\mathcal{P}, f) = \Omega_2$. We shall show now that $\mathcal{P}$ is a pseudo-Markov partition.

In order to prove this, we claim that for every $k \in \mathbb{N}$ the following statement is true: for every point $z \in \Omega_2$, the set $D_k(z)$ is an open arc of length $2\pi/2^k$, included in P_0 or in P_1. For $k = 1$ the statement is obviously true: if $z \in \Omega_2$ then the set P_{z_0} equals either P_0 or P_1, which is an open arc of length $2\pi/2$. Now assume the statement is true for a certain value of $k \in \mathbb{N}$. Then for any point $z \in \Omega_2$ one obviously has

$$D_{k+1}(z) = P_{z_0} \cap \bigcap_{n=1}^{k} (f^n)^{\leftarrow}[P_{z_n}] = P_{z_0} \cap f^{\leftarrow}[D_k(\sigma z)]\,. \tag{6.1-7}$$

By assumption, $D_k(\sigma z)$ is an open arc with length $2\pi/2^k$ which is included in P_0 or in P_1. We show that (6.1-7) implies that $D_{k+1}(z)$ is an open arc with length half the length of $D_k(\sigma z)$, i.e., with length $2\pi/2^{k+1}$, which is also included in P_0 or P_1. In fact, f stretches both P_0 and P_1 uniformly by a factor 2 over $\mathbb{S} \setminus \{[0]\}$. Consequently if J is an open arc included in P_0 or P_1 then $f^{\leftarrow}[J]$ consists of two open arcs, each with length half the length of J, one of which is included in P_0 and the other in P_1. It follows that $P_i \cap f^{\leftarrow}[J]$ is an open arc with length half the length of J, obviously included in P_i $(i = 0, 1)$. This concludes the proof of the claim.

It follows from the claim that for every point $z \in \Omega_2$ the set $\overline{D_k(z)}$ has diameter at most $2\pi/2^k$ $(k \in \mathbb{N})$. This implies that the intersection of these sets can have not more than one point, which was to be proved.

Proposition 6.1.8. *Let $\mathcal{P}$ be a pseudo-Markov partition of X. Then the mapping ψ defined above is a morphism of dynamical systems from (Z, σ_Z) to (X, f). Consequently, $\psi[Z]$ is a closed invariant subset of X. Moreover, the subspace $\psi[Z]$ of X is metrizable.*

Proof. First, we show that ψ is continuous. Consider any point $z \in Z$ and an open neighbourhood U of $\psi(z)$ in X. Because the set $\{\psi(z)\}$ is the intersection of the nested sequence of compact sets $\overline{D_k(z)}$ with $k \in \mathbb{N}$, it follows from Appendix A.2.2 that there exists $k \in \mathbb{N}$ such that $\overline{D_k(z)} \subseteq U$. Consider the neighbourhood $V := Z \cap \widetilde{B}_k(z)$ of the point z in Z. It will be sufficient to show that $\psi[V] \subseteq U$. If $y \in V$ then $y \in Z$ and $y_n = z_n$ for $n = 0, \ldots, k-1$. This obviously implies that $D_k(y) = D_k(z)$ which, in turn, implies that $\psi(y) \in \bigcap_{l=1}^{\infty} \overline{D_l(y)} \subseteq \overline{D_k(y)} = \overline{D_k(z)} \subseteq U$. This completes the proof that $\psi[V] \subseteq U$.

In order to show that $f \circ \psi = \psi \circ \sigma$, consider an arbitrary point $z \in Z$. Note that equality (6.1-7) – which holds for every topological partition in any system – (with k replaced by $k-1$) implies that for $k \geq 2$

$$f[D_k(z)] = f[P_{z_0} \cap f^{\leftarrow}[D_{k-1}(\sigma z)]] = f[P_{z_0}] \cap D_{k-1}(\sigma z),$$

which implies that $f[D_k(z)] \subseteq D_{k-1}(\sigma z)$. Using this, one finds

$$f(\psi(z)) \in f\Big[\bigcap_{k=1}^{\infty} \overline{D_k(z)}\Big] \subseteq \bigcap_{k=2}^{\infty} f[\,\overline{D_k(z)}\,] \subseteq \bigcap_{k=2}^{\infty} \overline{D_{k-1}(\sigma z)}.$$

Since the right-hand side of this inclusion is the singleton set $\{\psi(\sigma z)\}$, it follows that $f(\psi(z)) = \psi(\sigma z)$. This completes the proof that $f \circ \psi = \psi \circ \sigma$.

The final statement of the proposition follows from Proposition 1.5.4 (3) and the fact that Z is compact: $\psi[Z]$ is compact, hence closed in X. Moreover, it is well-known that the continuous image of a compact metric space (in this case, the shift space Z) is metrizable: see Appendix A.7.8. □

Remark. The final statement of the proposition implies that it is not a limitation of generality if we consider only dynamical systems on compact metric spaces as the possible candidates for symbolic representation.

Corollary 6.1.9. *Let $\mathcal{P}$ be a pseudo-Markov partition.*

(1) *For every point $z \in Z$ and for every $n \in \mathbb{Z}^+$ one has $f^n(\psi(z)) \in \overline{P}_{z_n}$. Consequently, if $f^n(\psi(z)) \in P_\alpha$ for some $n \in \mathbb{Z}^+$ and $\alpha \in \mathbf{S}$ then $z_n = \alpha$. In particular, if b is a $(\mathcal{P}, f)$-allowed k-block and $\psi(z) \in D_k(b)$ then $b = z_{[0\,;\,k)}$.*

(2) *If $\psi\colon Z \to X$ is a surjection then ψ is semi-open.*

Proof. (1) If $n \in \mathbb{Z}^+$ then $f^n(\psi(z)) = \psi(\sigma^n(z)) \in \overline{D_1(\sigma^n z)} = \overline{P}_{z_n}$. For the second statement in 1, note that if $\alpha \neq z_n$ then the sets P_α and P_{z_n} are disjoint, so that P_α and $\overline{P}_{z_n}$ are disjoint as well, because P_α is open.

(2) It is sufficient to show that for every point $z \in Z$ and every $k \in \mathbb{N}$ the image under ψ of the basic neighbourhood $\widetilde{B}_k(z) \cap Z$ of z in Z incudes a non-empty open subset of X.

Claim. If $z \in Z$ and $k \in \mathbb{N}$ then $D_k(z) \subseteq \psi[\widetilde{B}_k(z) \cap Z]$.

This claim implies the desired result: if $z \in Z$ then the block $z_{[0\,;\,k)}$ is admissible, so $D_k(z)$ is non-empty – and open, of course. To prove the claim, consider a point $x' \in D_k(z)$. Since ψ is surjective there exists $z' \in Z$ such that $x' = \psi(z')$, hence $f^i(\psi(z')) \in P_{z_i}$ for $i = 0, \dots, k-1$. Then the conclusion of 1 above implies that $z'_i = z_i$ for these values of i, which means that $z' \in \widetilde{B}_k(z)$. This proves the claim. □

Remarks.

(1) By stretching the definition of itinerary, we might express the first statement in 1 above by saying that z is an (not 'the') itinerary of the point $\psi(z)$ with respect to the cover $\{\,\overline{P}_\alpha \,\vdots\, \alpha \in \mathbf{S}\,\}$ of X.

(2) In general, ψ is not an open mapping: see Remark 1 following Theorem 6.1.11 below.

Proposition 6.1.10. *Let $\mathcal{P}$ be a pseudo-Markov partition of X and assume that $X^* \neq \emptyset$. Then:*

(1) *$\psi \circ \iota = \mathrm{id}_{X^*}$. Consequently, ι is injective and ψ is injective on $\iota[X^*]$. In particular, the mapping $\iota\colon X^* \to Z$ is a topological embedding with inverse $\psi|_{\iota[X^*]}$, and $X^* \subseteq \psi[Z]$.*

(2) *$\forall x \in X^* : \psi^{\leftarrow}[x] = \{\iota(x)\}$. It follows that $\psi^{\leftarrow}[X^*] = \iota[X^*]$ and that ψ maps $Z \setminus \iota[X^*]$ into $X \setminus X^*$. Moreover, $\iota[X^*]$ is a G_δ-set in Z.*

Proof. (1) Let $x \in X^*$. Then by formula (6.1-6) we have $x \in D_k(\iota(x))$ for every $k \in \mathbb{N}$, so $x \in \bigcap_{k=1}^{\infty} \overline{D_k(\iota(x))} = \{\psi(\iota(x))\}$, i.e., $x = \psi(\iota(x))$. This shows that $\psi \circ \iota = \mathrm{id}_{X^*}$. This, in turn, implies that the mapping $\iota\colon X^* \to Z$ is injective and that $X^* \subseteq \psi[Z]$. It also implies that $\psi|_{\iota[X^*]}$ is the inverse of ι. Since this inverse is continuous, it follows that ι is a topological embedding of X^* into Z.

(2) Let $x \in X^*$ and let $z := \iota(x)$. In view of 1 it remains to show that the fibre $\psi^{\leftarrow}[x]$ contains no points of Z different from z. Consider an arbitrary point $z' \in Z$ such that $\psi(z') = x$. Then $f^j(\psi(z')) = f^j(x) \in P_{\iota(x)_j} = P_{z_j}$ for every $j \in \mathbb{Z}^+$, so Corollary 6.1.9 (1) implies that $z'_j = z_j$ for all $j \in \mathbb{Z}^+$. This shows that $z' = z$, which completes the proof

that $\psi^{\leftarrow}[x] = \{z\}$. This clearly implies that $\iota[X^*] = \psi^{\leftarrow}[X^*]$. Because the inverse image of a G_δ-set under a continuous mapping is a G_δ-set as well, it follows from 6.1.3 (2) that $\iota[X^*]$ is a G_δ-set in Z. □

Remark. The conditions '$\mathcal{P}$ is pseudo-Markov' and '$X^* \neq \emptyset$' are mutually independent. In Example A in 6.3.5 below, $X^* \neq \emptyset$ but $\mathcal{P}$ is not a pseudo-Markov partition. In Example B in 6.3.5, we have a pseudo-Markov partition but X^* is empty.

Theorem 6.1.11. *Let $\mathcal{P}$ be a pseudo-Markov partition of X and let X^* be dense in X. Then $\psi : (Z, \sigma) \to (X, f)$ is a surjective morphism of dynamical systems which is almost 1-to-1, hence irreducible and semi-open.*

Proof. If X^* is dense in X then Proposition 6.1.10 (1) implies that $\psi[Z]$ is dense in X. But $\psi[Z]$ is closed, so $\psi[Z] = X$, that is, ψ is surjective.

In addition, Proposition 6.1.10 (2) implies that $\psi^{\leftarrow}[\psi(z)] = \{z\}$ for every $z \in \iota[X^*]$. Since by Proposition 6.1.5 (2) and the final conclusion in Proposition 6.1.10 the set $\iota[X^*]$ is a dense G_δ-set in Z it follows that the mapping ψ is almost 1-to-1. Finally, by the theorem in Appendix A.9.4, an almost 1-to-1 continuous surjection from a compact Hausdorff space to another compact Hausdorff space is irreducible and semi-open. □

Remarks.
(1) If in the situation of the above theorem, ψ would be an open mapping then Proposition A.9.3 would imply that ψ is a homeomorphism, hence a conjugation. This is not always the case (e.g., not for the argument-doubling transformation, as $\mathbb{S}$ is not 0-dimensional).
(2) In view of Corollary 6.1.9 (2), the conclusion that ψ is semi-open would follow already from the fact that ψ is surjective.

Corollary 6.1.12. *If $\mathcal{P}$ is a pseudo-Markov partition of X and f is semi-open then the conclusions of the previous theorem hold. In particular, this is true if f is a homeomorphism.*

Proof. Use Theorem 6.1.6. □

Example. We revisit, again, the argument-doubling system $(\mathbb{S}, f)$ with the topological partition $\mathcal{P} := \{P_0, P_1\}$, where P_0 and P_1 are the open arcs into which $\mathbb{S}$ is divided by the points $[0]$ and $[1/2]$. We have seen already that $\mathcal{P}$ is a Markov partition and that $Z(\mathcal{P}, f) = \Omega_2$. Since the phase mapping f of the argument-doubling system is semi-open, the morphism $\psi_{\mathcal{P},f} : (\Omega_2, \sigma) \to (\mathbb{S}, f)$ is surjective and almost 1-to-1. Now it follows from Corollary 5.2.6 (1) and Proposition 1.5.4 (5) that the system $(\mathbb{S}, f)$ has transitive points; moreover, by Corollary 5.2.4 (2) and Proposition 1.5.4 (2), it has a dense set of periodic points. (We know this already by other arguments, but this illustrates what can be done with symbolic representations.)

Corollary 6.1.13.
(1) *If $\mathcal{P}$ is a Markov partition then the following conditions are equivalent:*
  (i) $\psi: Z \to X$ *is surjective.*
  (ii) f *is a semi-open mapping.*
  (iii) X^* *is dense in* X.
(2) *If $\mathcal{P}$ is a clopen pseudo-Markov partition then ψ is a conjugation.*

Proof. (1) Assume (i). Since $\mathcal{P}$ is a Markov partition, the symbolic representation Z is an SFT, hence by Proposition 5.3.12 the restricted shift $\sigma_Z : Z \to Z$ is an open mapping. Moreover, by Corollary 6.1.9 (2), the mapping $\psi: Z \to X$ is a semi-open. Consequently, the mapping $\psi \circ \sigma_Z$ is semi-open. Since f is surjective, any non-empty open subset U of X can be written as $U = \psi[V]$ with $V := \psi^{\leftarrow}[U]$ a non-empty open subset of Z. Then $f[U] = (\psi \circ \sigma_Z)[V]$ has a non-empty interior. If (ii) holds then Theorem 6.1.6 implies that X^* is dense in X (recall that X is a Baire space). Finally, if (iii) holds then Theorem 6.1.11 implies that ψ is surjective.
(2) In this case, $\mathcal{P}$ is a covering of X, hence every point of X has a full itinerary. So $X^* = X$, and Proposition 6.1.10 implies that ψ is surjective and injective. Because Z is compact and X is Hausdorff this implies that ψ is a homeomorphism. (Of course, we could also refer to Lemma 6.1.3 (3).) □

Remarks.
(1) *$\mathcal{P}$ is a clopen partition iff $X^* = X$.* "Only if": obvious (see the proof of part 2 of the above corollary). "If": as points in $X \setminus \bigcup \mathcal{P}$ have no itinerary, the equality $X^* = X$ implies that $\bigcup \mathcal{P} = X$, i.e., that $\mathcal{P}$ is a partition. Moreover, each member of $\mathcal{P}$ is the complement of the (open) union of the other members, hence closed.
(2) A clopen partition is not necessarily a pseudo-Markov partition. As an example, consider a system with phase space X that is the disjoint union of two non-empty compact sets P_0 and P_1, one of which has at least two points, let $\mathcal{P} := \{P_0, P_1\}$ and let f leave P_0 and P_1 invariant under f. Then $X^* = X$ and $Z = \{0^\infty, 1^\infty\}$, so ι is not injective. Hence by Proposition 6.1.10 (1), $\mathcal{P}$ is not pseudo-Markov.

By the preceding results, the system (X, f) has a representation by a shift system whenever it admits a pseudo-Markov partition such that X^* is dense in X. So the art of constructing symbolic representations consists of finding (or proving the existence of) a suitable topological partition. In the next section we shall say more about the existence of pseudo-Markov partitions. As to the condition that X^* is dense in X, a sufficient condition is that the phase mapping f is semi-open, and by Corollary 6.1.13 (1) above this is also a necessary condition for $\mathcal{P}$ to be a Markov partition. We present another necessary and sufficient condition for mappings of a compact interval into itself.

Lemma 6.1.14. *Let $\mathcal{P}$ be a pseudo-Markov partition of X and let $\alpha \in \mathbf{S}$. If $x \in X^* \cap P_\alpha$, $y \in P_\alpha$ and $f(x) = f(y)$ then $x = y$. In particular, the mapping $f|_{X^* \cap P_\alpha}$ is injective.*

Proof. If $x \in X^*$, $y \in X$ and $f(y) = f(x)$ then $f^n(y) = f^n(x) \in P_{\iota(x)_n}$ for all $n \geq 1$. If, in addition, the points x and y both belong to the same member P_α of $\mathcal{P}$ then, obviously, $f^0(y) \in P_{\iota(x)_0}$. It follows that $y \in X^*$ and that $\iota(x) = \iota(y)$. Hence $x = y$, because ι is injective by Proposition 6.1.10 (1). □

Proposition 6.1.15. *Let (X, f) be a dynamical system on a compact interval in $\mathbb{R}$ admitting a pseudo-Markov partition $\mathcal{P}$ consisting of intervals*[2]*. Then the set X^* is dense in X iff f is strictly monotonous on each member of $\mathcal{P}$. In that case, f is piecewise monotonous and each member of $\mathcal{P}$ is included in an interval of monotonicity of f.*

Proof. “If”: If f is strictly monotonous on each member of $\mathcal{P}$ then f is easily seen to be semi-open. Now apply Theorem 6.1.6.

“Only if”: Let X^* be dense in X and assume that f is not strictly monotonous on an interval P_α for some $\alpha \in \mathbf{S}$. Then either f is constant on a non-degenerate subinterval J of P_α, or else f has a strict local extremum at an interior point x_0 of P_α. We show that none of these possibilities is compatible with the conclusions of Lemma 6.1.14.

In the first case this is obvious: J contains at least two different points of X^* and Lemma 6.1.14 implies that f should have different values at these points. In the second case, the intermediate value theorem implies that every value of f on a sufficiently small left neighbourhood J_1 of the point x_0 in P_α is also assumed on a right neighbourhood J_2 of x_0 in P_α. Obviously, J_1 contains a point x of X^* different from x_0, and if $y \in J_2$ is such that $f(y) = f(x)$ then clearly $y \neq x$. But Lemma 6.1.14 implies that $x = y$, a contradiction. □

We shall not discuss the ‘standard’ applications of symbolic representations occurring in the literature; but see Note 9 at the end of this chapter. However, the results of the present chapter will be used in Chapter 8 for the calculation of the topological entropy of certain systems.

The construction and usefulness of a symbolic representation of a system require a subtle balance between its simplicity and its complexity. If it is very complex it may be quite difficult to find a suitable symbolic representation, but it may be very rewarding to stumble upon one. On the other hand, for very simple systems a symbolic representation may give less information than is needed for its construction. See e.g., the Examples B and C in 6.3.5 below. In these examples the information about the nature of the invariant points (attracting or repelling) is lost; all properties of the system can be seen straight away, and a symbolic representation adds nothing to our knowledge.

---

2 Open in X, not necessarily open in $\mathbb{R}$. So the left (right) end point of X is allowed to be the left (right) end point of the left-most (right-most) member of $\mathcal{P}$.

6.2 Expansive systems

Notation is as before. In particular, (X, f) is a dynamical system with a compact Hausdorff phase space X and $\mathcal{P} = \{P_0, \dots, P_{s-1}\}$ is a topological partition of X. In order that $\mathcal{P}$ be pseudo-Markov (intersections of) the inverse images of its members have to shrink. So it may be expected that an f-adapted topological partition is pseudo-Markov if f is 'expanding' in some sense. With a few exceptions, the results of this section apply only to metric spaces. The following definition has proved to be useful.

A dynamical system (X, f) on a *metric space* (X, ρ) is said to be *expansive* whenever there exists a real number $\eta > 0$ such that

$$\forall x, y \in X: \ x \neq y \Rightarrow \exists n \in \mathbb{Z}^+ : \rho(f^n(x), f^n(y)) \geq \eta . \tag{6.2-1}$$

In this case η is called an *expansive coefficient* of (X, f). This condition means that points that are close to each other will have a distance of at least η at a certain moment in the future (if the points themselves have already a distance of at least η then (6.2-1) is certainly fulfilled).

Obviously, every subsystem of an expansive system is expansive with the same expansive coefficient. In general, the property of being expansive and the value of the coefficient depend on the particular metric used. See Exercise 6.8 (4). Consequently, expansiveness is not a dynamical property. On the other hand, being a *compact* expansive system turns out to be a dynamical property. See Exercises 6.8 (2) and 6.8 (3).

The standard examples of expansive systems are the shift spaces:

Proposition 6.2.1. *The full shift system (Ω_S, σ) is expansive with expansive coefficient* 1, *hence every shift system is expansive with coefficient* 1.

Proof. Let $x, y \in \Omega_S$. If $x \neq y$ then there exists $n \in \mathbb{Z}^+$ such that $x_n \neq y_n$, that is, $(\sigma^n x)_0 \neq (\sigma^n y)_0$. Hence $d(\sigma^n x, \sigma^n y) = 1$. □

Example. The mapping $x \mapsto ax : [0; \infty) \to [0; \infty)$ with $|a| > 1$ is expansive. However, a compact interval in $\mathbb{R}$ admits no expansive mappings; see Exercise 6.8 (6).

The following example shows that the 'natural' clopen partition of a shift space is a pseudo-Markov partition.

Example. Consider a shift space $X \subseteq \Omega_S$ and let $f := \sigma_X$. For every $\alpha \in S$, let $R_\alpha := X \cap C_0[\alpha] = \{x \in X : x_0 = \alpha\}$. We may assume that $R_\alpha \neq \emptyset$ for every α (after a suitable relabelling of S one finds $s' \leq s$ such that $R_\alpha \neq \emptyset$ iff $\alpha = 0, \dots, s'-1$, hence the symbols $s', \dots, s-1$ can be neglected). Then $\mathcal{R} := \{R_0, \dots, R_{s-1}\}$ is a clopen partition of X, the *natural partition* of X.

For every block b of length k one has $D_k(b) = X \cap C_0[b]$. This set is non-empty iff b is the initial block of a point of X, iff b occurs in a point of X. Consequently, the set of allowed blocks for the symbolic representation Z of X coincides with the set of the X-present blocks. It follows that $Z = X$. Moreover, if $z \in Z$ then $D_k(z) = X \cap \widetilde{B}_k(z)$. The intersection of these sets is the singleton set $\{z\}$, so $\mathcal{R}$ is a pseudo-Markov partition (with $\psi: Z \to X$ the identity mapping).

Lemma 6.2.2. *Let X be a compact metric space. If the mapping f is expansive with expansive coefficient η and* diam $(P_\alpha) < \eta$ *for every $\alpha \in \mathbf{S}$, then*

$$\lim_{k \rightsquigarrow \infty} \operatorname{diam}\left(\overline{D_k(z)}\right) = 0 \text{ uniformly in } z \in Z\,. \tag{6.2-2}$$

If $\mathcal{P}$ is f-adapted and satisfies (6.2-2) *then $\mathcal{P}$ is a pseudo-Markov partition.*

Proof. Suppose that $Z \neq \emptyset$ and that formula (6.2-2) holds. Then it is clear that for every point $z \in Z$ the set $\bigcap_{k=1}^{\infty} \overline{D_k(z)}$ contains at most one element. So if $\mathcal{P}$ is f-adapted then it is a pseudo-Markov partition.

In order to prove formula (6.2-2) under the condition that diam $(P_\alpha) < \eta$ for every $\alpha \in \mathbf{S}$ we may assume that $Z \neq \emptyset$, otherwise there is nothing to prove. Moreover, if $\eta' := \max\{\operatorname{diam}(P_\alpha) \vdots \alpha \in \mathbf{S}\}$ then $0 < \eta' < \eta$.

Assume that (6.2-2) is false: then there exists $\varepsilon > 0$ such that, for every $k \in \mathbb{N}$, there are a point $z^{(k)} \in Z$ and a natural number $n_k \geq k$ for which $\operatorname{diam}\left(\overline{D_{n_k}(z^{(k)})}\right) \geq \varepsilon$. Consequently, for every $k \in \mathbb{N}$ there are points $x_k, y_k \in D_{n_k}(z^{(k)})$ with $d(x_k, y_k) \geq \frac{1}{2}\varepsilon$. The definition of the set $D_{n_k}(z^{(k)})$ implies that for $i = 0, \dots, k-1$ (even up to $n_k - 1$) the points $f^i(x_k)$ and $f^i(y_k)$ both belong to the same set $P_{z_i^{(k)}}$, so that $d(f^i(x_k), f^i(y_k)) \leq \eta'$ for these values of i. By considering convergent subsequences of the sequences $(x_k)_{k\in\mathbb{N}}$ and $(y_k)_{k\in\mathbb{N}}$ in X, with limits x and y, respectively, one finds points $x, y \in X$ such that $d(x, y) \geq \frac{1}{2}\varepsilon$ and $d(f^i(x), f^i(y)) < \eta$ for every $i \geq 0$. By expansiveness, the latter inequality implies that $x = y$, contradicting the former inequality. □

Lemma 6.2.3. *Let X be a compact metric space and let $\eta > 0$. Then X has a topological partition $\mathcal{P}$ all of whose members have a diameter less than η. If X is 0-dimensional then $\mathcal{P}$ may be assumed to be a clopen partition of X.*

Proof. By compactness, X has a finite cover $\{U_0, \dots, U_k\}$ consisting of open balls with diameter less than η. Define inductively open sets P_i for $i = 0, \dots, k$ by

$$P_0 := U_0 \quad \text{and} \quad P_i := U_i \setminus \overline{P_0 \cup \ldots \cup P_{i-1}} \quad \text{for } i = 1, \dots, k\,.$$

These sets are mutually disjoint and have a diameter less than η. We show that the union of their closures is X, as follows: Clearly, $\overline{P}_0 = \overline{U}_0 \supseteq U_0$, and if $\overline{P_0 \cup \ldots \cup P_i} \supseteq U_0 \cup \ldots \cup U_i$ for some $i \in \{0, \dots, k-1\}$ then

$$\begin{aligned}
\overline{P_0 \cup \ldots \cup P_i \cup P_{i+1}} &= \overline{P_0 \cup \ldots \cup P_i} \cup \overline{P}_{i+1} \\
&\supseteq \overline{P_0 \cup \ldots \cup P_i} \cup \left(U_{i+1} \setminus \overline{P_0 \cup \ldots \cup P_i}\right) \\
&= \overline{P_0 \cup \ldots \cup P_i} \cup U_{i+1} \\
&\supseteq U_0 \cup \ldots \cup U_i \cup U_{i+1}\,.
\end{aligned}$$

So by induction one gets:

$$\overline{P}_0 \cup \ldots \cup \overline{P}_k = \overline{P_0 \cup \ldots \cup P_k} \supseteq U_0 \cup \ldots \cup U_k = X\,.$$

Thus, if one leaves out the empty sets from the collection $\{P_0, \dots, P_k\}$ then one gets the desired topological partition.

Finally, if X is 0-dimensional then for the sets U_i above one may take clopen subsets of balls with diameter less than η, in which case the sets P_i are clopen as well. In that case the sets P_i cover all of X and $\mathcal{P}$ is a clopen partition of X. □

Remark. A topological partition $\mathcal{P}$ of X for which (6.2-2) holds is called a *generator* of (X, f). The final part of Lemma 6.2.2 shows that an f-adapted generator is pseudo-Markov. The first part of Lemma 6.2.2 states that if f is expansive then every topological partition consisting of sufficiently small sets is a generator. So Lemma 6.2.3 implies that if f is expansive then (X, f) actually has a generator.

Corollary 6.2.4. *Let (X, f) be a dynamical system on a compact metric space. The following statements are equivalent:*
(i) *(X, f) is conjugate to a shift system.*
(ii) *X is 0-dimensional and X has a metric with respect to which f is expansive.*

Proof. (i)⇒(ii): by 5.1.4, a full shift space is 0-dimensional, hence every shift space is 0-dimensional. In addition, by 6.2.1 above, a full shift system is expansive, hence every subshift is expansive as well. These properties are carried over to (X, f) by the conjugation.

(ii)⇒(i): By Lemma 6.2.3 there is a partition $\mathcal{P}$ of X consisting of clopen subsets of X with diameter less than the expansive coefficient of f. It follows from Remark 1 after 6.1.13 that $X^* = X$. Hence 6.1.5 (1) implies that $\mathcal{P}$ is f-adapted. Consequently, Lemma 6.2.2 implies that $\mathcal{P}$ is a pseudo-Markov partition, hence we infer from Theorem 6.1.11 that (X, f) is a factor of a shift system. Finally, Corollary 6.1.13 (2) implies that the factor mapping is a conjugation. □

Remarks.
(1) Exercise 6.8 (2) implies that if (ii) holds then f is expansive with respect to every compatible metric on X.
(2) The proof of the implication (ii)⇒(i) above does not need the full machinery of symbolic representations. See Exercise 6.9.

Corollary 6.2.5. *Let (X, f) be an expansive dynamical system on a compact metric space and assume that f is a semi-open mapping. Then there exists an almost 1-to-1 factor mapping $\psi: (Z, \sigma_Z) \to (X, f)$ of a shift system (Z, σ_Z) onto (X, f).*

Proof. By Theorem 6.1.6, the topological partition obtained in Lemma 6.2.3 is f-adapted. Hence by Lemma 6.2.2 it is a pseudo-Markov partition. Now apply Theorem 6.1.12. □

Remark. The properties of being semi-open and of being expansive are unrelated. For example, the mapping $f: x \mapsto x^2: [0; 1] \to [0; 1]$ is semi-open but not expansive.

Conversely, the shift on the system defined in Exercise 5.11 (3) is expansive but not semi-open.

So every dynamical systems on a compact metric space X with a semi-open expansive phase mapping has an almost 1-to-1 surjective symbolic representation $\psi: Z \to X$. It is a different – and usually quite hard – problem to actually construct a suitable pseudo-Markov partition for a particular system with these properties. We shall pay hardly any attention to this problem; see Note 10 at the end of this chapter.

If a system has an almost 1-to-1 symbolic representation $\psi: Z \to X$ then there remain two questions:

(1) What can be said about the size of fibres of points of X outside of the set T_ψ of points with singleton fibres (see Section A.9 for the definition of T_ψ)? For the applications of symbolic dynamics in Chapter 8 it would be desirable if all fibres were finite with uniformly bounded sizes.
(2) Often the set T_ψ is topologically large: $T_\psi \supseteq X^*$ and if X^* is dense (which is often the reason that ψ is almost 1-to-1) then T_ψ includes a dense G_δ-set. Is it also 'dynamically large' in the sense that it includes transitive points (provided that the system (X, f) is transitive)?

In Theorem 6.2.11 below we show that under rather natural conditions the answers to the questions stated in (1) and (2) is affirmative.

We say that a factor map $\varphi: (X, f) \to (Y, g)$ of dynamical systems is *finite-to-one* whenever every fibre $\varphi^{\leftarrow}[y]$ for $y \in Y$ is a finite set. We say that φ is *boundedly finite-to-one* whenever there is a natural number $K > 0$ such that for every $y \in Y$ the number of elements in the fibre $\varphi^{\leftarrow}[y]$ is at most K.

We start with a couple of lemmas needed for the proof of Theorem 6.2.11.

Lemma 6.2.6. *Let A be a subset of Ω_S containing more than s^2 points. Then there are $z, z' \in A$ with the following property:*

$$\exists\, l, m \in \mathbb{Z}^+,\ 0 < l < m : z_0 = z'_0,\ z_l \neq z'_l,\ z_m = z'_m\,. \tag{D}$$

Proof. Under the assumptions of the lemma we can choose $s^2 + 1$ different points in A. For sufficiently large m, the initial $(m+1)$-blocks of these points are mutually different. Among the $s^2 + 1$ pairs of symbols, formed by the initial and final coordinates of each these blocks there must be at least two equal ones, for there are at most s^2 different pairs of symbols. Consequently, there are two points $z, z' \in A$ such that $z_0 = z'_0$ and $z_m = z'_m$. Since the blocks $z_{[0;m]}$ and $z'_{[0;m]}$ are different there exists l between 0 and m such that $z_l \neq z'_l$. □

Remark. The following picture explains why a pair of points with property (D) is also known as a 'diamond':

$$\begin{array}{ccccc} & & z_l & & \\ & \cdot^{\cdot^{\cdot}} & & \ddots & \\ z_0 = z_0' & & & & z_m = z_m' \\ & \ddots & & \cdot^{\cdot^{\cdot}} & \\ & & z_l' & & \end{array}$$

Lemma 6.2.7. *Suppose $\mathcal{P}$ is a pseudo-Markov partition. If $x \in U_{\mathcal{P}}$ is a transitive point and $\psi^{\leftarrow}[x]$ contains more than two points then every pair of distinct points $z, z' \in \psi^{\leftarrow}[x]$ has property* (D).

Proof. Assume that $x \in P_\alpha$ ($\alpha \in \mathbf{S}$) and let $z, z' \in \psi^{\leftarrow}[x]$ with $z \neq z'$. Then $\psi(z) \in P_\alpha$, hence $z_0 = \alpha$ by Corollary 6.1.9 (1). Similarly, $z_0' = \alpha$, so $z_0 = z_0'$. Moreover, because $z \neq z'$ there is $l \in \mathbb{N}$ such that $z_l \neq z_l'$. Finally, by transitivity of the point x there exists $m > l$ such that $f^m(x) \in P_\alpha$ so, again by Corollary 6.1.9 (1), the points z and z' have equal m-th coordinates. □

Lemma 6.2.8. *Let X be a compact metric space and let f be an expansive mapping with expansive coefficient η. In addition, assume that $\mathcal{P}$ is a Markov partition, that Z is an SFT of order 2, and that $\psi\colon Z \to X$ is a surjection. If* $\operatorname{diam}(P_\alpha) < \frac{1}{2}\eta$ *for all $\alpha \in \mathbf{S}$ then there are no pairs of mutually distinct points $z, z' \in Z$ with property* (D) *such that* $\psi(z) = \psi(z')$.

Proof. Let $z, z' \in Z$ such that $\psi(z) = \psi(z') =: x$ and such that, in addition, $z_0 = z_0' =: \alpha$ and $z_m = z_m' =: \beta$ for some $m \geq 2$:

$$\begin{aligned} z &= \alpha b_0 \dots b_{m-2} \beta z_{m+1}\, z_{m+2} \dots, \\ z' &= \alpha c_0 \dots c_{m-2} \beta z_{m+1}'\, z_{m+2}' \dots . \end{aligned}$$

We want to show that the $(m-1)$-blocks b and c are equal to each other. To this end, first note that, by Corollary 6.1.9 (1), for every $i \in \{1, \dots, m-1\}$ the point $f^i(x)$ belongs to the set $\overline{P}_{z_i} = \overline{P}_{b_{i-1}}$ as well as to the set $\overline{P}_{z_i'} = \overline{P}_{c_{i-1}}$. So for every $i \in \{1, \dots, m-1\}$ the sets $\overline{P}_{b_{i-1}}$ and $\overline{P}_{c_{i-1}}$ have a point in common. Hence the triangle equality implies that for all points $x' \in \overline{P}_{b_{i-1}}$ and $x'' \in \overline{P}_{c_{i-1}}$ one has

$$d(x', x'') \leq \operatorname{diam}(P_{b_{i-1}}) + \operatorname{diam}(P_{c_{i-1}}) < \eta . \tag{6.2-3}$$

Because the block $\alpha b \beta$ is allowed, the set $D_{m+1}(\alpha b \beta)$ is not empty. Select a point y in it: $y \in P_\alpha$, $f^i(y) \in P_{b_{i-1}}$ for $i = 1, \dots, m-1$ and $f^m(y) \in P_\beta$. As ψ is surjective there is a point $u \in Z$ such that $\psi(u) = y$, and the final statement in Corollary 6.1.9 (1) implies that the initial m-block of u equals $\alpha b \beta$, i.e.,

$$u = \alpha\, b_0 \dots b_{m-2} \beta u_{m+1}\, u_{m+2} \dots .$$

Next, consider the point of Ω_S that is obtained from u by replacing each coordinate b_i by c_i ($i = 0, \dots, m-2$):

$$v := \alpha\, c_0 \dots c_{m-2} \beta u_{m+1}\, u_{m+2} \dots .$$

Every 2-block in this sequence is either in z' or in u, hence is allowed. As Z is an SFT of order 2 it follows that $v \in Z$. Let $y' := \psi(v)$.

By the choice of y we have $f^i(y) \in P_{b_{i-1}}$ for $i = 1, \dots, m-1$. In addition, Corollary 6.1.9 (1) implies that $f^i(y') \in \overline{P}_{c_{i-1}}$ for these values of i. So by inequality (6.2-3), $d((f^i(y), f^i(y')) < \eta$ for $i = 1, \dots, m-1$. For all other values of i there exists $\gamma \in \mathbf{S}$ such that both points $f^i(y)$ and $f^i(y')$ belong to $\overline{P}_\gamma$ (namely, $\gamma = \alpha$ if $i = 0$, $\gamma = \beta$ if $i = m$ and $\gamma = u_i$ if $i > m$), hence have a distance less than η as well. Conclusion: $d((f^i(y), f^i(y')) < \eta$ for all $i \in \mathbb{Z}^+$. Since f is expansive with coefficient η, it follows that $y = y'$. In particular, for $i = 1, \dots, m-1$ one has $f^i(y') = f^i(y) \in P_{b_{i-1}}$. Using the second statement in Corollary 6.1.9 (1) once again one easily shows that this implies that $c_{i-1} = b_{i-1}$ for $i = 1, \dots, m-1$, that is, the blocks b and c are equal to each other. □

Lemma 6.2.9. *Assume that X is a compact metric space and that f is an expansive mapping with expansive coefficient η. In addition, assume that $\mathcal{P}$ is a Markov partition such that Z is an SFT of order 2 and that* diam $(P_\alpha) < \frac{1}{2}\eta$ *for all $\alpha \in \mathbf{S}$. Finally, assume that $\psi\colon Z \to X$ is surjective. Then:*
(1) *ψ is boundedly finite-to-one.*
(2) *If (X, f) is transitive then ψ is an almost 1-to-1 mapping and, consequently, (Z, σ_Z) is transitive as well. In addition, there is a dense G_δ-set of transitive points in X that have a singleton fibre under ψ. In point of fact, every transitive point x_0 in $U_\mathcal{P}$ has a unique pre-image under ψ.*

Proof. (1) By Lemmas 6.2.8 and 6.2.6, no fibre of ψ contains more than s^2 points.
(2) Denote the set of transitive points in (X, f) by T. By Remark 1 after Theorem 1.3.5, T includes a dense G_δ-set T_0. Then $T' := U_\mathcal{P} \cap T_0$ is also a G_δ-set, and because the set $U_\mathcal{P}$ is open and dense in X it is easily seen that the set T' is dense in X as well. Now recall from Corollary 6.1.9 (2) that ψ is a semi-open mapping. Hence Appendix A.3.6 implies that $S' := \psi^{\leftarrow}[T']$ is a dense subset of Z. Moreover, it obviously is a G_δ-set.
If $z \in S'$ then $\psi(z)$ is a transitive point in $U_\mathcal{P}$. By the Lemma's 6.2.7 and 6.2.8, every transitive point in $U_\mathcal{P}$ has a unique pre-image under ψ so, in particular, the fibre $\psi^{\leftarrow}[\psi(z)]$ consists of only one point. Since S' is a dense G_δ-set in Z, it follows that ψ is an almost 1-to-1 mapping.
Finally, Theorem A.9.4 implies that the mapping ψ is irreducible, hence it lifts transitivity – see Exercise 1.8 (2). This completes the proof. □

In our final result we want to relax the condition that the diameters of the members of $\mathcal{P}$ are at most $\frac{1}{2}\eta$ to the condition that they are at most η, and we want to get rid of the restriction on the order of Z. To this end, we need a lemma that has some interest of its own. We could formulate it for systems with an arbitrary compact Hausdorff phase space X, but as X is the image of the compact metric shift space Z, X is metrizable – see Appendix A.7.8. But we do not need that f is expansive.

Lemma 6.2.10. *Assume that $\mathcal{P}$ is a pseudo-Markov partition and that the symbolic representation $\psi\colon (Z, \sigma_Z) \to (X, f)$ is a surjection. Let $k \in \mathbb{N}$ and let $\mathcal{P}_k^*$ be the family of all sets $D_k(b)$ for all $(\mathcal{P}, f)$-allowed blocks b of length k. Then $\mathcal{P}_k^*$ is a pseudo-Markov partition. Let $\psi_k^* : (Z_k^*, \sigma) \to (X, f)$ be the symbolic representation generated by $\mathcal{P}_k^*$. Then $Z_k^* = Z^{(k)}$, the k-th higher block representation of Z, and $\psi_k^* \circ \mathrm{I}_\infty^{(k)} = \psi$. In particular, it follows that ψ_k^* is a surjection of Z_k^* onto X.*

Proof. The proof consists of a straightforward but slightly dull verification of all statements of the lemma. For the sake of completeness we spell out the details. Denote the set of all $(\mathcal{P}, f)$-allowed k-blocks by $\mathbf{S}_k^*$. Hence we can write $\mathcal{P}_k^* = \{ D_k(b) \,\vdots\, b \in \mathbf{S}_k^* \}$. First, we show that $\mathcal{P}_k^*$ is a topological partition.

Clearly, the sets $D_k(b)$ for $b \in \mathbf{S}_k^*$ are mutually disjoint and open. Moreover, if $z \in Z$ then (by definition) $\psi(z) \in \overline{D_k(z)}$, where $D_k(z) = D_k(z_{[0;k)})$, with $z_{[0;k)} \in \mathbf{S}_k^*$. So the union of the closures of the sets $D_k(b)$ for $b \in \mathbf{S}_k^*$ includes $\psi[Z]$, which is given to be equal to X. This completes the proof that $\mathcal{P}_k^*$ is a topological partition[3].

In accordance with definition (6.1-4), for any n-word $b^{(0)} \dots b^{(n-1)}$ over the symbol set $\mathbf{S}_k^*$ we define

$$D_n^*(b^{(0)} \dots b^{(n-1)}) := \bigcap_{j=0}^{n-1} (f^j)^{\leftarrow}[D_k(b^{(j)})] \quad (n \geq 1)\,.$$

An n-word $b^{(0)} \dots b^{(n-1)}$ over $\mathbf{S}_k^*$ is $(\mathcal{P}_k^*, f)$-allowed iff $D_n^*(b^{(0)} \dots b^{(n-1)})$ is not empty and, by definition, Z_k^* is the set of all points in $\Omega_{\mathbf{S}_k^*}$ in which all finite blocks are $(\mathcal{P}_k^*, f)$-allowed.

We first show that $X^{(k)} \subseteq Z_k^*$, which implies that $Z_k^* \neq \emptyset$, i.e., that the topological partition $\mathcal{P}_k^*$ is f-adapted. Consider a point of $Z^{(k)}$, that is, let $z \in Z$ and consider the point $z^* := \mathrm{I}_\infty^{(k)}(z)$. Using the definition of the mapping $\mathrm{I}_\infty^{(k)}$ in (5.4-3) one gets

$$\begin{aligned} D_{n+1}^*(z^*) &= \bigcap_{j=0}^{n} (f^j)^{\leftarrow}[D_k(z_{[j;j+k)})] = \bigcap_{j=0}^{n} \Big(\bigcap_{i=j}^{j+k-1} (f^i)^{\leftarrow}[P_{z_i}] \Big) \\ &= \bigcap_{i=0}^{n+k-1} \Big(\bigcap_{j \in I(n,i)} (f^i)^{\leftarrow}[P_{z_i}] \Big) = \bigcap_{i=0}^{n+k-1} (f^i)^{\leftarrow}[P_{z_i}] \\ &= D_{n+k}(z)\,, \end{aligned}$$

3 This argument shows that we can replace $\mathbf{S}_k^*$ by the set of all Z-present k-blocks. Usually, this is a smaller set, but in this case it is the same: see Exercise 6.2 (5) (fortunately, for otherwise we would have a family of mutually disjoint non-empty open sets with a dense proper subfamily). So the symbol set for Z_k^* is $\mathcal{L}_k(X)$, the same as for $X^{(k)}$.

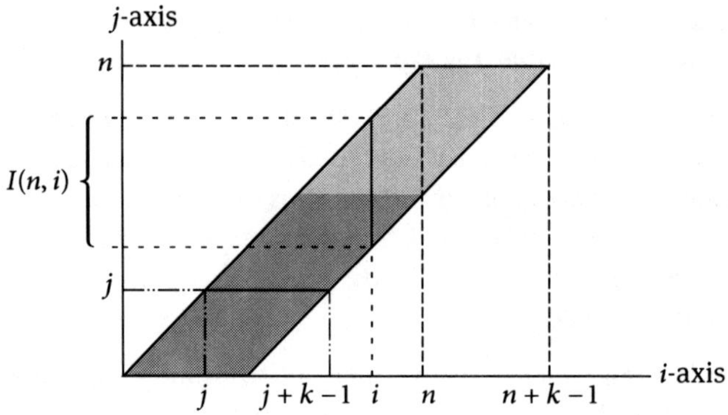

Fig. 6.2. Illustrating the proof of Lemma 6.2.10.

where $I(n,i) := \{ j \in \mathbb{Z}^+ : \max\{0, i-k+1\} \le j \le \min\{i,n\} \}$. See the schematic picture in Figure 6.2 above (where we assume for convenience that $n > k-1$). As $D_{n+k}(z) \neq \emptyset$ for every $n \in \mathbb{Z}^+$, it follows that every initial block of z^* is $(\mathcal{P}_k^*, f)$-allowed. Hence $z^* \in Z_k^*$, which completes the proof that $X^{(k)} \subseteq Z_k^*$ and that $\mathcal{P}_k^*$ is f-adapted.

Next, we show that, conversely, $Z_k^* \subseteq X^{(k)}$. To this end, consider an arbitrary point $z^* = b^{(0)}b^{(1)} \dots b^{(j)} \dots$ in Z_k^*, where $b^{(j)} = b_0^{(j)} \dots b_{k-1}^{(j)}$ is a $(\mathcal{P}, f)$-allowed k-block over **S** for every $j \in \mathbb{Z}^+$. Then a similar computation as above shows that for arbitrary $n \in \mathbb{Z}^+$

$$\begin{aligned} D_{n+1}^*(z^*) &= \bigcap_{j=0}^{n} (f^j)^{\leftarrow}[D_k(b^{(j)})] = \bigcap_{j=0}^{n} \Big(\bigcap_{i=j}^{j+k-1} (f^i)^{\leftarrow}[P_{b_{i-j}^{(j)}}] \Big) \\ &= \bigcap_{i=0}^{n+k-1} \Big(\bigcap_{j \in I(n,i)} (f^i)^{\leftarrow}[P_{b_{i-j}^{(j)}}] \Big) = \bigcap_{i=0}^{n+k-1} (f^i)^{\leftarrow}\Big[\bigcap_{j \in I(n,i)} P_{b_{i-j}^{(j)}} \Big], \end{aligned} \tag{6.2-4}$$

where $I(n,i)$ is as above. The fact that $z^* \in Z_k^*$ means that every initial block of z^* is $(\mathcal{P}_k^*, f)$-allowed, i.e., that $D_{n+1}^*(z^*) \neq \emptyset$ for every $n \in \mathbb{Z}^+$. By the above computation, this obviously implies that

$$\forall n \in \mathbb{Z}^+ \forall i \in \{0, \dots, n+k-1\} : \bigcap_{j \in I(n,i)} P_{b_{i-j}^{(j)}} \neq \emptyset . \tag{6.2-5}$$

That the intersection in formula (6.2-5) is not empty means that the symbols $b_{i-j}^{(j)}$ involved in this intersection are equal to each other (recall that the sets P_α are mutually disjoint for different values of $\alpha \in$ **S**). If $i \le n$ then $i \in I(n,i)$, so one of those symbols is $b_0^{(i)}$ (which occurs for $j = i$). It follows that (we silently exchange quantors) for every $i \ge 0$ and every $n \ge i$ we have $b_{i-j}^{(j)} = b_0^{(i)}$ for all $j \in I(n,i)$. However, if we consider even larger values of n, namely, $n \ge i+k-1$, then $I(n,k) = \mathbb{Z}^+ \cap [i-k+1; i]$ (which is independent of n). It follows that $b_{i-j}^{(j)} = b_0^{(i)}$ for every $j \in \mathbb{Z}^+ \cap [i-k+1; i]$. This can also be formulated as

$$\forall j \ge 0 : b_{i-j}^{(j)} = b_0^{(i)} \quad \text{for } i = j, \dots, j+k-1 . \tag{6.2-6}$$

For every $j \in \mathbb{Z}^+$ put $z_i := b_0^{(i)}$. This defines a point $z \in \Omega_{\mathbf{S}}$ and it follows from (6.2-6) that $z_i = b_{i-j}^{(j)}$ for *all* pairs $(i, j) \in \mathbb{Z}^+ \times \mathbb{Z}^+$ with $0 \le i - j \le k-1$. So if we substitute z_i

for $b_{i-j}^{(j)}$ in the right-hand side of formula (6.2-4) then we get $D_{n+1}^*(z^*) = D_{n+k}(z)$. The left-hand side of this equality is assumed to be non-empty. Since this holds for every $n \in \mathbb{Z}^+$, every initial block (and therefore, every subblock) of z is $(\mathcal{P}, f)$-allowed, which means that $z \in Z$. Next, note that (6.2-6) implies that

$$b^{(j)} = b_0^{(j)} b_1^{(j)} \dots b_{k-1}^{(j)} = b_0^{(j)} b_0^{(j+1)} \dots b_0^{(j+k-1)} = z_{[j\,;\,j+k-1)}$$

for every $j \in \mathbb{Z}^+$. This means that $z^* = \mathrm{I}_\infty^{(k)}(z)$; see, again, the definition of the mapping $\mathrm{I}_\infty^{(k)}$ in (5.4-3). Conclusion: $Z_k^* \subseteq \mathrm{I}_\infty^{(k)}[Z] = Z^{(k)}$.

Finally, consider an arbitrary point $z \in Z$, or, what amounts to the same, consider an arbitrary point $\mathrm{I}_\infty^{(k)}(z)$ in Z_k^*. Since $D_{n+1}^*(\mathrm{I}_\infty^{(k)}(z)) = D_{n+k}(z)$ for all $n \in \mathbb{Z}$ it is clear that

$$\bigcap_{n\in\mathbb{N}} \overline{D_n^*(\mathrm{I}_\infty^{(k)}(z))} = \bigcap_{n\in\mathbb{N}} \overline{D_{n+k-1}(z)} = \{\psi(z)\}\,.$$

This implies that $\mathcal{P}_k^*$ is a pseudo-Markov partition (the left-hand intersection consists of one point) and that $\psi_k^*(\mathrm{I}_\infty^{(k)}(z)) = \psi(z)$ for all $z \in Z$. □

Theorem 6.2.11. *Let (X, f) be an expansive dynamical system on a compact metric space X and let $\mathcal{P}$ be a topological partition of X all of whose members have a diameter less than the expansive coefficient η of f. Finally, assume that $\mathcal{P}$ is a Markov partition, generating a symbolic representation $\psi\colon (Z, \sigma_Z) \to (X, f)$ in which ψ is a surjection. Then:*

(1) *The factor mapping ψ is boundedly finite-to-one.*
(2) *If, in addition, (X, f) is transitive then ψ is almost 1-to-1 and (Z, σ_Z) is transitive as well. In addition, there is a dense G_δ-set of transitive points in X that have a singleton fibre under ψ.*

Proof. We want to apply Lemma 6.2.9, but to do so we need a topological partition whose members have a diameter of at most $\frac{1}{2}\eta$. This is not the case for $\mathcal{P}$ and therefore we shall consider a refinement of $\mathcal{P}$.

By Lemma 6.2.10, for every $k \in \mathbb{N}$ the family $\mathcal{P}_k^*$ of all sets $D_k(b)$ with b a $(\mathcal{P}, f)$-allowed block of length k is a Markov-partition. It defines the symbolic representation $\psi_k^*\colon Z^{(k)} \to X$ such that $\psi_k^* \circ \mathrm{I}_\infty^{(k)} = \psi$. By Lemma 6.2.2, all members of $\mathcal{P}_k^*$ have a diameter less than $\frac{1}{2}\eta$ for almost all k. Moreover, Corollary 5.4.7 implies that $X^{(k)}$ is an SFT of order 2 for almost all k.

So there exists k so large that Lemma 6.2.9 above can be applied to $\mathcal{P}_k^*$ and ψ_k^*. It follows that ψ_k^* is boundedly finite-to-one, say, every fibre of ψ_k^* has at most N points. Now observe that, again by Lemma 6.2.10, $\mathrm{I}_\infty^{(k)}$ maps fibres of ψ bijectively onto those of ψ_k^*. Consequently, also all fibres of ψ have at most N elements. So ψ is boundedly finite-to-one.

If (X, f) is transitive then Lemma 6.2.9 also implies that ψ_k^* is an almost 1-to-1 mapping and that the shift on $Z^{(k)}$ is transitive. By a similar reasoning as above it follows that ψ is almost 1-to-1 and that the shift on Z is transitive. Moreover, the proof of Lemma 6.2.9 shows: if $T' := U_{\mathcal{P}_k^*} \cap T_0$, where T_0 is a dense G_δ-set of transitive points in X, then T' is a dense G_δ-set consisting of transitive points in X and all points of T' have singleton fibres (for ψ_k^*, hence for ψ). □

Remark. In the above theorem the conditions that X^* is dense or that f is semi-open are not explicitly mentioned. However, $\mathcal{P}$ is assumed to be a Markov partition, so by Corollary 6.1.13 the condition that ψ is surjective is equivalent to the condition that f is semi-open, which by itself already implies that the representation mapping ψ is almost 1-to-1: the points of the dense G_δ-set X^* have singleton ψ-fibres; see Corollary 6.1.12. So if (X, f) is transitive and f is semi-open then under the conditions of Theorem 6.2.11 a dense G_δ-set of transitive points is added to X^* as the set of points with singleton ψ-fibres.

6.3 Applications

Let notation be as before. In particular, (X, f) is a dynamical system with a compact Hausdorff phase space X and $\mathcal{P} = \{P_0, \dots, P_{s-1}\}$ is a topological partition of X. We can assign to $\mathcal{P}$ a vertex-labelled directed graph G in the following way: the set of vertices of G is labelled faithfully by the symbol set $\mathbf{S}$, and there is an edge from the vertex with label α to the vertex with label β iff $P_\alpha \cap f^\leftarrow[P_\beta] \neq \emptyset$, iff $f[P_\alpha] \cap P_\beta \neq \emptyset$, iff $\alpha\beta$ is a partial itinerary.

Obviously, every $(\mathcal{P}, f)$-allowed allowed block is represented by a finite path in G, but in general, the converse is not true. Stated otherwise, the shift space $Z = Z(\mathcal{P}, f)$ defined by $\mathcal{P}$ is included in $\mathfrak{M}_v(G)$ and the inclusion may be proper (see the example below). We shall present a condition which implies that the shift space Z is equal to the shift space $\mathfrak{M}_v(G)$.

We say that a topological partition $\mathcal{P}$ of X has property **(M)** whenever it is f-adapted and, in addition, the following condition holds: if b is a finite block such that $P_{b_i} \cap f^\leftarrow[P_{b_{i+1}}] \neq \emptyset$ for $i = 0, \dots, |b| - 2$ then $D_{|b|}(b) \neq \emptyset$. Stated otherwise, an f-adapted topological partition $\mathcal{P}$ has property **(M)** whenever a block is $(\mathcal{P}, f)$-allowed iff all of its 2-blocks are $(\mathcal{P}, f)$-allowed. The latter condition means that the allowed blocks correspond to finite walks in the graph G defined above. By the definition of $\mathfrak{M}_v(G)$ in Section 5.5 this implies that $Z = \mathfrak{M}_v(G)$; see also Lemma 5.5.1. So if $\mathcal{P}$ has property **(M)** then $Z = \mathfrak{M}_v(G)$ and, consequently, Z is an SFT of order 2.

Example. In Figure 6.3 (a) we have sketched the graph of a continuous mapping f of the interval $[0; 1]$ into itself with a topological partition $\mathcal{P} = \{P_0, P_1, P_2\}$. Then $\mathcal{P}$ is f-adapted: the shift space Z consists of the orbits of the periodic points $(02)^\infty$ and $(12)^\infty$ (all points of $[0; 1]$ are periodic under f). The 2-blocks 02 and 21 are allowed, but the block 021 is not: if $x \in P_0$ then $f(x) \in P_2$ and $f^2(x) \in P_0$. So $\mathcal{P}$ does not have property **(M)**. The graph G is shown below. The shift space $\mathfrak{M}_v(G)$ contains, among others, the point $(021)^\infty$, which is not in Z because the block 021 is not allowed. So Z is a proper subset of $\mathfrak{M}_v(G)$.

Lemma 6.3.1. *Let $\mathcal{P}$ be an f-adapted topological partition. Consider the following properties:*
(i) *$\mathcal{P}$ has property* (**M**).
(ii) $Z = \mathfrak{M}_v(G)$.
Then always (i)⇒(ii) *and Z is an SFT of order 2. If G has no stranded edges then also the implication* (ii)⇒(i) *holds.*

Proof. "(i)⇒(ii)": Clear from the discussion preceding the above example.

"(ii)⇒(i) provided G has no stranded edges": Let b be a finite block over $\mathbf{S}$ and suppose all its 2-blocks are $(\mathcal{P}, f)$-allowed. Then b is represented by a finite path in G. Assuming that G has no stranded edges, b can be extended to an infinite path in G. Consequently, the block b occurs in a point of $\mathfrak{M}_v(G)$. If (ii) holds this means that b occurs in a point of Z, i.e., b is Z-present, hence certainly $(\mathcal{P}, f)$-allowed. □

Example. In Figure 6.3 (b) we have sketched the graph of a continuous mapping f of the interval $[0; 1]$ into itself with a topological partition $\mathcal{P} = \{P_0, P_1, P_2\}$. Then $\mathcal{P}$ is f-adapted, and Z includes the full shift Ω_2 (considered in the obvious way as a subshift of $\Omega_{\{0,1,2\}}$). In fact, the subsystem on $Y := [0; 1/2]$ (together with the topological partition $\mathcal{P} \cap Y$) is conjugate to the system of the tent map, hence the set of all partial itineraries of points of Y under f is the same as for the tent map. According to Example E in 6.3.5 below, this set consists of all finite blocks of 0's and 1's. Moreover, partial itineraries of points $x \in (1/2; 1]$ all have the form 1^n with $n \in \mathbb{N}$ or 1^n2 with $n \in \mathbb{Z}^+$. Obviously, the latter blocks are not Z-present: the symbol 2 cannot be followed by any symbol.

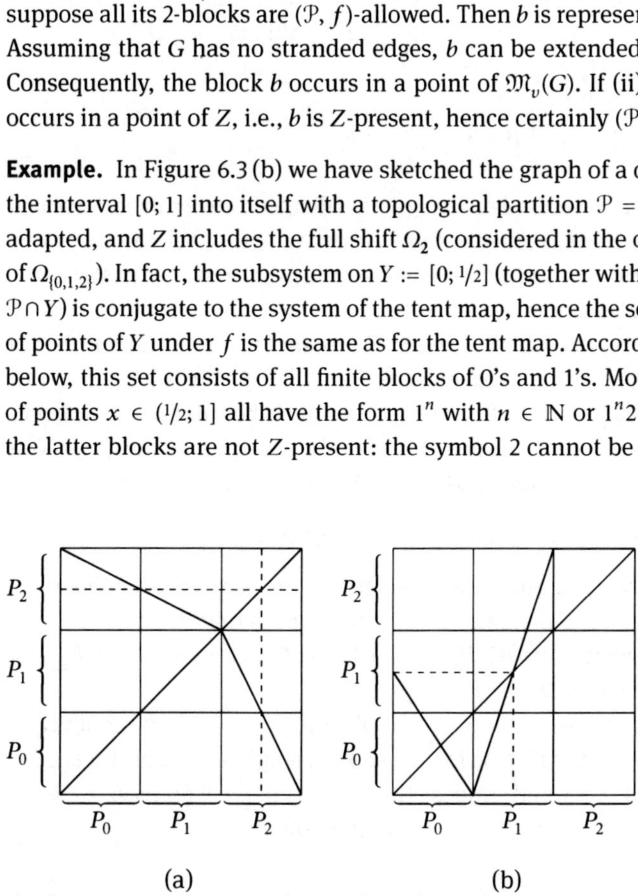

Fig. 6.3. Two examples of an f-adapted partition $\mathcal{P}$ not having property (**M**). In (a), $Z \neq \mathfrak{M}_v(G)$ and in (b), $Z = \mathfrak{M}_v(G)$.

It follows that $Z = \Omega_2$. The graph G for this topological partition is sketched below. There is one stranded edge – which does not contribute to $\mathfrak{M}_v(G)$ – so G defines the full shift on the two symbols 0 and 1. Consequently, $Z = \mathfrak{M}_v(G)$. However, $\mathcal{P}$ does not have property (**M**): the blocks 01 and 12 are allowed, but 012 is not: if $x \in P_0$ then $f^2(x) \notin P_2$. Finally, note that $\mathcal{P}$ is a pseudo-Markov partition (hence a Markov partition). To prove this, use that $\mathcal{P} \cap Y$ is a Markov partition for $f|_Y$ on Y (see Example E in 6.3.5 below). Moreover, blocks of the form $1^k 2$ are obtained as partial itineraries of points in the interval $[^1/_2; {}^1/_2 + {}^2/_{2\cdot 3^k}]$, which shrink with increasing k.

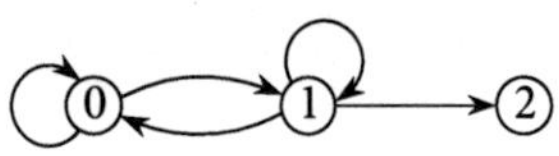

The following condition is more restrictive, but easier to check, than property (**M**). We say that $\mathcal{P}$ has property (**M***) whenever the following conditions are fulfilled:

(1) $\forall\,(\alpha,\beta) \in \mathbf{S}\times\mathbf{S} : f[P_\alpha] \supseteq P_\beta$ or $f[P_\alpha] \cap P_\beta = \emptyset$;

(2) $\exists\,(\alpha,\beta) \in \mathbf{S}\times\mathbf{S} \vdots\; f[P_\alpha] \supseteq P_\beta$;

(3) $\forall\,(\kappa,\alpha) \in \mathbf{S}\times\mathbf{S} : f[P_\kappa] \supseteq P_\alpha \Rightarrow \exists\,\beta \in \mathbf{S} : f[P_\alpha] \supseteq P_\beta$.

If $\mathcal{P}$ has property (**M***) then the vertex-labelled graph G defined above can be described more precisely as follows: there is an edge from the vertex with label α to the vertex with label β iff $f[P_\alpha] \supseteq P_\beta$. So condition (2) just requires that $\mathcal{E}_2(G) \neq \emptyset$. Moreover, condition (3) means that G has no stranded edges.

If $\mathcal{P}$ satisfies condition (1) *and X^* is dense in X then the conditions* (2) *and* (3) *are satisfied.* For in that case for every $\alpha \in \mathbf{S}$ one has $P_\alpha \cap X^* \neq \emptyset$. If x is a point of this intersection then obviously $f(x) \in P_\beta$ for some $\beta \in \mathbf{S}$ (this is because $x \in X^*$), hence $f[P_\alpha] \cap P_\beta \neq \emptyset$ (this is because $x \in P_\alpha$), Consequently, $f[P_\alpha] \supseteq P_\beta$ by condition (1). This shows that every vertex has an outgoing edge, so (2) and (3) hold.

Lemma 6.3.2. *If $\mathcal{P}$ has property* (**M***) *then it has property* (**M**).

Proof. Let $\mathcal{P}$ have property (**M***). The properties (2) and (3) together imply that there are infinite walks in the graph G defined above, which means that $Z \neq \emptyset$, i.e., $\mathcal{P}$ is f-adapted. In addition, we want to prove: if b is a block over $\mathbf{S}$ such that $f[P_{b_i}] \cap P_{b_{i+1}} \neq \emptyset$ for every $i \in \{0, \dots, |b| - 2\}$ then $D_{|b|}(b) \neq \emptyset$.

The proof is by induction on the length k of the block b. For $k = 2$ there is nothing to prove. Next, assume that the claim is true for all blocks of length k for some integer $k \geq 2$. Let b be a block of length $k+1$ and suppose that $f[P_{b_i}] \cap P_{b_{i+1}} \neq \emptyset$ for $i = 0, \dots, k-1$. Since $D_{k+1}(b) = P_{b_0} \cap f^{\leftarrow}[D_k(b')]$ with $b' := b_{[1\,;\,k+1)}$, it follows that

$$f[D_{k+1}(b)] = f[P_{b_0}] \cap D_k(b') = D_k(b')\,. \tag{6.3-1}$$

Here the final equality is justified by the fact that $D_k(b') \subseteq P_{b_1}$ whereas the assumptions imply that $P_{b_1} \subseteq f[P_{b_0}]$. The induction hypothesis applies to the block b', so $D_k(b') \neq \emptyset$. It follows that $D_{k+1}(b) \neq \emptyset$ as well. □

Remarks.

(1) The proof of the lemma shows: If $\mathcal{P}$ is f-adapted and only condition (1) is satisfied then $\mathcal{P}$ already has condition (**M**).

(2) Let $\mathcal{P}$ have property (**M***) and let $z \in Z$. Then formula (6.3-1) implies

$$f[D_{k+1}(z)] = f[P_{z_0}] \cap D_k(\sigma z)) = D_k(\sigma z)\,. \tag{6.3-2}$$

See also formula (6.1-7). It follows easily by induction from (6.3-1) that for $j = 1, \ldots, k$ we have $f^j[D_{k+1}(b)] = D_{k+1-j}(b_{[j;k+1)})$. In particular,

$$f^k[D_{k+1}(b)] = P_{b_k} \quad \text{and} \quad f^{k+1}[D_{k+1}(b)] = f[P_{b_k}]\,. \tag{6.3-3}$$

If $z \in Z$ then in (6.3-3) one may substitute z for b, and the equality so obtained holds for all $k \in \mathbb{Z}^+$.

Corollary 6.3.3. *Let (X, f) be a dynamical system on a compact Hausdorff space and assume that f is semi-open. Moreover, let $\mathcal{P}$ be a pseudo-Markov partition with property* (**M***)*. Then $\mathcal{P}$ is a Markov partition, Z is an SFT of order 2 and $\psi\colon (Z, \sigma_z) \to (X, f)$ is an almost 1,1-factor mapping.*

Proof. Use Corollary 6.1.12, Lemma 6.3.2 and Lemma 6.3.1. □

Remark. By Lemma 6.3.1, if $\mathcal{P}$ is a pseudo-Markov partition with property (**M***) then $\mathcal{P}$ is a Markov partition. So by Corollary 6.1.13 (1) it is also necessary that f is semi-open for ψ to be surjective.

We apply the above to a dynamical system (X, f) on a non-degenerate bounded closed interval $X = [a; b]$. Let $s \in \mathbb{N}$ and $\mathcal{P} = \{P_0, \ldots, P_{s-1}\}$ with $P_i = (a_i; a_{i+1})$ for $i = 0, \ldots, s-1$, where $a = a_0 < a_1 < \cdots < a_s = b$. Motivated by Proposition 6.1.15, we assume that f is strictly monotonous on each of the intervals $\overline{P}_i$, which is necessary and sufficient for X^* to be dense in X. In this situation, property (**M***) is easily seen to be equivalent to the condition $f[\{a_0, \ldots, a_s\}] \subseteq \{a_0, \ldots, a_s\}$.

Proposition 6.3.4. *Let (X, f) and $\mathcal{P}$ be as above.*

(1) *For every $k \in \mathbb{N}$ and every $(\mathcal{P}, f)$-allowed k-block b, $D_k(b)$ is an open interval and the function f^k is strictly monotonous on this interval, mapping it into the interval $f[P_{b_{k-1}}]$. If $\mathcal{P}$ has property* (**M***) *then f^k maps $D_k(b)$ onto the interval $f[P_{b_{k-1}}]$.*

(2) *If $\mathcal{P}$ is a pseudo-Markov partition then $\psi\colon Z \to X$ is at most 2-to-1.*

Proof. (1) First, we show that for every $k \in \mathbb{N}$ and every $(\mathcal{P}, f)$-allowed k-block b the set $D_k(b)$ is an open interval. The proof is by induction in k. For $k = 1$ this statement is true: $D_1(b) = P_{b_0}$ is given to be an open interval for every 1-block $b = b_0$. Next, let

$k \in \mathbb{N}$ and assume that for every $(\mathcal{P}, f)$-allowed k-block b' the set $D_k(b')$ is an open interval. Let b be a $(\mathcal{P}, f)$-allowed $(k+1)$-block and recall that $D_{k+1}(b) = P_{b_0} \cap f^{\leftarrow}[D_k(b')]$, where $b' := b_{[1;k]}$. So the set $D_{k+1}(b)$ is the inverse image in P_{b_0} of the set $D_k(b')$ under the restriction of f to P_{b_0}, which is a strictly monotonous continuous function. By the induction hypothesis, $D_k(b')$ is an open interval. Consequently, the set $D_{k+1}(b)$ is an open interval as well. This concludes the proof that $D_k(b)$ is an open interval for every $k \in \mathbb{N}$ and every $(\mathcal{P}, f)$-allowed k-block b.
Next, we show that the function f^k is strictly monotonous on $D_k(b)$ for every $k \in \mathbb{N}$ and every $(\mathcal{P}, f)$-allowed k-block b. The proof is also by induction in k. For $k = 1$ the statement is true: f is assumed to be strictly monotonous on the interval $D_1(b) = P_{b_0}$ for every 1-block $b = b_0$. Assume that the statement is true for $k \in \mathbb{N}$ and for every $(\mathcal{P}, f)$-allowed k-block. Let b be any $(\mathcal{P}, f)$-allowed $(k+1)$-block. First, note that $D_{k+1}(b) \subseteq P_{b_0}$ so that f is strictly monotonous on P_{b_0}, hence on $D_{k+1}(b)$. Moreover,

$$f[D_{k+1}(b)] = f[P_{b_0}] \cap D_k(b') \tag{6.3-4}$$

with $b' := b_{[1;k+1)}$; see the first equality in (6.3-1). Hence $f[D_{k+1}(b)]$ is a subset of $D_k(b')$, on which the mapping f^k is assumed to be strictly monotonous. Consequently, $f^{k+1}|_{D_{k+1}(b)} = f^k|_{D_k(b')} \circ f|_{D_{k+1}(b)}$ is the composition of two strictly monotonous functions, hence is strictly monotonous as well.
Finally, it follows from equation (6.3-4) that $f[D_k(b)] \subseteq D_{k-1}(b_{[1;k)})$ for every $k \geq 2$ and every $(\mathcal{P}, f)$-allowed k-block b. It follows easily by induction that $f^k[D_k(b)] \subseteq f[D_1(b_{[k-1,k)})] = f[P_{b_{k-1}}]$. That we have equality here if $\mathcal{P}$ has property (**M***) follows from formula (6.3-3).
(2) Consider two different points $z, z' \in Z$ such that $\psi(z) = \psi(z') =: x$. Let k be the smallest non-negative integer such that $z_k \neq z'_k$. Then it is clear that $z_{[0;k)} = z'_{[0;k)}$, so that $D_k(z) = D_k(z')$. On the other hand, the (open) sets $D_{k+1}(z) = D_k(z) \cap (f^k)^{\leftarrow}[P_{z_k}]$ and $D_{k+1}(z') = D_k(z') \cap (f^k)^{\leftarrow}[P_{z'_k}]$ are disjoint, because the intervals P_{z_k} and $P_{z'_k}$ are different, hence disjoint. Next, recall from 1 that $D_{k+1}(z)$ and $D_{k+1}(z')$ are open intervals. Moreover, $\psi(z)$ is in the closure of the first of these intervals, and $\psi(z')$ is in the closure of the second one. So the point x is a common end point of these intervals. It follows easily that for all $l \geq k+1$ the open intervals $D_l(z)$ and $D_l(z')$ are disjoint as well and that they have x as a common end point. Now suppose there is a third point z'' in Z which is different from z and from z' and is for which $\psi(z'') = x$. Then for sufficiently large l the open interval $D_l(z'')$ must be disjoint from the intervals $D_l(z)$ and $D_l(z')$, but it must have x as an end point. This is impossible in a space like $\mathbb{R}$; see Figure 6.4 below[4]. □

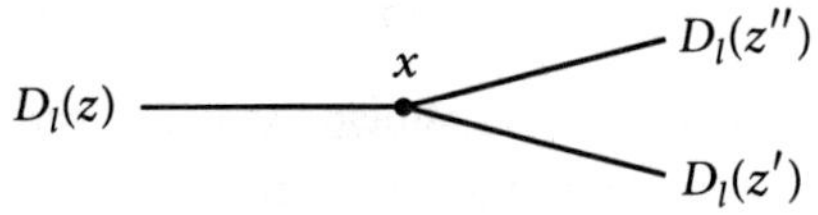

Fig. 6.4. This situation is impossible in $\mathbb{R}$.

4 This suggests other spaces where a similar result holds, e.g., $\mathbb{S}$.

6.3.5. We present a few simple examples (in no particular order), illustrating the preceding theory. Admittedly, most of these examples are not very exciting, but they may clarify some issues. In particular, they show that, in general, one has to choose a topological partition carefully, adapted to the system under consideration. Actually, constructing a (pseudo-)Markov partition is, if possible at all, not always easy. In all examples we consider a dynamical system (X, f) with $X = [0; 1]$, where f is given by its graph. The partition $\mathcal{P}$ is indicated in the obvious way by horizontal and vertical separation lines in the unit square. In each case we describe X^*, ι and the symbolic representation $\psi\colon (Z, \sigma_Z) \to (X, f)$ and we investigate whether $\mathcal{P}$ is a pseudo-Markov partition. The proofs are rather sketchy. Everything boils down to the search for partial itineraries. Unless stated otherwise (as in Example D) we assume that $0, 1 \notin U_{\mathcal{P}} := \bigcup \mathcal{P}$.

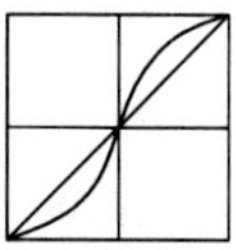

Example A.

- $X^* = P_0 \cup P_1$: the sets P_0 and P_1 are invariant. That X^* is dense in X is in accordance with the fact that f is semi-open.
- $Z = \{0^\infty, 1^\infty\}$: the only partial itineraries are blocks of 0's and blocks of 1's of any length, which can be realized by starting anywhere in P_0 or P_1, respectively. This is in accordance with the Lemma's 6.3.1 and 6.3.2, as $\mathcal{P}$ has property $(\mathbf{M}^*)$ and the graph of $\mathcal{P}$ is equal to the 5th example in the table on page 246.
- $\mathcal{P}$ is not a pseudo-Markov partition: $\iota[P_0] = \{0^\infty\}$, $\iota[P_1] = \{1^\infty\}$, so the mapping ι is not injective; now use Proposition 6.1.10 (1).

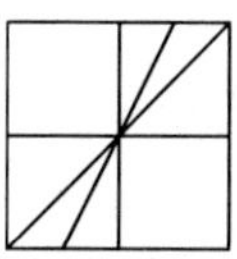

Example B.

- $X^* = \emptyset$: every point in $U_{\mathcal{P}}$ is ultimately 0 or 1, hence no orbit remains in $U_{\mathcal{P}}$.
- $Z = \{0^\infty, 1^\infty\}$: by taking an initial point sufficiently close to the invariant point $1/2$ one gets partial itineraries of prescribed length, consisting of either 0's (left of $1/2$) or 1's (right of $1/2$), and there are no other partial itineraries. So $Z = \mathfrak{M}_v(G)$ with G as in Example A.
- $\mathcal{P}$ is a pseudo-Markov-partition (even a Markov partition, because Z is an SFT) and $\psi(0^\infty) = \psi(1^\infty) = 1/2$: for every $k \in \mathbb{N}$, the set $D_k(0^\infty)$ is an open interval of the form $(a_k; 1/2)$, where a_k increases to $1/2$ if k tends to infinity. It follows that the closure of the set $D_k(0^\infty)$ shrinks to the point $1/2$. Similarly, the closure of $D_k(1^\infty)$ shrinks to the point $1/2$.

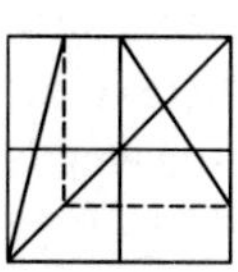

Example C.

- $X^* = \{a_k : k \in \mathbb{Z}^+\}$, where a_k is the unique point in X such that $f^k(a_k)$ equals the invariant point $a_0 := 7/10$, that is, $a_k := 7 \cdot (10 \cdot 4^k)^{-1}$ for $k \in \mathbb{Z}^+$. All other points get trapped in the periodic orbit

$\{1/4, 1\}$, which is not included in $U_{\mathcal{P}}$. Note that X^* is a sequence that converges to the point 0; it is neither closed nor dense in X.
- $Z = \{0^\infty\} \cup \{0^k 1^\infty : k \in \mathbb{Z}^+\}$: by taking an initial point sufficiently close to the invariant point 0 in X and sufficiently close to one of the points a_k one can get partial itineraries consisting of blocks of 0's of prescribed length, followed by blocks of 1's of arbitrary length. So Z is the SFT of order 2, defined by the forbidden block 10 (see the second graph on page 246). However, the graph G of $\mathcal{P}$ defines the full shift on $\{0, 1\}$, and therefore $Z \neq \mathfrak{M}_v(G)$. So the topological partition $\mathcal{P}$ cannot have property (**M**).
- $\mathcal{P}$ is pseudo-Markov (hence Markov): the reasoning is similar as in the previous example. One finds that $\psi(0^\infty) = 0$ and $\psi(0^k 1^\infty) = a_k$ for $k \in \mathbb{Z}^+$.
- $\iota[X^*] = Z \setminus \{0^\infty\}$ is dense in Z. Accordingly, X^* is dense in $\psi[Z]$.

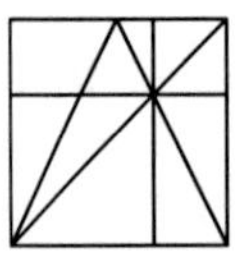

Example D (the tent map).
- with $P_0 := [0; 2/3)$, $P_1 := (2/3; 1]$
- $X^* = X \setminus A$, where A is the set of the preimages of the point $2/3$ under all iterates of f (countably many points); note that the points 0 and 1 are in $U_{\mathcal{P}}$. Obviously, X^* is dense in X, which is in accordance with the fact that f is semi-open.
- Z is the golden mean shift (with 0 and 1 interchanged as in Example (3) in Section 5.4): clear from Lemma 6.3.1, because $\mathcal{P}$ has property (**M***) with $f[P_0] \supseteq P_0 \cup P_1$ and $f[P_1] \supseteq P_0$.
- $\mathcal{P}$ is not a pseudo-Markov partition: clear from Proposition 6.1.15, as f is not monotonous on P_0. Of course, Proposition 6.1.10 (1) can also be used, as ι is not injective on X^*: select a point y in X^* between $1/2$ and $2/3$, then the points y and $1 - y$ have the same itinerary because $f(1 - y) = f(y)$.

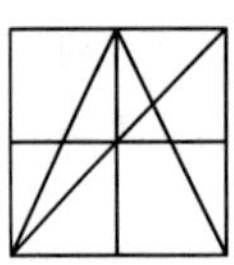

Example E (the tent map).
- $X^* = (P_0 \cup P_1) \setminus A$, where A is the set of the preimages of the point $1/2$ under all iterates of f (countably many points). Obviously, X^* is a dense subset of X.
- Z is the SFT defined by the complete directed graph with two vertices 0 and 1 and, consequently, $Z = \Omega_2$: use the Lemma's 6.3.2 and 6.3.1.
- $\mathcal{P}$ is a pseudo-Markov partition: by formula (6.3-2), if $z \in Z$ and $k \in \mathbb{N}$ then $f[D_{k+1}(z)] = D_k(\sigma z)$. Taking into account that $D_k(z) \subseteq P_{z_0} = P_i$ for $i = 0$ or 1 and that f stretches $\overline{P}_i$ by a factor 2 over the unit interval, one easily shows that the preimage in $\overline{P}_i$ under f of any subset of the unit interval is a copy of that subset, scaled by a factor $1/2$. By induction it follows that, for every $k \in \mathbb{N}$, $D_k(z)$ is an interval – see also Proposition 6.3.4 (1 –) and that its length, hence also the length of its closure, is 2^{-k}.
- Some values of ψ: $\psi(0^\infty) = 0$, $\psi(1^\infty) = \frac{2}{3}$, $\psi(010^\infty) = \psi(110^\infty) = \frac{1}{2}$, $\psi(01^\infty = \frac{1}{3}$ (preimage of $\frac{2}{3}$ in P_0 under T), $\psi(001^\infty) = \frac{1}{6}$ (preimage of $\frac{1}{3}$ in P_0 under T),

$\psi(101^\infty) = \frac{5}{6}$ (preimage of $\frac{1}{3}$ in P_1 under T) and $\psi(10^\infty) = 1$ (in order to prove this, determine the sets $D_k(z)$ ($k \in \mathbb{N}$) for these points, or use that $T \circ \psi = \psi \circ \sigma$).

NB. The preimages in Z under ψ of the points 0 and 1 of $X \setminus X^*$ are singleton sets. This can be explained as follows: replace $\mathcal{P}$ by the topological partition $\{ P_0 \cup \{0\}, P_1 \cup \{1\} \}$; then essentially nothing changes, except that the points 0 and 1 now belong to X^*.

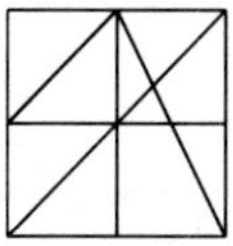

Example F.

- $X^* = U_{\mathcal{P}} \setminus B = (P_0 \cup P_1) \setminus B$, where B is the set of all preimages of the point $1/2$ under all iterates of f.
- Z is the golden mean shift: by Lemma 6.3.1 it is the shift space determined by the 7th graph in the table on page 246.
- $\mathcal{P}$ is a pseudo-Markov partition: As in Example E, use formula (6.3-2), but now use that f stretches $\overline{P}_1$ uniformly over the full unit interval by a factor 2 but that it maps $\overline{P}_0$ isometrically onto $\overline{P}_1$. By induction this implies that each set $D_k(z)$ for $k \in \mathbb{N}$ and $z \in Z$ is an open interval of length $2^{-\kappa(k,z)}$, where $\kappa(k, z)$ is the number of 1's in the block $z_{[0;k)}$. Since the 0's are isolated in z, this number as at least $\frac{1}{2}(k-1)$: the worst possible cases are $10\,10\ldots 10$ and $01\,01\ldots 01$ for even k, and $01\,01\ldots 01\,0$ for odd k. Consequently, the closures of the sets $D_k(z)$ shrink to singletons.

NB. The point 0 is periodic with period 3. This example illustrates a particular case of the situation used in the proof of the Li–Yorke Theorem: see (2.2-2) and (2.2-3). In addition, the golden mean shift is transitive and it has a dense set of periodic points because it is irreducible: see the Example after Proposition 5.3.7. So (X, f) has these properties as well.

6.3.6 (The Cantor set revisited). Let $f\colon \mathbb{R} \to \mathbb{R}$ be defined by

$$f(x) := \tfrac{3}{2}\left(1 - |2x - 1|\right) \quad \text{for } x \in \mathbb{R}\,.$$

Let C be the Cantor set and recall from 1.7.5 that C is the largest f-invariant subset of $\mathbb{R}$. In point of fact, $C = \Lambda$, where

$$\Lambda := \bigcap_{n=0}^{\infty} (f^n)^{\leftarrow}[0;1]\,.$$

Recall also from 1.7.5 that Λ is completely invariant under f. In what follows we use the notation of 1.7.5. In particular, the reader may re-read the definition and properties of the intervals I_b^n for $n \in \mathbb{N}$ and $b \in \{0,1\}^n$ (i.e., b an n-block).

For $i = 0, 1$, let $P_i := \Lambda \cap I_i^1$. Then $\mathcal{P} := \{P_0, P_1\}$ is a topological partition of Λ; in fact, it is a clopen partition. Consequently, every point of Λ has a full itinerary with respect to $\mathcal{P}$, i.e., $\Lambda^* = \Lambda$. So $\mathcal{P}$ is f-adapted and the symbolic representation $Z := Z(\mathcal{P}, f|_\Lambda)$ is a shift space.

(1) *If b is a finite block with length $k \geq 1$ over the symbol set $\{0,1\}$ then*

$$D_k(b) = \Lambda \cap \bigcap_{n=0}^{k-1} (f^n)^{\leftarrow}[I^1_{b_n}]\,. \tag{6.3-5}$$

It follows that $D_k(b) = \Lambda \cap I^k_b$ for every k-block b.

Proof. For the proof of (6.3-5), use the definition of $D_k(b)$ and the equalities

$$(f^n)^{\leftarrow}[P_{b_n}] = (f^n)^{\leftarrow}[\Lambda \cap I^1_{b_n}] = \Lambda \cap (f^n)^{\leftarrow}[I^1_{b_n}]\,.$$

For the proof of the final statement we introduce the following notation: if $k \in \mathbb{N}$ and $b \in \{0,1\}^k$ then $D^*_k(b) := \bigcap_{n=0}^{k-1}(f^n)^{\leftarrow}[I^1_{b_n}]$. Clearly, (6.3-5) means that $D_k(b) = \Lambda \cap D^*_k(b)$, so we have to prove that $D^*_k(b) = I^k_b$ for every $k \in \mathbb{N}$ and every k-block b over $\{0,1\}$. As an intermediate step we first prove that

$$I^{k+1}_{bj} = I^k_b \cap (f^k)^{\leftarrow}[I^1_j] \tag{6.3-6}$$

for all $k \in \mathbb{N}$, all $b \in \{0,1\}^k$ and all $j \in \{0,1\}$. The proof is by induction. First, note that for every $n \in \mathbb{N}$, for every n-block d and for every $i \in \{0,1\}$ the interval I^{n+1}_{di} is is defined as the subinterval of I^n_d that is mapped by f onto $I^n_{d_1 \dots d_{n-1} i}$, which implies that

$$I^{n+1}_{di} = I^n_d \cap f^{\leftarrow}[I^n_{d_1 \dots d_{n-1} i}]\,. \tag{6.3-7}$$

If $n = 1$ then this proves (6.3-6) for $k = 1$, every 1-block b and $j \in \{0,1\}$. Now assume that (6.3-6) holds for a certain value of k, for all k-blocks b and $j \in \{0,1\}$. Then for any $(k+1)$-block $b = b_0 \dots b_k$ and $j \in \{0,1\}$ we have, by (6.3-7) for $n = k+1$:

$$I^{k+2}_{bj} = I^{k+1}_b \cap f^{\leftarrow}[I^{k+1}_{b_1 \dots b_k j}]\,.$$

By the induction hypothesis, the second term in right-hand side in this equality is equal to $f^{\leftarrow}[I^k_{b_1 \dots b_k}] \cap (f^{k+1})^{\leftarrow}[I^1_j]$. Since $I^{k+1}_b \subseteq f^{\leftarrow}[I^k_{b_1 \dots b_k}]$ by (6.3-7) (with $n = k$, $c = b_{[0;k)}$ and $j = b_k$), we get

$$I^{k+2}_{bj} = I^{k+1}_b \cap (f^{k+1})^{\leftarrow}[I^1_j]\,.$$

This completes the proof of (6.3-6).

The proof of the desired equality $D^*(b) = I^k_b$ ($k \in \mathbb{N}$, b a k-block) is by induction as well. For $k = 1$ the equality is obviously true. Suppose it is true for some $k \in \mathbb{N}$ and every k-block b. Then for any $(k+1)$-block bj with b an arbitrary k-block and $j \in \{0,1\}^n$ we have

$$D^*_{k+1}(bj) = D^*_k(b) \cap (f^k)^{\leftarrow}[J^1_j] = I^k_b \cap (f^k)^{\leftarrow}[I^1_j]\,.$$

By (6.3-6), the right-hand side is equal to I^{k+1}_{bj}. This completes the proof. □

(2) *$\mathcal{P}$ is a Markov partition and $Z = \Omega_2$. Consequently, $(\Lambda, f|_\Lambda)$ is conjugate to the full shift (Ω_2, σ).*

Proof. Recall that for $k \in \mathbb{N}$ and $b \in \{0, 1\}^k$ the intervals I_b^k have a length of 3^{-k}. Hence 1 implies that or every $z \in Z$ the diameter of $D_k(z)$ tends to zero if k tends to infinity. This implies that $\mathcal{P}$ is a pseudo-Markov partition. Moreover, f is injective on I_i^1 for $i = 0, 1$ and $f[\Lambda] = \Lambda$, so

$$f[P_i] = f[\Lambda \cap I_i^1] = f[\Lambda] \cap f[I_i^1] = \Lambda \cap [0; 1] = P_0 \cup P_1 \,.$$

Consequently, $\mathcal{P}$ has property $(\mathbf{M}^*)$ and $Z = \Omega_2$: the graph G defined by the partition $\mathcal{P}$ is the full graph with two vertices. As $\mathcal{P}$ is a clopen partition of Λ, Corollary 6.1.13 (2) implies that the representation mapping ψ is a conjugation. □

Remarks.

(1) There is a slightly different approach which avoids the need to identify the sets $D_k(b)$ as in 1 above: show that there is a metric on Λ such that f is expansive with coefficient larger than $1/3 = \operatorname{diam} P_0 = \operatorname{diam} P_1$ and apply 6.2.2 to show that $\mathcal{P}$ is a pseudo-Markov partition. In order to prove that f is expansive, note that if x and y are two distinct points in Λ then there exists $n \in \mathbb{Z}^+$ such that $f^n(x)$ and $f^n(y)$ do not both belong to $I_0^{(1)}$ or to $I_1^{(1)}$ (we leave the proof to the reader). So there exists $n \in \mathbb{Z}^+$ such that the points $f^n(x)$ and $f^n(y)$ have distance at least $1/3$. This shows that the mapping f is expansive with expansive coefficient $1/3$. This is not sufficient to apply Lemma 6.2.2, but the following trick does the job: change the metric on $\mathbb{R}$ by doubling all distances within the interval $[\frac{1}{3}; \frac{2}{3}]$. The metric so obtained is compatible with the topology of Λ (on Λ nothing changes), but it makes f expansive with coefficient $2/3$.

(2) Replace the mapping f above by a function satisfying the conditions mentioned in 1.7.5 (3). Then one gets the same results as above: a system on a Cantor space which is conjugate to the full shift system (Ω_2, σ).

6.3.7 (Semi-Sturmian systems[5]).

(1) Consider the rigid rotation $(\mathbb{S}, \varphi_a)$ of the circle with $a \in [0; 1) \setminus \mathbb{Q}$. Select a point b in the open unit interval, let $P_0 := \{\, [t] \,\vdots\, 0 < t < b \,\}$ and let $P_1 := \{\, [t] \,\vdots\, b < t < 1 \,\}$. Then $\mathcal{P} := \{P_0, P_1\}$ is a topological partition of $\mathbb{S}$. Since φ_a is semi-open (it is even a homeomorphism), Theorem 6.1.6 implies that $Z := Z(\mathcal{P}, \varphi_a)$ is a shift space. We shall call it the *semi-Sturmian system of type* (a, b).

Claim. $\mathcal{P}$ is a pseudo-Markov partition. To prove this, it is sufficient to show that for every $z \in Z$ and for every $\varepsilon > 0$ there exists $k \in \mathbb{N}$ such that the diameter of $D_k(z)$ (hence the diameter of $\overline{D_k(z)}$ as well) is at most ε.

So let $\varepsilon > 0$; without limitation of generality we may assume that $\varepsilon < b$. Recall from the proof of Case 2 in Example 0.4.3 in the Introduction that there exists $k \in \mathbb{N}$ such that the partial orbit $\{\, [0], \varphi_a([0]), \ldots, \varphi_a^{k-1}([0]) \,\}$ of the point $[0]$ is ε-dense in $\mathbb{S}$ (has

5 See Note 8 at the end of this chapter.

a point in every arc of length ε). If $x \in \mathbb{S}$ is arbitrary then, using the isometry φ_x of $\mathbb{S}$ onto itself, one easily shows that partial orbit $\{x, \varphi_a(x), \dots, \varphi_a^{k-1}(x)\}$ of the point x is ε-dense as well. Now let z be an arbitrary point of Z and consider two points $x, y \in D_k(z)$. Then for every $n \in \{0, 1, \dots, k-1\}$ the points $\varphi_a^n(x)$ and $\varphi_a^n(y)$ are both in the same set P_0 or P_1. Let $L_\varepsilon := \{[t] \vdots 0 < t < \varepsilon/2\pi\}$, the counter-clockwise open arc of length ε that starts in the point $[0]$. Since $\varepsilon < b$, it is clear that $L_\varepsilon \subseteq P_0$. Now assume that in the counter-clockwise orientation of $\mathbb{S}$ the point y comes before the point x, that is, assume that $x = [t]$ and $y = [s]$ with $0 \le s < t < 1$. By the choice of k there exists $n \in \{0, 1, \dots, k-1\}$ such that $\varphi_a^n(x) \in L_\varepsilon$. In particular, $\varphi_a^n(x) \in P_0$, hence $\varphi_a^n(y) \in P_0$ as well. Because φ_a^n is orientation preserving and because of the assumption that in counter-clockwise orientation y comes before x, it follows that $\varphi_a^n(y)$ is on the counter-clockwise arc from $[0]$ to $\varphi_a^n(x)$. In particular, the distance between $\varphi_a^n(x)$ and $\varphi_a^n(y)$ is at most ε. But φ_a^n is an isometry, so the distance between x and y is at most ε as well. This completes the proof that $\mathcal{P}$ is a pseudo-Markov partition.

It follows from Corollary 6.1.12 and Proposition 6.1.10 (2) that there is a factor mapping $\psi: (Z, \sigma_Z) \to (\mathbb{S}, \varphi_a)$ such that $\psi^{\leftarrow}[t] = \{\iota[t]\}$ for all $[t]$ in the set $\mathbb{S}^*$ of all points that have a full itinerary, i.e., all points that do not have $[0]$ or $[b]$ in their orbit and that, consequently, ψ is almost 1-to-1. By Proposition 1.5.6 this implies that Z is minimal. Finally, the proof of Proposition 6.3.4 (2) can easily be adapted so as to show that the mapping ψ is at most 2-to-1.

Resuming. *For every choice of $a \in \mathbb{R} \setminus \mathbb{Q}$ and $b \in (0; 1)$ the semi-Sturmian system of type (a, b) is minimal. It is an almost 1-to-1 extension of the rigid rotation such that all points of $\mathbb{S} \setminus \mathbb{S}^*$ have fibres consisting of two points.*

Proof. Above we observed that the fibre of a point from $\mathbb{S} \setminus \mathbb{S}^*$ has at most two points. We have yet to prove that it also has at least two points. First, we consider the point $[0] \in \mathbb{S} \setminus \mathbb{S}^*$. As $[0]$ is a common end point of the open arcs P_0 and P_1 there are two sequences, one in P_0 and the other in P_1, that both converge in $\mathbb{S}$ to the point $[0]$. Since $\psi: Z \to \mathbb{S}$ is a surjection, one can find two sequences, $(z^{(0,n)})_n$ and $(z^{(1,n)})_n$, in Z such that $\psi(z^{(i,n)}) \in P_i$ for all n and $\psi(z^{(i,n)}) \rightsquigarrow [0]$ if n tends to infinity ($i = 0, 1$). By passing to convergent subsequences – recall that Z is a compact metric space – it may be assumed that these sequences in Z converge, say with limits $z^{(0)}$ and $z^{(1)}$. As ψ is continuous it follows that $\psi(z^{(i)}) = [0]$ ($i = 0, 1$), so the points $z^{(0)}$ and $z^{(1)}$ both belong to $\psi^{\leftarrow}[0]$. In order to show that the points $z^{(0)}$ and $z^{(1)}$ are different from each other, note that for $i = 0, 1$ one has $\psi(z^{(i,n)}) \in P_i$ for all n, so the final conclusion in Corollary 6.1.9 (1) implies that $(z^{(i,n)})_0 = i$. Since the projection of Z onto its zero coordinate is continuous, it follows that the limit $z^{(i)}$ of the sequence $(z^{(i,n)})_n$ has zero coordinate i as well, that is, $(z^{(0)})_0 = 0$ and $(z^{(1)})_0 = 1$. Consequently, $z^{(0)} \neq z^{(1)}$. This concludes the proof that $\psi^{\leftarrow}[0]$ contains at least, hence precisely, two different points.

Next, note that $\psi^{\leftarrow}[(\varphi_a^k)^{\leftarrow}[0]] = (\sigma^k)^{\leftarrow}[\psi^{\leftarrow}[0]]$ for all $k \in \mathbb{Z}^+$. Using that σ^k is surjective (because Z is minimal) and φ_a^k is injective (it is an isometry), it easily follows that every point $[t] \in \mathbb{S}$ with $[0]$ in its orbit has at least two points in $\psi^{\leftarrow}[t]$. Similar arguments show that every point $[t] \in \mathbb{S}$ with $[b]$ in its orbit has at least two points in $\psi^{\leftarrow}[t]$. □

Remarks.

(1) Obviously, a semi-Sturmian systems of type (a, b) is infinite, so it is a non-trivial minimal system. If $0 < b \leq a < 1/2$ then P_0 cannot contain two successive points of an orbit. So the block 00 does not occur in any point of Z, i.e., Z is a subshift of the golden mean shift. Consequently, the golden mean shift has *non-trivial* minimal subsets. Note also that the system has no isolated points (otherwise it would be finite), so its phase space is a Cantor space.

(2) Remark 3 after Proposition 5.3.12 implies that Z is not a subshift of finite type. So the pseudo-Markov partition $\mathcal{P}$ of $\mathbb{S}$ is not a Markov partition.

(3) The set $\mathbb{S}^*$ is equal to the set of all points that have neither $[0]$ nor $[b]$ in their orbit. Hence $\mathbb{S} \setminus \mathbb{S}^* = \mathcal{O}_{\varphi_c}[0] \cup \mathcal{O}_{\varphi_c}[b]$ with $c := -a$.

(2) *Suppose that* $[b] \notin \mathcal{O}_{\varphi_a}[0]$ *and* $[0] \notin \mathcal{O}_{\varphi_a}[b]$ *(so the orbits of* $[0]$ *and* $[b]$ *in* $\mathbb{S}$ *under* φ_a *are disjoint). If* $z, z' \in Z$ *and* $\psi(z) = \psi(z')$ *then there exists* $k \in \mathbb{N}$ *such that* $z_n = z'_n$ *for all* $n \neq k$*, hence* $\sigma^n z = \sigma^n z'$ *for all* $n \geq k$.

Proof. The proof is based on Corollary 6.1.9 (1): if $\varphi_a^n(\psi z)$ and $\varphi_a^n(\psi z')$ both belong to the same set P_i ($n \in \mathbb{Z}^+$, $i \in \{0, 1\}$) then $z_n = i = z'_n$.
Now consider $z, z' \in Z$ with $\psi(z) = \psi(z')$. If $z = z'$ then there is nothing to prove, so we assume that $z \neq z'$. Then $[t] := \psi(z) = \psi(z') \in \mathbb{S} \setminus \mathbb{S}^*$, so by Remark 3 above there exists $k \in \mathbb{Z}^+$ such that $\varphi_a^k[t] \in \{[0], [b]\}$. Since $a \notin \mathbb{Q}$ and $[b] \notin \mathcal{O}_{\varphi_a}[0]$ it follows easily that $\varphi_a^n([t]) \notin \{[0], [b]\}$ for all $n \geq k + 1$. Moreover, a moment's reflection shows that it is also impossible that $\varphi_a^n([t]) \in \{[0], [b]\}$ for any $n < k$, because, in addition, $[0] \notin \mathcal{O}_{\varphi_a}[b]$. Consequently, for every $n \neq k$ the points $\varphi_a^n(\psi z)$ and $\varphi_a^n(\psi z')$ both belong to $P_0 \cup P_1$, so (as they coincide) both points are in P_0 or in P_1. So the initial observation in this proof implies that $z_n = z'_n$ for every $n \neq k$. Finally, note that this implies that $\sigma^n z = \sigma^n z'$ for all $n \geq k + 1$. □

NB 1. Inspection of the proof shows that if $\psi(z) = \psi(z') = [0]$ and $z \neq z'$ then $z_0 \neq z'_0$ and $z_n = z'_n$ for all $n \geq 1$, hence $\sigma^n z = \sigma^n z'$ for all $n \geq 1$.

NB 2. The (unique) value of k in the above statement is the unique value of k for which $\varphi_a^k(\psi z) \in \{[0], [b]\}$.

(3) *The system* (Z, σ) *satisfies the conditions of* 1.6.13 (3)*, so the construction of* 1.6.13 – *the introduction of a delay – produces a minimal weakly mixing system* (Z^*, σ^*) *on a compact metric space.*

Proof. Let $Z_1 := \{ z \in Z : z_0 = 1 \}$ and $Z_2 := \{ z \in Z : z_0 = 0 \}$. Then Z_1 and Z_2 are mutually disjoint non-empty clopen subsets of Z whose union is all of Z. Let $z^{(1)}$ and $z^{(2)}$ be the two points of $\psi^{\leftarrow}[0]$. It follows from 2 above that $z_0^{(1)} \neq z_0^{(2)}$, so we may assume that the numbering of the points is such that $z^{(i)} \in Z_i$ for $i = 1, 2$. In addition, $\sigma^n z^{(1)} = \sigma^n z^{(2)}$ for all $n \geq 1$. Hence for $n \geq 1$ the points $\sigma^n z^{(1)}$ and $\sigma^n z^{(2)}$ belong both to Z_1 or they belong both to Z_2. Moreover, as $(\sigma^n z^{(1)}, \sigma^n z^{(2)}) \in \Delta_Z$ for all $n \geq 1$, it is clear that $\overline{\mathcal{O}_{\sigma\times\sigma}(z^{(1)}, z^{(2)})} \cap \Delta_Z \neq \emptyset$.
Next, we look at complete pasts of the points $z^{(1)}$ and $z^{(2)}$. If $k \in \mathbb{N}$ then the set $(\sigma^k)^{\leftarrow}[(z^{(1)}, z^{(2)})]$ consists of just the two points of $\psi^{\leftarrow}[\varphi_c^k[0]]$ with $c := -a$ (for clarity: recall that in this discussion σ stands for $\sigma|_Z$). Label these points as $z^{(1,k)}$ and $z^{(2,k)}$ in such a way that $\sigma^k z^{(i,k)} = z^{(i)}$ for $i = 1, 2$. So $P := \{(z^{(1,k)}, z^{(2,k)}) : k \in \mathbb{N}\}$ is a complete past in $Z \times Z$ under $\sigma \times \sigma$ of the point $(z^{(1)}, z^{(2)})$.
Let $k \in \mathbb{N}$. Since $\varphi_a^k(\psi z^{(i,k)}) = [0]$ it follows from 2 above that all corresponding coordinates of the points $z^{(1,k)}$ and $z^{(2,k)}$ are equal to each other, except their k-th coordinates. In particular, their 0-coordinates are equal, so both points belong to Z_1 or both belong to Z_2. Moreover, since $(z^{(1,k)})_j = (z^{(2,k)})_j$ for $j = 0, \dots, k-1$ and $(z^{(1,k)})_k \neq (z^{(2,k)})_k$ it is clear that $d(z^{(1,k)}, z^{(2,k)}) = 1/(k+1)$ (here d is the metric that Z inherits from the full shift space Ω_2). By Appendix A.7.6, the collection of all sets $\alpha_\varepsilon := \{(z, z') \in Z \times Z : d(z, z') < \varepsilon\}$ for $\varepsilon > 0$ is a neighbourhood base of the diagonal Δ_Z, so it follows from the above that P meets every neighbourhood of Δ_Z. Consequently, $\overline{P} \cap \Delta_Z \neq \emptyset$. □

6.3.8 (The generalized tent map). For $0 \leq s \leq 2$ we shall consider the 'generalized' tent map $T_s : [0; 1] \to [0; 1]$, defined by

$$T_s(x) := \tfrac{1}{2} s \left(1 - |2x - 1| \right) \quad \text{for } x \in \mathbb{R}$$

(not to be confused with the truncated tent map defined in Section 2.3). Clearly, for $s = 2$ we get the ordinary tent map; for $s = 0$ the mapping T_s is identically equal to 0. In what follows we assume that $0 < s \leq 2$.

Let $J_0 := (0; \frac{1}{2})$ and let $J_1 := (\frac{1}{2}; 1)$. Then $\mathcal{P} := \{J_0, J_1\}$ is a topological partition of the interval $[0; 1]$. Note that T_s is not expansive (there are arbitrarily close points with the same image under T_s), so the results of Section 6.2 do not apply. On the other hand, T_s is monotonous on the closures of the members of the topological partition $\mathcal{P}$, so Proposition 6.3.4 applies. However, if $s \neq 2$ then $\mathcal{P}$ does not have property (**M***). In fact, in that case $\mathcal{P}$ does not even have property (**M**): see Exercise 6.11.

According to Proposition 6.3.4 (1), for every $k \in \mathbb{N}$ the function T_s^k is strictly monotonous on every interval $\overline{D_k(z)}$ with $z \in Z$. We shall prove a slightly stronger result: these intervals are the maximal intervals on which the function T_s^k is monotonous. For every $k \in \mathbb{N}$, let

$$\Gamma_k := \bigcup \{ (T_s^n)^{\leftarrow}[c] : 0 \leq n \leq k - 1 \},$$

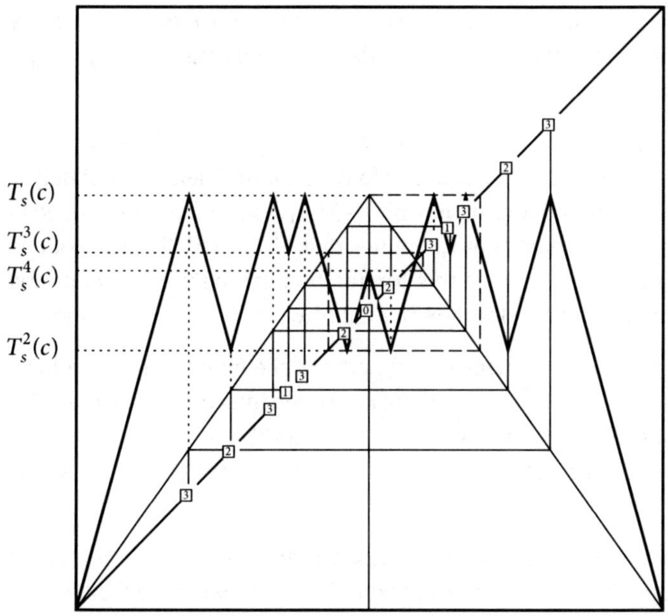

Fig. 6.5. The graph of T_s^4. The local extrema are at the points of Γ_4. The points of $(T_s^n)^{\leftarrow}[c]$ are labelled $\boxed{n}$ for $n = 0, 1, 2, 3$ (along the diagonal). The value of T_s^4 in a point with label $\boxed{n}$ is $T_s^{4-n}(c)$ $(n = 0, 1, 2, 3)$.

where $c := \frac{1}{2}$. Obviously, if $k, l \in \mathbb{N}$ and $l > k$ then $\Gamma_l \supseteq \Gamma_k$. Experimenting with graphical iteration suggest that the points of Γ_k are the local extrema of T_s^k; see Figure 6.5. In the next lemmas this will be rigorously proved.

Lemma 6.3.9. *For every $k \in \mathbb{N}$, the function T_s^k is strictly monotonous on every non-degenerate interval that has no points of Γ_k in its interior.*

Proof. The proof is by induction in k. For $k = 1$ the statement in the lemma is certainly true. Suppose the statement is true for a certain value of $k \in \mathbb{N}$, and consider a non-degenerate interval J whose interior contains no points of Γ_{k+1}. Then that interior contains no points of Γ_k either, so by hypothesis, T_s^k is strictly monotonous on the interval J, mapping it homeomorphically onto $T_s^k[J]$. It follows that $T_s^k[J]$ is a non-degenerate interval. In addition, as T_s^k maps the interior of J onto the interior of $T_s^k[J]$, the interior of $T_s^k[J]$ does not contain the point c, for otherwise the interior of J would contain a point from $(T_s^k)^{\leftarrow}[c] \subseteq \Gamma_{k+1}$, contradicting the hypothesis. Hence by the case for $k = 1$, T_s is strictly monotonous on $T_s^k[J]$. It follows that the composition $T_s \circ T_s^k$ is strictly monotonous on the interval J. □

Lemma 6.3.10. *For every $k \in \mathbb{N}$, the function T_s^k has a strict local extremum in the points of Γ_k.*

Proof. The proof is by induction in k. For $k = 1$ the statement in the lemma is certainly true. Suppose the statement is true for a certain value of k, and consider a point $x_0 \in \Gamma_{k+1}$. We distinguish two cases:

Case 1: $x_0 \in (T_s^k)^{\leftarrow}[c]$. Then $T_s^k(x_0) = c$, so the composition $T_s \circ T_s^k = T_s^{k+1}$ assumes its largest possible value $T_s(c) = \frac{1}{2}s$ at the point x_0. Moreover, a sufficiently small neighbourhood of x_0 contains no other points of the finite set Γ_{k+1}, so Lemma 6.3.9 applied to suitable left and right neighbourhoods of x_0 implies that T_s^{k+1} is strictly monotonous on each of them (x_0 cannot be 0 or 1, so this formulation can be used without any difficulties). It follows that T_s^{k+1} has a local extremum at x_0.

Case 2: $x_0 \notin (T_s^k)^{\leftarrow}[c]$. Then $x_0 \in \Gamma_k$, hence by hypothesis the function T_s^k has a local extremum at x_0. In addition, since $T_s^k(x_0) \neq c$ there is an open interval around the point $T_s^k(x_0)$ that does not contain the point c. Hence the function T_s is strictly monotonous on this interval. So the composition $T_s \circ T_s^k = T_s^{k+1}$ has a local extremum at the point x_0. □

Corollary 6.3.11. *For every $k \in \mathbb{N}$, the successive closed intervals into which the interval $[0; 1]$ is divided by the points of Γ_k are just the maximal intervals on which the function T_s^k is (strictly) monotonous.*

Proof. By Lemma 6.3.9, the function T_c^k is strictly monotonous on each of these intervals. By Lemma 6.3.10, T_c^k cannot be monotonous on any interval with a point of Γ_k in its interior, which implies that none of the intervals mentioned above can be extended in such a way that T_s^k remains monotonous on it. □

Lemma 6.3.12. *Let $k \in \mathbb{N}$. The collection of sets $\overline{D_k(b)}$ for all $(\mathcal{P}, T_s)$-allowed k-blocks b coincides with the set of intervals mentioned in the previous lemma, i.e., the set of maximal intervals of monotonicity of T_s^k.*

Proof. By Proposition 6.3.4 (1), the set $\overline{D_k(b)}$ for any $(\mathcal{P}, T_s)$-allowed k-block b is an interval on which T_s is strictly monotonous, hence it is included in one of the maximal intervals of monotonicity of T_s^k. So it is sufficient to prove that the end points of each set $D_k(b)$ are in the set $\{0, 1\} \cup \Gamma_k$.

Suppose this is false. Then there is a point x different from 0 and 1 which is an end point of $D_k(b)$ for some $(\mathcal{P}, T_s)$-allowed k-block b and which is not in Γ_k. The point x has a partial itinerary of length k because none of the points $T_s^i(x)$ for $0 \leq i \leq k-1$ is c (hence they are neither 1 nor 0). Hence there is a $(\mathcal{P}, f)$-allowed k-block b' such that $x \in D_k(b')$. As x is an end point of the interval $D_k(b)$ and $D_k(b')$ is open, it follows that $D_k(b') \cap D_k(b) \neq \emptyset$. This would imply that $b = b'$, which is impossible, because $x \notin D_k(b)$. This contradiction completes the proof. □

Corollary 6.3.13. *If $s > 1$ then the topological partition $\mathcal{P}$ of $[0; 1]$ is pseudo-Markov.*

Proof. By Lemma 6.3.12 we have to show that for $k \rightsquigarrow \infty$ the lengths of the maximal intervals of monotonicity of T_s^k tend to 0. To do so, observe that for a point x in such an interval the points $T_s^i(x)$ for $0 \le i \le k-1$ are either in J_0 or in J_1 (never equal to 0, 1 or c), hence T_s^k is differentiable at the point x with derivative $(T_s^k)'(x)$ satisfying

$$(T_s^k)'(x) = \prod_{i=0}^{k-1} T_s'(T_s^i(x)) = \pm s^k .$$

Since the values of T_s^k are between 0 and 1, it follows that the length of such an interval is at most $1/s^k$. For $s > 1$ this tends to 0 for $k \rightsquigarrow \infty$. □

6.3.14 (Conclusion). *For $1 < s \le 2$ the system $([0; 1], T_s)$ is factor of a shift system (Z, σ_Z) under a factor mapping $\psi\colon (Z, \sigma_Z) \to ([0; 1], T_s)$ which is at most 2-to-1 (and, in addition, almost 1-to-1).*

Remarks.

(1) In the case that $s = 2$, Example E in 6.3.5 above shows that $Z = \Omega_2$.

(2) Every orbit under T_s in included in the interval $[0; \frac{1}{2}s]$, so if $0 < s < 2$ then *no orbit is dense. and the system* $([0; 1], T_s)$ *is not transitive*. A similar argument shows that *the periodic points cannot form a dense subset of the interval* $[0; 1]$. If $s > 1$ then by Example (3) following Corollary 8.6.9 ahead the system has periodic points. Now a simple finiteness argument shows that if $x \in [0; 1]$ is periodic under T_s then $\psi^{\leftarrow}[x]$ includes an eventually periodic point. So the shift system (Z, σ_Z) has periodic points.

(3) If $0 < s \le 1$ then it is easily checked that $Z = \{0^\infty, 10^\infty\}$ and that for every $k \in \mathbb{N}$ and every $(\mathcal{P}, T_s)$-allowed k-block b the set $\overline{D_k(b)}$ equals either $\overline{P_0}$ or $\overline{P_1}$ (this is in accordance with Lemma 6.3.12, as $T_s^k = T_{s^k}$). So in this case, $\mathcal{P}$ is not a pseudo-Markov partition.

Exercises

6.1.

(1) Consider the argument-doubling system $(\mathbb{S}, \psi)$, let $P_0 := \{[t] ⋮ 0 \le t < 1/2\}$ and $P_1 := \{[t] ⋮ 1/2 \le t < 1\}$. Since $\{P_0, P_1\}$ is a genuine partition of $\mathbb{S}$, every point $[t]$ of $\mathbb{S}$ has an itinerary $\iota([t]) \in \Omega_2$. Show that for $0 \le t < 1$ the sequence $\iota([t])$ is the unique binary expansion of t that does not end with the half-sequence 1^∞.

(2) In 1 above, the set $\iota[\mathbb{S}]$ is not closed in Ω_2.

(3) Let (X, f) be an arbitrary dynamical system and let $\mathcal{P} = \{P_0, \dots, P_{s-1}\}$ be a partition of X. The mapping $\iota\colon X \to \Omega_{\mathbf{S}}$ is continuous iff the sets P_α for $\alpha \in \mathbf{S} := \{0, \dots, s-1\}$ are clopen.

Unless specified differently, in the following exercises (X, f) is an arbitrary dynamical system, $\mathcal{P} = \{P_0, \dots, P_{s-1}\}$ is a topological partition of X and $\mathbf{S} := \{0, \dots, s-1\}$.

6.2.

(1) Show that the set of all $(\mathcal{P}, f)$-allowed blocks coincides with the language of $Z(\mathcal{P}, f)$ iff the following condition is fulfilled: for every $(\mathcal{P}, f)$-allowed word b there exists a symbol $\alpha \in \mathbf{S}$ such that the word $b\alpha$ is $(\mathcal{P}, f)$-allowed. If this condition is fulfilled then $Z(\mathcal{P}, f) \neq \emptyset$.

(2) The system of the example illustrated by Figure 6.1 (b) satisfies the condition of 1 above. Moreover, $Z(\mathcal{P}, f)$ is the golden mean shift, represented as the subset of $\Omega_{\{0,1,2\}}$ in which the symbol 1 does not occur and the symbol 2 occurs isolated.

(3) Recall that $U_{\mathcal{P}} := \bigcup \mathcal{P}$. Show that if for every $k \in \mathbb{N}$ the set $(f^k)^{\leftarrow}[U_{\mathcal{P}}]$ is dense, then the condition of 1 above is fulfilled, so that $\mathcal{P}$ is f-adapted.

(4) If f is a semi-open mapping then every topological partition $\mathcal{P}$ of X is f-adapted and its language coincides with the set of $(\mathcal{P}, f)$-allowed blocks.

(5) Let $Z := Z(\mathcal{P}, f)$. Assume that $\mathcal{P}$ is a pseudo-Markov partition and that the mapping $\psi : Z \to X$ is surjective, i.e., that X^* is dense in X – see Corollary 6.1.13 (1). Then the set of all $(\mathcal{P}, f)$-allowed blocks coincides with the language of Z.

6.3. For every $k \in \mathbb{N}$, let $X_k := \bigcup \{ D_k(z) \,\vdots\, z \in Z \}$.

(1) Show that $X^* = \bigcap_{k\geq 1} X_k$.

(2) If $\mathcal{P}$ is a pseudo-Markov partition then $\psi[Z] = \bigcap_{k\geq 1} \overline{X}_k$.

(3) Identify the sets X_k in the example in 6.3.6 and conclude that the general construction of a symbolic representation can be carried out completely analogous to the construction of the Cantor set.

6.4.

(1) Let $\mathcal{P} \vee \mathcal{Q} := \{P \cap Q \,\vdots\, P \in \mathcal{P} \,\&\, Q \in \mathcal{Q}\}$, where $\mathcal{Q}$ is yet another topological partition of X. Then $\mathcal{P} \vee \mathcal{Q}$ is a topological partition of X.

(2) If f is semi-open then $f^{\leftarrow}\mathcal{P} := \{ f^{\leftarrow}[P] \,\vdots\, P \in \mathcal{P} \}$ is a topological partition.

(3) Assume that f is semi-open and let $k \in \mathbb{N}$. Then $\mathcal{P} \vee f^{\leftarrow}\mathcal{P} \vee \cdots \vee (f^{k-1})^{\leftarrow}\mathcal{P}$ is a topological partition.

NB. Clearly, $\mathcal{P} \vee f^{\leftarrow}\mathcal{P} \vee \cdots \vee (f^{k-1})^{\leftarrow}\mathcal{P}$ coincides with the collection $\mathcal{P}_k^*$ of sets of the form $D_k(b)$ for all $(\mathcal{P}, f)$-allowed k-blocks as defined in Lemma 6.2.10.

6.5. Assume that $\mathcal{P}$ is a pseudo-Markov partition. Show that the following statements are equivalent:

(i) X^* is dense in X.

(ii) $\iota[X^*]$ is dense in Z and ψ is a surjection of Z onto X.

(iii) Every partial itinerary can be extended to a full itinerary.

NB. Condition (iii) is just the condition mentioned in Exercise 6.2 (1). It does not imply that every point (which defines a partial itinerary) is in X^* (defines a full itinerary): in the example illustrated in Figure 6.1 (b) the partial itinerary – admissible block – 0020 can be extended to the full itinerary $(002)^\infty$. However, the point $11/144$ has partial itinerary 0020 but except the first four points its orbit is not in $P_0 \cup P_1 \cup P_2$.

6.6. Assume that X is a compact Hausdorff space and that $\mathcal{P}$ is a pseudo-Markov partition with $\psi\colon Z \to X$ the corresponding symbolic representation. Assume that ψ is a surjection. Then

$$\forall z \in Z, \forall k \in \mathbb{N} : \ \psi[\widetilde{B}_k(z) \cap Z] = \overline{D_k(z)}\,.$$

NB. In particular, $\overline{P}_\alpha = \psi[C_0[\alpha] \cap Z]$ for every $\alpha \in \mathbf{S}$. So if the members of $\mathcal{P}$ are regular open sets (i.e., each of them is the interior of its closure) then they can be retrieved from the symbolic representation: $\forall \alpha \in \mathbf{S} : P_\alpha = \operatorname{int}_X \psi[C_0[\alpha] \cap Z]$.

6.7. Let (X, σ_X) be a subshift and let $k \in \mathbb{N}$. Let $\mathbf{T} := \mathcal{L}_k(X)$ and let $\mathcal{P}$ be the clopen partition of X into sets of the form $X \cap C_0[b]$ with $b \in \mathbf{T}$. Show that $\mathcal{P}$ is a pseudo-Markov partition; by Corollary 6.1.13 (2), the symbolic representation $\psi\colon (Z, \sigma_Z) \to (X, \sigma_X)$ is a conjugation. Show that $Z = X^{(k)}$ and that $\psi = (\mathrm{I}_\infty^{(k)})^{-1}$.

6.8.

(1) Let (X, f) be an expansive dynamical system on a compact metric space X. Then for every $n \in \mathbb{N}$ there are only finitely many periodic points with period n.

(2) Let $\varphi\colon (X, f) \to (Y, g)$ be a factor map and assume that X and Y are compact metric spaces. If (Y, g) is expansive then so is (X, f). Consequently, if φ is a conjugation then (X, f) is expansive iff (Y, g) is expansive.
NB. The rigid rotation $(\mathbb{S}, \varphi_a)$ with $a \in \mathbb{R}\setminus\mathbb{Q}$ is a factor of a the semi-Sturmian shift system: see 6.3.7. So expansiveness is not preserved by factor maps.

(3) Prove the following topological characterization of expansiveness of a dynamical system (X, f) on a compact metric space X:
The mapping f is expansive iff there is a neighbourhood U of the diagonal Δ_X in $X \times X$ such that Δ_X is the maximal $f \times f$-invariant subset of U.
NB. This characterization is independent of a metric on X. So it implies the final conclusion of 2 above. Moreover, it enables one to define expansiveness of arbitrary compact systems.

(4) Let $X := (0; 1)$ and let $f\colon X \to X$ be given by $f(x) := x^2$ for $x \in X$. Consider the two metrics $\rho\colon (x_1, x_2) \mapsto |x_1 - x_2|$ and $\tau\colon (x_1, x_2) \mapsto |1/x_1 - 1/x_2|$ on X (both compatible with the topology of X). Prove: with respect to the metric τ the system (X, f) is expansive and with respect to the metric ρ it is not expansive.

(5) Prove: if (X, f) is an expansive system on a compact metric space X then for every $k \in \mathbb{N}$ the system (X, f^k) is expansive.

(6) Show that if X is a non-degenerate compact interval then no continuous mapping $f\colon X \to X$ is expansive with respect to the ordinary metric inherited from $\mathbb{R}$.

6.9. Let X be a compact metric 0-dimensional space and let $f\colon X \to X$ be an expansive mapping. Use the following steps the prove that (X, f) is conjugate to a shift space:

(a) X admits a clopen partition $\mathcal{P}$ consisting of sets with diameter less than the expansive coefficient of f [this is Lemma 6.2.3];

(b) the itinerary-mapping ι with respect to $\mathcal{P}$ is defined on all of X (i.e., $X^* = X$), continuous [this is Lemma 6.1.3 (3)] and injective [use (a)].

6.10.

(1) Let $a \in \mathbb{R} \setminus \mathbb{Q}$ and let (X, f) be the corresponding Ellis minimal system; for the notation, see 1.7.6. Show that (X, f) admits no symbolic representation.

(2) Consider the subsets $P_0 := \{([t], 1) \vdots 0 \leq t < 1/2\} \cup \{([t], 2) \vdots 0 < t \leq 1/2\}$ and $P_1 := \{([t], 1) \vdots 1/2 \leq t < 1\} \cup \{([t], 2) \vdots 1/2 < t \leq 1\}$ of X. Then $\{P_0, P_1\}$ is a clopen partition of X, so every point $x \in X$ has a full itinerary $\iota_X(x)$ with respect to this partition. Note that $\iota_X : X \to \Omega_2$ is a continuous mapping – Exercise 6.1 (3) – such that $\sigma \circ \iota_X = \iota_X \circ f$. Show that $\iota_X[X]$ is the semi-Sturmian system of type $(a, 1/2)$.

(3) Show that $\psi \circ \iota_X$ is the canonical projection of X onto $\mathbb{S}$.
NB. This agrees with the following straightforward observation: for $[t] \in \mathbb{S} \setminus \mathbb{S}^*$ the points $\iota_X([t], 1)$ and $\iota_X([t], 2)$ differ in just one coordinate (namely, the k-th coordinate if $\varphi_a^k[t] \in \{[0], [1/2]\}$). Apparently, in this case these two points are the two points of $\psi^{\leftarrow}[t]$.

6.11. Consider the system $([0; 1], T_s)$ discussed in 6.3.8, $1 < s < 2$, and let $\mathcal{P}$ be the topological partition of the unit interval considered there. Show that blocks of the form $0^k 1^l$ for arbitrary $k, l \in \mathbb{Z}^+$ are allowed, but that a block of the form 10^n is allowed iff $n < -\log(2 - s)/\log s$. Conclude that $\mathcal{P}$ does not have property **(M)**.

6.12. Let $p \in \mathbb{N}$, $p \geq 5$ and p odd (for the case $p = 3$ we refer to Example F in 6.3.5). Let $X_p := [0; p-1]$ and let $g_p : X_p \to X_p$ be the mapping whose definition is described in the final example of Section 2.5 (note the change in notation: we use p instead of k), and for $i = 0, \ldots, p-2$, let P_i be the interior of the closed interval J_{i+1} (see Proposition 2.5.3 for the intervals J_i). Show that $\mathcal{P} := \{P_0, \ldots, P_{p-2}\}$ is a Markov partition and that the shift space Z representing the system (X_p, g_p) is irreducible and has a dense set of periodic points. Consequently, the system (X_p, g_p) is transitive and it has – in agreement with Theorem 2.6.2 – a dense set of periodic points as well.

Notes

1 At first sight the coding by means of itineraries seems quite natural, but it follows from the results in Section 6.1 that useful results are to be expected only for 0-dimensional spaces; see also the Exercises 6.1 (3) and 6.9. In Chapter 7 we shall deal with a coding defined by itineraries: see the construction in 7.4.2 and the remark following Proposition 7.4.3. However, we have to add the remark that itineraries sometimes provide a splendid tool for book-keeping. In this respect, the so-called kneading theory for interval maps comes to mind; see J. Milnor & W. Thurston [1988]. Another example can be found in the by now classical results by Marston Morse (1921) concerning geodesic flows on surfaces of negative curvature.
2 There are various results on embeddings of certain systems in a 'universal' system. One of the earliest results in this direction (for systems with continuous time) is Bebutov's Theorem, generalized later by Kakutani and Hajek; see O. Hajek [1971]. For systems with discrete time the following construction appears several times in the literature: for references, see P. C. Baayen [1964]. For variations on this

theme, see J. de Vries [1983]. In general, one can say that embedding a system in some 'universal' system (even a linear one) gives no additional information on the dynamical properties of the system under consideration; it only shows that such a universal system must be very complicated.

3 Much of the present chapter is adapted from D. Lind & B. Marcus [1995]. But some ideas – in particular, the results in 6.2.6 up to 6.2.11 – are borrowed from the paper R. L. Adler [1998]. A notable difference with existing literature is that we consider systems (X, f) in which f is not a homeomorphism, so that images of open sets under f need not be open, or preimages of dense sets need not be dense. But in Section 6.1 we show that much of the theory for homeomorphism on compact spaces also works in this context if f is semi-open. See Note 9 below for more remarks on this topic. In addition, for expansive systems we use the concept of a pseudo-Markov partition instead of that of a generator; see the Remark following Lemma 6.2.3. Thus, Corollary 6.2.5 might be formulated as follows: if f is semi-open and $\mathcal{P}$ is a generator then $\psi: Z \to X$ is an almost 1-to-1 surjection. Moreover, in Theorem 6.2.11 the conditions that (X, f) is expansive and that all members of $\mathcal{P}$ have a diameter of at most the expansive coefficient of f can be replaced by the condition that $\mathcal{P}$ is a generator. We leave the straightforward details to the reader. Finally, we pay no attention at all to the 'standard' examples of symbolic dynamics, i.e., constructing Markov partitions for (two-dimensional) toral automorphisms or Axiom A diffeomorphisms. For the former, see Note 10 below; the latter is extensively treated in R. Bowen [1975].

4 By Proposition 6.1.10 (2), ψ maps the remainder $Z \setminus \iota[X^*]$ of $\iota[X^*]$ in Z into the remainder $\psi[Z] \setminus X^*$ of X^* (or equivalently, into the remainder $X \setminus X^*$ of X^* in X). If $\iota[X^*]$ is dense in Z then this is in accordance with Lemma 6.11 in L. Gillman & M. Jerison [1960].

5 According to Adler in the paper mentioned in Note 3 above, symbolic representation of exceptions, the results of this section apply only to metric spacesdynamical systems by SFT's, while possible for certain non-expansive systems, only seems natural for expansive ones.

6 Condition (**M**) is taken from Adler's publication mentioned above, where it is called the condition that $\mathcal{P}$ has *the n-fold intersection property for all* $n \geq 3$. A topological partition satisfying condition ($\mathbf{M}^*$) is sometimes called a exceptions, the results of this section apply only to metric spaces*distillation*. An interval map as meant in Proposition 6.3.4 – strictly monotonous on the intervals of the topological partition – with property ($\mathbf{M}^*$) is also called a *Markov mapping*.

7 The result of 6.3.6 (2) holds, in particular, for the mapping f_μ on $\mathbb{R}$ for all $\mu > 2 + \sqrt{5}$. Actually, it holds for all $\mu > 4$: exceptions, the results of this section apply only to metric spacessee Note 8 in Chapter 1.

8 Traditionally, *Sturmian sequences* are defined as the two-sided infinite sequences $(z_n)_{n\in\mathbb{Z}} \in \{0,1\}^{\mathbb{Z}}$ obtained as $z_n := \chi_0(\varphi_a^n([0]))$, where $\chi_0 : \mathbb{S} \to \{0,1\}$ is the characteristic function of the half-open arc $\{\,[t] \in \mathbb{S} : 0 \leq t < b\,\} = P_0 \cup \{[0]\}$, and P_0 is as in 6.3.7. Moreover, a *Sturmian system* is defined as the orbit closure in $\{0,1\}^{\mathbb{Z}}$ under the shift mapping of such a Sturmian sequence. In the context of the present book, where we work with one-sided infinite sequences instead of two-sided infinite ones this amounts to the following.

Let $E_0 := P_0 \cup \{[0]\}$ and $E_1 := P_1 \cup \{[b]\}$. Then $\{E_0, E_1\}$ is a genuine partition of $\mathbb{S}$, so every point $[t] \in \mathbb{S}$ has a full itinerary $\iota^*([t]) \in \Omega_2$. Recall that, by definition, $\iota^*([t])_n = i$ iff $\varphi_a^n([t]) \in E_i$ $(i = 0, 1)$, so $\iota^*([t])_n = \chi_1(\varphi_a^n([t]))$, where χ_1 is the characteristic function of E_1 (in order to agree with the traditional definition given above exceptions, the results of this section apply only to metric spaceswithout interchanging the symbols 0 and 1 we have to consider χ_1 instead of χ_0). In analogy with the two-sided case mentioned above, we may call $x := \iota^*([0])$ a *semi-Sturmian sequence*, and its orbit closure X in Ω_2 under the shift mapping a *semi-Sturmian system*. Note that $\iota^*|_{\mathbb{S}^*} = \iota$, where ι is the itinerary-mapping used in 6.3.7. We shall show now that the orbit closure X of the semi-Sturmian sequence agrees with the minimal shift space Z obtained in 6.3.7, provided the point $[b]$ is not in the orbit of $[0]$ under φ_a.

Proof. If the point $[b]$ is not in the orbit of $[0]$ under φ_a then $[a] = \varphi_a[0] \in \mathbb{S}^*$: recall from 6.3.7 that $\mathbb{S}^*$ consists of all points with $[0]$ and $[b]$ not in their orbits, and it is easily seen that $[a]$ is such a point (note that the point $[0]$ is not periodic). As $\iota^*|_{\mathbb{S}^*} = \iota$, it follows that $\iota^*([a]) = \iota([a]) \in Z$. (This already implies that $X \cap Z \neq \emptyset$, hence that $Z \subseteq X$ by minimality of Z, but this is exceptions, the results of this section apply only to metric spacesnot sufficient, unless we prove directly that X is minimal under the condition that $[b]$ is not in the orbit of $[0]$; however, we want to avoid such a proof.) Next, observe that $\sigma(x) = \sigma(\iota^*[0]) = \iota^*(\varphi_a[0]) = \iota^*[a] = \iota[a]$, which implies that the sequence x consists of the sequence $\iota[a]$ – an element of Z – preceded by a 0 or a 1. However, it was shown exceptions, the results of this section apply only to metric spacesin 6.3.7 that the fibre of the point $[0]$ under the factor map $\psi: (Z, \sigma) \to (\mathbb{S}, \varphi_a)$ consists of two points, and that the fibre of the point $[a]$ consists of the single point $\iota[a]$ (recall that $[a] \in \mathbb{S}^*$ but that $[0] \notin \mathbb{S}^*$). As $\varphi_a \circ \psi = \psi \circ \sigma$ (on Z) it follows that the fibre of the point $\iota[a]$ under σ in Z consists of two points. Obviously, these exceptions, the results of this section apply only to metric spacespoints are the sequences $0\,\iota[a]$ and $1\,\iota[a]$. Stated otherwise, thesesequences belong both to Z. It was argued above that the semi-Sturmian sequence x is one of these sequences. Consequently, $x \in Z$ and therefore, as Z is minimal under σ, it follows that $X = Z$. □

So in the situation that $[b]$ is not in the φ_a-orbit of $[0]$ – this condition figures also in the traditional theory – our approach in 6.3.7 is the one-sided counterpart of the classical approach.

9 Invertible vs. non-invertible systems. Without any problems the results of this chapter can be applied to invertible dynamical systems, i.e., systems (X, f) in which f is a homeomorphism. In that case one would like to get symbolic representations on invertible shift spaces, that is, on subshifts of (Σ_S, σ) (see Note 8 in Chapter 5). We briefly indicate how the constructions are to be modified so as to achieve this.

Consider a dynamical system (X, f) with a compact Hausdorff phase space X and assume that f is a homeomorphism. Let $\mathcal{P} = \{P_0, \dots, P_{s-1}\}$ be a topological partition of X and let $S := \{0, \dots, s-1\}$. If $x \in X$ is such that $f^n(x) \in U_{\mathcal{P}} := P_0 \cup \dots \cup P_{s-1}$ for all $n \in \mathbb{Z}$ – equivalently, if $x \in X^* := \bigcap_{n \in \mathbb{Z}} f^{-n}[U_{\mathcal{P}}]$ – then the full itinerary of x is defined as the element $\iota(x)$ of Σ_S such that $f^n(x) \in P_{\iota(x)_n}$ for all $n \in \mathbb{Z}$. As f is a homeomorphism, X^* is a two-sided invariant dense G_δ-subset of X, and $\iota: X^* \to \Sigma_S$ is continuous, hence a morphism of dynamical systems; see Lemma 6.1.3 and Theorem 6.1.6.

Next, one defines $(\mathcal{P}, f)$-forbidden and $(\mathcal{P}, f)$-allowed blocks as in 6.1.4, and the shift space $Z = Z(\mathcal{P}, f)$ is defined as the set of all points in Σ_S in which every block is $(\mathcal{P}, f)$-allowed. By Theorem 6.1.6, $Z \neq \emptyset$ and $\iota[X^*]$ is dense in Z; so the condition that $\mathcal{P}$ be f-adapted is always fulfilled (hence superfluous) for invertible systems.

The criterion for a topological partition $\mathcal{P}$ to be pseudo-Markov (or Markov) is, similar to the definition in 6.1.7, that $\bigcap_{k \in \mathbb{N}} \overline{D_k(z)}$ is a singleton-set for every $z \in Z$ (and that Z is an SFT, respectively); in that case, the single point in this set is denoted $\psi(z)$. However, the definition of the sets $D_k(z)$ is more 'symmetric' than in 6.1.2, namely,

$$\forall\, k \in \mathbb{N}: \; D_k(z) := \bigcap_{n=-k+1}^{k-1} f^{-n}[P_{z_n}]$$

(this can be defined for all $z \in \Sigma_S$, but will be non-empty only for $z \in Z$). In that case we get a morphism $\psi: (Z, \sigma_Z) \to (X, f)$, which is almost 1,1 because $\psi \circ \iota = \mathrm{id}_{X^*}$.

The application of the theory to mappings of intervals into itself is not very exciting if that mapping is a homeomorphism: in that case the mapping is a monotonous surjection and the dynamics is rather trivial.

For homeomorphisms, expansiveness is defined similarly to our definition in Section 6.2, but then the n for which $\rho(f^n(x), f^n(y)) \geq \eta$ is to be taken from all of $\mathbb{Z}$. In that case, the notion defined by us in Section 6.2 should be called *positive* expansiveness. For homeomorphisms on compact metric spaces, positive expansiveness is too strong to be really useful: an old result by S. Schwartzman (1952) states that a compact metric space that admits a positively expansive homeomorphism is finite. See also [GH], 10.30. A simple proof is in E. M. Coven & M. Keane [2006]. Our proof of Proposition 5.3.2 is adapted from this paper. Exercise 6.8 (5) is adapted from W. R. Utz [1950].

The definition of an expansive system has been generalized to dynamical systems on uniform spaces (see also Exercise 6.8 (3)). The following variant of Corollary 6.2.5 holds: let X be a compact

Hausdorff space, let f be a homeomorphism of X onto itself and assume that the dynamical system (X, f) is expansive (with respect to the unique uniformity compatible with the topology of X). Then (X, f) is a factor of a shift system (the proofs of Lemmas 6.2.2 and 6.2.3 can easily be adapted to this situation) and, consequently, X is metrizable. In particular, a compact Hausdorff space admitting an expansive homeomorphism is metrizable; this result is originally due to B. F. Bryant [1960]. For generalizations see J. de Vries [1972].

Corollary 6.2.4 also holds for expansive homeomorphisms. In this form, it can be found in W. L. Reddy [1968], and in H. B. Keynes & J. B. Robertson [1969]. Also Corollary 6.2.5 holds: if f is an expansive homeomorphism on a compact metric space then(X, f) is an almost 1,1 factor of a (two-sided) shift system.

With the modifications mentioned above the proof of Theorem 6.2.11 can also be given in the invertible case, provided 'transitive' is replaced by 'bilaterally transitive' and X has no isolated points; see the paper R. L. Adler [1998]. In this case it follows easily that if all bilaterally transitive points in an open set like $U_{\mathcal{P}_k^*}$ have singleton fibres then *all* transitive points in X have singleton fibres.

10 In constructing a symbolic representation one hopes that the topological properties of X and f induce special properties in the representing shift space and that (dynamical) consequences of these properties are preserved by ψ. By this *detour* one might then be able to derive properties of the system (X, f) which would otherwise be hard to prove. See the Example following Corollary 6.1.12 above for an idea of how this might work. In addition, symbolic representations may be used for classification purposes.

One of the motivating examples of a symbolic representation of a dynamical system using a Markov partition concerns so-called hyperbolic automorphisms of the torus $\mathbb{S}^2$. Let $A = \left(\begin{smallmatrix} a & b \\ c & d \end{smallmatrix}\right) \in GL(2, \mathbb{Z})$, i.e., A is a 2×2-matrix with integer entries and determinant ± 1. Then the linear mapping

$$L_A : (s, t) \mapsto A \left(\begin{smallmatrix} s \\ t \end{smallmatrix}\right) = (as + ct, bs + dt) : \mathbb{R}^2 \to \mathbb{R}^2$$

induces a well-defined continuous mapping

$$T_A : ([s], [t]) \mapsto ([as + ct], [bs + dt]) : \mathbb{S}^2 \to \mathbb{S}^2 .$$

In fact, T_A is a bijection with inverse defined in a similar way by $A^{-1} \in GL(2, \mathbb{Z})$. In addition, for every $n \in \mathbb{Z}$ we have $A^n \in GL(2, \mathbb{Z})$, and $(T_A)^n = T_{A^n}$. It is not too difficult to show that the set of mappings defined in this way coincides with set of the continuous automorphisms of the topological group $\mathbb{S}^2 \approx \mathbb{R}^2/\mathbb{Z}^2$, hence the name: *toral automorphisms*. If a point $([s], [t]) \in \mathbb{S}^2$ is invariant under T_A then it is straightforward to show that $s, t \in \mathbb{Q}$. Consequently, if a point is periodic under T_A – that is, invariant under T_{A^n} – then it has rational coordinates. Conversely, if a point $([s], [t])$ of $\mathbb{S}^2$ has rational coordinates then the denominators of the coordinates of the points of the orbit of this point, i.e., of the points $T_A([s], [t])$, $T_A^2([s], [t])$, …, are bounded by the product of the denominators of s and t. Hence the orbit of the point $([s], [t])$ is finite, which means that this point is periodic (not just eventually periodic, for T_A is invertible). *It follows that the set of periodic points of T_A is dense in $\mathbb{S}^2$.*

Let $A \in GL(2, \mathbb{Z})$ and let λ and μ be the eigenvalues of A. Then there are two, mutually exclusive, possibilities:

(a) λ and μ are complex conjugate and $|\lambda| = |\mu| = 1$;
(b) $\lambda, \mu \in \mathbb{R} \setminus \mathbb{Q}$ and $0 < |\mu| < 1 < |\lambda|$ or $0 < |\lambda| < 1 < |\mu|$.

[Proof: use that $\lambda + \mu = \mathrm{Trace}(A) \in \mathbb{Z}$ and that $\lambda\mu = \det(A) = \pm 1$.] The toral automorphism defined by A is said to be *hyperbolic* whenever case (b) applies.

Consider a hyperbolic toral automorphism T_A. Let λ and μ be the eigenvalues of A and let L and M be the corresponding eigenspaces of A. Assume that $|\mu| < 1$ and that, consequently, $|\lambda| > 1$. Then the action of the linear operator L_A on $\mathbb{R}^2$ consists of an expansion by a factor $|\lambda|$ in the direction of L and a contraction by a factor $|\mu|$ in the direction of M. These actions may also involve direction reversals along L or M if λ or μ is negative. In order to visualize the action of T_A on $\mathbb{S}^2$ this description has to be

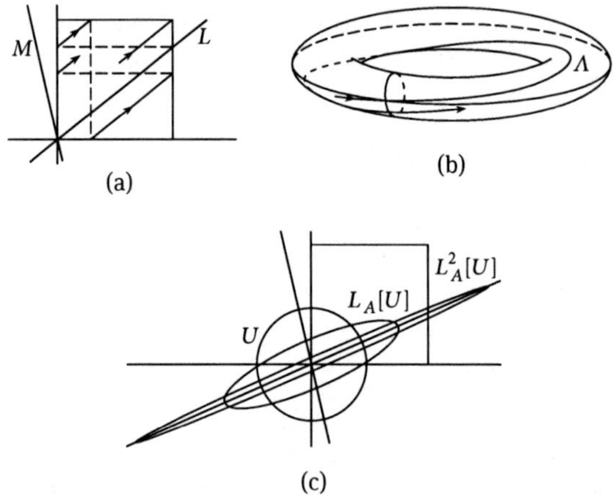

Fig. 6.6. (a) The eigenspaces L and M of T_A and the image Λ of L under the mapping $(s,t) \mapsto ([s],[t]) : \mathbb{R}^2 \to \mathbb{S}^2$ ($\mathbb{S}^2$ represented as the closed unit square). (b) The torus with a segment of Λ. (c) The images of a disc U around the origin under L_A and $(L_A)^2$.

reduced modulo 1. The image Λ of L in $\mathbb{S}^2$ under this reduction is dense in $\mathbb{S}^2$, because L has irrational slope (this is a consequence of Case 2 of the example in 0.4.3 in the Introduction). So Λ meets every non-empty open subset of $\mathbb{S}^2$.

The following argument makes it plausible that the system $(\mathbb{S}^2, T_A)$ is topologically ergodic – even strongly mixing – if T_A is hyperbolic. Let λ, μ, L and M be as above. Consider two non-empty open sets U and V in $\mathbb{R}^2$ and suppose that U is a neighbourhood of the origin. For large n the set $(L_A)^n[U]$ consists of a long but narrow strip around a segment of L (in both directions), the longer and narrower according as n is larger. Now reduce everything modulo 1 so as to see what happens in $\mathbb{S}^2$ under T_A. By what has been said above, Λ meets V. Consequently, for sufficiently large n the set $(T_A)^n[U]$ is a strip (very narrow, but that does not matter) around such a large segment of Λ that it meets V as well. If U is not a neighbourhood of the origin then a similar argument can be used: now for large n the set $(T_A)^n[U]$ is a large narrow strip close to a segment of Λ. This completes the proof that T_A acts ergodically on $\mathbb{S}^2$. Similarly, $\mathbb{S}^2$ is topologically ergodic under T_A^{-1}. As $\mathbb{S}^2$ is a 2nd countable Baire space, Remark 1 after Theorem 1.3.5 implies that the set of transitive points under T_A and the set of transitive points under T_A^{-1} both include a dense G_δ-set. By Baire's Theorem, the intersection of two dense G_δ-sets is also a dense G_δ-set. Conclusion: *the set of bilaterally transitive points under T_A is not empty and is, in fact, residual in $\mathbb{S}^2$.*

By similar geometric arguments it can be shown that *every hyperbolic toral automorphism T_A is expansive*, that is, expansive as used in the literature for homeomorphisms: see Note 9 above. So the results of Section 6.2 do not apply without modification. But they are relatively easy to modify so as to work for homeomorphisms like hyperbolic toral automorphisms: see, again, Note 9 above. For details, see the paper R. L. Adler [1998].

In 1967, K. Berg in his Ph. D. thesis constructed Markov partitions of $\mathbb{S}^2$ for hyperbolic automorphisms. Shortly after that, Adler and Weiss proved that such automorphisms are measure theoretically isomorphic if and only if they have the same (measure theoretic) entropy. Their proof was based on two ideas:

(1) symbolic representations of dynamical systems by means of Markov partitions;
(2) the shift spaces occurring as symbolic representations of hyperbolic toral automorphisms with equal entropy are measure theoretically conjugate.

Each of these two aspects has undergone extensive development since. In this chapter we spend some attention to (1). Aspect (2) falls completely outside the scope of this book.

7 Erratic behaviour

Abstract. In this section we discuss a number of possible definitions that are meant to capture the notion of erratic[1] behaviour (chaos) in a dynamical system. The first one is 'sensitive dependence on initial conditions' which, roughly, means that initial states that are close together may evolve far apart, the metaphor being that the wing-beat of a butterfly in the Amazon woods may cause a hurricane in Texas. The second notion of chaos that we study was defined in the influential paper "Period three implies chaos" by Li and Yorke; it involves the notion of a 'scrambled set'. These notions are most useful for topologically ergodic systems. Moreover, 'weak mixing' turns out to be a rather strong notion, implying both forms of chaos just mentioned.

In the final section of this chapter we show that a dynamical system on an interval with a periodic point whose primitive period is not a power of 2 has a scrambled set. In the next chapter it will be shown that this condition on the system is equivalent to 'positive topological entropy'. In point of fact, positive topological entropy is often considered as yet another form of chaos.

7.1 Stability revisited

Notation. As before, (X, f) will denote a dynamical system. Unlike in previous chapters, X will be assumed to be a *metric space*, with metric d.

A point $x \in X$ (not necessarily invariant) is said to be *stable* (*in the sense of Lyapunov*), or alternatively, the system (X, f) is said to be *stable at* x, whenever the set of mappings $\{ f^n : n \in \mathbb{Z}^+ \}$ is equicontinuous at x. Recall from Appendix A.7.2 that this means: for every $\varepsilon > 0$ there is a neighbourhood U of x such that

$$d(f^n(x), f^n(y)) < \varepsilon \quad \text{for all } y \in U \text{ and all } n \in \mathbb{Z}^+ . \tag{7.1-1}$$

This can also be written as $f^n[U] \subseteq B_\varepsilon(f^n(x))$ for all $n \in \mathbb{Z}^+$. In that case we also say that *the system (X, f) is equicontinuous at the point x*. A stable point is also called an *equicontinuity point*. This notion of stability of points extends the notion of stability of invariant points (see Section 3.2) to non-invariant points: for an invariant point the two notions obviously coincide. Though the term 'stable point' is unambiguous, we prefer to use the term 'equicontinuity point' when it is not clear whether the point is invariant or not. The set of all equicontinuity points in X will be denoted by $\mathbf{Eq}(X, f)$.

Unlike Section 1.6, the present section pays mainly attention to individual points of equicontinuity. In the literature the term '(Lyapunov) stable point' is still often used to denote an equicontinuity point. In this connection, see Exercise 7.1.

1 The word 'erratic' is used only in an informal, not mathematically defined, sense.

In (7.1-1) it may be assumed that $U = B_\delta(x)$ for some $\delta > 0$. So an equivalent formulation is: for every $\varepsilon > 0$ there exists $\delta > 0$ such that

$$f^n[B_\delta(x)] \subseteq B_\varepsilon(f^n(x)) \quad \text{for all } n \in \mathbb{Z}^+ . \tag{7.1-2}$$

If $\varepsilon > 0$ then a point $x \in X$ is said to be an *ε-stable point* or an *ε-equicontinuity point* whenever there exists a neighbourhood U of x such that (7.1-1) holds for this particular value of ε. In other words, if $\varepsilon > 0$ then a point $x \in X$ is ε-stable under f iff the set

$$U_f(x, \varepsilon) := \{\, y \in X : d\,(f^n(x), f^n(y)) < \varepsilon \text{ for all } n \in \mathbb{Z}^+ \,\}$$

is a neighbourhood of x. The set of all ε-stable points of (X, f) will be denoted by $S_\varepsilon(X, f)$. Obviously, if a point is an ε-stable then it is ε'-stable for all ε' with $\varepsilon' \geq \varepsilon$, i.e., if $0 \leq \varepsilon \leq \varepsilon'$ then $S_\varepsilon(X, f) \subseteq S_{\varepsilon'}(X, f)$. Note that a point in X is an equicontinuity point iff it is ε-stable for every $\varepsilon > 0$. So

$$\mathbf{Eq}(X, f) = \bigcap_{\varepsilon > 0} S_\varepsilon(X, f) . \tag{7.1-3}$$

Examples.

(1) Every isolated point is an equicontinuity point.
(2) Consider the tent map T on $[0; 1]$ and let $0 < \varepsilon \leq 1/2$. Claim: no point of $[0; 1]$ is ε-stable under T. For let $x \in [0; 1]$ and let U be an arbitrary neighbourhood of x in $[0; 1]$. We may assume that U is an interval, hence by the argument used in Example (1) after Theorem 1.3.5 there is $n \in \mathbb{N}$ such that $T^n[U] = [0; 1]$. One of the end points of $[0; 1]$ has distance of at least $1/2$ to $T^n(x)$. Then for the point $y \in U$ for which $T^n(y)$ is just that end point one has $|T^n(x) - T^n(y)| \geq 1/2$.

Lemma 7.1.1. *Let $\varepsilon > 0$. Then:*

(1) $S_\varepsilon(X, f) \subseteq \operatorname{int}_X S_{2\varepsilon}(X, f)$.
(2) $f^{\leftarrow}[S_\varepsilon(X, f)] \subseteq S_\varepsilon(X, f)$, *hence* $f^{\leftarrow}[\mathbf{Eq}(X, f)] \subseteq \mathbf{Eq}(X, f)$.

Proof. (1) Consider an arbitrary point $x \in S_\varepsilon(X, f)$ and let U be an open neighbourhood of x such that $U \subseteq U_f(x, \varepsilon)$. So if $y \in U$ then for every point $y' \in U$ the triangle inequality implies that $d(f^n(y), f^n(y')) < 2\varepsilon$ for all $n \in \mathbb{Z}^+$. This shows that every point y of U is 2ε-stable, because U, being open, is a neighbourhood of the point y. Thus, $U \subseteq S_{2\varepsilon}(X, f)$. It follows that x is an interior point of $S_{2\varepsilon}(X, f)$.
(2) Let $x \in X$ such that $f(x) \in S_\varepsilon(X, f)$. Then $W := f^{\leftarrow}[U_f(f(x), \varepsilon)]$ is a neighbourhood of the point x. If $y \in W$ then for every $n \in \mathbb{N}$ one has

$$d(\, f^n(y), f^n(x)\,) = d(\, f^{n-1}(f(y)), f^{n-1}(f(x))\,) < \varepsilon ,$$

which implies that $W \cap B_\varepsilon(x) \subseteq U_f(x, \varepsilon)$ – the intersection with $B_\varepsilon(x)$ is needed to account for $n = 0$. It follows that $U_f(x, \varepsilon)$ is a neighbourhood of the point x, i.e., $x \in S_\varepsilon(X, f)$. This concludes the proof of the first statement. The second statement now follows easily from (7.1-3). □

Remarks.

(1) By induction, statement 2 implies that, for all $n \in \mathbb{N}$,

$$(f^n)^{\leftarrow}[S_\varepsilon(X,f)] \subseteq S_\varepsilon(X,f) \quad \text{and} \quad (f^n)^{\leftarrow}[\mathbf{Eq}(X,f)] \subseteq \mathbf{Eq}(X,f).$$

(2) If $\varepsilon > 0$ and $S_{\frac{1}{2}\varepsilon}(X,f) \neq \emptyset$ then the set $\mathrm{int}_X S_\varepsilon(X,f)$ is not empty. In particular, if $\mathbf{Eq}(X,f) \neq \emptyset$ then for every $\varepsilon > 0$ the set $S_\varepsilon(X,f)$ has a non-empty interior. Moreover, from formula (7.1-3), taking into account that the sets $S_\varepsilon(X,f)$ for $\varepsilon > 0$ form a decreasing family, it follows easily that $\mathbf{Eq}(X,f) \supseteq \bigcap_{k\geq 0} \mathrm{int}_X S_{1/k}(X,f) \supseteq \bigcap_{k\geq 0} S_{1/2k}(X,f) = \mathbf{Eq}(X,f)$, which implies that $\mathbf{Eq}(X,f) = \bigcap_{k\geq 0} \mathrm{int}_X S_{1/k}(X,f)$ is a G_δ-set.

Proposition 7.1.2. *Let A be a non-empty subset of X. If $A \subseteq \mathbf{Eq}(X,f)$ then also $\mathcal{B}(A) \subseteq \mathbf{Eq}(X,f)$.*

Proof. Let $\varepsilon > 0$. If $A \subseteq \mathbf{Eq}(X,f)$ then $A \subseteq S_\varepsilon(X,f)$, Hence Lemma 7.1.1 (1) implies that $S_{2\varepsilon}(X,f)$ is a neighbourhood of A. So for any point $z \in \mathcal{B}(A)$ there exists $n \in \mathbb{Z}$ such that $f^n(z) \in S_{2\varepsilon}(X,f)$ (actually, there are infinitely many of such values of n), hence $z \in (f^n)^{\leftarrow}[S_{2\varepsilon}(X,f)] \subseteq S_{2\varepsilon}(X,f)$, where we have used Remark 2 above. This shows that $\mathcal{B}(A) \subseteq S_{2\varepsilon}(X,f)$. As this holds for every $\varepsilon > 0$, the desired result follows from formula (7.1-3). □

Remarks.

(1) In particular, all points in the basins of stable invariant points and of stable periodic orbits (which are, by Exercise 7.1 (3) , included in $\mathbf{Eq}(X,f)$) are equicontinuity points.

(2) *The basin of an arbitrary stable set is not necessarily included in the set* $\mathbf{Eq}(X,f)$. Example: consider the map $f\colon (x_1,x_2) \mapsto (T(x_1), x_2^2)$ on $X := [0;1] \times [0;1]$, where T denotes the tent map. Clearly, the invariant set $A := [0;1] \times \{0\}$ is stable, and $\mathcal{B}_f(A) = [0;1] \times [0;1)$. Using Example (2) above one easily shows that no point of $\mathcal{B}_f(A)$ (actually, no point of X) is an equicontinuity point.

7.1.3. Let $\varepsilon > 0$. If a point $x \in X$ is not ε-stable then it is said to be *ε-unstable*, and the system is said to be *ε-sensitive at the point x*. Thus, the point x is ε-unstable iff $x \notin S_\varepsilon(X,f)$, iff

$$\forall U \in \mathcal{N}_x \; \exists y \in U : \; d(f^n(x), f^n(y)) \geq \varepsilon \text{ for some } n \in \mathbb{Z}^+ . \tag{7.1-4}$$

A point x is said to be *unstable*, and the system is said to be *sensitive at the point x*, whenever it is ε-unstable for some $\varepsilon > 0$. So by (7.1-3), the point x is unstable iff $x \notin \mathbf{Eq}(X,f)$. Stated otherwise, the system is sensitive at x iff there exists $\varepsilon > 0$ such that the system is ε-sensitive at x.

It is possible that a system has no equicontinuity points at all, i.e., that it is unstable at every point (in that case it is called *pointwise unstable*) in such a way that there exists no 'universal' $\varepsilon > 0$ such that all points are ε-unstable: see Example (8) below.

If there is such a universal ε then the system (X, f) is said to be *ε-sensitive to initial conditions* (or more briefly, it is called *ε-sensitive*). Thus, (X, f) is ε-sensitive iff there exists $\varepsilon > 0$ such that each point of X is ε-unstable, i.e., iff $S_\varepsilon(X, f) = \emptyset$. If a system is ε-sensitive then it is ε'-sensitive for every ε' with $0 < \varepsilon' \leq \varepsilon$.

The dynamical system (X, f) is said to be *sensitive to initial conditions* (or more briefly, just *sensitive*) whenever it is ε-sensitive for some $\varepsilon > 0$. Thus, (X, f) is sensitive iff there exists $\varepsilon > 0$ such that (7.1-4) holds for all points $x \in X$ (with the same value of ε for all points x). Note that in a sensitive system there are no equicontinuity points. As observed above, Example (8) below shows that the converse is not true in general. But if the system is transitive then it turns out to be true, by Corollary 7.1.5 below.

If the system has no equicontinuity points, i.e., if it is pointwise unstable, then every point x is ε_x-unstable with its own ε_x. Therefore, sometimes a sensitive system is called *uniformly unstable*: all points are ε-unstable with a common ε.

Let us summarize the above definitions and their negations in terms of the sets $S_\varepsilon(X, f)$ of ε-stable point:

$$(X, f) \text{ is sensitive} \Leftrightarrow \exists\, \varepsilon > 0 \,\vdots\, S_\varepsilon(X, f) = \emptyset\,. \tag{7.1-5}$$

$$(X, f) \text{ has no equicont. pts} \Leftrightarrow \forall\, x \in X \exists \varepsilon_x > 0 \,\vdots\, x \notin S_{\varepsilon_x}(X, f)\,. \tag{7.1-6}$$

Equivalently:

$$(X, f) \text{ is non-sensitive} \Leftrightarrow \forall\, \varepsilon > 0 \;:\; S_\varepsilon(X, f) \neq \emptyset\,. \tag{7.1-7}$$

$$(X, f) \text{ has an equicont. pt} \Leftrightarrow \bigcap_{\varepsilon>0} S_\varepsilon(X, f) \neq \emptyset\,. \tag{7.1-8}$$

Recall that an isolated point is an equicontinuity point. Consequently, *a sensitive system has no isolated points.*

Examples.

(1) Every expansive system, say, with expansive coefficient η, and without isolated points, is η-sensitive to initial conditions. Recall from Proposition 6.2.1 that every shift system is expansive with expansive coefficient 1. Hence all shift systems are 1-sensitive.

(2) By Example (2) at the beginning of this section, the tent map $T\colon [0; 1] \to [0; 1]$ is $1/2$-sensitive.

(3) Consider the *generalized tent map* $T_s\colon [0; 1] \to [0; 1]$, defined for $0 < s \leq 2$ by

$$T_s(x) := \tfrac{1}{2}\, s \left(1 - |2x - 1| \right) \quad \text{for } x \in \mathbb{R}\,.$$

We shall show that for $1 < s \leq 2$ the system $([0; 1], T_s)$ is sensitive. The idea of the proof is the same as in the preceding example: find a finite set A such that for every interval J there exists n such that $T_s^n[J]$ contains at least two points of A (in the preceding example, $A = \{0, 1\}$). If we apply this to an arbitrary neighbourhood of a point of $[0; 1]$ then this implies that the system $([0; 1], T_s)$ is $\frac{1}{2}a$-sensitive, where a is the minimal distance between two different points of A. The graph of a higher iterate of the generalized tent map in Figure 6.4 in the previous chapter – local maxima and minima which lie more dense if the iterate is higher – suggests that a good candidate for A might be the set $\{ c, T_s(c), \dots, T_s^k(c) \}$ for some fixed $k \in \mathbb{N}$, where $c := 1/2$.

One might be tempted to argue as follows: It is easy to show that for every non-degenerate subinterval J of $[0; 1]$ there are two values $m, n \in \mathbb{N}$ with $m < n$ such that $c \in T_s^n[J] \cap T_s^m[J]$ (see Exercise 7.2 (1)). Then $T_s^{n-m}(c) \in T_s^{n-m}[T_s^m[J]] = T_s^n[J]$ and $c \in T_s^n[J]$, so $T_s^n[J]$ contains two points of the orbit of c. This is not sufficient for our purpose: $n - m$ may depend on J – so the points may not come from a *fixed* initial segment of the orbit of c – and perhaps these points are not distinct (in the case that c is eventually periodic). For hints to a complete proof, see Exercise 7.2.

(4) The argument-doubling system $(\mathbb{S}, \psi)$ with the metric d_c on $\mathbb{S}$ is π-sensitive. (Recall that d_c is the metric that assigns to any two points of $\mathbb{S}$ the length of the shortest arc in $\mathbb{S}$ with those points as end points.) The proof is similar to that of Example (2) preceding Lemma 7.1.1. Let $x \in \mathbb{S}$ and let U be a neighbourhood of x in $\mathbb{S}$. By the argument in Example (2) after Theorem 1.3.5 there exists $n \in \mathbb{N}$ such that $\psi^n[U] = \mathbb{S}$. Consequently, there exists $y \in U$ such that the point $\psi^n(y)$ is diametrically opposed to the point $\psi^n(x)$, that is, $d_c(\psi^n(x), \psi^n(y)) = \pi$.

(5) Let $(X, f) := (\Omega_2 \times \mathbb{S}, \sigma \times \varphi_a)$ with $a \in \mathbb{R}$. Thus, $f(x, [t]) := (\sigma x, \varphi_a[t])$ for $x \in \Omega_2$ and $t \in \mathbb{R}$. Then X is a compact metric space with metric D given by (for example)

$$D((x, [s]), (y, [t])) := \max\{ d(x, y), d_c([s], [t]) \} \quad \text{for } x, y \in \Omega_2 \text{ and } [s], [t] \in \mathbb{S},$$

where d is the usual metric on Ω_2 and d_c is the metric on $\mathbb{S}$ (any metric on $\mathbb{S}$ would do). We claim that (X, f) is 1-sensitive with respect to the metric D. In order to prove this, observe that $D((x, [s]), (y, [t])) \geq d(x, y)$ for every pair of points $(x, [s])$ and $(y, [t])$ in X, so that

$$D(f^k(x, [s]), f^k(y, [t])) \geq d(\sigma^k x, \sigma^k y) \quad \text{for all } k \in \mathbb{Z}^+ .$$

Using this it is easy to show that the 1-sensitivity of (Ω_2, σ) – see 1 above – implies that the system (X, f) is 1-sensitive.

NB. In the second example in the beginning of Section 7.2 it will be shown that this system is transitive.

(6) It follows from the definitions that if a system (X, f) is equicontinuous in at least one point then the system is not sensitive. Examples: the rigid rotation $(\mathbb{S}, \varphi_a)$ for $a \in \mathbb{R}$ and the adding machine (G, f) are equicontinuous at all points of their phase spaces, because their phase mappings are isometries (for the adding machine, see Exercise 5.2 (2)). Hence these systems are not sensitive.

(7) Let $X := (0; 1/2]$ and $f(x) := x^2$ for $x \in X$. With respect to the usual metric the system (X, f) is easily seen to be equicontinuous on X: the mean value theorem

implies that $|f^n(x) - f^n(y)| \leq |x - y|$ for all $x, y \in X$ with $x, y < 1/2$ and all $n \in \mathbb{Z}^+$. Next, consider the metric on X defined by $\rho(x, y) := |\frac{1}{x} - \frac{1}{y}|$ for $x, y \in X$. This metric is compatible with the topology of X. Then

$$\lim_{n \rightsquigarrow \infty} \rho(f^n(x), f^n(y)) = \lim_{n \rightsquigarrow \infty} |x^{-2^n} - y^{-2^n}| = \infty$$

for all $x, y \in X$, $x \neq y$. So with this metric the system (X, f) is expansive with expansive coefficient equal to, for example, 1081. Consequently, with respect to the metric ρ the system (X, f) is sensitive to initial conditions.

(8) Denote the point $r(\cos 2\pi t + \mathrm{i} \sin 2\pi t)$ of $\mathbb{C}$ by $[r, t]$ ($r \geq 0$ and $0 \leq t < 1$). Let X be the closed unit disk in $\mathbb{C}$ with the origin removed and let $f : [r, t] \mapsto [r, \psi([t])] : X \to X$, where ψ is the argument-doubling transformation. On each circle $r = const.$ the mapping f acts as a scaled-down version of the argument-doubling transformation. So for every point x of X there is an $\varepsilon_x > 0$ such that x is ε_x-unstable. But there is no value of $\varepsilon > 0$ such that all points of X are ε-unstable. For if $\varepsilon > 0$ then for all n the Euclidean distance between points $f^n([r, t])$ and $f^n([r', t'])$ is at most $r + r'$, which is less than ε if $r, r' < \varepsilon/2$. Hence points $[r, t]$ with $r < \varepsilon/2$ are not ε-unstable. Stated otherwise, the system (X, f) is not sensitive. To get a compact example, think of $\mathbb{C}$ as situated in the x, y-plane and form the union of X with the unit interval of the z-axis. Define f on this interval as the tent map. On this extended space the system still is not sensitive and yet all points are unstable.

Example (7) shows that equicontinuity and sensitivity of a system may depend on the metric used. If the phase space is compact this is not the case; the proof will be postponed to the next section: see Corollary 7.2.5 ahead.

In Example (5) above, the set $\{0^\infty\} \times \mathbb{S}$ is invariant; the subsystem on this set is conjugate to the equicontinuous system $(\mathbb{S}, \varphi_a)$. Thus, *a subsystem of a sensitive system is not necessarily sensitive* (but an open subsystem is).

The following result reveals an important connection between transitivity and equicontinuity/sensitivity:

Theorem 7.1.4.
(1) *If* (X, f) *is non-sensitive then* $\mathbf{Trans}\,(X, f) \subseteq \mathbf{Eq}\,(X, f)$.
(2) *If* (X, f) *is topologically ergodic then* $\mathbf{Eq}\,(X, f) \subseteq \mathbf{Trans}\,(X, f)$.

Proof. (1) Let the point x have a dense orbit in X (this will be sufficient). As (X, f) is non-sensitive, formula (7.1-7) and Lemma 7.1.1 (1) imply that for every $\varepsilon > 0$ the set $S_\varepsilon(X, f)$ has a non-empty interior. It follows that $f^n(x) \in S_\varepsilon(X, f)$ for some $n \in \mathbb{Z}^+$ and, consequently, that $x \in S_\varepsilon(X, f)$ by Lemma 7.1.1 (2). Since this holds for every $\varepsilon > 0$, the desired conclusion follows from formula (7.1-3).

(2) Let $x \in \mathbf{Eq}\,(X, f)$ and let V be an arbitrary non-empty open subset of X. Without restriction of generality we may assume that $V = B_{2\varepsilon}(z)$ for some point $z \in X$ and some $\varepsilon > 0$. Because x as an equicontinuity point, the set $U_f(x, \varepsilon)$ is a neighbourhood of x. Since the system is topologically ergodic, there are a point $y \in U_f(x, \varepsilon)$ and infinitely

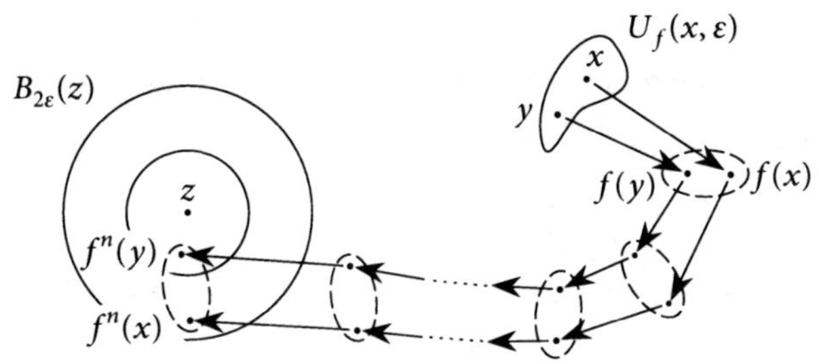

Fig. 7.1. The point $x \in \mathbf{Eq}(X, f)$ is dragged along with the point y into the vicinity of the point z.

many values of $n \in \mathbb{N}$ such that $f^n(y) \in B_\varepsilon(z)$. By the choice of y, for each of these values of n we have $d(f^n(x), f^n(y)) < \varepsilon$, hence $f^n(c) \in B_{2\varepsilon}(z) = V$ by the triangle inequality. This completes the proof that $x \in \mathbf{Trans}(X, f)$. □

Remarks.

(1) By part 2 of the theorem, in a system with equicontinuity points ergodicity implies topological transitivity. In contrast with Theorem 1.3.5 the phase space X is not required to be a 2nd countable Baire space.

(2) Non-sensitivity cannot be omitted from Theorem 7.1.4 (1): the tent map is sensitive, so $\mathbf{Eq}(X, f) = \emptyset$, but the set $\mathbf{Trans}(X, f)$ is not empty. See also Theorem 7.1.13 below. Similarly, topological ergodicity of the system cannot be omitted from the conditions in Theorem 7.1.4 (2). Example: the rigid rotation $(\mathbb{S}, \varphi_a)$ with $a \in \mathbb{Q}$ has no transitive points, but φ_a is equicontinuous at every point of $\mathbb{S}$.

Corollary 7.1.5. *If the system (X, f) includes an unstable point with a dense orbit then it is sensitive (hence all points are unstable) and transitive. Consequently, if a transitive system is pointwise unstable then it is sensitive.*

Proof. An unstable point is not isolated, so when it has a dense orbit then it is transitive: see Corollary 1.3.3. If the system is not sensitive then Theorem 7.1.4 (1) implies that such a point cannot be unstable: a contradiction. This proves the first statement. The second statement is an obvious consequence of the first. □

A dynamical system (X, f) is said to be *almost equicontinuous* whenever it is transitive and has at least one equicontinuity point, that is, whenever $\mathbf{Trans}(X, f) \neq \emptyset$ and $\mathbf{Eq}(X, f) \neq \emptyset$. As the existence of an equicontinuity point implies non-sensitivity, Theorem 7.1.4 (1) implies that a system is almost equicontinuous iff it has a transitive equicontinuity point. One might also apply Theorem 7.1.4 (2): transitivity implies topological ergodicity, hence the equicontinuity points are transitive. Resuming, we have:

Proposition 7.1.6. *The following conditions are equivalent:*

(i) *(X, f) is almost equicontinuous.*

(ii) *(X, f) has a transitive equicontinuity point.*

(iii) *(X, f) is transitive and not sensitive.*

(iv) $\mathbf{Trans}(X, f) = \mathbf{Eq}(X, f) \neq \emptyset$.
(v) *(X, f) is transitive and $\mathbf{Eq}(X, f)$ is a residual set (actually, it is a dense G_δ-set).*

Proof. "(i)⇒(ii)": Clear from Theorem 7.1.4 (1) or from Theorem 7.1.4 (2).
"(ii)⇒(iii)": A system with an equicontinuity point is not sensitive.
"(iii)⇒(iv)": Clear from Theorem 7.1.4 (1), (2).
"(iv)⇒(v)": The set $\mathbf{Trans}(X, f)$ is dense and by Remark 2 following Lemma 7.1.1, $\mathbf{Eq}(X, f)$ is a G_δ-set.
"(v)⇒(i)": Obvious. □

Corollary 7.1.7 (the Auslander–Yorke Dichotomy Theorem).
(1) *A transitive system is either almost equicontinuous or sensitive.*
(2) *A minimal system is either equicontinuous or sensitive.*

Proof. (1) Clear from Proposition 7.1.6 (i)⇔(iii).
(2) If the system is minimal then $\mathbf{Trans}(X, f) = X$, so if the system is not sensitive then by Proposition 7.1.6 (iii)⇔(iv) it is equicontinuous. □

Corollary 7.1.8.
(1) *A transitive equicontinuous system is minimal.*
(2) *A minimal almost equicontinuous system is equicontinuous.*

Proof. One of the two sets $\mathbf{Trans}(X, f)$ and $\mathbf{Eq}(X, f)$ is equal to X, hence the other is equal to X as well. □

Examples.
(1) The Morse–Thue system and the semi-Sturmian systems are sensitive (even expansive) minimal systems.
(2) The rigid rotations $(\mathbb{S}, \varphi_a)$ with $a \in \mathbb{R} \setminus \mathbb{Q}$ and the adding machine are examples of equicontinuous minimal systems (in accordance with Theorem 7.1.11 below the phase mappings in these systems are invertible). For an example of a non-minimal almost equicontinuous (hence not equicontinuous) system, see Example (4) below.
(3) Examples of non-minimal transitive systems that are sensitive are the tent map, the argument-doubling system and the non-minimal transitive shift systems mentioned in the Example in Proposition 5.3.7; see the Examples (1), (2), (4) and (5) in 7.1.3.
(4) For an example of a non-minimal almost equicontinuous system we refer to Theorem 4.2 in the paper E. Akin, J. Auslander & K. Berg [1996]. Another example: there exists a system (X, f) with a compact metric phase space such that
   - the product of (X, f) with every minimal system on a compact metric space is topologically ergodic;
   - (X, f) is not weakly mixing;
   - (X, f) is not sensitive;

(for an existence proof, see E. Akin & S. Glasner [2001]; an explicit example is in W. Huang & X. Ye [2002b]; systems with the first property are called *scattering*). Since (X, f) is conjugate to the product of itself with the trivial one-point system (which is minimal), the first property implies that (X, f) is topologically ergodic, hence transitive. Moreover, (X, f) is not minimal, otherwise the product with itself would be topologically ergodic, contradicting the second property. Consequently, (X, f) is a non-minimal transitive system that is, by the third property and Corollary 7.1.7 (1), almost equicontinuous but, by 7.1.8 (1), not equicontinuous.

Almost equicontinuous systems have an important property: they are uniformly rigid, see Theorem 7.1.11 below. Here is the definition: a non-empty subset K of X is said to be *uniformly rigid*, and the dynamical system (X, f) is said to be *uniformly rigid on K*, whenever

$$\forall \varepsilon > 0 \exists n \in \mathbb{N} : d(x, f^n(x)) < \varepsilon \quad \text{for all } x \in K \tag{7.1-9}$$

(K is not required to be invariant). If this is the case then we also say that f is uniformly rigid on K. The *dynamical system* (X, f) is said to be *uniformly rigid* whenever f is uniformly rigid on X.

Example. The rigid rotation $(\mathbb{S}, \varphi_a)$ with $a \in \mathbb{R}$ is uniformly rigid. If $a \in \mathbb{Q}$ this is obvious: there exists $p \in \mathbb{N}$ such that $\varphi_a^p = \mathrm{id}_{\mathbb{S}}$. In the case that $a \notin \mathbb{Q}$ then proceed as follows: For every $\varepsilon > 0$, let U_ε be the symmetric open arc around $[0]$ with length 2ε. Then for every $n \in D([0], U_\varepsilon)$ we have $d_c([0], \varphi_a^n[0]) < \varepsilon$. As each rotation of $\mathbb{S}$ is an isometry that commutes with every φ_a^n, it follows that, for every $n \in D([0], U_\varepsilon)$, we have $d_c([t], \varphi_a^n[t]) < \varepsilon$ for all $[t] \in \mathbb{S}$.

Lemma 7.1.9. *Let K be a non-empty subset of X. Then f is uniformly rigid on K iff for every $\varepsilon > 0$ there are infinitely many values of n in $\mathbb{N}$ such that $d(x, f^n(x)) < \varepsilon$ for all $x \in K$.*

Proof. "If": Obvious. "Only if": If there exist $n \in \mathbb{N}$ such that $f^n|_K = \mathrm{id}_K$ then $f^{ni}|_K = \mathrm{id}_K$ for all $i \in \mathbb{N}$. This proves our statement in this particular case. Next, assume that $f^n|_K \neq \mathrm{id}_K$ for every $n \in \mathbb{N}$. Then for every $n \in \mathbb{N}$ the real number $M_n := \sup_{x \in K}\{d(f^n(x), x)\}$ is strictly positive (possibly infinite). So if $\varepsilon > 0$ and $m \in \mathbb{N}$ then $\varepsilon' := \frac{1}{2}\min\{\varepsilon, M_1, \dots, M_m\}$ is strictly positive as well. By (7.1-9) there exists $n \in \mathbb{N}$ such that $d(f^n(x), x) < \varepsilon' < \varepsilon$, and by the choice of ε' it is easily seen that $n > m$. Since m can be chosen arbitrarily large, this completes the proof. □

Remark. It follows from the lemma that f is uniformly rigid on K iff there is a subsequence $(n_i)_{i \in \mathbb{N}}$ of $\mathbb{Z}^+$ such that $f^{n_i} \rightsquigarrow \mathrm{id}_K$ for $i \rightsquigarrow \infty$, uniformly on K.

Lemma 7.1.10. *Assume that X is compact and that f is uniformly rigid on X. Then:*
(1) *f is a homeomorphism.*
(2) *f^{-1} is uniformly rigid on X.*
(3) *For all $x \in X$, $\omega_{f^{-1}}(x) = \omega_f(x)$, hence $\mathbf{Trans}(X, f) = \mathbf{Trans}(X, f^{-1})$.*

Proof. (1) Let $x_1, x_2 \in X$, $x_1 \neq x_2$. Then $\varepsilon := \frac{1}{3}d(x_1, x_2) > 0$, so according to (7.1-9) there exists $n \in \mathbb{N}$ such that $d(f^n(x_i), x_i) < \varepsilon$ for $i = 1, 2$. Then by the triangle equality, $d(f^n(x_1), f^n(x_2)) \geq \varepsilon$, hence $f^n(x_1) \neq f^n(x_2)$. It follows that $f(x_1) \neq f(x_2)$. This shows that f is injective.
Moreover, it follows immediately from (7.1-9) that every point of X is recurrent. So if U is a non-empty open subset of X then $U \cap f[X] \neq \emptyset$, for otherwise no point of U would be recurrent. It follows that $f[X]$ is dense in X. As $f[X]$ is closed (X is assumed to be compact) it follows that f is surjective. Consequently, f, being a bijection of the compact space X onto the Hausdorff space X, is a homeomorphism.
(2) Clear from (7.1-9) (replace $f^n(x)$ by y).
(3) Let $x \in X$; then $x \in \omega_f(x)$, because the point x is recurrent. By Exercise 3.2, the set $\omega_f(x)$ is completely invariant under f. Because f is a bijection this implies that this set is invariant under f^{-1}. Since this set is closed, it follows that it includes $\omega_{f^{-1}}(x)$; thus, $\omega_{f^{-1}}(x) \subseteq \omega_f(x)$. The same reasoning can be applied to f^{-1}, because f^{-1} is uniformly rigid on X. So $\omega_f(x) \subseteq \omega_{f^{-1}}(x)$ and, consequently, $\omega_{f^{-1}}(x) = \omega_f(x)$.
Now the final equality follows easily, taking into account that a point is transitive under f or f^{-1} iff its limit set under f or f^{-1}, respectively, is equal to X. □

Theorem 7.1.11. *Let the system (X, f) be almost equicontinuous. Then the system (X, f) is uniformly rigid. Moreover, if X is compact then f is a homeomorphism and the system (X, f^{-1}) is almost equicontinuous.*

Proof. Select a equicontinuity point x_0 in X and let $\varepsilon > 0$. Then x_0 has an open neighbourhood U such that

$$\forall x \in U: \ d(f^k(x), f^k(x_0)) < \varepsilon \quad \text{for all } k \in \mathbb{Z}^+ . \tag{7.1-10}$$

by Proposition 7.1.6 (iv), the point x_0, being an equicontinuity point, is also a transitive point, so there exist $n \in \mathbb{N}$ such that $f^n(x_0) \in U$. Then (7.1-10) implies that $d(f^{k+n}(x_0), f^k(x_0)) < \varepsilon$ for all $k \in \mathbb{Z}^+$, that is, the continuous function $x \mapsto d(f^n(x), x) : X \to \mathbb{R}$ is at most ε at all points $f^k(x_0)$ of the dense orbit of x_0. Consequently, this function is at most ε on all of X, that is, $d(f^n(x), x) \leq \varepsilon$ for all $x \in X$. This completes the proof that (X, f) is uniformly rigid.

It follows from Lemma 7.1.10 (1) that f is a homeomorphism. As (X, f) is transitive, it follows from Lemma 7.1.10 (3) that the system (X, f^{-1}) is transitive as well. It remains to show that there exists an equicontinuity point for f^{-1}. We shall show that x_0 is such an equicontinuity point. To this end, note that in the above we have actually shown

$$\forall n \in \mathbb{Z}^+ : \ f^n(x_0) \in U \Rightarrow d(x, f^n(x)) \leq \varepsilon \quad \text{for all } x \in X .$$

Replace in the right-hand side the arbitrary element $x \in X$ by $f^{-k}(x_0)$ for arbitrary $k \in \mathbb{Z}^+$ and get

$$\forall n \in \mathbb{N} : \ f^n(x_0) \in U \Rightarrow d(f^{-k}(x_0), f^{n-k}(x_0)) \leq \varepsilon \quad \text{for all } k \in \mathbb{Z}^+ .$$

This means that, for every $k \in \mathbb{Z}^+$, the function $y \mapsto d(f^{-k}(x_0), f^{-k}(y))$ is at most ε on the set $U \cap \mathcal{O}_f(x_0)$. Hence this function is at most ε on the closure of this set. As $\mathcal{O}_f(x_0)$ is dense in X it follows that U is included in that closure. Consequently, $d(f^{-k}(x_0), f^{-k}(y)) \leq \varepsilon$ for all $y \in U$. As this holds for every $k \in \mathbb{Z}^+$, this completes the proof that x_0 is an equicontinuity point for f^{-1}. □

Remarks.

(1) In particular, if X is a compact metric space and (X, f) is a minimal equicontinuous system then f is a homeomorphism. In that case, Lemma 7.1.10 (3) and Corollary 7.1.8 (2) imply that the system (X, f^{-1}) is minimal and equicontinuous. Thus, both sets $\{ f^n : n \in \mathbb{Z}^+ \}$ and $\{ f^{-n} : n \in \mathbb{Z}^+ \}$ are equicontinuous, hence their union $\{ f^n : n \in \mathbb{Z} \}$ is equicontinuous. This can also be concluded from Theorem 1.6.9.

(2) In general, the converse of the theorem is not true: see Note 2 at the end of this chapter.

Corollary 7.1.12. *A non-invertible transitive system on a compact metric space is sensitive.*

Proof. Clear from Theorem 7.1.11 and Corollary 7.1.7 (1). □

The next application of Theorem 7.1.4 (1) explains why in previous chapters we paid much attention to examples of transitive systems with a dense set of periodic points (the tent map, the argument-doubling transformation, shift systems). It is yet another answer to the question: when is a transitive system chaotic?

Theorem 7.1.13. *Let X be an infinite metric space. If the system (X, f) is transitive and has a dense set of periodic points then it is sensitive.*

Proof. Assume that (X, f) is not sensitive. Let x_0 be a transitive point. Then by Theorem 7.1.4 (1), x_0 is an equicontinuity point. We claim that this implies that the point x_0 is almost periodic. Once this has been proved, proceed as follows: by Theorem 4.2.2 (1), the orbit closure $\overline{\mathcal{O}(x_0)}$ is minimal. As this orbit closure equals all of X – recall that the point x_0 is transitive – this means that the system (X, f) is minimal. This is not compatible with the existence of periodic points in X, unless X consists of one single periodic orbit. As X is infinite this is not the case. This contradiction shows that (X, f) is sensitive.

It remains to prove our claim, namely, that the point x_0 is almost periodic. Let U be an arbitrary neighbourhood of x_0 and let $\varepsilon > 0$ be such that $B_{2\varepsilon}(x_0) \subseteq U$. Select $\delta > 0$ according to (7.1-2); without limitation of generality we may assume that $\delta \leq \varepsilon$. Since the periodic points are dense in X there is a periodic point $z \in B_\delta(x_0)$. By the choice of δ we then have, for all $n \in \mathbb{N}$, $d(f^n(x_0), f^n(z)) < \varepsilon$. For every $n \in D(z, z)$ – the set of periods of the point z – we have the equality $f^n(z) = z$, hence

$$d(f^n(x_0), x_0) \leq d(f^n(x_0), f^n(z)) + d(z, x_0) < \varepsilon + \delta \leq 2\varepsilon,$$

that is, $f^n(x_0) \in U$. Thus, $D(z,z) \subseteq D(x_0, U) := \{n \in \mathbb{Z}^+ : f^n(x_0) \in U\}$. By Lemma 1.1.2 (c), the set $D(z,z)$ has bounded gaps, hence the set $D(x_0, U)$ has bounded gaps as well. So according to the definition in Section 4.2, the point x_0 is almost periodic. □

Note the similarity with the proof of Theorem 7.1.4 (2): the equicontinuity point is dragged along with a (now not transitive, but) periodic point in an orbit that is 'almost' periodic. The above proof shows that the point x_0 has a slightly stronger form of almost periodicity: for every neighbourhood U of x_0 the set $D(x_0, U)$ includes a set of the form $p\mathbb{Z}^+$ for some $p \in \mathbb{N}$: the point x_0 is *regularly almost periodic*.

Example (7) in 7.1.3 shows that pointwise equicontinuity is not a dynamical property in the class of metrizable dynamical systems. But Exercise 7.8 (1) implies that it is a dynamical property in the class of of compact metric systems. The following examples show that, in general, equicontinuity at single points is not preserved or lifted by factor maps (but a factor of a compact equicontinuous system is equicontinuous: see Exercise 1.12 (2)):

Examples.

(1) The Example after Proposition 3.2.5 shows that an equicontinuity point is not preserved by a factor mapping. However, an open factor mapping preserves equicontinuity: see Exercise 7.8 (1).

(2) Let $X := [-1;1]$, $f(x) := 0$ for $-1 \leq x \leq 0$ and $f(x) := \sqrt{x}$ for $0 \leq x \leq 1$. Then the set $\mathbf{Eq}(X,f)$ is not invariant under f, for $[-1;0) \subseteq \mathbf{Eq}(X,f)$ but $0 \notin \mathbf{Eq}(X,f)$.

(3) Equicontinuity points are not lifted by factor maps (points that move far apart can be identified by the factor map). Trivial example: the factor map of a system without equicontinuity points (e.g., any shift system) onto the trivial one-point system. Also, in Example (5) in 7.1.3 the projection onto the second factor does not lift equicontinuity.

Since equicontinuity is not lifted it is clear that, in general, unstability and sensitivity are not preserved by factor maps (see Example (3) above). As we shall see in the next section, it behaves slightly better with respect to lifting.

7.2 Chaos(1): sensitive systems

As in the previous section we consider only dynamical systems (X, f) with a metric phase space (X, d).

Based on observations in many examples, the following notion of 'chaotic' behaviour has emerged: a system (or invariant set in a system) is called chaotic whenever it is sensitive and, in addition, it behaves such that 'every part comes everywhere'. The latter condition can be formalized as: the system is topologically ergodic or, which is almost the same, transitive.

Recall that every transitive system is topologically ergodic and that, conversely, every topologically ergodic system on a 2nd countable Baire space is transitive. In particular, for systems with compact metric phase spaces the two notions are equivalent.

In itself, transitivity does not necessarily imply erratic behaviour (example: equicontinuous minimal systems): it is the combination with sensitivity that makes the behaviour chaotic everywhere. Similarly, in non-compact spaces sensitivity may not imply erratic behaviour either: there is no reason to say that the system $(\mathbb{R}^+, f)$ with $f(t) := 2t$ has erratic behaviour, yet the system is sensitive (even expansive with expansive coefficient, say, 1081).

Many authors require also that a chaotic system has a dense set of periodic points. Therefore, the following properties will play a role in this section:

(1) (X, f) is transitive.
(2) (X, f) is sensitive to initial conditions.
(3) The periodic points of (X, f) are dense in X.

These conditions are not completely independent of each other: we have seen in Theorem 7.1.13 that (1) & (3) $\Rightarrow$ (2). The following examples show that there are no other generally valid implications between these properties.

Examples.

(2) & (3) $\not\Rightarrow$ (1): The disjoint union of any two systems having the properties (2) and (3) – e.g., two copies of the tent map T – will satisfy. For a connected example, glue two intervals, each with the tent map T, together by identifying their invariant end points: the system $([0; 1], S^2)$, where S is the mapping described in the example in 2.6.3. A second example: the system $([0; 1] \times [0; 1], \mathrm{id}_{[0;1]} \times T)$ is sensitive and it has a dense set of periodic points, but it is not transitive.

(1) & (2) $\not\Rightarrow$ (3): Any infinite minimal shift space, like the Morse–Thue system or a semi-Sturmian system (see Example (1) after Corollary 7.1.8 above) is sensitive and transitive but, due to minimality, it has no periodic points.
Example (5) in 7.1.3 with $a \notin \mathbb{Q}$ provides another example: this system has no periodic points and it is sensitive. In addition, by Corollary 1.6.3 this system is topologically ergodic, hence by Theorem 1.3.5 it is transitive.
It is easy to give a direct proof of its transitivity. Consider two non-empty basic open sets $U := U_1 \times U_2$ and $V := V_1 \times V_2$ in $\Omega_2 \times \mathbb{S}$ with U_1 and V_1 open in Ω_2 and U_2 and V_2 open in $\mathbb{S}$. We may assume that U_1 and V_1 are cylinder sets, i.e., $U_1 = C_0[b]$ and $V_1 = C_0[c]$ for finite blocks b and c. Minimality of the system $(\mathbb{S}, \varphi_a)$ implies that the set $D(U_2, V_2) = \{n \in \mathbb{Z}^+ : \varphi_a^n[U_2] \cap V_2 \neq \emptyset\}$ is infinite: some point (actually, every point) of U_2 is transitive, hence visits V_2 infinitely often. Select any member n of this set with $n > |b|$ and consider a point $x \in \Omega_2$ of the following form:

$$x := b_0 \ldots b_{|b|-1} \cdots\cdots \underset{\uparrow \text{ position } n}{c_0} \ldots c_{|c|-1} \cdots\cdots$$

Then $x \in U_1$, as x starts with the block b and $\sigma^n(x) \in V_1$, as $\sigma^n(x)$ starts with the block c. Consequently, $\sigma^n[U_1] \cap V_1 \neq \emptyset$. Now it is clear that for this value of n we have $f^n[U_1 \times U_2] \cap (V_1 \times V_2) \neq \emptyset$. This shows that (X, f) is topologically ergodic.

(2) $\not\Rightarrow$ (1), (3) $\not\Rightarrow$ (1), (1) $\not\Rightarrow$ (3) and (2) $\not\Rightarrow$ (3): Clear from the above.

(1) $\not\Rightarrow$ (2): Consider any minimal equicontinuous system like $(\mathbb{S}, \varphi_a)$ with $a \notin \mathbb{Q}$.

(3) $\not\Rightarrow$ (2): In the non-sensitive system $(\mathbb{S}, \varphi_a)$ with $a \in \mathbb{Q}$ all points are periodic.

We now come to the definition(s) of chaos. A dynamical system (X, f) is said to be *chaotic in the sense of*

- *Auslander–Yorke* whenever it has (1) and (2), i.e., whenever it is transitive and sensitive,
- *Devaney* whenever it has (1), (2) and (3), i.e., whenever it is transitive and sensitive and it has a dense set of periodic points.

A system that is chaotic in the sense of Auslander–Yorke or Devaney will also be called *AY-chaotic* or *D-chaotic*, respectively. By Theorem 7.1.13, the definition of D-chaos is slightly redundant if X is infinite. However, usually condition (2) is included in the definition in order to emphasize that Devaney-chaos is Auslander–Yorke-chaos plus dense periodic points.

Theorem 7.2.1. *A minimal dynamical system is either equicontinuous or AY-chaotic.*

Proof. This is Corollary 7.1.7 (2) (the original form of the Auslander–Yorke Dichotomy Theorem). □

Theorem 7.2.2. *A transitive dynamical system on a non-degenerate interval is D-chaotic.*

Proof. Clear from the Theorems 7.1.13 and 2.6.2. □

Examples.

(1) See "(1) & (2) $\not\Rightarrow$ (3)" above for systems that are AY-chaotic but not D-chaotic.

(2) Every full shift is D-chaotic, as are the golden mean shift, the even shift, the prime gap shift, the (1,3) run-length limited shift and the context free shift; see the Example after Proposition 5.3.7 and Exercise 5.9-(4),(5). Also D-chaotic are the system of the tent map and the argument-doubling system – see the Examples (2) and 4 in 7.1.3 – and the systems (X_p, g_p) for odd $p \geq 3$ considered in Exercise 6.12. By the Exercises 5.13 (1) and 2.13 these systems are (strongly, hence) weakly mixing, so the fact that they are at least AY-chaotic would also follow from Exercise 7.7.

(3) Let $\mu > 2 + \sqrt{5}$ and consider the mapping $f_\mu : x \mapsto \mu x(1 - x) : \mathbb{R} \to \mathbb{R}$. By 1.7.5 the system $(\mathbb{R}, f_\mu)$ has an invariant Cantor set Λ and by 6.3.6 (2) the subsystem on Λ is conjugate to the shift (Ω_2, σ), which is D-chaotic. So $(\mathbb{R}, f_\mu)$ has a D-chaotic subsystem (here we use Corollary 7.2.5 below). However, for these values of μ the full system $(\mathbb{R}, f_\mu)$ is not AY-chaotic (it is not transitive), even though it is sensitive (see Exercise 7.5).

By Example (3) after Theorem 7.1.13, AY-chaos is not, in general, preserved by factor maps. But D-chaos is preserved:

Proposition 7.2.3. *An infinite factor of a Devaney chaotic system is Devaney chaotic.*

Proof. A factor of a transitive system is transitive and the image of a dense set of periodic points under a factor map is a dense set of periodic points. Now use Theorem 7.1.13. □

Example. The system considered in Example F of 6.3.5 is a factor of the golden mean shift, hence Devaney-chaotic.

Lifting of chaos is more interesting. In point of fact, an often used method to prove that a system is sensitive is to show that is has a sensitive factor. A first result on lifting is considered in Exercise 7.8. Another one follows here:

Proposition 7.2.4. *Let $\varphi : (X, f) \to (Y, g)$ be a factor map of dynamical systems with X (hence[2] Y) a compact metric space, and assume that φ is semi-open. If (Y, g) is sensitive then (X, f) is sensitive as well.*

Proof. There is a metric ρ on Y with respect to which the system (Y, g) sensitive; say, it is η-sensitive ($\eta > 0$). Since X is a compact space, the mapping $\varphi : X \to Y$ is uniformly continuous with respect to the metrics ρ and d. Consequently, there is a constant $\zeta > 0$ such that

$$\forall x_1, x_2 \in X : d(x_1, x_2) < \zeta \Rightarrow \rho(\varphi(x_1), \varphi(x_2)) < \eta . \tag{7.2-1}$$

Claim. The system (X, f) is $\frac{1}{2}\zeta$-sensitive with respect the metric d on X. Suppose it is not: there are a point $x_0 \in X$ and a neighbourhood U of x_0 such that $d(f^n(x_0), f^n(x)) < \frac{1}{2}\zeta$ for all points $x \in U$ and for all $n \in \mathbb{Z}^+$, hence by the triangle inequality,

$$d(f^n(x), f^n(x')) < \zeta \quad \text{for all } x, x' \in U \text{ and all } n \in \mathbb{Z}^+ . \tag{7.2-2}$$

Since φ is semi-open, the set $\varphi[U]$ includes a non-empty open set W. For any two points $y, y' \in W$, select $x, x' \in U$ such that $\varphi(x) = y$ and $\varphi(x') = y'$. Then (7.2-1) and (7.2-2) imply that

$$\begin{aligned} \rho(g^n(y), g^n(y')) &= \rho(g^n(\varphi(x)), g^n(\varphi(x'))) \\ &= \rho(\varphi(f^n(x)), \varphi(f^n(x'))) < \eta . \end{aligned}$$

In particular, if $y \in W$ and W' is an arbitrary neighbourhood of y in Y then this inequality holds for every point $y' \in W \cap W'$, contradicting the fact that the point y is η-unstable with respect to the metric ρ. □

2 Compactness of Y is obvious, but metrizability is not so trivial: see Appendix A.7.8.

Examples.

(1) The canonical projection π_1 of the product $\Omega_2 \times \mathbb{S}$ onto Ω_2 is an open mapping and it is easily seen to be a factor mapping. As (Ω_2, σ) is sensitive it follows that the system $(\Omega_2 \times \mathbb{S}, \sigma \times \varphi_a)$ is sensitive. This accounts for Example (5) in 7.1.3.

(2) Consider the system $(\mathbb{S} \times \mathbb{S}, f)$, where $f([s],[t]) := (\psi[s], g[t])$ for $s, t \in [0;1)$, where g is any mapping from $\mathbb{S}$ onto itself (ψ is the argument-doubling transformation). As in the example above, the projection $\pi_1 : (\mathbb{S} \times \mathbb{S}, f) \to (\mathbb{S}, \psi)$ is an open factor mapping. As $(\mathbb{S}, \psi)$ is sensitive it follows that $(\mathbb{S} \times \mathbb{S}, f)$ is sensitive as well.

Corollary 7.2.5. *Sensitivity is a dynamical property within the class of dynamical systems on compact metric spaces. In particular, if a dynamical system on a compact metric space is sensitive with respect to one compatible metric then it is sensitive with respect to all compatible metrics. Similar statements hold for AY-chaos and D-chaos.*

Proof. Clear from the Propositions 7.2.4 and 1.5.2 (2), (5). □

Remark. In compact systems, the (supremum of the) set of values of ε for which the system is ε-sensitive may depend on the metric used. For example, for the argument-doubling system this supremum is equal to the diameter of $\mathbb{S}$ – see Example (4) in 7.1.3 – which is 2 with respect to the metric d_{eucl} and π with respect to the metric d_c.

Corollary 7.2.6. *Let $\varphi: (X, f) \to (Y, g)$ be a factor mapping of minimal dynamical systems on compact metric spaces. If (Y, g) is sensitive then so is (X, f). Stated otherwise: if (Y, g) is AY-chaotic then so is (X, f).*

Proof. By Theorem 1.5.7, φ is semi-open, so Proposition 7.2.4 can be applied. □

We now remove the condition that φ is semi-open:

Proposition 7.2.7. *Let $\varphi : (X, f) \to (Y, g)$ be a factor map of dynamical systems with compact metric phase spaces (X, d) and (Y, ρ). Then for every $\varepsilon > 0$ there exists $\delta > 0$ with the following property: if the point $y \in Y$ is ε-unstable (with respect to the metric ρ) then there exists a δ-unstable point (with respect to the metric d) in $\varphi^{\leftarrow}[y]$.*

Proof. Let $\varepsilon > 0$. As φ is uniformly continuous there is a $\delta > 0$ such that

$$\rho(\varphi(x), \varphi(x')) < \varepsilon \quad \text{for all } x, x' \in X \text{ with } d(x, x') < \delta\,. \tag{7.2-3}$$

Let $y \in Y$ an ε-unstable point and assume that every point of $\varphi^{\leftarrow}[y]$ is δ-stable. Then every point $x \in \varphi^{\leftarrow}[y]$ has an open neighbourhood U_x in X such that

$$\forall x' \in U_x : \; d(f^n(x), f^n(x')) < \delta \quad \text{for all } n \in \mathbb{Z}^+\,. \tag{7.2-4}$$

By the Lemma A.3.3 in Appendix A there is a neighbourhood V of y with

$$\varphi^{\leftarrow}[V] \subseteq \bigcup \{U_x : x \in \varphi^{\leftarrow}[y]\} =: W_y\,.$$

In particular, $V \subseteq \varphi[W_y]$, because φ is surjective. So if $y' \in V$ then there are $x \in \varphi^{\leftarrow}[y]$ and $x' \in U_x$ such that $y' = \varphi(x')$, hence

$$\rho(g^n(y), g^n(y')) = \rho(\varphi(f^n x), \varphi(f^n x')) < \varepsilon \quad \text{for all } n \in \mathbb{Z}^+$$

by (7.2-3) and (7.2-4). This would be true for every $y' \in V$, contradicting the fact that the point y is ε-unstable. □

Remarks.

(1) By mapping an equicontinuous system onto an unstable invariant point in any system one sees that surjectivity of φ is essential in the above proposition. Example (1) after Theorem 7.1.13 shows that not all points of the fibre of an unstable point of Y are necessarily unstable in X.

(2) Application of Proposition 7.2.7 to a conjugation gives us an alternative proof of Corollary 7.2.5. Another consequence of Proposition 7.2.7 is that a factor of a compact equicontinuous system can have no unstable points, so that it is equicontinuous as well: an alternative proof of Exercise 1.12 (2).

Corollary 7.2.8. *Let $\varphi : (X, f) \to (Y, g)$ be a factor map of dynamical systems with compact metric phase spaces and assume that there is a transitive point x_0 in X such that $\varphi^{\leftarrow}[\varphi(x_0)] = \{x_0\}$. Under these conditions, if the system (Y, g) is sensitive then so is (X, f). Stated otherwise: if (Y, g) is AY-chaotic then (X, f) is AY-chaotic.*

Proof. A straightforward consequence of Proposition 7.2.7 and Theorem 7.1.4 (1). □

Proposition 7.2.9. *Let $\varphi: (X, f) \to (Y, g)$ be a factor mapping of dynamical systems on compact metric spaces. If (Y, g) is AY-chaotic then there is a closed f-invariant subset X_0 of X such that the subsystem (X_0, f) is AY-chaotic and $\varphi[X_0] = Y$.*

Proof. By Proposition 1.5.5 (1), there is a closed invariant subset X_0 of X such that $\varphi[X_0] = Y$ and which has the following property: if $x \in X_0$ and the point $\varphi(x)$ is transitive in (Y, g) then x is transitive in (X_0, f). In particular, the system (X_0, f) is transitive.

Next, we show that the system (X_0, f) is sensitive. Let d and ρ be the metrics for X and Y, respectively and assume that (Y, g) is ε-sensitive. Select $\delta > 0$ in accordance with Proposition 7.2.7 and let y_0 be a transitive point in Y. Then the point y_0 is ε-unstable under g, and by Proposition 7.2.7 there exist a point $x_0 \in \varphi^{\leftarrow}[y_0]$ which is δ-unstable under f. By the property of X_0 mentioned above, the point x_0 is transitive in (X_0, f), so Theorem 7.1.4 (1) implies that (X_0, f) is sensitive. □

Remarks.

(1) If X is minimal under f then in the situation of the theorem above one has $X_0 = X$, hence the system (X, f) is AY-chaotic. So we get an alternative proof of Corollary 7.2.6 above.

(2) For later reference, recall that the proof of Proposition 1.5.5 (1) shows that the set X_0 in the above proposition is minimal with respect to the property of being a closed invariant subset of X that is mapped onto Y by φ.

7.3 Chaos(2): scrambled sets

7.3.1. The following notion is more complicated than sensitivity: it involves pairs of points that not only drift apart (not just once, but infinitely often as in Devaney chaotic systems: see Exercise 7.6-(1)) but also come arbitrarily close to each other infinitely often. The setting is, as before, a dynamical system (X, f) with a metric phase space X with metric d. A pair of points $(x, y) \in X^2 := X \times X$ is called a *Li–Yorke pair* whenever

$$\limsup_{n\rightsquigarrow\infty} d(f^n(x), f^n(y)) > 0\,, \quad \liminf_{n\rightsquigarrow\infty} d(f^n(x), f^n(y)) = 0\,. \tag{7.3-1}$$

Obviously, if (x, y) is a Li–Yorke pair then $x \neq y$.

Example.

(1) Consider the system (X^*, f^*) discussed in the Example after Theorem 3.1.1. Recall that X^* is a compact Hausdorff space and observe that the topology of X^* has a countable base. So X^* is metrizable: see Appendix A.7.7. Let d^* be a metric on X^*. We could obtain an explicit metric on X^* by constructing a homeomorphic image of X^* in $\mathbb{R}^2$, but that is not really necessary. The only facts that are relevant are the following: points in an ε-ball about the point $Q := (0, \infty)$ have a distance of at most ε to Q and they have a mutual distance of at most ε. Moreover, the compact sets $B := \mathbb{Z}^* \times \{1\}$ and $X^* \setminus B$ have a positive distance, say, δ.
It is easy to see that the orbit under f^* of the point $P := (0, 1)$ of X^* has a subsequence that converges to the point $Q := (0, \infty)$ and a subsequence that converges to the point $R := (\infty, 1)$. As the point Q is invariant, it follows that

$$\liminf_{n\rightsquigarrow\infty} d^*((f^*)^n(P), (f^*)^n(Q)) = 0$$

and that

$$\limsup_{n\rightsquigarrow\infty} d^*((f^*)^n(P), (f^*)^n(Q)) \geq d^*(R, Q)) \geq \delta\,.$$

So (P, Q) is a Lie–Yorke pair. By a similar argument, (P', Q) is a Lie–Yorke pair for every point P' in the orbit of P. With a little more effort one shows that any two distinct points P' and P'' in the orbit of P form a Lie–Yorke pair. Briefly: every neighbourhood of Q includes both $(f^*)^n(P')$ and $(f^*)^n(P'')$ for infinitely many values of n, hence for every $\varepsilon > 0$ one has $d^*((f^*)^n(P'), (f^*)^n(P'')) < \varepsilon$ for infinitely many values of n. Thus, $\liminf_{n\rightsquigarrow\infty} d^*((f^*)^n(P'), (f^*)^n(P'')) = 0$. Moreover, supposing that P'' is in the orbit of P', say, $P'' = (f^*)^m(P')$, if n is such that $f^n(P'') = (1, k)$ with $k \geq 1$ and k is sufficiently large, then $f^n(P')$ is straight above $(1, k)$, namely,

$f^k(P') = (m+1, k)$, hence $d^*((f^*)^n(P'), (f^*)^n(P'')) \geq \frac{1}{2}$. This happens for infinitely many values of n, hence $\limsup_{n\rightsquigarrow\infty} d^*((f^*)^n(P'), (f^*)^n(P'')) \geq \frac{1}{2}$. So all pairs of distinct points in the subset X of X^* – see the Example after Theorem 3.1.1 for notation – are Li–Yorke.

A subset $S \subseteq X$ is said to be *scrambled* whenever every pair of distinct points of S is a Li–Yorke pair. A subset S of X is said to be *δ-scrambled* (with $\delta > 0$) if it is scrambled and the first condition of (7.3-1) reads

$$\limsup_{n\rightsquigarrow\infty} d(f^n(x), f^n(y)) \geq \delta \tag{7.3-1*}$$

for every pair of points $x, y \in S$ with $x \neq y$.

The first condition of (7.3-1) is easily seen to hold for every $(f \times f)$-recurrent point $(x, y) \in X^2$ such that $x \neq y$: then condition (7.3-1*) is fulfilled with $\delta := d(x, y) > 0$, because for every $\beta > 0$ there are infinitely many values of n such that the point $(f^n(x), f^n(y))$ is so close to the point (x, y) that $d(f^n(x), f^n(y)) > d(x, y) - \beta$.

A pair $(x, y) \in X^2 \setminus \Delta_X$ is called a *strong Li–Yorke pair* whenever it is recurrent in the system $(X^2, f \times f)$ and the second condition of (7.3-1) is fulfilled as well, which means that $(f^n(x), f^n(y))$ not only approaches the point (x, y) arbitrarily close, but also the diagonal Δ_X of X^2. If a point of X^2 is transitive in X^2 under $f \times f$ then it is a strong Li–Yorke pair: see Exercise 7.10 (3). Clearly, a strong Li–Yorke pair is a Li–Yorke pair.

A subset S of X is said to be *strongly scrambled* whenever every pair of distinct points in S is a strong Li–Yorke pair. Obviously, a strongly scrambled set is scrambled.

Examples.

(2) The system $([0; 1], f)$ with $f(x) := x^2$ for $0 \leq x \leq 1$ has no scrambled subsets: see the Exercises 7.11 (1),(2).

(3) Consider the shift system (Ω_2, σ). For every point $x \in \Omega_2$, define $x^* \in \Omega_2$ by

$$x^* := x_0\ 1\ x_{[0;2)}\ 11 \ldots x_{[0;m)}\ 1^m\ \ldots\ldots$$

Claim. the set $S := \{x^* : x \in \Omega_2\}$ is 1-scrambled. To prove this claim, consider two points x^* and y^* in S with $x^* \neq y^*$ (i.e., $x, y \in \Omega_2$ and $x \neq y$). Then there is $k \in \mathbb{Z}^+$ such that $x_k \neq y_k$, hence $x_{[0;m)} \neq y_{[0;m)}$ for all $m > k$. It follows that there are infinitely many positions where x^* differs from y^*. Consequently, $d(\sigma^n x^*, \sigma^n y^*) = 1$ for infinitely many values of n, hence $\limsup_{n\rightsquigarrow\infty} d(\sigma^n x^*, \sigma^n y^*) = 1$. Moreover, for every $k \in \mathbb{N}$, let $i_k := \sum_{n=1}^{k}(n-1+n) = k^2$, so that the sequences of coordinates of the points $\sigma^{i_k} x^*$ and $\sigma^{i_k} y^*$ both start with the block 1^k. Then $d(\sigma^{i_k} x^*, \sigma^{i_k} y^*) = 1/(1+k)$. It follows that $\liminf_{n\rightsquigarrow\infty} d(\sigma^n x^*, \sigma^n y^*) = 0$.

The mapping $x \mapsto x^* : \Omega_2 \to \Omega_2$ is injective, so S is uncountable. As this mapping is also continuous (all coordinates of x^* depend continuously on x), S is homeomorphic with Ω_2, hence it is a Cantor space. So Ω_2 includes a scrambled Cantor set.

A system (X, f) is said to be *chaotic in the sense of Li–Yorke*, or *Li–Yorke chaotic* – abbreviation: *LY-chaotic* – whenever there exists an uncountable scrambled set in X. If this set is strongly scrambled then the system is said to be *strongly Li–Yorke chaotic* and if this set is dense in X then we say that the system is *densely Li–Yorke chaotic*.

In contrast to AY-chaos, which has everywhere chaotic behaviour, LY-chaos can be restricted to a small region in the phase space (see Example (5) below). On the other hand, dense LY-chaos is spread all over the phase space. Note that the initial example in 7.3.1 has a dense δ-scrambled set; however, the system is not densely LY-chaotic, as this set is not uncountable.

By Example (3) above, the shift system (Ω_2, σ) is LY-chaotic. It does not follow from the proof above that it is densely LY-chaotic, but it is, as will follow immediately from Corollary 7.3.7 (a) below. For a number of LY-chaotic subshifts, see Exercise 7.11 (3). A famous example: a system on an interval with a periodic point of period 3 is LY-chaotic: see Corollary 7.5.6.

The definitions of a scrambled set and of LY-chaos originate from the paper *Period three implies chaos* by Li and Yorke; see T. Y. Li & J. A. Yorke [1975], where it is shown that an interval map with a periodic point of period three has periodic points of all periods – see Theorem 2.2.2 – and has an uncountable scrambled set. They (and some other authors as well) imposed also the following condition on a scrambled set S: for every point $x \in S$ and every periodic point $z \in X$,

$$\limsup_{n\rightsquigarrow\infty} d(f^n(x), f^n(z)) > 0 . \tag{7.3-2}$$

Thus, a scrambled set contains no points asymptotic to (i.e., with the same long-run behaviour as) periodic points. This condition turns out to make no difference for the definition of LY-chaos: it can be satisfied by removal of at most one point from a set satisfying condition (7.3-1):

Lemma. *Let (X, f) be a dynamical system on a metric space and let S be a subset of X such that for all points $x, y \in S$, $x \neq y$,*

$$\limsup_{n\rightsquigarrow\infty} d(f^n(x), f^n(y)) > 0 , \quad \liminf_{n\rightsquigarrow\infty} d(f^n(x), f^n(y)) = 0 . \tag{7.3-3}$$

Then there is at most one point $x \in S$ for which there exists a periodic point $z \in X$ such that $\limsup_{n\rightsquigarrow\infty} d(f^n(x), f^n(z)) = 0$, *or, equivalently,* $\lim_{n\rightsquigarrow\infty} d(f^n(x), f^n(z)) = 0$.

Proof. Assume the contrary: there are two points $x_1, x_2 \in S$, $x_1 \neq x_2$, and two periodic points $z_1, z_2 \in X$ such that $\lim_{n\rightsquigarrow\infty} d(f^n(x_i), f^n(z_i)) = 0$ for $i = 1, 2$. Let

$$\varepsilon := \limsup_{n\rightsquigarrow\infty} d(f^n(x_1), f^n(x_2)) . \tag{7.3-4}$$

Then in view of (7.3-3) one has $\varepsilon > 0$. Let p be a common multiple of the periods (hence a common period) of the points z_1 and z_2.

The finitely many continuous functions f^k for $0 \leq k \leq p$ form an equicontinuous set on X, and as the set $\mathcal{O}(z_1)$ is finite there exists $\eta > 0$ such that $d(f^k(x), f^k(y)) < \frac{1}{3}\varepsilon$ for $0 \leq k \leq p$ and for all points $x, y \in X$ with $x \in \mathcal{O}(z_1)$ and $d(x, y) < \eta$. In addition, we may assume that $\eta \leq \frac{1}{2}\varepsilon$. By our assumptions and the second equality in (7.3-3) there exists $N \in \mathbb{N}$ such that

$$\forall n \geq N : d(f^n(x_i), f^n(z_i)) < \tfrac{1}{3}\eta \quad (i = 1, 2) \tag{7.3-5}$$
$$\text{and} \quad d(f^N(x_1), f^N(x_2)) < \tfrac{1}{3}\eta \tag{7.3-6}$$

By the triangle inequality it follows that $d(f^N(z_1), f^N(z_2)) < \eta$. The choice of η then implies that $d(f^{N+k}(z_1), f^{N+k}(z_2)) < \frac{1}{3}\varepsilon$ for $0 \leq k \leq p$. As both points z_1 and z_2 are periodic with period p, it follows that the inequality $d(f^n(z_1), f^n(z_2)) < \frac{1}{3}\varepsilon$ holds for all $n \geq N$, hence by (7.3-5), $d(f^n(x_1), f^n(x_2)) < \frac{1}{3}\varepsilon + \frac{2}{3}\eta < \frac{2}{3}\varepsilon$. This contradicts the definition of ε in (7.3-4). □

If the subsystem on an invariant subset Y of (X, f) is LY-chaotic then the system (X, f) is Li–Yorke chaotic as well: an uncountable scrambled subset of Y is, obviously, an uncountable scrambled subset of X. Moreover, if (Y, f) is strongly LY-chaotic then so is (X, f), because 'recurrent in $Y \times Y$' implies 'recurrent in $X \times X$'. But if Y is not dense in X then dense LY-chaos of (Y, f) does not imply *dense* LY-chaos of (X, f).

Finally, the definitions of lim sup and lim inf imply that, if for some $q \in \mathbb{N}$ the system (X, f^q) is LY-chaotic then so is (X, f). Similarly, if (X, f^q) is densely or strongly LY-chaotic then (X, f) has the same property.

Examples.

(4) For every finite symbol set **S** the shift system (Ω_S, σ) includes a copy of Ω_2. So Example (3) above implies that *every full shift system is LY-chaotic*. It will follow from Corollary 7.3.7 (a) below that these systems are strongly and densely LY-chaotic.

(5) The system $(\Omega_2 \times [0; 1], \sigma \times f)$ with $f(t) := t^2$ for $0 \leq t \leq 1$ is LY-chaotic, because the subsystem on $\Omega_2 \times \{0\}$ is isometrically isomorphic to the LY-chaotic system (Ω_2, σ). But the subset $\{0^\infty\} \times [0; 1]$ includes no scrambled subsets – see Example (2) above. So an uncountable subsystem of a LY-chaotic system is not necessarily LY-chaotic. This example also shows that a factor – in this case, $([0; 1]), f)$ – of an LY-chaotic system is not necessarily LY-chaotic.

7.3.2. It will be convenient to introduce two new notions that will enable us to give an alternative definition of a Li–Yorke pair. The first one is fairly classical: two points $x, y \in X$ are said to be *asymptotic to each other*, or *the pair* (x, y) is called *asymptotic*, whenever $\lim_{n \rightsquigarrow \infty} d(f^n(x), f^n(y)) = 0$.

Examples.

(1) A trivial (though important) example of an asymptotic pair is a pair (x, y) in a system such that y is an invariant point and $f^n(x) \rightsquigarrow y$ for $n \rightsquigarrow \infty$.

(2) By Exercise 3.3 (2) , if the limit set of a point $x \in X$ is a periodic orbit then x is asymptotic with a point from that periodic orbit.

(3) It is easy to show that two points $x, y \in \Omega_2$ are asymptotic iff $x_n = y_n$ for almost all n. So in Example (3) in 7.3.1, the points x^* and y^* for $x \neq y$ are not asymptotic. Note that the set of asymptotic pairs is dense in $\Omega_2 \times \Omega_2$.
(4) Consider a semi-Sturmian system (Z, σ) of type (a, b) with $a \notin \mathbb{Q}$ and $[b]$ not in the orbit of $[0]$. in 6.3.7 (2) it is shown that two points in a fibre of the 'canonical' map $\psi: Z \to \mathbb{S}$ are asymptotic. The converse is true as well: if two points are asymptotic then their ψ-images in $\mathbb{S}$ are asymptotic under φ_a, hence equal to each other, so the two points are in the same fibre of ψ. A similar example is in Exercise 7.9 (5).

The discussion in 7.3.1 immediately after (7.3-1*) can be interpreted as follows: if a point $(x, y) \in X^2 \setminus \Delta_X$ is recurrent under the mapping $f \times f$ then (x, y) is not an asymptotic pair. Moreover, it follows easily from the definition of a uniformly rigid set in formula (7.1-9) that if K is a uniformly rigid set and $x, y \in K$ then $(x, y) \in X \times X$ is recurrent under $f \times f$. Hence *no pair of mutually distinct points in a uniformly rigid set is asymptotic.*

Two points $x, y \in X$ are said to be *proximal to each other* or, alternatively, the pair $(x, y) \in X \times X$ is called a *proximal pair*, whenever for every $\varepsilon > 0$ there exists $n \in \mathbb{Z}^+$ such that $d(f^n(x), f^n(y)) < \varepsilon$. Obviously, an asymptotic pair is proximal, but the converse is not generally true. In point of fact (and here comes our alternative description) *a pair of points in X is a Li–Yorke pair iff it is proximal but not asymptotic.*

A convenient characterization of proximal pairs can be given using the following notation: for every $\varepsilon > 0$, let

$$\alpha_\varepsilon := \{(x', y') \in X^2 : d(x', y') < \varepsilon\}$$

(an open neighbourhood of the diagonal Δ_X in X^2). So a pair $(x, y) \in X^2$ is proximal iff $\mathcal{O}_{f\times f}(x, y) \cap \alpha_\varepsilon \neq \emptyset$ for every $\varepsilon > 0$.

Lemma 7.3.3. *Consider the following conditions for $x, y \in X$:*
(i) *(x, y) is a proximal pair.*
(ii) *For every $\varepsilon > 0$ there are infinitely many values of $n \in \mathbb{Z}^+$ with $d(f^n(x), f^n(y)) < \varepsilon$.*
(iii) $\liminf_{n\rightsquigarrow\infty} d(f^n(x), f^n(y)) = 0$.
(iv) $\overline{\mathcal{O}_{f\times f}(x, y)} \cap \Delta_X \neq \emptyset$.
(v) *For every neighbourhood W of Δ_X in $X \times X$ there exists $n \in \mathbb{Z}^+$ such that $(f^n(x), f^n(y)) \in W$.*
Then always (i)⇔(ii)⇔(iii) *and* (iv)⇔(v)⇒(i). *If X is compact then all five conditions are mutually equivalent.*

Proof. The implications (i)⇐(ii)⇔(iii) are trivial.

"(i)⇒(ii)": If there exists $n \in \mathbb{Z}^+$ such that $f^n(x) = f^n(y)$ then, obviously, (ii) holds. So assume that $f^n(x) \neq f^n(y)$ for all $n \in \mathbb{Z}^+$. Let $m \in \mathbb{Z}^+$ be arbitrary. By (i), for $\varepsilon' := \min\{\varepsilon, \frac{1}{2}\min\{d(f^k(x), f^k(y)) : 0 \leq k \leq m\}\}$ there exists $n \in \mathbb{Z}^+$ such that $d(f^n(x), f^n(y)) < \varepsilon' < \varepsilon$. Obviously, $n > m$.

"(iv)⇒(v)": If $(z,z) \in \overline{\mathcal{O}_{f\times f}(x,y)} \cap \Delta_X$ and W is a neighbourhood of Δ_X then W is a neighbourhood of the point (z,z), hence $W \cap \mathcal{O}_{f\times f}(x,y) \neq \emptyset$.

"(v)⇒(iv)": If (iv) does not hold then (v) fails for the open neighbourhood $W := (X \times X) \setminus \overline{\mathcal{O}_{f\times f}(x,y)}$ of Δ_X.

"(v)⇒(i)": If $\varepsilon > 0$ then α_ε is a neighbourhood of Δ_X.

"(i)⇒(v) if X is compact": Use that, by Appendix A.7.6, in this case the sets α_ε with $\varepsilon > 0$ form a neighbourhood base of Δ_X in X^2. □

Examples.

(1) Let $X := (0;\infty)$ with its usual metric and let $f(t) := \frac{1}{2}t$ for $t \in X$. Then every pair of points in X is proximal (even asymptotic), but condition (v) of the above lemma is not fulfilled, because $(0,0)$ does not belong to X^2.

(2) Two points in Ω_2 are proximal under the shift σ iff they have arbitrarily large common blocks in the same position. So in Example (3) in 7.3.1, the points x^* and y^* for $x, y \in \Omega_2$ are proximal. Note that the set of proximal pairs is dense in $\Omega_2 \times \Omega_2$.

(3) If f is an isometry then the only proximal pairs are (x,x) with $x \in X$.

A subset K of X is said to be a *proximal subset of* X whenever for every selection of finitely many elements $x_1, \dots, x_n \in K$ and for every $\varepsilon > 0$ there exists $k \in \mathbb{Z}^+$ such that $\operatorname{diam}\{f^k(x_1), \dots, f^k(x_n)\} < \varepsilon$. Obviously, if K is a proximal set then every pair of points in K is proximal. In this connection, the following notion is convenient: for every $n \in \mathbb{N}$, put

$$\mathbf{Prox}_n(X,f) := \big\{(x_1, x_2, \dots, x_n) \in X^n \,\vdots\, \forall\,\varepsilon > 0 \exists k \in \mathbb{N} \quad \text{such that} \\ \operatorname{diam}(\{f^k(x_1), f^k(x_2), \dots, f^k(x_n)\}) < \varepsilon\big\}.$$

In this notation a subset K of X is proximal iff for every $n \in \mathbb{N}$, every n-tuple $(x_1, \dots, x_n) \in K^n$ is in $\mathbf{Prox}_n(X,f)$, i.e., $K^n \subseteq \mathbf{Prox}_n(X,f)$.

Example.

(4) Let $a, b \in [0;1)$, $a \notin \mathbb{Q}$, and let (Z,σ) be the semi-Sturmian system of type (a,b). Recall that there are a factor map $\psi: (Z,\sigma) \to (\mathbb{S}, \varphi_a)$ and a subset $\mathbb{S}^*$ of $\mathbb{S}$ such that $\psi^{\leftarrow}[\psi(z)] = \{z\}$ if $\psi(z) \in \mathbb{S}^*$, whereas $\psi^{\leftarrow}[\psi(z)]$ consists of two points if $\psi(z) \in \mathbb{S} \setminus \mathbb{S}^*$. Moreover, if $[b]$ is not in the orbit of $[0]$ then every pair $(z,z') \in Z^2$ with $\psi(z) = \psi(z')$ is asymptotic. see 6.3.7 (2) and Example (4) in 7.3.2.
Claim: every proximal pair in Z is asymptotic. To prove this, consider any proximal pair $(z,z') \in Z^2$. Uniform continuity of ψ implies that $(\psi(z), \psi(z'))$ is a proximal pair in $\mathbb{S}$ under φ_a. But φ_a is an isometry, hence $\psi(z) = \psi(z')$ – see Example (3) above. It follows that the pair (z,z') is asymptotic.
Consequently, the system has no Li–Yorke pairs, hence it has no scrambled subsets: it is not LY-chaotic. Yet, being a minimal shift system, it is AY-chaotic.

The following lemma is a modification of the construction mentioned in Appendix B.3.3.

Lemma 7.3.4. *Let (X, f) be a transitive dynamical system on a compact metric space X with no isolated points. Then there is a sequence of uniformly rigid Cantor sets $C_1 \subseteq C_2 \subseteq C_3 \subseteq \dots$ such that $K := \bigcup_{n=1}^{\infty} C_n$ is a dense subset of X, all of whose points have a dense orbit (hence are transitive) in X. If, in addition, for each $n \in \mathbb{N}$ the set $\mathbf{Prox}_n(X, f)$ is dense in X^n then K may be assumed to be a proximal set.*

Proof. The proof consists of three parts. First, we prove the existence of the rigid Cantor sets whose union K is dense. Then we show how to adapt the proof in such a way that all points of K have dense orbits. Finally, if it is given that $\mathbf{Prox}_n(X, f)$ is dense in X^n for every n we show how to adapt the construction in such a way that K is, in addition, a proximal set.

I. Recall from Section 1.3 that the system is transitive iff it is topologically ergodic; this is because X has no isolated points and is a second countable Baire space. Let Y be a countable dense subset of X and let $Y = \{y_1, y_2, y_3, \dots\}$ be an enumeration of Y. Moreover, for each $n \in \mathbb{N}$, let $Y_n := \{y_1, \dots, y_n\}$ and, finally, let $a_0 := 0$ and $V_{0,1} := X$.

We claim that there are sequences $(a_n)_{n\in\mathbb{N}}$ and $(b_n)_{n\in\mathbb{N}}$ in $\mathbb{N}$ and that for every $n \in \mathbb{N}$ there are non-empty open sets $V_{n,1}, V_{n,2}, \dots, V_{n,a_n}$ in X with the following properties:

(1) $2a_{n-1} \leq a_n \leq 2a_{n-1} + n$.
(2) $\mathrm{diam}(V_{n,i}) < \frac{1}{n}$ for $i = 1, 2, \dots, a_n$.
(3) The sets $\overline{V}_{n,i}$ for $i = 1, 2, \dots, a_n$ are mutually disjoint.
(4) $\overline{V}_{n,2i-1} \cup \overline{V}_{n,2i} \subseteq V_{n-1,i}$ for $i = 1, 2, \dots, a_{n-1}$.
(5) $Y_n \subseteq B_{\frac{1}{n}}(\bigcup_{i=1}^{a_n} V_{n,i})$.
(6) $f^{b_n}[V_{n,2i-1} \cup V_{n,2i}] \subseteq V_{n-1,i}$ for $i = 1, 2, \dots, a_{n-1}$.

Proof of the claim (by induction): For $n = 1$, take $a_1 := 2$, $b_1 := 1$ and take for $V_{1,1}$ and $V_{1,2}$ two non-empty open subsets of $X = V_{0,1}$ with disjoint closures and with diameter at most 1 such that $y_1 \in V_{1,1} \cup V_{1,2}$. Then the conditions (1) through (6) are satisfied for $n = 1$.

Let $k \in \mathbb{N}$ and suppose that for $n = 1, \dots, k-1$ we have $a_n, b_n \in \mathbb{N}$ and non-empty open sets $V_{n,1}, V_{n,2}, \dots, V_{n,a_n}$ such that the conditions (1) through (6) are satisfied. For $i = 1, 2, \dots, 2a_{k-1}$ select non-empty open subsets $V_{k,i}^{(0)}$ of X such that

(a) $\mathrm{diam}(V_{k,i}^{(0)}) < \frac{1}{2k}$ for $i = 1, 2, \dots, 2a_{k-1}$.
(b) The sets $V_{k,i}^{(0)}$ for $i = 1, 2, \dots, 2a_{k-1}$ have mutually disjoint closures.
(c) $\overline{V_{k,2i-1}^{(0)}} \cup \overline{V_{k,2i}^{(0)}} \subseteq V_{k-1,i}$ for $i = 1, 2, \dots, a_{k-1}$.

As X has no isolated points, this can easily be done by selecting $V_{k,2i-1}^{(0)}$ and $V_{k,2i}^{(0)}$ as suitable subsets of the set $V_{k-1,i}$ $(i = 1, \dots, a_{k-1})$. Some of the points of Y_k may be included in the set $B_{\frac{1}{2k}}(\bigcup_{i=1}^{2a_{k-1}} V_{k,i}^{(0)})$; let $a_k \in \mathbb{N}$ be such that $a_k - 2a_{k-1}$ is equal to the number of points of Y_k not yet covered (which is at most the cardinality k of Y_k). So $2a_{k-1} \leq a_k \leq 2a_{k-1} + k$. Select for each of the remaining points of Y_k a sufficiently small open set containing that point and get in this way non-empty open sets $V_{k,i}^{(0)}$ for

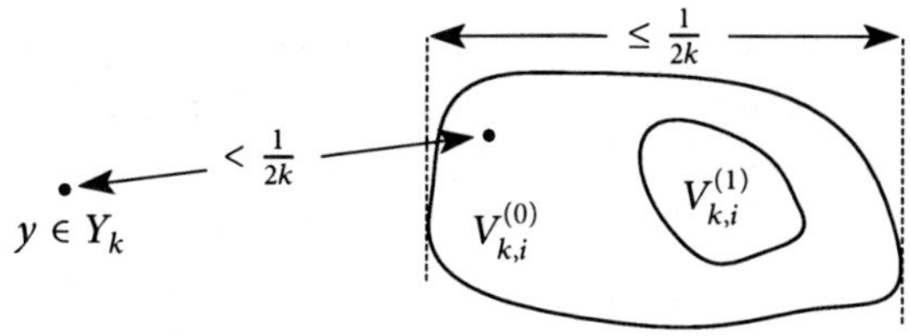

Fig. 7.2. Illustration of the induction step in the proof of condition (5).

$2a_{k-1} < i \leq a_k$ such that the conditions (a) and (b) are fulfilled for $1 \leq i \leq a_k$. In addition, it should be clear that we also have

(d) $Y_k \subseteq B_{\frac{1}{2k}}\big(\bigcup_{i=1}^{a_k} V^{(0)}_{k,i}\big)$.

Recall that $D(A,B) := \{m \in \mathbb{Z}^+ : A \cap (f^m)^{\leftarrow}[B] \neq \emptyset\}$ for any two subsets A and B of X. Since the system (X,f) is topologically ergodic, every set $D(U,U)$ for U a non-empty open subset of X is not empty and, in addition, the collection of all these sets is a filter base, i.e., the intersection of two of these sets includes another such set: see Exercise 1.6 (5). By induction, it follows that each finite intersection of such sets includes another one; consequently, finite intersections of such sets are not empty. In particular, the intersection of the sets $D(V^{(0)}_{k,i}, V^{(0)}_{k,i})$ for $i = 1,2,\dots,a_k$ is not empty. Select $b_k \in \bigcap_{i=1}^{a_k} D(V^{(0)}_{k,i}, V^{(0)}_{k,i})$; then the set $V^{(1)}_{k,i} := V^{(0)}_{k,i} \cap (f^{b_k})^{\leftarrow}[V^{(0)}_{k,i}]$ is a non-empty open subset of $V^{(0)}_{k,i}$ for $i = 1,2,\dots,a_k$. As $f^{b_k}[V^{(1)}_{k,i}] \subseteq V^{(0)}_{k,i}$, it follows that

(e) $f^{b_k}[V^{(1)}_{k,2i-1} \cup V^{(1)}_{k,2i}] \subseteq V^{(0)}_{k,2i-1} \cup V^{(0)}_{k,2i} \subseteq V_{k-1,i}$ for $i = 1,2,\dots,a_{k-1}$.

Since $V^{(1)}_{k,i} \subseteq V^{(0)}_{k,i}$ for $i = 1,2,\dots,a_k$ the conditions (a), (b) and (c) are also fulfilled for the sets $V^{(1)}_{k,i}$ instead of the sets $V^{(0)}_{k,i}$; the conditions (a) and (b) hold even for $i = 1,2,\dots,a_k$. Finally, condition (d) implies that for each point $y \in Y_k$ there exists $i \in \{1,2,\dots,a_k\}$ such that this point has distance less than $1/2k$ to at least one point of the set $V^{(0)}_{k,i}$, which point has, in turn, a distance of at most $1/2k$ to all points the set $V^{(1)}_{k,i}$. Hence y has a distance of at most $1/k$ to all points of $V^{(1)}_{k,i}$, that is, $y \in B_{\frac{1}{k}}(V^{(1)}_{k,i})$. Conclusion: if we put $V_{k,i} := V^{(1)}_{k,i}$ for $i = 1,2,\dots,a_k$ then the conditions (1) through (6) hold for $n = k$. This completes the proof by induction of the claim.

Similar to the argument in Appendix B.3.3, it follows from the conditions (1) through (4) that for every $N \in \mathbb{N}$ the set

$$C_N := \bigcap_{n=N}^{\infty}\Big(\bigcup_{i=1}^{2^{n-N}a_N} \overline{V}_{n,i}\Big)$$

is a Cantor set. Obviously, $C_1 \subseteq C_2 \subseteq C_3 \subseteq \dots$ and condition (5) implies that $K := \bigcup_{N=1}^{\infty} C_N$ is dense in X. In point of fact, if U is a non-empty open subset of X then U includes a point y of Y, and we can choose $n \in \mathbb{N}$ so large that $y \in Y_n$ and such that, in addition, the open ball with radius $2/n$ about y is included in U. Then it follows from condition (5) that $V_{n,i} \subseteq U$ for some $i \in \{1,2,\dots,a_n\}$. However, $V_{n,i}$ includes $\overline{V}_{n+1,2i}$ which, in turn, includes a point of C_N for every $N \geq n+1$. Consequently, U includes a point of K. This shows that K is dense in X.

Next, we show that for every $N \in \mathbb{N}$ the set C_N is uniformly rigid by showing that $f^{b_n}|_{C_N} \rightsquigarrow \mathrm{id}_{C_N}$ uniformly on C_N. By the definition of C_N, for every $n \geq N$ and every point $x \in C_N$ there exists $i \in \{1, 2, \dots, a_n\}$ such that $x \in \overline{V}_{n,i}$. Then either $x \in \overline{V}_{n+1,2i-1}$ or $x \in \overline{V}_{n+1,2i}$. By (6), this implies that $f^{b_n}(x) \in \overline{V}_{n,i}$. So both x and $f^{b_n}(x)$ belong to the set $\overline{V}_{n,i}$. Now it follows from condition (2) that $d(x, f^{b_n}(x)) < 1/n$. This holds for all $x \in C_N$ and for every $n \geq N$, so $\lim_{n\rightsquigarrow\infty} f^{b_n}(x) = x$ uniformly on C_N. This completes the proof that the set C_N is uniformly rigid.

Note that in the above proof the sets $V_{k,i}^{(1)}$ may be replaced by non-empty open subsets of itself: then condition (e) and the proof of the properties (1) through (6) remain valid, though for (6) we may have to choose other values of b_n. In particular, the conditions (a) and (d) guarantee that condition (5) holds. This observation will be used in the following steps of the proof.

II. In order to modify the construction in such a way that the points of K have dense orbits in X under f, we claim that the sets $V_{n,i}$ can be chosen such that, in addition to the conditions (1) through (6), also the following condition is satisfied for every $n \in \mathbb{N}$:

(7) $Y_n \subseteq B_{\frac{1}{n}}(\mathcal{O}(x))$ for every $x \in \bigcup_{i=1}^{a_n} \overline{V}_{n,i}$.

The proof is, again, by induction. For $n = 1$, rename the sets $V_{1,1}$ and $V_{1,2}$ above as $V_{1,1}^{(1)}$ and $V_{1,2}^{(1)}$, respectively. The set of transitive points is dense, so for $i = 1, 2$ there is a transitive point z_i in $V_{1,i}^{(1)}$. Then there are $n_1, n_2 \in \mathbb{Z}^+$ such that $f^{n_i}(z_i) \in B_1(y_1)$. Continuity of the mapping f^{n_i} implies that there is an open neighbourhood $V_{1,i}^{(2)}$ of z_i such that $f^{n_i}[\overline{V}_{1,i}^{(2)}] \in B_1(y_1)$; without limitation of generality we may assume that $V_{1,i}^{(2)} \subseteq V_{1,i}^{(1)}$. So if we put $V_{1,i} := V_{1,i}^{(2)}$ for $i = 1, 2$ then the conditions (1) through (6) are satisfied for $n = 1$; in particular, (5) holds because $y_1 \in V_{1,i}^{(1)}$ and $V_{1,i}^{(1)}$ has diameter at most 1. Also condition (7) holds for $n = 1$, because (by construction) for every point $x \in \overline{V}_{1,i} = \overline{V}_{1,i}^{(2)}$ the point y_1 has a distance of at most 1 to the point $f^{n_i}(x)$ of $\mathcal{O}(x)$ $(i = 1, 2)$.

Now assume that (1) through (7) are true for $n = 1, \dots, k-1$. Select the sets $V_{k,i}^{(1)}$ for $i = 1, 2, \dots, a_k$ as above, namely, by renaming each set $V_{k,i}$ as $V_{k,i}^{(1)}$. For each value of i under consideration, the set $V_{k,i}^{(1)}$ includes a transitive point z_i. Under the iterates of f, this point approaches the points of Y_k, so for $j = 1, \dots, k$ there exist $n_{ij} \in \mathbb{Z}^+$ such that $f^{n_{ij}}(z_i) \in B_{\frac{1}{k}}(y_j)$. By continuity of the (finitely many) mappings $f^{n_{ij}}$, there is an open neighbourhood $V_{k,i}^{(2)}$ of z_i such that $f^{n_{ij}}[\overline{V}_{k,i}^{(2)}] \subseteq B_{\frac{1}{k}}(y_j)$ for $j = 1, \dots, k$. We may assume that $V_{k,i}^{(2)} \subseteq V_{k,i}^{(1)}$ for $i = 1, 2, \dots, a_k$, so if we put $V_{k,i} := V_{k,i}^{(2)}$ then (1) through (6) hold for $n = k$. In addition, for every $j \in \{1, \dots, k\}$ and every $x \in \overline{V}_{k,i}$ the point y_j of Y_k has a distance of at most $1/k$ to the point $f^{n_{ij}}(x)$ of $\mathcal{O}(x)$ $(i \in \{1, 2, \dots, a_k\})$. So condition (7) is satisfied for $n = k$. This completes the proof by induction that the conditions (1) through (7) hold for every $n \in \mathbb{N}$.

Let $x \in K$; then there exists $N \geq 1$ such that $x \in C_N$. Then by the definition of C_N[3], $x \in \bigcup_{i=1}^{a_n} \overline{V}_{n,i}$ for every $n \in \mathbb{N}$, hence $Y_n \subseteq B_{\frac{1}{n}}(\mathcal{O}(x))$, i.e., every point of Y_n has

3 Here we use that $2^{n-N} a_N \leq a_n$ for all $N \geq 1$ and $n \geq N$.

distance less than $1/n$ to some point of $\mathcal{O}(x)$. If U is any non-empty open subset of X then for sufficiently large n there is a point $y \in Y_n$ such that $B_{\frac{1}{n}}(y) \subseteq U$. It follows that $\mathcal{O}(x)$ meets U. This completes the proof that $\mathcal{O}(x)$ is dense in X.

III. Finally, assume that for every $n \in \mathbb{N}$ the set $\mathbf{Prox}_n(X, f)$ is dense in the product space X^n. Consider, for every $n \in \mathbb{N}$ and $i \in \{1, 2, \dots, a_n\}$ the set $V_{n,i}^{(2)}$ constructed above. The set $V_{n,1}^{(2)} \times \dots \times V_{n,a_n}^{(2)}$ is non-empty and open in X^{a_n}, hence it contains a point of $\mathbf{Prox}_{a_n}(X, f)$. It follows that for $i = 1, 2, \dots, a_n$ there is a point $x_i \in V_{n,i}^{(2)}$ and there exists $t_n \in \mathbb{Z}^+$ such that the set $\{f^{t_n}(x_1), \dots, f^{t_n}(x_{a_n})\}$ has diameter less than $1/2n$. By continuity of the mapping f^{t_n} each of the points x_i has an open neighbourhood $V_{n,i}^{(3)}$ such that for every choice of $x_i' \in V_{n,i}^{(3)}$ the set $\{f^{t_n}(x_1'), \dots, f^{t_n}(x_{a_n}')\}$ has diameter less than $1/n$. Stated otherwise, the set $f^{t_n}[V_{n,1}^{(3)} \cup \dots \cup V_{n,a_n}^{(3)}]$ has diameter at most $1/n$. Of course, we may assume that $\overline{V}_{n,i}^{(3)} \subseteq V_{n,i}^{(2)}$ for every i. Put $V_{n,i} := V_{n,i}^{(3)}$ for $i = 1, 2, \dots, a_n$. By our earlier remark, as $V_{n,i}^{(3)}$ is a subset of $V_{n,i}^{(2)}$ for all $n \geq 1$ and $i = 1, \dots, a_n$, the conditions (1) through (6) remain valid (for suitable choice of b_n), and it is obvious that (7) remains valid as well. Moreover, the closure of a set has the same diameter as that set, so $\operatorname{diam}(f^{t_n}[\bigcup_{i=1}^{a_n} \overline{V}_{n,i}]) \leq 1/n$. Since $C_N \subseteq \bigcup_{i=1}^{a_n} \overline{V}_{n,i}$ for every $n \geq N$ it follows that $\operatorname{diam}(f^{t_n}[C_N]) \leq 1/n$ for almost all n. This obviously implies that the Cantor set C_N is proximal. It follows easily that the set K is proximal as well. □

Remarks.

(1) If $x, y \in K$ then there exists $N \in \mathbb{N}$ such that both x and y belong to the uniformly rigid set C_N. So if $x \neq y$ then the pair (x, y) is not asymptotic.

(2) Part III of the proof of the lemma shows that, if $\mathbf{Prox}_n(X, f)$ is dense in X^n for every $n \in \mathbb{N}$, then for every $N \in \mathbb{N}$ the Cantor set C_N is *uniformly proximal*, which means (by definition) that for every $\varepsilon > 0$ there exists $n \in \mathbb{Z}^+$ such that $\operatorname{diam}(f^n[C_N]) < \varepsilon$.

Corollary 7.3.5. *Let (X, f) be a transitive dynamical system on a compact metric space with no isolated points. If for every $n \in \mathbb{N}$ the set $\mathbf{Prox}_n(X, f)$ is dense in X^n then (X, f) is densely and strongly LY-chaotic.*

Proof. By Lemma 7.3.4, there is a dense subset of X such that all pairs of mutually different points from K are proximal but, by Remark 1 above not asymptotic. □

Theorem 7.3.6. *Let (X, f) be a transitive system on a compact metric space without isolated points. If there is an invariant subset Y of X such that the system $(X \times Y, f \times f|_Y)$ is transitive then (X, f) is densely and strongly LY-chaotic.*

Proof. By Corollary 7.3.5, it remains to show that for every $n \in \mathbb{N}$ the set $\mathbf{Prox}_n(X, f)$ is dense in X^n. Fix any $n \in \mathbb{N}$ and for every $\varepsilon > 0$, let

$$P_n(\varepsilon) := \left\{ (x_1, \dots, x_n) \,\vdots\, \exists l \in \mathbb{N} \text{ such that } \operatorname{diam}(f^l[\{x_1, \dots x_n\}]) < \varepsilon \right\}.$$

Then $\mathbf{Prox}_n(X, f) = \bigcap_{m=1}^{\infty} P_n(\frac{1}{m})$. For every $\varepsilon > 0$ the set $P_n(\varepsilon)$ is open in X^n, so if we can show that $P_n(\varepsilon)$ is dense in X^n for every $\varepsilon > 0$ then Baire's Theorem implies that $\mathbf{Prox}_n(X, f)$ is dense as well.

So let $\varepsilon > 0$ and consider an arbitrary non-empty open subset O of X^n. We may and shall assume that $O = U_1 \times \cdots \times U_n$, with U_j a non-empty open subset of X for $j = 1, \ldots, n$. In addition, let W be a non-empty open subset of Y with $\mathrm{diam}(W) < \varepsilon$. As $X \times Y$ is assumed to be transitive under the mapping $f \times f|_Y =: g$, it follows that $D_g(U_1 \times W, U_2 \times W) \neq \emptyset$. So the equality $D_g(U_1 \times W, U_2 \times W) = D_f(U_1, U_2) \cap D_{f|_Y}(W, W)$ implies that that there exists $m_2 \in \mathbb{Z}^+$ such that

$$U_1 \cap (f^{m_2})^{\leftarrow}[U_2] \neq \emptyset \quad \text{and} \quad W \cap (f^{m_2})^{\leftarrow}[W] \neq \emptyset\,.$$

Then find $m_3 \in \mathbb{Z}^+$ such that the non-empty open sets $U_1 \cap (f^{m_2})^{\leftarrow}[U_2]$ and $W \cap (f^{m_2})^{\leftarrow}[W]$ have non-empty intersections with the open sets $(f^{m_3})^{\leftarrow}[U_3]$ and $(f^{m_3})^{\leftarrow}[W]$, respectively. Etc.: there are $m_2, \ldots, m_n \in \mathbb{Z}^+$ such that

$$U^* := U_1 \cap \bigcap_{i=2}^{n} (f^{m_i})^{\leftarrow}[U_i] \neq \emptyset \quad \text{and} \quad W^* := W \cap \bigcap_{i=2}^{n} (f^{m_i})^{\leftarrow}[W] \neq \emptyset\,.$$

So the two sets U^* and W^* are non-empty and open. Now recall that the system (X, f) is transitive, hence topologically ergodic: there exists $l \in \mathbb{Z}^+$ such that $U^* \cap (f^l)^{\leftarrow}[W^*] \neq \emptyset$. Select a point x in this set. As $x \in U^*$ it is clear that $(x, f^{m_1}(x), \ldots, f^{m_n}(x)) \in U_1 \times U_2 \times \cdots \times U_n = O$. On the other hand $f^l(x) \in W^*$, so for $i = 1, \ldots, m_n$ we have $f^{m_i}(f^l(x)) \in W$ (for convenience, we let $m_1 := 0$). As W has diameter at most ε it follows that

$$\mathrm{diam}\Big(\big\{f^l(x), f^l(f^{m_2}(x)), \ldots, f^l(f^{m_n}(x))\big\}\Big) < \varepsilon\,.$$

It follows that $(x, f^{m_2}(x), \ldots, f^{m_n}(x)) \in P_n(\varepsilon)$. Conclusion: $P_n(\varepsilon) \cap O \neq \emptyset$. This completes the proof that $P_n(\varepsilon)$ is a dense subset of X^n. □

Corollary 7.3.7. *Let (X, f) be a dynamical system on a compact metric space without isolated points. In the following cases (X, f) is densely and strongly LY-chaotic:*
(a) *(X, f) is transitive and has an invariant point;*
(b) *(X, f) is totally transitive and has a periodic point*
(c) *(X, f) is weakly mixing;*
(d) *(X, f) is scattering, that is, for every minimal system (Y, g) with Y a compact metric space the product system $(X \times Y, f \times g)$ is transitive.*

Proof. The statements (a) and (c) are clear from Theorem 7.3.6, with Y equal to the singleton set consisting of the invariant point or equal to X, respectively. As to (b), if p is a period of a periodic point then (X, f^p) is a transitive system with an invariant point, to which case (a) can be applied. So (X, f^p) is densely and strongly LY-chaotic; therefore, (X, f) is densely and strongly LY-chaotic as well. For (d), Apply Theorem 7.3.6 with for Y a minimal subset of X, which exists in view of Theorem 1.2.7. □

Note that (c) follows from (d): by Corollary 1.6.3, every weakly mixing system is scattering; see also the small print after Corollary 1.6.3.

Though by Exercise 1.10 (5) a weakly mixing system on a compact metric space is totally transitive, (c) does not follow from (b) or (a): such a system may have no periodic points – it can be minimal: see Theorem 5.6.12 or 6.3.7 (3).

Examples.

(1) The tent map and the argument-doubling system are strongly LY-chaotic.

(2) The converse of (c) does not hold. Consider the system $(X, f) = (\Omega_2 \times \mathbb{S}, \sigma \times \varphi_a)$ with $a \notin \mathbb{Q}$, defined in Example (5) in 7.1.3, which was shown to be transitive in the second example in the beginning of Section 7.2. Let $A := \{0^\infty\} \times \mathbb{S}$ and let Y be the quotient of the space $\Omega_2 \times \mathbb{S}$, obtained by collapsing A to a point: $Y := X/R$, where R is the equivalence relation $(A \times A) \cup \Delta_X$. If $((x, [t]), (y, [s])) \in R$ then either $x = y = 0^\infty$ or $x = y$ and $[t] = [s]$; in both cases $((\sigma x, \varphi_a[t]), (\sigma y, \varphi_a[s])) \in R$. Hence there is an unique continuous mapping $g: Y \to Y$ such that $q \circ (\sigma \times \varphi_a) = g \circ q$; here $q := R[\cdot]: X \to Y$ is the quotient map. As the system (X, f) is transitive, its factor (Y, g) is transitive as well. As (Y, g) has an invariant point, namely, the point $q[A]$, it follows that (Y, g) is strongly and densely LY-chaotic.
On the other hand, (Y, g) is not weakly mixing. In order to prove this, consider open subsets $W_i := U_i \times V_i$ of X $(i = 1, 2, 3, 4)$, where the U_i are open subsets of Ω_2 not containing the point 0^∞ and where the V_i are small open arcs in $\mathbb{S}$ such that the distance of V_1 to V_2 is much larger than the distance between V_3 to V_4, implying that $D_{\varphi_a}(V_1, V_2) \cap D_{\varphi_a}(V_3, V_4) = \emptyset$. Then $D_f(W_1, W_2) \cap D_f(W_3, W_4) = \emptyset$. As $A \cap W_i = \emptyset$ it is clear that $q^{\leftarrow} q[W_i] = W_i$, so the subsets $W_i' := q[W_i]$ of Y are open in Y $(i = 1, 2, 3, 4)$. It also follows easily that $D_g(W_1', W_2') \cap D_g(W_3', W_4') = \emptyset$. This completes the proof that (Y, g) is not weakly mixing.

Corollary 7.3.8. *A transitive dynamical system on a non-degenerate compact interval is densely and strongly LY-chaotic.*

Proof. By Lemma 2.6.4 (1), such a system has an invariant point. □

Corollary 7.3.9. *Let (X, f) be a dynamical system on a compact metric space without isolated points. If (X, f) is transitive and has a periodic orbit then (X, f) is strongly (but not necessarily densely) LY-chaotic.*

Proof. Let $x_0 \in X$ be a point with dense orbit under f and let $y_0 \in X$ be an f-periodic point, say, with primitive period p. Moreover, let $g := f^p$. Lemma 3.3.10 (1) and the choice of x_0 as a f-transitive point imply that $X = \omega_f(x_0) = \bigcup_{i=0}^{p-1} \omega_g(f^i(x_0))$. In particular, there exists r with $0 \leq r \leq p - 1$ such that $y_0 \in \omega_g(f^r(x_0)) =: X_0$

Since $\omega_{g|_{X_0}}(f^r(x_0)) = \omega_g(f^r(x_0))$ – see Exercise 3.1 (1) – it follows that the point $f^r(x_0)$ is transitive in X_0 under g. So the system (X_0, g) is transitive. Since X_0 contains the invariant point y_0, it follows from the first case in Corollary 7.3.7 that X_0 includes a dense strongly scrambled subset K (scrambled under g, that is). Then it is obvious that K is strongly scrambled in X under f (but not necessarily dense in X). □

Remark. Using Proposition 3.1.2 (2), applied to the endomorphism f^i of the (compact) system (X, g), and taking into account that that $\omega_g(g(x)) = \omega_g(x)$ for all $x \in X$, one easily shows that $f^i[X_0] = \omega_g(f^{r+i \pmod p}(x_0))$, which implies that $X = \bigcup_{i=0}^{p-1} f^i[X_0]$. The strongly scrambled set K in the above proof is dense in X_0. It follows that $\bigcup_{i=0}^{p-1} f^i[K]$ is dense in X.

Corollary 7.3.10. *Let (X, f) be a dynamical system on a compact metric space. If this system is AY-chaotic and has at least one periodic orbit then it is strongly LY-chaotic.*

Proof. Clear from Corollary 7.3.9. (Note that sensitivity of (X, f) is only needed to guarantee that X has no isolated points.) □

Corollary 7.3.11. *Let (X, f) be a dynamical system on a compact metric space. If a subsystem on a closed invariant subset is D-chaotic then the system is strongly LY-chaotic.*

Proof. Apply Corollary 7.3.10 to the D-chaotic subsystem. □

Remark. In general, AY-chaos (without periodic points) does not imply LY-chaos: every semi-Sturmian system is AY-chaotic and (being minimal) without periodic points, but if it has type (a, b) such that $[b]$ is not in the φ_a-orbit of $[0]$ then the system is not LY-chaotic: see Example (4) after Lemma 7.3.3. So the assumption in Corollary 7.3.10 that there is a periodic orbit is essential. However, on a compact interval AY-chaos implies LY-chaos: see S. Ruette [2005].

Also the converse implication is not true in general: LY-chaos does not imply AY-chaos: see Note 6. However, for compact minimal systems it does: see Exercise 7.13.

We may summarize some of the above results in the following implications:

$$
\begin{array}{ccccc}
\text{weakly mixing} & & \Longrightarrow & & \text{scattering} \\
\Downarrow & & & & \Downarrow \\
\text{AY-chaotic} & \Leftarrow & h(f) > 0 & \Rightarrow & \text{LY-chaotic} \\
\Uparrow & & & & \Uparrow \\
\text{AY-chaotic + periodic pt.} & & \Longrightarrow & &
\end{array}
$$

For the meaning of '$h(f) > 0$' (positive topological entropy) we refer to the next chapter; for the proof that this property implies LY-chaos – a long standing open problem – we refer to F. Blanchard, E. Glasner, S. Kolyada & A. Maass [2002]. See also Note 6 at the end of this chapter. For the implication 'weakly mixing ⇒ scattering', see the small print after Corollary 1.6.3. In the case of *minimal* systems on compact metric spaces the following can be shown:

$$
\begin{array}{ccccc}
 & & \text{weakly mixing} & \Leftrightarrow & \text{scattering} \\
 & & & & \Updownarrow \\
\text{AY-chaotic} & \Leftarrow & \text{LY-chaotic} & \Leftarrow & \text{densely LY-chaotic} \\
 & & \Uparrow & & \\
 & & h(f) > 0 & &
\end{array}
$$

For the 'up'-implication from dense Li–Yorke chaos to scattering we have to refer to the literature: see Corollary 3.6 in W. Huang & X. Ye [2002].

7.4 Horseshoes for interval maps

In this section, (X, f) will always be a dynamical system on a subinterval X of $\mathbb{R}$ (not necessarily bounded or closed). We shall show that if such a system has a periodic point whose primitive period is not a power of 2 then it is LY-chaotic. Our tool will be the one-dimensional variant of a 'horseshoe', which has much the same effect as the original two-dimensional definition by Smale. If the presence of a horseshoe would enable us to construct an f-invariant Cantor set in X then in view of Example (4) in 7.3.1 and Example (2) after Theorem 7.2.2 we might conclude that the system is LY-chaotic and that there is a D-chaotic subsystem. I general, this is asking too much, but a horseshoe will enable us to construct an invariant set of which (Ω_2, σ) is an almost 1,1 factor, and this turns out to be sufficient to obtain a scrambled Cantor set in the interval and, additionally, a D-chaotic subsystem.

Suppose $J_1, \dots, J_n$ $(n \geq 1)$ are closed non-degenerate subintervals of X with pairwise disjoint interiors such that $f: J_i \overset{\circ}{\rightarrow} J_1 \cup \cdots \cup J_n$ for $i = 1, \dots, n$:

$$f[J_i] \supseteq J_1 \cup \cdots \cup J_n \quad \text{for } i = 1, \dots, n .$$

(see Section 2.2 for the notation $\overset{\circ}{\rightarrow}$). Then $(J_1, \dots, J_n)$ is called an *n-horseshoe* for f. If $n = 2$ then (J_1, J_2) will simply be called a *horseshoe*. If there is an n-horseshoe in X then we say that the system (X, f) *admits* (or *has*) an n-horseshoe. Clearly, if (X, f) admits an n-horseshoe with $n \geq 2$ then it also admits m-horseshoes for $m = 1, \dots, n-1$: select any m of the intervals that form the n-horseshoe.

If $(J_1, \dots J_n)$ is an n-horseshoe such that the successive intervals J_i have common end points, then the interval $J := J_1 \cup \cdots \cup J_n$ is folded over itself at least n times: every value in J, except possibly the end points of J, is assumed (at least) n times. See also Figure 7.3.

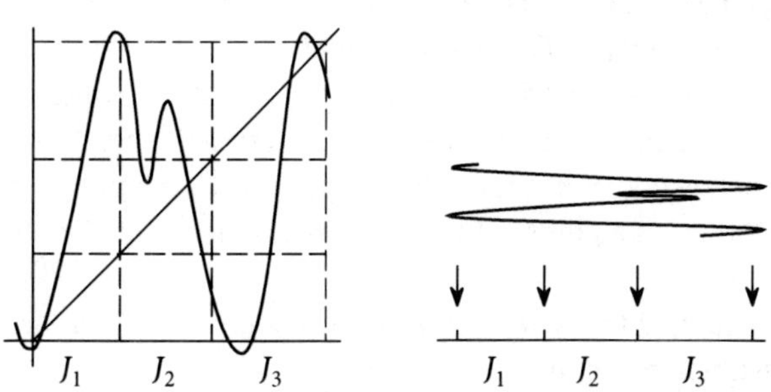

Fig. 7.3. (J_1, J_3) and (J_2, J_3) are 2-horseshoes, but (J_1, J_2) is not (J_2 is not mapped over J_1), neither is (J_1, J_2, J_3) a 3-horseshoe.

Examples.

(1) The tent map $x \mapsto 1 - |1 - 2x| : \mathbb{R} \to \mathbb{R}$ has a 2-horseshoe, namely, the pair of intervals $([0; 1/2], [1/2; 1])$. The mapping $x \mapsto 3/2 - |3/2 - 3x| : \mathbb{R} \to \mathbb{R}$ has a horseshoe consisting of the two disjoint closed intervals $[0; 1/3]$ and $[2/3; 1])$.

(2) Let f be a mapping of an interval into itself which has a point with period 3. Using the notation of the proof of Theorem 2.2.2, there are intervals I_0 and I_1 with a common end point and disjoint interiors such that $f : I_0 \overset{\circ}{\rightarrow} I_0 \cup I_1$ and $f : I_1 \overset{\circ}{\rightarrow} I_0$. Obviously, (I_0, I_1) is a horseshoe for f^2.

Proposition 7.4.1. *If the system (X, f) has a horseshoe consisting of two compact intervals then for every $n \in \mathbb{N}$ there is a periodic point in X with primitive period n.*

Proof. Let (J, K) be a horseshoe for (X, f) with J and K compact. Then

$$f : J \overset{\circ}{\rightarrow} J \cup K \quad \text{and} \quad f : K \overset{\circ}{\rightarrow} J \cup K .$$

First, assume that J and K are disjoint. Then proceed in the following way: if $n \in \mathbb{N}$ then

$$f^n : J \overset{\circ}{\rightarrow} \underbrace{K \overset{\circ}{\rightarrow} K \overset{\circ}{\rightarrow} \dots \overset{\circ}{\rightarrow} K}_{n-2 \text{ arrows}} \overset{\circ}{\rightarrow} J .$$

By Lemma 2.2.1 (3), this implies that there are a closed subinterval J_0 of J and closed subintervals K_i of K $(i = 1, \dots, n - 1)$ such that

$$f^n : J_0 \twoheadrightarrow K_1 \twoheadrightarrow \cdots \twoheadrightarrow K_{n-1} \twoheadrightarrow J \supseteq J_0 . \tag{7.4-1}$$

Thus, $f^n : J_0 \overset{\circ}{\rightarrow} J_0$. Hence there is an invariant point x of f^n in J_0, i.e, a periodic point of f with period n. Now by (7.4-1), $f^i(x) \in K_i$ and, by the assumption that J and K are disjoint, $K_i \cap J_0 = \emptyset$ for $i = 1, \dots, n - 1$. Hence it is clear that x cannot have a period less than n: otherwise some of the points $f^i(x)$ for $0 \le i \le n - 1$ would coincide with each other; these coinciding points must be situated in K, but then all points of this periodic orbit would be situated in K. So n is the primitive period of x.

Next assume that J and K have a common end point, say, $J = [a; b]$ and $K = [b; c]$ with $a < b < c$. Assume first that b is an invariant point. Since $f[\,[b; c]\,] \supseteq [a; c]$ there are points in $[b; c]$ that are mapped onto the points a and c, but the point b is not among them, because b is invariant. By continuity of f, there is a neighbourhood $(b - \delta; b + \delta)$ of b on which the values a and c are nor assumed. So the interval $f[\,[b + \delta; c]\,]$ contains both points a and c, hence it includes the complete interval $[a; c]$. From this it follows easily that $([a; b], [b + \delta; c])$ is a horseshoe. Since it has disjoint intervals, the previous case implies that for every $n \in \mathbb{N}$ there is a point in the interval $[a; c]$, hence in X, with primitive period n.

Finally, let $J = [a; b]$ and $K = [b; c]$ with $a < b < c$, but suppose b is not an invariant point. Consider the following sequence of maps:

$$J \overset{\circ}{\rightarrow} K \overset{\circ}{\rightarrow} K \overset{\circ}{\rightarrow} J .$$

By Lemma 2.2.1 (3), there are closed intervals $J_0 \subseteq J$ and $K_1, K_2 \subseteq K$ such that

$$J_0 \twoheadrightarrow K_1 \twoheadrightarrow K_2 \twoheadrightarrow J \supseteq J_0 \,.$$

This gives rise to a point $z \in J_0$ with period 3, that is, an invariant point or a periodic point with primitive period 3. The first case cannot occur, for otherwise z, being a common point of J, K_1 and K_2, coincides with b, which is supposed to be not invariant. So we have a point with primitive period 3. By Li and Yorke's Theorem 2.2.2 this implies that for every $n \in \mathbb{N}$ there is a point in $[a; b]$ with primitive period n. □

Example. The converse of the above proposition is not true. Let $f : [0; 1] \to [0; 1]$ be the function whose graph is shown in Example F in 6.3.5. Then 0 is a point with period 3, hence the system $([0; 1], f)$ has periodic points of all primitive periods. But it is clear that f has no horseshoe: if a subinterval J of $([0; 1], f)$ is mapped over itself, then the (unique) invariant point c of f belongs to J. If there are two of such intervals then c will belong to both intervals. So if (J, K) is a horseshoe then c is a common end point of J and K. Inspection of the graph of f then shows that such a horseshoe cannot exist. In Example (3) after Corollary 8.6.3 ahead an alternative proof will be given that f has no horseshoe. However, in accordance with Proposition 7.5.1 ahead, the mapping f^2 admits a horseshoe. See Figure 7.4.

We shall show now how a horseshoe consisting of two mutually disjoint compact intervals induces a coding of a set of orbits by symbolic sequences (elements of Ω_2): we find a non-empty compact invariant set Z of X and a factor mapping $\varphi\colon (Z, f) \to (\Omega_2, \sigma)$ which is 1-to-1 outside of a countable subset E of Ω_2 (i.e., has singleton fibres at points not in E). This is not an example of a symbolic representation, as the factor mapping is in the wrong direction. The construction is quite similar to the construction of the Cantor set in 1.7.5 (where we also have a horseshoe!).

7.4.2 (Construction of the coding). Let (J_0, J_1) be a horseshoe consisting of two disjoint compact intervals. Similar to the construction of the Cantor set we define a decreasing nest of non-empty compact sets D_n ($n \in \mathbb{N}$), each of which is a union of 2^n mutually disjoint non-degenerate compact intervals.

For D_1 we take $J_0 \cup J_1$. In order to define the intervals that form D_2, note that by assumption we have $f\colon J_0 \overset{\circ}{\rightarrow} J_0 \cup J_1$ and $f\colon J_1 \overset{\circ}{\rightarrow} J_0 \cup J_1$. Consequently, Lemma 2.2.1 (2) implies that there are closed intervals J_{ij} such that $J_{ij} \subseteq J_i$ and $f[J_{ij}] = J_j$ for $i, j \in \{0, 1\}$. By the Remark following the proof of Lemma 2.2.1 we may assume that f maps the

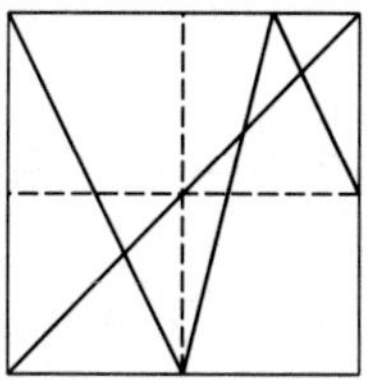

Fig. 7.4. The graph of f^2 for f as in Example F in 6.3.5. The two closed intervals $[0; 1/2]$ and $[1/2; 1]$ form a horseshoe for f^2.

interior of the interval J_{ij} onto the interior of the interval J_j and that f maps the set of end points of J_{ij} bijectively onto the set of end points of J_j. The two intervals J_{00} and J_{01} in J_0 are easily seen to be mutually disjoint, as are the two intervals J_{10} and J_{11} in J_1 (a formal proof is in 1 below). Now let $D_2 := \bigcup_{i,j\in\{0,1\}} J_{ij}$.

Next, as $f : J_{ij} \twoheadrightarrow J_j \supseteq J_{jk}$ for $k \in \{0, 1\}$, a similar argument shows that there are closed subintervals J_{ijk} of J_{ij} such that $f[J_{ijk}] = J_{jk}$, in such a way that f maps the interior of J_{ijk} onto the interior of J_{jk}, hence it maps the set of the end points of J_{ijk} onto the set of the end points of J_{jk}. See Figure 7.5. Put $D_3 := \bigcup_{i,j,k\in\{0,1\}} J_{ijk}$. Repeat this procedure inductively and find for every $n \in \mathbb{N}$ and every $(n + 1)$-block $a_0 \dots a_n$ over the symbol set $\{0, 1\}$ a closed interval $J_{a_0\dots a_n}$ such that

$$J_{a_0\dots a_n} \subseteq J_{a_0\dots a_{n-1}} \quad \text{and} \quad f[J_{a_0\dots a_n}] = J_{a_1\dots a_n} \,. \tag{7.4-2}$$

Moreover, we may assume that f maps the interior of $J_{a_0\dots a_n}$ onto the interior of $J_{a_1\dots a_n}$, hence it maps the two end points of $J_{a_0\dots a_n}$ onto the two end points of $J_{a_1\dots a_n}$. For every $n \in \mathbb{N}$, let

$$D_n := \bigcup \{ J_{a_0\dots a_{n-1}} \vdots a_i \in \{0, 1\} \text{ for } 0 \le i < n \} = \bigcup_{a\in\Omega_2} J_{a_{[0;n)}} \,.$$

Here $\{ a_{[0;n)} \vdots a \in \Omega_2 \}$ is a convenient way to denote the set $\{0, 1\}^n$ of all n-blocks over the symbol set $\{0, 1\}$. Obviously, the set D_n, being the union of finitely many compact non-empty intervals, is compact and non-empty.

The reader may have noticed that the above construction is similar to the construction in 1.7.5. However, in the present situation there is no question of monotonicity of f on J_0 or J_1. Neither is there any reason that the length of the intervals $J_{a_{[0;n)}}$ tends to 0 if n tends to infinity.

(1) *For every $n \in \mathbb{Z}^+$ the 2^{n+1} intervals $J_{a_0\dots a_n}$ for $a \in \{0, 1\}^{n+1}$ are mutually disjoint. In point of fact, if $k, n \in \mathbb{Z}^+$, $k \le n$, and $a_0 \dots a_k$ and $b_0 \dots b_n$ are two finite blocks over*

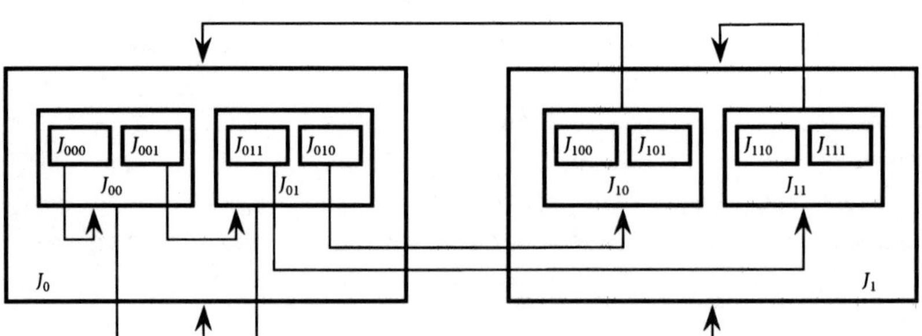

Fig. 7.5. Schematic view of the construction of the intervals (for clarity represented as rectangles) J_{ij} and J_{ijk} for $i, j, k \in \{0, 1\}$.

the symbol set $\{0, 1\}$, *then*

$$J_{a_0\dots a_k} \cap J_{b_0\dots b_n} = \begin{cases} J_{b_0\dots b_n} & \text{if } a_0 \dots a_k \text{ is an initial block of } b_0 \dots b_n, \\ \emptyset & \text{otherwise.} \end{cases}$$

Proof.[4] First, we prove the statement for the case that $k = n$. The proof is by induction. For $n = 0$ the statement is obviously true: the intervals J_0 and J_1 are given to be disjoint. Next, suppose the statement is true for a given value of $n \in \mathbb{Z}^+$ and consider two different $(n + 1)$-blocks $a_0a_1 \dots a_n$ and $b_0b_1 \dots b_n$. If $a_0 \neq b_0$ then the first part of equation (7.4-2) and the induction hypothesis imply that the intervals $J_{a_0a_1\dots a_n}$ and $J_{b_0b_1\dots b_n}$ are disjoint. And if $a_0 = b_0$ then $a_1 \dots a_n \neq b_1 \dots b_n$, in which case the second part of (7.4-2) and the induction hypothesis imply that $J_{a_0a_1\dots a_n}$ and $J_{b_0b_1\dots b_n}$ have disjoint images under f, hence are disjoint. This completes the proof for the case that $k = n$.

The general statement with $k \leq n$ follows from what was just proved and the first inclusion in (7.4-2), as follows: If $a_0 \dots a_k$ is an initial block of $b_0 \dots b_n$ then $J_{b_0\dots b_n} \subseteq J_{b_0\dots b_k} = J_{a_0\dots a_k}$, hence $J_{a_0\dots a_k} \cap J_{b_0\dots b_n} = J_{b_0\dots b_n}$. If $a_0 \dots a_k$ is not an initial block of $b_0 \dots b_n$ then by the first part of the proof, $J_{b_0\dots b_k}$ is disjoint from $J_{a_0\dots a_k}$, so the possibly smaller set $J_{b_0\dots b_n}$ is certainly disjoint from $J_{a_0\dots a_k}$. □

(2) $\forall n \in \mathbb{N} : D_{n+1} \subseteq D_n$ *and* $f[D_{n+1}] \subseteq D_n$.

Proof. The set D_{n+1} can be seen as arising from D_n by replacing each interval of the form $J_{a_0\dots a_{n-1}}$ by the intervals $J_{a_0\dots a_{n-1}0}$ and $J_{a_0\dots a_{n-1}1}$. By the inclusion in (7.4-2), the latter are subintervals of the former. It follows that $D_{n+1} \subseteq D_n$.

By the equality in (7.4-2), $f[J_{a_0a_1\dots a_n}] = J_{a_1\dots a_n} \subseteq D_n$ for every $(n+1)$-block $a_0a_1 \dots a_n$. As D_{n+1} is the union of the intervals $J_{a_0\dots a_n}$, it follows that $f[D_{n+1}] \subseteq D_n$. □

Comparison of the above construction with the construction of the Cantor set in Appendix B and with the procedure followed in 1.7.5 might tempt the reader to consider the set $D := \bigcap_{n\in\mathbb{N}} D_n$, which is non-empty and compact, being the intersection of a descending chain of non-empty compact sets. Though D is invariant under f (this follows easily from the second part of statement 2), we shall see below that (and why) D is not the subspace Z of X we are looking for.

For every point $a \in \Omega_2$ the intervals $J_{a_{[0;n)}}$ for $n \in \mathbb{N}$ form a decreasing family of non-empty compact intervals. Hence

$$J_a := \bigcap_{n=1}^{\infty} J_{a_{[0;n)}} \tag{7.4-3}$$

is a non-empty compact interval, possibly a point.

4 For later use in Lemma 8.6.1 ahead, note that if J_0 and J_1 are not disjoint – but still have disjoint interiors – this proof shows that the intervals $J_{a_0\dots a_n}$ have mutually disjoint interiors.

(3) *If $a, b \in \Omega_2$, $a \neq b$, then $J_a \cap J_b = \emptyset$.*

Proof. If $a, b \in \Omega_2$, $a \neq b$, then there is $n \in \mathbb{N}$ such that $a_{[0;n)} \neq b_{[0;n)}$. Since $J_a \subseteq J_{a_{[0;n)}}$ and $J_b \subseteq J_{b_{[0;n)}}$ the result follows from statement 1. □

(4) *Let $a \in \Omega_2$ and let p and q be the left and right end points of the (possibly degenerate) interval J_a, respectively, so that $J_a = [p; q]$ (possibly, $p = q$). Similarly, for every $n \in \mathbb{Z}^+$, let p_n and q_n be the left and right end points of the interval $J_{a_0 \dots a_n}$, so that $J_{a_0 \dots a_n} = [p_n; q_n]$. Then $p_n \nearrow p$ and $q_n \searrow q$ for $n \rightsquigarrow \infty$.*

Proof. By the definition of J_a, the intervals $[p_n; q_n]$ for $n \in \mathbb{Z}^+$ form a descending chain with intersection the interval $[p; q]$. It follows that the sequences $(p_n)_{n \in \mathbb{Z}^+}$ and $(q_n)_{n \in \mathbb{Z}^+}$ are monotonously increasing and decreasing, respectively (not necessarily strictly so). Clearly, $p_n \leq p \leq q \leq q_n$ for all n, so the sequences are bounded, hence have limits l and r, respectively, with $l \leq p \leq q \leq r$. Obviously, if $l < p$ or $q < r$ we would get a contradiction with the fact that $[p; q] = \bigcap_{n \in \mathbb{Z}^+} [p_n; q_n]$. (Alternatively, we might refer to Appendix A.2.2.) □

(5) *Let $a \in \Omega_2$. Then for every $k \in \mathbb{N}$, f^k maps the set of the two end points*[5] *of J_a onto the set of the two end points of $J_{\sigma^k a}$.*

Proof. We prove this statement only for the case that $k = 1$. The other cases follow easily by induction.
As in statement 4 above, let $J_{a_0 \dots a_n} = [p_n; q_n]$ and $J_a = [p; q]$. Similarly, let $J_{a_1 \dots a_n} = [p'_n; q'_n]$ and $J_{\sigma a} = [p'; q']$. Then by statement 4 above one has $p_n \nearrow p$ and $q_n \searrow q$; similarly, as $J_{\sigma a}$ is the intersection of the intervals $J_{a_1 \dots a_n}$ for $n \in \mathbb{N}$, one also has $p'_n \nearrow p'$ and $q'_n \searrow q'$ for $n \rightsquigarrow \infty$.
By construction, f maps for every $n \in \mathbb{Z}^+$ the end points of $J_{a_0 \dots a_n}$ onto those of $J_{a_1 \dots a_n}$, i.e., $f(p_n) = p'_n$ and $f(q_n) = q'_n$, or just the other way round, $f(p_n) = q'_n$ and $f(q_n) = p'_n$, depending on n. Since there are only two possibilities for each $n \in \mathbb{Z}^+$, one of these possibilities must hold for infinitely many values of $n \in \mathbb{Z}^+$. Assume that the first possibility holds infinitely often, say for n belonging to a subsequence $(n_i)_{i \in \mathbb{N}}$ of $\mathbb{N}$. Then by the continuity of f and the limits mentioned above one gets

$$f(p) = \lim_{i \rightsquigarrow \infty} f(p_{n_i}) = \lim_{i \rightsquigarrow \infty} p'_{n_i} = p'$$

and, similarly, $f(q) = q'$. By the same reasoning, if the second possibility holds for infinitely many values of n one gets $f(p) = q'$ and $f(q) = p'$. □

---

5 If J_a consists of a single point then this point is considered both as the left and right end point of J_a. A similar remark applies to $J_{\sigma a}$.

As observed earlier, it is possible that some of the sets J_a are proper intervals, not just points. To take care of this, consider the set

$$E := \{ a \in \Omega_2 : \text{the interval } J_a \text{ is non-degenerate} \}$$

and let Z be the set D from which all interiors of intervals J_a for $a \in E$ are omitted:

$$Z := D \setminus \bigcup_{a \in E} \operatorname{int} J_a = \left(\bigcap_{n=1}^{\infty} \bigcup_{a \in \Omega_2} J_{a_{[0;n)}} \right) \setminus \bigcup_{a \in E} \operatorname{int} J_a \,. \tag{7.4-4}$$

(6) *The set E is at most countable.*

Proof. By 3 above, the set $\{ J_a : a \in E \}$ is a collection of mutually disjoint non-degenerate intervals. By a well-known argument, namely, that each member of this collection contains an element of the countable set $\mathbb{Q}$, the set E is at most countable. □

Remark. *The complement of E is dense in* Ω_2: every basic open set in Ω_2 is uncountable, hence cannot be included in the countable set E. Moreover, it follows easily from 5 above that *the complement of E in* Ω_2 *is invariant under* σ.

(7) *Z is a non-empty compact invariant subset of* $J_0 \cup J_1$.

Proof. Recall that the set D is compact. From this compact set we omit the union of the open interiors of the intervals J_a for $a \in E$, which union is also open. So we end up with a closed, hence compact, subset of D.
In order to show that $Z \neq \emptyset$, we claim that in the right-hand side of (7.4-4) we may interchange the intersection over $n \in \mathbb{N}$ and the union over $a \in \Omega_2$, that is, we claim that

$$\bigcap_{n=1}^{\infty} \bigcup_{a \in \Omega_2} J_{a_{[0;n)}} = \bigcup_{a \in \Omega_2} J_a \,. \tag{7.4-5}$$

The inclusion "⊇" follows easily from the fact that $J_{a_{[0;n)}} \supseteq J_a$ for all $a \in \Omega_2$ and $n \in \mathbb{N}$. Conversely, if x belongs to the left-hand side of (7.4-5) then for every $n \in \mathbb{N}$ there is an n-block a^n such that $x \in J_{a^{(n)}}$. In view of 1 above, for every $n \in \mathbb{N}$ the block a^{n+1} starts with the block a^n. Hence there is a point $a \in \Omega_2$ which, for every $n \in \mathbb{N}$, starts with a^n, that is, $a_{[0\,;\,n)} = a^n$. Consequently, $x \in J_{a_{[0;n)}}$ for every $n \in \mathbb{N}$, that is, $x \in J_a$. This concludes the proof of (7.4-5). Using (7.4-5) and taking into account that $\operatorname{int} J_a = \emptyset$ if $a \notin E$, we get

$$Z = \bigcup_{a \in \Omega_2} (J_a \setminus \operatorname{int} J_a) \,. \tag{7.4-6}$$

Here for every $a \in \Omega_2$ the set $J_a \setminus \operatorname{int} J_a$ consists of the end points of the (possibly degenerate) interval J_a, which is a 1-point set if $a \notin E$ and a 2-point set if $a \in E$. This shows that $Z \neq \emptyset$.

Finally, we show that Z is invariant under f. To this end, consider any point $z \in Z$ and let, according to (7.4-6), $a \in \Omega_2$ be such that $z \in J_a \setminus \operatorname{int} J_a$. By statement 5 above, f maps the end points of the interval J_a onto the end points of the interval $J_{\sigma a}$. This means that

$$f(z) \in J_{\sigma a} \setminus \operatorname{int} J_{\sigma a} . \tag{7.4-7}$$

So by (7.4-6) again, $f(z) \in Z$. □

Finally, we define the mapping $\varphi: Z \to \Omega_2$. If $z \in Z$ then by (7.4-6) there exists $a \in \Omega_2$ such that $z \in J_a$. In view of statement 3 above, this a is unique: call it $\varphi(z)$. Thus, $\varphi: Z \to \Omega_2$ is defined such that

$$\forall\, z \in Z: \quad z \in J_{\varphi(z)} . \tag{7.4-8}$$

(8) *The mapping $\varphi: Z \to \Omega_2$ is continuous and surjective. If $a \in \Omega_2$ then the fibre $\varphi^{\leftarrow}[a]$ consists of two points if $a \in E$ and of one point if $a \notin E$. Moreover,*

$$\varphi \circ f|_Z = \sigma \circ \varphi . \tag{7.4-9}$$

Proof. It is obvious that $\varphi: Z \to \Omega_2$ is surjective. Actually, it follows from the definition of E that for every $a \in \Omega_2$, the fibre $\varphi^{\leftarrow}[a]$ consists of one point if $a \in \Omega_2 \setminus E$ and that $\varphi^{\leftarrow}[a]$ consists of two points if $a \in E$.
Next, we prove equation (7.4-9). Let $z \in Z$ and let $a := \varphi(z)$, so $z \in J_a$. By formula (7.4-7), $f(z) \in J_{\sigma a}$, so the definition of φ – see also (7.4-8) – implies that $\varphi(f(z)) = \sigma a = \sigma\varphi(z)$.
Finally, we show that φ is continuous. For every $n \in \mathbb{N}$, let δ_n be the minimum distance between the (finitely many) mutually disjoint compact intervals $J_{a_{[0;n)}}$ for $a \in \Omega_2$. Consider two arbitrary points y and z in Z, and let $b := \varphi(y)$ and $c := \varphi(z)$. Then by definition, $y \in J_b \subseteq J_{b_{[0;n)}}$ and $z \in J_c \subseteq J_{c_{[0;n)}}$. Consequently, if $|y - z| < \delta_n$ then the intervals $J_{b_{[0;n)}}$ and $J_{c_{[0;n)}}$ must coincide, which implies that $b_{[0;n)} = c_{[0;n)}$. This obviously means that $d(\varphi(y), \varphi(z)) < 1/(n+1)$ (where d is the metric on Ω_2). This completes the proof that φ is continuous. □

Proposition 7.4.3. *Let (J_0, J_1) be a horseshoe consisting of two disjoint compact intervals. Then there are a compact 0-dimensional invariant subset Z of X and a continuous surjection $\varphi: Z \to \Omega_2$ such that:*
(1) *$\varphi \circ f|_Z = \sigma \circ \varphi$, that is, $\varphi: (Z, f) \to (\Omega_2, \sigma)$ is a factor mapping.*
(2) *For all points $z \in Z$: $\varphi(z)_0 = i$ iff $z \in J_i$ $(i \in \{0, 1\})$.*
(3) *There is an at most countable subset E of Ω_2 such that, for all points $a \in \Omega_2$, the fibre $\varphi^{\leftarrow}[a]$ consists of two points if $a \in E$ and it consists of one point if $a \notin E$.*
(4) *For every point $a \in \Omega_2 \setminus E$ the number*

$$\operatorname{diam}\{\, z \in Z \,\vdots\, \varphi(z) \textit{ begins with the block } a_{[0;n]} \,\}$$

tends to 0 for $n \rightsquigarrow \infty$.

Proof. Let Z and φ be as before and recall from 7 and 8 above that Z is a closed invariant subset of X having the properties (1) and (3). In order to prove (2), note that[6], for every $n \in \mathbb{N}$,

$$J_{a_{[0;n)}} \supseteq \{ z \in Z : \varphi(z) \text{ begins with the block } a_{[0;n)} \}, \tag{7.4-10}$$

Indeed, if $n \in \mathbb{N}$, $z \in Z$ and $\varphi(z)$ begins with the block $a_{[0;n)}$ then it follows from the definitions that $z \in J_{\varphi(z)} \subseteq J_{\varphi(z)_{[0;n)}} = J_{a_{[0;n)}}$. In particular, for $n = 1$ we get $z \in J_{a_0}$. This is equivalent to statement (2).

In order to show that the set Z is 0-dimensional, we consider a connected subset L of Z. Then $\varphi[L]$ is a connected subset of Ω_2, and because Ω_2 is 0-dimensional, it follows that $\varphi[L]$ consists of a single point. So it follows from (3) that L consists of at most two points. But in a Hausdorff space a set consisting of two distinct points is not connected, so L consists of a single point only. Thus, all connected subsets of Z are singleton sets. It follows from Proposition A.6.2 that Z is 0-dimensional.

It remains to prove property (4). To do so, we again use that the fact that the intervals $J_{a_{[0;n)}}$ for $n \in \mathbb{N}$ form a decreasing family of compact sets. If $a \notin E$ then this intersection is a single point $z' \in X$. It follows from Appendix A.2.2 that for every $\varepsilon > 0$ the open neighbourhood $B_{\varepsilon/2}(z')$ of z' includes almost all intervals $J_{a_{[0;n)}}$, hence diam $(J_{a_{[0;n)}}) < \varepsilon$ for almost all $n \in \mathbb{N}$. The desired result now follows from (7.4-10). □

Remark. By statements (1) and (2) above, for all $z \in Z$ and $k \in \mathbb{Z}^+$ we have $\varphi(z)_k = (\sigma^k \varphi(z))_0 = \varphi(f^k(z))_0 = i$ iff $f^k(z) \in J_i$ $(i = 0, 1)$, hence

$$\forall z \in Z \; \forall k \in \mathbb{Z}^+ : \; f^k(z) \in J_{\varphi(z)_k} . \tag{7.4-11}$$

This means that $\varphi(z)$ *is the itinerary of the point z with respect to the partition* $(Z \cap J_0, Z \cap J_1)$ *of* Z. Since the intervals J_a may be non-degenerate, this partition is not necessarily a pseudo-Markov partition.

Theorem 7.4.4. *Assume that (X, f) has a horseshoe (J_0, J_1) consisting of two disjoint compact intervals. Then there exists $\delta > 0$ such that f admits a δ-scrambled Cantor set.*

Proof. Let Z, φ and E be as in Proposition 7.4.3. The proof of the present theorem consists of lifting a scrambled subset of Ω_2 to Z by means of φ. Select a point $a \in \Omega_2 \setminus E$. For every point $b \in \Omega_2$, put

$$b^* := a_0 \, b_0 \, a_{[0;2)} \, b_{[0;2)} \ldots a_{[0;n)} \, b_{[0;n)} \ldots\ldots$$

(the main difference with Example (3) in 7.3.1 is that we now use the initial blocks of the selected point a instead of those of the point 1^∞). For every point $b \in \Omega_2$ select a point $z_b \in \varphi^{\leftarrow}[b^*]$; so $z_b \in Z$ and $\varphi(z_b) = b^*$. Put $S := \{ z_b : b \in \Omega_2 \}$ and let $\delta > 0$ be the distance between the intervals J_0 and J_1. Obviously, S has the same cardinality as Ω_2 (for $b^* \neq c^*$ if $b \neq c$), hence is uncountable. We shall show that S is a δ-scrambled set in X.

---

6 In point of fact, the set in the right-hand side is equal to $Z \cap J_{a_{[0;n)}}$.

Consider two distinct elements z_b and z_c in S with $b, c \in \Omega_2$. Then obviously $b \neq c$, and $b_{[0;k)} \neq c_{[0;k)}$ for all sufficiently large $k \in \mathbb{N}$. Hence there are infinitely many values of n such that b^* and c^* have different coordinates at position n, that is, $\varphi(z_b)_n \neq \varphi(z_c)_n$. By (7.4-11), this implies that there are infinitely many values of n such that the points $f^n(z_b)$ and $f^n(z_c)$ are not both in the same interval J_0 or J_1. Consequently, for these values of n we have $|f^n(z_b)-f^n(z_c)| \geq \delta$. This shows that $\limsup_{n\rightsquigarrow\infty} |f^n(z_b)-f^n(z_c| \geq \delta$.

Next, in order to show that $\liminf_{n\rightsquigarrow\infty} |f^n(z_b) - f^n(z_c)| = 0$, let $\varepsilon > 0$ and select $k \in \mathbb{N}$ such that the set mentioned in property (4) of Proposition 7.4.3 has diameter less than ε (recall that $a \notin E$). Since the block $a_{[0;k)}$ occurs in b^* and c^* infinitely often in the same position it follows that there are infinitely many values of n such that $\sigma^n b^*$ and $\sigma^n c^*$ both begin with the block $a_{[0;k)}$. Then the choice of k implies that $|f^n(z_b) - f^n(z_c)| < \varepsilon$ for infinitely many values of n. This completes the proof that $\liminf_{n\rightsquigarrow\infty} |f^n(z_b) - f^n(z_c)| = 0$. So S is a δ-scrambled set in X.

But S need not be a Cantor set (there may be isolated points). Since every subset of S is a δ-scrambled set as well, it is sufficient to show that S includes a Cantor set. To this end, recall the definition of S. If $b^* \notin E$ then there is a unique choice for $z_b \in \varphi^{\leftarrow}[b^*]$. If $b^* \in E$ then we make one of the two possible choices for $z_b \in \varphi^{\leftarrow}[b^*]$, so one point is not chosen. Hence S equals the uncountable the set $\varphi^{\leftarrow}[\{b^* \vdots b \in \Omega_2\}]$ minus a countable set. As the mapping $b \mapsto b^* : \Omega_2 \to \Omega_2$ is continuous – all coordinates of b^* (are constant or) depend continuously on b – it follows that $\{b^* \vdots b \in \Omega_2\}$ is compact, hence closed in Ω_2. Consequently, the above description of S implies that S is an uncountable closed set in Z minus a countable set: an uncountable Borel set. A well-known result by Aleksandrov and Hausdorff[7] states that every uncountable Borel set in a complete separable metric space includes a Cantor set. So S includes a Cantor set (take into account that Z is a compact subset of $\mathbb{R}$, hence a complete separable metric space). □

Corollary 7.4.5. *If the system (X, f) has a horseshoe consisting of two disjoint compact intervals then it is LY-chaotic.*

Let notation be as above. In particular, Z is the compact 0-dimensional invariant subset of X defined just before 7.4.2 (6) that contains the scrambled set S. We want to show that Z also has a D-chaotic subset. This means, among others, that there exist periodic points in Z.

Lemma 7.4.6. *Let $z \in Z$ and assume that $a := \varphi(z)$ is a periodic point in (Ω_2, σ). Then the point z is periodic in (Z, f). If $a \in E$ then both end points of the interval J_a (one of which is the point z) are periodic under f.*

Proof. Let k be a period of a. Then $\sigma^k a = a$, hence $f^k(z) \in \varphi^{\leftarrow}[a]$. We distinguish the cases that $a \in E$ and that $a \notin E$.

If $a \notin E$ then $\varphi^{\leftarrow}[a] = \{z\}$, so in this case $f^k(z) = z$: the point z is periodic with period k.

7 See C. Kuratowski [1966], Ch. III, §37, Theorem 3.

If $a \in E$ then $\varphi^{\leftarrow}[a]$ consists of the two end points of the interval J_a. One of these points is z; denote the other by z'. Statement 5 of 7.4.2 implies that $\{f^k(z), f^k(z')\} = \{z, z'\}$. So there are two possibilities. The first one is that $f^k(z) = z$ and $f^k(z') = z'$. So in this case the points z and z' are periodic and the proof is completed. The other possibility is that $f^k(z) = z'$ and $f^k(z') = z$. In that case the point z is periodic with period $2k$ and the point z belongs to the orbit of z, hence z' is periodic as well. □

Remark. If one of the end points of an interval J_a with $a \in E$ is periodic then (a is periodic, hence) the other end point is periodic as well. The above proof shows that these points have equal primitive periods.

Theorem 7.4.7. *Assume that (X, f) has a horseshoe consisting of two disjoint compact intervals. Then X has a compact invariant subset Z_0 such that the subsystem (Z_0, f) is D-chaotic.*

Proof. Let Z, φ and E be as in Proposition 7.4.3. By Proposition 7.2.9, there is a closed invariant subset Z_0 of Z such that the subsystem (Z_0, f) of (Z, f) – which is, consequently, a subsystem of (X, f) – is transitive and sensitive. In order to prove the theorem it is sufficient to show that the set P of all periodic points of (Z_0, f) is dense in Z_0. In order to do so, recall from the proof of Proposition 7.2.9 that φ maps Z_0 onto Ω_2 and that, among the closed invariant subsets of Z, Z_0 is minimal with respect to the property that $\varphi[Z_0] = \Omega_2$: see Remark 2 after Proposition 7.2.9.

It is easy to see that Lemma 7.4.6 also holds with Z_0 instead of Z: a periodic point of (Z, f) that happens to be in Z_0 has its orbit in Z_0, hence is periodic in (Z_0, f). Since φ maps Z_0 onto Ω_2, Lemma 7.4.6 implies that the set P of periodic points of (Z_0, f) is not empty and that, in fact, $\varphi[P]$ is equal to the set of periodic points of Ω_2, which is dense in Ω_2. Consequently, we have $\varphi[\overline{P}] = \overline{\varphi[P]} = \Omega_2$; note that the first equality holds because the space Z_0 is compact. Thus, φ maps the set $\overline{P}$ onto Ω_2. Since $\overline{P}$, being the closure of an invariant set, is a compact and invariant subset of Z_0, it follows from the minimality property of Z_0 that $\overline{P} = Z_0$. □

Proposition 7.4.3 has yet another consequence, which is interesting in the context of the search for non-trivial minimal sets. See Exercise 7.14 (1).

7.5 Existence of a horseshoe

In this section we consider, again, a dynamical system (X, f) on an interval X in $\mathbb{R}$. Though X will not be assumed to be compact, all results in this section concern a bounded subset of X. We shall use the notation, the terminology and the conventions of Section 2.5. In particular, recall from Proposition 2.5.3 the two possible orderings for the orbit of a periodic point with the smallest possible odd primitive period greater than 1.

Proposition 7.5.1. *Let (X, f) be a dynamical system on an interval which has a periodic point of odd period larger than 1. Then there exists a horseshoe (J, K) for f^2 consisting of mutually disjoint compact intervals which contain no end points of X (if there are any).*

Proof. Let p be the smallest odd natural number, $p > 1$, for which there is a periodic point x with primitive period p. Let c the middle one of the points of the orbit of x and put $c_i := f^i(c)$ for $i = 0, \dots, p-1$. According to Proposition 2.5.3, the points c_i are ordered in the following way:

$$c_{p-1} < c_{p-3} < \cdots < c_2 < c_0 < c_1 < \cdots < c_{p-2}\,,$$

or just the other way round (in which case the proof will be similar to the proof we present here). Clearly, the mapping f^2 acts on these points in the following way:

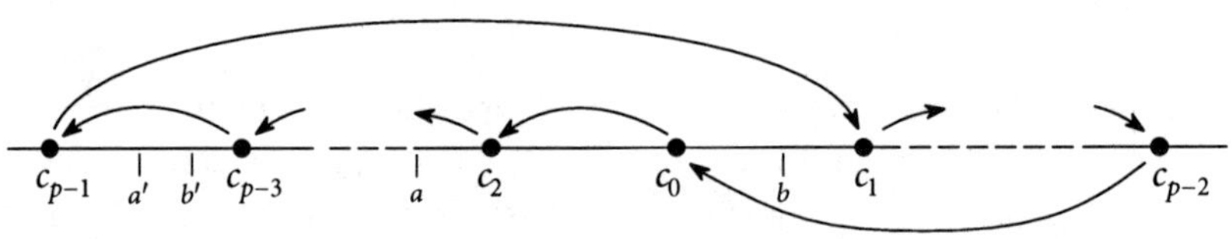

(in the case $p = 3$ there are no points c_i with $i > 2$, and one only has the arrows $c_0 \to c_2 \to c_1 \to c_0$). We shall find points $a' < b' < a < b$ in the interval $[c_{p-1}; c_1]$ such that $f^2(b'), f^2(a) \le a'$ and $f^2(a'), f^2(b) \ge b$, which implies that the two intervals $[a'; b']$ and $[a; b]$ form a horseshoe for f^2. We start with the point b, then we find, successively, the points a', b' and a.

In order to find the point b, recall that $f\colon [c_0; c_1] \xrightarrow{\circ} [c_2; c_1] \supseteq [c_0; c_1]$, hence $f^2\colon [c_0; c_1] \xrightarrow{\circ} [c_0; c_1]$. Consequently, there is a point $b \in [c_0; c_1]$ such that[8] $f^2(b) = c_1 \ge b$. Note that the $b \neq c_1$, for otherwise the point c_1 would be invariant under f^2, which is not the case: it has odd period $p \ge 3$. Thus, we have a point b with $b < c_1$ and $f^2(b) = c_1 > b$.

Next, observe that $f^2\colon [c_{p-1}; c_{p-3}] \xrightarrow{\circ} [c_{p-1}; c_1]$. So there is a point $a' \in [c_{p-1}; c_{p-3}]$ with $f^2(a')$ strictly between b and c_1 (here it is, of course, essential that $b < c_1$). In particular, $f^2(a') > b$.

Note that $f^2(a') > a'$, so $f^2\colon [a'; c_{p-3}] \xrightarrow{\circ} [c_{p-1}; a']$. It follows that there is a point $b' \in [a'; c_{p-3}]$ such that $f^2(b') < a'$.

Finally, $f^2\colon [c_{p-3}; b] \xrightarrow{\circ} [c_{p-1}; c_1]$, hence there is a point $a \in [c_{p-3}; b]$ with $f^2(a) < a'$.

So we have points a', b', a and b such that $c_{p-1} \le a' < b' < a < b < c_1$ for which $[a'; b] \subseteq \langle f^2(a'); f^2(b') \rangle$ and $[a'; b] \subseteq \langle f^2(a); f^2(b) \rangle$. Consequently, if we put $J := [a'; b']$ and $K := [a; b]$ then $J \cap K = \emptyset$ and (J, K) is a horseshoe for f^2.

Finally, we show that the intervals J and K contain no end points of the interval X. To this end, it is sufficient to show that J and K are included in the interior of the

8 If $p \ge 5$ this is also clear from the fact that $f^2\colon [c_0; c_1] \xrightarrow{\circ} [c_2; c_3]$ with $c_3 > c_1$. For the case that $p = 3$ (where $c_3 = c_0$) this would be not sufficient for our purpose.

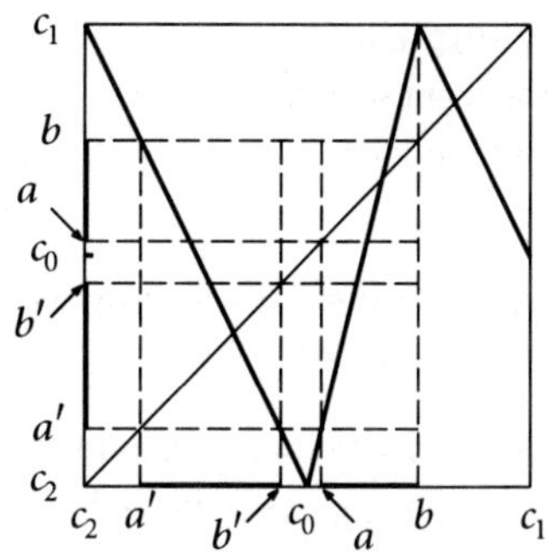

Fig. 7.6. Illustration of the example following Proposition 7.5.1.

interval $[c_{p-1}; c_1]$. As we know already that $b < c_1$, it remains to prove that $c_{p-1} < a'$. Recall that the point a' has been chosen such that $f^2(a')$ is an interior point of the interval $[b; c_1]$; in particular, this means that $f^2(a') < c_1$. Then $a' \neq c_{p-1}$, for otherwise $f^2(a')$ would be equal to c_1. This completes the proof that the intervals J and K contain no end points of X. □

Example. Follow the procedure outlined in the above proof for the function f defined in Example F in 6.3.5; for the graph of f^2, see Figure 7.4. We find $b := 3/4$. Then we have to find $a' \in [c_2; c_0]$ with $f^2(a') > b$; thus, let a' be slightly less than $1/8$. In addition, we need $b' \in [a'; c_0]$ and $a \in [c_0; b]$ with f^2-values both less than a': take these points sufficiently close to c_0, situated right and left of c_0, respectively. See Figure 7.6 (for convenience, in this figure we have drawn the points a', a and b' so that $f^2(a') = b$ and $f^2(b') = f^2(a) = a'$; in this case, this also produces the desired horseshoe for f^2).

Theorem 7.5.2. *Let (X, f) be a dynamical system on an interval. The following conditions are equivalent:*
(i) *f has a periodic point whose primitive period is not a power of 2, i.e., f has a periodic point with primitive period $q \cdot 2^m$ with $m \geq 0$, $q > 1$ and q odd.*
(ii) *There exists $n \in \mathbb{N}$ such that f^n has a horseshoe consisting of two disjoint closed intervals.*

Proof. (i)⇒(ii): Assume (i) and let $g := f^{2^m}$. Then by Proposition 1.1.4 (1), g has a periodic point with primitive period q, where $q > 1$ and q is odd. By Proposition 7.5.1, there is a horseshoe for g^2 consisting of two disjoint intervals. So condition (ii) is fulfilled with $n := 2 \cdot 2^m$.

(ii)⇒(i): If (ii) holds then Proposition 7.4.1 implies that f^n has a periodic point with period 3. Then by Proposition 1.1.4 (2), f has a periodic point with primitive period $3n/b$ with $b = 1$, hence condition (i) is fulfilled. □

Remark. Since Proposition 7.4.1 does not require disjointness of the intervals of the horseshoe, condition (ii) above can be weakened to:
(iii) There exists $k \in \mathbb{N}$ such that f^k has a 2-horseshoe.

see the proof of the implication (ii)⇒(i). This is not very surprising: by Lemma 8.6.1 below, if f^k has a 2-horseshoe then f^{2k} has a 4-horseshoe from which one can select two disjoint members which will form a 2-horseshoe for f^{2k} consisting of two disjoint intervals.

The above theorem applies to transitive maps on a compact interval: see Proposition 7.5.4 below.

Lemma 7.5.3. *Let $([a; b], g)$ be a transitive system on a non-degenerate compact interval and assume that g has at least two distinct invariant points*[9]. *Then g has a horseshoe.*

Proof. Below, we shall repeatedly use the fact that a closed invariant subset of the phase space $[a; b]$ has empty interior: see Exercise 1.6 (3) in combination with the first part of Theorem 1.3.5. In particular, a non-degenerate closed subinterval of $[a; b]$ cannot be invariant.

By Lemma 2.6.4 (1) there exists an invariant point strictly between a and b. We consider two cases.

Case 1. there are at least two invariant points strictly between a and b. The complement in $[a; b]$ of the set of invariant points is a dense open set, so it is a union of mutually disjoint open intervals (open in $[a; b]$), one of which is situated between the two invariant points just mentioned (otherwise the set would not be dense). We may assume that this is a maximal open interval in the complement of the set of invariant points (a connected component of this complement). Consequently, there are two invariant points x_1 and x_2 such that $a < x_1 < x_2 < b$ and such that there are no invariant points strictly between x_1 and x_2. The mean value theorem now implies that either $g(x) > x$ for all $x \in (x_1; x_2)$ or $g(x) < x$ for all $x \in (x_1; x_2)$. Assume the former case.

By transitivity, the interval $[x_1; b]$ is not invariant. Hence there exists a point $x \in [x_1; b]$ such that $g(x) < x_1$. By assumption, all values of g on the interval $(x_1; x_2]$ are larger than x_1, so $x > x_2$. As $g(x_2) \geq x_2 > x_1$, the mean value theorem implies that there is a point between x_2 and x where g assumes the value x_1. Let z be the minimum of such points, i.e., the minimum of the compact set $g^{\leftarrow}[x_1] \cap [x_2; b]$. This choice of the point z clearly implies that $g(x) \neq x_1$ for all x with $x_1 < x < z$. Since there are points in the interval $[x_1; z]$ where g assumes a value larger than x_1 it clearly follows that $g(x) \geq x_1$ for all $x \in [x_1; z]$. But the interval $[x_1; z]$ is not invariant, so there is a point in this interval where g has a value greater than z. By the mean value theorem this implies that there is a point $y \in [x_1; z]$ such that $g(y) = z$. Consequently, $g: [x_1; y] \xrightarrow{\circ} [x_1; z]$ (recall that $g(x_1) = x_1$) and $g: [y; z] \xrightarrow{\circ} [x_1; z]$, which means that the intervals $[x_1; y]$ and $[y; z]$ form a horseshoe. See Figure 7.7 (a).

9 By Lemma 2.6.4 (1), at least one of the invariant points is strictly between a and b.

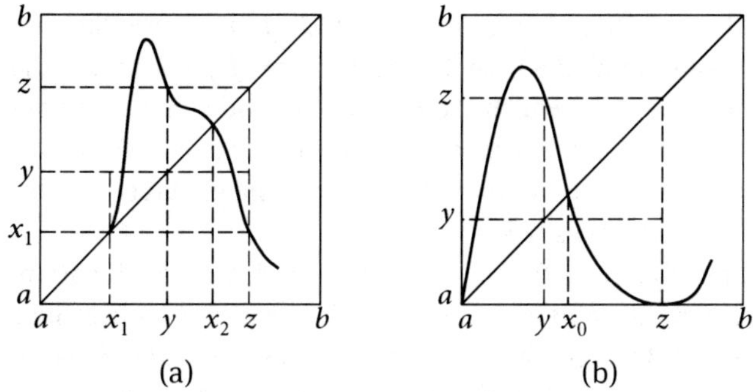

Fig. 7.7. (a) Illustrating Case 1 of the proof of Lemma 7.5.3 with $g(x) > x$ for $x_1 < x < x_2$. (b) Illustrating Case 2 of the proof of Lemma 7.5.3.

In the case that $g(x) < x$ for all $x \in (x_1; x_2)$, let z be the maximal point in $[a; x_1]$ with $g(z) = x_2$ and find y between z and x_2 such that $g(y) = z$. Then the intervals $[z; y]$ and $[y; x_2]$ form a horseshoe.

Case 2. g has only one invariant point strictly between a and b. Call it x_0. The other invariant point of g must be an end point of $[a; b]$; say, the point a is invariant under g. Let $l := (b - a)/2$. For every $n \in \mathbb{N}$ none of the two intervals $[a; a + l/n]$ or $[a + l/n; b]$ is invariant, so there are points a_n and b_n such that

$$a < a_n \le a + l/n \quad \text{and} \quad g(a_n) > a + l/n \ge a_n\,;$$
$$a + l/n \le b_n \le b \quad \text{and} \quad g(b_n) < a + l/n \le b_n\,.$$

Then between a_n and b_n there is an invariant point of g, which can only be the point x_0. Since $a_n < x_0$ for almost all $n \in \mathbb{N}$ it follows that $b_n > x_0$ for almost all $n \in \mathbb{N}$. The sequence $(b_n)_{n\in\mathbb{N}}$ has a convergent subsequence with limit z, say. Obviously, $g(z) = a$ and $z > x_0$ (equality is excluded because $g(x_0) = x_0$). As the interval $[a; z]$ is not invariant there exist a point $y \in (a; z)$ with $g(y) = z$ (y cannot coincide with a or z because $g(a) = g(z) = a$). Then the intervals $[a; y]$ and $[y; z]$ form a horseshoe. See Figure 7.7 (b) (note that $g(x) > x$ for $a < x < x_0$, for otherwise the interval $[a; x_0]$ would be invariant). □

Remark. The horseshoe for g does not necessarily consist of mutually disjoint intervals: the tent map on the unit interval is transitive and has two invariant points, yet it has no horseshoe consisting of disjoint closed intervals. See Exercise 7.14 (3).

Proposition 7.5.4. *Let (X, f) be a transitive system on a compact interval. Then f^2 has a horseshoe. Moreover:*

(a) *If f is totally transitive then it has Šarkovskij type q for some odd integer $q > 1$.*

(b) *If f is not totally transitive then it is of type 6.*

In both cases f has a periodic point of period 6.

Proof. Assume that f is totally transitive: then by Theorem 2.6.8, f has a periodic point with odd primitive period greater than 1. So the Šarkovskij type of f is equal to the smallest odd integer q greater than 1 for which there is an f-periodic point with period q. Then Šarkovskij's Theorem 2.2.3 implies that f has a periodic point of primitive period 6. In addition, Proposition 7.5.1 implies that f^2 has a horseshoe. This completes the proof in the case that f is totally transitive.

If f is not totally transitive then X is the union of two non-degenerate invariant closed subintervals I_0 and I_1 with disjoint interiors, on each of which f^2 is transitive and which are interchanged by f. Their common end point is invariant under f, hence it is an invariant point for f^2. So in view of Lemma 2.6.4 (1), Lemma 7.5.3 can be applied to $g := f^2$ on I_0 and on I_1. Consequently, f^2 has a horseshoe.

Moreover, Proposition 7.4.1 now implies that f^2 has a periodic point with primitive period 3. Therefore, Proposition 1.1.4 (2) implies that f has a periodic point with primitive period 6. Obviously, f has no periodic points with a primitive period that precedes 6 in the Šarkovskij order (i.e., f has type 6): otherwise such a point would have a primitive period q for some odd integer $q > 1$, which is incompatible with the fact that all non-invariant f-periodic points have even periods (because f interchanges the intervals I_0 and I_1). □

Next, we combine Theorem 7.5.2 with one of our previous results about the existence of scrambled sets:

Theorem 7.5.5. *Let (X, f) be a dynamical system on an interval which has a periodic point with a primitive period which is not a power of 2. Then there exists $\delta > 0$ such that f admits a δ-scrambled Cantor set.*

Proof. By Theorem 7.5.2, there is $n \in \mathbb{N}$ such that f^n admits a horseshoe consisting of two disjoint intervals. Then by Theorem 7.4.4 there exists a δ-scrambled Cantor set K for f^n with $\delta > 0$. Then K is also a δ-scrambled Cantor set for f. □

Corollary 7.5.6. *A system on an interval with a periodic point whose primitive period is not a power of 2 is LY-chaotic.*

Proof. Clear from Theorem 7.5.5. □

Remark. By Proposition 7.5.4, the corollary applies to any transitive system on a compact interval, so such a system is LY-chaotic. See Corollary 7.3.8 for a stronger statement: such a system is strongly and densely LY-chaotic.

Theorem 7.5.7. *A dynamical system on an interval which has a periodic point whose primitive period is not a power of 2 has a compact Devaney chaotic subsystem.*

Proof. By Theorem 7.5.2, there is an $n \in \mathbb{N}$ such that f^n admits a horseshoe consisting of two disjoint intervals. Then Theorem 7.4.7 implies that (X, f^n) has a D-chaotic compact subsystem (Z_0, f^n). Note that Z_0 is invariant under f^n, not necessarily under f, so we have yet to find a suitable f-invariant subset of Z_0.

Let z be a transitive point in Z_0 under f^n and let $Z_1 := \overline{\mathcal{O}_f(z)}$, the orbit closure of the orbit of the point z under f. Then Z_1 is a closed f-invariant subset of X. We first show that Z_1 is compact and that the set of periodic points of (Z_1, f) is dense in Z_1. To this end, note that

$$\mathcal{O}_f(z) = \bigcup_{i=0}^{n-1} \mathcal{O}_{f^n}(f^i(z)).$$

Here the inclusion "$\supseteq$" is trivial. In order to prove "$\subseteq$", write every $m \in \mathbb{Z}^+$ as $m = nk+i$ with $k \in \mathbb{Z}^+$ and $0 \le i \le n-1$, so that $f^m(z) = (f^n)^k(f^i(z))$; this implies the desired inclusion.

For $i = 0, \dots, n-1$ put $Y_i := \overline{\mathcal{O}_{f^n}(f^i(z))}$. Since the closure of a finite union is the union of the closures, it follows easily from the equality just proved that $Z_1 = \bigcup_{i=0}^{n-1} Y_i$.

As the point z is chosen to be transitive in Z_0 under f^n, it is clear that $Y_0 = Z_0$. Moreover, if $0 \le i \le n-1$ then f^i maps $\mathcal{O}_{f^n}(z)$ onto $\mathcal{O}_{f^n}(f^i(z))$, hence by compactness of the set $\overline{\mathcal{O}_{f^n}(z)} = Z_0$, f^i maps Z_0 onto Y_i. Consequently, the sets Y_i are compact and therefore their union Z_1 is compact as well. Moreover, each of the sets Y_i, being an orbit closure under f^n, is invariant under f^n, hence defines a subsystem of (X, f^n). As f^i commutes with f^n, it follows that $f^i : (Z_0, f^n) \to (Y_i, f^n)$ is a factor map (recall that we have observed above that f^i maps Z_0 onto Y_i). As the system (Z_0, f^n) has a dense set of periodic points, it follows that every system (Y_i, f^n) has a dense set of periodic points. The union of these sets of periodic points is dense in Z_1 (again, the closure of a finite union is the union of the closures). Obviously, these points are, as points in Z_1, periodic under f. Conclusion: the set of periodic points of (Z_1, f) is dense in Z_1.

Next, recall that the system (Z_0, f^n) is sensitive, which implies that the point z is unstable in the system (Z_0, f^n). Consequently, the point z is unstable in Z_1 under f^n, hence it is unstable under f as well. As the point z has a dense orbit in Z_1, (the proof of) Corollary 7.1.5 shows that the point z is transitive in Z_1 under f and that the system (Z_1, f) is sensitive. This completes the proof. □

Remarks.

(1) The converse statement is also true: see Note 9 below.

(2) By Corollary 7.3.11, a dynamical system as in the above theorem is LY-chaotic, a conclusion which is in accordance with Corollary 7.5.6.

(3) As $Z_0 \subseteq Z$ and φ maps Z_0 onto Ω_2, it is clear that Z_0 meets every fibre of φ in Z, but that it includes no complete fibres. Consequently, the difference $Z \setminus Z_0$ is included in $(\varphi|_Z)^{\leftarrow}[E]$, hence is at most countable.

Exercises

7.1. Let (X, f) be a dynamical system on a metric space, let A be a non-empty compact invariant subset of X and let $x_0 \in X$ be a periodic point with primitive period p.

(1) If $A \subseteq \mathbf{Eq}(X, f)$ then condition (3.2-1) is satisfied (so if A is completely invariant then A is stable).

(2) Conversely, if $A := \mathcal{O}_f(x_0)$ and A is a stable set then $A \subseteq \mathbf{Eq}(X, f)$.
NB. This statement does not hold for arbitrary compact completely invariant sets: for an example, use Remark 2 after Proposition 7.1.2.

(3) The following statements are equivalent:
 (i) $x_0 \in \mathbf{Eq}(X, f)$.
 (ii) $\mathcal{O}_f(x_0) \subseteq \mathbf{Eq}(X, f)$.
 (iii) $\mathcal{O}_f(x_0)$ is stable under f.
 (iv) $x_0 \in \mathbf{Eq}(X, f^p)$.
 (v) $\mathcal{O}_f(x_0) \subseteq \mathbf{Eq}(X, f^p)$.

7.2. Consider the generalized tent map $T_s\colon [0; 1] \to [0; 1]$, defined for $0 < s \leq 2$ by

$$T_s(x) := \tfrac{1}{2}s\,(1 - |2x - 1|) \quad \text{for } x \in \mathbb{R}.$$

Let $c := 1/2$ and assume that $1 < s \leq 2$; note that the point c is not invariant under the mapping T_s. The following exercises form steps in the proof that there is a partial orbit $A := \{c, T_s(c), \dots, T_s^k(c)\}$ of the point c such that for every non-degenerate subinterval J of $[0; 1]$ there exists $n \in \mathbb{Z}^+$ such that $T_s^n[J]$ contains at least two different points of A. In point of fact, if the point c is not periodic then any $k \in \mathbb{N}$ with $\frac{1}{2}s^k > 1$ suffices; if the point c is periodic with primitive period p then one has to take $k = p - 1$. Note that in both cases the points of A are mutually distinct.

(1) Denote the length of an interval L by $|L|$. Let J be an arbitrary non-degenerate subinterval of $[0; 1]$. Note that

$$|T_s[J]| = s|J| \quad \text{if } c \text{ is not an interior point of } J,$$
$$|T_s[J]| \geq \tfrac{1}{2}s|J| \quad \text{if } c \text{ is an interior point of } J.$$

Use this to show that there are infinitely many values of $n \in \mathbb{N}$ such that $c \in T_s^n[J]$.

(2) Assume that the point $c := 1/2$ is not periodic. Fix $k \in \mathbb{N}$ such that $\frac{1}{2}s^k > 1$ and let J be an arbitrary non-degenerate subinterval of $[0; 1]$.
 (a) Show that there are $m, j \in \mathbb{N}$ such that $0 < j \leq k$ and $c \in T_s^{m+j}[J] \cap T_s^m[J]$.
 (b) Let $A := \{c, T_s(c), \dots, T_s^k(c)\}$. With m and j as in (a), $T_s^{m+j}[J]$ includes the two points c and $T_s^j(c)$ of A.

(3) Assume that the point c is periodic with primitive period $p > 1$ ($p = 1$ is excluded because $s > 1$[10]) and let J be a non-degenerate subinterval of $[0; 1]$. By 1 one can select $m \in \mathbb{Z}^+$ such that $c \in T_s^m[J] =: J'$.

---

10 Actually, $p \geq 3$: one easily computes that if c has period 2, then $s = 1$. A straightforward computation shows that if $s = \frac{1}{2}(1 + \sqrt{5}) \approx 1.618\dots$ then $T_s^3(c) = c$.

(a) There exists $n \in \mathbb{N} \setminus p\mathbb{N}$ such that $c \in T_s^n[J']$.
(b) Let $A := \{c, T_s(c), \ldots, T_s^{p-1}(c)\}$; then the points of A are mutually different and two different points c and $T_s^j(c)$ with $1 \leq j \leq p-1$ of A belong to $T_s^{n+m}[J]$ for some $n \in \mathbb{N}$.

7.3. Let $X := [0;1]$ and define $f : X \to X$ by:

$$f(x) = \begin{cases} \frac{3}{2}x & \text{for } 0 \leq x \leq \frac{2}{3}\,, \\ \frac{3}{2}(\frac{4}{3} - x) & \text{for } \frac{2}{3} \leq x \leq 1\,. \end{cases}$$

(1) Show that the set of periodic points is not dense in $[0;1]$.
NB. Similarly: the system is not transitive. This is in accordance with Theorem 2.6.2.
(2) Show that the system (X, f) is sensitive.

7.4. Prove the following refinement of Theorem 7.1.13: *Let (X, f) be a dynamical system on a metric space and assume that (X, f) is not minimal. If the system is transitive and has a dense set of almost periodic points then it is sensitive.*
NB. A non-minimal transitive system with a dense set of almost periodic points is sometimes called an M-system. Thus, an M-system is AY-chaotic.

7.5. Show that the quadratic system $(\mathbb{R}, f_\mu)$ with $\mu > 2 + \sqrt{5}$ is sensitive.
NB. By Note 8 in Chapter 1 this holds for all $\mu > 4$.

7.6.
(1) Let (X, f) be a dynamical system on an arbitrary metric space with metric d and assume that the system is sensitive and has a dense set of periodic points. Show that the system that (X, f) is *strongly sensitive* in the following sense: there exists $\varepsilon > 0$ such that for every point $x \in X$ and every neighbourhood U of x there is a point $y \in U$ such that $d(f^n(x), f^n(y)) \geq \varepsilon$ for infinitely many $n \in \mathbb{N}$.
(2) Let $Z := \{x \in \Omega_2 : x_n = 0 \text{ for almost all } n \in \mathbb{Z}^+\}$. Then Z is σ-invariant (Z is not closed: it is dense) and the system (Z, σ) is 1-sensitive, but not strongly sensitive.

NB. Sensitivity means that an error in selecting the initial state (however small) inevitably gives rise to a deviation of at least ε at some moment in the future. But perhaps this deviation occurs only once, not precluding the possibility that in the long run the deviation goes to zero. In strong sensitivity this is not the case.

7.7. Show that a weakly mixing system is Auslander–Yorke chaotic.

7.8. Let $\varphi\colon (X, f) \to (Y, g)$ be a factor mapping of dynamical systems on compact metric spaces X and Y.
(1) If $x \in X$ is an equicontinuity point under f and φ is open at x then $\varphi(x) \in Y$ is an equicontinuity point under g.
(2) If (X, f) is transitive, (Y, g) is Auslander–Yorke chaotic and φ is open at a transitive point x of X then (X, f) is Auslander–Yorke chaotic.

7.9. Let (T, σ) be the Toeplitz system defined in 5.6.13. Recall that there is a factor map $\varphi: (T, \sigma) \to (G, f)$, where (G, f) is the adding machine, defined and completely determined by $\varphi(\tau) = \underline{0} := 0^\infty$.

(1) As a preliminary result, show that for every $n \in \mathbb{N}$ the sequence of coordinates of the point $\underline{n} := f^n(\underline{0})$ consists of the binary expansion of n in reverse order followed by 0^∞. Example: $f^{40}(\underline{0}) = 000101\,0^\infty$.
Show that for every $n \in \mathbb{N}$ the point $\underline{n}$ starts with k 0's iff $n \in 2^k\mathbb{Z}^+$. (In the example: $40 = 2^3 \cdot 5$.)

(2) For every point x in $\mathcal{O}_\sigma(\tau)$ one has $\varphi^\leftarrow[\varphi(x)] = \{x\}$.

NB 1. As $\mathcal{O}_\sigma(\tau)$ is dense in T, it follows from Corollary A.9.6 that φ *is an almost 1-to-1 mapping*. Below it will be shown that for every $x \in T$ the fibre $\varphi^\leftarrow[\varphi(x)]$ has at most two points.

NB 2. The factor mapping φ is not injective: by part (c) in the proof of Theorem 5.6.14 the point $x := 0\tau$ is in T and, by a similar argument, the point $y := 1\tau$ is in T; moreover, $\varphi(x) = \varphi(y)$ is the unique point in G that is mapped by f onto the point $\underline{0}$, i.e., $\varphi(x) = \varphi(y) = 1^\infty$. More generally: if $x, y \in T$, $x \neq y$, and there exists $k \in \mathbb{N}$ such that $\sigma^k x = \sigma^k y$ then $\varphi(x) = \varphi(y)$.

(3) Let $x \in \Omega_2$ and $n \in \mathbb{Z}^+$. A *$t(n)$-representation* of x is a parsing of the sequence of coordinates of x as a concatenation of blocks of the form $0t(n)$ and $1t(n)$, preceded by a 'tail' of $t(n)$:

$$x = t_i(n) * b(n) * b(n) * b(n) * b(n) * \ldots\ldots$$

where each $*$ (called a *separator*) represents a 0 or a 1 and where $t_i(n)$ is the block consisting of the final i coordinates of $t(n)$, $0 \le i \le 2^{n+1} - 1$ (this includes the possibilities that $t_i(n)$ is $t(n)$ or the empty block). One might say that it is the pattern of the occurrences of the $*$'s that makes up the $t(n)$-representation, hence such a representation is completely determined by the value of i. In particular, it makes sense to say that two distinct points have the same $t(n)$-representation: the positions of separators is the same for both points, but at least one of their values differ. Show that *every element of T has, for every $n \in \mathbb{Z}^+$, a unique $t(n)$-representation.*

(4) If $x \in T$ then $\varphi^\leftarrow[\varphi(x)]$ consists of at most two points. Moreover, if $y \in \varphi^\leftarrow[\varphi(x)]$ and $y \neq x$ then x and y differ in just one coordinate from each other, hence (x, y) is an asymptotic pair.

(5) The Toeplitz system has no Li–Yorke pairs, hence it is not LY-chaotic.

7.10. Let (X, f) be a dynamical system on a compact metric space X.

(1) Show that the set $LY_\delta(X, f)$ of all δ-scrambled pairs of points of X is a G_δ-subset of $X \times X$.

(2) If X has no isolated points and X has a dense δ-scrambled subset then $LY_\delta(X, f)$ is dense in $X \times X$ and (X, f) is ε-sensitive for all $\varepsilon < \delta/2$.

(3) If the point (x, y) is transitive in X^2 under $f \times f$ (so it cannot belong to the closed invariant set Δ_X) then (x, y) is a strong Li–Yorke pair that belongs to the set $LY_\delta(X, f)$ with $\delta = \operatorname{diam}(X)$.
(4) If the system (X, f) is weakly mixing – which implies that $(X \times X, f \times f)$ is transitive – then $LY_\delta(X, f)$ is a dense G_δ-subset of $X \times X$ with $\delta = \operatorname{diam}(X)$.

NB. This elementary result does not directly imply that X has an uncountable scrambled subset; see, however, Corollary 7.3.7 (c).

7.11.
(1) A pointwise stable (i.e., equicontinuous) system on a compact metric space has no scrambled subsets.
(2) The basin of a stable invariant point includes no points of any scrambled set. Hence the basin of a stable periodic orbit contains no points of any scrambled set.
(3) Show that the golden mean shift, the even shift, the prime gap shift, the context free shift and the (1,3) run-length limited shift are LY-chaotic.
(4) If the phase space of a dynamical system is a compact metrizable space then the property of being LY-chaotic is independent of the metric used.

7.12. Let X be a compact metric space. If the dynamical system (X, f) is transitive and not minimal then (X, f) has a LY-chaotic factor.

7.13. Let (X, f) be a minimal system with a compact metric phase space. If the system is LY-chaotic then it is AY-chaotic.

7.14. Let (X, f) be a dynamical system on an interval.
(1) If the system (X, f) has a periodic point whose period is not a power of 2 then it contains a non-trivial minimal set.
(2) Let (X, f) have a periodic point with odd primitive period greater than 1 and let p be the smallest of such periods. Using the notation of Proposition 2.5.3, show that if $p \geq 5$ then (J_2, J_4) is a horseshoe under f^p (consisting of disjoint intervals).
NB. Compare this with the result of Proposition 7.5.1.
(3) Show that the tent map T on the interval $[0; 1]$ admits no horseshoe consisting of two disjoint intervals.

Notes

1 In this chapter 'chaos' means *deterministic* chaos. In general, this term is used to indicate that a certain process is unpredictable. In a predictable system, all future states can be determined up to a given accuracy by following an initial state for a (sufficiently long) finite stretch of time: roughly, there are two aspects to this notion: (a) Lagrange stability – the complete orbit of a point is approximated by a sufficiently long initial segment of that orbit (see Note 1 in Chapter 3), and (b) Lyapunov stability – the evolution of a given state approximates the evolution of all neighbouring states.

Many physical processes, which are believed to be deterministic (e.g., turbulence of a liquid flow past a solid object, convection in the earth's atmosphere), are inherently unpredictable: small differences in initial conditions may cause large deviations. This 'sensitive dependence on initial conditions' was already known for a long time to physicists and mathematicians (be it under different names), e.g., to Maxwell, Hadamard and Poincaré. The first to discover a system of differential equations with chaotic behaviour was the meteorologist E. N. Lorenz – not to be confused with the Dutch physicist H. A. Lorentz – who found the 'Lorenz attractor' in 1963 while studying a model of convection in the earths atmosphere. Over time, the ideas about how exactly chaos should be defined (the mathematical concepts to describe those processes) have evolved. For a brief overview of the relevant literature we refer to the introduction of J. Auslander & J. A. Yorke [1980] (in this paper the concept that we call Auslander–Yorke chaos is attributed to Takens and Ruelle).

Most important for the description of chaos is ergodic theory. However, this book is about the topology of dynamical systems, so we completely neglect that approach to chaos. A second approach to chaos is to discuss turbulent systems in geometrical or topological terms: capture the flavour of turbulence with the concept of a 'strange attractor', i.e., a compact attracting set which is not a point, a circle or an n-torus (or any other 'nice' manifold). A third approach is to emphasize the complicated nature of the dynamics rather than the shape of the set. Thus, in the paper T. Y. Li & J. A. Yorke [1975] the notion of a scrambled set was introduced as an attempt to describe chaos. Other approaches stress sensitive dependence on initial conditions: see the paper by Auslander and Yorke mentioned above and the book R. L. Devaney [1989]. But there are also other definitions in circulation. For example, systems with positive entropy have exponentially diverging orbits (see Chapter 8), reason why many authors call such systems (topologically) chaotic. In this context, it is instructive to compare Corollary 8.6.9 below with the results of Corollary 7.5.6 and Theorem 7.5.7, and to compare Corollary 8.6.10 below with Theorem 7.2.2 and Corollary 7.3.8. See also Note 8 in Chapter 8. Moreover, strongly and weakly mixing systems have a very erratic behaviour, so such systems are also often called chaotic. See the summary after Corollary 7.3.11 and Exercise 7.7.

The existence of chaos has important consequences for modelling real-world systems. Even though a system exhibits random behaviour, modelling strategies need not be restricted to stochastic models but can consider deterministic ones as well. Most properly, chaos is a concept that is applied to models, not to natural or experimental systems. The question is often asked of a system exhibiting complicated motion, whether the system *is* random or chaotic-deterministic. A more useful question is whether the system is better approximated using a deterministic model that allows chaotic dynamics, or alternatively by a stochastic, or a mixed deterministic/stochastic model.

2 Several results from Section 1 (in particular, those on lifting unstability) as well as Exercise 7.14 (1) ('period three implies there are non-trivial minimal sets') can be found in the paper J. Auslander & J. A. Yorke [1980].

The converse of Theorem 7.1.11 does not hold, i.e., a uniformly rigid system on a compact metric space is not necessarily almost equicontinuous. In E. Akin, J. Auslander & K. Berg [1996] is an example of a non-minimal transitive uniformly rigid system that is sensitive, hence not almost equicontinuous. In S. Glasner & D. Maon [1989] it is shown that there exists a minimal uniformly rigid system that is weakly mixing, hence sensitive (see Exercise 7.7), so certainly not almost equicontinuous.

3 A well-known and much-studied type of extensions are the so-called *group extensions*. A group extension $\varphi\colon (X, f) \to (Y, g)$ has the property that there is a group G of automorphisms of the system (X, f) such that for every two elements of X that are in the same φ-fibre there is a member of G mapping the one element onto the other – for precise definitions, see [deV], VI(4.1). A group extension is rather easily seen to be an open mapping (briefly: Y is the orbit space of X under the action of G), so if (Y, g) is sensitive then Proposition 7.2.4 implies that (X, f) is sensitive as well.

4 The problem of whether there exist asymptotic pairs in a given system has been well-studied. For example, in B. F. Bryant & P. Walters [1969] it was shown that in an expansive invertible dynamical system on a compact metric space, every infinite closed invariant set includes a positively and a negatively asymptotic pair of points (we leave the, obvious, definitions of 'positively asymptotic' and 'negatively asymptotic' to the reader). See also [GH], 10.36.

5 Much is known about proximality in compact dynamical systems. For a start, cf. [deV], Section IV.2. A well-known result, due to J. Auslander is: *In a dynamical system in a compact space every point is proximal to an almost periodic point.* For an elementary proof, see J. Auslander [1988], 5.3.
6 The main results of Section 7.3 are in the paper E. Akin, E. Glasner, W. Huang, S. Shao & X. Ye [2010], which extends earlier results from W. Huang & X. Ye [2002a].

In E. Akin & S. Glasner [2001] it is shown that there is a scattering system on an infinite compact metric space that is not sensitive to initial conditions: an example showing that, in general, Li–Yorke chaos does not imply Auslander–Yorke chaos. See also Example (4) in Corollary 7.1.8.
7 Corollary 7.5.6 implies that a mapping of an interval into itself which has a point of period three is LY-chaotic. This is one of the main results – the other is the result that we have included as Theorem 2.2.2 – of the paper T. Y. Li & J. A. Yorke [1975].
8 For the existence of a period 6 orbit in Proposition 7.5.4 see L. Block & E. M. Coven [1987]. Theorem 7.5.5 comes, essentially, from K. Janková & J. Smítal [1986] (see also Y. Oono [1978] and G. J. Butler & G. Pianigiani [1978]). It should be noticed that in this paper, instead of the condition 'there is a point whose primitive period is not a power of 2' the (for intervals equivalent) condition '(X, f) has positive topological entropy' is used (see Corollary 8.6.9 ahead). Thus, positive entropy implies LY-chaos (on an interval). In this form, the result is generalized to arbitrary compact metric spaces in F. Blanchard, E. Glasner, S. Kolyada & A. Maass [2002].

The proof that $h(f) > 0$ implies AY-chaos is also outside of the scope of this book, but it can easily be outlined here, as follows: Let (X, f) be a transitive system and assume that it is not sensitive. Then by the Auslander–Yorke Dichotomy Theorem and Theorem 7.1.11, f is a uniformly rigid homeomorphism. It can be shown that this implies $h(f) = 0$: it implies that for every f-invariant probability measure μ on X the (measure-theoretic) entropy $h_\mu(f) = 0$; then the so-called Variational Principle – the topological entropy $h(f)$ is the supremum of the numbers $h_\mu(f)$, where μ ranges over all invariant probability measures on X – implies that $h(f) = 0$.
9 As explained in Note 8 above, positive entropy implies Li–Yorke chaos. The converse is not true, not even for interval maps: it is possible that a system on a compact interval with entropy zero is Li–Yorke chaotic. A necessary and sufficient condition (in terms of limit sets) for such a system to have a δ-scrambled Cantor set is in J. Smítal [1986].

For interval maps a characterization of the existence of a Devaney chaotic subsystem is given by Theorem 7.5.7 together with its converse, which turns out to be true as well: *Devaney chaos on some compact subset is necessary and sufficient for the existence of a periodic point whose period is not a power of* 2 (i.e., necessary and sufficient for positive entropy). See Shihai Li [1993]. Outline of proof of sufficiency: let X be a Devaney-chaotic subset of the interval under consideration. By sensitivity, X has no isolated points, so X is infinite. Moreover, X has a transitive point x, so $\omega(x) = X$. If the mapping under consideration has entropy zero then by a theorem of Šarkovskij's $\omega(x)$ contains no periodic points, contradicting that X has a dense set of periodic points.
10 If (X, f) is a system on a compact interval then the existence of one single Li–Yorke pair is already sufficient for the existence of a scrambled Cantor set. See M. Kuchta & J. Smítal [1987]. In view of Theorem 7.5.5 there remains only something to prove in the case that all periods of periodic points are a power of two.

In this context we mention some variants of the notion of Li–Yorke chaos. One may read the usual definition of this type of chaos as follows: Let $LY(X, f) := \bigcup_{\delta>0} LY_\delta(X, f)$ be the set of all scrambled pairs in X (see Exercise 7.10 (1) for the definition of $LY_\delta(X, f)$). Then Li–Yorke chaos means that $LY(X, f)$ is 'large' in the sense that there is an uncountable set S in X such that $(S \times S) \setminus \Delta_X \subseteq LY(X, f)$. But 'large' may be defined in another way, e.g., by requiring that the set $LY(X, f)$ is a dense subset of $X \times X$ (*dense chaos*, not to be confused with dense Li–Yorke chaos discussed in the main text, which means the existence of a dense uncountable scrambled set) or that it is a dense G_δ-subset of $X \times X$ (*generic chaos*). For interval maps the variety in these notions is rather restricted: if X is a compact interval then (X, f) is generically chaotic iff for some $\delta > 0$ it is densely δ-chaotic. See L. Snoha [1990].

8 Topological entropy

Abstract. In this chapter we introduce the reader to the notion of topological entropy. Our discussion of topological entropy is far from complete: our main goal is to characterize the dynamical systems on intervals that play such an important role in the previous chapter, i.e., the systems in which not all primitive periods of periodic points are a power of 2.

8.1 The definition

Like in the previous chapter, (X, f) will always be a dynamical system on an arbitrary metric space (however, for many results, X will be required to be compact as well). The metric on X will be denoted by d.

'Topological entropy' is meant to be a measure of the complexity of a system. It measures the exponential growth rate of the number of distinguishable orbits as time advances. We shall explain the various parts of this description.

First, the *exponential growth rate* of a sequence $(s_n)_{n\in\mathbb{N}}$ of positive real numbers is defined as the real number[1] $\limsup_{n\rightsquigarrow\infty} \frac{1}{n} \log s_n$. For example, a bounded sequence has exponential growth rate 0. More generally, if there is a polynomial P such that $0 < s_n \leq P(n)$ for almost all $n \in \mathbb{N}$ then the sequence $(s_n)_{n\in\mathbb{N}}$ has exponential growth rate 0, because $\lim_{n\rightsquigarrow\infty} \frac{1}{n} \log n = 0$. An exponential sequence has positive exponential growth rate: if $s_n = cb^n$ with $b > 1$ then the growth rate of $(s_n)_{n\in\mathbb{N}}$ is $\log b$.

To illustrate the phrase 'the number of distinguishable orbits as time advances' we first give an example. Consider the full shift (Ω_2, σ) on the two symbols 0 and 1. Assume that we can distinguish two points $x, y \in \Omega_2$ only if their distance is at least 1, i.e., if their first coordinates x_0 and y_0 are different. Looking at points x and y 'as time advances' means 'looking at the orbits starting with the coordinates $x, \sigma x, \dots, \sigma^{n-1}x$ and $y, \sigma y, \dots, \sigma^{n-1}y$, respectively, for increasing n'. The corresponding orbits can be distinguished whenever among these first n corresponding coordinates of x and y there is a difference. Since there are 2^n possible n-blocks this means that during the time-interval $[0; n-1]$ we are able to distinguish just 2^n different orbits in Ω_2. So according to the above definition the topological entropy of the shift system is $\log 2$, the exponential growth rate of the sequence $(2^n)_{n\in\mathbb{N}}$.

The reason for looking at 'the number of distinguishable orbits as time advances' is that it is expected to be a measure for the *complexity* of the system. From this point of view a system is considered 'simple' if pairs of points that are close to each other

---

1 For the base of the logarithm one usually takes 2, e or 10, in which case $\exp(x)$ means 2^x, e^x or 10^x, respectively ($x \in \mathbb{R}$). Either choice is fine (but once made one should stick to it).

always remain close if time advances. On the other hand, if nearby points strongly diverge (become distinguishable in the course of time) then the system is considered complex.

A general (non-mathematical) definition is that 'the complexity' of a particular system is the degree of difficulty in predicting the properties of the system if the properties of the system's parts are given. Thus, complexity of a system comes from the lack of correlation between elements in the system. In mathematics there are several notions of complexity, depending on the theory where they are used. The definition of topological entropy comes closest to the notion of complexity used in information processing: a measure of the amount of information included in a message. The more predictable a message, the lower the information content – the most random signal contains the most information. Therefore, the 'amount of information' is defined as follows: if a device is equally likely to send one of N messages then 'the information produced when one message is chosen from the set' is defined as $\log N$.
Why '$\log N$' and not another function of N? Without being rigorous, this can be motivated as follows: First of all, this function has to be non-negative and increasing. In addition, if we have two lists of messages, one of length M and one of length N, then the information obtained by choosing one of the MN possible combinations should be the sum of the separate amounts of information. This additivity of the information function suggests the choice of the logarithm.
Resuming, if a situation is observed by selecting it from N possible situations then $\log N$ is a measure of the complexity of the system. In the definition of 'topological entropy' his idea is applied with for N the number of distinguishable orbits at any moment.

If the number distinguishable orbits at time n is $N(n)$ then the quantity $\frac{1}{n}\log N(n)$ can be interpreted as the time-average of the complexity of the system during the time-interval $[0; n-1]$. Then $\limsup_{n\rightsquigarrow\infty} \frac{1}{n}\log N(n)$ is a time-independent measure of the complexity of the system.

Finally, we give a more precise description of 'the number of distinguishable orbits at time n'. Actually we shall encounter two definitions of this notion. The first one can be given for any system with a metric phase space, the second one for any system with a compact phase space. In this way we get two notions of 'topological entropy', one for systems on metric spaces and one for systems on compact spaces. For systems on a compact metric space these two notions coincide. We start with the approach for metric spaces. The approach for compact spaces will be postponed to Section 8.4.

Let us say that for given $\varepsilon > 0$ and $n \in \mathbb{N}$ the orbits of two points x and x' are '(n, ε)-different' whenever there exists $i \in \{0, 1, \dots, n-1\}$ such that the distance between the corresponding points $f^i(x)$ and $f^i(x')$ is at least ε. Thus, we distinguish only orbits up to accuracy ε, and we consider only initial parts of orbits (the first n elements). As to counting 'the number of distinguishable orbits', we can consider minimal sets of orbits such that every orbit is indistinguishable from at least one orbit from such a set (this will lead to the notion of *(n, ε)-spanning sets*) or we can consider maximal sets of orbits that can can be distinguished from each other (leading to the equivalent approach via *(n, ε)-separated sets*). The result thus obtained may depend on the value of ε (the 'resolution' of the observation of the orbits). Therefore, at the end we considers the supremum over ε of all values obtained. Then the entropy may depend on the metric used, but this turns out to be not be the case if the phase space is compact.

Notation. The number of points in a finite set E will be denoted by $\#E$. If E is not finite then $\#E := \infty$.

8.1.1. For every $n \in \mathbb{N}$ we define a mapping $d_n^f : X \times X \to \mathbb{R}^+$ by

$$d_n^f(x, y) := \max_{0 \le i \le n-1} d(f^i(x), f^i(y)) \quad \text{for } x, y \in X .$$

It is straightforward to check that this is a metric on X. We claim that, for every $n \in \mathbb{N}$, the metric d_n^f is equivalent with the original metric d on X, that is: d_n^f and d generate the same topology on X. In order to prove this, consider an arbitrary point $x \in X$. Then for every point $y \in X$ we obviously have $d(x, y) = d_0^f(x, y) \le d_n^f(x, y)$. Conversely, if $\varepsilon > 0$ and $n \in \mathbb{N}$ then the continuity of the functions $f^0, f, \dots f^{n-1}$ at the point x implies that there exists a number $\delta(x, n, \varepsilon) > 0$ such that for every point $y \in X$ with $d(x, y) < \delta(x, n, \varepsilon)$ one has $d(f^i(x), f^i(y)) < \varepsilon$ for all $i \in \{0, \dots, n-1\}$ It follows that $d_n^f(x, y) < \varepsilon$ if $d(x, y) < \delta(x, n, \varepsilon)$. So we have

$$B_{\delta(x,n,\varepsilon)}(x, d) \subseteq B_\varepsilon(x, d_n^f) \subseteq B_\varepsilon(x, d)$$

for every point $x \in X$. This shows that the metrics d en d_n^f are equivalent. It follows, among others, that a subset K of X that is compact in the original topology of X is also compact in the topology generated by d_n^f.

8.1.2. Let K be a non-empty compact subset of X. If $n \in \mathbb{N}$ and $\varepsilon > 0$ then we say that a subset S of X is an *(n, ε)-spanning set for K*, or that it *(n, ε)-spans K*, whenever

$$\forall x \in K \; \exists y \in S : \; d_n^f(x, y) < \varepsilon .$$

Stated otherwise: S (n, ε)-spans K whenever the open (d_n^f, ε)-balls $B_\varepsilon(s, d_n^f)$ with $s \in S$ cover K (note that we do not require that $S \subseteq K$). Since K is compact, there is a finite cover of K by open (d_n^f, ε)-balls (centred at points of K). Thus, K admits at least one finite (n, ε)-spanning set. Consequently, the following definition is meaningful:

$$\text{minspan}_n(\varepsilon, K, f) := \min\{ \#S : S \subseteq X \text{ is an } (n, \varepsilon)\text{-spanning set for } K \} .$$

In terms of the introductory remarks of this section: $\text{minspan}_n(\varepsilon, K, f)$ is the minimal number of points in X needed to describe all orbits starting in points of K (K need not be invariant) with (n, ε)-accuracy.

8.1.3. If $n \in \mathbb{N}$ and $\varepsilon > 0$ then a subset S of X is said to be *(n, ε)-separated* whenever all pairs of distinct points of S are spaced at least ε apart under d_n^f:

$$\forall s, s' \in S : s \ne s' \Rightarrow d_n^f(s, s') \ge \varepsilon .$$

Stated otherwise, no open (d_n^f, ε)-ball centred at a point of S contains other points of S.

If K is a non-empty compact subset of X then K has a finite cover by open $(d_n^f, \frac{1}{2}\varepsilon)$-balls. If S is an (n, ε)-separated subset of K then, by the triangle inequality, each of these balls contains at most one point of S. Thus, an (n, ε)-separated subset of K must be finite. Of course, the family of (n, ε)-separated subsets of K is not empty (every singleton subset of K is a member). This shows that the following definition is meaningful:

$$\mathrm{maxsep}_n(\varepsilon, K, f) := \max\{\, \#S \,\vdots\, S \subseteq K \mathrm{is} (n, \varepsilon)\text{-separated}\,\}\,.$$

In terms of the introductory remarks of this section: $\mathrm{maxsep}_n(\varepsilon, K, f)$ is the greatest number of (n, ε)-different orbits emanating from K.

Lemma 8.1.4. *Let K be a non-empty compact subset of X, let $n \in \mathbb{N}$ and let $\varepsilon > 0$.*
(1) *If $0 < \varepsilon' < \varepsilon$ then*[2]

$$\mathrm{minspan}_n(\varepsilon', K, f) \geq \mathrm{minspan}_n(\varepsilon, K, f)\,,$$
$$\mathrm{maxsep}_n(\varepsilon', K, f) \geq \mathrm{maxsep}_n(\varepsilon, K, f)\,.$$

(2) *Always:* $\mathrm{minspan}_n\,(\varepsilon, K, f) \leq \mathrm{maxsep}_n(\varepsilon, K, f) \leq \mathrm{minspan}_n\,(\frac{1}{2}\varepsilon, K, f)$.

Proof. (1) Take into account that every (n, ε')-spanning set for K is (n, ε)-spanning, and that every (n, ε)-separated set is (n, ε')-separated.
(2) The second inequality follows from the argument used in 8.1.2 above: the number $\mathrm{maxsep}_n(\varepsilon, K, f)$ is at most equal to $\#\mathcal{C}$ for any open cover $\mathcal{C}$ of K by open $(d_n^f, \frac{1}{2}\varepsilon)$-balls. However, by definition, there is such a cover that has precisely $\mathrm{minspan}_n(\frac{1}{2}\varepsilon, K, f)$ members.
In order to prove the first inequality, consider an (n, ε)-separated subset S of K with maximal cardinality $\mathrm{maxsep}_n(\varepsilon, K, f)$. We claim that S is an (n, ε)-spanning set for K: if this is true then, by definition, the first inequality holds. To prove the claim, consider an arbitrary point $x \in K$. Then the set $S \cup \{x\}$ cannot be (n, ε)-separated (S has maximal cardinality), hence there exists $s \in S$ such that $d_n^f(s, x) < \varepsilon$. This proves our claim. □

For every non-empty compact subset K of X and every $\varepsilon > 0$ we shall consider the exponential growth rate of the sequences $(\mathrm{minspan}_n(\varepsilon, K, f))_{n\in\mathbb{N}}$ and $(\mathrm{maxsep}_n(\varepsilon, K, f))_{n\in\mathbb{N}}$:

$$r(\varepsilon, K, f) := \limsup_{n\rightsquigarrow\infty} \frac{1}{n} \log\Big(\mathrm{minspan}_n(\varepsilon, K, f)\Big)\,,$$
$$s(\varepsilon, K, f) := \limsup_{n\rightsquigarrow\infty} \frac{1}{n} \log\Big(\mathrm{maxsep}_n(\varepsilon, K, f)\Big)\,.$$

The base of the logarithms does not really matter, provided we use always the same. In examples we shall assume that it is e.

2 Look sharper and you see more.

Lemma 8.1.5. *Let K be a non-empty compact subset of X.*

(1) *If $0 < \varepsilon' < \varepsilon$ then*

$$r(\varepsilon', K, f) \geq r(\varepsilon, K, f) \quad \textit{and} \quad s(\varepsilon', K, f) \geq s(\varepsilon, K, f)\,.$$

(2) *For all $\varepsilon > 0$ we have*

$$r(\varepsilon, K, f) \leq s(\varepsilon, K, f) \leq r(\tfrac{1}{2}\varepsilon, K, f)\,.$$

(3) $r(\varepsilon, K, f) \geq 0$ *for all* $\varepsilon > 0$.

Proof. Statements 1 and 2 are clear from the definitions and Lemma 8.1.4. Statement 3 follows from the trivial observation that $\mathrm{minspan}_n(\varepsilon, K, f) \geq 1$ for all $n \in \mathbb{N}$ and all $\varepsilon > 0$. □

8.1.6. Let K be a non-empty compact subset of X. It follows from Lemma 8.1.5 (1) that both $\lim_{\varepsilon \rightsquigarrow 0} r(\varepsilon, K, f)$ and $\lim_{\varepsilon \rightsquigarrow \infty} s(\varepsilon, K, f)$ exist (they need not be finite) and are equal to $\sup_{\varepsilon>0} r(\varepsilon, K, f)$ and $\sup_{\varepsilon>0} s(\varepsilon, K, f)$, respectively. Moreover, Lemma 8.1.5 (2) implies that these limits are equal to each other. This common limit will be denoted by $h(K, f)$:

$$h(K, f) := \lim_{\varepsilon \rightsquigarrow 0} r(\varepsilon, K, f) = \lim_{\varepsilon \rightsquigarrow 0} s(\varepsilon, K, f)\,.$$

It is called the *topological entropy of K with respect to f*. Note that, by Lemma 8.1.5, we have $h(K, f) \geq s(\varepsilon, K, f) \geq r(\varepsilon, K, f) \geq 0$ for all $\varepsilon > 0$.

The *topological entropy of the system* (X, f), also called the *topological entropy* (or just 'the entropy') *of the mapping* f, is the quantity

$$h(f) := \sup\{\, h(K, f) \,\vdots\, K \in \mathcal{K} \,\}\,,$$

where $\mathcal{K}$ is the set of all non-empty compact subsets of X. Since $h(K, f) \geq 0$ for all $K \in \mathcal{K}$, it follows that $h(f)$ is a non-negative real number or $+\infty$.

In general, the entropy depends on the metric d used on X (see the initial example in Section 8.2 below). If we want to stress this dependence of the entropy on the metric d we shall write $h_d(K, f)$ and $h_d(f)$.

As said before, the topological entropy is a measure for the dispersion of orbits. Consequently, it is to be expected that if the orbits do not diverge then the entropy is zero. In particular, it is immediately clear that, if f is an isometry or a *weak contraction* (which means, by definition, that $d(f(x), f(y)) \leq d(x, y)$ for all $x, y \in X$) then $d_n^f = d$ for all n. Hence for every $\varepsilon > 0$, $n \in \mathbb{N}$ and every non-empty compact subset K of X the number $\mathrm{minspan}_n(\varepsilon, K, f) = \mathrm{minspan}_1(\varepsilon, K, f)$ is independent of n and the sequence $(\mathrm{minspan}_n(\varepsilon, K, f))_{n\in\mathbb{N}}$ has exponential growth rate 0, i.e., $r(\varepsilon, K, f) = 0$. Therefore, $h(K, f) = 0$ as well. This holds for every compact subset K of X, so $h(f) = 0$.

If, on the contrary, for all points x, y in X the distance $d(f^n(x), f^n(y))$ grows with n then d_n^f will grow as well and, therefore, d_n^f-balls will shrink. As a consequence, for given $\varepsilon > 0$ and $K \subseteq X$ the quantities $\mathrm{minspan}_n(\varepsilon, K, f)$ and $\mathrm{maxsep}_n(\varepsilon, K, f)$ will grow with n. If their growth rate is sufficiently large then, at the end, $h(f)$ may be positive or infinite. This discussion also suggests that the precise value of $h(f)$ is not important: it only matters whether it is zero, positive or infinite. Note that this distinction is independent of the choice of the base of the logarithms in the definitions above.

Not surprisingly, the entropy does not shrink if we enlarge the space:

Proposition 8.1.7. *If Y is an invariant subset of X then $h(f|_Y) \leq h(f)$.*

Proof. If K is a non-empty compact subset of Y then K is compact in X as well, and it is obvious[3] that $\mathrm{maxsep}_n(\varepsilon, K, f|_Y) = \mathrm{maxsep}_n(\varepsilon, K, f)$ for all $n \in \mathbb{N}$ and every $\varepsilon > 0$. It follows that $h(K, f|_Y) = h(K, f)$. Since the set of compact subsets of Y is a subset of the set of compact subsets of X (those that are included in Y), the desired inequality follows easily. □

Lemma 8.1.8. *Let K_1 and K_2 be non-empty compact subsets of X such that $K_1 \subseteq K_2$. Then for all $n \in \mathbb{N}$ and $\varepsilon > 0$,*

$$\mathrm{minspan}_n(\varepsilon, K_1, f) \leq \mathrm{minspan}_n(\varepsilon, K_2, f)\,,$$
$$\mathrm{maxsep}_n(\varepsilon, K_1, f) \leq \mathrm{maxsep}_n(\varepsilon, K_2, f)\,.$$

Consequently, $r(\varepsilon, K_1, f) \leq r(\varepsilon, K_2, f)$, $s(\varepsilon, K_1, f) \leq s(\varepsilon, K_2, f)$ and $h(K_1, f) \leq h(K_2, f)$.

Proof. Straightforward application of the definitions. □

Corollary 8.1.9. *If X is compact then $h(f) = h(X, f)$.* □

Thus, in the case that the phase space X is a *compact* metric space, we do not need to compute $h(K, f)$ for compact subsets of X other than X itself:

$$\begin{aligned} h(f) &= \lim_{\varepsilon \rightsquigarrow 0} \limsup_{n \rightsquigarrow \infty} \frac{1}{n} \log \mathrm{minspan}_n(\varepsilon, X, f) \\ &= \lim_{\varepsilon \rightsquigarrow 0} \limsup_{n \rightsquigarrow \infty} \frac{1}{n} \log \mathrm{maxsep}_n(\varepsilon, X, f)\,. \end{aligned}$$

In Corollary 8.4.10 ahead we shall see that in these formulas "lim sup" may be replaced by "lim inf" if X is a compact metric space.

Examples.

(1) The Adding Machine (G, f) – see 4.2.8 – has entropy $h(f) = 0$. This follows from the discussion in 8.1.6, because the mapping f is an isometry with respect to the metric of G: see Exercise 5.2 (2). Similarly, every rigid rotation $(\mathbb{S}, \varphi_a)$ has entropy 0 ($a \in \mathbb{R}$).

3 The equality $\mathrm{minspan}_n(\varepsilon, K, f|_Y) = \mathrm{minspan}_n(\varepsilon, K, f)$ is not obvious at all: it may even be false (for small values of n and large values of ε), because (d_n^f, ε)-balls with centre in $X \setminus Y$ cannot be used for the computation of $\mathrm{minspan}_n(\varepsilon, K, f|_Y)$.

(2) Let $\mathbf{S}$ be a finite set with $s := \#\mathbf{S}$ elements, $s \geq 2$, and consider the full shift system $(\Omega_{\mathbf{S}}, \sigma)$. We shall show that $h(\sigma) = \log s$. By Corollary 8.1.9, we have to prove that $h(\Omega_{\mathbf{S}}, \sigma) = \log s$.
Recall that for every $n \in \mathbb{N}$ the cylinders based on the s^n different initial n-blocks of points of $\Omega_{\mathbf{S}}$ form a partition of $\Omega_{\mathbf{S}}$. Two points x and y in different cylinders differ in their initial n-blocks, hence there is $i \in \{0, \dots, n-1\}$ such that $d(\sigma^i x, \sigma^i y) = 1$, and therefore $d_n^\sigma(x, y) = 1 \geq \varepsilon$ for all $\varepsilon > 0$ with $\varepsilon \leq 1$. So if we select in each cylinder just one point we get an (n, ε)-separated set with cardinality s^n. This shows that $\text{maxsep}_n(\varepsilon, \Omega_{\mathbf{S}}, \sigma) \geq s^n$ (provided $\varepsilon \leq 1$). As the exponential growth rate of the sequence $(s^n)_{n\in\mathbb{N}}$ is equal to $\log s$, we get $s(\varepsilon, \Omega_{\mathbf{S}}, \sigma) \geq \log s$. Taking the limit for $\varepsilon \rightsquigarrow 0$ then gives $h(\Omega_{\mathbf{S}}, \sigma) \geq \log s$.
Next, let $n, k \in \mathbb{N}$ and consider the partition of $\Omega_{\mathbf{S}}$ in cylinders based on the initial $(n+k)$-blocks of points. Note that if the points x and y are situated in the same cylinder of this partition, then for $0 \leq i \leq n$ the points $\sigma^i x$ and $\sigma^i y$ agree in their initial k-blocks, so for these values of i their distance is at most $1/(k+1)$. Stated otherwise, $d_n^\sigma(x, y) \leq 1/(k+1)$ for two points x and y in the same cylinder. So if $0 < \varepsilon < 1$ and $k \in \mathbb{N}$ is such that $1/(k+1) < \varepsilon$, and we choose in each cylinder one point, we get an (n, ε)-spanning set in $\Omega_{\mathbf{S}}$ with s^{n+k} elements. It follows that $\text{minspan}_n(\varepsilon, \Omega_{\mathbf{S}}, \sigma) \leq s^{n+k}$. Taking into account that $\frac{1}{n}\log s^{n+k} = \frac{n+k}{n}\log s$ tends to $\log s$ for $n \rightsquigarrow \infty$, we find that $r(\varepsilon, \Omega_{\mathbf{S}}, \sigma) \leq \log s$. Taking the limit for $\varepsilon \rightsquigarrow 0$ now shows that $h(\Omega_{\mathbf{S}}, \sigma) \leq \log s$.

(3) Let X be a shift space and let, for every $n \in \mathbb{N}$, $\theta_n(X) := \#\mathcal{L}_n(X)$, the number of X-present n-blocks. Then

$$h(\sigma|_X) = \lim_{n\rightsquigarrow\infty} \frac{1}{n}\log\theta_n(X) = \lim_{n\rightsquigarrow\infty} \log\sqrt[n]{\theta_n(X)}\,. \tag{8.1-1}$$

First, we show that the limit in the right-hand side of (8.1-1) exists and is finite. This will be a consequence of the following inequality:

$$\forall\, m, n \in \mathbb{N}:\ \theta_{m+n}(X) \leq \theta_m(X)\,\theta_n(X)\,. \tag{8.1-2}$$

To prove (8.1-2), note that an X-present $(m+n)$-block consists of an X-present m-block followed by an X-present n-block. There are $\theta_m(X)\,\theta_n(X)$ possible combinations of such blocks, but it is possible that not every concatenation of words of length m and n from the language of X is again in the language of X, so this product may be larger than $\theta_{m+n}(X)$. This completes the proof of inequality (8.1-2). Obviously, this inequality implies

$$\log\theta_{m+n}(X) \leq \log\theta_m(X) + \log\theta_n(X)\,.$$

Now Lemma 8.1.10 below with $a_n := \log\theta_n(X)$ for $n \in \mathbb{N}$ implies that $\frac{1}{n}\log\theta_n(X)$ has a finite limit for $n \rightsquigarrow \infty$.
For every $n \in \mathbb{N}$ the shift space X has a partition in precisely $\theta_n(X)$ non-empty intersections of X with cylinders based on words of length n in the language of X

(take into account that every member of the language of X occurs as initial block in an element of X). Just as in the previous example, with $\theta_n(X)$ instead of s^n, we get $\text{maxsep}_n(\varepsilon, X, \sigma|_X) \geq \theta_n(X)$ (provided $\varepsilon \leq 1$), which results, finally, in

$$h(X, \sigma|_X) \geq \limsup_{n\rightsquigarrow\infty} \frac{1}{n} \log \theta_n(X) = \lim_{n\rightsquigarrow\infty} \frac{1}{n} \log \theta_n(X)$$

Similarly, by considering the partition of X in the intersections of X with cylinders based on words of length $(n + k)$ in the language of X we find that

$$\text{minspan}_n\left(\frac{1}{k}, X, \sigma|_X\right) \leq \theta_{n+k}(X) \leq \theta_n(X)\,\theta_k(X)\,.$$

Since the right-hand side has exponential growth rate $\lim_{n\rightsquigarrow\infty} \frac{1}{n} \log \theta_n(X)$ this eventually implies that $h(X, \sigma|_X) \leq \lim_{n\rightsquigarrow\infty} \frac{1}{n} \log \theta_n(X)$. This completes the proof.

(4) The topological entropy of the Golden Mean shift is equal to

$$\log\left(\frac{1+\sqrt{5}}{2}\right) = \log(1+\sqrt{5}) - \log 2 \approx 0.48121\ldots > 0$$

(in base e). The proof is an application of Example (3) above. Let $n \in \mathbb{N}$; in order to compute the quantity $\theta_n(X)$ in this case, let

$$\theta_n^0 := \text{the number of } X\text{-present } n\text{-blocks ending with } 0\,,$$
$$\theta_n^1 := \text{the number of } X\text{-present } n\text{-blocks ending with } 1\,.$$

Then $\theta_n(X) = \theta_n^0 + \theta_n^1$. Any $(n + 1)$-block can be considered as an initial n-block followed by the symbol 0 or 1. If it is a 0 then the initial n block must end with a 1, if it is a 1 then there is no restriction on the final symbol of the initial n-block. So $\theta_{n+1}^0 = \theta_n^1$ and $\theta_{n+1}^1 = \theta_n(X)$, whence

$$\theta_{n+1}(X) = \theta_{n+1}^0 + \theta_{n+1}^1 = \theta_n^1 + \theta_n(X) = \theta_{n-1}(X) + \theta_n(X)\,.$$

Because $\theta_1(X) = 2$ and $\theta_2(X) = 3$, it follows that the sequence $(\theta_n(X))_{n\in\mathbb{N}}$ is just the well-known Fibonacci sequence. For this sequence it is known that

$$\theta_n(X) = \frac{(1+\sqrt{5})^n - (1-\sqrt{5})^n}{2^n\sqrt{5}}\,.$$

By elementary calculus, this implies that $\lim_{n\rightsquigarrow\infty} \sqrt[n]{\theta_n(X)} = \frac{1}{2}(1+\sqrt{5})$. This proves our claim.

(5) Consider the recurrent point $x \in \Omega_2$ defined in 5.6.1 (3):

$$x := 010\,111\,010\;\;111\,111\,111\;\;010\,111\,010\ldots\,.$$

Recall that the initial n-blocks $b^{(n)}$ of x for $n \in \mathbb{N}$ are inductively given by $b^{(0)} = 0$ and $b^{(n+1)} = b^{(n)} 1^{b^{(n)}} b^{(n)}$. Let X be the orbit closure of x in Ω_2.

In order to estimate $\theta_n(X)$, first recall that, according to Exercise 5.3 (3), the language of X coincides with the set of all finite blocks occurring in x. Now take into account that, for every $k \in \mathbb{N}$, the sequence x consists of blocks $b^{(k)}$, separated by blocks of 1's of length $|b^{(k)}|$ or more. Thus, a block of length $|b^{(k)}|$ in x can coincide with an occurrence of $b^{(k)}$ in x, it can consist of a block of 1's followed by an initial block of $b^{(k)}$, it can start with a final block of $b^{(k)}$ followed by a block of 1's, or it is a block of only 1's. Consequently, there occur at most $2 \cdot |b^{(k)}|$ different blocks of length $|b^{(k)}|$ in x. Next, if $n \in \mathbb{N}$ then any subblock of x of length n occurs as the initial block of a subblock of length $|b^{(k)}|$ of x for all $k \in \mathbb{N}$ such that $|b^{(k)}| = 3^k \geq n$; this block is uniquely determined by the n-block if we choose for k the smallest value satisfying this inequality, that is, if $3^{k-1} < n \leq 3^k$. Since that subblock of x of length $|b^{(k)}|$ (for this value of k) starting with the n-subblock under consideration is one of a collection of at most $2 \cdot |b^{(k)}| = 2 \cdot 3^k$ different blocks, there cannot be more than $2 \cdot 3^k$ different n-subblocks of x. Since $3^{k-1} < n$, this implies that $\theta_n(X) \leq 6n$. As $\frac{1}{n} \log n \rightsquigarrow 0$ if n tends to infinity, it follows that $h(\sigma|_X) = 0$.

NB. The system considered in Example (5) above is transitive (it is the orbit closure of a recurrent point) and it is sensitive (it is expansive), so it is AY-chaotic. Nevertheless, it has entropy 0. For other examples of transitive (even minimal) subshifts with entropy 0, see Example (4) after Theorem 8.2.7 and Exercise 8.2 (1).

Lemma 8.1.10. *Let $(a_n)_{n\in\mathbb{N}}$ be a sequence of positive real numbers and assume that*

$$\forall\, m, n \in \mathbb{N} : \ a_{m+n} \leq a_m + a_n \,. \tag{8.1-3}$$

Then $\lim_{n\rightsquigarrow\infty} \frac{a_n}{n}$ exists, and is equal to $\inf_{n\in\mathbb{N}} \frac{a_n}{n}$.

Proof. Let $\alpha := \inf_{n\in\mathbb{N}} \frac{a_n}{n}$. For every $\varepsilon > 0$ there exists $k \in \mathbb{N}$ such that

$$\frac{a_k}{k} < \alpha + \frac{1}{2}\varepsilon \,.$$

For every j with $0 \leq j \leq k-1$ and every $m \in \mathbb{Z}^+$ it follows that

$$\begin{aligned} \frac{a_{mk+j}}{mk+j} &\overset{(1)}{\leq} \frac{a_{mk}}{mk+j} + \frac{a_j}{mk+j} \leq \frac{a_{mk}}{mk} + \frac{a_j}{mk} \\ &\overset{(2)}{\leq} \frac{ma_k}{mk} + \frac{ja_1}{mk} \leq \frac{a_k}{k} + \frac{a_1}{m} < \alpha + \frac{1}{2}\varepsilon + \frac{a_1}{m} \,, \end{aligned}$$

where for the inequalities (1) and (2) we have used (8.1-3). If $n = mk + j$ is sufficiently large (keeping k fixed, so equivalently: if m is sufficiently large) then $\frac{a_1}{m} < \frac{\varepsilon}{2}$, hence $\alpha \leq \frac{a_n}{n} < \alpha + \varepsilon$. □

Proposition 8.1.11. $\forall\, m \in \mathbb{N} : \ h(f^m) \leq mh(f)$.

Proof. Let K be a non-empty compact subset of X and let $\varepsilon > 0$. Then for every $n \in \mathbb{N}$ and all points $x, y \in X$,

$$\max_{0\leq i\leq n-1} d(f^{mi}(x), f^{mi}(y)) \leq \max_{0\leq j\leq mn-1} d(f^j(x), f^j(y)) \,.$$

It follows that every (n, ε)-separated subset E of K under f^m, in particular, the one with maximal cardinality, is an (mn, ε)-separated subset of K under f (but in that capacity possibly not of maximal cardinality). Consequently, $\text{maxsep}_n(\varepsilon, K, f^m) \leq \text{maxsep}_{mn}(\varepsilon, K, f)$ and it follows that

$$\begin{aligned}\frac{1}{n}\log \text{maxsep}_n(\varepsilon, K, f^m) &\leq \frac{1}{n}\log \text{maxsep}_{mn}(\varepsilon, K, f)\\ &= m\left(\frac{1}{mn}\log \text{maxsep}_{mn}(\varepsilon, K, f)\right).\end{aligned}$$

Take the lim sup for $n \rightsquigarrow \infty$ and take into account that, for any sequence $(c_n)_{n\in\mathbb{N}}$ of real numbers, $\limsup_{n\rightsquigarrow\infty} c_{mn} \leq \limsup_{k\rightsquigarrow\infty} c_k$. Therefore, the lim sup of the right hand side is at most $ms(\varepsilon, K, f)$. After taking the limit for $\varepsilon \rightsquigarrow 0$ we get $h(K, f^m) \leq mh(K, f)$. Finally, take the supremum over all non-empty compact subsets of X. □

Remark. In the above, X is only assumed to be a metric space. In Proposition 8.5.2 below it will be shown that if, in addition, X is compact then $h(f^m) = mh(f)$.

8.2 Independence of the metric; factor maps

The next example shows that, in general, the entropy of a mapping may depend on the metric of the phase space under consideration. In particular, it follows that topological entropy is not a dynamical property.

Example. Let $X := (0;\infty)$ and let $f(x) := 2x$ for $x \in X$. Moreover, let ρ be the usual metric on X. For $x, y \in K := [1;2]$ one readily shows that $\rho_n^f(x, y) = 2^n|x - y|$, so if $\varepsilon > 0$ then an (n, ε)-spanning set for K has at least $2^n/2\varepsilon$ points. Consequently, $\text{minspan}_n(\varepsilon, F, f) \geq 2^{n-1}/\varepsilon$. It follows that $h_\rho(f) \geq h(K, f) \geq \log 2$.

Next, consider the mapping $d:(x, y) \mapsto |\log x - \log y| : X \times X \to \mathbb{R}^+$. It is straightforward to check that this is a metric which generates the Euclidean topology on X and that f is an isometry with respect to d. By the discussion in 8.1.6, the entropy of X with respect to that metric is 0, hence different from $h_\rho(f)$.

It will turn out that on compact metric spaces the entropy does not depend on the metric used.

Lemma 8.2.1. *Let $\varphi : (X, f) \to (Y, g)$ be a factor mapping and assume that the phase spaces X and Y are metrizable with metrics d and ρ, respectively. In addition, assume that φ is uniformly continuous with respect to these metrics, and that every compact subset of Y is image of a compact subset of X. Then is $h_\rho(g) \leq h_d(f)$.*

Proof. Let $\varepsilon > 0$, let $n \in \mathbb{N}$ and let K be a non-empty compact subset of Y. Let E be an (n, ε)-separated subset of K with cardinality $\text{maxsep}_n(\varepsilon, K, g)$. By the assumptions on φ there is a $\delta > 0$ such that

$$\forall x_1, x_2 \in X:\ d(x_1, x_2) < \delta \Rightarrow \rho(\varphi(x_1), \varphi(x_2)) < \varepsilon. \tag{8.2-1}$$

In addition, there is a compact subset K' of X such that $\varphi[K'] = K$. For every $y \in E$, select one point in the fibre $\varphi^{\leftarrow}[y] \cap K'$. In this way we get a subset E' of K' with cardinality $\text{maxsep}_n(\varepsilon, K, g)$. It is straightforward to see that E' is an (n, δ)-separated set: if it is not then there are $x', x'' \in E'$ such that $d(f^i(x'), f^i(x'')) < \delta$ for $i = 0, \dots, n-1$. Using (8.2-1), taking into account that $\varphi \circ f^i = g^i \circ \varphi$ for all i, we get $\rho(g^i(\varphi(x')), g^i(\varphi(x''))) < \varepsilon$ for $i = 0, \dots, n-1$. This contradicts the choice of E as an (n, ε)-separated set, so the subset E' of K' is, indeed, (n, δ)-separated.

It follows that $\text{maxsep}_n(\delta, K', f) \geq \#E' = \text{maxsep}_n(\varepsilon, K, g)$. This implies that $s(\delta, K', f) \geq s(\varepsilon, K, g)$. Since $h_d(f) \geq h_d(K', f) \geq s(\delta, K', f)$ this, in turn, implies that $h_d(f) \geq s(\varepsilon, K, g)$. Now take the limit of the right-hand side for $\varepsilon \rightsquigarrow 0$; we get $h_d(f) \geq h_\rho(K, g)$. This holds for every non-empty compact subset K of Y, hence $h_d(f) \geq h_\rho(g)$. □

Examples.

(1) In the example preceding Lemma 8.2.1 the factor mapping $\text{id}_X : ((X, d), f) \to ((X, \rho), f)$ satisfies only one of the two conditions of Lemma 8.2.1: the preimage of a compact set is compact (obvious, as the two metrics generate the same topology), but the mapping $\text{id}_X : (X, d) \to (X, \rho)$ is not uniformly continuous. And, indeed, we have seen in the example that $h_\rho(f) = \log 2 \not\leq 0 = h_d(f)$. As $\text{id}_X : (X, f) \to (X, f)$ is a conjugation, this example also shows that topological entropy is not a dynamical property.

(2) Let (X, f) be a dynamical system on a compact metric space (X, ρ) and assume that $h_\rho(f) > 0$. As ρ is bounded on $X \times X$, the mapping

$$d : (x, y) \mapsto \sup\{\rho(f^n(x), f^n(y)) : n \in \mathbb{Z}^+\} : X \times X \to \mathbb{R}^+$$

is well-defined. It is easy to show that d is a metric on X and that $\rho(x, y) \leq d(x, y)$ for all $x, y \in X$. It follows that the mapping $\varphi := \text{id}_X : (X, d) \to (X, \rho)$ is uniformly continuous. As $d(f(x), f(y)) \leq d(x, y)$ for all $x, y \in X$, it is clear that f is continuous with respect to the topology generated by d and that $h_d(f) = 0$ in view of the discussion in 8.1.6. Thus, we have a uniformly continuous factor map, but the conclusion of Lemma 8.2.1 does not hold. Apparently, the second condition on φ mentioned in Lemma 8.2.1 is not fulfilled: not every compact set in (X, ρ) is (the image under φ of) a compact set in (X, d).

NB. The metric d introduced above will be used in the proof of Corollary 8.2.6 (2) below (with the roles of ρ and d interchanged).

Corollary 8.2.2. *Let (X, f) be a dynamical system on a metric phase space (X, d) and let ρ be a metric on X which is uniformly equivalent*[4] *with d. Then $h_d(f) = h_\rho(f)$.*

Proof. Clear from Lemma 8.2.1. □

4 This means that both $\text{id}_X : (X, d) \to (X, \rho)$ and $\text{id}_X : (X, \rho) \to (X, d)$ are uniformly continuous.

Corollary 8.2.3. *The topological entropy of a dynamical system on a compact metric phase space is independent of the metric used (provided it is compatible with the topology).* □

Corollary 8.2.4. *Let (X, f) and (Y, g) be dynamical systems on compact metric spaces (X, d) and (Y, ρ), and let $\varphi\colon (X, f) \to (Y, g)$ be a factor mapping. Then $h_\rho(g) \le h_d(f)$.*

Proof. Clear from Lemma 8.2.1. □

Corollary 8.2.5. *Conjugate dynamical systems on compact metric phase spaces have the same topological entropy. Thus, topological entropy is a dynamical property on the class of compact metric systems.* □

Corollary 8.2.6. *Let (X, f) be a dynamical system on a compact metric space.*

(1) *If f is semi-open and expansive then $h(f)$ is finite. In particular, an expansive homeomorphism on a compact metric space has finite entropy.*
(2) *If $\{f^n\}_{n\in\mathbb{Z}^+}$ is equicontinuous on X then $h(f) = 0$.*

Proof. (1) By Corollary 6.2.5, the system (X, f) is a factor of a shift system (Z, σ_Z). Hence $h(f) \le h(\sigma_Z) < \infty$ (recall from Proposition 8.1.7 and Example (2) after Corollary 8.1.9 that every subshift has finite entropy).
(2) Define a mapping $\rho\colon X \times X \to \mathbb{R}^+$ by

$$\rho(x, y) := \sup\left\{d(f^n(x), f^n(y)) \,\vdots\, n \in \mathbb{Z}^+\right\} \quad \text{for } x, y \in X\,.$$

Because d is continuous, hence bounded on $X \times X$, it is clear that ρ is finite on $X \times X$. Moreover, is easily seen that ρ is a metric on X.
Since X is compact, the set of mappings $\{f^n\}_{n\in\mathbb{Z}^+}$ is *uniformly* equicontinuous on X. It is straightforward to show that this implies that the mapping $\mathrm{id}_X : (X, d) \to (X, \rho)$ is (uniformly) continuous. As X is compact, this implies that this mapping is a homeomorphism; so the metrics d en ρ generate the same topology. Consequently, $h_d(f) = h_\rho(f)$ by Corollary 8.2.3. But f is easily seen to be a weak contraction with respect to the metric ρ, so the discussion in 8.1.6 implies that $h_\rho(f) = 0$. It follows that $h_d(f) = 0$ as well. □

Remark. The proof of part 2 of this Corollary shows: *for an equicontinuous system (X, f) on a compact metric space X there is a compatible metric on X with respect to which f is a weak contraction.* If, in addition, f is a homeomorphism then there is a compatible metric making f an isometry (replace $\mathbb{Z}^+$ by $\mathbb{Z}$ in the definition of ρ).

The following result extends Corollary 8.2.4. It is a special case of a more general result: see Note 3 at the end of this chapter.

Theorem 8.2.7. *Let $\varphi : (X, f) \to (Y, g)$ be a factor map of dynamical systems with compact metric phase spaces (X, d) and (Y, ρ). Assume that the mapping φ is boundedly finite-to-one: there exists $N > 0$ such that every fibre $\varphi^{\leftarrow}[y]$ for $y \in Y$ has at most N points. Then $h_d(f) = h_\rho(g)$.*

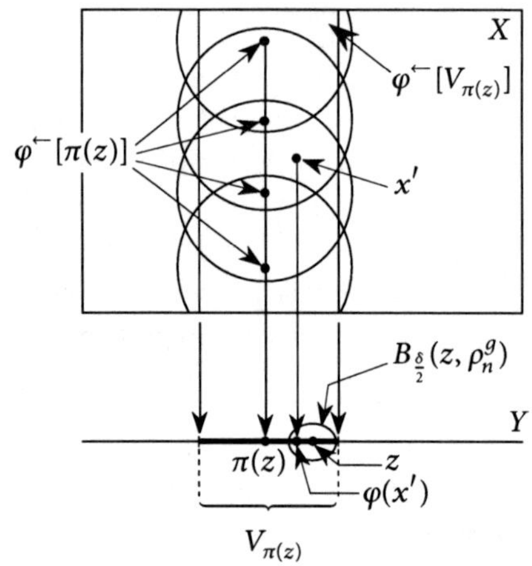

Fig. 8.1. Illustrating the proof of Theorem 8.2.7 with $y = \pi(z)$.

Proof. In view of Corollary 8.2.4 it remains to show that $h_d(f) \geq h_\rho(g)$. Let $\varepsilon > 0$ and $n \in \mathbb{N}$. By Lemma A.3.3 in Appendix A, for every $y \in Y$ there exists an open neighbourhood V_y of y in Y such that

$$\varphi^{\leftarrow}[V_y] \subseteq \bigcup_{x \in \varphi^{\leftarrow}[y]} B_\varepsilon(x, d_n^f) =: U_y . \tag{8.2-2}$$

See also Figure 8.1 with $y := \pi(z)$.

The open sets V_y with $y \in Y$ cover the compact space Y, hence there is a finite subset P of Y such that $\{V_y : y \in P\}$ is a covering of Y. Let δ be a Lebesgue number of this covering with respect to the metric ρ_n^g on Y (see Appendix A.10.1 for the definition).

Now let S be an $(n, \frac{1}{2}\delta)$-spanning set for Y with minimal cardinality, which means that $\#S = \text{minspan}_n(\frac{1}{2}\delta, Y, g)$. By the definition of δ it is possible to select for every point $z \in S$ a point $\pi(z) \in P$ such that the open $(\rho_n^g, \frac{1}{2}\delta)$-ball centred at z is included in $V_{\pi(z)}$. Let $S' := \varphi^{\leftarrow}[\pi[S]]$. Then S' is a subset of X with at most $N \cdot (\#\pi[S])$ points, so $\#S' \leq N \cdot \text{minspan}_n(\frac{1}{2}\delta, Y, g)$. We claim that S' is an (n, ε)-span of X. If this is true, then

$$\text{minspan}_n(\varepsilon, X, f) \leq \#S' \leq N_S \cdot \text{minspan}_n(\tfrac{1}{2}\delta, Y, g) ,$$

which implies that $r(\varepsilon, X, f) \leq r(\frac{1}{2}\delta, Y, g) \leq h(Y, g) = h(g)$. As this holds for every $\varepsilon > 0$, we get $h(X, f) \leq h(g)$, which is just what we want to prove.

It remains to show that S' is an (n, ε)-span of X. Consider an arbitrary point $x' \in X$. Then the point $\varphi(x')$ is situated in one of the open $(\rho_n^g, \frac{1}{2}\delta)$-balls centred at a point of S, say, centred at $z \in S$. As that open ball is included in $V_{\pi(z)}$, it follows that $\varphi(x') \in V_{\pi(z)}$, hence (see also Figure 8.1)

$$x' \in \varphi^{\leftarrow}[V_{\pi(z)}] \subseteq U_{\pi(z)} = \bigcup_{x \in \varphi^{\leftarrow}[\pi(z)]} B_\varepsilon(x, d_n^f) \subseteq \bigcup_{x \in S'} B_\varepsilon(x, d_n^f) .$$

Thus, the open (d_n^f, ε)-balls centred at the points of S' cover X. □

Examples.

(1) The argument-doubling system $(\mathbb{S}, \psi)$ is a factor of the full shift system (Ω_2, σ) under a factor map with fibres that each consist of at most two points: see the Example in Corollary 6.1.12 and note that the conclusion of Proposition 6.3.4 (2) also holds in this case. Using Example (2) after Corollary 8.1.9 above one finds $h(\psi) = \log 2$.

(2) The tent map is a factor of the full shift system (Ω_2, σ) under a factor map with fibres consisting of at most two points: see Example E in 6.3.5, taking into account Proposition 6.3.4 (2). Hence the tent map has entropy log 2. The quadratic mapping $f_4 : [0; 1] \to [0; 1]$, which is conjugate to the tent map, has entropy log 2 as well.

(3) Similarly, the system in Example F of 6.3.5 has the same entropy as the golden mean shift, which is, by Example (4) after Corollary 8.1.9, equal to $\log(1+\sqrt{5}) - \log 2$.

(4) The semi-Sturmian systems, defined in 6.3.7, have entropy equal to the entropy of their factor $(\mathbb{S}, \varphi_a)$, which is zero.

8.3 Maps on intervals and circles

As before, the length of a bounded interval J in $\mathbb{R}$ will be denoted by $|J|$.

Lemma 8.3.1. *Let X be an interval in $\mathbb{R}$ (not necessarily bounded, open or closed), let $f : X \to X$ be an injective continuous mapping and assume that for every compact subinterval J of X and every $n \in \mathbb{N}$ there is positive real number $L(J, n)$ such that*

(1) $|f^i[J]| \le L(J, n)$ for $i = 0, \dots, n-1$, and

(2) $\limsup_{n\rightsquigarrow\infty} \frac{1}{n} \log L(J, n) = 0$.

Then $h(f) = 0$.

Proof. Let $n \in \mathbb{N}$, let $\varepsilon > 0$ and let K be a non-empty compact subset of X. Select a compact subinterval J of X containing K. In addition, select for every $i \in \{0, \dots, n-1\}$ finitely many points in the interval $f^i[J]$ containing the end points of this interval and such that the distance between successive points is strictly less than ε. Let S_i be the set of these points minus the end points of $f^i[J]$. It is clear that S_i can be chosen such that $\#S_i \le |f^i[J]|/\varepsilon \le L(J, n)/\varepsilon$. Now let $S := \bigcup_{i=0}^{n-1} (f^i)^{\leftarrow}[S_i]$. Because for every $i \in \mathbb{Z}^+$ the mapping f^i is injective, the set S has at most $nL(J, n)/\varepsilon$ points. We claim that S is an (n, ε)-span of J (hence of K). If this claim is true then $\text{minspan}_n(\varepsilon, K, f) \le \#S \le nL(J, n)/\varepsilon$ and

$$r(\varepsilon, K, f) \le \limsup_{n\rightsquigarrow\infty} \frac{1}{n}\Big(\log n + \log L(J, n) - \log \varepsilon\Big).$$

The right-hand side of the above inequality is 0, hence $h(K, f) = 0$ for every non-empty compact subset K of X. This implies that $h(f) = 0$ as well.

It remains to show that S is an (n, ε)-span of J. Let $x \in J$ and let y be a point in S with minimal distance to x (there are at most two candidates: select one of them). We shall show that

$$d_n^f(x, y) := \max\{|f^i(x) - f^i(y)| : 0 \le i \le n-1\} < \varepsilon\,,$$

which will complete the proof. Suppose that there is an $i \in \{0, \ldots, n-1\}$ such that $|f^i(x) - f^i(y)| \ge \varepsilon$. Since the points of S_i are spaced less than ε apart, and both points $f^i(x)$ and $f^i(y)$ are in the interval $f^i[J]$ (recall that not only $x \in J$ but also $y \in S \subseteq J$), there is a point $z \in S_i$ situated strictly between $f^i(x)$ and $f^i(y)$. Since the mapping f^i is monotonous, it follows that the point $(f^i)^{-1}(z)$, which is in $(f^i)^{\leftarrow}[S_i] \subseteq S$, is strictly between the points x and y. This contradicts the choice of y as a point in S with minimal distance to x. □

Corollary 8.3.2. *Let f be an injective mapping of a bounded interval into itself. Then $h(f) = 0$. In particular, the topological entropy of a homeomorphism of a compact interval onto itself is zero.*

Proof. Apply Lemma 8.3.1 with $L(J, n)$ equal to the length of the interval on which f is defined, which is independent of $n \in \mathbb{N}$. (Alternative proof: by Theorem 8.6.6 ahead, if $h(f) > 0$ then there exists an $n \in \mathbb{N}$ such that f^n has a horseshoe. This contradicts injectivity of f.) □

Example. Examples (2) and (3) at the end of Section 8.2 show that the entropy need not be zero if f is not a homeomorphism. The initial example in Section 8.2 shows that boundedness cannot be left out from the conditions in Corollary 8.3.2: there we consider the homeomorphism $f : x \mapsto 2x : (0; \infty) \to (0; \infty)$, which has $h(f) \ge \log 2 > 0$.
NB. The method used in the proof of Lemma 8.3.1 can easily be applied to this example: take $L(J, n) := 2^n|J|$ and find for every $n \in \mathbb{N}$, $\varepsilon > 0$ and any compact subinterval J of $(0; \infty)$ that $\text{minspan}_n(\varepsilon, J, f) \le n2^n|J|/\varepsilon$. Hence $h(J, f) \le \log 2$. This, in turn, implies that $h(f) \le \log 2$. Conclusion: $h(f) = \log 2$.

Theorem 8.3.3. *Let $f : \mathbb{S} \to \mathbb{S}$ be a homeomorphism. Then $h(f) = 0$.*

Proof. The proof of Lemma 8.3.1 will be adapted so as to work in the context of the system $(\mathbb{S}, f)$, replacing 'length of an interval' by 'length of an arc'. Thus, let $n \in \mathbb{N}$ and $\varepsilon > 0$. In order to avoid ambiguities we assume that $\varepsilon \le \pi/2$ (a quarter of the size of the circumference of $\mathbb{S}$). Select a finite subset S_ε of $\mathbb{S}$ such that the distance between successive points of S_ε is strictly less than ε in the usual 'arc-length' metric d_c of $\mathbb{S}$. It is clear that S_ε can be chosen such that $\#S_i \le 2\pi/\varepsilon + 1 \le 3\pi/\varepsilon$. Now let $S^{(n)} := \bigcup_{i=0}^{n-1} (f^i)^{\leftarrow}[S_\varepsilon]$. Obviously, the set $S^{(n)}$ has at most $3n\pi/\varepsilon$ points, because the mappings f^i for $0 \le i \le n-1$ are injective.

We claim that $S^{(n)}$ is an (n, ε)-span of $\mathbb{S}$. If this claim is true then it follows that $\text{minspan}_n(\varepsilon, \mathbb{S}, f) \le \#S^{(n)} \le 3n\pi/\varepsilon$, so that

$$r(\varepsilon, \mathbb{S}, f) \le \limsup_{n \rightsquigarrow \infty} \frac{1}{n}(\log n + \log 3\pi - \log \varepsilon) = 0\,.$$

Consequently, $h(f) = h(\mathbb{S}, f) = 0$.

It remains to show that $S^{(n)}$ is an (n, ε)-span of $\mathbb{S}$. Let $x \in \mathbb{S}$ and let y be a point in $S^{(n)}$ with minimal distance to x (there are at most two candidates: select one of them). Let J be the shortest arc in $\mathbb{S}$ with end points x and y. We shall show that

$$(d_c)_n^f(x, y) := \max\{d_c(f^i(x), f^i(y)) : 0 \le i \le n-1\} < \varepsilon,$$

which will complete the proof. Suppose there is an $i \in \{0, \dots, n-1\}$ such that $d_c(f^i(x), f^i(y)) \ge \varepsilon$, that is, the shortest arc in $\mathbb{S}$ with $f^i(x)$ and $f^i(y)$ as end points has length at least ε. Consequently, the arc $f^i[J]$ – which is one of the two arcs in $\mathbb{S}$ connecting $f^i(x)$ and $f^i(y)$ – certainly has length at least ε. Since the points of S_ε are spaced less than ε apart, there must be a point $z \in S_\varepsilon$ situated on the arc $f^i[J]$, lying strictly between the points $f^i(x)$ and $f^i(y)$. Since the mapping f^i is injective, it follows that the point $(f^i)^{-1}(z)$, which is in $(f^i)^{\leftarrow}[S_\varepsilon] \subseteq S^{(n)}$, is situated in the arc J strictly between the points x and y. This contradicts the choice of y as a point in $S^{(n)}$ with minimal distance to x. □

8.3.4 (The generalized tent map). This mapping was defined in 6.3.8, as follows: for $0 \le s \le 2$, let

$$T_s(x) := \begin{cases} sx & \text{for } 0 \le x \le \frac{1}{2}, \\ s(1-x) & \text{for } \frac{1}{2} \le x \le 1. \end{cases}$$

If $0 < s \le 1$ then T_s is a weak contraction, hence $h(T_s) = 0$ by the discussion in 8.1.6. Next, we consider the case that $1 < s \le 2$. It follows from 6.3.14 that for these values of s the system $([0;1], T_s)$ is a factor of some shift system (Z, σ_Z) over the symbol set $\{0, 1\}$ under an at most 2-to-1 factor map. So Theorem 8.2.7 implies that $h(T_s) = h(\sigma_Z)$, and by Example (3) after Corollary 8.1.9 one gets

$$h(\sigma_Z) = \lim_{k \rightsquigarrow \infty} \log \sqrt[k]{\theta_k(Z)},$$

where $\theta_k(Z)$ is the number of Z-present k-blocks. It follows from Lemma 6.3.12 that $\theta_k(Z)$ is equal to the number of maximal intervals of monotonicity of T_s^k. Taking into account that each such interval has length at most s^{-k} – see the proof of Corollary 6.3.13 – it follows that there are at least s^k of such intervals. Consequently, $\theta_k(Z) \ge s^k$, hence $h(T_s) = h(\sigma|_Z) \ge \log s$.

In order to prove the reverse inequality, consider for every $n \in \mathbb{N}$ and every $\varepsilon > 0$ a suitable (n, ε)-spanning subset of $[0;1]$, as follows: There is a smallest number of points $0 =: x_0 < x_1 < \cdots < x_p := 1$ in $[0;1]$ such that $|x_i - x_{i-1}| < \varepsilon/s^n$ for all $i \in \{1, \dots, p\}$. Then $s^n/\varepsilon < p \le s^n/\varepsilon + 1$. We claim that the set $S := \{x_1, \dots, x_{p-1}\}$ is an (n, ε)-span of $[0;1]$. If this is true, then $\mathrm{minspan}_n(\varepsilon, [0;1], T_s) \le \#S = p - 1 \le s^n/\varepsilon$. This implies that $r(\varepsilon, [0;1], T_s) \le \log s$, hence $h(T_s) \le \log s$.

It remains to prove the above claim, namely, that S is an (n, ε)-span of $[0;1]$. Observe that, for all $i \in \{0, \dots, n\}$ and all pairs of points x and y in $[0;1]$ one has

$|T_s^i(x) - T_s^i(y)| \leq s^i|x - y| \leq s^n|x - y|$ (recall that $s > 1$). For if x and y are in the same maximal interval of monotonicity of T_s^i then this inequality follows from the mean value theorem, taking into account that the absolute value of the derivative of T_s^i on such an interval is s^i. And if x and y are not in the same maximal interval of monotonicity of T_s^i then this inequality is easily seen to hold as well. Since any point $x \in [0; 1]$ has a distance of less than ε/s^n to one of the points of S it follows easily that our claim is true.

Conclusion. *If* $1 < s \leq 2$ *then* $h(T_s) = \log s$.

8.4 The definition with covers

In this section we present an alternative definition of 'topological entropy' and we show that for systems on compact metric spaces this definition agrees with the one given in Section 8.1. In what follows, X will be an arbitrary Hausdorff space; metrizability or compactness of X is not (yet) assumed.

The idea of this definition is the same as explained in the beginning of Section 8.1: topological entropy measures the exponential growth rate of the number of distinguishable orbits as time advances. But now we use another method to distinguish orbits from each other.

Let $\mathfrak{U} = \{U_0, \ldots, U_{s-1}\}$ be an open cover of X ($s \geq 2$). Then one can try to keep track of points of X by looking at their partial itineraries with respect to this cover. In fact, we consider the collection of all non-empty sets of the form $U_{i_0} \cap f^{\leftarrow}[U_{i_1}] \cap \cdots \cap (f^{n-1})^{\leftarrow}[U_{i_{n-1}}]$ with $n \in \mathbb{N}$ and $i_j \in \{0, \ldots, s-1\}$ for $j = 0, \ldots, n-1$ (recall that a point x is in such a set iff $f^j(x) \in U_{i_j}$ for all j under consideration, which means that the sequence $(U_{i_0}, \ldots, U_{i_{n-1}})$ is a partial itinerary of the point x). Then the partial orbits (of length n) of two points will be said to be distinguishable (with respect to this cover of X) whenever they belong to different members of this collection. In that case we shall also say that the orbits of those points are distinguishable at time n.

As an example, consider the 2-shift (Ω_2, σ) with the natural partition $\{C_0[0], C_0[1]\}$ of Ω_2 – which is an open cover. It is easily seen that the orbits of the points x and y in Ω_2 are distinguishable at time n iff the initial n-blocks of x and y are different.

As before, if $N(n)$ is the number of distinguishable orbits at time n then the real number $\log N(n)$ is a measure for the complexity of the system and the quantity $\frac{1}{n} \log N(n)$ is the time-average of that complexity.

There is another argument why $N(n)$ measures the complexity of the system: the number of sets of the form $U_{i_0} \cap f^{\leftarrow}[U_{i_1}] \cap \cdots \cap (f^{n-1})^{\leftarrow}[U_{i_{n-1}}]$ needed to cover X represents the number of words of length n needed to encode the points of X according to their behaviour under the first $n-1$ iterates of f. (Of course, we will be interested in the minimum of that number, that is why the definition preceding Lemma 8.4.2 below involves the minimal size of a subcover.) The more words are needed, the more complex the system is.

In order to get a time-independent measure of the complexity we consider, as before, $\limsup_{n\rightsquigarrow\infty}\frac{1}{n}\log N(n)$. The result so obtained depends on the cover of X used – the 'resolution' of our observations – so we have to consider all possible open covers of X and take the supremum of the outcomes. The formalism developed below reflects these ideas.

Recall that a cover of a space is a collection of *non-empty* subsets of that space whose union is all of the space. Traditionally, the following definitions are given only for open covers of compact spaces, but they are useful for other purposes as well: therefore, we shall consider *special covers* of X, i.e., covers having a finite subcover. Note that *if X is compact then every open cover of X is special.*

8.4.1. A cover $\mathcal{B}$ of X is said to be a *refinement* of a cover $\mathcal{A}$ of X whenever every member of $\mathcal{B}$ is included in a member of $\mathcal{A}$:

$$\forall B \in \mathcal{B}\ \exists A \in \mathcal{A} \,\vdots\, B \subseteq A\,.$$

Notation: $\mathcal{A} < \mathcal{B}$. In that case, $\mathcal{B}$ is said to be *finer than* $\mathcal{A}$ and $\mathcal{A}$ is said to be *coarser than* $\mathcal{B}$. Note the following (at first sight perhaps slightly counter intuitive) fact: if $\mathcal{B}$ is a subcover of $\mathcal{A}$ then $\mathcal{A} < \mathcal{B}$.

Let $\mathcal{A}$ and $\mathcal{B}$ be covers of X; *the join* of $\mathcal{A}$ and $\mathcal{B}$, denoted $\mathcal{A} \vee \mathcal{B}$, is the cover of X consisting of all non-empty sets $A \cap B$ with $A \in \mathcal{A}$ and $B \in \mathcal{B}$. Clearly, $\mathcal{A} < \mathcal{A} \vee \mathcal{B}$ and $\mathcal{B} < \mathcal{A} \vee \mathcal{B}$, so $\mathcal{A} \vee \mathcal{B}$ is a common refinement of $\mathcal{A}$ and $\mathcal{B}$ (in fact, it is their coarsest common refinement; see Exercise 8.3).

If $f: X \to X$ is a mapping and $\mathcal{A}$ is a cover of X then the sets $f^{\leftarrow}[A]$ for $A \in \mathcal{A}$ cover X. Stated otherwise,

$$f^{\leftarrow}(\mathcal{A}) := \{\, f^{\leftarrow}[A] \,\vdots\, A \in \mathcal{A}\,\}$$

is a cover of X; we shall often write $f^{\leftarrow}\mathcal{A}$ for $f^{\leftarrow}(\mathcal{A})$. If $\mathcal{A}$ is a cover of X then the cover

$$\bigvee_{i=0}^{n-1}(f^i)^{\leftarrow}\mathcal{A} = \mathcal{A} \vee f^{\leftarrow}\mathcal{A} \vee \cdots \vee (f^{n-1})^{\leftarrow}\mathcal{A}$$

for $n \in \mathbb{N}$ will be denoted by $\mathcal{A}^{f,n}$. Stated otherwise, the cover $\mathcal{A}^{f,n}$ consists of all sets of the form $A_0 \cap f^{\leftarrow}[A_1] \cap \cdots \cap (f^{n-1})^{\leftarrow}[A_{n-1}]$ with $A_i \in \mathcal{A}$ for $i = 0, \ldots, n-1$ (with the silent understanding that the empty sets of this form are omitted). Thus, using the terminology from Section 6.1, an element $x \in X$ belongs to a member of $\mathcal{A}^{f,n}$ iff there are $A_i \in \mathcal{A}$ such that the sequence $(A_0, A_1, \ldots, A_{n-1})$ is a partial itinerary of x.

We collect some trivial observations regarding these definitions.

Let $\mathcal{A}$ and $\mathcal{B}$ be covers of X. Then:

(1) *If $\mathcal{A} < \mathcal{B}$, and $\mathcal{B}$ is special then $\mathcal{A}$ is special as well.*
(2) *$\mathcal{A} \vee \mathcal{B}$ is special iff $\mathcal{A}$ and $\mathcal{B}$ are special. In addition, $\mathcal{A} \vee \mathcal{B}$ is an open cover iff $\mathcal{A}$ and $\mathcal{B}$ are open covers.*
(3) *If $\mathcal{A}$ is special then so is $f^{\leftarrow}(\mathcal{A})$. Moreover, if $\mathcal{A}$ is an open cover and f is continuous then $f^{\leftarrow}(\mathcal{A})$ is an open cover.*
(4) *$f^{\leftarrow}(\mathcal{A} \vee \mathcal{B}) = f^{\leftarrow}(\mathcal{A}) \vee f^{\leftarrow}(\mathcal{B})$ and if $\mathcal{A} < \mathcal{B}$ then $f^{\leftarrow}(\mathcal{A}) < f^{\leftarrow}(\mathcal{B})$.*

Proof. (1) Select for every member of a finite subcover of $\mathcal{B}$ a member of $\mathcal{A}$ containing it; this defines a finite subcover of $\mathcal{A}$.
(2) First, we prove the statement about 'special'. "Only if": Clear from 1. "If": For (finite) subcovers $\mathcal{A}'$ and $\mathcal{B}'$ of $\mathcal{A}$ and $\mathcal{B}$, respectively, $\mathcal{A}' \vee \mathcal{B}'$ is a (finite) subcover of $\mathcal{A} \vee \mathcal{B}$.
To prove the statement about 'open', recall that $\mathcal{A} \vee \mathcal{B}$ is the collection of all non-empty sets $A \cap B$ with $A \in \mathcal{A}$ and $B \in \mathcal{B}$. "If": If A and B are open then so is $A \cap B$. "Only if": Take into account that $A = \bigcup\{ A \cap B \vdots B \in \mathcal{B} \}$ for every subset A of X.
(3), (4) The straightforward proofs are left for the reader. □

If $\mathcal{A}$ is a special cover of X then it makes sense to define $N(\mathcal{A})$ as the smallest cardinality of a finite subcover of $\mathcal{A}$. Obviously, $N(\mathcal{A})$ is a finite integer, and $N(\mathcal{A}) \geq 1$. The *entropy* of the special cover $\mathcal{A}$ of X is defined as

$$H(\mathcal{A}) := \log N(\mathcal{A}) .$$

Lemma 8.4.2. *Let $\mathcal{A}$ and $\mathcal{B}$ be special covers of X and let $f: X \to X$ be any mapping. Then:*
(1) $H(\mathcal{A}) \geq 0$ *and* $H(\mathcal{A}) = 0$ *iff* $X \in \mathcal{A}$.
(2) *If* $\mathcal{A} < \mathcal{B}$ *then* $N(\mathcal{A}) \leq N(\mathcal{B})$, *hence* $H(\mathcal{A}) \leq H(\mathcal{B})$.
(3) $H(\mathcal{A} \vee \mathcal{B}) \leq H(\mathcal{A}) + H(\mathcal{B})$.
(4) $H(f^{\leftarrow}\mathcal{A}) \leq H(\mathcal{A})$ *and if f is a surjection then* $H(f^{\leftarrow}\mathcal{A}) = H(\mathcal{A})$.

Proof. (1) That $H(\mathcal{A}) \geq 0$ is clear from the observation that $N(\mathcal{A}) \geq 1$. Moreover, it is obvious that $N(\mathcal{A}) = 1$ iff $X \in \mathcal{A}$.
(2) Let $\{ B_1, \dots, B_{N(\mathcal{B})} \}$ be a subcover of $\mathcal{B}$ with minimal cardinality. For every $i \in \{1, \dots, N(\mathcal{B})\}$, select a member A_i of $\mathcal{A}$ such that $B_i \subseteq A_i$. Obviously, $\{ A_1, \dots, A_{N(\mathcal{B})} \}$ covers X, hence is a subcover of $\mathcal{A}$. It has cardinality $N(\mathcal{B})$, hence $N(\mathcal{A}) \leq N(\mathcal{B})$. Consequently, $H(\mathcal{A}) \leq H(\mathcal{B})$.
(3) Taking intersections of the members of a minimal subcover of $\mathcal{A}$ with the members of a minimal subcover of $\mathcal{B}$ and deleting the empty intersections, we find a subcover of $\mathcal{A} \vee \mathcal{B}$ with at most $N(\mathcal{A})N(\mathcal{B})$ elements. It follows that $N(\mathcal{A} \vee \mathcal{B}) \leq N(\mathcal{A})N(\mathcal{B})$.
(4) Let $\{ A_1, \dots, A_{N(\mathcal{A})} \}$ be a subcover of $\mathcal{A}$ with minimal cardinality. Then it is easy to see that $\{ f^{\leftarrow}[A_1], \dots, f^{\leftarrow}[A_{N(\mathcal{A})}] \}$ is (finite) cover of X. It is a subcover of $f^{\leftarrow}\mathcal{A}$, so $N(f^{\leftarrow}\mathcal{A}) \leq N(\mathcal{A})$.
Now assume that f is a surjection and let $\{f^{\leftarrow}[A_1], \dots, f^{\leftarrow}[A_{N(f^{\leftarrow}\mathcal{A})}]\}$ be a finite subcover of $f^{\leftarrow}\mathcal{A}$ of minimal cardinality. Taking into account that surjectivity of f implies that $f[f^{\leftarrow}[A]] = A \cap f[X] = A$ for every subset A of X, it follows that the collection $\{A_1, \dots, A_{N(f^{\leftarrow}\mathcal{A})}\}$ covers X. Consequently, $N(\mathcal{A}) \leq N(f^{\leftarrow}\mathcal{A})$. □

If $\mathcal{A}$ is a special cover then $\mathcal{A}^{f,n} < \mathcal{A}^{f,m}$ for all $n, m \in \mathbb{N}$ with $n < m$, hence by statement 2 in the lemma above, $N(\mathcal{A}^{f,n}) \leq N(\mathcal{A}^{f,m})$. Stated otherwise, the function $n \mapsto N(\mathcal{A}^{f,n}) : \mathbb{N} \to \mathbb{N}$ is non-decreasing. It is called the *complexity function* of the special cover $\mathcal{A}$. The next proposition is about the exponential growth rate of this function.

Proposition 8.4.3. *Let $\mathcal{A}$ be a special cover of X and let $f: X \to X$ be any mapping. Then*

$$h(f,\mathcal{A}) := \lim_{n\rightsquigarrow\infty} \frac{1}{n} H(\mathcal{A}^{f,n}) = \lim_{n\rightsquigarrow\infty} \frac{1}{n} \log N(\mathcal{A}^{f,n}) \tag{8.4-1}$$

exists and is finite.

Proof. For every $n \in \mathbb{N}$, let $a_n := H(\mathcal{A}^{f,n})$. Since $a_n > 0$ for every $n \in \mathbb{N}$, it is, in view of Lemma 8.1.10, sufficient to show that

$$\forall\, m, n \in \mathbb{N}: \; a_{m+n} \leq a_m + a_n\,.$$

Let $m, n \in \mathbb{N}$; then

$$\begin{aligned}
a_{m+n} &= H\left(\bigvee_{i=0}^{m+n-1} (f^i)^{\leftarrow}\mathcal{A}\right) \\
&= H\left(\left(\bigvee_{i=0}^{n-1} (f^i)^{\leftarrow}\mathcal{A}\right) \vee \left(\bigvee_{i=n}^{n+m-1} (f^i)^{\leftarrow}\mathcal{A}\right)\right) \\
&\overset{(1)}{\leq} H\left(\bigvee_{i=0}^{n-1} (f^i)^{\leftarrow}\mathcal{A}\right) + H\left((f^n)^{\leftarrow}\left(\bigvee_{j=0}^{m-1} (f^j)^{\leftarrow}\mathcal{A}\right)\right) \\
&\overset{(2)}{\leq} a_n + a_m\,,
\end{aligned}$$

where (1) follows from Lemma 8.4.2 (3) and (2) from Lemma 8.4.2 (4) applied to f^n. □

Let $f: X \to X$ be a mapping and let $\mathcal{A}$ be a special cover of X. The real number $h(f,\mathcal{A})$ defined in (8.4-1) is called the *entropy of f with respect to $\mathcal{A}$*.

Lemma 8.4.4. *Let $f: X \to X$ be a mapping and let $\mathcal{A}$ and $\mathcal{B}$ be special covers of X. Then:*

(1) $h(f,\mathcal{A}) \geq 0$.
(2) *If $\mathcal{A} < \mathcal{B}$ then $h(f,\mathcal{A}) \leq h(f,\mathcal{B})$. In particular, this inequality holds if $\mathcal{B}$ is a subcover of $\mathcal{A}$.*
(3) $h(f,\mathcal{A}) \leq H(\mathcal{A})$.

Proof. (1) Clear from Lemma 8.4.2 (1).

(2) If $\mathcal{A} < \mathcal{B}$ and $n \in \mathbb{N}$ then, by statement 4 in 8.4.1, $(f^i)^{\leftarrow}\mathcal{A} < (f^i)^{\leftarrow}\mathcal{B}$ for $i = 0, \ldots, n-1$, hence $\mathcal{A}^{f,n} < \mathcal{B}^{f,n}$. Now Lemma 8.4.2 (2) immediately implies that $H(\mathcal{A}^{f,n}) \leq H(\mathcal{B}^{f,n})$, from which the desired result easily follows.

(3) It follows easily from the inequalities in Lemma 8.4.2 (3) and Lemma 8.4.2 (4) that

$$H\left(\bigvee_{i=0}^{n-1} (f^i)^{\leftarrow}\mathcal{A}\right) \leq \sum_{i=0}^{n-1} H((f^i)^{\leftarrow}\mathcal{A}) \leq n \cdot H(\mathcal{A})\,.$$

Now apply the definition of $h(f,\mathcal{A})$. □

8.4.5. Let (X, f) be a dynamical system with X a compact Hausdorff space. The *topological entropy* of f is the real number

$$h(f) := \sup\{h(f, \mathcal{A}) \,\vdots\, \mathcal{A} \text{ is an open cover of } X\}. \tag{8.4-2}$$

By Lemma 8.4.4 (1), always $h(f) \geq 0$. Note that, without limitation of generality, we may restrict ourselves to *finite* open covers of X in the definition of $h(f)$. Though in (8.4-2) we would perhaps get a smaller number by considering only finite open covers, Lemma 8.4.4 (2) above implies that this is not the case.

Examples.

(1) We shall use the definition with coverings to show that the rigid rotation φ_a on the circle has entropy 0, which we know already: see Example (1) after Corollary 8.1.9). Notation concerning arcs will be as in 1.7.6.
Let $\mathcal{A}$ be an open cover of $\mathbb{S}$ and let δ be a Lebesgue number of $\mathcal{A}$. Select a natural number $k > 2\pi/\delta$, and divide $\mathbb{S}$ in k equal arcs by k points $\xi_0, \ldots, \xi_{k-1}$; it will be convenient to define $\xi_k := \xi_0$. Let $\mathcal{B}$ be the cover of $\mathbb{S}$ consisting of the half-open arcs $[\xi_i; \xi_{i+1})$ for $i = 0, \ldots, k-1$. Each of these arcs has length at most δ, so $\mathcal{A} < \mathcal{B}$. Since φ_a shifts the points ξ_i in a rigid way, the covering $\varphi_a^{\leftarrow}\mathcal{B}$ is just a shifted copy of the covering $\mathcal{B}$. Since all arcs of $\mathcal{B}$ and those of $\varphi_a^{\leftarrow}\mathcal{B}$ have the same length, it should be clear that the way the division points of $\varphi_a^{\leftarrow}\mathcal{B}$ occur in members of $\mathcal{B}$ is the same for all members of $\mathcal{B}$: either the two sets of division points coincide or the interior of each member of $\mathcal{B}$ contains precisely one division point of $\varphi_a^{\leftarrow}\mathcal{B}$. Consequently, each member of $\mathcal{B}$ meets at most two members of $\varphi_a^{\leftarrow}\mathcal{B}$. So the covering $\mathcal{B} \vee \varphi_a^{\leftarrow}\mathcal{B}$ has at most twice the number of elements of $\mathcal{B}$, i.e., $\#\mathcal{B}^{\varphi_a,2} \leq 2\cdot(\#\mathcal{B})$. See also Figure 8.2.
Similarly, $(\varphi_a^2)^{\leftarrow}\mathcal{B}$ has at most one division point in any member of $\mathcal{B}$. So of the (at most) two members of $\mathcal{B} \vee (\varphi_a^{\leftarrow}\mathcal{B})$ that are included in a given member of $\mathcal{B}$ at most one is 'split' into two members of $\mathcal{B} \vee (\varphi_a^{\leftarrow}\mathcal{B}) \vee (\varphi_a^2)^{\leftarrow}\mathcal{B}$. This obviously implies that $\#\mathcal{B}^{\varphi_a,3} \leq 3\cdot(\#\mathcal{B})$. Proceeding in this way, we find $\#\mathcal{B}^{\varphi_a,n} \leq n\cdot(\#\mathcal{B}) = nk$ for all $n \in \mathbb{N}$, so $N(\mathcal{B}^{\varphi_a,n}) \leq nk$ for all $n \in \mathbb{N}$. Since $\frac{1}{n}\log n$ tends to 0 for $n \rightsquigarrow \infty$ it follows easily that $h(\varphi_a, \mathcal{B}) = 0$. Now recall that $\mathcal{A} < \mathcal{B}$, so that $h(\varphi_a, \mathcal{A}) = 0$ by Lemma 8.4.4 (2). Since this holds for every open cover $\mathcal{A}$ of $\mathbb{S}$, it follows that $h(\varphi_a) = 0$.

(2) Let (X, f) be Ellis' minimal system. We shall show that $h(f) = 0$. Recall that X is a compact Hausdorff space, but that X is not metrizable: see 1.7.6. So we have to use the covering definition of topological entropy. The proof is very similar to the one in the previous example. To this end, we have to define something like a Lebesgue number for open coverings of X. Notation will be as in 1.7.6.

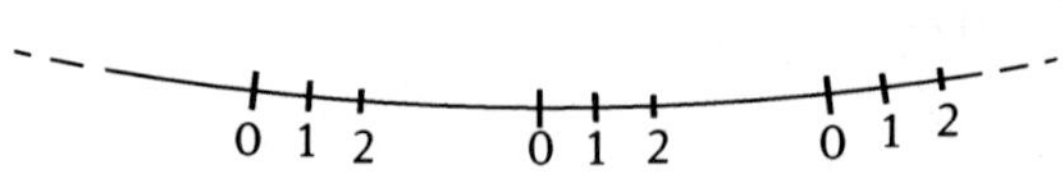

Fig. 8.2. Illustrating Example (1). The point labelled by 0, 1 and 2 denote division points of $\mathcal{B}$, $\varphi_a^{\leftarrow}\mathcal{B}$ and $(\varphi_a^2)^{\leftarrow}\mathcal{B}$, respectively.

For every point $\xi \in \mathbb{S}$ and every $\delta > 0$, let $\xi + \delta$ denote the unique point of $\mathbb{S}$ such that the arc $[\xi; \xi + \delta]$ has length δ. Similarly, $\xi - \delta$ will denote the unique point of S$\mathbb{S}$ such that the arc $[\xi - \delta; \xi]$ has length δ. In addition, let

$$\begin{aligned} V_{(\xi,1)}(\delta) &:= \Big([\xi; \xi + \delta) \times \{1\}\Big) \cup \Big((\xi; \xi + \delta) \times \{2\}\Big), \\ V_{(\xi,2)}(\delta) &:= \Big((\xi - \delta; \xi) \times \{1\}\Big) \cup \Big((\xi - \delta; \xi] \times \{2\}\Big), \end{aligned}$$

basic neighbourhoods of the points $(\xi, 1)$ and $(\xi, 2)$ of X, respectively.
Consider an arbitrary open cover $\mathcal{A}$ of X. Then for every $\xi \in \mathbb{S}$ there exists a real number $\delta_\xi > 0$ such that the neighbourhood $V_{(\xi,i)}(\delta_\xi)$ of the point (ξ, i) of X is entirely included in a member of $\mathcal{A}$ (without restriction of generality we may assume that we have the same δ_ξ for the two points $(\xi, 1)$ and $(\xi, 2)$ of X). Cover X by finitely many of the sets $V_{(\xi,i)}(\delta_\xi/2)$ with $\xi \in \mathbb{S}$ and $i = 1, 2$, and let $\delta_{\mathcal{A}}$ be the minimum of the values of $\delta_\xi/2$ involved in that finite cover. Then every set of the form $V_{(\xi,i)}(\delta_{\mathcal{A}})$ with $\xi \in \mathbb{S}$ and $i \in \{1, 2\}$ is included in a member of $\mathcal{A}$.
Select a natural number $k > 2\pi/\delta_{\mathcal{A}}$, and divide $\mathbb{S}$ in k equal arcs by k equidistant points $\xi_0, \dots, \xi_{k-1}$; it will be convenient to define $\xi_k := \xi_0$. Let $\mathcal{C}$ be the (open) cover of $\mathbb{S}$ consisting of the sets $V_{(\xi_j,i)}(1/k)$ for $j = 0, \dots, k - 1$ and $i = 1, 2$. Since each of these sets is included in a set of the form $V_{(\xi,i)}(\delta_{\mathcal{A}})$ with $\xi \in \mathbb{S}$ and $i \in \{1, 2\}$ it follows from the definition of $\delta_{\mathcal{A}}$ that $\mathcal{A} < \mathcal{C}$.
Similar to the method used in Example (1) above one shows that $\#\mathcal{C}^{f,n} \leq nk$, which implies that $h(f, \mathcal{C}) = 0$. Since $\mathcal{A} < \mathcal{C}$, it follows that $h(f, \mathcal{A}) = 0$ as well. This holds for every open cover $\mathcal{A}$ of X, so $h(f) = 0$.

We shall show now that in the case of a compact metric space the above definition agrees with the one given in Section 8.1. To avoid confusion, we shall denote the topological entropy defined in Section 8.1 by $h(X, f)$.

The *diameter of the collection* $\mathcal{A}$ of subsets of a metric space (X, d) is defined as

$$\operatorname{diam}(\mathcal{A}) := \sup\{\operatorname{diam}(A) : A \in \mathcal{A}\},$$

where $\operatorname{diam}(A) := \sup_{x,y \in A} d(x, y)$ is the usual diameter of a set A. If X is a compact metric space, $\mathcal{A}$ and $\mathcal{C}$ are open covers of X, and γ is a Lebesgue number of $\mathcal{C}$, then it follows immediately from the definitions that

$$\operatorname{diam}(\mathcal{A}) \leq \gamma \Rightarrow \mathcal{C} < \mathcal{A}. \tag{8.4-3}$$

Theorem 8.4.6. *Let (X, f) be a dynamical system on a compact metric space (X, d) and let $(\mathcal{A}_n)_{n\in\mathbb{N}}$ be a sequence of open covers of X such that* $\operatorname{diam}(\mathcal{A}_n) \rightsquigarrow 0$ *for $n \rightsquigarrow \infty$. Then* $\lim_{n\rightsquigarrow\infty} h(f, \mathcal{A}_n)$ *exists and is equal to $h(f)$ (which may be ∞).*

Proof. First, assume that $h(f) < \infty$. Let $\varepsilon > 0$ and let $\mathcal{C}$ be an open cover of X such that $h(f, \mathcal{C}) > h(f) - \varepsilon$. If $\mathcal{C}$ has Lebesgue number δ then there exists $n_0 \in \mathbb{N}$ such that $\operatorname{diam}(\mathcal{A}_n) < \delta$ for all $n \geq n_0$. For these values of n we have, by (8.4-3), $\mathcal{C} < \mathcal{A}_n$,

Hence Lemma 8.4.4 (2) implies that $h(f, \mathcal{C}) \leq h(f, \mathcal{A}_n)$. Consequently, $h(f) \geq h(f, \mathcal{A}_n) > h(f) - \varepsilon$ for all $n \geq n_0$, where the first inequality follows from (8.4-2). This shows that $\lim_{n \rightsquigarrow \infty} h(f, \mathcal{A}_n) = h(f)$.

If $h(f) = \infty$, consider an arbitrary positive real number a and select an open cover $\mathcal{C}$ of X with $h(f, \mathcal{C}) > a$. As before, there exists $n_0 \in \mathbb{N}$ such that $h(f, \mathcal{A}_n) > a$ for all $n \geq n_0$. □

Example. Consider the shift system (Ω_S, σ) with $s := \#S$ symbols and let $\mathcal{A}_n$ be the clopen partition of Ω_S into cylinders based on initial n-blocks of points ($n \in \mathbb{N}$). It is straightforward to check that, for every $k \in \mathbb{N}$, $\mathcal{A}_n^{\sigma,k} = \mathcal{A}_n \vee \sigma^{\leftarrow}\mathcal{A}_n \vee \cdots \vee (\sigma^{k-1})^{\leftarrow}\mathcal{A}_n$ is the clopen partition of Ω_S into cylinders based on the initial $(n + k)$-blocks of points. This partition has no proper subcovers, so it follows that $N(\mathcal{A}_n^{f,k}) = s^{n+k}$. Consequently, $h(\sigma, \mathcal{A}_n) = \lim_k \frac{1}{k} \log s^{n+k} = \log s$ for every $n \in \mathbb{N}$. Finally, it is clear that $\operatorname{diam} \mathcal{A}_n = 1/(n + 1)$, and since $1/(n + 1) \rightsquigarrow 0$ for $n \rightsquigarrow \infty$, Theorem 8.4.6 implies that $h(f) = \log s$.

Lemma 8.4.7. *Let (X, f) be a dynamical system on a compact metric space (X, d).*
(1) *If $\mathcal{A}$ is an open cover of X with Lebesgue number δ then*

$$\forall\, n \in \mathbb{N}: \quad N(\mathcal{A}^{f,n}) \leq \operatorname{minspan}_n(\tfrac{1}{2}\delta, X, f) \leq \operatorname{maxsep}_n(\tfrac{1}{2}\delta, X, f)\,.$$

(2) *If $\varepsilon > 0$ and $\mathcal{C}$ is an open cover of X with* $\operatorname{diam}(\mathcal{C}) < \varepsilon$ *then we have*

$$\forall\, n \in \mathbb{N}: \quad \operatorname{minspan}_n(\varepsilon, X, f) \leq \operatorname{maxsep}_n(\varepsilon, X, f) \leq N(\mathcal{C}^{f,n})\,.$$

Proof. In view of Lemma 8.1.4 (2), we need only prove the first inequality in 1 and the second inequality in 2.
(1) Let S be an $(n, \frac{1}{2}\delta)$-span of X with cardinality $\operatorname{minspan}_n(\frac{1}{2}\delta, X, f)$. Then the open balls $B_{\delta/2}(x, d_n^f)$ with $x \in S$ form a finite cover $\mathcal{B}$ of X which has no proper subcover, so that $N(\mathcal{B}) = \#S = \operatorname{minspan}_n(\frac{1}{2}\delta, X, f)$. Now

$$\begin{aligned} B_{\delta/2}(x, d_n^f) &= \left\{x' \in X : d(f^i(x), f^i(x')) < \tfrac{1}{2}\delta \text{ for } 0 \leq i \leq n-1\right\} \\ &= \bigcap_{i=0}^{n-1} (f^i)^{\leftarrow}[B_{\delta/2}(f^i(x), d)]\,. \end{aligned}$$

For every $i \in \{0, \ldots, n-1\}$, the ball $B_{\delta/2}(f^i(x), d)$ has diameter at most δ, hence is included in some member A_i of $\mathcal{A}$. Consequently, for every $x \in S$, the open ball $B_{\delta/2}(x, d_n^f)$ is included in a set of the form $\bigcap_{i=0}^{n-1}(f^i)^{\leftarrow}[A_i]$ with $A_i \in \mathcal{A}$ for $i = 0, \ldots, n-1$, which is a member of $\bigvee_{i=0}^{n-1}(f^i)^{\leftarrow}\mathcal{A} = \mathcal{A}^{f,n}$. Hence the open cover $\mathcal{B} = \{B_{\delta/2}(x, d_n^f) : x \in S\}$ is a refinement of $\mathcal{A}^{f,n}$. Consequently, Lemma 8.4.2 (2) implies that $N(\mathcal{A}^{f,n}) \leq N(\mathcal{B}) = \operatorname{minspan}_n(\frac{1}{2}\delta, X, f)$.
(2) Let $n \in \mathbb{N}$ and consider two points $x, x' \in X$ which are situated in the same member of the cover $\bigvee_{i=0}^{n-1}(f^i)^{\leftarrow}\mathcal{C}$. Then for every $i \in \{0, \ldots, n-1\}$ the points $f^i(x)$ and $f^i(x')$ are situated in the same member of $\mathcal{C}$ (depending on i), hence they have a distance

less than ε. This means that $d_n^f(x, x') < \varepsilon$. It follows that every member of $\mathcal{C}^{f,n}$ contains at most one member of an (n, ε)-separated set. Applying this to any finite subcover $\mathcal{D}$ of $\mathcal{C}^{f,n}$, we see that any (n, ε)-separated set has at most $\#\mathcal{D}$ elements. It follows that $\text{maxsep}_n(\varepsilon, X, f) \leq N(\mathcal{C}^{f,n})$. □

Remark. For later reference we note that the proof of statement 2 actually shows: if $\mathcal{C}$ is any *finite* cover of X with $\text{diam}(\mathcal{C}) < \varepsilon$ then $\text{maxsep}_n(\varepsilon, X, f) \leq \#\mathcal{C}^{f,n}$.

Corollary 8.4.8. *Let (X, f) be a dynamical system on a compact metric space (X, d) and let $0 < \varepsilon' < \varepsilon$.*

(1) *If $\mathcal{A}_\varepsilon$ is the open cover of X consisting of all open balls with radius 2ε then*

$$\forall n \in \mathbb{N}: \quad N(\mathcal{A}_\varepsilon^{f,n}) \leq \text{minspan}_n(\varepsilon', X, f).$$

(2) *If $\mathcal{C}_{\varepsilon'}$ is an arbitrary open cover of X consisting of open balls with radius $\varepsilon'/2$ then*

$$\forall n \in \mathbb{N}: \quad \text{maxsep}_n(\varepsilon, X, f) \leq N(\mathcal{C}_{\varepsilon'}^{f,n}).$$

Proof. (1) Obviously, $2\varepsilon'$ is a Lebesgue number for the open cover $\mathcal{A}_\varepsilon$. So Lemma 8.4.7 (1) with $\delta := 2\varepsilon'$ implies that $N\left(\mathcal{A}_\varepsilon^{f,n}\right) \leq \text{minspan}_n(\varepsilon', X, f)$.
(2) Clearly, the open cover $\mathcal{C}_{\varepsilon'}$ has diameter less than or equal to ε', so that $\text{diam}(\mathcal{C}_{\varepsilon'}) < \varepsilon$. Now apply Lemma 8.4.7 (2). □

Theorem 8.4.9. *Let (X, f) be a dynamical system on a compact metric space (X, d). Then $h(X, f) = h(f)$, so the two definitions of topological entropy coincide.*

Proof. For every $k \in \mathbb{N}$, let $\mathcal{A}_k$ be the open cover of X consisting of all open balls with radius $2/k$ (in order to avoid the clumsy $\mathcal{A}_{1/k}$, the notation deviates slightly from that in Corollary 8.4.8 above). In addition, let $0 < \varepsilon' < 1/k$. Then Corollary 8.4.8 (1) with $\varepsilon = 1/k$ implies that

$$\forall n \in \mathbb{N}: \quad N(\mathcal{A}_k^{f,n}) \leq \text{minspan}_n(\varepsilon', X, f). \tag{8.4-4}$$

Take logarithms, divide by n and take the lim sup of both sides for $n \rightsquigarrow \infty$:

$$h(f, \mathcal{A}_k) \leq r(\varepsilon', X, f)(0 < \varepsilon' < 1/k).$$

Here $r(\varepsilon', X, f) \leq h(X, f)$ for all $\varepsilon' > 0$. As Theorem 8.4.6 implies that $h(f, \mathcal{A}_k) \rightsquigarrow h(f)$ if $k \rightsquigarrow \infty$, it follows that $h(f) \leq h(X, f)$.

Conversely, if $\varepsilon > 0$ then let $k \in \mathbb{N}$ be such that $1/k < \varepsilon$ and let $\mathcal{C}_k$ be any open cover of X consisting of open balls with radius $1/2k$. Then Corollary 8.4.8 (2) with $\varepsilon' = 1/k$ implies that

$$\forall n \in \mathbb{N}: \quad \text{maxsep}_n(\varepsilon, X, f) \leq N(\mathcal{C}_k^{f,n}). \tag{8.4-5}$$

Take logarithms, divide by n and take the lim sup of both sides for $n \rightsquigarrow \infty$:

$$s(\varepsilon, X, f) \leq h(f, \mathcal{C}_k) \quad (0 < 1/k < \varepsilon).$$

Taking the limit for $k \rightsquigarrow \infty$ shows that $s(\varepsilon, X, f) \leq h(f)$ for all $\varepsilon > 0$. Finally, let $\varepsilon \rightsquigarrow 0$ and get $h(X, f) \leq h(f)$. □

Corollary 8.4.10. *Let (X, f) be a dynamical system on a compact metric space (X, d). Then*[5]

$$\begin{aligned} h(f) &= \lim_{\varepsilon \rightsquigarrow 0} \liminf_{n \rightsquigarrow \infty} \frac{1}{n} \log \text{minspan}_n(\varepsilon, X, f) \\ &= \lim_{\varepsilon \rightsquigarrow 0} \liminf_{n \rightsquigarrow \infty} \frac{1}{n} \log \text{maxsep}_n(\varepsilon, X, f) . \end{aligned}$$

Proof. Let notation be as in the proof of Theorem 8.4.9. Let $k \in \mathbb{N}$ and let $0 < \varepsilon' < 1/k$; then for every $n \in \mathbb{N}$ we have inequality (8.4-4). Take logarithms, divide by n, but instead of taking the lim sup take the lim inf for $n \rightsquigarrow \infty$. In view of Proposition 8.4.3 we get

$$h(f, \mathcal{A}_k) \le \liminf_{n \rightsquigarrow \infty} \frac{1}{n} \log \text{minspan}_n(\varepsilon', X, f) . \tag{8.4-6}$$

Using Lemma 8.1.4 (1), it is easily seen that the right-hand side increases if ε' decreases, hence it has a limit for $\varepsilon' \rightsquigarrow 0$ (not necessarily finite). After taking this limit in the right-hand side of (8.4-6), let $k \rightsquigarrow \infty$ in the left-hand side of the resulting inequality, taking into account Theorem 8.4.6: we get

$$h(f) \le \lim_{\varepsilon' \rightsquigarrow 0} \liminf_{n \rightsquigarrow \infty} \frac{1}{n} \log \text{minspan}_n(\varepsilon', X, f) . \tag{8.4-7}$$

If $\varepsilon > 0$ and $k \in \mathbb{N}$ is such that $0 < 1/k < \varepsilon$ then we get in a similar way, starting with (8.4-5),

$$\liminf_{n \rightsquigarrow \infty} \frac{1}{n} \log \text{maxsep}_n(\varepsilon, X, f) \le h(f, \mathcal{C}_k) .$$

Taking the limit for $k \rightsquigarrow \infty$ and subsequently for $\varepsilon \rightsquigarrow 0$, we get

$$\lim_{\varepsilon \rightsquigarrow 0} \liminf_{n \rightsquigarrow \infty} \frac{1}{n} \log \text{maxsep}_n(\varepsilon, X, f) \le h(f) \tag{8.4-8}$$

Finally, by Lemma 8.1.4 (2) the right-hand side of (8.4-7) is less than or equal to the left-hand side of (8.4-8). Using this, we get the desired equalities. □

8.5 Miscellaneous results

In this section some results are collected that use either Theorem 8.4.9 or Corollary 8.4.10. In all cases, (X, f) is a dynamical system on a compact metric space (though in Proposition 8.5.1 metrizability is not needed).

Proposition 8.5.1. *Let f be a homeomorphism of X onto itself. Then $h(f^{-1}) = h(f)$.*

---

5 According to Theorem 8.4.9 the quantity $h(f)$ can be interpreted either according to the definition in the present section or according to the one in Section 8.1. Recall that we have similar formulas with 'lim inf' replaced by 'lim sup'. see Corollary 8.1.9.

Proof. Let $\mathcal{A}$ be an open cover of X and let $n \in \mathbb{N}$. For any bijection g and any set B in the range of g one has $g^{\leftarrow}[B] = g^{-1}[B]$, hence[6]

$$\begin{aligned} \mathcal{A}^{f^{-1},n} &= \bigvee_{j=0}^{n-1}((f^{-1})^j)^{\leftarrow}\mathcal{A} = \bigvee_{j=0}^{n-1} f^j\mathcal{A} = f^{n-1}\Big[\bigvee_{i=0}^{n-1} f^{-i}\mathcal{A}\Big] \\ &= f^{n-1}\Big[\bigvee_{i=0}^{n-1}(f^i)^{\leftarrow}\mathcal{A}\Big] = (f^{1-n})^{\leftarrow}\mathcal{A}^{f,n}\,. \end{aligned}$$

Recall from Lemma 8.4.2 (4) that $H((f^{1-n})^{\leftarrow}\mathcal{A}^{f,n}) = H(\mathcal{A}^{f,n})$, so by the above, $H(\mathcal{A}^{f^{-1},n}) = H(\mathcal{A}^{f,n})$. Now apply Proposition 8.4.3. □

Example. In general, this result does not hold for non-compact spaces. For example, consider the system (X, f) with $X := (0;\infty)$ and $f(x) := 2x$ for $x \in X$. In the Example following Corollary 8.3.2 it is shown that $h(f) = \log 2$. On the other hand, f^{-1} is the mapping $x \mapsto x/2 : X \to X$ (a contraction) which has entropy zero.

Proposition 8.5.2. *For every $m \in \mathbb{N}$, $h(f^m) = mh(f)$.*

Proof. Consider any $m \in \mathbb{N}$. In view of Proposition 8.1.11 we need only show that $h(f^m) \geq mh(f)$. Let $n \in \mathbb{N}$. Since the metric space X is compact, the finite family $\{ f^i : i = 0, \dots, n-1 \}$ of continuous functions is uniformly equicontinuous, hence for every $\varepsilon > 0$ there exists $\delta(n, \varepsilon) > 0$ such that

$$\left.\begin{aligned} d(f^i(x), f^i(y)) < \varepsilon \quad &\text{for all } i \in \{0, \dots, n-1\} \text{ and all} \\ &x, y \in X \text{ with } d(x, y) < \delta(n, \varepsilon) \end{aligned}\right\} \tag{8.5-1}$$

Let $\delta := \delta(m, \varepsilon)$. Then for every pair of points $x, y \in X$ and every $j \in \mathbb{Z}^+$ it follows from (8.5-1) with $n = m$ and $f^{mj}(x)$ and $f^{mj}(y)$ instead of x and y, respectively: if $d(f^{mj}(x), f^{mj}(y)) < \delta$ then $d(f^{mj+i}(x), f^{mj+i}(y)) < \varepsilon$ for every $i \in \{0, , \dots, m-1\}$. This implies for every $n \in \mathbb{N}$:

$$\max_{0\leq j\leq n-1} d(f^{mj}(x), f^{mj}(y)) < \delta \;\Rightarrow\; \max_{0\leq k\leq mn-1} d(f^k(x), f^k(y)) < \varepsilon$$

Hence every (n, δ)-span of X under f^m (in particular, the one with minimal cardinality) is an (mn, ε)-span of X under f. Consequently, for every $\varepsilon > 0$ there exists $\delta = \delta(m, \varepsilon)$ such that

$$\text{minspan}_{mn}(\varepsilon, X, f) \leq \text{minspan}_n(\delta, X, f^m)\,.$$

6 Formally we did not define $f^j\mathcal{A}$ for $j \geq 0$, but its meaning should be clear: the collection of all sets $f^j[A]$ for $A \in \mathcal{A}$. As f is a bijection this is equal to the collection of sets $(g^j)^{\leftarrow}[A]$ for $A \in \mathcal{A}$ with $g := f^{-1}$.

Using this, one gets

$$\begin{aligned}\liminf_{k\rightsquigarrow\infty}\frac{1}{k}\log \text{minspan}_k(\varepsilon, X, f) &\leq \liminf_{n\rightsquigarrow\infty}\frac{1}{mn}\log \text{minspan}_{mn}(\varepsilon, X, f)\\ &\leq \liminf_{n\rightsquigarrow\infty}\frac{1}{mn}\log \text{minspan}_n(\delta, X, f^m)\\ &\leq \limsup_{n\rightsquigarrow\infty}\frac{1}{mn}\log \text{minspan}_n(\delta, X, f^m)\end{aligned}$$

(with 'lim sup' in stead of 'lim inf' the first inequality would be the other way round). by Lemma 8.1.5 (1) and the definition of $h(X, f)$ in 8.1.6, the right-hand side of this inequality is at most $\frac{1}{m}h(f^m)$. Now take the limit of the left-hand side for $\varepsilon \rightsquigarrow 0$. Then Corollary 8.4.10 implies that $h(f) \leq \frac{1}{m}h(f^m)$. □

The following result is in concordance with the fact that the interesting behaviour of the dynamical system (X, f) takes place in the non-wandering set $\Omega(X, f)$: see Corollary 4.3.4–Corollary 4.3.9 and Exercise 4.5.

Theorem 8.5.3. $h(f) = h(f|_{\Omega(X,f)})$.

Proof. By Proposition 8.1.7, it remains to show that $h(f) \leq h(f|_{\Omega(X,f)})$. Let $\varepsilon > 0$, $n \in \mathbb{N}$ and let S be an (n, ε)-span of the compact set $\Omega(X, f)$ with minimal cardinality, i.e., $\#S = \text{minspan}_n(\varepsilon, \Omega(X, f), f)$. Let $\mathcal{U}$ be the open cover of $\Omega(X, f)$ consisting of the open balls $B_\varepsilon(x, d_n^f)$ with $x \in S$ and let U be the union of the members of $\mathcal{U}$. Then $Y := X \setminus U$ is a compact subset of X disjoint from $\Omega(X, f)$. Every point $y \in Y$ is wandering, hence has an open neighbourhood V_y in X such that $V_y \subseteq B_\varepsilon(y, d_n^f)$ and $f^i[V_y] \cap V_y = \emptyset$ for all $i \in \mathbb{N}$. The sets V_y with $y \in Y$ form an open cover of Y; let $\mathcal{V}$ be a finite subcover. Then $\mathcal{W} := U \cup V$ is a finite open cover of X with sets that all have the property that two points in the same set have d_n^f-distance less than 2ε. The members of $\mathcal{V}$ have the additional property that $f^i[V] \cap V = \emptyset$ for every $V \in \mathcal{V}$ and every $i \in \mathbb{N}$. Let $M := \#\mathcal{V}$; note that M depends on the choices of n and of ε.

Consider arbitrary $k \in \mathbb{N}$ with $k > M$ and let E be an $(nk, 2\varepsilon)$-separated subset of X. In order to find an upper bound for $\#E$, consider for every point $x \in E$ the partial orbit $\{f^{ni}(x) \,⋮\, i = 0, \dots, k-1\}$ of x under f^n (recall that, for the time being, n is fixed). For every $i \in \{0, \dots, k-1\}$ we select a set $W_i(x) \in \mathcal{W}$ such that $f^{ni}(x) \in W_i(x)$ – such a set exists, but may be not unique. The finite sequences of sets $\{W_i(x) \,⋮\, i = 0, \dots, k-1\}$ with $x \in E$ have the following properties:

(a) If $x, y \in E$ and $W_i(x) = W_i(y)$ for $i = 0, \dots, k-1$ then $x = y$.

Proof. The d_n^f-distance of two points in the same member of $\mathcal{W}$ is less than 2ε, so the assumption implies that for $i = 0, \dots, k-1$ we have $d_n^f(f^{ni}(x), f^{ni}(y)) < 2\varepsilon$, and therefore $d(f^{ni+j}(x), f^{ni+j}(y)) < 2\varepsilon$ for $j = 0, \dots, n-1$. Stated otherwise, $d_{nk}^f(x, y) < 2\varepsilon$. Since E is an $(nk, 2\varepsilon)$-separated set, this is only possible if $x = y$.

(b) Let $x \in E$ and let $0 \leq i < j \leq k-1$ be such that both sets $W_i(x)$ and $W_j(x)$ are a member of $\mathcal{V}$; then $W_i(x) \neq W_j(x)$.

Proof. Suppose the contrary: $W_i(x) = W_j(x)$ with $0 \leq i < j \leq k-1$. Then $f^{j-i}(f^i(x)) = f^j(x) \in W_j(x) = W_i(x)$ which would imply that $f^{j-i}[W_i(x)] \cap W_i(x) \neq \emptyset$, contradicting the choice of $\mathcal{V}$.

It follows from (a) and (b) that $\#E$ is bounded from above by the number of different sequences of k elements from $\mathcal{W}$ that have the property that every element of $\mathcal{V}$ occurs at most once in such a sequence (thus, repetitions in such a sequence can occur only for elements of $\mathcal{U}$, the cover of the part of X with the complicated dynamics). In particular, the number of elements from $\mathcal{V}$ in such a sequence is at most M. Let us call this the 'admissible sequences'. If $0 \leq j \leq M$ then the number of admissible sequences with j elements from $\mathcal{V}$ is equal to the product of the following natural numbers: the number M^j of possibilities to select j elements from $\mathcal{V}$ (mutually different or not, this is not essential anymore), the number $k(k-1)\dots(k-j+1)$ of possible ways that these j members of $\mathcal{V}$ can be placed in a sequence of length k (here we use that $j \leq k$, which follows from the inequalities $j \leq M$ and $k > M$), and, finally, the number $(\#S)^{k-j}$ of ways the remaining $k-j$ elements of the admissible sequence can be chosen from $\mathcal{U}$. Hence

$$\#E \leq \sum_{j=0}^{M} M^j \cdot k(k-1)\dots(k-j+1)\cdot(\#S)^{k-j}\,.$$

Using the estimations $M^j \leq M^M$, $k(k-1)\dots(k-j+1) \leq k^j \leq k^M$ and, finally, $(\#S)^{k-j} \leq (\#S)^k$, we get $\#E \leq (M+1)M^M k^M (\#S)^k$. This holds for every $(nk, 2\varepsilon)$-separated subset E of X. Consequently,

$$\operatorname{maxsep}_{kn}(2\varepsilon, X, f) \leq (M+1)M^M k^M \cdot \left(\operatorname{minspan}_n(\varepsilon, \Omega(X,f), f)\right)^k.$$

Take logarithms, divide by kn and take the lim inf for $k \rightsquigarrow \infty$. Since the numbers kn form a subsequence of the sequence of all natural numbers, we get an inequality, the left-hand side of which is bounded from below by $\liminf_{i\rightsquigarrow\infty} \frac{1}{i}\log \operatorname{maxsep}_i(2\varepsilon, X, f)$. Hence

$$\liminf_{i\rightsquigarrow\infty} \frac{1}{i}\log \operatorname{maxsep}_i(2\varepsilon, X, f) \leq \frac{1}{n}\log \operatorname{minspan}_n(\varepsilon, \Omega(X,f), f)\,.$$

The lim sup for $n \rightsquigarrow \infty$ of the right-hand side is equal to $r(\varepsilon, \Omega(X,f), f)$, which is less than or equal to $h(\Omega(X,f), f)$. Now the desired result follows from Corollary 8.4.10. □

Corollary 8.5.4. *Suppose that the non-wandering set $\Omega(X)$ of a dynamical system (X, f) on a compact metric space is finite; then $h(f) = 0$.*

Proof. A finite system has entropy zero. □

Examples.

(1) Consider the quadratic system $([0;1], f_\mu)$ for $0 < \mu \leq 3$. It follows from the results in 2.1.5 that $\Omega([0;1], f_\mu) = \{0, p_\mu\}$, hence $h(f_\mu) = 0$ for these values of μ. For $3 < \mu < 1+\sqrt{6}$ we have $\Omega([0;1], f_\mu) = \{0, p_\mu, x_1^{(\mu)}, x_2^{(\mu)}\}$, where $\{x_1^{(\mu)}, x_1^{(\mu)}\}$ is an orbit with period 2 (see Exercise 2.2). So for these values of μ, $h(f_\mu) = 0$ as well. See also Example (1) in Corollary 8.6.9 ahead.

(2) Compactness of the phase space is essential in Theorem 8.5.3 and Corollary 8.5.4. Consider the quadratic system $(\mathbb{R}, f_\mu)$ for $\mu > 0$. If $K := [-2; -1]$ then similar to Exercise 8.1 one shows that $h(K, f_\mu) = \infty$, so that $h(f_\mu) = \infty$. On the other hand, since the orbit of every point outside of the unit interval remains outside of it and decreases monotonously to $-\infty$, it is easily seen that $\Omega(\mathbb{R}, f_\mu) = \Omega([0; 1], f_\mu)$. Hence for all values of μ mentioned in Example (1) above we have $h(f_\mu|_{\Omega(\mathbb{R}, f_\mu)}) = 0 < h(f_\mu) = \infty$. Note that for $\mu = 4$ one has $\Omega(\mathbb{R}, f_\mu) = [0; 1]$: see Example (2) after Proposition 4.3.10. Moreover, for $\mu > 2 + \sqrt{5}$ the non-wandering set of $(\mathbb{R}, f_\mu)$ is a Cantor set Λ: see Example (3) following Proposition 4.3.10. Recall that on Λ the mapping f_μ is conjugate to the full 2-shift: see Remark 2 Following 6.3.6 (2). So also in these cases $h(f_\mu|_{\Omega(\mathbb{R}, f_\mu)} = \log 2 < h(f_\mu) = \infty$. (By Note 7 in Chapter 6, this is even true for all $\mu > 4$.)

8.6 Positive entropy and horseshoes for interval maps

We consider a dynamical system (X, f) on a compact interval X in $\mathbb{R}$. See Section 7.4 for the definition of horseshoes.

Lemma 8.6.1. *If f has an s-horseshoe $(J_0, \dots, J_{s-1})$ consisting of compact intervals then for every $n \in \mathbb{N}$ the mapping f^n has an s^n-horseshoe consisting of compact intervals, each of which is included in $J := J_0 \cup \dots \cup J_{s-1}$ and which is mapped over J by f^n.*

Proof. Let $S := \{1, \dots, s\}$. Similar to the initial part of 7.4.2 one shows that for every $n \in \mathbb{N}$ and every n-block $a_0 \dots a_{n-1} \in S^n$ one can select a closed interval $J_{a_0 \dots a_{n-1}}$ such that for $n \geq 2$:

$$J_{a_0 \dots a_{n-1}} \subseteq J_{a_0 \dots a_{n-2}} \quad \text{and} \quad f[J_{a_0 \dots a_{n-1}}] = J_{a_1 \dots a_{n-1}} .$$

Obviously, it will be sufficent to show that for every $n \in \mathbb{N}$ the intervals $J_{a_0 \dots a_{n-1}}$ for $a_0 \dots a_{n-1} \in S^n$ form an s^n-horseshoe for f^n. To prove this, recall that f maps the interior of $J_{a_0 \dots a_{n-1}}$ onto the interior of $J_{a_1 \dots a_{n-1}}$. Using this, one easily shows by induction that, for every $n \in \mathbb{N}$ and for every choice of two distinct n-tuples a and b of symbols from S, the intervals $J_{a_0 \dots a_{n-1}}$ and $J_{b_0 \dots b_{n-1}}$ have disjoint interiors: see the footnote to the proof of 7.4.2 (1). Moreover, it should be clear that the inclusion above implies that $J_{a_0 \dots a_{n-1}} \subseteq J$ for every $n \in \mathbb{N}$ and $a = a_0 \dots a_{n-1} \in S^n$. Finally, one also easily shows by induction that $f^{n-1}[J_{a_0 \dots a_{n-1}}] = J_{a_{n-1}}$. The assumption that $(J_0, \dots, J_{s-1})$ is an s-horseshoe implies that $f[J_{a_{n-1}}] \supseteq J$, hence $f^n[J_{a_0 \dots a_{n-1}}] \supseteq J$. This completes the proof. □

Proposition 8.6.2. *Let (X, f) be a dynamical system on a compact interval X. If f has an s-horseshoe consisting of compact intervals ($s \geq 2$) then $h(f) \geq \log s$.*

Proof. First, assume that f admits an s-horseshoe $(J_1, \dots, J_s)$ consisting of mutually disjoint intervals. Let ε be the minimum of the distances between these intervals and let K be their union, $K := J_1 \cup \dots \cup J_s$. Then K is compact, and we shall show that

$\text{maxsep}_n(\varepsilon, K, f) \geq s^n$ for all $n \in \mathbb{N}$. Assuming that this has been shown, we conclude that

$$h(f) \geq h(K, f) \geq s(\varepsilon, K, f) \geq \limsup_{n \rightsquigarrow \infty} \frac{1}{n} \log s^n = \log s\,.$$

In order to prove the desired estimation for $\text{maxsep}_n(\varepsilon, K, f)$, consider for every $n \in \mathbb{N}$ and every word $J := J_{p(0)}J_{p(1)} \dots J_{p(n-1)}$ of length n over the symbol set $\mathcal{H} := \{J_1, \dots, J_s\}$ (so $1 \leq p(i) \leq s$ for all i) the set

$$\begin{aligned} D_n(J) &= \bigcap_{i=0}^{n-1} (f^i)^{\leftarrow}[J_{p(i)}] \\ &= \left\{x \in J_{p(0)} : f^i(x) \in J_{p(i)} \text{ for } i = 1, \dots, n-1\right\} \end{aligned}$$

(the reader will recognize this terminology and notation from Chapter 6, where it is used in connection with symbolic representations).

We shall show by induction that $D_n(J) \neq \emptyset$ for every $n \in \mathbb{N}$ and every word J of length n over $\mathcal{H}$. For $n = 1$ we have $D_1(J_{p(0)}) = J_{p(0)} \neq \emptyset$. Moreover, let $n \in \mathbb{N}$ and assume that $D_n(J') \neq \emptyset$ for every word J' of length n over $\mathcal{H}$. For a word $J = J_{p(0)}J_{p(1)} \cdots J_{p(n)}$ of length $(n + 1)$ it is easily seen that $D_{n+1}(J) = J_{p(0)} \cap f^{\leftarrow}[D_k(J')]$ with $J' := J_{p(1)} \cdots J_{p(n)}$ (see, e.g., formula (6.1-7) in Chapter 6). So $f[D_{n+1}(J)] = f[J_{p(0)}] \cap D_k(J')$. By the defining property of an n-horseshoe we have $f[J_{p(0)}] \supseteq K$, which implies that $f[J_{p(0)}] \supseteq J_{p(1)} \supseteq D_k(J')$. Consequently, $f[D_{n+1}(J)] = D_k(J')$, which is non-empty by the induction hypothesis. So $D_{n+1}(J) \neq \emptyset$ as well.

Select for every word J of length n over $\mathcal{H}$ an element $x_J \in D_n(J)$. Let J and J' be different words of length n over $\mathcal{H}$, say $J = J_{p(0)}J_{p(1)} \cdots J_{p(n)}$ and $J' = J_{q(0)}J_{q(1)} \cdots J_{q(n)}$ with $J_{p(i)} \neq J_{q(i)}$ for some $i \in \{0, \dots, n-1\}$. Since $f^i(x_J) \in J_{p(i)}$ and $f^i(x_{J'}) \in J_{q(i)}$, it follows that the points $f^i(x_J)$ and $f^i(x_{J'})$, being situated in different members of the horseshoe, have a distance of at least ε. This implies that $d_n^f(x_J, x_{J'}) \geq \varepsilon$. Consequently, the subset $\{x_J : J \text{ a word of length } n \text{ over } \mathcal{H}\}$ of K is (n, ε)-separated. It has s^n elements, which shows that $\text{maxsep}_n(\varepsilon, K, f) \geq s^n$. This completes the proof of the theorem for this case.

Now assume that f admits an s-horseshoe, the members of which are not all disjoint (there can be common end points). By Lemma 8.6.1, for every $k \geq 2$ the mapping f^k has an s^k-horseshoe; here $s^k \geq 4$ if $k \geq 2$. Number the intervals of such an s^k-horseshoe from left to right and consider only the intervals with odd number. Obviously we obtain an $s^k/2$-horseshoe if s^k is even, or an $(s^k+1)/2$-horseshoe if s^k is odd. In both cases, we have disjoint intervals, so by what was proved above for that case, we know that $h(f^k) \geq \log(s^k/2)$. By Proposition 8.5.2 – note that Proposition 8.1.11 will not do – this implies that $h(f) \geq \frac{1}{k}(k \log s - \log 2)$ for all $k \geq 2$. Taking the limit for $k \rightsquigarrow \infty$ completes the proof. □

Corollary 8.6.3. *If there exists $n \in \mathbb{N}$ such that f^n has an s-horseshoe for some $s \geq 2$ then $h(f) > 0$.*

Proof. If f^n has an s-horseshoe then the Propositions 8.5.2 and 8.6.2 imply that $h(f) = \frac{1}{n}h(f^n) \geq \frac{1}{n}\log s > 0$. □

Examples.

(1) Suppose (X, f) has a periodic point with period 3. By Example (2) at the beginning of Section 7.4, f^2 has a 2-horseshoe. It follows that $h(f^2) \geq \log 2$, hence $h(f) \geq \frac{1}{2}\log 2$.

(2) In Example (5) just before Proposition 1.1.1, f^2 maps the intervals $[2; 3]$ and $[3; 4]$ over the interval $[2; 4]$. So f^2 admits a 2-horseshoe. It follows that $h(f) \geq \frac{1}{2}\log 2$.

(3) Consider the system defined in Example F in 6.3.5. In Example (3) after Theorem 8.2.7 it was shown that $h(f) = \log\frac{1}{2}(1+\sqrt{5})$, which is less than $\log 2$. Hence this system has no s-horseshoe with $s \geq 2$. This is in accordance with the Example in Proposition 7.4.1.

We shall prove now the converse of the above corollary. First, we present two lemmas concerning lim sup's.

Lemma 8.6.4. *For $i = 1, \ldots, k$, let $(a_n^{(i)})_{n\in\mathbb{N}}$ be a sequence of positive real numbers. Then*

$$\limsup_{n\to\infty} \frac{1}{n}\log(a_n^{(1)} + \cdots + a_n^{(k)}) = \max_{1\leq i\leq k}\left\{\limsup_{n\to\infty}\frac{1}{n}\log a_n^{(i)}\right\}.$$

Proof. For every $i \in \{1, \ldots, k\}$ we have $a_n^{(1)} + \cdots + a_n^{(k)} \geq a_n^{(i)}$, so the left-hand side of the equality is greater than or equal to the right-hand side. Conversely, let R denote the right-hand side of the equality. If $R = \infty$ then there remains nothing to prove. If R is finite then for every $i \in \{1, \ldots, k\}$ one has $\limsup_{n\to\infty}\frac{1}{n}\log a_n^{(i)} \leq R$, so if $\varepsilon > 0$ then $a_n^{(i)} \leq \exp(n(R+\varepsilon))$ for almost all n (independently of i, as we have to deal with only finitely many values of i) and, consequently,

$$a_n^{(1)} + \cdots + a_n^{(k)} \leq n \cdot \exp(n(R+\varepsilon)).$$

It follows that, for almost all n,

$$\frac{1}{n}\log(a_n^{(1)} + \cdots + a_n^{(k)}) \leq \frac{1}{n}\log n + R + \varepsilon.$$

Thus, the left-hand side of the equality above is less than or equal to $R + \varepsilon$. Since this holds for every $\varepsilon > 0$, the desired result follows. □

Lemma 8.6.5. *Let $(a_n)_{n\in\mathbb{N}}$ and $(b_n)_{n\in\mathbb{N}}$ be two sequences of non-negative real numbers and let $a_0 := b_0 := 0$. Then*

$$\limsup_{n\to\infty}\frac{1}{n}\log\Big(\sum_{k=0}^{n}\exp(a_k + b_{n-k})\Big) \leq \max\Big\{\limsup_{n\to\infty}\frac{a_n}{n}, \limsup_{n\to\infty}\frac{b_n}{n}\Big\}.$$

Proof. Denote the right-hand side of the above inequality by R. If $R = \infty$ then the inequality is obviously true, so we shall assume that R is finite. Obviously, $R \geq 0$. If

$\varepsilon > 0$ then there exists $n_\varepsilon \in \mathbb{N}$ such that both a_n/n and b_n/n are less than $R + \varepsilon$ for all $n \geq n_\varepsilon$. Stated otherwise,

$$\forall n \geq n_\varepsilon : \quad a_n < n(R+\varepsilon) \quad \text{and} \quad b_n < n(R+\varepsilon)\,. \tag{$*$}$$

In addition, let $M_\varepsilon := \max_{1\leq n<n_\varepsilon}\left\{\frac{a_n}{n}, \frac{b_n}{n}\right\} \geq 0$. Then

$$\forall n \in \{0, 1, \dots, n_\varepsilon - 1\} : \quad a_n \leq nM_\varepsilon \quad \text{and} \quad b_n \leq nM_\varepsilon \tag{$**$}$$

(for $1 \leq n \leq n_\varepsilon - 1$ this follows from the definition of M_ε; for $n = 0$ we have equalities). Now consider $n, k \in \mathbb{N}$ such that $n \geq 2n_\varepsilon$ and $0 \leq k \leq n$. In order to get an estimate of the sum $a_k + b_{n-k}$ for these values of k and n we distinguish three cases (the fourth case would be $k < n_\varepsilon$ and $n - k < n_\varepsilon$, but this is not compatible with the choice of $n \geq 2n_\varepsilon$):

– $k \geq n_\varepsilon$ and $n - k \geq n_\varepsilon$. Then $(*)$ implies

$$a_k + b_{n-k} < k(R+\varepsilon) + (n-k)(R+\varepsilon) = n(R+\varepsilon)\,.$$

– $k \geq n_\varepsilon$ and $n - k < n_\varepsilon$. Then $(*)$ and $(**)$ imply

$$a_k + b_{n-k} < k(R+\varepsilon) + (n-k)M_\varepsilon \leq n(R+\varepsilon) + n_\varepsilon M_\varepsilon\,.$$

– $k < n_\varepsilon$ and $n - k \geq n_\varepsilon$. Then $(*)$ and $(**)$ imply

$$a_k + b_{n-k} < kM_\varepsilon + (n-k)(R+\varepsilon) \leq n_\varepsilon M_\varepsilon + n(R+\varepsilon)\,.$$

In all cases we have $a_k + b_{n-k} < n(R+\varepsilon) + n_\varepsilon M_\varepsilon$. This obviously implies that

$$\frac{1}{n}\log\Big(\sum_{k=0}^{n}\exp(a_k + b_{n-k})\Big) \;\leq\; \frac{1}{n}\log(n+1) + (R+\varepsilon) + \frac{n_\varepsilon M_\varepsilon}{n}$$

for all $n \geq 2n_\varepsilon$. By taking the lim sup of this expression for $n \rightsquigarrow \infty$ one sees that the left-hand side of the inequality to prove is at most $R + \varepsilon$. As this holds for every $\varepsilon > 0$ this completes the proof. □

Remark. It is just for symmetry that we start the summation in the formula above with $k = 0$. The proof works equally well if we start the summation at $k = 1$. Actually, that is what we really need in the proof of Theorem 8.6.6 below.

It will be convenient to adapt some of the notation and techniques used in Section 6.1 to the present context. If $\mathcal{B}$ is a collection of subsets of X (not necessarily a cover) then for every $n \in \mathbb{N}$ the n-tuples $B_0 \dots B_{n-1}$ of elements of $\mathcal{B}$, that is, the elements of $\mathcal{B}^n = \mathcal{B} \times \cdots \times \mathcal{B}$ (n-times), will be denoted by Greek letters. If $\beta = B_0 \dots B_{n-1} \in \mathcal{B}^n$ then let

$$D_n(\beta) := \bigcap_{i=0}^{n-1}(f^i)^{\leftarrow}[B_i]\,.$$

If $n \in \mathbb{N}$ then an element $\beta \in \mathcal{B}^n$ is said to be an *allowed n-string over* $\mathcal{B}$ whenever $D_n(\beta) \neq \emptyset$. The set of all allowed n-strings over $\mathcal{B}$ will be denoted by $\mathcal{B}^{(n)}$:

$$\mathcal{B}^{(n)} := \{\, \beta \in \mathcal{B}^n \;\vdots\; D_n(\beta) \neq \emptyset \,\}\,.$$

If $B \in \mathcal{B}$ then $\mathcal{B}^{(n)}|_B$ denotes the set of allowed n-strings $B_0 \dots B_{n-1}$ over $\mathcal{B}$ with $B_0 = B$. Note that $\mathcal{B}^{(n)}|_B$ can be empty for certain $B \in \mathcal{B}$, even if $\mathcal{B}^{(n)}$ is not empty; obviously, in that case $\mathcal{B}^{(n)}|_B$ cannot be empty for all $B \in \mathcal{B}$.

If $x \in X$ and $\beta = B_0 \dots B_{n-1} \in \mathcal{B}^{(n)}$ then $x \in D_n(\beta)$ iff $f^i(x) \in B_i$ for $i = 0, \dots, n-1$. Consequently, if the members of $\mathcal{B}$ are mutually disjoint then the sets $D_n(\beta)$ for $\beta \in \mathcal{B}^{(n)}$ are mutually disjoint as well.

We hope the notational burden is not too overwhelming. Let us recapitulate:
- $\mathcal{B}^n = \mathcal{B} \times \dots \times \mathcal{B}$ (n times),
- $\mathcal{B}^{(n)} = \{ \beta \in \mathcal{B}^n : D_n(\beta) \neq \emptyset \}$, and
- $\mathcal{B}^{f,n} = \mathcal{B} \vee \dots \vee (f^{n-1})^{\leftarrow}\mathcal{B} = \{ D_n(\beta) : \beta \in \mathcal{B}^{(n)} \}$.

In particular, it should be clear that in the case that $\mathcal{B}$ consists of mutually disjoint sets then the collections $\mathcal{B}^{(n)}$ and $\mathcal{B}^{f,n}$ have the same number of elements: if β_1 and β_2 are distinct elements of $\mathcal{B}^{(n)}$ then $D_n(\beta_1)$ and $D_n(\beta_2)$ are mutually disjoint non-empty sets, hence mutually different.

Theorem 8.6.6. *Assume that X is a compact interval. If $h(f) > 0$ then there exists $n \in \mathbb{N}$ such that f^n has a 2-horseshoe.*

Proof. Our goal is to find a (closed) interval J such that f^n maps two subintervals of J with disjoint interiors over J for some $n \in \mathbb{N}$. The proof is rather long and complicated. We break it up in four parts, as follows: In Part I we define, after some preparations, a suitable partition $\mathcal{F}$ of X in intervals and in Part II we select a subfamily $\mathcal{E}$ of $\mathcal{F}$ such that the exponential growth rate of $\#\mathcal{E}^{(n)}|_A$ is sufficiently large for every $A \in \mathcal{E}$. In Part III a counting argument is used to show that for every $A \in \mathcal{E}$ there are $n \in \mathbb{N}$ and two subsets of A of the special form $D_n(\alpha)$ for $\alpha \in \mathcal{E}^{(n)}|_A$ which are mapped over some member B of $\mathcal{E}$. In Part IV it is shown that this member A of $\mathcal{E}$ can be selected such that the corresponding set B is equal to A. Finally, this configuration is modified so as to obtain a horseshoe for f^n.

Part I. Without restriction of generality we may assume that $h(f) > \log 3$. If necessary, replace f by f^r with $r \in \mathbb{N}$, $r > \log 3/h(f)$, and take into account that, by Proposition 8.5.2, $h(f^r) = rh(f)$. Let $c \in \mathbb{R}$, $\log 3 < c < h(f)$. By the definition of topological entropy, the interval X has an open cover $\mathcal{A}$ such that $h(\mathcal{A}, f) > c$. Let δ be a Lebesgue number of $\mathcal{A}$. Select division points $a = \xi_0 < \xi_1 < \dots < \xi_k = b$ in X which are spaced less than δ apart, where a and b are the left and right end points of X, respectively. Let $A_i := [\xi_{i-1}; \xi_i)$ for $1 \leq i \leq k-1$ and $A_k := [\xi_{k-1}; b]$. Then $\mathcal{F} := \{A_1, \dots, A_k\}$ is a finite cover (not an open cover) of X consisting of mutually disjoint intervals. By the choice of the members of $\mathcal{F}$ and the definition of δ one has $\mathcal{A} < \mathcal{F}$, so $h(f, \mathcal{F}) \geq h(f, \mathcal{A}) > c$ by Lemma 8.4.4 (2), hence

$$h(f, \mathcal{F}) > c > \log 3\,. \tag{8.6-1}$$

Recall that $h(f, \mathcal{F}) = \limsup \frac{1}{n} \log N(\mathcal{F}^{f,n})$. The cover $\mathcal{F}^{f,n}$ consists of all sets of the form $D_n(\alpha)$ with $\alpha = A_{p(0)} \cdots A_{p(n-1)} \in \mathcal{F}^{(n)}$ with $1 \leq p(i) \leq k$ for $i = 0, \dots, n-1$. These sets are mutually disjoint, so the cover $\mathcal{F}^{f,n}$ has no proper subcover and, consequently, $N(\mathcal{F}^{f,n}) = \#\mathcal{F}^{f,n}$. Moreover, $\mathcal{F}^{f,n}$ and $\mathcal{F}^{(n)}$ have the same number of elements, i.e., $\#\mathcal{F}^{f,n} = \#\mathcal{F}^{(n)}$. Consequently, the inequalities in (8.6-1) imply that

$$\limsup_{n\rightsquigarrow\infty} \frac{1}{n}\log(\#\mathcal{F}^{(n)}) > c > \log 3\,. \tag{8.6-2}$$

Part II. Obviously, $\mathcal{F}^{(n)} = \bigcup_{A\in\mathcal{F}} \mathcal{F}^{(n)}|_A$, where the sets $\mathcal{F}^{(n)}|_A$ for $A \in \mathcal{F}$ are mutually disjoint. It follows that $\#\mathcal{F}^{(n)} = \sum_{A\in\mathcal{F}} \#(\mathcal{F}^{(n)}|_A)$. So Lemma 8.6.4 implies that the largest of the numbers $\limsup \frac{1}{n}\log\#(\mathcal{F}^{(n)}|_A)$ with $A \in \mathcal{F}$ is larger than c. Consequently, the set

$$\mathcal{E} := \{\, A \in \mathcal{F} : \limsup_{n\rightsquigarrow\infty} \frac{1}{n}\log\#(\mathcal{F}^{(n)}|_A) > c \,\}$$

is not empty.

Claim.

$$\forall\, A \in \mathcal{E} :\ \ \limsup_{n\rightsquigarrow\infty} \frac{1}{n}\log\#(\mathcal{E}^{(n)}|_A) > c\,. \tag{8.6-3}$$

Fix $A \in \mathcal{E}$ and let $A_{p(0)}\dots A_{p(n-1)} \in \mathcal{F}^{(n)}|_A$. Then $A_{p(0)} = A \in \mathcal{E}$, so there is a greatest integer m in $\{1,\dots,n\}$ such that $A_{p(i)} \in \mathcal{E}$ for $0 \le i \le m-1$. In the case that $m = n$ we obviously have $A_{p(0)}\dots A_{p(n-1)} \in \mathcal{E}^{(n)}|_A$, and in the case that $1 \le m \le n-1$ we have $A_{p(0)}\dots A_{p(m-1)} \in \mathcal{E}^{(m)}|_A$ and $A_{p(m)}\dots A_{p(n-1)} \in \mathcal{F}^{(n-m)}|_B$ for some $B \in \mathcal{F}\setminus\mathcal{E}$ (namely, $B := A_{p(m)}$). Consequently, every allowed n-string over $\mathcal{F}$ starting with a member A of $\mathcal{E}$ either is an allowed n-string over $\mathcal{E}$ starting with A (followed by the empty string) or there is an $m \in \{1,\dots,n-1\}$ such that it consists of an allowed m-string over $\mathcal{E}$ starting with A, followed by an allowed $(n-m)$-string over $\mathcal{F}$ starting with an element from $\mathcal{F}\setminus\mathcal{E}$. It follows that

$$\#(\mathcal{F}^{(n)}|_A) \le \sum_{m=1}^{n}\Big(\#(\mathcal{E}^{(m)}|_A)\sum_{B\in\mathcal{F}\setminus\mathcal{E}}\#(\mathcal{F}^{(n-m)}|_B)\Big) \tag{8.6-4}$$

where we use the convention that (for $m = n$) $\sum_{B\in\mathcal{F}\setminus\mathcal{E}}\#(\mathcal{F}^{(0)}|_B) = 1$.

Let $a_0 := 0$, and for $j = 1,\dots,n$, let $a_j := \log\#(\mathcal{E}^{(j)}|_A)$. In addition, for $j = 0,\dots,n-1$ let $b_j := \log(\sum_{B\in\mathcal{F}\setminus\mathcal{E}}\#(\mathcal{F}^{(j)}|_B))$ (so by the above convention, $b_0 = 0$). Then (8.6-4) can be rewritten as[7]

$$\#(\mathcal{F}^{(n)}|_A) \le \sum_{m=1}^{n}\exp(a_m + b_{n-m}) \le \sum_{m=0}^{n}\exp(a_m + b_{n-m})\,.$$

Since $A \in \mathcal{E}$, it follows from the definition of $\mathcal{E}$ that

$$\limsup_{n\rightsquigarrow\infty}\frac{1}{n}\log\Big(\sum_{m=0}^{n}\exp(a_m + b_{n-m})\Big) > c\,.$$

7 See the Remark after Lemma 8.6.5. If the summation in Lemma 8.6.5 would start at $k = 1$ then in the following we would not need the second inequality.

Now apply Lemma 8.6.5 above. We get

$$\max\left\{\limsup_{n\rightsquigarrow\infty}\frac{a_n}{n}\,,\ \limsup_{n\rightsquigarrow\infty}\frac{b_n}{n}\right\} > c\,. \tag{8.6-5}$$

However, by the definition of $\mathcal{E}$ we have $\limsup_{n\rightsquigarrow\infty}\frac{1}{n}\log\#(\mathcal{F}^{(n)}|_B) \le c$ for all $B \in \mathcal{F}\setminus\mathcal{E}$, which implies that

$$\begin{aligned}\limsup_{n\rightsquigarrow\infty}\frac{b_n}{n} &= \limsup_{n\rightsquigarrow\infty}\frac{1}{n}\log\Big(\sum_{B\in\mathcal{F}\setminus\mathcal{E}}\#(\mathcal{F}^{(n)}|_B)\Big)\\ &\overset{(*)}{=} \max_{B\in\mathcal{F}\setminus\mathcal{E}}\left\{\limsup_{n\rightsquigarrow\infty}\frac{1}{n}\log\#(\mathcal{F}^{(n)}|_B)\right\} \le c\,,\end{aligned}$$

where $\overset{(*)}{=}$ is justified by Lemma 8.6.4. In combination with (8.6-5) this shows that $\limsup_{n\rightsquigarrow\infty}\frac{a_n}{n} > c$. This completes the proof of claim (8.6-3).

For the time being, we consider a fixed (but arbitrary) member A of $\mathcal{E}$. Let $e_n := \#\mathcal{E}^{(n)}|_A$ ($n \in \mathbb{N}$) and put $d := \exp(c)$. Then $d > 3$, and (8.6-3) implies

$$e_n > d^n \text{ for infinitely many values of } n \in \mathbb{N}\,. \tag{8.6-6}$$

This implies one additional property of the numbers e_n, namely, that

$$e_{n+1} \ge 3e_n \text{ for infinitely many values of } n \in \mathbb{N}\,. \tag{8.6-7}$$

Suppose this is not true: then there exists an $r \in \mathbb{N}$ such that $e_{n+1} < 3e_n$ for all $n \ge r$. By induction, it would follow that $e_n < 3^n(e_r/3^r)$ for all $n \ge r$, which contradicts (8.6-6), because $d > 3$. This contradiction proves (8.6-7).

There is no *a priori* reason why there should be infinitely many values of n for which both (8.6-6) and (8.6-7) hold. Yet that is almost the case, namely, with 3 instead of d in (8.6-6) – see (8.6-8) below. To prove this, we need one more inequality:

$$\forall\, n \in \mathbb{N}: \quad e_{n+1} \le ke_n\,.$$

This inequality follows easily from the fact that every member of $\mathcal{E}^{(n+1)}|_A$, that is, every allowed $(n+1)$-string $A_{p(0)}\dots A_{p(n-1)}A_{p(n)}$, consists of the (allowed) n-string $A_{p(0)}\dots A_{p(n-1)}$, which is in $\mathcal{E}^{(n)}|_A$ (e_n possibilities), followed by the final element $A_{p(n)}$ (at most k possibilities).

We are now able to show that there infinitely many values of $n \in \mathbb{N}$ such that both $e_n > 3^n$ and $e_{n+1} > 3e_n$. Suppose this is not true. Then there exists $r \in \mathbb{N}$ such that

$$\forall\, n \ge r: \ e_n > 3^n \Rightarrow e_{n+1} \le 3e_n\,. \tag{$*$}$$

Let $C := \max\{\,k, e_r\,\}$. We shall show that $(*)$ implies that $e_n \le C3^n$ for all $n \ge r$. Together with (8.6-6) this implies that $d^n \le C3^n$ for infinitely many values of n, which contradicts the choice of $d > 3$. So in order to complete the proof that $(*)$ is impossible it remains to show that it implies that $e_n \le C3^n$ for all $n \ge r$. We do this with induction in n.

For $n = r$ the claim is true, because $C \geq e_r$. Suppose the claim is true for some $n \geq r$. If $e_n \leq 3^n$ then $e_{n+1} \leq ke_n \leq k3^n \leq C3^n < C3^{n+1}$. On the other hand, if $e_n > 3^n$ then $(*)$ implies that $e_{n+1} \leq 3e_n \leq 3 \cdot C3^n = C3^{n+1}$. This completes the proof. Resuming, we have

$$\left.\begin{aligned} e_n &> 3^n \\ e_{n+1} &> 3e_n \end{aligned}\right\} \text{ for infinitely many values of } n \in \mathbb{N}\,. \tag{8.6-8}$$

Part III. Define for every $B \in \mathcal{E}$ and $n \in \mathbb{N}$ the integer

$$\gamma(A, B, n) := \#\{\alpha \in \mathcal{E}^{(n)}|_A : f^n[D_n(\alpha)] \supseteq B\}$$

(recall that above we fixed some $A \in \mathcal{E}$). In Lemma 8.6.8 below we shall show that, for every value of $n \in \mathbb{N}$ and every allowed n-string α over $\mathcal{E}$, the set $f^n[D_n(\alpha)]$ is an interval. Consequently, if for some $n \in \mathbb{N}$ and some allowed n-string $\alpha \in \mathcal{E}^{(n)}|_A$ we know that $f^n[D_n(\alpha)]$ has a non-empty intersection with s members of $\mathcal{E}$ then it covers at least $(s-2)$ members[8] of $\mathcal{E}$. This accounts for the inequality in the following computation:

$$\begin{aligned} \sum_{B\in\mathcal{E}} \gamma(A, B, n) &= \sum_{B\in\mathcal{E}} \#\{\alpha \in \mathcal{E}^{(n)}|_A : f^n[D_n(\alpha)] \supseteq B\} \\ &= \sum_{\alpha\in\mathcal{E}^{(n)}|_A} \#\{B \in \mathcal{E} : f^n[D_n(\alpha)] \supseteq B\} \\ &\geq \sum_{\alpha\in\mathcal{E}^{(n)}|_A} \Big(\#\{B \in \mathcal{E} : f^n[D_n(\alpha)] \cap B \neq \emptyset\} - 2\Big) \\ &= \#\{(\alpha, B) \in \mathcal{E}^{(n)}|_A \times \mathcal{E} : f^n[D_n(\alpha)] \cap B \neq \emptyset\} - 2 \cdot (\#\mathcal{E}^{(n)}|_A)\,. \end{aligned}$$

Here the second term of the right-hand side is equal to $2e_n$. The first term turns out to be equal to $\#\mathcal{E}^{(n+1)}|_A = e_{n+1}$. To prove this, note that every $(n+1)$-string over $\mathcal{E}$ starting with A can be seen as an n-string α over $\mathcal{E}$ starting with A, followed by any member $B \in \mathcal{E}$. The condition that the concatenation αB is allowed means that $D_n(\alpha) \cap (f^n)^{\leftarrow}[B] \neq \emptyset$, which is equivalent with the condition that $f^n[D_n(\alpha)] \cap B \neq \emptyset$. This completes the proof that the term under consideration equals $\#\mathcal{E}^{(n+1)}|_A$. Resuming:

$$\sum_{B\in\mathcal{E}} \gamma(A, B, n) \geq e_{n+1} - 2e_n\,.$$

Now recall the result stated in (8.6-8): for infinitely many values of $n \in \mathbb{N}$ we not only have $e_{n+1} > 3e_n$, so that $e_{n+1} - 2e_n > e_n$, but also $e_n > 3^n$. Consequently, there is a value of n such that $e_{n+1} - 2e_n > k$, i.e., such that

$$\sum_{B\in\mathcal{E}} \gamma(A, B, n) > k\,.$$

8 Obviously, this is also true for $s = 0$, $s = 1$ and $s = 2$.

Recall that $k = \#\mathcal{F}$. Since $\mathcal{E} \subseteq \mathcal{F}$ it is clear that $\mathcal{E}$ has at most k elements, so the above inequality implies that there exists (at least one) $B \in \mathcal{E}$ such that $\gamma(A, B, n) \geq 2$. This conclusion holds for every member A of $\mathcal{E}$, so we can select for every $A \in \mathcal{E}$ an element $n_A \in \mathbb{N}$ and an element $\varphi(A) \in \mathcal{E}$ with the property that $\gamma(A, \varphi(A), n_A) \geq 2$.

Conclusion. There is a mapping $\varphi: \mathcal{E} \to \mathcal{E}$ with the property that for every $A \in \mathcal{E}$ there exists an $n_A \in \mathbb{N}$ such that $\gamma(A, \varphi(A), n_A) \geq 2$. This mapping can be iterated, and we shall show now:

$$\forall A \in \mathcal{E}\ \forall i \in \mathbb{N}\ \exists n_{A,i} \in \mathbb{N} \vdots\ \gamma(A, \varphi^i(A), n_{A,i}) \geq 2 . \tag{8.6-9}$$

The straightforward (but tedious) proof is by induction in i. Let $A \in \mathcal{E}$. If $i = 1$ then, by the very definition of φ, we have $\gamma(A, \varphi^1(A), n_{A,1}) \geq 2$ with $n_{A,1} := n_A$. Next, suppose that for some $i \in \mathbb{N}$ there exists $n_i \in \mathbb{N}$ such that $\gamma(A, \varphi^i(A), n_i) \geq 2$ (we suppress the 'A' in $n_{A,i}$). This means that there are $\alpha, \alpha' \in \mathcal{E}^{(n_i)}|_A$ such that $\alpha \neq \alpha'$, $f^{n_i}[D_{n_i}(\alpha)] \supseteq \varphi^i(A)$ and $f^{n_i}[D_{n_i}(\alpha')] \supseteq \varphi^i(A)$. In addition, the definition of φ implies that we also have $\gamma(\varphi^i(A), \varphi^{i+1}(A), l) \geq 2 \geq 1$ for $l := n_{\varphi^i(A),1} \in \mathbb{N}$, so at least one subset of $\varphi^i(A)$ of the form $D_l(\beta)$ for some l-string $\beta \in \mathcal{E}^{(l)}|_{\varphi^i(A)}$ satisfies the condition $f^l[D_l(\beta)] \supseteq \varphi^{i+1}(A)$. If $\alpha = A_{p(0)} \dots A_{p(n_i-1)}$ and $\beta = A_{q(0)} \dots A_{q(l-1)}$ with $p(j), q(j) \in \{1, \dots, k\}$ for all j, then it is easily seen that

$$D_{n_i+l}(\alpha\beta) = \bigcap_{j=0}^{n_i-1} (f^j)^{\leftarrow}[A_{p(j)}] \cap (f^{n_i})^{\leftarrow}\Big[\bigcap_{j=0}^{l-1} (f^j)^{\leftarrow}[B_{p(j)}]\Big],$$

hence

$$\begin{aligned} f^{n_i}[D_{n_i+l}(\alpha\beta)] &= f^{n_i}[\,D_{n_i}(\alpha) \cap (f^{n_i})^{\leftarrow}[D_l(\beta)]\,] \\ &= f^{n_i}[D_{n_i}(\alpha)] \cap D_l(\beta) \\ &\supseteq \varphi^i(A) \cap D_l(\beta) = D_l(\beta) \neq \emptyset, \end{aligned}$$

so $D_{n_i+l}(\alpha\beta) \neq \emptyset$ as well. Stated otherwise, the $(n_i + l)$-string $\alpha\beta$ is allowed, that is, $\alpha\beta \in \mathcal{E}^{(n_i+l)}|_A$. It follows also from the above computation that

$$f^{n_i+l}[D_{n_i+l}(\alpha\beta)] \supseteq f^l[D_l(\beta)] \supseteq \varphi^{i+1}(A) .$$

Similarly, we get $\alpha'\beta \in \mathcal{E}^{(n_i+l)}|_A$ and $f^{n_i+l}[D_{n_i+l}(\alpha'\beta)] \supseteq \varphi^{i+1}(A)$. This shows that for $n_{i+1} := n_i + l$ we have $\gamma(A, \varphi^{i+1}(A), n_{i+1}) \geq 2$. This completes the proof of (8.6-9).

Part IV. The set $\mathcal{E}$ is finite, so every member A of $\mathcal{E}$ is ultimately periodic under φ. Consequently, there is a φ-periodic element $A_0 \in \mathcal{E}$, say, with period $m \geq 1$, that is, $\varphi^m(A_0) = A_0$. For $A := A_0$ and $i := m$, formula (8.6-9) becomes:

$$\exists n_m \in \mathbb{N} \vdots\ \gamma(A_0, A_0, n_m) \geq 2 . \tag{8.6-10}$$

Thus, there are $\alpha^{(1)}, \alpha^{(2)} \in \mathcal{E}^{(n_m)}|_{A_0}$ such that $\alpha^{(1)} \neq \alpha^{(2)}$ and

$$f^{n_m}[D_{n_m}(\alpha^{(j)})] \supseteq A_0 \supseteq D_{n_m}(\alpha^{(1)}) \cup D_{n_m}(\alpha^{(2)}) \tag{8.6-11}$$

for $j = 1, 2$. This looks like a horseshoe for f^{n_m}, except that we don't know if the sets $D_{n_m}(\alpha^{(1)})$ and $D_{n_m}(\alpha^{(2)})$ are intervals.

However, suppose $\alpha^{(j)} = A^{(j)}_{p_j(0)} \dots A^{(j)}_{p_j(n_m-1)}$ with $p_j(i) \in \{1, \dots, k\}$ for $i = 0, \dots, n_m - 1$ and $j = 1, 2$. Then it follows from Lemma 8.6.8 (2) below that for $j = 1, 2$ there exists an interval $K_j \subseteq D_{n_m}(\alpha^{(j)})$ such that

$$f^{n_m}[K_j] = f^{n_m}[D_{n_m}(\alpha^{(j)})] \quad \text{and} \quad f^i[K_j] \subseteq A^{(j)}_{p_j(i)} \quad \text{for } i = 0, \dots, n_m - 1 .$$

The latter condition implies that the two intervals K_1 and K_2 are disjoint: since $\alpha^{(1)} \neq \alpha^{(2)}$ there is an $i \in \{0, \dots, n_m - 1\}$ such that the sets $A^{(1)}_{p_1(i)}$ and $A^{(2)}_{p_2(i)}$ are different, hence disjoint. Then $f^i[K_1]$ and $f^i[K_2]$ are disjoint, hence the intervals K_1 and K_2 are disjoint as well. It follows that $\overline{K}_1$ and $\overline{K}_2$ are intervals with disjoint interiors.

The first condition for K_1 and K_2 and the first inclusion in (8.6-11) together imply that, for $j = 1, 2$, $f^{n_m}[K_j] \supseteq A_0$, where A_0 is a non-degenerate interval. Hence the interval K_j is non-degenerate as well. Moreover,

$$f^{n_m}[\,\overline{K_j}\,] \supseteq \overline{A_0} \supseteq \overline{D_{n_m}(\alpha^{(1)})} \cup \overline{D_{n_m}(\alpha^{(2)})} \supseteq \overline{K}_1 \cup \overline{K}_1$$

for $j = 1, 2$. So $(\overline{K}_1, \overline{K}_2)$ is a 2-horseshoe for f^{n_m}. □

Remark. Recall from the proof of Corollary 8.6.3: if f^n has an s-horseshoe then $\frac{1}{n}\log s \leq h(f)$. By refining the arguments in the proof above one can show that this number $\frac{1}{n}\log s$ can be arbitrarily close to $h(f)$: for every $\varepsilon > 0$ there are $n \in \mathbb{N}$ and an s-horseshoe for f^n such that $h(f) - \varepsilon < \frac{1}{n}\log s \leq h(f)$. See Note 7 at the end of this chapter.

For the proof of Lemma 8.6.8 below we need Lemma 2.2.1 (2), but with some modification, because in Lemma 2.2.1 we consider compact intervals, while below we need a similar result for arbitrary bounded intervals (due to the fact that not all members of $\mathcal{F}$ can be compact[9]). The modification is as follows:

Lemma 8.6.7. *Let Z be a non-empty bounded subset of $\mathbb{R}$, let $g\colon Z \to \mathbb{R}$ be a continuous mapping and let I and J be non-empty intervals such that $\overline{I} \subseteq Z$ and $\overline{J} \subseteq g[Z]$. If $g\colon I \overset{\circ}{\rightarrow} J$ then there is a subinterval K of I such that $f[K] = J$. If J is non-degenerate then K is non-degenerate.*

Proof. First, assume that both I and J are non-degenerate. Then the closures $\overline{I}$ and $\overline{J}$ of I and J are compact non-degenerate intervals and, by compactness of $\overline{I}$, we have $g[\,\overline{I}\,] = \overline{g[I]} \supseteq \overline{J}$. So by Theorem 2.2.2 (2), there is a closed subinterval K' of $\overline{I}$ such that $g[K'] = \overline{J}$. In addition, by the Remark following Lemma 2.2.1, we may assume that g maps the interior of K' onto the interior of $\overline{J}$ (so K' cannot be degenerate) and that g maps the set of the two end points of K' onto the set of the two end points of $\overline{J}$.

9 Other choices can be made for $\mathcal{F}$, but X has no partition existing only of compact intervals.

Now take into account that J equals $\overline{J}$ minus one or two end points. By deleting the corresponding end point(s) from K' we get the desired interval K.

If I is degenerate then I is a singleton set, and it follows that J is a singleton set as well. So $f[I] = J$, and we can take $K = I$. If J is degenerate we can take for K any singleton subset or any closed interval included in the non-empty set $g^{\leftarrow}[J] \cap I$ (possibly, the only such intervals are degenerate). This completes the proof of the first statement. The second statement is an easy consequence of the first: if K is a singleton set then so is $J = f[K]$. □

Let notation be as in the proof of Theorem 8.6.6. In particular, $\mathcal{F}$ is a partition of X into k intervals A_i ($1 \le i \le k$), and if $\alpha = A_{i_0} \dots A_{i_{n-1}} \in \mathcal{F}^n$ is an n-tuple ($n \in \mathbb{N}$) then $D_n(\alpha) = \bigcap_{j=0}^{n-1} (f^j)^{\leftarrow}[A_{i_j}]$. Moreover, $\mathcal{F}^{(n)}$ is the set of elements $\alpha \in \mathcal{F}^n$ such that $D_n(\alpha) \neq \emptyset$ (the allowed n-strings) and $\mathcal{F}^{(n)}|_A$ is the set of all allowed n-strings that start with a given element A of $\mathcal{F}$.

Lemma 8.6.8. *For every $n \in \mathbb{N}$ and every $\alpha = A_{i_0} \dots A_{i_{n-1}} \in \mathcal{F}^{(n)}$ the following statements hold:*
1. *$f^n[D_n(\alpha)]$ is an interval.*
2. *There is an interval $K \subseteq D_n(\alpha)$ such that $f^n[K] = f^n[D_n(\alpha)]$ and $f^j[K] \subseteq A_{i_j}$ for $j = 0, \dots, n-1$.*

In particular, if the interior of $f^n[D_n(\alpha)]$ is not empty, i.e., if it is a non-degenerate interval, then the interval K in (2) is non-degenerate as well.

Proof. (1) Of course, statement (1) follows from statement (2), but (1) has simple direct proof, while that of (2) is more involved. So we prove statement (1) directly, using induction in n.

For $n = 1$ the statement is obviously true: for a 1-string $\alpha = A_{i_0}$ we have $f[D_1(\alpha)] = f[A_{i_0}]$, which is the continuous image of an interval, hence an interval. Suppose the statement is true for some $n \in \mathbb{N}$ and all allowed n-strings. For an allowed $(n+1)$-string $\alpha = A_{i_0} \dots A_{i_{n-1}} A_{i_n}$ we clearly have $D_{n+1}(\alpha) = D_n(\alpha') \cap (f^n)^{\leftarrow}[A_{i_n}]$ with $\alpha' := A_{i_0} \dots A_{i_{n-1}}$ an allowed n-string (if it were not allowed then $D_{n+1}(\alpha)$ would be empty). It follows that $f^n[D_{n+1}(\alpha)] = f^n[D_n(\alpha')] \cap A_{i_n}$. By the induction hypothesis, this is an intersection of two intervals; this intersection is not empty, because α is allowed. Hence $f^n[D_{n+1}(\alpha)]$ is an interval (possibly degenerate). Consequently, its image $f^{n+1}[D_{n+1}(\alpha)]$ under f is an interval as well. This completes the proof.

(2) The proof of statement (2) is by induction as well, and it starts like the proof of (1). For $n = 1$, the statement is obviously true: if $\alpha = A_{i_0} \in \mathcal{F}$ then $f[D_1(\alpha)] = f[A_{i_0}]$ and we can take $K = A_{i_0}$. Next, assume that statement (2) holds for some $n \in \mathbb{N}$ and all allowed n-strings over $\mathcal{F}$. Then for an arbitrary allowed $(n+1)$-string $\alpha = A_{i_0} \dots A_{i_{n-1}} A_{i_n}$ we obviously have $D_{n+1}(\alpha) = D_n(\alpha') \cap (f^n)^{\leftarrow}[A_{i_n}]$ with $\alpha' := A_{i_0} \dots A_{i_{n-1}}$ an allowed n-string over $\mathcal{F}$. It follows that

$$f^n[D_{n+1}(\alpha)] = f^n[D_n(\alpha')] \cap A_{i_n} \tag{8.6-12}$$

By the induction hypothesis there exists an interval $L \subseteq D_n(\alpha')$ such that $f^j[L] \subseteq A_{i_j}$ for $j = 0, \dots, n-1$ and $f^n[L] = f^n[D_n(\alpha')]$. By (8.6-12), the non-empty set $J := f^n[D_{n+1}(\alpha)]$ is the intersection of two intervals, hence an interval itself. It is included in the interval $f^n[D_n(\alpha')]$, i.e., it is included in $f^n[L]$. So we have $f^n : L \overset{\circ}{\to} J$, and by Lemma 8.6.7 above, there is an interval $K \subseteq L$ such that $f^n[K] = J = f^n[D_{n+1}(\alpha)]$. By applying f once more, we get $f^{n+1}[K] = f^{n+1}[D_{n+1}(\alpha)]$. Moreover, since $K \subseteq L$ it is clear that $f^j[K]$ is included in A_{i_j} for $j = 0, \dots, n-1$. For $j = n$ we get, in view of the choice of K and (8.6-12): $f^n[K] = f^n[D_{n+1}(\alpha)] \subseteq A_{i_n}$ This concludes the proof of (2).
Finally, the concluding statement of the lemma is obvious: if K is degenerate then $f^n[K] = f^n[D_n(\alpha)]$ consists of one point, so this interval cannot be non-degenerate. □

Remark. In the above lemma, it is possible that the intervals $f^n[D_n(\alpha)]$ for allowed n-strings α are degenerate. But in the argument in the proof of Part III of Theorem 8.6.6 where we used Lemma 8.6.8 (1) – if $f^n[D_n(\alpha)]$ meets s members of $\mathcal{F}$ then it covers at least $s - 2$ of them – this is irrelevant.

Corollary 8.6.9. *Let (X, f) be a dynamical system on a compact interval. The following conditions are equivalent:*
(i) $h(f) > 0$.
(ii) *There exists $n \in \mathbb{N}$ such that f^n has an s-horseshoe for some $s \geq 2$.*
(iii) *There exists $n \in \mathbb{N}$ such that f^n has an 2-horseshoe consisting of two disjoint closed intervals.*
(iv) *f has a periodic point whose primitive period is not a power of 2.*

Proof. "(i)⇔(ii)": Clear from Corollary 8.6.3 and Theorem 8.6.6.

"(ii)⇒(iii)": If f^n has an s-horseshoe with $s \geq 3$ then the left-most and right-most members of this s-horseshoe form a 2-horseshoe for f^n consisting of two disjoint closed intervals. If $s = 2$ and the two intervals of the horseshoe are not disjoint then apply Lemma 8.6.1 to f^n: the mapping f^{2n} has a 4-horseshoe. Now apply the previous case.

"(iii)⇒(ii)": Obvious.

"(iii)⇔(iv)": See Theorem 7.5.2. □

Examples.
(1) The quadratic mapping $f_\mu : x \mapsto \mu x(1 - x) : [0; 1] \to [0; 1]$ has for μ less than the Feigenbaum point μ_∞ only periodic points with primitive periods that are a power of 2 (for $0 < \mu < 1 + \sqrt{6}$ we have proved this rigorously: see 2.1.5 and Exercise 2.2. Hence for those values of μ, $h(f_\mu) = 0$. See also Note 9 at the end of this chapter.
(2) In 4.3.11 it is stated that the non-wandering set of the system $([0; 1], f_\infty)$ is equal to $C \cup P(f_\infty)$, where C is a minimal set (hence includes no periodic points) and $P(f_\infty)$ is the set of periodic points of f_∞, all of which have a period that is a power of 2. So Corollary 8.6.9 implies that $h(([0; 1], f_\infty)) = 0$.

(3) By 8.3.4, the generalized tent map with $1 < s \leq 2$ has positive entropy, hence it has periodic points with primitive period not a power of 2.

It may be useful to remind the reader at this point to the fact that the systems satisfying the conditions of the above corollary is LY-chaotic and that is has a D-chaotic subsystem: see Corollary 7.5.6 and Theorem 7.5.7. In the other direction we have:

Corollary 8.6.10. *Let (X, f) be a transitive dynamical system on a compact interval. Then $h(f) > 0$.*

Proof. Clear from Corollary 8.6.9 and Proposition 7.5.4. □

Remark. This statement does not hold for transitive (or even minimal) systems on arbitrary compact metric spaces, not even on the circle: by Example (1) after Corollary 8.1.9 – or Example (1) in 8.4.5 – the minimal system $(\mathbb{S}, \varphi_a)$ with $a \in \mathbb{R} \setminus \mathbb{Q}$ has entropy 0. See also Example (2) above (f_∞ has entropy 0 on the minimal set C) and Exercise 8.2.

Exercises

8.1. Let $X := \mathbb{R}$ with its usual metric $\rho: (x, y) \mapsto |x - y|: \mathbb{R}^2 \to \mathbb{R}^+$ and consider the mapping $f: x \mapsto x^2: \mathbb{R} \to \mathbb{R}$. Let $K := [3; 4]$. Show that

$$\text{minspan}_n(\varepsilon, K, f) \geq \frac{2^{n-2}\, 2^{2^{n-1}-1}}{\varepsilon}$$

for all $\varepsilon > 0$ and $n \in \mathbb{N}$, so that $r(\varepsilon, K, f) = \infty$.

NB.. It follows that the dynamical system $((0; \infty), f)$ has $h_\rho(f) = \infty$ (here ρ is the ordinary metric on $(0; \infty)$).

8.2.

(1) Show that the Morse–Thue system has entropy 0.
(2) Let (X, σ_X) be the even shift. Show that $h(\sigma_X) = \log \frac{1}{2}(1 + \sqrt{5})$.
(3) Let (X, σ_X) be a shift system. Recall from Corollary 5.2.4 (1) that for every $n \in \mathbb{N}$ the number $p_n(X)$ of periodic points with primitive period n is finite. Show that $h(\sigma_X) \geq \limsup_{n \rightsquigarrow \infty} \frac{1}{n} p_n(X)$.
(4) More generally, let (X, f) be an expansive system on a compact metric space. Recall from Exercise 6.8 (1) that for every $n \in \mathbb{N}$ the number $p_n(X)$ of periodic points with primitive period n is finite. Show that $h(f) \geq \limsup_{n \rightsquigarrow \infty} \frac{1}{n} p_n(X)$.

8.3. Prove: if $\mathcal{A}$ and $\mathcal{B}$ are covers of a space X then the join $\mathcal{A} \vee \mathcal{B}$ is the coarsest common refinement of $\mathcal{A}$ and $\mathcal{B}$, that is, if $\mathcal{C}$ is a cover of X such that both $\mathcal{A} < \mathcal{C}$ and $\mathcal{B} < \mathcal{C}$ then $\mathcal{A} \vee \mathcal{B} < \mathcal{C}$.

8.4. Let X be an interval and let $f: X \to X$ be a λ-Lipschitz mapping with $\lambda \geq 1$ (that is, $|f(x) - f(y)| \leq \lambda|x - y|$ for all $x, y \in X$). Then $h(f) \leq \log \lambda$.

8.5. Let (X, f) be a dynamical system on a compact metric space. For every $\delta > 0$ the real number $h(f)$ is equal to the supremum of the quantities $h(K, f)$, where K ranges over all compact subsets of X with diameter less than or equal to δ.

Notes

1 The original definition of topological entropy for a continuous selfmap of a compact metric space appeared in R. L. Adler, A. G. Konheim & M. H. McAndrew [1965]. In their definition they imitated the definition of the Kolmogorov–Sinaïentropy for measure preserving transformations in ergodic theory. This is the approach sketched in Section 8.4.

The approach sketched in the Sections 8.1, 8.2 and 8.3, based on the dispersion of orbits in metric spaces, comes from R. Bowen [1971/3] and E. I. Dinaburg [1970]. The proof that for compact metric spaces these two definitions yield the same quantity, which is an invariant of topological conjugacy in the class of compact metric spaces, was given in R. Bowen [1971]. When the space is not compact the situation is more complicated.

2 The definition of entropy in Section 8.1 is usually only given for a uniformly continuous self-map of a metric space. See, e.g., the standard exposition in P. Walters [1982]. The assumption of uniform continuity plays no role in this definition, although the examples motivating Bowen were all uniformly continuous. We have avoided the assumption of uniform continuity. However, uniform continuity is essential for factor maps to have useful properties, and when all phase spaces under consideration are compact this condition is automatically fulfilled. If the assumption of compactness of the phase space is dropped, a number of useful properties of entropy that hold for uniformly continuous maps can fail; see e.g. Example (1) after Lemma 8.2.1.

3 Theorem 8.2.7 is a special case of a result by Rufus Bowen: for a factor map $\varphi: (X, f) \to (Y, g)$, where X and Y are compact metric spaces, always

$$h(g) \leq h(f) \leq h(g) + \sup_{y \in Y} h(\varphi^{\leftarrow}[y], f)\,.$$

It is easy to see that Theorem 8.2.7 is a special case of this result: if K is a finite subset of X then $\mathrm{maxsep}_n(\varepsilon, K, f) \leq \#K$ for all $n \in \mathbb{N}$ and $\varepsilon > 0$, hence $h(K, f) = 0$. It follows that if the fibre $\varphi^{\leftarrow}[y]$ is finite, then $h(\varphi^{\leftarrow}[y], f) = 0$. Note that it follows from this general version that in Theorem 8.2.7 the uniform bound on the size of the fibres is superfluous.

4 The following result is in accordance with one of the remarks motivating the definition of topological entropy preceding 8.1.1: for $i = 1, 2$, let (X_i, f_i) be a dynamical system with metric phase space (X_i, ρ_i), and let d be the metric on $X_1 \times X_2$ defined by $d((x_1, x_2), (y_1, y_2)) := \max\{\rho(x_1, y_1), \rho(x_2, y_2)\}$ for $x_i, y_i \in X_i$. If one of the phase spaces under consideration is compact then $h(f_1 \times f_2) = h(f_1) + h(f_2)$; if none of the phase spaces is compact then only $h(f_1 \times f_2) \leq h(f_1) + h(f_2)$. See P. Walters [1982], Theorem 7.1.10(ii). For the proof one needs Corollary 8.4.10.

5 Recall that a function $f: J \to \mathbb{R}$ (J any interval in $\mathbb{R}$) is said to be *piecewise monotonous* whenever J is the union of finitely many intervals with disjoint interiors (so adjacent intervals have a common end point) on each of which f is strictly monotonous. In that case we can divide J in *maximal intervals* of monotonicity of f (sometimes called *laps*). Let $l(f)$ denote the number of maximal intervals of monotonicity of f (also called *the lap number of f*).

If f is piecewise monotonous then for every $n \in \mathbb{N}$ the function f^n is also piecewise monotonous, hence $l(f^n)$ is defined as well. It can be shown that the limit

$$\mathbf{s}(f) := \lim_{n\rightsquigarrow\infty} \sqrt[n]{l(f^n)}$$

always exists; this is rather easy: first show that $l(f^{m+n}) = l(f^m \circ f^n) \le l(f^m) \cdot l(f^n)$, then apply Lemma 8.1.10. It turns out that for every continuous, piecewise monotonous mapping of a compact interval into itself the equalities

$$h(f) = \log \mathbf{s}(f) := \lim_{n\rightsquigarrow\infty} \log \sqrt[n]{l(f^n)} = \lim_{n\rightsquigarrow\infty} \frac{1}{n} \log l(f^n)$$

hold. See M. Misiurewicz & W. Szlenk [1980].

6 Concerning Example (2) after Corollary 8.6.3, it can be shown that $h(f) \ge \log \lambda_5$, where λ_5 is the largest real root of the equation $x^5 - 2x^3 - 1 = 0$. This gives a better estimate for $h(f)$ than the one obtained in Example (2), because $\lambda_5 \simeq 1.51\cdots > 1.41\cdots \simeq \sqrt{2}$.

The idea behind the above statements is as follows: Let f be an interval map with a periodic point of period p with $p > 1$, p odd and p minimal with respect to these properties. Then the incidence matrix of the Markov graph of that periodic point (see Proposition 2.5.3 for what that graph looks like) has a maximal eigenvalue λ_p which is the unique positive root of $x^p - 2x^{p-2} - 1$, and $h(f) \ge \log \lambda_p$. Consequently, if f has a periodic point with period $2^d p$, $p > 1$ and p odd, then $h(f) \ge \frac{1}{2^d} \log \lambda_p$. See L. Block, J. Guckenheimer, M. Misiurewicz & L. S. Young [1980].

7 Proposition 8.6.2 can be found in L. S. Block & W. A. Coppel [1992]. It is a special case of a result in the paper mentioned at the end of the previous Note.

Theorem 8.6.6 is a simplification of a result from M. Misiurewicz [1979]. For the full result, see the Remark after Theorem 8.6.6. It is essentially a one-dimensional result. There are examples of homeomorphisms in dimension 2 (and of diffeomorphisms in dimension 4) that define non-trivial minimal systems with positive entropy. If positive entropy would always imply the existence of a horseshoe, there would be periodic points in such systems (Proposition 7.4.1 is the version for dimension 1 of this more general statement), which is impossible by minimality. For details, see the book J. Llibre & M. Misiurewicz [1993] (in particular, page 209).

8 In view of Corollary 8.6.9, the result of Corollary 7.4.5 can be formulated as follows: for a dynamical system on a compact interval, *positive entropy implies Li–Yorke chaos*. In F. Blanchard, E. Glasner, S. Kolyada & A. Maass [2002] it is shown that this conclusion is true for every dynamical system on a compact metric space (see also Note 8 to Chapter 7). The converse is not true: there are examples of systems on compact intervals with entropy 0 that are Li–Yorke chaotic; see J. Smítal [1986]. There are also such systems that are not Li–Yorke chaotic: see S. Ruette [2003].

9 For $\mu \le \mu_\infty$ the logistic system $([0;1], f_\mu)$ has only periodic points with period a power of 2, hence $h(f_\mu) = 0$. For larger values of μ there are points with other periods (see Note 6 in the Introduction), so then $h(f_\mu) > 0$. Finally, for $\mu = 4$ we get $h(f_4) = \log 2$; see Example (2) after Theorem 8.2.7. It can be shown that the mapping $\mu \mapsto h(f_\mu) : [1;4] \to \mathbb{R}^+$ is continuous and monotonously increasing, constant on non-degenerate closed intervals whose union is dense in $[1;4]$. So the graph of this mapping is a so-called *devil's staircase*. (The standard example of a devil's staircase is the Cantor function from $[0;1]$ onto itself, which is continuous, monotonously increasing and constant on the middle thirds deleted in the construction of the Cantor set.

10 In Corollary 8.6.10 the following estimation can be made about the value of the topological entropy of a transitive interval mapping f: $h(f) \ge \frac{1}{2} \log 2$. This result, originally due to A. M. Blokh, can be found in L. S. Block & E. M. Coven [1987]. This lower bound cannot be improved: for the transitive system $([0;1], S)$ of Figure 2.16 (a), S^2 is, essentially, the union of two tent maps, so that $h(S^2) = \log 2$ (see the hint to Exercise 8.5), hence $h(f) = \frac{1}{2} \log 2$. If f has two invariant points then there is a better estimate: then $h(f) \ge \log 2$ (the tent map shows that also this bound cannot improved).

The relationship between transitivity and positive entropy has attained much attention. Of course, the question whether of positive entropy implies transitivity is rather pointless: entropy is a local property (it can be determined by a small part of the system; see also Exercise 8.5), while transitivity is global. In fact, the disjoint union of two systems, one of which has positive entropy, is not transitive but has positive entropy. On the other hand, by Corollary 8.6.9 and Theorem 7.4.7, an interval map with positive entropy has an Auslander–Yorke chaotic subsystem which is, by definition, transitive. However this is a rather weak conclusion: it just means that there is an orbit closure without isolated points – see the Remark 2 after Proposition 1.3.3.

The question the other way round is much more interesting: has a compact transitive system positive entropy? According to Proposition 8.6.10 the answer is affirmative for systems on intervals in $\mathbb{R}$. Also for the circle the results are satisfactory: if $f: \mathbb{S} \to \mathbb{S}$ is transitive then $h(f) > 0$ or f is conjugate to an irrational rigid rotation (which has entropy zero).

Other results for one-dimensional spaces concern, for instance, trees (a *tree* is a connected space that is homeomorphic to the union of finitely many copies of the unit interval, but which does not contain a subset homeomorphic to a circle). Any transitive mapping on a tree has positive entropy. This follows from results in A. M. Blokh [1987].

In dimension two the situation is less satisfactory, For instance, there exists a transitive orientation preserving homeomorphism of the unit square with zero topological entropy. See Ll. Alsed'a, S. Kolyada, J. Llibre & L. Snoha [1999].

11 Most of this chapter is about topological entropy for interval maps. There is, of course, much more to say. The mere treatment of topological entropy for shift systems requires a book in itself: see, for instance, D. Lind & B. Marcus [1995] and B. Kitchens [1998]. We mention two typical results. First, for an irreducible SFT of order 2 represented by a graph G, the topological entropy is equal to $\log \lambda_G$, where λ_G is the unique positive eigenvalue of the adjacency matrix of G. Secondly, if (X, σ_X) is a sofic shift and for every $n \in \mathbb{N}$ the number of periodic points with primitive period n is denoted by $p_n(X)$, then $h(\sigma_X) = \limsup_{n\rightsquigarrow\infty} \frac{1}{n} p_n(X)$ (the inequality "≥" is easy: see Exercise 8.2 (3)).

It is interesting to compare the properties of topological entropy and measure theoretic entropy as invariants for a standard class of examples. In ergodic theory the standard models are Bernoulli shifts, and for this class Ornstein proved that measure theoretic entropy is a *complete* conjugacy invariant, meaning that two Bernoulli shifts are conjugate (by means of a measure preserving transformation) iff they have the same entropy (here the 'if' is the interesting and non-trivial part). In topological dynamics the standard models are aperiodic SFT's ('aperiodic' means that the gcd of all primitive periods of periodic points in the shift is equal to 1; it can be shown that the aperiodic transitive SFT's are just the strongly mixing SFT's). It is not too difficult to see that topological entropy is not a complete invariant for topological conjugacy in this class of systems, but in 1979 Adler and Marcus proved that topological entropy is a complete invariant under so-called *almost topological conjugacy*: see Theorem 9.3.2 in the book by Lind and Marcus mentioned above.

Finally, there is a strong connection between having positive entropy and sensitive behaviour: the intuitive feeling is that 'positive entropy' quantifies a rate for sensitive dependence and says that many nearby points have orbits that diverge exponentially fast. And indeed, in systems with compact metric phase spaces without isolated points, positive entropy implies both AY- and LY-chaos. See Note 8 in the previous chapter.

In this connection, let me mention scattering systems. A dynamical system (X, f) is said to be *scattering* whenever for every finite cover $\mathcal{U}$ of X by non-dense open sets the complexity function $n \mapsto N(\mathcal{U}^{f,n}): \mathbb{N} \to \mathbb{N}$ – see just above Proposition 8.4.3 for the definition – is unbounded. In F. Blanchard, B. Host & A. Maass [2000] it is shown that a system with a compact metric phase space is scattering iff its cartesian product with any compact minimal system is transitive This is the characterization used in Example (4) after Corollary 7.1.8 and in Section 7.3.

A. Topology

Abstract. The purpose of this Appendix is to give an overview of the basic notions of set theoretical topology. Most of the definitions and results included in this section can be found in every introductory book on topology. The proofs of most statements in this Appendix are omitted, but those who have followed a course in general topology will no doubt be able to provide most of the arguments. However, we give full proofs of Baire's theorem (because that's sometimes skipped in elementary courses) and of some less well-known results about connectedness and about semi-open and irreducible mappings.

A.1 Elementary notions

A.1.1. A *topological space* is an ordered pair $(X, \mathcal{O})$, where X is a set and $\mathcal{O}$ is a collection of subsets of X satisfying the following conditions:
(a) $\emptyset \in \mathcal{O}$ and $X \in \mathcal{O}$.
(b) If $\mathcal{U} \subseteq \mathcal{O}$ then also $\bigcup \mathcal{U} \in \mathcal{O}$.
(c) If O_1 en O_2 are members of $\mathcal{O}$ then also $O_1 \cap O_2 \in \mathcal{O}$.
A collection $\mathcal{O}$ of subsets of X that satisfies these conditions is called a *topology* on X. In that case the members of $\mathcal{O}$ are called the *open (sub)sets* of X. A subset F of X is said to be *closed* in X whenever the set $X \setminus F$ is open. A set which is both open and closed will be called *clopen*. The definition of a topology immediately implies that the subsets X and $\emptyset$ of X are closed in X. Also, the intersection of an arbitrary collection of closed sets is closed, as is the union of a finite number of closed sets. The intersection of a countable family of open sets (which need not be open) is called a G_δ-set.

A *base* for the topology $\mathcal{O}$ on X is a collection $\mathcal{B} \subseteq \mathcal{O}$ such that

$$\forall U \in \mathcal{O} : U = \bigcup \{B \,\vdots\, B \in \mathcal{B} \;\&\; B \subseteq U\} .$$

The space X is said to be *2nd countable* whenever it has a countable base. A *subbase* for the topology $\mathcal{O}$ is a family $\mathcal{B}' \subseteq \mathcal{O}$ such that the collection of all intersections of finitely many of its members is a base for $\mathcal{O}$.

Example. The collection of open intervals in $\mathbb{R}$ is a base for the 'usual' topology of $\mathbb{R}$. In addition, the family of all open half-lines $(a; \infty)$ and $(-\infty; b)$ with $a, b \in \mathbb{R}$ is a subbase for this topology. Note that one should consider this as a *definition* of the term 'usual topology' on $\mathbb{R}$. Indeed, the 'usual topology' on $\mathbb{R}$ is defined as the collection of all sets that are a union of a family of open intervals (not necessarily bounded). It is easy to see that this collection is a topology on $\mathbb{R}$.

In the same vain, the usual topology on $\mathbb{R}^2$ is the collection of all sets that can be written as a union of a family of open rectangles in $\mathbb{R}^2$ (i.e., rectangles without boundary). Clearly, the set of all open half-planes is a subbase for this topology.

In general, the following is true: if $\mathcal{B}$ is any non-empty collection of subsets of a set X then there is a *smallest* or *weakest* topology $\mathcal{O}_{\mathcal{B}}$ on X such that $\mathcal{B} \subseteq \mathcal{O}_{\mathcal{B}}$: if $\mathcal{O}$ is any topology such that $\mathcal{B} \subseteq \mathcal{O}$ then $\mathcal{O}_{\mathcal{B}} \subseteq \mathcal{O}$. In fact, $\mathcal{O}_{\mathcal{B}}$ is the intersection of all topologies that include $\mathcal{B}$. In that case we call $\mathcal{O}_{\mathcal{B}}$ the *topology generated by* $\mathcal{B}$. It turns out *that* $\mathcal{B}$ *is a subbase of the topology* $\mathcal{O}_{\mathcal{B}}$. These notions are important in connection with the definitions of *product* and *quotient* topologies; see A.5.2 and A.5.5 below.

A topology is a subset of the power set of X. Thus, one can form the union and intersection of sets of topologies; in most cases the former is not a topology, but the latter always is. Moreover, if $\mathcal{O}_1$ and $\mathcal{O}_2$ are topologies and $\mathcal{O}_1 \subseteq \mathcal{O}_2$ then we say that $\mathcal{O}_1$ is *weaker* (or *coarser*) than $\mathcal{O}_2$, and we say that $\mathcal{O}_2$ is *stronger* (or *finer*) than $\mathcal{O}_1$. Every set has a finest topology: the *discrete topology*, that is, the topology in which every set is open; in particular, every singleton set is open in the discrete topology. Similarly, every set has a coarsest topology, namely, $\{\emptyset, X\}$.

A.1.2. Let $(X, \mathcal{O})$ be a topological space and assume that X is non-empty. Consider an arbitrary subset A of X. The *interior* of A is the union of all open subsets of X that are included in A. Obviously, the interior of A is the largest open subset of A. Notation: $\operatorname{int}_X(A)$ or, when the space is understood, $\operatorname{int}(A)$ or A°.

The *closure* of A is the intersection of all closed subsets of X in which A is included. Obviously, the closure of A is the smallest closed subset in which A is included. Notation: $\operatorname{cl}_X(A)$, or $\operatorname{cl}(A)$, $\overline{A}$ or A^-.

Using the De Morgan rules it is easy to express the interior and closure of the complement of a set as the complements of the closure and the interior, respectively, of that set:

$$(X \setminus A)^\circ = X \setminus A^- \quad \text{and} \quad (X \setminus A)^- = X \setminus A^\circ . \tag{A.1-1}$$

Also, for two subsets A and B of X we always have

$$\overline{A \cup B} = \overline{A} \cup \overline{B} . \tag{A.1-2}$$

The *boundary* of a set A is defined as the set $\partial A := A^- \setminus A^\circ$. The boundary of a set is always closed and it has an empty interior.

Example. The boundary of a bounded interval in $\mathbb{R}$ with respect to the usual topology consists of the two end points of that interval.

Any subset B of X such that $A \subseteq B^\circ$ is called a *neighbourhood* (abbreviated to 'nbd') of A (in X). In particular, every open subset U of X which includes A is a nbd of A (an *open* nbd of A). When we apply this notion to a singleton set $\{x\}$ for $x \in X$ we get the notion of a *neighbourhood of a point*: any subset B of X such that $x \in B^\circ$. In particular, every open subset U of X such that $x \in U$ is a nbd of x (an *open* nbd of x). Notation: the set of all nbds of a subset A of X is denoted by $\mathcal{N}_A$; if $x \in X$ then instead of $\mathcal{N}_{\{x\}}$ we write $\mathcal{N}_x$.

A point $x \in X$ is said to be *isolated* whenever it has a nbd U such that x is the only point in U, that is, $U = \{x\}$. Clearly, a point x is isolated iff the singleton set $\{x\}$

is open. In this book we consider only spaces in which singleton sets are closed – see A.1.4 below – and if in such a space a point x is isolated then the set $\{x\}$ is clopen. A *local base* in (or: at) the point $x \in X$ is a collection $\mathcal{B}_x \subseteq \mathcal{N}_x$ of nbds of x with the following property:

$$\forall U \in \mathcal{N}_x \ \exists B \in \mathcal{B}_x \ \vdots \ B \subseteq U .$$

The notion of *local base of a set* is defined analogously. If $x \in X$ and $\mathcal{B}_x$ is a local base at x then for any subset A of X we have:

- $x \in A^\circ$ iff there exists $U \in \mathcal{B}_x$ such that $x \in U \subseteq A$.
- $x \in A^-$ iff for every $U \in \mathcal{B}_x$ we have $U \cap A \neq \emptyset$.

Of course, in these characterizations $\mathcal{B}_x$ can be replaced by $\mathcal{N}_x$.

A subset A of X is said to be *dense in* X whenever $A^- = X$. Hence A is dense in X iff every non-empty open subset of X contains a point of A, iff

$$\forall x \in X \ \forall U \in \mathcal{N}_x \ : \ U \cap A \neq \emptyset . \tag{A.2-3}$$

Obviously, in order to show that a set A is dense in X it is sufficient to show that every member of a base for the topology of X meets A, or that for every point x in X each member of a local base at x meets A.

A topological space is said to be *separable* whenever it contains a *countable* dense subset. Observe that if a space has a countable base then it is separable: select a point in each basic set.

A.1.3. Let $(X, \mathcal{O})$ be a topological space and let Y be a subset of X. The *relative topology* of Y in X, or *the topology that Y inherits from X*, is the collection $\mathcal{O} \cap Y := \{O \cap Y \ \vdots \ O \in \mathcal{O}\}$; so by definition, the open sets in Y are precisely all sets $O \cap Y$ where O is open in X. With this topology we call Y a (topological) *subspace* of X. Unless stated otherwise, we always assume that a subset of a topological space carries the relative topology. In the case that Y is an open subset of X the open sets of Y (in the relative topology) are just the open sets of X that are included in Y.

When Y is a subspace of X, the closed subsets of Y are precisely the sets of the form $F \cap Y$ with F a closed subset of X. Consequently, if Y is a closed subset of X then the closed subsets of Y (in the relative topology) are the closed subsets of X that are included in Y. It follows quite easily that for any subset A of Y the closure of A in Y equals $Y \cap A^-$, where A^- is the closure of A (as a subset of X) in X. In different notation: if $A \subseteq Y$ then $\mathrm{cl}_Y(A) = Y \cap \mathrm{cl}_X(A)$.

A.1.4. A topological space is said to satisfy the *Hausdorff separation axiom*, or to be a *Hausdorff space*, or a T_2-space, whenever any pair of distinct points can be separated by disjoint open sets, or equivalently:

$$\forall x, y \in X : \ x \neq y \ \Rightarrow \ \exists U \in \mathcal{N}_x \, \exists V \in \mathcal{N}_y \ \vdots \ U \cap V = \emptyset .$$

In a Hausdorff the complement of a single point is easily seen to be open. Consequently, every singleton set is closed[1].

A *regular space* is a topological space in which points and closed sets can be separated by open sets. Thus, a space X is regular iff

$$\begin{aligned}\forall x \in X\ \forall A \subseteq X :\ & A \text{ closed and } x \notin A \ \Rightarrow \\ & \exists U \in \mathcal{N}_x\ \exists V \in \mathcal{N}_A\ \vdots\ U \cap V = \emptyset .\end{aligned}$$

Equivalently: a space is regular iff every point has a local base consisting of closed nbds of that point. A regular T_2-space (regular Hausdorff space) is also called a T_3-space. The defining property of regularity immediately implies the following: if A is a non-empty closed subset of a regular space then

$$A = \bigcap\{V \vdots V \in \mathcal{N}_A\} = \bigcap\{\overline{V} \vdots V \in \mathcal{N}_A\} \tag{A.1-3}$$

Another, stronger, separation property is normality: a topological space is said to be *normal* whenever each pair of mutually disjoint closed subsets can be separated by disjoint open nbds. A normal Hausdorff space is also called a T_4-space. Using the definitions it is easy to see that a subspace of a T_i-space is a T_i-space ($i = 2, 3, 4$).

Example. With the usual topologies, both $\mathbb{R}$ and $\mathbb{R}^2$ are T_2-spaces. Using the fact that these spaces are *metrizable* (see Section A.7 below) it is not difficult to show that these spaces are T_3 and T_4 as well.

A.2 Compactness

A.2.1. A *cover* of a set X is a collection $\mathcal{U}$ of non-empty subsets of X whose union is all of X, i.e., such that $\bigcup \mathcal{U} = X$. An *open cover* is a cover consisting of open sets. A subcollection of a cover $\mathcal{U}$ which is also a cover is called a *subcover* of $\mathcal{U}$. A topological space is said to be *compact* whenever every open cover has a finite subcover. A collection of subsets of a set X is said to have the *finite intersection property* (abbreviated: FIP) whenever it has the following property:

$$\forall \mathcal{F}' \subseteq \mathcal{F} :\ \mathcal{F}' \text{ is finite } \Rightarrow \bigcap \mathcal{F}' \neq \emptyset .$$

By looking at complements and applying the De Morgan rules one gets the following characterization of compactness: *A topological space is compact iff every collection of closed subsets with FIP has a non-empty intersection.* It follows that in a compact space every descending chain of non-empty closed sets has a non-empty intersection (in fact, this characterizes compactness, but the proof is not elementary: it uses Zorn's Lemma).

1 A space with this property is called a T_1-space.

A *subset* of a topological space is called *compact* whenever this subset with its relative topology is a compact topological space. Hence a subset A of a topological space is compact iff

$$\forall \mathcal{U} \subseteq \mathcal{O} : \bigcup \mathcal{U} \supseteq A \;\; \Rightarrow \; \exists \mathcal{U}' \subseteq \mathcal{U} : \mathcal{U}' \text{ is finite and } \bigcup \mathcal{U}' \supseteq A\,.$$

Every closed subset of a compact space is compact. Conversely, in a Hausdorff space every compact subset is closed. This is because in a Hausdorff space X points and compact sets can be separated by disjoint open sets:

If A is a compact subset of X and $x \in X \setminus A$ then there are open sets U and V such that $x \in U$, $A \subseteq V$ and $U \cap V = \emptyset$.

Proof. Separate each point a of A from x by disjoint open nbds of a and x, respectively, cover A by the nbds of finitely many of those points a and take the intersection of the corresponding nbds of x. □

This implies:
Let A and B be disjoint compact sets in a Hausdorff space. Then A and B have mutually disjoint open nbds. Consequently, a compact Hausdorff space is normal (hence a T_4-space).

Proof. Repeat the proof of the previous statement, replacing x by B (and using the previous statement with A replaced by B). □

It follows that every compact Hausdorff space is regular (hence T_3). In particular, in a compact Hausdorff space every point has a local base consisting of closed, hence compact, neighbourhoods. Also, for every non-empty compact subset of a (not necessarily regular) Hausdorff space we have, similar to (A.1-3):

$$A = \bigcap \{ V : V \in \mathcal{N}_A \} = \bigcap \{ \overline{V} : V \in \mathcal{N}_A \} \tag{A.2-1}$$

A topological space $(X, \mathcal{O})$ is called *locally compact* whenever each of its points has a compact nbd. If X is a Hausdorff space then X is locally compact iff every point of X has a local base consisting of compact nbds. In particular, every compact Hausdorff space is locally compact.

Example. In $\mathbb{R}$ or $\mathbb{R}^2$ with their usual topologies: a subset is compact iff it is closed and bounded (Heine–Borel). It follows that both $\mathbb{R}$ and $\mathbb{R}^2$ are locally compact.

In a locally compact space, every non-empty compact subset A has a local base consisting of compact nbds of A.

Proof. Cover A by finitely many sufficiently small open nbds, each with a compact closure. □

The following lemma is included for easy reference: a *filter base* is a collection $\mathcal{B}$ of non-empty sets such that the intersection of any two members of $\mathcal{B}$ includes a member of $\mathcal{B}$. It follows easily that the intersection of finitely many members of $\mathcal{B}$ includes a member of $\mathcal{B}$ (use induction on the number of intersecting sets), so a filter base has FIP. A *descending sequence of sets* is a sequence $(B_n)_{n\in\mathbb{N}}$ of sets such that $B_n \supseteq B_{n+1}$ for all $n \in \mathbb{N}$. Obviously, a descending sequence of sets is a filter base.

Lemma A.2.2. *Let $\mathcal{B}$ be a filter base consisting of non-empty compact sets of a Hausdorff space X. Then $B_0 := \bigcap \mathcal{B} \neq \emptyset$ and for every nbd U of B_0 in X there exists $B \in \mathcal{B}$ such that $B \subseteq U$.*

In particular, if $\mathcal{B} = (B_n)_{n\in\mathbb{N}}$ is a descending sequence of non-empty compact sets then $B_0 := \bigcap\{B_n : n \geq 1\} \neq \emptyset$ and for every nbd U of B_0 in X there exists $N_U \in \mathbb{N}$ such that $B_n \subseteq U$ for all $n \geq N_U$.

Proof. The family $\mathcal{B}$ has FIP, hence $B_0 \neq \emptyset$. Let U be an arbitrary nbd of B_0 in X and assume that $B \setminus U \neq \emptyset$ for all $B \in \mathcal{B}$. Without limitation of generality U is open, hence $\{ B \setminus U : B \in \mathcal{B} \}$ is a family of non-empty compact sets, which is easily seen to have FIP: the intersection of finitely many members includes $B' \setminus U$ for some $B' \in \mathcal{B}$, hence is non-empty by assumption. Consequently, its intersection $\bigcap \{B\setminus U : B \in \mathcal{B}\} = B_0 \setminus U$ is non-empty. This contradicts the choice of U as a neighbourhood of B_0. □

A.3 Continuous mappings

A.3.1. Let $(X, \mathcal{O})$ and $(Y, \mathcal{U})$ be topological spaces. A function (often also called a mapping) $f : X \to Y$ is said to be *continuous* whenever $f^{\leftarrow}[U] \in \mathcal{O}$ for every $U \in \mathcal{U}$. This means that f is continuous iff for every point $x \in X$ and every $W \in \mathcal{N}_{f(x)}$ one has $f^{\leftarrow}[W] \in \mathcal{N}_x$. If this condition is only fulfilled for a particular point $x_0 \in X$ then we say that f is *continuous at* x_0. In order to show that a mapping $f : X \to Y$ is continuous it is sufficient to check openness only for the inverse images under f of sets taken from a (sub)base of Y or from a local base for each point of Y. By taking complements in the definition of continuity we get: f is continuous iff $f^{\leftarrow}[F]$ is closed in X for every closed subset F of Y. It follows that f is continuous iff

$$f[\overline{A}\,] \subseteq \overline{f[A]}\,. \tag{A.3-1}$$

for every subset A of X. In particular, if A is dense in X then $f[A]$ is dense in $f[X]$. Some additional elementary properties of continuous maps are:

(1) A composition of continuous mappings is continuous.
(2) If $f, g : X \to Y$ are continuous mappings and Y is a Hausdorff space, then the set $\{ x \in X : f(x) = g(x) \}$ is closed in X. In particular, if X is a Hausdorff space and $f : X \to X$ is a continuous mapping then the set $\{ x \in X : f(x) = x \}$ is closed in X.
(3) If $f : X \to Y$ is a continuous function and A is a compact subset of X then $f[A]$ is a compact subset of Y.
(4) If $f : X \to Y$ is a continuous mapping, Y a Hausdorff space, and A is a subset of X with a compact closure $\overline{A}$, then $f[\overline{A}\,] = \overline{f[A]}$.

A mapping $f: X \to Y$ is said to be *open* or *closed*, respectively, whenever the image of an open (respectively, closed) subset of X under f is open (respectively, closed) in Y. A *topological mapping* or *homeomorphism* from a topological space X to a topological space Y is a bijection $f : X \to Y$ for which the inverse mapping $f^{-1} : Y \to X$ is continuous as well. Thus, a mapping f is a homeomorphism iff f is a continuous bijection which is also open, iff f is a continuous bijection which is also closed. A *topological embedding* is a not-necessarily surjective mapping $f: X \to Y$ such that the corestriction $f: X \to f[Y]$ is a homeomorphism.

If X is a compact space and Y is a Hausdorff space then every continuous mapping $f : X \to Y$ is closed: if F is closed in X then F is compact, so $f[F]$ is compact, hence closed, in Y. If, in this situation, f is a bijection then f is a homeomorphism. This proves:

Theorem A.3.2. *A continuous bijection from a compact space onto a Hausdorff space is a homeomorphism.*

Lemma A.3.3. *Let X be compact, let Y be Hausdorff and let $f: X \to Y$ be a continuous mapping. If $y \in Y$ and U is an open nbd of the fibre $f^{\leftarrow}[y]$ then there is a nbd V of y in Y such that $f^{\leftarrow}[V] \subseteq U$.*

Proof. It is clear that $F := f[X \setminus U]$ is a compact, hence a closed, subset of Y and that $y \notin F$. Then $V := Y \setminus F$ is a neighbourhood of y such that $V \cap F = \emptyset$ and, consequently, $f^{\leftarrow}[V] \subseteq U$. □

Proposition A.3.4. *Let X be a compact Hausdorff space. Then for every pair A and B of disjoint non-empty closed subsets of X there is a continuous function $f: X \to [0;1]$ such that $f[A] = \{0\}$ and $f[B] = \{1\}$.*

Proof. This follows from the fact that X is a normal space (see A.2.1 above) and the Tietze-Urysohn Extension Theorem: see [Eng], 2.1.8. □

The following lemma is quite useful, but not easily found in this form in the standard text books on general topology: we include it for easy reference.

Lemma A.3.5. *Let X and Y be Hausdorff spaces and let $f: X \to Y$ be a continuous mapping. If $(B_n)_{n\in\mathbb{N}}$ be a descending sequence*[2] *of non-empty compact sets in X then*

$$f\Big[\bigcap_{n\in\mathbb{N}} B_n\Big] = \bigcap_{n\in\mathbb{N}} f[B_n] .$$

Proof. The inclusion "$\subseteq$" is trivial. In order to prove de inclusion "$\supseteq$", consider a point $y_0 \in Y \setminus f[\bigcap_{n\in\mathbb{N}} B_n]$. We shall show that $y_0 \notin \bigcap_{n\in\mathbb{N}} f[B_n]$. Let $U := Y \setminus \{y_0\}$, an open neighbourhood of the set $f[\bigcap_{n\in\mathbb{N}} B_n]$. By continuity of f, the set $V := f^{\leftarrow}[U]$ is open in X. Since it includes the set $\bigcap_{n\in\mathbb{N}} B_n$, it follows from Lemma A.2.2 that there exists

2 The conclusion holds (with a very similar proof) also for a filter base of compact sets.

$k \in \mathbb{N}$ such that $B_k \subseteq V = f^{\leftarrow}[U]$, and therefore $f[B_k] \subseteq U$. This obviously implies that $\bigcap_{n\in\mathbb{N}} f[B_n] \subseteq f[B_k] \subseteq U$. In particular, $y_0 \notin \bigcap_{n\in\mathbb{N}} f[B_n]$. □

A continuous mapping $f : X \to Y$ is said to be *semi-open* whenever the set $f[V]$ has a non-empty interior in Y for every non-empty open subset V of X. Every homeomorphism is semi-open, but the converse is not true, as can easily be concluded from the following example.

Example. Let I be an interval in $\mathbb{R}$. It is easily seen that *a continuous mapping $f : I \to \mathbb{R}$ is semi-open iff the image of every non-degenerate subinterval of I is a non-degenerate interval of* $\mathbb{R}$. Consequently, if $f : I \to \mathbb{R}$ is not constant on any subinterval of I then f is semi-open. In particular, a piecewise monotonous mapping is semi-open. (Recall that a mapping $f : I \to \mathbb{R}$ is said to be *piecewise monotonous* whenever there are points $a_0 < a_1 < \cdots < a_n$ such that a_0 and a_n are the end points of I and on each interval with end points a_i and a_{i+1} $(i = 0, \ldots, n-1)$ the mapping f is strictly monotonous.)

Lemma A.3.6. *Let $f : X \to Y$ be a continuous mapping. The following statements are equivalent:*
(i) *The mapping f is semi-open.*
(ii) *For every dense subset B of Y, the set $f^{\leftarrow}[B]$ is dense in X.*

Proof. “(i)⇒(ii)”: Let B be a dense subset of Y and consider a non-empty open set V in X. Since f is semi-open, $f[V]$ contains a non-empty open subset of Y, hence it contains a point of B. Consequently, $V \cap f^{\leftarrow}[B] \neq \emptyset$.

“(ii)⇒(i)”: Let U be a non-empty open set in X and assume that $f[U]$ has empty interior in Y, that is, $Y \setminus f[U]$ is dense in Y. Then $f^{\leftarrow}[Y \setminus f[U]]$ is dense in X. Since $f^{\leftarrow}[Y \setminus f[U]] \subseteq X \setminus U$, it would follow that $X \setminus U$ is dense in X, contradicting that U is non-empty and open. □

Lemma A.3.7. *Let $f : X \to X$ be a continuous mapping. If f is semi-open then for every $k \in \mathbb{N}$ the mapping f^k is semi-open as well.*

Proof. Use an obvious induction argument. □

A.4 Convergence

A *directed set* is a set Σ endowed with a relation $\leq$ such that

(1) $\forall \sigma \in \Sigma : \sigma \leq \sigma$;
(2) $\forall \sigma, \tau, \zeta \in \Sigma : \sigma \leq \tau \;\&\; \tau \leq \zeta \Rightarrow \sigma \leq \zeta$;
(3) $\forall \sigma, \tau \in \Sigma\, \exists \zeta \in \Sigma \vdots\, \sigma \leq \zeta \;\&\; \tau \leq \zeta$.

(Note condition (2) of a partial order as defined in A.10.4 ahead is missing.) By induction one easily shows that condition (3) implies that for every finite subset Φ of Σ there exists $\tau \in \Sigma$ such that $\tau \geq \sigma$ for every $\sigma \in \Phi$ (as is quite usual, $\tau \geq \sigma$ will be used as an alternative writing for $\sigma \leq \tau$). A subset Σ' of Σ is said to be *cofinal* in Σ whenever for every $\sigma \in \Sigma$ there exists $\sigma' \in \Sigma'$ such that $\sigma' \geq \sigma$.

Example. Every filter base $\mathcal{B}$ is a directed set with the partial ordering $\leq$ defined by $B_1 \leq B_2$ iff $B_1 \supseteq B_2$ for $B_1, B_2 \in \mathcal{B}$. In particular, if X is a topological space and $x \in X$ then $\mathcal{N}_x$ is a filter base, hence a directed set with respect to this partial ordering. A cofinal subset of $\mathcal{N}_x$ with respect to this ordering is nothing but a local base at the point x.

Let X be a topological space. A *net* in X is a subset $\{x_\sigma : \sigma \in \Sigma\}$ of X (also denoted as $\{x_\sigma\}_{\sigma \in \Sigma}$), where Σ is a directed set. A net $\{x_\sigma\}_{\sigma \in \Sigma}$ in X is said to *converge* to a point $x \in X$, in which case x is called a *limit* of the net, whenever every nbd U of x includes *almost all* elements of the net[3], that is, whenever for every $U \in \mathcal{N}_x$ there exists $\sigma \in \Sigma$ such that $x_\tau \in U$ for all $\tau \in \Sigma$ with $\tau \geq \sigma$. If X is a Hausdorff space then every net in X has at most one limit (which is, when it exists, called *the* limit of the net). For this reason we shall consider only convergence of nets Hausdorff spaces. If a net $\{x_\sigma\}_{\sigma \in \Sigma}$ in X converges to a point x we write $x_\sigma \rightsquigarrow x$ or $x = \lim_{\sigma \in \Sigma} x_\sigma$.

Lemma A.4.1. *If A is a subset of a topological space X and $x \in X$ then $x \in \overline{A}$ iff there is a net $\{x_\sigma\}_{\sigma \in \Sigma}$ in A such that $x_\sigma \rightsquigarrow x$.*

Proof. "If": For every $U \in \mathcal{N}_x$, select a point $x_U \in U \cap A$. Then $\{x_U\}_{U \in \mathcal{N}_x}$ is a net in A that converges to the point x.

"Only if": Obvious. □

Lemma A.4.2. *A function $f : X \to Y$ is continuous at a point $x \in X$ iff $f(x_\sigma) \rightsquigarrow f(x)$ in Y for every net $\{x_\sigma\}_{\sigma \in \Sigma}$ in X such that $x_\sigma \rightsquigarrow x$.*

Proof. "Only if": straightforward.

"If": if f is not continuous at x then there exists $U \in \mathcal{N}_{f(x)}$ such that every $V \in \mathcal{N}_x$ contains a point x_V such that $f(x_V) \notin U$; then $\{x_V\}_{V \in \mathcal{N}_x}$ is a net in X such that $x_V \rightsquigarrow x$, whereas the net $\{f(x_V)\}_{V \in \mathcal{N}_x}$ in Y does not converge to $f(x)$. □

A point $x \in X$ is called a *cluster point* of a net $\{x_\sigma : \sigma \in \Sigma\}$ in X whenever for every $U \in \mathcal{N}_x$ the set $\{\sigma \in \Sigma : x_\sigma \in U\}$ is cofinal in Σ. Clearly, a limit of a net is a cluster point, but not conversely.

Example. A sequence is a special case of a net (and convergence of a sequence is a special case of convergence of a net), namely, a net indexed by the directed set $\mathbb{N}$ with its natural order. The sequence $\{(-1)^n(1 - 1/n)\}_{n \in \mathbb{N}}$ has two cluster points, -1 and 1, but it has no limit.

---

3 In the literature this is often expressed by saying that the net is *eventually* in U.

Proposition A.4.3. *Every net $\{x_\sigma\}_{\sigma\in\Sigma}$ in a compact space has a cluster point.*

Proof. Clearly, the family of compact sets $A_\sigma := \overline{\{x_\tau \,\vdots\, \tau \in \Sigma \& \tau \geq \sigma\}}$ for $\sigma \in \Sigma$ has FIP. So the sets have a point in common, with is easily seen to be a cluster point. □

One can define the notion of a *subnet* (in [Eng] this is called a *finer* net). A point is a cluster point of a net iff that net has a subnet that converges to that point. So the statement above implies: in a compact space, every net has a convergent subnet. The converse is true as well: see [Eng], Theorem 3.1.23.

A.5 Subspaces, products and quotients

A.5.1 (Subspaces). Let Y be a subset of X. If Y is given the relative topology inherited from X then the embedding mapping $i_Y : Y \to X$ is continuous. Moreover, the relative topology on Y is the weakest topology on Y making the mapping $i_Y : Y \to X$ continuous. In addition: *if Z is a topological space then a mapping $f : Z \to Y$ is continuous with respect to the relative topology on Y iff the mapping $i_Y \circ f : Z \to X$ is continuous.* In particular, it makes no difference if one considers f as a mapping from Z into Y or as a mapping from Z into X. If $Z \subseteq Y \subseteq X$ then the relative topology of Z in Y (Y with the relative topology inherited from X) coincides with the relative topology that Z inherits from X as a subset of X.

A.5.2 (Product spaces). Let $\{(X_a, \mathcal{O}_a) \,\vdots\, a \in A\}$ be an indexed set of topological spaces. The *Cartesian product* $\prod_{a\in A} X_a$ is the set of all functions $x: a \mapsto x_a : A \to \bigcup_{a\in A} X_a$ such that $x_a \in X_a$ for all $a \in A$. We shall use without explicit reference the assumption that $\prod_{a\in A} X_a \neq \emptyset$ if $X_a \neq \emptyset$ for all $a \in A$ (the axiom of Choice; see A.10.4 below). The *topological product* of this set of spaces consists of the product set $X := \prod_{a\in A} X_a$ endowed with the *product topology*, that is, the weakest topology $\mathcal{O}$ making all projections $\pi_a : x \mapsto x_a : X \to X_a$ continuous ($a \in A$). This means that $\mathcal{O}$ is the intersection of all topologies on X for which every projection is continuous (such topologies exist – the discrete topology is an example – hence $\mathcal{O}$ is well-defined). The product topology $\mathcal{O}$ is generated by the collection of all sets of the form $\pi_a^{\leftarrow}[U]$ with $a \in A$ and $U \in \mathcal{O}_a$. The collection of these sets forms a subbase for the product topology (see the final paragraph in A.1.1 above). Note that one also gets a subbase for the product topology if the sets U are taken from a base or a subbase of the topology $\mathcal{O}_a$ on X_a.

From the ‘standard’ subbase for the product topology described above one gets a base by forming all intersections of finite subcollections of this subbase. This *standard base* consists of all subsets U of X of the form

$$U = \bigcap_{a\in A_U} \pi_a^{\leftarrow}[U_a] = \prod_{a\in A} U_a, \tag{A.5-1}$$

with A_U a finite subset of A, U_a an open set in X_a for every $a \in A$ and $U_a = X_a$ if $a \notin A_U$. If for every $a \in A$ the sets U_a are taken from a base for the topology of X_a then one obtains a base for the product topology as well. A local base at a point $x = (x_a)_{a \in A} \in X$ is formed by the collection of all sets of the form

$$U = \prod_{a \in A} U_a , \tag{A.5-2}$$

with U_a a nbd of x_a in X_a for every $a \in A$ and $U_a = X_a$ for all $a \in A \setminus A_U$, where A_U is a finite subset of A. The sets U_a may also be taken from a local base at x_a for every $a \in A$.

The following lemma is an easy consequence of the definitions: it provides the most convenient way to prove that a mapping into a product space is continuous. We shall sometimes refer to it as *the defining property of a product topology*.

Lemma A.5.3. *A mapping $f : Z \to \prod_{a \in A} X_a$ (Z any topological space) is continuous iff for every $a \in A$ the mapping $\pi_a \circ f : Z \to X_a$ is continuous.*

Proof. Only "if" needs a proof. It is sufficient the check that $f^{\leftarrow}[W]$ is open in Z for subbasic open sets W in X, that is, for sets of the form $\pi_a^{\leftarrow}[U]$ with $a \in A$ and $U \in \mathcal{O}_a$. But this follows from continuity of $\pi_a \circ f$. □

It is easy to see that the product of a set of Hausdorff spaces is a Hausdorff space. Less trivial (in fact, equivalent to the Axiom of Choice) is:

Theorem A.5.4 (Tychonov). *The product of compact spaces is compact.*

It follows easily that the topological product of a set of locally compact spaces is locally compact iff all but finitely many of these spaces are compact.

Example. If $n \in \mathbb{N}$ then $\mathbb{R}^n$ is the Cartesian product of n copies of $\mathbb{R}$. The product topology in $\mathbb{R}^n$ (each factor $\mathbb{R}$ with its usual topology) coincides with the usual topology on $\mathbb{R}^n$. So $\mathbb{R}^n$ with this topology is a locally compact Hausdorff space.

A.5.5 (Quotient spaces). Let $(X, \mathcal{O})$ be a topological space, let R be an equivalence relation on X and denote the set of all R-equivalence classes by X/R. Recall that the *quotient mapping* is the mapping $\varphi_R : X \to X/R$ that assigns to each element $x \in X$ the equivalence class $R[x]$ to which it belongs; thus, $\varphi_R(x) := R[x]$. The *finest* topology on X/R that makes the quotient mapping $\varphi_R : X \to X/R$ continuous is called the *quotient topology*, and the set X/R endowed with this topology is called the *quotient space* defined by the equivalence relation R. It is easy to see this topology is the collection of all subsets O of X/R for which $\varphi_R^{\leftarrow}[O] \in \mathcal{O}$: any topology for which φ_R is continuous consists of sets with this property, and these sets together form already a topology. So a subset of X/R is open in X/R iff its preimage under φ_R is open in X. It follows that a subset of X/R is closed in X/R iff its preimage under φ_R is closed in X.

A *partition* of a space is a collection of mutually disjoint subsets whose union is the whole space. A partition is called *open*, *closed* or *clopen* whenever all its members are open, closed or clopen sets, respectively[4]. If $\mathcal{P}$ is a partition of X then an equivalence relation R is defined by

$$\forall x, y \in X: \; xRy \Leftrightarrow x \text{ and } y \text{ belong to the same member of } \mathcal{P}$$

(if R is given this defines a partition of X: the partition in R-equivalence classes) and the quotient space $X/\mathcal{P}$ is defined as the quotient space X/R.

The property mentioned in the following lemma will be called the *defining property of quotient maps*:

Lemma A.5.6. *Let Y be a topological space and let $f : X/R \to Y$ be a mapping. Then f is continuous iff the composition $f \circ \varphi_R : X \to Y$ is continuous.*

Example. Let $X := [0; 1]$ and let R be the equivalence relation that identifies the points 0 and 1 of $[0; 1]$, that is, $R = \{ (t, t) \colon t \in X \} \cup \{(0, 1)\} \cup \{(1, 0)\}$. The quotient space X/R (a continuous image of a compact space) is compact. Now $g(\varphi_R(t)) := [t]$ for $t \in X$, where $[\,t\,] := \exp(2\pi i t)$, unambiguously defines a bijection $g : X/R \to \mathbb{S}$. The composition $g \circ \varphi_R = [\cdot]$ is continuous, hence g is continuous. Since X/R is a compact space and $\mathbb{S}$ is a Hausdorff space it follows that g is a homeomorphism.

A.6 Connectedness

A topological space $(X, \mathcal{O})$ is said to be *not connected* (or: *non-connected*, or: *disconnected*) whenever there are two mutually disjoint non-empty open subsets U and V of X such that $X = U \cup V$. In this situation the sets U and V, each one being the complement of the other, are closed: we have two clopen sets. The space X is said the be *connected* whenever it is not non-connected, that is, whenever the following property holds: if

$$X = U \cup V \quad \text{with } U \cap V = \emptyset \text{ and } U, V \in \mathcal{O}$$

then $U = \emptyset$ or $V = \emptyset$. A *subset* of a topological space X is said to be *connected* whenever as a subspace of X (with its relative topology) it is connected. A characterization of connectedness of a set in terms of the ambient space is as follows: *A subset C of X is connected iff for every pair of subsets X_1 and X_2 of C such that $C = X_1 \cup X_2$ and $\overline{X_1} \cap X_2 = \emptyset = X_1 \cap \overline{X_2}$ we have either $X_1 = \emptyset$ or $X_2 = \emptyset$* (the closures are in X).

The closure and the continuous image of a connected set are connected as well. Moreover, if $\{A_i\}_{i \in I}$ is a family of connected sets such that any two members of the family have a non-empty intersection then $\bigcup_{i \in I} A_i$ is connected. In particular, this is the case if $\bigcap_{i \in I} A_i \neq \emptyset$.

4 A *topological* partition as defined in Section 6.1 is *not* a partition as defined here.

Example.

(1) The connected subsets of $\mathbb{R}$ are the singleton sets and the intervals (open, closed, half open, not necessarily bounded). The continuous image of an interval under a real–valued function is connected, hence it is a one-point set or an interval.

The *connected component*, or just *component*, of a point in a topological space X is the union of all connected subsets of X that contain that point. The component of a point is the largest connected subset containing that point: it is connected because it is the union of connected sets with non-empty intersection. Using the fact that the closure of a connected set is connected, it follows easily that *every component is closed.* In addition, different components are disjoint: if not then their union would be a larger connected set. Hence *the components of a space form a partition of the space into non-empty closed subsets.*

If the space has the additional property that every point has a connected nbd (e.g., because it is *locally connected*, that is, each point has a local base consisting of connected nbds) then every component is an open set. *So in this case the components form a partition of the space into clopen subsets.*

Example.

(2) Every open subset X of $\mathbb{R}$ is locally connected. So its components form a clopen partition of X into connected subsets of $\mathbb{R}$ that are also open in $\mathbb{R}$ (open in X, but X is open in $\mathbb{R}$): open intervals. In $\mathbb{R}$ a disjoint collection of non-empty open intervals is at most countably infinite (each of them contains a point of the countable set $\mathbb{Q}$). Hence X is the union of a finite or a countably infinite family of mutually disjoint open intervals.

A topological space is said to be *totally disconnected* whenever all its connected subsets are singleton sets. Hence a space is totally disconnected iff the component of any point consists of that point alone. Since components are closed, it follows that a totally disconnected space is a T_1-space.

Example.

(3) *Let R be the equivalence relation on a topological space X defined by the partition of X in components. Then the quotient space X/R with its quotient topology is totally disconnected.* For if F is a closed connected subset of X/R then it is rather easy to show that $q^{\leftarrow}[F]$ is connected, hence it is included in the component of any of its points, x, say. Then $F = q[q^{\leftarrow}[F]] = \{q(x)\}$, a singleton set in X/R.

The *quasi-component* of a point is the intersection of all its clopen neighbourhoods. Since a connected subset is included in every clopen set with which it has a non-empty intersection it should be clear that *the component of a point is always included in the quasi-component of that point.* So if all quasi-components are singletons then the space is totally disconnected.

Warning. In [Eng], totally disconnected spaces are called *hereditarily* disconnected, whereas the term totally disconnected is applied to a space in which all quasi-components consist of one point. We shall not follow this terminology.

In general, the converse statement is not true: quasi-components need not be connected, hence may be strictly larger than components. However:

Theorem A.6.1. *In a compact Hausdorff space X the component of any point coincides with the quasi-component of that point.*

Proof. Let Q be the quasi-component of some point $x_0 \in X$. We want to show that Q is connected. Assume it is not: it is the union of two non-empty closed, hence compact, sets A and B; assume that $x_0 \in A$. By A.2.1, the compact sets A and B have disjoint neighbourhoods U and V, respectively. Then $U \cup V$ is a neighbourhood of Q, so by Lemma A.2.2, applied to the collection $\mathcal{B}$ of clopen nbds of x_0 – of which Q is the intersection and which are compact – there is a member W of $\mathcal{B}$ such that $Q \subseteq W \subseteq U \cup V$. Then the set $U' := U \cap W = W \setminus V$ is a clopen nbd of x_0. As $Q \nsubseteq U'$, this contradicts the definition of Q. □

A T_1-space is said to be 0-*dimensional* whenever each of its points has a local base consisting of clopen sets. Thus, a space is 0-dimensional iff for every open subset U and every point $x \in U$ there is a clopen set V such that $x \in V \subseteq U$. Note that a 0-dimensional space is a Hausdorff space (even a *Tychonov space*: a closed set and a point outside of that set can be separated by a continuous function, namely, the characteristic function of a sufficiently small clopen neighbourhood of the point).

Obviously, in a 0-dimensional the quasi-components are singleton sets. Conversely, if in a locally compact space X all quasi-components are singleton sets then the space is 0-dimensional: any point of X is the intersection of its clopen neighbourhoods, so by Lemma A.2.2 applied to to the family $\mathcal{B}$ of all clopen neighbourhoods of x (which may assumed to be compact), $\mathcal{B}$ is a local base.

Proposition A.6.2. *A compact Hausdorff space is totally disconnected iff it is 0-dimensional.*

Proof. "If": Obviously, in a 0-dimensional space the components are singletons.

"Only if": By the observations above we have to show that if the components are singletons then the quasi-components are singletons as well. This is a direct consequence of the theorem in 2 above. □

Let R be the equivalence relation on the Hausdorff space X defined by the partition of X into its connected components. Then the space X/R, the *component space* of X, is totally disconnected. As X/R need not be a Hausdorff space – see the picture in 2 above – it may be not 0-dimensional. However, if X is compact and X/R turns out to be a Hausdorff space then the proposition above implies that X/R is 0-dimensional.

Theorem A.6.3. *Let X be a compact Hausdorff space and let R be the equivalence relation on X defined by the partition of X into its components. Then the quotient space X/R is a 0-dimensional compact Hausdorff space.*

Proof. By the discussion above it is sufficient to show that X/R is a Hausdorff space, i.e., that different points of X/R have disjoint nbds.

Let $q: X \to X/R$ be the quotient map. So for every point $x \in X$ the q-fibre $C_x := q^{\leftarrow}[q(x)]$ is the component of x in X. Though we do not know yet that X/R is a Hausdorff space we claim that the conclusion of Lemma A.3.3 is correct in the present setting, namely: if $x \in X$ and U is a nbd of C_x then there is a nbd U' of $q(x)$ in X/R such that $q^{\leftarrow}[U'] \subseteq U$.

In order to prove this, note that by the Theorem in 2 above the component C_x of x is equal to the quasi-component of x, that is, C_x is the intersection of all its clopen nbds. Hence Lemma A.2.2 implies that there is a clopen subset W of X such that $C_x \subseteq W \subseteq U$. Now observe that for every point $z \in W$ the component C_z is connected, hence C_z is completely included in the clopen set W. It follows that W is the union of the components of its points, i.e., $W = q^{\leftarrow}[q[W]]$. This means that $U' := q[W]$ is open (cf. the definition of a quotient topology), so it is a nbd in X/R of the point $q(x)$. Moreover, by the choice of U' we have $q^{\leftarrow}[W] \subseteq U$. This completes the proof of the claim.

Now consider two different points x' and y' of X/R, say, $x' = q(x)$ and $y' = q(y)$ with $x, y \in X$. Then the components of x and y are different, so the compact sets C_x and C_y are mutually disjoint. It follows that C_x and C_y have disjoint open neighbourhoods U and V, respectively. By what we have shown above, there are nbds U' and V' of the points x' and y', respectively, such that $q^{\leftarrow}[U'] \subseteq U$ and $q^{\leftarrow}[V'] \subseteq V$. Clearly, the sets $q^{\leftarrow}[U']$ and $q^{\leftarrow}[V']$ are disjoint, hence U' and V' are disjoint as well. □

A.7 Metric spaces

A.7.1. A *metric* on a set X is a mapping $d : X \times X \to \mathbb{R}$ such that:
(a) $\forall x, y \in X : d(x, y) \geq 0$ and $d(x, y) = 0 \Leftrightarrow x = y$.
(b) $\forall x, y \in X : d(x, y) = d(y, x)$.
(c) $\forall x, y, z \in X : d(x, z) \leq d(x, y) + d(y, z)$ (triangle inequality).
If d is a metric on X then the ordered pair (X, d) is called a *metric space* and d is called the *distance function* of this metric space. If $x, y \in X$ then the non-negative real number $d(x, y)$ is called the *distance of x and y*. In addition to the axioms above there is the following consequence of the triangle inequality (which is often formulated as a part of it):
(c′) $\forall x, y, z \in X : |d(x, y) - d(x, z)| \leq d(y, z)$.
Let (X, d) be a metric space. For every $x \in X$ and every $r > 0$ we call the set $B_r(x, d) := \{ y \in X : d(x, y) < r \}$ the ***(open) ball*** about its *centre* x (or: *centred at* x) and with *radius* r. If it is clear which metric is being used we write $B_r(x)$ instead of $B_r(x, d)$. The

set $S_r(x, d) := \{ y \in X \,\vdots\, d(x, y) \leq r \}$ is called the *closed ball*[5] about x with radius r; here we also write, for convenience, $S_r(x)$ if the metric under consideration is clear.

The *topology generated by a metric d* is the topology $\mathcal{O}_d$ generated by the collection of all open balls $B_r(x, d)$ with $x \in X$ and $r > 0$. It is easy to see that *the collection of all open balls is a base for this topology*. So a set U is open with respect to this topology iff it is a union of open balls, iff for every point $x \in U$ there exists an $\varepsilon > 0$ such that $B_\varepsilon(x) \subseteq U$. Thus, if $x \in X$ and A is a subset of X then $x \in A^\circ$ iff there exists $\varepsilon > 0$ such that $B_\varepsilon(x) \subseteq A$. Also, a subset F of X is closed iff for every $x \in X \setminus F$ there exists $\varepsilon > 0$ such that $B_\varepsilon(x) \cap F = \emptyset$. Consequently, for an arbitrary subset F of X and any point $x \in X$ we have:

$$x \in \overline{F} \iff \forall \varepsilon > 0 \exists y \in F \,\vdots\, d(x, y) < \varepsilon \,.$$

If there is a metric on a topological space that generates the topology of that space then that space is said to be *metrizable*. See 7 below for one of the simplest metrization theorems (i.e., theorems that give (necessary and) sufficient conditions for a space to be metrizable).

If Y is a non-empty subset of a metric space (X, d) then the restriction of d to $Y \times Y$ is a metric on Y. The topology on Y generated by this restriction turns out to coincide with the relative topology that Y inherits from the topology on X generated by d. In the sequel we shall denote the restriction of d to Y also by d.

Example. The metric $d : (s, t) \mapsto |s - t| : \mathbb{R} \times \mathbb{R} \to \mathbb{R}$ generates the usual topology on $\mathbb{R}$. The usual topology on $\mathbb{R}^n$ ($n \geq 1$) is generated, by the Euclidean metric $d_{\text{eucl}} : (x, y) \mapsto \|x - y\| : \mathbb{R}^n \times \mathbb{R}^n \to \mathbb{R}^n$, where $\|\,.\,\| : x \mapsto \sqrt{x_1^2 + \cdots + x_n^2} : \mathbb{R}^n \to \mathbb{R}^+$ is the de euclidean norm on $\mathbb{R}^n$. Another metric on $\mathbb{R}^n$ that generates this topology is the 'Manhattan' metric $d_m : (x, y) \mapsto \max\{ |x_i - y_i| \,\vdots\, i = 1, \ldots, n \}$.

A.7.2. Continuity of mappings between metric spaces is easily characterized in terms of the metrics of these spaces. Let (X, d) and (Y, ρ) be metric spaces. Then a mapping $f : X \to Y$ is continuous (with respect to the topologies on X and Y generated by their respective metrics) iff for every $x \in X$ the following condition is fulfilled

$$\forall \varepsilon > 0 \; \exists \delta > 0 \,\vdots\; \rho(f(x), f(x')) < \varepsilon \quad \text{for all } x' \in X \text{ with } d(x, x') < \delta$$

(in general, δ depends on f, x and ε). The mapping $f : X \to Y$ is said to be *uniformly continuous* whenever

$$\forall \varepsilon > 0 \; \exists \delta > 0 \,\vdots\; \rho(f(x), f(x')) < \varepsilon \quad \text{for all } x, x' \in X \text{ with } d(x, x') < \delta$$

(δ depends only on f and ε). Obviously, a uniformly continuous mapping is continuous. If the space X is compact then the converse holds: *If (X, d) is a compact metric*

5 One should not (yet) interpret this in terms of a topology. Rather, one should think about a ball including its 'skin'. See the footnote preceding formula (A.7-4) below.

space and (Y, ρ) is any metric space then every continuous mapping $f : X \to Y$ is uniformly continuous. There is a slightly more subtle statement, as follows: Assume that (X, d) and (Y, ρ) are arbitrary metric spaces, let $f : X \to Y$ be a continuous function and let A be a compact subset of X. Then for every $\varepsilon > 0$ there exists $\delta > 0$ such that

$$\rho(f(x), f(x')) < \varepsilon \quad \text{for all } x \in A \text{ and } x' \in X \text{ with } d(x, x') < \delta \tag{A.7-1}$$

(note that x' is not necessarily in A).

A set $\mathcal{F}$ of mapping from X to Y is said to be *equicontinuous* at the point $x \in X$ whenever the following condition is satisfied:

$$\forall\, \varepsilon > 0\ \exists\, \delta > 0 \ \vdots\ \rho(f(x), f(x')) < \varepsilon \quad \text{for all } f \in \mathcal{F} \text{ and all } x' \in X \text{ with } d(x, x') < \delta$$

(δ depends only on x and ε, but not on the particular choice of f). If this condition is fulfilled for every point $x \in X$ then $\mathcal{F}$ is said to be *equicontinuous on* X. If in that case the value of δ can be chosen independently of the choice of the point x then the set is said to be *uniformly equicontinuous*:

$$\forall\, \varepsilon > 0\ \exists\, \delta > 0 \ \vdots\ \rho(f(x_1), f(x_2)) < \varepsilon \quad \text{for all } f \in \mathcal{F} \text{ and all } x_1, x_2 \in X \text{ with } d(x_1, x_2) < \delta\,.$$

Also here a simple compactness argument shows: *An equicontinuous family of functions on a compact metric space is uniformly equicontinuous.* Note that every finite set of mappings is equicontinuous, so a finite family of functions on a compact metric space is uniformly equicontinuous.

A.7.3. If A is a non-empty subset of a metric space (X, d) then the *diameter* of a non-empty set A in a metric space (X, d) is the (possibly infinite) non-negative real number $\mathrm{diam}(A) := \sup\{\, d(x, y) \vdots x, y \in A\,\}$. For example, for any point x in X and every $r > 0$ we have[6]

$$0 \le \mathrm{diam}(B_r(x)) = \mathrm{diam}(\,\overline{B_r(x)}\,) \le \mathrm{diam}(S_r(x)) \le 2r\,.$$

For every non-empty subset A of X we define the distance of a point x to A and the r-neighbourhood ($r > 0$) of A by

$$\begin{aligned} d(x, A) &:= \inf\{\, d(x, y) \vdots y \in A\,\}, \\ B_r(A, d) &:= \{\, y \in X \vdots d(y, A) < r\,\} = \bigcup \{\, B_r(x, d) \vdots x \in A\,\}. \end{aligned}$$

If d is understood we write $B_r(A)$ for $B_r(A, d)$. If $x \in A$ then $d(x, A) = 0$; the converse implication need not be true. In addition, $B_r(A)$ is an open nbd of A if $r > 0$, but in

6 Examples in the shift space Ω_2 with the metric defined in 5.1.5 show that the second and third inequalities can be strict. In a discrete metric space the first inequality is an equality for small r.

general, it is not true that every open nbd of A includes a nbd of this form (it does if A is compact: see below).

The mapping $x \mapsto d(x, A): X \to \mathbb{R}$ is continuous. This implies that for every subset A of X the set $\{x \in X \vdots d(x, A) = 0\}$ is closed. This proves the "$\subseteq$"–part of the following equality, the other inclusion being a direct consequence of the characterization of a closure given earlier in 1 above:

$$\overline{A} = \{x \in X \vdots d(x, A) = 0\}. \tag{A.7-2}$$

If A is a compact subset of X and F is a closed subset of X, then the continuous mapping $x \mapsto d(x, F) : X \to \mathbb{R}$ assumes its minimum on A. So if the sets A and F are disjoint then this minimum is strictly positive. Consequently, there exists $\varepsilon > 0$ such that $B_\varepsilon(A) \cap F = \emptyset$. In particular, if U is an arbitrary open nbd of A then an application of this conclusion with $F := X \setminus U$ shows: there exists $\varepsilon > 0$ such that $B_\varepsilon(A) \subseteq U$. This proves the following statement: *Let A be a compact subset of a metric space. Then its open neighbourhoods $B_r(A)$ with $r > 0$ form a neighbourhood base for A.*

The *distance* of two non-empty subsets A and B of a metric space is the non-negative real number dist(A, B) defined by

$$\begin{aligned} \operatorname{dist}(A, B) \ &:= \ \inf\{d(a, b) \vdots a \in A \text{ and } b \in B\} \\ &= \ \inf\{d(a, B) \vdots a \in A\}. \end{aligned} \tag{A.7-3}$$

Obviously, $\operatorname{dist}(A, B) = \operatorname{dist}(B, A)$. The discussion following (A.7-2) above can be reformulated as follows: *if A is compact, B is closed and $A \cap B = \emptyset$ then* $\operatorname{dist}(A, B) > 0$.

The continuity of the mapping $y \mapsto d(x, y) : X \to \mathbb{R}$ implies that every closed ball is a closed subset of X which, in turn, is included in every open ball about the same point with larger radius: if $x \in X$ then[7]

$$B_r(x) \subseteq \overline{B_r(x)} \subseteq S_r(x) \subseteq B_{r'}(x) \quad \text{for all } r' > r > 0. \tag{A.7-4}$$

It follows that every point has a local base consisting of closed balls; in addition, each point has a local base consisting of closures of open balls about that point.

Every metric space is a Hausdorff space. By the preceding paragraph, every metric space is regular, hence a T_3-space. A metric space is even normal, hence a T_4-space: if A and B are disjoint closed subsets of a metric space X then the sets $U := \{x \in X \vdots d(x, A) - d(x, B) < 0\}$ and $V := \{x \in X \vdots d(x, A) - d(x, B) > 0\}$ are mutually disjoint open nbds of A and B, respectively.

A.7.4. The following definition is a special case of the definition of converge of nets, given in Section A.4: a sequence is nothing but a net, indexed by the set $\mathbb{N}$ (with its natural order this is a directed set). A sequence $(x_n)_{n\in\mathbb{N}}$ in X is said *to be convergent in X*

7 The second inclusion can be proper. For example, if Ω_2 is the *shift space* with the metric defined in 5.1.5 then it is easily checked that $B_1(x)^- = B_1(x) \neq \Omega_2 = S_1(x)$ for every point $x \in \Omega_2$.

with limit $x \in X$, or to *converge* in X *to* the point x, whenever the sequence $(d(x, x_n))_{n\in\mathbb{N}}$ converges in $\mathbb{R}$ to 0. Notation: $x_n \rightsquigarrow x$. Thus, a sequence in X converges to the point $x \in X$ iff every nbd of x contains almost all (that is, all but finitely many) members of the sequence. Obviously, the limit of a sequence is unique.

A cluster point of a sequence is characterized by the property that each of its nbds contains infinitely many terms of the sequence; such a point is the limit of a suitable subsequence and will therefore also be called a *limit point* of that sequence.

A number of topological concepts can be reformulated for a metric spaces in terms of sequences. Let (X, d) be a metric space.

- Let A be a subset of X and let $x \in X$. Then $x \in A^-$ iff there is a sequence $(x_n)_{n\in\mathbb{N}}$ in A such that $x_n \rightsquigarrow x$.
- Let (Y, ρ) be also a metric space and let $f : X \to Y$ be a mapping. Then f is continuous with respect to the topologies generated by the metrics on X and Y iff for every convergent sequence $(x_n)_{n\in\mathbb{N}}$ in X the sequence $(f(x_n))_{n\in\mathbb{N}}$ converges in Y with $\lim_{n\rightsquigarrow\infty} f(x_n) = f(\lim_{n\rightsquigarrow\infty} x_n)$.
- *A subset A of X is compact iff every sequence in A has a convergent subsequence with limit in A.*

A metric space is said to be *totally bounded* whenever for every $\varepsilon > 0$ it can be covered by a finite number of open ε-balls: for every $\varepsilon > 0$ there is a finite subset K of X such that

$$\forall\, x \in X \exists y \in K : d(x, y) < \varepsilon\,.$$

A metric space (X, d) is said to be *complete* whenever every Cauchy–sequence converges. Recall that a *Cauchy–sequence* is a sequence $(x_n)_{n\in\mathbb{N}}$ in X with the following property:

$$\forall\, \varepsilon > 0\, \exists n_\varepsilon \in \mathbb{N} : d(x_k, x_m) < \varepsilon \quad \text{for all } k, m \geq n_\varepsilon$$

(in general, this is a necessary condition for convergence of a sequence in a metric space). A complete subset of a metric space is always closed. Conversely, a closed subset Y of a complete metric space is complete with respect to the metric restricted to Y.

Theorem A.7.5. *A metric space is compact iff it is complete and totally bounded.*

Examples.

(1) The Euclidean space $\mathbb{R}^n$ ($n \in \mathbb{N}$) with its usual metric is complete (Cauchy's Theorem).

(2) Let $X := C([0; 1])$, the set of all continuous functions $f: [0; 1] \to \mathbb{R}$. For $f \in X$, let $\|f\| := \sup\{\,|f(s)| : s \in [0; 1]\,\}$. Then $\rho: (f, g) \mapsto \|f - g\| : X \times X \to \mathbb{R}^+$ is a metric on X. A sequence converges in X with respect to this metric iff the sequence converges uniformly on $[0; 1]$. *With this metric X is a complete metric space.*

(3) More generally, if X is a compact space and Y is a metric space with metric d then the mapping

$$d_u : (f, g) \mapsto \sup\{ d(f(x), g(x)) \,\vdots\, x \in X \} : C(X, Y) \times C(X, Y) \to \mathbb{R}^+$$

is a metric in $C(X, Y)$. The topology defined by this metric is called the *topology of uniform convergence* and d_u will be called the *uniform metric*. The space $C(X, Y)$ endowed with this topology will be denoted by $C_u(X, Y)$. If Y is a complete metric space then $C_u(X, Y)$ is complete as well.

A.7.6. Let for every $i \in \mathbb{N}$ a metric space (X_i, d_i) be given. Since the two metrics d_i and $d_i' : (x, y) \mapsto \min\{ 1, d(x, y) \} : X_i^2 \to \mathbb{R}^+$ generate the same topology on X_i, we may assume that each of the metrics d_i is bounded by 1. Consider the space $X := \prod_{i=1}^{\infty} X_i$ and for every pair of points $x = (x_i)_{i\in\mathbb{N}}$ and $y = (y_i)_{i\in\mathbb{N}}$ of points of X let

$$\rho(x, y) := \sum_{i=1}^{\infty} \frac{1}{2^i} d_i(x_i, y_i) \,. \tag{A.7-5}$$

It is easily verified that ρ is a metric on X and it is not too difficult to show that the topology generated by ρ is the product topology on X. (Note that in formula (A.7-5) the real numbers $1/2^i$ for $i \in \mathbb{N}$ can be replaced by positive real numbers a_i provided the series $\sum_i a_i$ converges.) In particular, it follows that *a countable product of metric spaces is metrizable*.

In the case that we consider a product of two metric spaces – now the condition that the metrics are bounded by 1 can be omitted – a metric on the product space X^2 is also given by

$$\rho'((x_1, x_2), (y_1, y_2)) := \max\{ d_1(x_1, y_1), d_2(x_2, y_2) \} \tag{A.7-6}$$

for $(x_1, x_2), (x_2, y_2) \in X^2$. In particular, if (X, d) is a metric space then X^2 is a metric space with metric ρ given by (A.7-6) with $d_1 = d_2 = d$. In that case, for every $\varepsilon > 0$ the set $B_\varepsilon(\Delta_X, \rho)$ for $\varepsilon > 0$ is an open neighbourhood of the diagonal $\Delta_X := \{(x, x) \,\vdots\, x \in X\}$ in X^2. *If X is compact then the sets $B_\varepsilon(\Delta_X, \rho)$ with $\varepsilon > 0$ form a neighbourhood base of the set Δ_X in X^2*; see the final statement in the paragraph preceding formula (A.7-3) above. Finally, for every $\varepsilon > 0$ let

$$\alpha_\varepsilon := \{ (x, y) \in X^2 \,\vdots\, d(x, y) < \varepsilon \} \,. \tag{A.7-7}$$

Then α_ε is an open neighbourhood of Δ_X in X^2 as well. Using the triangle inequality it is easily shown that for every $\varepsilon > 0$ one has

$$\alpha_\varepsilon \subseteq B_\varepsilon(\Delta_X, \rho) \subseteq \alpha_{2\varepsilon} \,.$$

So in a compact metric space the sets α_ε with $\varepsilon > 0$ form a neighbourhood base of the set Δ_X in X^2 as well.

Proposition A.7.7. *A compact Hausdorff space admits a metric compatible with its topology iff it has a countable base.*

Proof. "Only if": For every $n \in \mathbb{N}$ the compact space X has a finite covering $\mathcal{B}_n$ by open balls with radius $1/n$. It is easily checked that the (countable!) set $\bigcup_n \mathcal{B}_n$ is a base for the topology of X.

"If": Let $\mathcal{B}$ be a countable base for the topology of X and let A be the set of all pairs (B_1, B_2) of elements of $\mathcal{B}$ with the property that $\overline{B}_1 \subset B_2$. For every $a = (B_1, B_2) \in A$ select a continuous function $f_a : X \to [0; 1]$ satisfying the conditions

$$f_a[B_1] = \{1\} \quad \text{and} \quad f_a[X \setminus B_2] = \{0\} .$$

This is possible in view of Proposition A.3.4 above. Then the mapping $\varphi : x \mapsto (f_a(x))_{a \in A} : X \to [0; 1]^A$ is continuous: this follows from the defining property of a product topology – see Lemma A.5.3 – since the mapping $\pi_a \circ \varphi = f_a$ is continuous for every $a \in A$ (here π_a is a projection). It is routine to check that φ is injective: if $x_1, x_2 \in X$ and $x_1 \neq x_2$ then there are $B_1, B_2 \in \mathcal{B}$ such that $x_1 \in B_1$ and $x_2 \notin B_2$ (this is because X is regular), hence $f_a(x_1) = 1$ and $f_a(x_2) = 0$ for $a := (B_1, B_2)$. As X is compact and $[0; 1]^A$ is a Hausdorff space it follows that f is a topological embedding. Finally, note that the set A is countable, so that $[0; 1]^A$ is metrizable. Hence X, being homeomorphic with a subspace of a metrizable space, is metrizable as well. □

Consequently, by the final remark in A.1.2: every compact metric space is separable, i.e., has a countable dense subset. Note that a quotient space of a space with a countable base has a countable base as well. Hence (see also [Eng], Theorem 4.4.15):

Theorem A.7.8. *Let X be a compact metric space, let Y be a Hausdorff space and let $f : X \to Y$ be a continuous surjection. Then Y is compact and metrizable.*

The following result has many applications.

Theorem A.7.9 (Banach). *Let (X, d) be a complete metric space and consider a contraction $f : X \to X$, i. e., a mapping such that:*

$$\exists c \in (0; 1) \;\vdots\; \forall x, y \in X : \; d(f(x), f(y)) \leq c\, d(x, y) . \tag{A.7-8}$$

Then there is a unique fixed point $z \in X$, i.e., a point z such that $f(z) = z$; moreover, $f^n(x) \rightsquigarrow z$ for every point $x \in X$.

Proof. "Unicity": If z_1 and z_2 are two fixed points then (A.7-8) implies that $d(z_1, z_2) = d(f(z_1), f(z_2)) \lneq d(z_1, z_2)$. Hence $d(z_1, z_2) = 0$ and $z_1 = z_2$.

"Existence": Let $x_0 \in X$ be arbitrary and define $x_n := f^n(x_0)$ for $n \in \mathbb{N}$. Then for every $k, m \in \mathbb{N}$ with $k > m$ we have $x_k = f^m(x_{k-m})$, hence

$$d(x_m, x_k) = d(f^m(x_0), f^m(x_{k-m})) \leq c^m\, d(x_0, x_{k-m}) .$$

If X is not bounded we can estimate $d(x_0, x_{k-m})$ in the following way: note that $d(x_0, x_{k-m}) \leq d(x_0, x_1) + d(x_1, x_2) + \cdots + d(x_{k-m-1}, x_{k-m})$, where, for every

$n \in \mathbb{N}$, $d(x_n, x_{n+1}) = d(f^n(x_0), f^n(x_1)) \le c^n\, d(x_0, x_1)$. This implies that $d(x_0, x_{k-m}) \le d(x_0, x_1)/(1-c) =: K$, hence $d(x_m, x_k) \le c^m K$, which implies that $(x_n)_{n\in\mathbb{N}}$ is a Cauchy-sequence. Therefore, the sequence converges, say with limit z. As f is continuous (this follows easily from the assumption on f) we get

$$f(z) = f(\lim_{n\rightsquigarrow\infty} x_n) = \lim_{n\rightsquigarrow\infty} f(x_n) = \lim_{n\rightsquigarrow\infty} x_{n+1} = z\,. \qquad \square$$

A.8 Baire category

A topological space X is said to be a *Baire space* whenever the intersection of a countable family of dense open subsets of X is dense in X. An equivalent formulation of the Baire property is as follows: A subset A of X is said to be *nowhere dense* whenever its closure has empty interior. Consequently, a subset A of X is nowhere dense iff the interior of its complement is dense in X; this follows from the equalities in (A.1-1). It is easily seen that the union of finitely many nowhere dense sets is nowhere dense. A set which is the union of a countable family of nowhere dense sets is called *meagre*, or is said to be a set of the *first category* in X.

A subset of X is said to be *residual* whenever it is the complement of a meagre set, i.e., the complement of a countable union of nowhere dense sets. Thus, *a subset B of X is residual iff B is the intersection of a countable family of sets each of which has a dense interior.* For example, a dense G_δ-set is residual (use that if $\bigcap_n U_n$ is dense then each U_n is dense).

Lemma A.8.1. *Let X be a topological space. The following conditions are equivalent:*
(i) *X is a Baire space.*
(ii) *Every residual subset of X is dense in X.*
(iii) *Every meagre subset of X has empty interior in X.*
It follows that in a Baire space a residual set includes a dense G_δ-set, and that a G_δ-set is residual iff it is dense.

Proof. "(i)⇒(ii)": Suppose X is a Baire space and let $\{\, A_n \,\vdots\, n \in \mathbb{N}\,\}$ be a family of sets with dense interiors. Then $\{\, A_n^\circ \,\vdots\, n \in \mathbb{N}\,\}$ is a countable family of dense open sets, so by the Baire property of X the set $\bigcap_n A_n^\circ$ is dense in X. Hence the larger set $\bigcap_n A_n$ is dense as well.

"(ii)⇒(i)": Clear from the definitions.

"(ii)⇔(iii)": Take complements, using De Morgan's laws and taking into account the equalities in (A.1-1).

Final statement: observe that the dense set $\bigcap_n A_n^\circ$ obtained in the proof of (i)⇒(ii) is a G_δ-set. $\square$

In topology the residual sets are considered to be the 'big' sets. Often a property is said to hold *almost everywhere* whenever it holds on a residual set. Hence by the above observations, if X is a Baire space then a property holds almost everywhere on X iff the set on which the property holds includes a dense G_δ-set.

We shall show now that every complete metric space is a Baire space. First a lemma (due to Cantor).

Lemma A.8.2. *Let (X, d) be a complete metric space and let $\{F_n\}_{n\in\mathbb{N}}$ be a descending sequence of closed subsets such that $\lim_{n\rightsquigarrow\infty} \operatorname{diam}(F_n) = 0$. Then the set $F := \bigcap_{n\in\mathbb{N}} F_n$ consists of exactly one point. Moreover, if for every $n \in \mathbb{N}$ an element $x_n \in F_n$ is selected then $(x_n)_{n\in\mathbb{N}}$ is a Cauchy sequence that converges to the unique element of F.*

Proof. First, we show that F contains at most one point. Consider two points z' and z'' in F. Then $z', z'' \in F_n$, hence $d(z', z'') \leq \operatorname{diam}(F_n)$ for every $n \in \mathbb{N}$. It follows that $d(z', z'') = 0$, hence $z' = z''$.

Next, select for every $n \in \mathbb{N}$ a point $x_n \in F_n$. Then for all $n \in \mathbb{N}$ and for all $k, m \in \mathbb{N}$ with $k \geq n$ and $m \geq n$ we have $d(x_k, x_m) \leq \operatorname{diam}(F_n)$, because both x_k, $x_m \in F_n$. As $\operatorname{diam}(F_n) \rightsquigarrow 0$ as $n \rightsquigarrow \infty$, this implies that $(x_n)_{n\in\mathbb{N}}$ is a Cauchy sequence. So this sequence converges, say, with limit $z \in X$. Finally, if $n \in \mathbb{N}$ then for all $k \geq n$ we have $x_k \in F_k \subseteq F_n$, and because the set F_n is closed, it follows that also the limit z of the sequence $(x_n)_{n\in\mathbb{N}}$ belongs to F_n. This holds for every $n \in \mathbb{N}$, hence $z \in F$. □

Theorem A.8.3 (Baire). *A complete metric space is a Baire space.*

Proof. Let (X, d) be a complete metric space and consider a countable collection $\{U_n \vdots n \in \mathbb{N}\}$ of open sets each of which is dense in X. Let V be an arbitrary non-empty open subset of X; we want to show that V has a non-empty intersection with $\bigcap_{n\in\mathbb{N}} U_n$.

The set $V \cap U_1$ is not empty (recall that U_1 is dense) and open. Since closures of open balls form local bases it follows that there is an open ball V_1 with radius at most $1/2$ such that $\overline{V}_1 \subseteq V \cap U_1$. Now proceed by induction: there is a sequence of open balls $\{ V_n \vdots n \in \mathbb{N} \}$ such that, for every $n \in \mathbb{N}$, the radius of V_n is at most $1/2^n$ and $\overline{V}_n \subseteq V_{n-1} \cap U_n$ (to make notation consistent, let $V_0 := V$). In point of fact, if we have $V_1, \dots, V_n$ satisfying these conditions then, as U_{n+1} is dense, the set $V_n \cap U_{n+1}$ is not empty and open, hence it includes the closure of an open ball with radius at most $1/2^{n+1}$. Now the lemma above implies that $F := \bigcap_{n\in\mathbb{N}} \overline{V}_n \neq \emptyset$. Next, observe that $F \subseteq \overline{V}_1 \subseteq V$ and that $F \subseteq \overline{V}_n \subseteq U_n$ for every $n \in \mathbb{N}$, hence $F \subseteq V \cap \bigcap_{n\in\mathbb{N}} U_n$. This shows that $V \cap \bigcap_{n\in\mathbb{N}} U_n \neq \emptyset$. □

Theorem A.8.4 (Baire). *A locally compact Hausdorff space is a Baire space.*

Proof. The proof is similar to the proof of Theorem A.8.3 above, the difference being that now each V_n is selected as a non-empty open set such that $\overline{V}_n$ is compact (and, as before, $\overline{V}_n \subseteq V_{n-1} \cap U_n$). Now $\{ \overline{V}_n \vdots n \in \mathbb{N} \}$ is a descending sequence of non-empty compact sets. From compactness we infer that $F := \bigcap_{n\in\mathbb{N}} \overline{V}_n \neq \emptyset$, and the proof is completed as in Theorem A.8.3. □

Example. For every $n \in \mathbb{N}$ the space $\mathbb{R}^n$ with its usual topology and metric is both a complete metric space and a locally compact Hausdorff space. Hence it is a Baire space. Similarly, the unit circle $\mathbb{S}$ in the complex plane (a compact Hausdorff space) is a Baire space, as are all closed subsets of $\mathbb{R}^n$, all open subsets of $\mathbb{R}^n$ and all arbitrary intervals in $\mathbb{R}$. By the observation in 1 above, the space $\mathbb{Q}$ (with its topology inherited from $\mathbb{R}$) is not a Baire space: the union of its countably many nowhere dense singletons (a meagre set) is all of $\mathbb{Q}$, hence has *not* empty interior.

In general, a closed subspace of a Baire space is not a Baire space (standard counter example: $\mathbb{R}^2$ from which all irrational points on the x-axis are removed is a Baire space, but its closed subspace $\mathbb{Q}\times\{0\}$ is not). A class of Baire spaces with the property that every closed subspace is a Baire space is formed by the so-called Čech-complete spaces. A Tychonov space X is said to be *Čech-complete* whenever it is a dense subspace of a compact Hausdorff space Z such that the remainder $Z \setminus X$ is an F_σ-set[8]. It can be shown that

- Closed subspaces and arbitrary products of Čech-complete spaces are Čech-complete (see [Eng], Theorems 3.9.6 and 3.9.8).
- Every Čech-complete space is a Baire space (see [Eng], Theorem 3.9.3).

Consequently, closed subspaces of products of Čech-complete spaces are Baire spaces.

Example. Every locally compact Hausdorff space X is Čech-complete: it has a one-point remainder in its one-point compactification $X \cup \{\infty\}$ (obtained by adding one point ∞ to X with as a local base at ∞ the family of all complements of compact subsets of X). Also, every complete metric space is Čech-complete: see [Eng], Theorem 4.3.26. An example of a non-locally compact and not completely metrizable Čech-complete space is the space $\mathbb{R} \setminus \mathbb{Q}$ of irrationals (it has a countable complement in the one-point compactification of $\mathbb{R}$).

A.9 Irreducible mappings

A continuous mapping $f: X \to Y$ is said to be *irreducible* whenever it is surjective and it maps no proper closed subset of X onto Y. In what follows we shall employ the following notation: if $A \subseteq X$ then let

$$F(A) := \{ y \in Y : f^{\leftarrow}[y] \subseteq A \} \quad \text{and} \quad A_f := f^{\leftarrow}[F(A)] .$$

(possibly, these sets are empty). Clearly, $A_f = \{x \in X : f^{\leftarrow}[f(x)] \subseteq A\}$, hence $f^{\leftarrow}[f[A_f]] = A_f \subseteq A$. It is also easy to see that

$$F(A) = Y \setminus f[X \setminus A] . \tag{A.9-1}$$

Consequently, for a continuous and closed mapping f we have: if A is open in X then $F(A)$ is open in Y and A_f is open in X.

8 An F_σ-set is a countable union of closed sets. Every Tychonov space can be densely embedded in a compact Hausdorff space, but the remainder need not be an F_σ-set.

Lemma A.9.1. *Let $f: X \to Y$ be a continuous closed surjection. Then the following conditions are equivalent:*
(i) *f is irreducible.*
(ii) *Every non-empty open subset of X contains a full fibre of f.*
(iii) *Every non-empty open subset U of X includes a non-empty open subset V such that $f^{\leftarrow}[f[V]] = V$ and $f[V]$ is an open set in Y.*
and they imply that f is semi-open.

Proof. First observe that condition (ii) means precisely that for every non-empty open subset U of X the set $F(U)$ defined above is not empty.

(i)⇒(ii): Let U be a non-empty open set in X. Then $X \setminus U$ is a proper closed subset of X, so $f[X \setminus U] \neq Y$, hence $F(U) \neq \emptyset$.

(ii)⇒(i): Let F be a closed proper subset of X. Then $X \setminus F$ is a non-empty open subset of X, hence there is a point $y \in Y$ such that $f^{\leftarrow}[y] \subseteq X \setminus F$. Then obviously $y \notin f[F]$, so $f[F] \neq Y$.

(iii)⇒(ii): Trivial (also if V or $f[V]$ are not assumed to be open). Moreover, it is obvious that (iii) implies that f is semi-open.

(ii)⇒(iii): Let U be a non-empty open subset of X and put $V := U_f$. Since $F(U) \neq \emptyset$ it follows that $V \neq \emptyset$ and in view of the remarks preceding the lemma, V satisfies the requirements of (iii). □

Corollary A.9.2 (See [deV], Appendix (A.4)). *Let $f: X \to Y$ be an irreducible mapping and assume that X is compact and that Y is a Hausdorff space. Then f is semi-open.*

Proof. In this situation f is a closed mapping, so the corollary follows from the implication (i)⇒(iii) in the lemma above. □

Proposition A.9.3 (See A. Wilansky [1970], 14.2.3). *Let X and Y be compact Hausdorff spaces and let $f: X \to Y$ be a continuous surjection. The following are equivalent:*
(i) *f is irreducible and open.*
(ii) *f is a homeomorphism.*

Proof. Only (i)⇒(ii) needs a proof. So assume that f is irreducible and open. It is sufficient to show that f is injective. Suppose it is not: there are points $x_1, x_2 \in X$ such that $x_1 \neq x_2$ and $f(x_1) = f(x_2) =: y$. Select open nbds U_1 of x_1 and U_2 of x_2 such that $U_1 \cap U_2 = \emptyset$. As f is open, $f[U_1]$ is an open nbd of y, hence there is an open nbd V of x_2 such that $V \subseteq U_2$ and $f[V] \subseteq f[U_1]$. So V cannot contain a full fibre: for every point $x \in V$ the fibre $f^{\leftarrow}[f(x)]$ intersects U_1. By the lemma above, this contradicts the irreducibility of f. □

If $f: X \to Y$ is continuous mapping then the (possibly empty) sets $T_f \subseteq Y$ and $S_f \subseteq X$ are defined by

$$T_f := \{\, y \in Y \,\vdots\, \operatorname{card}(f^{\leftarrow}[y]) = 1 \,\},$$

where card(A) denotes the cardinality of the set A, and

$$S_f := f^{\leftarrow}[T_f] = \{\, x \in X : f^{\leftarrow}[f(x)] = \{x\}\,\}.$$

A continuous mapping $f: X \to Y$ is said to be *weakly almost 1-to-1* or *weakly almost 1,1* whenever the set S_f is dense in X. Note that $T_f = f[S_f]$, so if f is a weakly almost 1,1 surjection then the set T_f is dense in $f[X]$. Conversely, if T_f is dense in Y and f is semi-open then the set S_f is dense in X by Lemma A.3.6, hence f is weakly almost 1-to-1.

A 'really' almost 1-to-1 mapping[9] is a mapping that is injective 'almost everywhere', i.e., on a residual set (as opposed to a set which is just dense). So a mapping $f: X \to Y$ is said to be *almost 1-to-1* whenever S_f includes a residual set or, equivalently, whenever S_f incudes a dense G_δ-set. If T_f includes a dense G_δ-set and f is semi-open then S_f includes a dense G_δ-set, so in this case f is almost 1-to-1. Clearly, an almost 1,1-mapping is a weakly almost 1,1-mapping. We shall see below that the two notions coincide if X is a metric space.

Theorem A.9.4. *Let $f: X \to Y$ be a weakly almost 1-to-1 continuous surjection. Then f is irreducible. If, in addition, f is a assumed to be a closed mapping then f is semi-open.*

Proof. Let U be a non-empty open subset of X. As the set S_f is dense in X, it follows that $S_f \cap U \neq \emptyset$. For any point $x \in S_f \cap U$ the fibre $f^{\leftarrow}[f(x)]$ is equal to the singleton set $\{x\}$, which is included in U. So condition (ii) of Lemma A.9.1 is fulfilled. It follows that f is irreducible. The final statement follows from the implication (ii)⇒(iii) in that lemma. □

We shall formulate and prove now a converse.

Lemma A.9.5. *Let $f: X \to Y$ be a continuous mapping. If X is a metric space, say, with metric d, then*

$$T_f = \bigcap_{n\in\mathbb{N}} \Big(\bigcup_{x\in X} F(B_{\frac{1}{n}}(x,d)) \Big).$$

Proof. For every $x \in X$ and $n \in \mathbb{N}$, let

$$G_n(x) := F(B_{\frac{1}{n}}(x,d)) = \{\, y \in Y : f^{\leftarrow}[y] \subseteq B_{\frac{1}{n}}(x,d)\,\},$$

and let $G_n := \bigcup_{x\in X} G_n(x)$. If $n \in \mathbb{N}$ then for every $y \in G_{2n}$ the diameter of $f^{\leftarrow}[y]$ is less than $1/n$ and, conversely, if the diameter of $f^{\leftarrow}[y]$ is less than $1/n$ then $y \in G_n$. As T_f is the set of all points $y \in Y$ for which the fibre $f^{\leftarrow}[y]$ has diameter 0, this implies that $T_f = \bigcap_{n=1}^{\infty} G_n$. □

9 This conflicts with usage in the theory of dynamical system; see the final remark in Note 12 at the end of Chapter 1.

Corollary A.9.6. *If X is a metric space and f is a closed continuous mapping then T_f and S_f are G_δ-sets. So in that case S_f is dense iff S_f is residual, that is, f is weakly almost 1-to-1 iff f is almost 1-to-1.* □

Theorem A.9.7. *Let $f: X \to Y$ be an irreducible closed mapping and assume that X is a metric Baire space. Then f is almost 1-to-1.*

Proof. Define the sets $G_n(x)$ and G_n ($n \in \mathbb{N}, x \in X$) as in the proof of the lemma above. By the observations just after (A.9-1), for every $n \in \mathbb{N}$ the set $G_n(x)$ is open in Y for every $x \in X$, hence G_n is open in Y as well. Consequently, the set $f^{\leftarrow}[G_n]$ is open in X for every $n \in \mathbb{N}$. We claim that it is dense in X. This will imply that the set $\bigcap_{n=0}^{\infty} f^{\leftarrow}[G_n]$ is dense in the Baire space X, that is, that S_f is dense in X.

Suppose that for some $n \in \mathbb{N}$ the set $f^{\leftarrow}[G_n]$ is not dense in X. Then there are $x \in X$ and $k \in \mathbb{N}$ such that the open ball $B_{\frac{1}{k}}(x)$ is disjoint from the set $f^{\leftarrow}[G_n]$. If we enlarge k or n then the corresponding sets shrink, so these sets remain disjoint. Consequently, we may assume that $k = n$. Because f is irreducible, the open ball $B_{\frac{1}{n}}(x)$ includes a full fibre of f, i.e., there exists a point $y \in Y$ with $f^{\leftarrow}[y] \subseteq B_{\frac{1}{n}}(x)$. This means, by definition, that $y \in G_n(x) \subseteq G_n$, which contradicts the fact that $f^{\leftarrow}[y]$, being a subset of $B_{\frac{1}{n}}(x)$, is disjoint from $f^{\leftarrow}[G_n]$. □

The hypotheses of the theorem are certainly fulfilled if X is a compact metric space and Y is a Hausdorff space: then f is a closed mapping and X is a Baire space. So a continuous surjection $f: X \to Y$ with X and Y compact metric spaces is almost 1-to-1 iff it is irreducible (in which case f is semi-open).

A.10 Miscellaneous results

Lemma A.10.1. *Let (X, d) be a compact metric space and let $\mathfrak{U}$ be an open cover of X. Then there is a real number $\delta > 0$ such that every subset of X with diameter less than or equal to δ is completely included in a member of the family $\mathfrak{U}$.*

Proof. Select for every $x \in X$ an open ball $B_{r(x)}(x)$ such that the closed ball $S_{2r(x)}(x)$ is included in a member of $\mathfrak{U}$. In this way we get an open cover of X, which has a finite subcover $\{ B_{r(x_1)}(x_1), \dots, B_{r(x_k)}(x_k) \}$. Now let $\delta := \min\{ r(x_1), \dots, r(x_k) \}$. □

The real number δ whose existence is proved in the above lemma is called a *Lebesgue number* of the open cover $\mathfrak{U}$. Also in a non-compact space it is possible that a covering has a Lebesgue number, but there its existence is not *a priori* guaranteed.

Example. Let $\mathfrak{U}$ be the covering of the metric space X by the collection of *all* open balls of radius ε. Then every real number $\delta > 0$ such that $0 < \delta < \varepsilon$ is a Lebesgue number for $\mathfrak{U}$.

A.10.2 (Function spaces).

(1) If X and Y are topological spaces then the product topology on the set Y^X of all mappings from X to Y is often called the *topology of pointwise convergence* or just the *pointwise topology*. This terminology is also used for the relative product topology on subsets of Y^X. For example, the set of all continuous mappings from X to Y will be denoted by $C(X, Y)$; if this set is endowed with the pointwise topology then we write $C_p(X, Y)$.
Recall that the pointwise topology on a subset A of Y^X is the weakest topology making all evaluation mappings $\delta_x : A \to Y$ for $x \in X$ continuous. Hence the defining property of a product topology (see Lemma A.5.3 above) implies that a mapping $f : Z \to A$ from a topological space Z to A is continuous iff the composition $\delta_x \circ f : Z \to Y$ is continuous for every $x \in X$.

(2) Let X and Y be as above. The *compact-open topology* in $C(X, Y)$ is the topology generated by the subbase consisting of all sets of the form

$$\mathcal{W}(K, O) := \{ g \in C(X, X) \,\vdots\, g[K] \subseteq O \}$$

with K a compact subset of X and O an open subset of Y. If Y is a metric space with metric d, say, then the compact-open topology coincides with the *topology of uniform convergence on compacta*, in which an element f of $C(X, Y)$ has a neighbourhood basis consisting of all sets

$$\mathcal{W}_\varepsilon(K, f) := \{ g \in C(X, Y) \,\vdots\, d(g(x), f(x)) < \varepsilon \text{ for all } x \in K \}$$

with K a compact subset of X and $\varepsilon > 0$. See [Eng], Theorem 8.2.6. The space $C(X, Y)$ endowed with this topology will be denoted by $C_c(X, Y)$. Obviously, if X is a compact metric space then the compact-open topology on $C(X, Y)$ coincides with the topology of uniform convergence on $C(X, Y)$; so in that case $C_u(X, Y) = C_c(X, Y)$. (For the meaning of $C_u(X, Y)$, see Example (3) after Theorem A.7.5 above.)

(3) Assume that X is a metric space with metric d, and let F be a set of mappings from X into itself, i.e., $F \subseteq X^X$. Then:
   - If F is equicontinuous on X then its closure in X^X is equicontinuous.
   - If F is equicontinuous on X then on F the topology of pointwise convergence coincides with the compact-open topology on F.

   For the straightforward proofs, see A. Wilansky [1970], Section 13.3. These two statements form the basis for the Arzela-Ascoli Theorem, of which the following is a special case:

Theorem A.10.3. *Let X be a compact metric space and let $F \subseteq X^X$. Then the following conditions are equivalent:*

(i) *F is equicontinuous on X;*

(ii) *$F \subseteq C(X, X)$ and F has a compact closure in $C_u(X, X)$.*

(iii) *For every $\varepsilon > 0$ there is a finite subset K of F with the following property:*

$$\forall f \in F\; \exists g \in K \,\vdots\, d(f(x), g(x)) < \varepsilon \quad \textit{for all } x \in X\,.$$

Proof. "(i)⇒(ii)": The space X^X is compact, hence the closure $\mathrm{cl}_p F$ of F in X^X is compact and, by the first statement above, it is an equicontinuous subset of $C(X, X)$. By the second statements above, on $\mathrm{cl}_p F$ the topology of uniform convergence and the pointwise topology coincide, hence F is dense in $\mathrm{cl}_p F$ with the uniform topology. Consequently, $\mathrm{cl}_p F$ is the closure of F in $C_u(X, X)$.

"(ii)⇒(iii)": If $F \subseteq C(X, X)$ and F is compact with respect to the uniform metric in $C(X, X)$ then it is totally bounded. This is precisely what is expressed by the formula in condition (iii).

"(iii)⇒(i)": Let $\varepsilon > 0$ and select K according to (iii). As the set K is finite, it is equicontinuous on X: if $x_0 \in X$ then there is a nbd U of x_0 such that $d(g(x), g(x_0)) < \varepsilon$ for all $x \in U$ and all $g \in K$. If $f \in F$ then select g according to (iii); then for every $x \in U$ one has

$$d(f(x), f(x_0)) \leq d(f(x), g(x)) + d(g(x), g(x_0)) + d(g(x_0), f(x_0)) \leq 3\varepsilon .$$

So F is equicontinuous at the (arbitrary) point x_0. □

A.10.4 (Zorn's Lemma and the Axiom of Choice). The following set-theoretical axiom is widely used in the development of mathematical theories. This axiom is taken as an unstated hypothesis in this book.

Axiom of Choice. *If X_α is a non-empty set for every member α of an index set A then the product set $\prod_{\alpha \in A} X_\alpha$ is non-empty.*

We use this axiom a couple of times in an equivalent form, which is known as *Zorn's Lemma*. To formulate it, recall that a *partial order* on a set X is a relation on that set which is reflexive, anti-symmetric and transitive. Thus, if we denote the relation by $\leq$ then the following conditions characterize it as a partial order:

(1) $\forall x \in X: \quad x \leq x;$

(2) $\forall x, y \in X: \quad x \leq y \ \&\ y \leq x \Rightarrow x = y;$

(3) $\forall x, y, z \in X: \quad x \leq y \ \&\ y \leq z \Rightarrow x \leq z.$

The partial order $\leq$ is said to be *linear* or *total* whenever

(4) $\forall x, y \in X:$ either $x \leq y$ or $y \leq x$.

A linearly ordered set is called a *chain*. In particular, a chain in a a partially ordered set is a totally ordered subset. If A is a subset of a partially ordered set X then an *upper bound* for A is an element $x \in X$ such that $a \leq x$ for all $a \in A$, and a *least upper bound* or *supremum* of A is an upper bound x of A that is less than every other upper bound, that is, if $y \in X$ is any upper bound of A then $x \leq y$. A *maximal element* (or *maximum*) of a subset A of X is an element $m \in A$ such that for every $a \in A$ the relation $m \leq a$ implies $m = a$ (if the order on A is not total this does not necessarily imply that $a \leq m$ for all $a \in A$). Finally, a partial order $\leq$ on a set X is said to be *inductive*, and X is said to be *inductively ordered* whenever each chain in X has an upper bound in X. We are now in a position the formulate Zorn's Lemma:

Zorn's Lemma. *Every inductively ordered set has a maximal element.*

If $x \in X$ then an application of this Lemma to the subset $\{a \in X \,\vdots\, x \leq a\}$ of X gives the following, equivalent, statement:

Zorn's Lemma. *If X is an inductively ordered set then for every $x \in X$ there is a maximal element*[10] *m of X such that $x \leq m$.*

It can be shown that the Axiom of Choice and Zorn's Lemma are logically equivalent. For a proof and for a more complete treatment of this topic we refer to Section I.4 in [Eng].

Examples.

(1) Let X be a subset of the power set of a set U. Let $\leq$ be the partial order on X defined by

$$\forall x, y \in X: \; x \leq y \iff x \subseteq y\,.$$

In this situation a chain in X is called a *nest* in X. Now Zorn's Lemma implies: If for each nest in X there is a member of X that includes all members of the nest then each member of X is included in a maximal element of X (the Maximal Principle).

(2) Let X be a subset of the power set of a set U. Let $\leq$ be the partial order on X defined by

$$\forall x, y \in X: \; x \leq y \iff x \supseteq y\,.$$

Also in this situation a totally ordered subset of X is called a nest in X. However, in this setting a maximal element is usually called 'minimal', and Zorn's Lemma implies: If for each nest in X there is a member of X included in all members of the nest then each member of X includes a minimal element of X (the Minimal Principle).

10 A maximal element in the subset $\{a \in X \,\vdots\, x \leq a\}$ of X is easily seen to be a maximal element of X.

B. The Cantor set

Abstract. For easy reference and in order to agree on terminology we give here short introduction to the Cantor set and its topology. In particular, we prove the theorem by L. E. J. Brouwer that characterizes this space.

B.1 The construction

In what follows, certain intervals will be indexed by n-tuples (also called *words*) of 0's and 1's ($n \in \mathbb{N}$), as in J_0 ($n = 1$) or J_{0110} ($n = 4$). We denote such n-tuples as elements $b = b_0 \dots b_{n-1}$ of $\{0, 1\}^n$, written without parentheses and separating comma's. We shall use *concatenation* to get larger words from shorter ones, as follows: if $b \in \{0, 1\}^n$ and $c \in \{0, 1\}^m$ then $bc := b_0 \dots b_{n-1} c_0 \dots c_{m-1}$. For example, if $b = 0101$ then $b0 = 01010$ and $b1 = 01011$.

Let $J := [p; q]$ be a closed interval in $\mathbb{R}$ ($p < q$). By definition, the *open middle third (part)* of J is the open interval $(\frac{2}{3}p + \frac{1}{3}q; \frac{1}{3}p + \frac{2}{3}q)$. The closed interval $[p; \frac{2}{3}p + \frac{1}{3}q]$ will be called *left third (part)* of J and the closed interval $[\frac{1}{3}p + \frac{2}{3}q; q]$ the *right third (part)* of J; both intervals together will be called the *extreme thirds* of J. Note that the union of the left and right third parts contain the end points p and q of J. Define by induction a sequence $(C_n)_{n \in \mathbb{Z}^+}$ of closed subsets of the closed unit interval such that

(1) $C_0 := [0; 1]$.
(2) For every $n \in \mathbb{N}$, the set C_n is the union of 2^n mutually disjoint closed intervals J^n_b, $b \in \{0, 1\}^n$, each of which has length 3^{-n}.
(3) For every $n \in \mathbb{Z}^+$ the intervals forming C_{n+1} are obtained from the intervals whose union is C_n by omitting from each of them its open middle third: if $b \in \{0, 1\}^n$ then the intervals J^{n+1}_{b0} and J^{n+1}_{b1} are the left and right thirds of the interval J^n_b.

For every $n \in \mathbb{N}$ and $b \in \{0, 1\}^n$, let and M^n_b be defined as the open middle third of the interval J^n_b. Let M_0 be the middlethird of $[0; 1] = C_0$ and for every $n \in \mathbb{N}$, let $M_n := \bigcup \{M^{(n)}_b : b \in \{0, 1\}^n\}$. Then the above definition can be summarized as

$$
\begin{aligned}
C_1 &= [0; 1] \setminus M_0 \,, \\
C_2 &= C_1 \setminus M_1 = [0; 1] \setminus (M_0 \cup M_1) \,, \\
&\dots\dots\dots\dots\dots\dots\dots\dots \\
C_{n+1} &= C_n \setminus M_n = [0; 1] \setminus (M_1 \cup M_2 \cup \dots \cup M_n) \,.
\end{aligned}
$$

It is easy to see that, for every $n \in \mathbb{N}$, the set C_n consists of all real numbers that *can be* written as

$$x = \sum_{i=0}^{\infty} a_i \, 3^{-i-1} \tag{B.1-1}$$

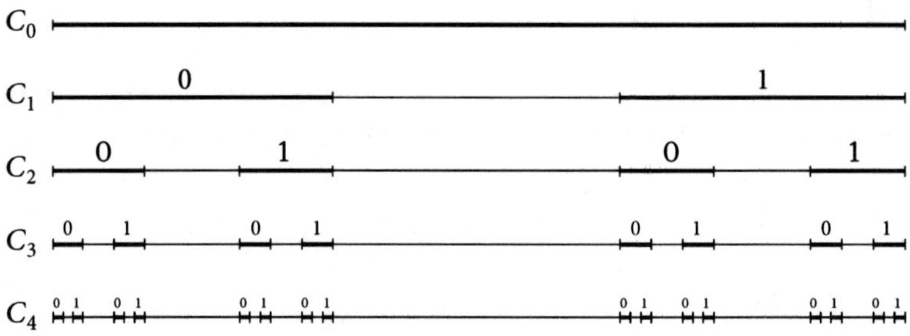

Fig. B..1. The construction of the Cantor set. The labels 0 and 1 above an interval of the form J_b^n ($b \in \{0,1\}^n$) indicates the common value of $\frac{1}{2}a_{n-1}$ in the ternary expansion $\sum_{i=0}^{\infty} a_i 3^{-i-1}$ of the points in that interval. The values of $\frac{1}{2}a_0$ through $\frac{1}{2}a_{n-2}$ for such points can be read off from the higher levels. These values together form the ordered n-tuple b for the interval J_b^n.

with $a_i \in \{0,1,2\}$ for all $i \in \mathbb{Z}^+$ and $a_i \neq 1$ for $0 \leq i \leq n-1$. In point of fact, if $n \in \mathbb{N}$ and $b \in \{0,1\}^n$ then J_b^n is the set of all points in $[0;1]$ for which the expansion (B.1-1) has $a_i = 2b_i$ for $0 \leq i \leq n-1$ and a_i arbitrary in $\{0,1,2\}$ for all $i \geq n$. The straightforward proof by induction is left to the reader. The basic ingredient is the observation that the sum $\sum_{i=n}^{\infty} a_i 3^{-i-1}$ with $a_i \in \{0,1,2\}$ can assume every value between 0 and 3^{-n} (0 and 3^{-n} included).

If a point $x \in \mathbb{R}$ has ternary expansion (B.1-1) then we may write $x \equiv 0\,.\,a_0\,a_1\,a_2\,\ldots$. Just as for ordinary decimal expansions, there is a certain ambiguity in this representation: the real numbers represented by $0\,.\,a_0\ldots a_{n-1}a_n 222\ldots$ with $a_n \neq 2$ and by $0\,.\,a_0\ldots a_{n-1}\,a_n'\,000\ldots$ with $a_n' := a_n + 1$ are equal. Thus, the points $0.01222\ldots$ and $0.01000\ldots$ belong to C_3: they can also be represented as $0.02000\ldots$ and $0.00222\ldots$, respectively. But $0.010\,\xi\ldots$ with $\xi = 0, 1$ or 2 does not belong to C_3.

If $n \in \mathbb{N}$ and $b \in \{0,1\}^n$ then the end points of the intervals J_b^n belong to the set C_n: they have ternary expansion with $a_i = 0$ or $a_i = 2$ for all $i \geq n$ (and, of course, $a_i \neq 1$ for $0 \leq i \leq n-1$).

The sets C_n with $n \in \mathbb{Z}^+$ form a descending sequence of subsets of $[0;1]$ (each of which is a union of closed intervals). We call their intersection C *the Cantor set* (also: the *Cantor discontinuum*):

$$C := \bigcap_{n=0}^{\infty} C_n\,.$$

Proposition B.1.1. *The Cantor set C is a non-empty compact subset of the unit interval. It consists of all real numbers x that can be written as $x = \sum_{i=0}^{\infty} a_i 3^{-i-1}$ with $a_i = 0$ or 2 for all $i \geq 0$. In particular, all end points of the intervals J_b^n belong to C ($n \in \mathbb{N}$ and $b \in \{0,1\}^n$).*

Proof. It is obvious that the points 0 and 1 belong to C_n for every $n \in \mathbb{Z}^+$, hence to C. So $C \neq \emptyset$. It is also obvious that for fixed $m \in \mathbb{N}$ the end points of the intervals J_b^m

($b \in \{0,1\}^m$) belong to C_m, hence to C_n for every $n \in \mathbb{N}$: for $n \le m$ this is because then $C_n \supseteq C_m$, for $n = m+1$ this is trivial and for larger n use induction. Hence these end points belong to C. Moreover, being the intersection of compact sets, C is compact. Finally, it follows easily from (B.1-1) that the points of C are characterised as the points that have a ternary expansion without 1's. □

Recall that a subset of a topological space is said to be *clopen* whenever it is both closed and open and that a topological space is said to be 0*-dimensional* whenever each of its points has a local base consisting of clopen sets. Thus, a space is 0-dimensional iff for every open subset U and every point $x \in U$ there is a clopen set V such that $x \in V \subseteq U$.

Proposition B.1.2. *With the relative topology inherited from* $\mathbb{R}$, *the Cantor set* C *has the following topological properties:*
(a) *C is a compact metric space.*
(b) *C is 0-dimensional.*
(c) *C has no isolated points.*

Proof. (a) Clear.
(b) Let $x \in C$ and consider the neighbourhood $U := C \cap (x-\varepsilon; x+\varepsilon)$ of x in C with arbitrary $\varepsilon > 0$. Select $n \in \mathbb{N}$ such that $3^{-n} < \varepsilon$. Then $x \in C_n$ and the (unique) interval J_b^n with $b \in \{0,1\}^n$ such that $x \in J_b^n$ – which has length 3^{-n} – is entirely included in the interval $(x-\varepsilon; x+\varepsilon)$. Then there are points p and q in the set $(x-\varepsilon; x+\varepsilon) \setminus C_n$ such that $p < x < q$: select p and q on either side of the interval J_b^n in the 'holes' of C_n. Then the intersection $V := C \cap (p; q)$ is an open neighbourhood of x in C which is included in U. But we also have $V = C \cap [p; q]$, because neither p nor q belongs to C. Hence V is closed in C. This completes the proof.
(c) Let $x \in C$ and let $U := C \cap (x-\varepsilon; x+\varepsilon)$ be an arbitrary neighbourhood of x in C. As in the proof of (b) we find an interval of the form J_b^n within the interval $(x-\varepsilon; x+\varepsilon)$. The two end points of J_b^n both belong to C and at least one of them is different from x. Hence $U \setminus \{x\} \neq \emptyset$. □

Remarks.
(1) In part (b) of the above proof the points p and q can be chosen so that $V = C \cap J_b^n$. This implies the following result: if $x \in C$ and if for every $n \in \mathbb{N}$ the unique n-tuple b in $\{0,1\}^n$ for which $x \in J_b^n$ is denoted by $b(n,x)$, then the collection of all sets $C \cap J_{b(n,x)}^n$ ($n \in \mathbb{N}$) is a local base at x (consisting of clopen subsets of C).
(2) The proof of (b) also shows that no interval is included in C. As C is closed, this means *that C is a nowhere dense subset of* $\mathbb{R}$. A set with no isolated points is sometimes called *dense-in-itself* and a set which is closed and dense-in-itself is said to be *perfect*. So C is a perfect nowhere dense subset of $\mathbb{R}$.

A topological space that has the three properties attributed to C in the proposition above is called a *Cantor space*. Thus, C is a Cantor space. The really interesting thing is that C is essentially the only Cantor space:

Theorem B.1.3 (Brouwer). *Every Cantor space is homeomorphic with C.*

Proof. A rigorous formal proof requires a quite complicated notation. In order to avoid this we present in the next section a rather informal proof. □

Example. The following construction produces a subset C^* of the unit interval which is formally different from C but which is a Cantor spaces, hence is homeomorphic to C (there are many variations of this construction). Define inductively a sequence $(C_n^*)_{n\in\mathbb{Z}^+}$ of closed subsets of the closed unit interval such that

(1) $C_0^* := [0; 1]$.

(2) $\forall n \in \mathbb{N}$: the set C_n^* is the union of 2^n mutually disjoint closed intervals I_b^n with $b \in \{0, 1\}^n$.

(3) $\forall n \in \mathbb{Z}^+$: the intervals whose union is C_{n+1}^* are obtained from the intervals whose union is C_n^* by selecting in each of them two disjoint closed subintervals. The subintervals selected in I_b^n ($b \in \{0, 1\}^n$) are denoted I_{b0}^{n+1} and I_{b1}^{n+1}, and enumeration is such that the first subinterval lies left of the second.

(4) If $r_n := \max\{\text{length of the interval } I_b^n : b \in \{0, 1\}^n\}$ then $\lim_{n\rightsquigarrow\infty} r_n = 0$.

The sets C_n^* for $n \in \mathbb{Z}^+$ form a descending chain of non-empty compact sets, hence the compact set $C^* := \bigcap_{n=0}^{\infty} C_n^*$ is not empty. The end points of the intervals I_b^n do not necessarily belong to C^*, but if $n \in \mathbb{N}$ and $b \in \{0, 1\}^n$ then it is easily shown that $I_b^n \cap C^* \neq \emptyset$: select one of the intervals I_{bi}^{n+1} included in I_b^n, then select one of the intervals I_{bij}^{n+2} included in I_{bi}^{n+1}, etc.; in this way one gets a descending sequence of intervals which has a non-empty intersection, obviously belonging to C^*.

Similar to the proof of Proposition B.1.2 (b) one shows that C^* is 0-dimensional; just replace the real number 3^{-n} by r_n. The proof that C^* has no isolated points is slightly different from the proof of Proposition B.1.2 (c), because the end points of an interval I_b^n do not necessarily belong to C^*. However, using the notation of the proof of Proposition B.1.2 (c), if $I_b^n \subseteq (x - \varepsilon; x + \varepsilon)$ then both I_{b0}^{n+1} and I_{b1}^{n+1} are included in $(x - \varepsilon; x + \varepsilon)$. Since both of these intervals have a non-empty intersection with C^*, we find two different points in U, at least one of which is different from x.

Conclusion. *C^* is a Cantor space.*

B.2 Proof of Brouwer's Theorem

Recall that a *partition* of a set is a family of mutually disjoint subsets whose union is the whole set. A partition of a topological space into open (or closed, or clopen) sets is called an open (or closed, or clopen) partition.

Let $\mathcal{A}$ and $\mathcal{A}'$ be two partitions of the same set. Then the partition $\mathcal{A}'$ is called a *refinement* of the partition $\mathcal{A}$ whenever for every $F' \in \mathcal{A}'$ there exists $F \in \mathcal{A}$ such that $F' \subseteq F$ (of course, this F is unique). If $\mathcal{A}'$ is a refinement of $\mathcal{A}$ then for every $F' \in \mathcal{A}'$ and every $F \in \mathcal{A}$ either $F \cap F' = \emptyset$ or $F' \subseteq F$, or, equivalently, if $F \cap F' \neq \emptyset$ then $F' \subseteq F$.

Also, it is clear that in this situation for every $F \in \mathcal{A}$ the collection of sets

$$\mathcal{A}' \restriction F := \{F' \in \mathcal{A}' : F' \subseteq F\}$$

is a partition of F.

Let X be a topological space. A sequence $\mathcal{F} = (\mathcal{F}_n)_{n\in\mathbb{Z}^+}$ of partitions of X is called a *calibration*[1] of X whenever it has the following properties:

(a) For every $n \in \mathbb{Z}^+$, $\mathcal{F}_n$ is a finite partition of X all of whose members are non-empty.
(b) For every $n \in \mathbb{Z}^+$, $\mathcal{F}_{n+1}$ is a refinement of $\mathcal{F}_n$.
(c) For every sequence $(A_n)_{n\in\mathbb{Z}^+}$ of subsets of X such that $A_n \in \mathcal{F}_n$ and $A_{n+1} \subseteq A_n$ for every $n \in \mathbb{Z}^+$ the intersection $\bigcap_n A_n$ consists of exactly one point.

We shall refer to the partition $\mathcal{F}_n$ ($n \in \mathbb{N}$) as the partition *at level n*. A calibration is said to be *open* (or *closed*, or *clopen*) whenever each level is an open (or closed, or clopen, respectively) partition of X.

A sequence satisfying the two conditions mentioned in (c) will be called an $\mathcal{F}$-*chain* (also if the condition that its intersection consists of just one point is perhaps not satisfied). If the intersection of an $\mathcal{F}$-chain consists of just one point x (as required in condition (c) above) then we say that x is the point *determined by* that $\mathcal{F}$-chain.

Examples.

(1) An ordinary measuring staff has a scale, say, in meters, each divided in ten decimetres, each of which is divided in ten centimetres, each of which is divided in ten millimetres, etc. On each level, this defines a partition (consider the decimetre segments, the centimetre segments, etc., as left-closed, right-open intervals) which refines the partition on the preceding level. Moreover, each nested sequence of sets, one from each level, defines a unique point, and vice versa. Indeed, $0.273\ldots$ is the point in the first meter-segment, the third decimetre-segment, the eights centimetre-segment, the fourth millimetre-segment, etc. Thus, we have a calibration as defined above and the chains for this calibration correspond to the decimal expansions of points.
This particular calibration has the property that at each level every set is divided in precisely ten sets of the next level. In an arbitrary calibration we do not require that all sets at the same level are split up in the same number of sets of the next level, let alone that these numbers are the same across the different levels.

(2) For every $n \in \mathbb{Z}^+$, let $\mathcal{F}_n$ be the partition of the Cantor set C consisting of the sets $C \cap J^n_b$ with $b \in \{0,1\}^n$. We have seen above that each $\mathcal{F}_n$ is a clopen partition of C. It is straightforward to show that $\mathcal{F} := (\mathcal{F}_n)_{n\in\mathbb{Z}^+}$ is a calibration of C As to condition (c), compactness implies that the intersection of an $\mathcal{F}$-chain is non-empty, and as the diameters of the subsequent sets in an $\mathcal{F}$-chain tend to zero, the intersection cannot have more than one point. This calibration of C will be called the *natural calibration* of C.

1 Warning: this term has a different meaning in other parts of topology. We have chosen this term because the definition reminds one of the calibration of a measuring staff: see Example (1).

Lemma B.2.1. *Let X be a topological space and let $\mathcal{F} = (\mathcal{F}_n)_{n\in\mathbb{Z}^+}$ be a calibration of X.*
(1) *If $m, n \in \mathbb{Z}^+$ and $m > n$, then $\mathcal{F}_m$ is a refinement of $\mathcal{F}_n$.*
(2) *Different $\mathcal{F}$-chains determine different points.*
(3) *For every point $x \in X$ there is a unique $\mathcal{F}$-chain $(A_n(x))_{n\in\mathbb{Z}^+}$ that determines the point x.*
(4) *Let $k \in \mathbb{Z}^+$, let $F \in \mathcal{F}_k$, let $x \in X$ and let $(A_n(x))_{n\in\mathbb{Z}^+}$ be the $\mathcal{F}$-chain that determines x. Then $x \in F$ iff $A_k(x) = F$. Consequently, the points of F are determined by $\mathcal{F}$-chains that all have the same initial segment up to and including level k, with F at level k.*

Proof. (1) The proof (by induction on m) is left to the reader.
(2) Let $(A_n)_{n\in\mathbb{Z}^+}$ and $(B_n)_{n\in\mathbb{Z}^+}$ be two $\mathcal{F}$-chains, determining points x and y, respectively. If these $\mathcal{F}$-chains are different then there is an $n \in \mathbb{Z}^+$ such that $A_n \neq B_n$, hence $A_n \cap B_n = \emptyset$ (both of these sets belong to the same partition $\mathcal{F}_n$). As $x \in A_n$ and $y \in B_n$ it follows that $x \neq y$.
(3) Unicity: clear from 2 above.
Existence: For every $n \in \mathbb{Z}^+$ there exists a unique member $A_n(x)$ of $\mathcal{F}_n$ such that $x \in A_n(x)$. As $A_n(x) \cap A_{n+1}(x) \neq \emptyset$ for all n, it follows from condition (b) in the definition of a calibration that $A_{n+1}(x) \subseteq A_n(x)$. So $(A_n(x))_{n\in\mathbb{Z}^+}$ is an $\mathcal{F}$-chain. Obviously, it determines the point x.
(4) The proof is clear from the definition of the $\mathcal{F}$-chain $(A_n(x))_{n\in\mathbb{Z}^+}$ as described in 3 above. □

Let X and Y be topological spaces with calibrations $\mathcal{F} = (\mathcal{F}_n)_{n\in\mathbb{Z}^+}$ and $\mathcal{G} = (\mathcal{G}_n)_{n\in\mathbb{Z}^+}$, respectively. We call these calibrations *similar* whenever there exists for every $n \in \mathbb{Z}^+$ a bijection $\Phi_n : \mathcal{F}_n \to \mathcal{G}_n$ such that

$$\forall F \in \mathcal{F}_n,\ F' \in \mathcal{F}_{n+1} \,\vdots\, F' \subseteq F \iff \Phi_{n+1}(F') \subseteq \Phi_n(F) \tag{B.2-1}$$

(or, equivalently: $F' \cap F = \emptyset$ iff $\Phi_{n+1}(F') \cap \Phi_n(F) = \emptyset$). In that case we call the sequence of bijections $\Phi := (\Phi_n)_{n\in\mathbb{Z}^+}$ a *similarity* of $\mathcal{F}$ and $\mathcal{G}$; notation: $\Phi : \mathcal{F} \to \mathcal{G}$. Also, if $\Phi : \mathcal{F} \to \mathcal{G}$ is a similarity, then $\Phi^{\leftarrow} := (\Phi_n^{\leftarrow})_{n\in\mathbb{Z}^+}$ is a similarity of $\mathcal{G}$ and $\mathcal{F}$: the notion of being similar is symmetric.

It is easy to see that the relation of being similar is reflexive and transitive. So 'similarity' defines an equivalence relation on the collection of all calibrations on all possible spaces.

Let X and Y be topological spaces with calibrations $\mathcal{F} = (\mathcal{F}_n)_{n\in\mathbb{Z}^+}$ and $\mathcal{G} = (\mathcal{G}_n)_{n\in\mathbb{Z}^+}$, respectively, and let $\Phi : \mathcal{F} \to \mathcal{G}$ be a similarity. Suppose we have an $\mathcal{F}$-chain $(A_n)_{n\in\mathbb{Z}^+}$ in X. Then $(\Phi_n(A_n))_{n\in\mathbb{Z}^+}$ is a $\mathcal{G}$-chain in Y. Conversely, if we have a $\mathcal{G}$-chain in Y then in a similar way we can form, using the similarity $\Phi^{\leftarrow}$, an $\mathcal{F}$-chain in X which is mapped by Φ onto the original $\mathcal{G}$-chain in Y. Thus, in this way Φ defines a 1,1-correspondence between $\mathcal{F}$-chains in X and $\mathcal{G}$-chains in Y.

In particular, let $x \in X$ and let $(A_n(x))_{n\in\mathbb{Z}^+}$ be the $\mathcal{F}$-chain that determines x. Then the point determined by the $\mathcal{G}$-chain $(\Phi_n(A_n(x)))_{n\in\mathbb{Z}^+}$ will be denoted by $\varphi(x)$. Thus, the mapping $\varphi\colon X \to Y$ is defined by

$$\{\varphi(x)\} = \bigcap_{n\in\mathbb{Z}^+} \Phi_n(A_n(x)) \quad \text{for all } x \in X\,.$$

It is not difficult to show that φ is a bijection of X onto Y. Indeed, this is a consequence of the fact that Φ defines a bijection of the set of $\mathcal{F}$-chains in X onto the set of $\mathcal{G}$-chains in Y.

Lemma B.2.2. *Let X and Y be topological spaces and let* $\mathcal{F} = (\mathcal{F}_n)_{n\in\mathbb{Z}^+}$ *and* $\mathcal{G} = (\mathcal{G}_n)_{n\in\mathbb{Z}^+}$ *be calibrations of X and Y, respectively. Assume that there is a similarity* $\Phi : \mathcal{F} \to \mathcal{G}$ *and let* $\varphi\colon X \to Y$ *be the bijection defined by* Φ.

(1) *For every* $k \in \mathbb{Z}^+$ *and every* $F \in \mathcal{F}_k$ *we have* $\varphi[F] = \Phi_k(F)$.

(2) *Let Y be a metric space and assume that for every $\mathcal{G}$-chain in Y the corresponding sequence of diameters converges to 0. If, in addition, the calibration $\mathcal{F}$ is open then φ is continuous.*

(3) *Let X and Y be metric spaces and assume that for every $\mathcal{F}$-chain in X and every $\mathcal{G}$-chain in Y the corresponding sequence of diameters converges to 0. If the calibrations $\mathcal{F}$ and $\mathcal{G}$ are open then φ is a homeomorphism.*

Proof. (1) Clear from Lemma B.2.1 (4).

(2) Let $x \in X$, let $\varepsilon > 0$ and let $(A_n(x))_{n\in\mathbb{Z}^+}$ be the $\mathcal{F}$-chain that determines x. Then there exists $k \in \mathbb{Z}^+$ such that $\Phi_k(A_k(x))$ has diameter less than ε, hence is included in the open ε-ball $B_\varepsilon(\varphi(x))$ centred at $\varphi(x)$. Then $A_k(x)$ is an open neighbourhood of x which is, by 1 above, included in $\varphi^{\leftarrow}[B_\varepsilon(\varphi(x))]$.

(3) Now φ^{-1} is continuous as well. □

Examples.

(1) Let $\nu \in \mathbb{R}$, $\nu > 1$. If $J = [a; b]$ is any non-degenerated interval in $\mathbb{R}$, then we call the open interval $(\frac{1}{2}(1+\frac{1}{\nu})a + \frac{1}{2}(1-\frac{1}{\nu})b; \frac{1}{2}(1-\frac{1}{\nu})a + \frac{1}{2}(1+\frac{1}{\nu})b)$ the open middle ν-th part of J; this interval is situated symmetrically around the middle of J and has length $\frac{1}{\nu}(b-a)$. Let D_ν be the subset of $[0; 1]$ that is constructed just like the Cantor set, with the difference that now the middle ν-th parts of intervals are deleted. For example, for $\nu = 3$ the ordinary Cantor set is obtained, i.e., $C = D_3$. There is a 'natural' calibration on D_ν and it is easily seen to be similar to the natural calibration on C: just let, at each level, intervals in similar positions (i.e., first or last remaining part after deleting the open middle part) correspond to each other. In addition, these calibrations satisfy the conditions of part 3 of the lemma. So the similarity induces a homeomorphism between C and D_ν. See also the Example after Theorem B.1.3.

(2) The above example gives an idea of how the proof of Brouwer's Theorem will proceed. The next example shows a difficulty: there can be 'natural' calibrations which are not similar.

Repeat the construction of the Cantor set, but with the following modification: at level n of the construction ($n \in \mathbb{Z}^+$) divide all remaining closed intervals in $2n + 1$ intervals of equal length and delete the interiors of the second, fourth, ..., $2n$-th of these intervals. So at level n there will remain $n!$ closed intervals, each of which has length $(3.5.7.\ldots.(2n-1))^{-1}$. The Cantor-like set obtained in this way is a Cantor space: the proof is completely similar to the proof that the ordinary Cantor set is a Cantor space. But the natural calibrations of the ordinary Cantor set and this Cantor-like set are not similar.

The next lemma will enable us to construct calibrations with at each level a prescribed number of sets (provided this number is not too small) of arbitrarily small diameter.

Lemma B.2.3. *Let X be a Cantor space. For every non-empty clopen subset O of X and for every $\varepsilon > 0$ there exists a natural number $P(\varepsilon, O)$ with the following property: if $n \geq P(\varepsilon, O)$ then O has a partition into exactly n non-empty clopen subsets, each with diameter less than ε.*

Proof. Because X is a 0-dimensional metric space and O is open, each point of O has a clopen neighbourhood with diameter less than ε. Since X is compact and O is closed, O can be covered by finitely many of such clopen neighbourhoods. By taking mutual intersections and leaving out all empty intersections we end up with a clopen partition of O all of whose members have a diameter less than ε. Let $P(\varepsilon, O)$ be the cardinality of this partition.

Claim. For every $n \geq P(\varepsilon, O)$ there is a clopen partition of O with n non-empty members that all have diameter less than ε. By the above, this claim is true for $n = P(\varepsilon, O)$. Suppose the claim is true for certain $n \geq P(\varepsilon, O)$: there is a partition $\{A_1, \ldots, A_n\}$ of O in n non-empty clopen sets, each of which has diameter less than ε. Because X has no isolated points, there are at least two different points x_1 and x_2 in A_n. Then x_1 has a clopen neighbourhood A'_n that does not contain x_2. Let $A''_n := A_n \setminus A'_n$. Then both sets A'_n and A''_n are non-empty, clopen, with diameter less than ε – they are included in A_n, which has diameter less than ε – and $\{A_1, \ldots, A'_n, A''_n\}$ is a partition of O with the required properties, having $n + 1$ members. Conclusion: the claim is true for all $n \geq P(\varepsilon, O)$. □

B.2.4 (Proof of Brouwer's Theorem). Let X and Y be Cantor spaces. We shall show that there exist similar clopen calibrations $\mathcal{F}$ on X and $\mathcal{G}$ on Y such that all members of $\mathcal{F}_n$ and $\mathcal{G}_n$ have a diameter less than $1/2^n$ ($n \in \mathbb{N}$). Then Lemma B.2.2 (3) implies that X and Y are homeomorphic. As this holds for any two Cantor spaces, they are all homeomorphic to C.

The proof that there are calibrations as required above is with an inductive construction. We shall construct that for every $n \in \mathbb{Z}^+$ clopen partitions $\mathcal{F}_n$ of X and $\mathcal{G}_n$ of Y and a bijection $\Phi_n : \mathcal{F}_n \to \mathcal{G}_n$ such that

P(n). $\mathcal{F}_{n+1}$ is a refinement of $\mathcal{F}_n$ and $\mathcal{G}_{n+1}$ is a refinement of $\mathcal{G}_n$, condition (B.2-1) in the definition of similarity is satisfied, all members of $\mathcal{F}_{n+1}$ and $\mathcal{G}_{n+1}$ have a diameter less than $1/2^{n+1}$.

The basis of the construction is straightforward: $\mathcal{F}_0 := \{X\}$, $\mathcal{G}_0 := \{Y\}$ (the trivial partitions of X and Y) and $\Phi_0 : \mathcal{F}_0 \to \mathcal{G}_0$ is defined by $\Phi_0(X) := Y$.

Now let $n \in \mathbb{Z}^+$ and suppose we have clopen partitions $\mathcal{F}_n$ of X and $\mathcal{G}_n$ of Y and a bijection $\Phi_n : \mathcal{F}_n \to \mathcal{G}_n$. We shall indicate how to obtain clopen partitions $\mathcal{F}_{n+1}$ of X and $\mathcal{G}_{n+1}$ of Y and a bijection $\Phi_{n+1} : \mathcal{F}_{n+1} \to \mathcal{G}_{n+1}$ such that condition **P**(n) is fulfilled. By Lemma B.2.3 there is a natural number M_n such that every $F \in \mathcal{F}_n$ and every $G \in \mathcal{G}_n$ has a clopen partition into M_n non-empty clopen sets, each with diameter less than $1/2^{n+1}$: take for M_n the maximum of the numbers $P(1/2^{n+1}, H)$ with H a member of $\mathcal{F}_n$ or $\mathcal{G}_n$. These partitions of the sets $F \in \mathcal{F}_n$ together form a clopen partition $\mathcal{F}_{n+1}$ of the space X and it is obvious that $\mathcal{F}_{n+1}$ is a refinement of $\mathcal{F}_n$. Similarly, the partitions of the sets $G \in \mathcal{G}_n$ together form a clopen partition $\mathcal{G}_{n+1}$ of Y, which is a refinement of $\mathcal{G}_n$. By construction, all members of $\mathcal{F}_{n+1}$ and $\mathcal{G}_{n+1}$ have a diameter less than $1/2^{n+1}$. Now define Φ_{n+1} on $\mathcal{F}_{n+1}$ in the following way: by the construction, for any $F \in \mathcal{F}_n$ the sets F and $\Phi_n(F)$ include the same number M_n of members of $\mathcal{F}_{n+1}$ and $\mathcal{G}_{n+1}$, respectively. Bring these into a 1,1-correspondence, and do this for every $F \in \mathcal{F}_n$. In this way one gets a bijection $\Phi_{n+1} : \mathcal{F}_{n+1} \to \mathcal{G}_{n+1}$ that satisfies condition (B.2-1).

Using this procedure it is easy construct by induction for every $n \in \mathbb{Z}^+$ clopen partitions $\mathcal{F}_n$ of X and $\mathcal{G}_n$ of Y and a bijection $\Phi_n : \mathcal{F}_n \to \mathcal{G}_n$ such that condition **P**(n) holds for every $n \in \mathbb{Z}^+$.

In order to show that $\mathcal{F} := (\mathcal{F}_n)_{n\in\mathbb{Z}^+}$ and $\mathcal{G} := (\mathcal{G}_n)_{n\in\mathbb{Z}^+}$ are calibrations and that $\Phi := (\Phi_n)_{n\in\mathbb{Z}^+}$ defines a similarity between these calibrations it remains only to verify that condition (c) in the beginning of Section B.2 is satisfied for $\mathcal{F}$ and $\mathcal{G}$. So consider an $\mathcal{F}$-chain (for $\mathcal{G}$-chains the proof is similar). Because this is a nested sequence of compact sets, their intersection is non-empty. However, the diameters of the sets in this chain tend to 0 because the diameter of any set in $\mathcal{F}_n$ is at most $1/2^n$ ($n \in \mathbb{N}$). Hence the intersection of the $\mathcal{F}$-chain cannot contain more than one point. □

Remarks. Thus, essentially, there is only one Cantor space: the Cantor set C. In order to avoid cumbersome formulations like 'consider a subset K of X which in the relative topology is a Cantor space' we shall say in such cases 'let K be a Cantor set in X'. So terminology becomes somewhat ambiguous: *a* Cantor set K is a subset of a space such that in the relative topology K is a Cantor space, and *the* Cantor set is the set C introduced in Section B.1. Usually, this sloppiness will cause no problems.

B.3 Cantor spaces

As explained in Chapter 5, for every $n \in \mathbb{N}$ the space $\{0, 1, \dots, n\}^{\mathbb{N}}$ with its product topology is a Cantor space. The proof can easily be adapted so as to show the following: *Let $(a_n)_{n\in\mathbb{N}}$ be a sequence of positive integers; then $K_0 := \prod_{n\in\mathbb{N}}\{1, 2, \dots, a_n\}$ with its product topology is a Cantor space.* This has the following interesting consequence:

Theorem B.3.1 (Alexandroff). *Every compact metric space is the continuous image of a closed subset of a Cantor space.*

Proof. Let X be a compact metric space. For every $n \in \mathbb{N}$ the cover of X with all open $1/n$-balls has a finite subcover $\mathfrak{U}_n$. Enumerate the cover $\mathfrak{U}_n$ as $\mathfrak{U}_n = \{U_{n,1}, \ldots, U_{n,a(n)}\}$. By the observation preceding this theorem, the set $K_0 := \prod_{n\in\mathbb{N}}\{1, 2, \ldots, a(n)\}$ endowed with the product topology is a Cantor space.

By definition, K_0 is the set of all sequences of the form $y = (y(n))_{n\in\mathbb{N}}$ with $y(n) \in \{1, \ldots, a(n)\}$ for every $n \in \mathbb{N}$. For every element y of K_0 the set $\bigcap_{n\in\mathbb{N}} \overline{U}_{n,y(n)}$ has diameter at most $1/n$ for every $n \in \mathbb{N}$, hence it has diameter 0, that is, it contains at most one element of X. Let

$$K := \left\{ y \in K_0 \,\vdots\, \bigcap_{n\in\mathbb{N}} \overline{U}_{n,y(n)} \neq \emptyset \right\}.$$

We claim that K is a non-empty closed subset of K_0. First, if x is an arbitrary element of X then for every $n \in \mathbb{N}$ there exists $\bar{x}(n) \in \{1, \ldots, a(n)\}$ such that $x \in U_{n,\bar{x}(n)}$. Then, obviously, $x \in \bigcap_{n\in\mathbb{N}} \overline{U}_{n,\bar{x}(n)}$, hence $\bar{x} := (\bar{x}(n))_{n\in\mathbb{N}} \in K$. This shows that K is not empty.

Next, assume that $z := (z(n))_{n\in\mathbb{N}} \in K_0$ belongs to the closure of K in K_0. Then every basic neighbourhood of the point z in K_0 contains an element of K. It follows[2] that for every $k \in \mathbb{N}$ there exists an point $y \in K$ such that $z(n) = y(n)$ for $n = 1, \ldots, k$. Consequently,

$$F_k := \bigcap_{n=1}^{k} \overline{U}_{n,z(n)} = \bigcap_{n=1}^{k} \overline{U}_{n,y(n)} \supseteq \bigcap_{n=1}^{\infty} \overline{U}_{n,y(n)} \neq \emptyset$$

So the sets F_k for $k \in \mathbb{N}$ form a decreasing sequence of non-empty closed sets in X. Therefore $\bigcap_{k\in\mathbb{N}} F_k \neq \emptyset$, which means that $\bigcap_{n\in\mathbb{N}} \overline{U}_{n,z(n)} \neq \emptyset$. Stated otherwise, $z \in K$. This completes the proof that K is closed in K_0.

For every $y \in K$ let $\varphi(y)$ denote the unique element of $\bigcap_{n\in\mathbb{N}} \overline{U}_{n,y(n)}$. Then φ maps K onto X: in the proof that K is not empty we constructed for every point $x \in X$ an element y of K such that $x = \varphi(y)$. It remains to show that φ is continuous.

Let $y \in K$ and let V be a neighbourhood of $\varphi(y)$ in X. We may assume that V is an open ball centred at the point $\varphi(y)$. Let $m \in \mathbb{N}$ be so large that $1/m$ is less than the radius of V. By definition, the point $\varphi(y)$ is situated in the set $\overline{U}_{m,y(m)}$. It follows from the choice of m that this set – which has diameter at most $1/m$ – is included in the ball V. Now consider the open subset $O := \{ z \in K \,\vdots\, z(m) = y(m) \}$ of K. If $z \in O$ then

$$\varphi(z) \in \bigcap_{n\in\mathbb{N}} \overline{U}_{n,z(n)} \subseteq \overline{U}_{m,z(m)} = \overline{U}_{m,y(m)} \subseteq V.$$

This completes the proof that φ is continuous. □

Remarks. By a result of Sierpiński's, every non-empty closed subset K of the Cantor space C is a retract of C, i.e., there exists a continuous mapping $\psi : C \to K$ such that $\psi|_K = \mathrm{id}_K$. For a (very elegant) proof, due to Halmos, see [Eng], 4.5.9 (a). Hence *every compact metric space is the continuous image of the full Cantor space.*

---

2 Use the description of the basic neighbourhoods of a point of K_0 given in the final conclusion of 5.1.3.

Finally, we mention a construction that produces a Cantor set in any compact metric space without isolated points.

Theorem B.3.2. *Let X be a compact metric space without isolated points. Then X includes a Cantor set.*

Proof. The following observation will be used without further reference: if U is a non-empty open subset of X then U cannot be finite, otherwise the points in U would be isolated in X: each point, being the complement in U of the (closed) union of the other points would be an open set. Select any finite number of mutually distinct points in U, let δ_1 be the minimum of their distances and let δ_2 be the minimum if their distances to the closed set $X \setminus U$; note that both δ_1 and δ_2 are positive. Now let $\varepsilon := \min\{\delta_1/3, \delta_2/2, 1/2n\}$. Then the ε-balls with centres at those points are non-empty open subsets of X with mutually disjoint closures included in U and with diameter at most $1/n$.

The proof of the theorem is by a construction which mimics the construction of the ordinary Cantor set. Let $V_{0,1} := X$, fix $a_1 \in \mathbb{N}$, $a_1 \geq 2$ and let $a_n := 2^{n-1}a_1$ for $n \in \mathbb{N}$. We shall show that for every $n \in \mathbb{N}$ there are non-empty open sets $V_{n,1}, V_{n,2}, \dots, V_{n,a_n}$ in X with the following properties:

(1) $\operatorname{diam}(V_{n,i}) < \frac{1}{n}$ for $i = 1, 2, \dots, a_n$.
(2) The sets $\overline{V}_{n,i}$ for $i = 1, 2, \dots, a_n$ are mutually disjoint.
(3) $\overline{V}_{n,2i-1} \cup \overline{V}_{n,2i} \subseteq V_{n-1,i}$ for $i = 1, 2, \dots, a_{n-1}$.

The proof is by induction and follows immediately from the initial observation above. For $n = 1$, select a_1 non-empty open subsets $V_{1,1}, \dots V_{1,a_1}$ of X with diameter at most 1 and with disjoint closures. Now consider $k \in \mathbb{N}$ and suppose that for $1 \leq n \leq k-1$ we have non-empty open sets $V_{n,1}, \dots, V_{n,a_n}$ satisfying the conditions (1), (2) and (3). Then one can select non-empty open sets $V_{k,1}, V_{k,2}, \dots, V_{k,a_k}$ in X with disjoint closures, with diameter at most $1/k$, and such that $\overline{V}_{k,2i-1} \cup \overline{V}_{k,2i} \subseteq V_{k-1,i}$ for $i = 1, 2, \dots, a_{k-1}$. So we have (1), (2) and (3) for $n = k$ as well.

For $X = [0; 1]$ and $a_1 = 2$ the above construction describes the situation in the example after Theorem B.1.3 above. The sets C_n^* in that example correspond to $\bigcup_{i=1}^{a_n} \overline{V}_{n,i}$

Now let $K := \bigcap_{n-1}^{\infty} \left(\bigcup_{i=1}^{a_n} \overline{V}_{n,i} \right)$. The proof that K is a Cantor space is similar to the proof of Proposition B.1.2: the sets $\overline{V}_{n,i} \cap K$ are clopen in K. Moreover, each of these sets includes at least two distinct points of K, namely, one in $\overline{V}_{n+1,2i-1}$ and one in $\overline{V}_{n+1,2i}$. In order to see this, note that the sets $\overline{V}_{k+1,2^{k-n}(2i-1)}$ for $k \geq n$ form a descending sequence of non-empty compact subsets of the set $\overline{V}_{n+1,2i-1}$, hence they have a non-empty intersection, which obviously is included in K. By applying the same reasoning to the set $\overline{V}_{n+1,2i}$ one infers that this set has a point in K as well. Further details are left to the reader. □

B.3.3. A modification of this construction shows that there are countably many Cantor sets $C_1 \subseteq C_2 \subseteq \dots$ in X such that $\bigcup_{n=1}^{\infty} C_n$ is dense in X. To this end, start with $a_0 := 0$ and $V_{0,1} := X$, and let Y be a countable dense subset of X (recall from the observation after Proposition A.7.7 that X, being a compact metric space, is separable). Let $Y = \{y_1, y_2, \dots\}$ be an enumeration of Y and for every $n \in \mathbb{N}$, let $Y_n := \{y_1, \dots, y_n\}$.

Claim: there is a sequence of natural numbers $(a_n)_{n\in\mathbb{N}}$ in $\mathbb{N}$ and for every $n \in \mathbb{N}$ there are a_n non-empty open subsets $V_{n,1}, V_{n,2}, \dots, V_{n,a_n}$ in X such that, in addition to the conditions (1), (2) and (3) in the above proof, also the following conditions are satisfied for each $n \in \mathbb{N}$:

(0) $2a_{n-1} \le a_n \le 2a_n + n$.

(4) $Y_n \subseteq \bigcup_{i=1}^{a_n} V_{n,i}$.

For $n = 1$, take $a_1 = 2$ and take for $V_{1,1}$ and $V_{1,2}$ any two non-empty open subsets of X with diameter at most 1 and with disjoint closures and such that, in addition, $y_1 \in V_{1,1} \cup V_{1,2}$ – e.g., select $V_{1,1}$ such that $y_1 \in V_{1,1}$. If for any $k \in \mathbb{N}$ and $1 \le n \le k-1$ we have natural numbers a_n and non-empty open sets $V_{n,1}, \dots, V_{n,a_n}$ satisfying the conditions (0), (1), (2), (3) and (4) then, as before, we can select non-empty open subsets $V_{k,1}, V_{k,2}, \dots, V_{k,2a_{k-1}}$ of X with mutually disjoint closures, with diameters at most $1/k$ and such that $\overline{V}_{k,2i-1} \cup \overline{V}_{k,2i} \subseteq V_{k-1,i}$ for $i = 1, 2, \dots, a_{k-1}$. This selection can be done in such a way that the points of Y_{k-1}, which are all in $\bigcup_{i=1}^{a_{k-1}} V_{k-1,i}$, are covered by the set $\bigcup_{i=1}^{2a_{k-1}} V_{k,i}$. Let a_k be such that $a_k - 2a_{k-1}$ is equal to the number of points of Y_k not yet covered. As Y_k has k points, it is clear that (0) holds for $n = k$. For each i with $2a_{k-1} + 1 \le i \le a_k$ we can select a non-empty open subset $V_{k,i}$ of X that includes one of the remaining points of Y_k, taking care that the closures of these sets are mutually disjoint and disjoint from the closures of the sets $V_{k,j}$ for $j = 1, \dots 2a_{k-1}$, and that its diameter is at most $1/k$. So we have (1), (2), (3) and (4) for $n = k$ as well.

Now let for each $N \in \mathbb{N}$

$$C_N := \bigcap_{n=N}^{\infty} \left(\bigcup_{i=1}^{2^{n-N} a_N} \overline{V}_{n,i} \right) .$$

Then C_N is a Cantor set: it is formed as in the proof of Theorem B.3.2, starting with a_N sets. It is clear that $C_1 \subseteq C_2 \subseteq \dots$ and it is easy to see that $K := \bigcup_{N=1}^{\infty} C_N$ is dense in X: for each $n \in \mathbb{N}$ the point y_n of Y, being situated in one of the sets $\overline{V}_{n,i}$, has distance at most $1/n$ to a point of[3] C_N for every $N \ge n$. As Y is dense it follows that K is dense.

---

3 By the end of the proof of Theorem B.3.2, for every $n \ge N$ and $i \in \{1, 2, \dots, a_n\}$ the set $\overline{V}_{n,i}$ includes a point of C_N.

C. Hints to the Exercises

We learn more by looking for the answer to a question and not finding it than we do from learning the answer itself.

Lloyd Alexander

Chapter 0

0.2.

(1) Don't derive formulas for $f^n(x)$, certainly not in case (e). All one needs is that $f(x) < x$ or $f(x) > x$ for all $x \in (0;1)$.]

(2) Distinguish the cases $a < -1$, $a = -1$, $-1 < a < 0$, $a = 0$, $0 < a < 1$, $a = 1$ and $a > 1$.

0.4. (c) Prove with induction on n that $f^n(x_0, y_0) = (x_n, y_n)$ with $x_n := 2^n x_0$ and $y_n := 2^n(\frac{1}{2}nx_0 + y_0)$ $(n \in \mathbb{N})$. Eliminate n and show that these points satisfy the equation $y = x(c_1 + \frac{1}{2\ln 2}(\ln x - c_2))$, with $c_1 := y_0/x_0$ and $c_2 := \ln|x_0|$, provided$x_0 \neq 0$. For $x_0 = 0$ these points are on the y-axis. See the following picture.

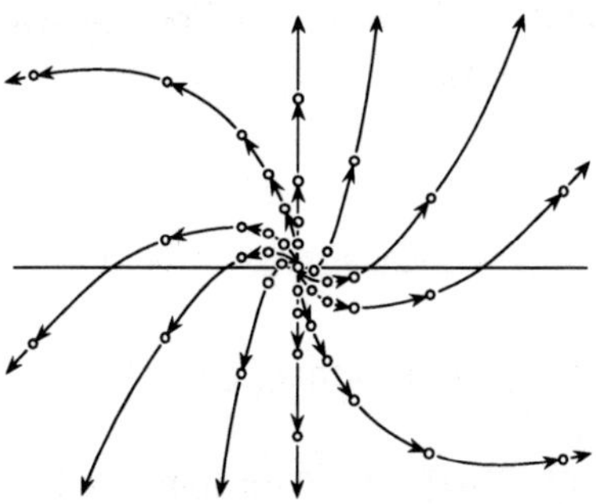

0.5. Use that the set of mappings $\{f^0, f, \ldots, f^N\}$ is equicontinuous at the point x.

Chapter 1

1.1. Inspect the proof of Theorem 1.1.5.

1.2.

(2) For $i = 0, 1$ and 3, the image under f^3 of the interval $[i; i+1]$ has at most an end point in common with its original, so only the interval $[2;3]$ can contain a fixed point of f^3. On this interval and its images under the mappings f and f^2, f is monotonously decreasing, so f^3 is monotonously decreasing here as well.

1.3.

(3) Suggestion: consider $\varphi_{1/3} \times g$ on $\mathbb{S} \times [-1;1]$, where $g(x) := -x$ for $-1 \leq x \leq 1$.

1.4. "Only if": Let $x \in \bigcap_{i=0}^{\infty} f^i[X]$. Then for every $n \in \mathbb{Z}^+$, $X_n := (f^n)^{\leftarrow}[x]$ is a non-empty compact subset of X. For $k, l \in \mathbb{Z}^+$ with $k \geq l$ let$\varphi_l^k := f^{k-l}|_{X_k} : X_k \to X_l$. Then the spaces X_n together with the mappings φ_l^k form an *inverse spectrum*. Such an inverse spectrum has a non-empty compact limit

$$L := \Big\{ (y_n)_{n\in\mathbb{Z}^+} \in \prod_{n\in\mathbb{Z}^+} X_n : \varphi_l^k(y_k) = y_l \text{ for all } k, l \in \mathbb{Z}^+ \text{ with } k \geq l \Big\} .$$

See Theorem 3.2.13 in [Eng] for the (straightforward) proof. Clearly, any point of L is a complete past of x.

"If": Obvious.

1.5.

(1) Note that there can be no invariant points.

(3) Consider the set $M \setminus \bigcup_{n=0}^{\infty} (f^n)^{\leftarrow}[\operatorname{int}_X(M)]$.

(5) There exists a non-empty open subset U of X with compact closure $F := \overline{U}$ such that $F \subseteq \operatorname{int}_X(\mathbf{Trans}\,(X, f))$. Every point $x \in F$ has an open neighbourhood U_x such that $f^{n_x}[U_x] \subseteq U$ for some $n_x \geq 1$. Then there is a finite open cover $\{U_{x_1}, \ldots, U_{x_m}\}$ of F. Let $N := \max\{n_{x_1}, \ldots, n_{x_m}\}$ and put $F_0 := \bigcup_{k=0}^{N} f^k[F]$. Then $f[f^N[F]] \subseteq F_0$; consequently, F_0 is invariant. Since F_0 includes points with a dense orbit it follows that $F_0 = X$. Hence X is compact. Moreover, $F \subseteq \mathbf{Trans}\,(X, f)$, hence $F_0 \subseteq \mathbf{Trans}\,(X, f)$. So all points of X are transitive.

1.6.

(1) Use that if $x_2 \notin \mathcal{O}(x_1)$ then $x_2 \in \overline{\mathcal{O}(x_1)} \setminus \mathcal{O}(x_1) \subseteq \omega(x_1)$.

(3) "If": Given V, let $A := X \setminus \bigcup_{n=0}^{\infty} (f^n)^{\leftarrow}[V]$. "Only if": Given A, take $U := A^\circ$ and $V := X \setminus A$.

(4) Let $x_0 \in X$ be isolated and let U be a non-empty open set with $x_0 \notin U$. Then the sets $D(x_0, U)$ and $D(U, x_0)$ are non-empty.

(5) If U, V are non-empty and open in X then $W := (f^k)^{\leftarrow}[U] \cap V \neq \emptyset$ for some $k \in \mathbb{Z}^+$. Then $D(U, U) \cap D(V, V) \supseteq D(W, W)$.

(6) Let $n \in D(U, V)$ and $U_1 := U \cap (f^n)^{\leftarrow}[V]$. Then for every $k \in D(U_1, U_1)$ one has $n{+}k \in D(U_1, V) \subseteq D(U, V)$, i.e., $n{+}D(U_1, U_1) \subseteq D(U, V)$. Finally, show that $D(U_1, U_1)$ is infinite (see also Proposition 4.3.2 ahead).

1.7.

(6) Let p_c be the largest invariant points of g_c. Then $\kappa_c : x \mapsto (-x + p_c)/2p_c : \mathbb{R} \to \mathbb{R}$ defines a conjugacy $\kappa_c : (\mathbb{R}, g_c) \simeq (\mathbb{R}, f_{\mu_c})$ with $\mu_c := 2p_c = 1 + \sqrt{4c + 1}$.

1.8.

(2) Use that φ maps orbit closures in X onto orbit closures in Y.

1.10.

(3) "(i)⇒(ii)": Assume that for some $k \geq 2$ the system (X, f^k) is not minimal and let A be an f^k-minimal proper closed subset X. Then for every $i \in \mathbb{Z}^+$ the subset $f^i[A]$ is f^k-minimal as well. If p is smallest positive integer such that $f^p[A] = A$

then $p \geq 2$, $f^i[A] = A$ iff $i \in \mathbb{Z}^+ p$ and $\bigcup_{i=0}^{p-1} f^i[A] = X$. Now put $A_i := f^i[A]$ for $i = 0, \ldots, p-1$.
"(ii)⇒(iii)" and "(iii)⇒(i)": Easy.

(5) Use Proposition 1.6.2.

(6) Use (3) above.

(7) Note that $X \times Y_n$ consists of n copies of X on each of which $(f \times g_n)^n$ acts like f^n on X. See the picture below. For the proof of "only if", take also into account that for $i = 0, \ldots n-1$ the point $f^i(x)$ is transitive in X.

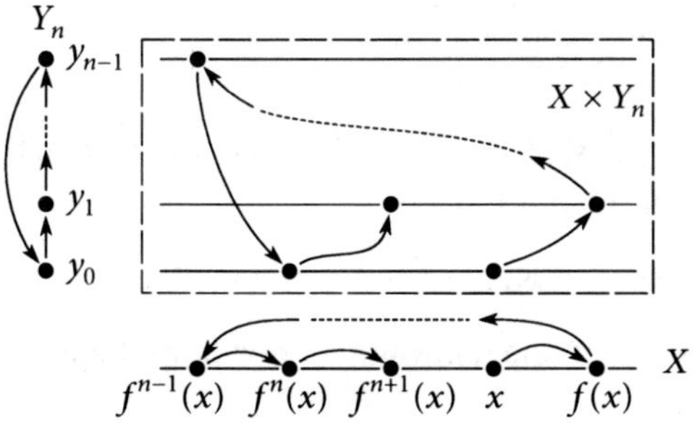

1.12.

(1) (i)⇔(ii): Standard: see Appendix A.7.2.
(ii)⇔(iii): See Appendix A.10.2 (3).

(2) Use (1) and the fact that φ is uniformly continuous.

1.13.

(1) The proof is by induction in n. For $n = 1$ and 2 the statement can easily be verified by direct inspection. The proof of the induction step follows easily from the following observation: let $b' := (b_1, \ldots, b_{n-1})$ and recallthat I_{bj}^{n+1} is the pre-image under f in I_b^n of the subinterval $I_{b'j}^n$ of $I_{b'}^{n-1}$ $(j = 0, 1)$. Hence the relative position of the intervals I_{b0}^{n+1} and I_{b1}^{n+1} in I_b^n is the same as (or opposite to) the relative position of the intervals $I_{b'0}^n$ and $I_{b'1}^n$ in $I_{b'}^{n-1}$ iff f is increasing(or decreasing) on I_b^n, iff $b_0 = 0$ (or $b_0 = 1$).

(2) Use the following observation: if the number of 1's among the coordinates of b is even (or odd, respectively) then by 1, $I_{b0}^{n+1} = J_{\kappa(b0)}^{n+1}$ and $I_{b1}^{n+1} = J_{\kappa(b1)}^{n+1}$ arethe left and right (or right and left, respectively) third parts of $I_b^n = J_{\kappa(b)}^n$. On the otherhand, the intervals $J_{\kappa(b)0}^{n+1}$ and $J_{\kappa(b)1}^{n+1}$ are always the left and right thirds of $J_{\kappa(b)}^n$.

Chapter 2

2.2.

(a) Take into account that $\alpha_\mu \le \hat{p}_\mu$ for $\mu \ge 2$.

(c) The equation $f_\mu^2(x) - x = 0$ has two known roots, namely, 0 and p_μ. Divide this equation by $x(x - p_\mu)$ to get an equation of degree two which can easily be analyzed.

(d) Using the hint in (c) above, find for $\mu > 3$ the real solutions

$$x_{1,2}^\mu = \frac{1}{2\mu}\Big(\mu + 1 \pm \sqrt{(\mu+1)(\mu-3)}\,\Big).$$

of the equation $f_\mu^2(x) - x = 0$. Then $(f_\mu^2)'(x_i^{(\mu)}) = f_\mu'(x_1^{(\mu)})f_\mu'(x_2^{(\mu)})$ is a polynomial in μ of degree two, with values between -1 and 1 for $3 < \mu < 1 + \sqrt{6}$.

2.6. For any $c \in \mathbb{R}$: $|c| < 1$ iff $|c|^k < 1$ for some $k \in \mathbb{N}$, iff $|c|^k < 1$ for all $k \in \mathbb{N}$.

2.7. The proof of Theorem 2.2.2 can be based on the following observation: there are two closed intervals I_0 and I_1 of X such that

(a) The formulas (2.2-2) and (2.2-3) hold,

(b) If the orbit of a common point of these intervals is included in their union then it is *not* completely included in I_0.

2.10.

(2) Take into account that $\mathbf{Per}(g_k) = \mathbf{S}(k)$ and $\{\, n \cdot 2^m \,\vdots\, n \in \mathbf{S}(k)\,\} = \mathbf{S}(k \cdot 2^m)$.

2.11.

(1) Let Q_n denote the set of all periodic points with primitive period at most n ($n \in \mathbb{N}$), and let Q_∞ denote the set of all periodic points. First, show that $Q_n \neq X$ for every $n \in \mathbb{N}$. Next, use the Exercises 1.3 (2) and 1.6 (3) to show that Q_n is a closed set with empty interior. As Q_∞ is dense, $Q_\infty \setminus Q_n$ is dense as well.

(2) Cf. 1 above. However, the case that $X = Q_n$ for some $n \in \mathbb{N}$ is less trivially dismissed. In that case, if there are $x, y \in X$ with disjoint orbits then consider nbds U of x and V of y such that $f^i[U]$ and V are disjoint for $i = 0, \dots, n-1$.

2.12. Assume Case 2 of Theorem 2.6.5 (otherwise there is something to prove). Let x_1 be a transitive point of (I_0, f^{2p}). Then the points $f^{2pn}(x_1)$ with $n \in \mathbb{Z}^+$ are dense in I_0. As f^p is a factor map of (I_0, f^{2p}) onto (I_1, f^{2p}), the points $f^{2pn}(f^p(x_1))$ are dense in I_1.

2.13. (i)⇒(iv): In the proof of the implication (iii)⇒(i) of Theorem 2.6.8, select x_1 and x_3 subject to the additional conditions $a < x_1 < a + \varepsilon$ and $b - \varepsilon < x_2 < b$.

(iv)⇒(iii)⇒(ii): Straightforward.

(ii)⇒(i): If condition (ii) holds then f is transitive, and the second possibility of Theorem 2.6.5 cannot hold. Alternatively: see Exercise 1.10 (5).

2.14. Write the points of the various intervals $J_b^{(n)}$ in ternary expansion.

Chapter 3

3.1.

(1) If B is a subset of Y then $\mathrm{cl}_Y(B) = Y \cap \overline{B}$; if Y is closed then $\overline{B} \subseteq Y$.

(2) The mapping f cyclically permutes the sets $\omega_{f^n}(f^i x)$ for $i = 0, \dots, n-1$. So either all sets $\omega_{f^n}(f^i x)$ for $0 \le i \le n-1$ are finite or all are infinite. Finally, apply 3.3.10 (1).

(3) Let $\emptyset \ne F \subset \omega(x)$ and let U and V be mutually disjoint compact neighbourhoods of F and $\omega(x) \setminus F$. Then there are infinitely many values of n such that $f^n(x) \in U$ and $f^{n+1}(x) \in V$. Then find a cluster point $y \in F$ with $f(y) \in \omega(x) \setminus F$.

3.2. Proof of the statement: Apply Proposition 3.1.2 with $\varphi := f$.
Counter example: Let X be the orbit of $[0]$ in $(\mathbb{S}, \varphi_a)$ with $a \notin \mathbb{Q}$ and let $x = [0]$.

3.3.

(1) By Theorem 3.1.9, $\omega(x)$ has this property. If K has has this property then Lemma 1.4.1 (2) implies that $d(z, K) = 0$ for every point $z \in \omega(x)$.

(2) Formula in (3.3-6) can be rewritten as $d(f^{np+j}(x), f^{np+j}(x_i)) \rightsquigarrow 0$ for $n \rightsquigarrow \infty$ and $j = 0, 1, \dots, p-1$.

3.4.

(a) Use Lemma 3.3.10 and observe that $\omega_g(f^i(x)) = f^i[\omega_g(x)]$.

(b) Now $f^i[\mathcal{B}_g(x_0)] = \mathcal{B}_g(x_i)$.

3.5. $(f^p)'(x_0) = \prod_{j=0}^{p-1} f'(x_j) = (f^p)'(x_i)$, where $x_i := f^i(x_0)$ for $i = 1, \dots, p-1$.

3.6.

(1) If $x \in A$ and $f(x) \notin A$ then let $U := X \setminus \{f(x)\}$ in (3.2-1).

(2) We may assume that U is compact. Let $V \in \mathcal{N}_A$ as in the definition of stability. Then V contains point x such that $\emptyset \ne \omega(x) \not\subseteq A$. Next, use Theorem 1.4.5 and Corollary 3.2.2 (2).

3.9. (i)⇒(ii): Let W be the interior of a nbd U of A satisfying condition (iii) of Theorem 3.4.2. (ii)⇒(i): Note that $A = \omega(\widetilde{W})$ and apply Theorem 3.4.2.

3.10.

(1) Stability: it is sufficient to prove that if U is an open neighbourhood of $A_1 \cap A_2$ then for $i = 1, 2$ there is an open neighbourhood W_i of A_i such that $W_1 \cap W_2 \subseteq U$.

3.11. Note that A is the largest completely invariant subset of X.

3.12.

(1) If $U_0 \supseteq A$ is strongly attracted by A then for every $U \in \mathcal{N}_A$ and sufficiently large $n \in \mathbb{N}$ one has $f[A] \subseteq f^n[A] \subseteq f^n[U_0] \subseteq U$.

(2) Assume that A strongly attracts a compact neighbourhood U of itself. Then for some $N \in \mathbb{N}$ one has $f^n[U] \subseteq U^\circ$ for all $n \ge N$. In particular, $f^N[U] \subseteq U^\circ$, so the set $B := \bigcap_{i\ge 0} f^{Ni}[U]$ is asymptotically stable under f^N, hence it is asymptotically

stable under f. Finally, if V is any neighbourhood of A then $f^n[U] \subseteq V$ for almost all n, hence $B \subseteq A$.

(3) Let $\mathcal{B}$ be a countable base for X. If A is an asymptotically stable subset of X then there is a finite subset $\mathcal{B}_A$ of $\mathcal{B}$ such that $A = \bigcap_{n\geq 0} f^n[\bigcup \mathcal{B}_A]$. So A is determined by a finite subset $\mathcal{B}_A$ of $\mathcal{B}$. Now use the fact that the set of all finite subsets of $\mathcal{B}$ is countable.

3.13.

(1) Use Corollary 3.5.5 and Exercise 1.10 (3).

Chapter 4

4.1.

(3) Apply Theorem 4.1.7 and 4.2.7, respectively, to the factor map $f: (X, f) \to (X, f)$.

4.2. Let $p_1 \geq 1$ be such that $f^{p_1}(x_0) \in U_0$. Continuity of f^{p_1} implies that there is a nbd U_1 of x_0 such that $f^{p_1}[U_1] \subseteq U_0$ and $U_1 \subseteq U_0$. There exists $p_2 \geq 1$ such that $f^{p_2}(x_0) \in U_1$ and observe that $f^{p_1+p_2}(x_0) \in f^{p_1}[U_1] \subseteq U_0$. Then find a nbd U_2 of x_0 such that $f^{p_2}[U_2] \subseteq U_1$ and $U_2 \subseteq U_1$, and find $p_3 \geq 1$ with $f^{p_3}(x_0) \in U_2$. Then $p_1, p_2, p_3 \in D(x_0, U_0)$ and also the sums $p_1+p_2+p_3$, p_1+p_2, p_1+p_3 and p_2+p_3 belong to $D(x_0, U_0)$. Now proceed by induction.

4.3.

(1) (iii)⇒(ii): If $\mathcal{O}(x) = \bigcup_{n\geq 0}\{f^n(x)\}$ is locally compact then there is an isolated point in $\mathcal{O}(x)$ (Baire). Now proceed similar to the proof of (iii)⇒(i) in Theorem 1.1.5.

(2) (i)⇒(iii): If $\mathcal{O}(f^n(x))$ is not a 1st-category space then some of its points would be isolated in $\mathcal{O}(f^n(x))$, and $\mathcal{O}(f^n(x))$ would be a discrete space. Now use 1 above.
(iii)⇒(ii): Obvious.
(ii)⇒(i): If $\mathcal{O}(f^m(x))$ is a 1st-category space then it is not locally compact (Baire); now use 1 above.

4.4.

(1) If X has an isolated point, then use Exercise 1.6 (4). If no point of X is isolated, use that every non-empty open set includes two disjoint open subsets.

(2) Let $W := (f^n)^{\leftarrow}[U] \cap V \neq \emptyset$. By (1) above and Proposition 4.3.2 the set $D(W, W)$ is infinite.

4.5. Let $x \notin f^N[X]$ for some $N \in \mathbb{N}$. Then $U := X \setminus f^N[X]$ is a neighbourhood of x and $f^n(y) \notin U$ for all $y \in X$ and $n \geq N$. The point x is not periodic, so there is a neighbourhood V of x such that $f^n(y) \notin V$ for all $y \in V$ and $n = 0, \ldots, N$. Then no point of $U \cap V$ can return to $U \cap V$.

4.6.

(1) by Lemma 2.6.1, if J is an open interval in X containing no periodic points, then for every point $x \in X$ the points $f^n(x)$ $(n \in \mathbb{Z}^+)$ that are in J form a (finite or infinite) monotonous sequence.

(2) "$\supseteq$": Easy (see also Corollary 4.3.9). "$\subseteq$": Let $x_0 \in \Omega(\Omega(X, f), f)$ and let J be an open neighbourhood of x_0 in X; we may assume that J is an interval. Suppose that J contains no periodic points. There exists $x \in J \cap \Omega(X, f)$ such that $f^n(x) \in J$ for some $n \geq 1$. Let W be a nbd of x in X such that $f^n[W] \subseteq J$ and $f^n[W] \cap W = \emptyset$. Then for every point $y \in W$ with $f^k(y) \in W$ for some $k \in \mathbb{N}$ the ordering of the points y, $f^n(y)$ and $f^k(y)$ and Lemma 2.6.1 imply that $k < n$, contradicting Proposition 4.3.2.

4.7.

(2) Suppose (∗) holds: for every $\varepsilon > 0$ there exists $l \in \mathbb{N}$ such that for every $k \in \mathbb{Z}^+$ there exists $i \in [0; l]$ with $x \in B_\varepsilon(f^{k+i}(x))$. This means that $k + i \in D(x, B_\varepsilon(x))$. So the point x is almost periodic. Conversely, assume that condition (∗) does not hold: then there exists $\varepsilon_0 > 0$ such that for every $l \in \mathbb{N}$ there are $n_l, k_l \in \mathbb{Z}^+$ such that $f^{n_l}(x) \notin B_{\varepsilon_0}(f^{k_l+i}(x))$ for $0 \leq i \leq l$. Passing to convergent subsequences of $(f^{n_l}(x))_l$ and $(f^{k_l}(x))_l$ with limits y and z in $\overline{\mathcal{O}(x)}$ we find $d(y, f^i(z)) \geq \varepsilon_0$ for all $i \in \mathbb{Z}^+$. It follows that y does not belong to the orbit closure of z, so $\overline{\mathcal{O}(x)}$ is not minimal.

(3) Condition (∗) implies that $\mathcal{O}(x)$ is totally bounded. Then $\overline{\mathcal{O}(x)}$ is totally bounded as well, hence compact. Now use (2) and Theorem 4.2.2 (2).

4.9. Without limitation of generality we may assume that for all sufficiently small δ there is a δ-chain from x_0 to itself of length precisely N. Now use Proposition 4.4.2.

4.10.

(2) Use the formulas (A.1-3) or (A.2-1), taking into account that $(x, f(x))$ is an ε-chain for every $\varepsilon > 0$ $(x \in A)$.

4.11.

(1) Use the hint for Exercise 3.10 (1).
(3) Example (4) in Lemma 3.3.6.

Chapter 5

5.1.

(4) Let R_{-1} be the given sequence and construct by induction for every $j \geq 0$ a subsequence R_j of R_{-1} such that R_j is a subsequence of R_{j-1} all of whose terms have the same j-th coordinate. Then take a suitable diagonal sequence.

5.3.

(2) If $x, y \in X$ and $k \in \mathbb{Z}^+$, then consider the point $(x_{[0;k)}y_{[0;k)})^\infty \in P(\sigma)$.

5.4. The equivalences of (i) and (ii) and of (iii) and (iv) are obvious.

(ii)⇒(iv): Clear from the definitions.

(iv)⇒(ii): Clearly, $\mathcal{B} \subseteq \mathcal{L}^c := \mathbf{S}^* \setminus \mathcal{L}$. Conversely, if $b \notin \mathcal{B}$ then a repeated application of (iv)(b) produces a sequence of symbols $a_0, a_1, a_2, \cdots$ such that the block $ba_0 \cdots a_k$ is not in $\mathcal{B}$ for any $k \in \mathbb{N}$. Therefore, the point $x := ba_0a_1a_2 \cdots$ of Ω_S has no initial block in $\mathcal{B}$. Consequently, by (iv)(a), no subblock of x belongs to $\mathcal{B}$. It follows that $x \in \mathfrak{X}(\mathcal{B})$, therefore $b \in \mathcal{L}$. This shows that $\mathcal{B} = \mathcal{L}^c$.

Final statement: start with $b_0 \in \mathbf{S} \setminus \mathcal{B}$ and use (iv)(b) to find a point in $\mathfrak{X}(\mathcal{B})$ with first coordinate b_0.

5.5.

(1) Denote the shift space in question by X. For every point $x \in X$ and $n \in \mathbb{N}$ find a point $x' \in X$ such that $x'_{[0;n)} = x_{[0;n)}$ but $x' \neq x$ by modifyingsome coordinates of x at suitable positions larger than n.

5.6.

(2) Any mapping $\Phi\colon \mathcal{L}_k(X) \to \mathbf{T}$ can be extended to a mapping $\Psi\colon \mathcal{L}_k(Y) \to \mathbf{T}$.

(3) *First proof*: If such an extension exists then by Remark 1 following Theorem 5.4.5 the even shift would be of finite type, which is not the case.
Second proof: Suppose there is such an extension ψ. by Theorem 5.4.3, it is a k-block code for some $k \in \mathbb{N}$, defined by $\Psi : \{0,1\}^k \to \{0,1\}$, say. Use that $\psi(x) = x$ if x is in the even shift in order to show that $\Psi(b) = b_0$ for every k-block b that is present in the even shift. Since every block of length k in $10^{2k+1}1^\infty$ is present in the even shift, it follows that $\psi(10^{2k+1}1^\infty) = 10^{2k+1}1^\infty$, a contradiction, as this is not an element of the even shift.

(4) The mapping $(\sigma_X)^{\leftarrow}\colon \sigma_X[X] \to X$ is an $(n-1)$-block code for some $n \geq 1$.

5.7. *First proof.* Copy and considerably simplify the proof of Theorem 5.4.4. (One may also apply Theorem 5.4.4 to the case that Y is the shift space consisting of only the invariant point 0^∞ and φ the unique mapping from X onto Y.)

Second proof. By Lemma 5.4.6, the SFT may be assumed to have order 2. Now use the fact that a finite graph admitting an infinite walk includes a cycle.

5.8.

(1) Recall that $(A^m)_{ij} = \sum_{k_1,k_2,\dots,k_{m-1}} A_{ik_1}A_{k_1k_2}\cdots A_{k_{m-1}j}$ and note that a term in this sum gives a positive contribution iff none of its factors is 0.

(2) There is a 1,1-correspondence between cycles of length m in G and periodic points of period m.

(3) Prove by induction that $A^m = \begin{bmatrix} a_m & a_{m+1} \\ a_{m+1} & a_{m+2} \end{bmatrix}$ if $A = \begin{bmatrix} 0 & 1 \\ 1 & 1 \end{bmatrix}$.

5.9.

(2) Use Exercise 5.8 (1).

(3) The shift space Ω_S is irreducible, but $\sigma : \Omega_S \to \Omega_S$ is not an irreducible mapping. The union X of two different periodic orbits is not an irreducible shift space but $\sigma_X : X \to X$ is an irreducible mapping.

(4) Without restriction of generality the SFT may be assumed to have order 2. Next, note that in a strongly connected graph without stranded edges every finite walk can be extended to a cycle.

(5) For any word b in the language of the shift under consideration, choose a block w such that the sequence $(bw)^\infty$ satisfies the *definition* of an element of the shift space in question: for the even shift and the prime gap shift, let $w = 0^m 10^n$ with suitable m and n; for the context free shift the proof is similar.

5.11.

(1) The shift space $\mathfrak{W}_e(G)$ consists of all sequences of 0's and 1's in which all blocks of consecutive 0's between two 1's have even length, except the first one, which can also have odd length.

(3) The shift maps the isolated point 10^∞ onto the non-isolated point 0^∞.

5.12.

(2) Observe that no block $b^{(n)}$ contains three equally spaced 0's.

5.13.

(1) For the (1,3) run-length limited shift: If the blocks p and q are present in that shift then there are $\alpha, \beta \in \{0, 1\}$ – depending on the final or first coordinate in p or q, respectively – such that the blocks $p\alpha 0$ and $1\beta q$ are present. Then $p(\alpha 0\,10\,1\beta)q$ is present: a bloc of length 6 can be inserted between p and q. By replacing here 101 by 1001 and 10101 one gets blocks of lengths 7 and 8. By repeating this procedure one gets blocks of all lengths greater than 6.
For the golden mean shift, the even shift and the prime gap shift: if the blocks p and q are present in the shift under consideration, then there are $m, n \in \mathbb{Z}^+$ such that $p\,0^m 1$ and $10^n q$ are present in the shift. Then for every $k \geq 1$ the block $p\,0^m 1^k 0^n q$ is present. For the context free shift the argument is similar (with 1 replaced by a, 0^m replaced by b^m and 0^n by c^n).

(2) (a) Let $U := X_P \cap C_0[1]$. Then $D(U, U) \subseteq P$.
(b) For $i = 1, 2, 3, 4$ let $b^{(i)} \in \mathcal{B}_P$, and for $i = 1, 3$, let F_i be the set of all distances between occurrences of 1's in $b^{(i)} b^{(i+1)}$. Then select k such that $k + (F_1 \cup F_3) \subseteq P$. Then for $i = 1, 3$ one has $b^{(i)} 0^k b^{(i+1)} \in \mathcal{B}_P$.

(3) Such a system has an equicontinuous factor (namely, $(\mathbb{S}, \varphi_a)$) which is not weakly mixing.

5.14.

(1) Use 5.6.2 (7) with $k = 1$ and Proposition 5.2.5, respectively.

(3) "Open": If $x \in M_1$ then select $n \geq 4$ so large that $x_{[0;n)}$ includes an occurrence of $q(2)$; then $M \cap C_0[x_{[0;n)}] \subseteq M_1$. "Closed": Use that M_1 is the intersection of the closed sets $\{x \in M : q(2)$ occurs at position $i\}$ for even i.

(4) Let $U_1 = U_2 = U_3 := M_1$ and $U_4 := M \setminus M_1 = \sigma[M_1]$.

5.15.

(1) Let B_U be the set of all points in Ω_2 with a complete history represented by a universally spaced concatenation. Then $B_U \subseteq B$. Moreover, B_U is closed, invariant, and includes the point β.

(2) This is, by 1 above, just what was shown in the second half of 5.6.11 (4).

(3) Let y be an non-exceptional point and let $x^{(0)}$ be the ordinary point to the complete history $\{x^{(n)} : n \in \mathbb{Z}\}$ of which the point y belongs. If $y = x^{(n)}$ with $n < 0$ then a point from any past of y is in the unique past of $x^{(0)}$. This implies that y has a unique complete past in B. If $y = x^{(n)}$ with $n > 0$ then any complete history of y includes a point y' with $\sigma^n y' = y = x^{(n)}$, thatis, $(y')_{[n;\infty)} = (x^{(0)})_{[n;\infty)}$. Then after some reflection one sees that 5.6.11 (4) implies that y' and $x^{(0)}$ have the same complete histories so that $y' = x^{(0)}$.

(4) β has at least two complete histories, $\tilde{\beta}\beta$ and $\tilde{\beta}1\beta$.

5.16.

(1) Initial observation: Let $n \in \mathbb{N}$ and let $0 < \varepsilon < 1/(n+1)$. If $x, y \in X$ and $d(y, \sigma(x)) < \varepsilon$ then the initial n-block of y is $x_1 \cdots x_{n-1}a_n$ with $a_n := x_n$ (differences in corresponding coordinates can only occur in position at least n). So if $a = a_0 \cdots a_{n-1}a_n$ is defined such that $x_{[0;n)} = a_{[0;n)}$ and $y_{[0;n)} = a_{[1;n]}$ then the two subblocks of a of length n are X-present.

(i)⇒(ii): If $n \in \mathbb{N}$ and u and v are X-present blocks of length n then there are points x and y starting with these blocks, and for $0 < \varepsilon < 1/(n+1)$ there is an ε-chain $(x = x^{(0)}, \dots, x^{(N)} = y)$. Then the initial observation above implies that there is an $(n+N)$-block a such thatevery subblock of length n is X-present, $x^{(0)}_{[0;n)} = a_{[0;n)}$ and $x^{(N)}_{[0;n)} = a_{[N;n+N-1]}$.Note that the block a isobtained as follows: $a_i := x^{(0)}_i$ for $0 \leq i \leq n$ and$a_{n+j} := x^{(j)}_n$ for $0 \leq j \leq N-1$.

(ii)⇒(i): Every n-subblock of a occurs as initial block in an element of X. The elements of X so obtained form an ε-chain if $\varepsilon < 1/(n+1)$.

(2) In view of Proposition 5.3.7 and Theorem 4.5.13 it remains to show that if X is chain-transitive then X is irreducible. Assume that the order of X is 2. Then the block a obtained in 1(ii) above is X-present because all its 2-subblocks are. For every pair u, v of X-present blocks, let $n := \max\{|u|, |v|\}$ and let $x, y \in X$ be such that u is the initial $|u|$-block of $x_{[0;n)}$ and v is the final $|v|$-block of $y_{[0;n)}$ (if $|v| < n$ then we need here the final statement of 1). By connecting the points x and y with an ε-chain of length $> n$ – concatenate with sufficiently many loops – statement 1 above implies that there is a block w such that the block uwv is X-present. If X has order larger than 2, use Corollary 5.3.7 and the fact that *chain-transitivity is an*

invariant for conjugations of compact metric spaces (the proof is virtually the same as that for Theorem 4.5.5).

(3) "Only if": Assume that X has order 2 and let ε and n be as in 1. The symbols a_i defined according to the hint in 1 form a sequence all of whose 2-blocks are X-present, i.e., they are the coordinates of an element a of X. Obviously, $d(\sigma^i(a), x^{(i)}) \le \varepsilon$ for all i. If X has order larger than 2, use Corollary 5.3.7 and the fact that *the pseudo-orbit tracing property is an invariant for conjugations of compact metric spaces.*
"If": Let $N \in \mathbb{N}$ be so large that every $1/(N-1)$-chain in X has a $1/2$-shadow. Then from a point x of Ω_S in which all N-blocks are in the alphabet of X one can easily construct an $1/(N-1)$-chain in X of points beginning with the successive N-blocks of x. If $z \in X$ is a $1/2$-shadow of this chain then $x = z$, so $x \in X$.

Chapter 6

6.1.

(1) An expansion that does not end with the half-sequence 1^∞ is uniquely determined by the condition that $2^n t \pmod 1 \in [0; \frac{1}{2}) + \frac{1}{2}t_n$, that is, $f^n([t]) \in P_{t_n}$.
(2) The points $1^n 0^\infty$ for $n \in \mathbb{N}$ converge in Ω_2 to the point 1^∞ for $n \rightsquigarrow \infty$.
(3) "If": apply Lemma 6.1.3 (3).
"Only if": Assume there is a point $x \in P_\alpha \cap \overline{P}_\beta$ for $\alpha \neq \beta$. Then $\iota(x)$ starts with the symbol α, but every neighbourhood of x contains a point of P_β, which has an itinerary starting with the symbol β, hence has a distance of at least 1 to $\iota(x)$.

6.2.

(1) Use Exercise 5.4.
(2) First statement: the proof is by induction on the length of the blocks; in the induction step, consider the final n-block of an $(n+1)$-block. Second statement: as the symbol 2 is isolated in every $(\mathcal{P}, f)$-allowed block – so $Z(\mathcal{P}, f)$ is included in the golden mean shift — it is sufficient to show that every block over $\{0, 2\}$ in which the symbol 2 occurs only isolated can be realized as a partial itinerary. The proof is by induction on the length of the block, using that every point in the union of P_0 and P_2 has a preimage in each of these sets. Alternative proof: use Lemma 6.3.1.
(4) Use 3.
(5) Let the k-block b be partial itinerary of the point $x \in X$. Then $x = \psi(z)$ for some point $z \in Z$. Now Corollary 6.1.9 (1) implies that $b = z_{[0;k-1)}$.

6.3.

(1) Let $X'_k := \bigcup\{D_k(b) : b \in \mathbf{S}^k\}$. Obviously, $X^* = \bigcap_{k\ge1} X'_k$, and for every $k \in \mathbb{N}$ one has $X^* \subseteq X_k \subseteq X'_k$.
(2) "$\subseteq$": Straightforward.
"$\supseteq$": Let $x \in X \setminus \psi[Z]$. For every $z \in Z$ there exists $k_z \in \mathbb{N}$ such that $x \notin \overline{D_{k_z}(z)}$. Then $x \notin \overline{D_{k_z}(z')}$ for every $z' \in \widetilde{B}_{k_z}(z)$, hence $x \notin \overline{D_k(z')}$ for all $k \ge k_z$. Cover Z by finitely

many sets $\widetilde{B}_{k_z}(z)$ with $z \in Z$ and let k_0 be the maximum of the corresponding values of k_z. Then $x \notin \overline{D_{k_0}(z')}$ for every $z' \in Z$, which implies that $x \notin \overline{X}_{k_0}$.

6.4.

(1) In order to show that $\bigcup(\mathcal{P} \vee \mathcal{Q})$ is dense in X, use that for every non-empty open subset W of X and every $P \in \mathcal{P}$ such that $W' := W \cap P \neq \emptyset$ the non-empty open set W' meets some $Q \in \mathcal{Q}$.

(3) Use 1 and 2 and induction on k (recall that all iterates of f are semi-open).

6.5. (ii)⇒(iii): Start with Exercise 6.2 (5). (iii)⇒(ii): See the Remark following Proposition 6.1.5.

6.6. "$\subseteq$": Clear from the definition of ψ. "$\supseteq$": See the proof of Corollary 6.1.9 (2).

6.8.

(1) Let η be the expansive coefficient. If $n \in \mathbb{N}$ then by uniform continuity of f there exists $\delta > 0$ such that $d(f^i(x), f^i(y)) < \eta$ for $i = 0, \dots, n-1$ for all points $x, y \in X$ with $d(x, y) < \delta$. Then two distinct periodic points with period n have a distance of at least δ.

(2) Straightforward, taking into account that φ is uniformly continuous.

(3) Observe: (a) A subset U of $X \times X$ is a neighbourhood of Δ_X iff there exists $\eta > 0$ such that $U_\eta := \{(x, y) \in X \times X : d(x, y) < \eta\} \subseteq U$, and (b) If $U \subseteq X \times X$ and $\Delta_X \subseteq U$ then Δ_X is the maximal invariant subset of U iff for every point $(x, y) \in U$ with $x \neq y$ one has $\mathcal{O}_{f\times f}(x, y) \nsubseteq U$.

(5) Let c be the expansive coefficient of f. The mappings f^i for $0 \leq i \leq k-1$ form a uniformly equicontinuous set, so there exists $c' > 0$ such that for all $x, y \in X$:

$$\exists i \in \{0, \dots, k-1\} : d(f^i(x), f^i(y)) > c \quad \Rightarrow \quad |x - y| > c' . \tag{$*$}$$

If $x \neq y$ and $d(f^m(x), f^m(y)) > c$ (expansiveness of f) then $f^m = f^i \circ (f^k)^n$ with $n \in \mathbb{Z}^+$ and $0 \leq i \leq k-1$. Now apply $(*)$.

(6) Suppose the continuous mapping $f : X \to X$ is expansive. First, show that f is strictly monotonous. If f is increasing then select a point $x \in X$ such that $f(x) \neq x$. The monotonous bounded sequence $(f^n(x))_n$ is a Cauchy sequence; use this to get a contradiction with expansiveness of f. If f is decreasing then f^2 is increasing and, by 4 above, expansive. By the above this is impossible.
Alternative proof: if f is surjective and monotonous then f is a homeomorphism, so expansiveness of f would imply that X is finite (see Note 9 in Chapter 6), hence degenerate.

6.9. (a) This is Lemma 6.2.3. (b) use Lemma 6.1.3 (3) and (a).

6.10.

(1) Use the final statement in Proposition 6.1.8.

(2) Consider rigid rotation $(\mathbb{S}, \varphi_a)$ with the topological partition considered in 6.3.7 for $b = 1/2$ and compare ι_X with the itinerary mapping ι on $\mathbb{S}^*$ in this system: for

every $[t] \in \mathbb{S}^*$ one has $\iota_X([t],1) = \iota_X([t],2) = \iota[t]$. So if $X' := \mathbb{S}^* \times \{1,2\}$ then X' is dense in X and $\iota_X[X'] = \iota[\mathbb{S}^*$. Hence $\mathrm{cl}_Z\iota_X[X'] = \mathrm{cl}_Z\iota[\mathbb{S}^*]$.

6.12. Note that $\mathcal{P}$ is g_p-adapted and has property (**M***). As $Z = Z(\mathcal{P}, g_p)$ is the SFT defined by the strongly connected graph shown in Figure 2.14, it follows that Z is irreducible. By Exercise 5.9 (4) the system (Z, σ_Z) has a dense set of periodic points. It remains to show that $\mathcal{P}$ is a pseudo-Markov partition.

For $i = 0, \dots, p-2$ let d_i denote the absolute (constant) value of the derivative of g_p on J_i. Thus, $d_0 = 2$, $d_{p-2} = \frac{1}{2}(p-1) \geq 2$ and all other d_i's are 1. Then formula (6.3-2) implies that $|D_{k+1}(z)| = d_{z_0}^{-1}|D_k(\sigma z)|$ for all $z \in Z$ (recall that $D_{k+1}(z) \subseteq J_{z_0}$), hence by induction $|D_{k+1}(z)| = (d_0 \cdots d_{k-1})^{-1}$. Consequently, $|D_{k+1}(z)| \leq 2^{-\gamma(k,z)}$, where $\gamma(k,z)$ is the total number of occurrences of the symbol 0 or the symbol $p-2$ in the block $z_{[0;k)}$. Inspection of Figure 2.14 shows that every infinite path in the graph includes infinitely many of these symbols. So $\gamma(k,z) \rightsquigarrow \infty$ if $k \rightsquigarrow \infty$, so $|D_{k+1}(z)| \rightsquigarrow 0$ for $k \rightsquigarrow \infty$.

Chapter 7

7.1.

(1) If U is an open nbd of A then $\varepsilon := \mathrm{dist}(A, X \setminus U) > 0$. For each $x \in A$, select a neighbourhood V_x of x according to (7.1-1) and consider $V := \bigcup_{x\in A} U_x$.

(2) By the implication Theorem 3.4.4 (i)⇒(iii), the points of A are stable under f^p. Hence the set $\{f^{pn} : n \in \mathbb{Z}^+\}$ is uniformly equicontinuous on the finite set A. Moreover, the set $\{f^i : 0 \leq i \leq p-1\}$ is equicontinuous at each point of A. Consequently, the set $\{f^m : m \in \mathbb{Z}^+\}$ is equicontinuous on A: write each f^m with $m \in \mathbb{Z}^+$ as $f^{np} \circ f^i$ with $n \in \mathbb{Z}^+$ and $0 \leq i \leq p-1$.

(3) (i)⇒(ii): For $i = 0, \dots, p-1$, let $x_i := f^i(x)$. As $x_i \in (f^{p-i})^{\leftarrow}(x_0)$ the result follows from Lemma 7.1.1 (2).
(ii)⇒(iii): Clear from 1.
(iii)⇒(ii): Clear from 2.
(ii)⇒(i): Obvious.
(iii)⇒(v): Use the Implication Theorem 3.3.4 (i)⇒(iii).
(v)⇒(iv): Obvious.
(iv⇒(iii): Use the Implication Corollary 3.4.4 (ii)⇒(iii).

7.2.

(2) (a) If not then for every $m \geq 0$ at most one of the intervals $T_s^m[J], \dots, T_s^{m+k}[J]$ contains the point c. So 1 would imply that $|T_s^{nk}[J]| \geq (\frac{1}{2}s^k)^n|J|$ for all n.

(3) (a) If not, then first show that c is an end point of $T_s^{ip}[J']$ for all $i \in \mathbb{N}$, as follows: Note that $T_s(c)$ is the maximal value of T_s, so $T_s(c)$ is an end point of $T_s[J']$. Using the fact that, by assumption, T_s is monotonous on $T_s^n[J']$ for all $n \in \mathbb{N}\setminus p\mathbb{N}$, it follows that $T_s^j(c)$ is an end point of $T_s^j[J']$ for $1 \leq j \leq p$. In particular,

$T^p(c) = c$ is an end point of $T_s^p[J']$. Repeat the procedure with J' replaced by $T_s^p[J']$. Etc. So, indeed, c is an end point of $T_s^{ip}[J']$ for all $i \in \mathbb{N}$. Together with the hypothesis this implies that c is not an interior point of $T_s^n[J']$ for all $n \geq 2$. So 1 would imply that $|T_s^n[J']| \geq \frac{1}{2}s^{n-1}$ for all $n \geq 2$.

(b) By (a), there exists $n \in \mathbb{N} \setminus p\mathbb{N}$ such that $c \in T_s^n[J']$. Recall that $c \in J'$.If one writes $n = ip + j$ with $0 < j < p$ then $T_s^j(c) = T_s^n(c) \in T_s^n[J']$.

7.3.

(2) Extend f over $[0; 4/3]$ by defining $f(x)$ also for $x \geq 1$ as $\frac{3}{2}(\frac{4}{3} - x)$, and apply Example (3) of 7.1.3 (i.e., Exercise 7.2).

7.4. If the system is not sensitive then any transitive point x_0 is almost periodic. This leads to a contradiction with the assumed non-minimality of the system.

7.6.

(1) If two points move far apart then there are two periodic points (close to these points) which move far apart as well – infinitely often.

7.7. A weakly mixing system is transitive, so only sensitivity needs a proof. Select two open sets U_3 and U_4 with a positive distance 2ε. If $x \in X$ and U is an open nbd of x then let $U_1 := U =: U_2$.

7.8.

(1) This is, essentially, shown in the proof of Proposition 7.2.4.
(2) Use (1) and apply Theorem 7.1.4 (1) to the transitive point x.

7.9.

(2) First show that it is sufficient to prove the statement only for the case that $x = \tau$ (use that σ maps T onto T and that $f: G \to G$ is injective). Next, let $y \in T$ and assume that $\varphi(y) = \varphi(\tau) = \underline{0}$. There is a sequence $(n_i)_{i\in\mathbb{N}}$ in $\mathbb{Z}^+$ such that $\sigma^{n_i}(\tau) \rightsquigarrow y$, hence $\underline{n_i} = f^{n_i}(\underline{0}) \rightsquigarrow \underline{0}$, for $i \rightsquigarrow \infty$. It follows that, for every $k \in \mathbb{N}$, the sequence of coordinates of $\underline{n_i}$ starts with k 0's, that is, $n_i \in 2^k\mathbb{Z}^+$, for almost all i. This implies that $\sigma^{n_i}(\tau)$ starts with the block $t(k-1)$ for almost all i. Consequently, $\sigma^{n_i}(\tau) \rightsquigarrow \tau$ for $i \rightsquigarrow \infty$, so $y = \tau$.

(3) Existence: Let T^* be the set of all points in Ω_2 that have, for every $n \in \mathbb{N}$, a $t(n)$-representation. Then T^* is invariant under σ, and it is straightforward to show that T^* is closed in Ω_2. Part (e) in the proof of Theorem 5.6.14 implies that $\tau \in T^*$, hence $T \subseteq T^*$.

Unicity: The proof is by induction. For $n = 0$: if x admids two $t(0)$-representations then $x = 0^\infty$, which does not belong to the infinite minimal set T. Let $n \in \mathbb{Z}^+$ and assume that x has a unique $t(n)$-representation. If there are two distinct $t(n+1)$-representations then these must induce the same $t(n)$-representation, so the blocks from the one $t(n+1)$-representation overlap with those of the other one as follows:

$$x = t_i(n) * \overbrace{\underbrace{t(n) * t(n)}_{t(n+1)} * t(n)}^{t(n+1)} * \overbrace{\underbrace{t(n) * t(n)}_{t(n+1)}}^{t(n+1)} * \cdots\cdots$$

If the central separator in $t(n+1)$ is a 0 (i.e., n is even) – if it is a 1 the proof is similar – then every $*$ in x, except perhaps the first one, is a 0. Then x is ultimately periodic, hence x does not belong to the infinite minimal set T.

(4) Let $x = \lim_{i\to\infty} \sigma^{n_i}\tau$ and $y = \lim_{i\to\infty} \sigma^{m_i}\tau$. Then $d(\underline{n_i}, \underline{m_i}) \to 0$ for $i \to \infty$, hence for every $k \in \mathbb{N}$ the sequences $\underline{n_i}$ and $\underline{m_i}$ start with the same k-block for almost all i. We may assume without limitation of generality that $n_i \leq m_i$ for all i, hence $m_i \in n_i + 2^k\mathbb{Z}^+$ for almost all i. Consequently, the sequence of coordinates of $\sigma^{m_i}\tau$ equals the sequence of coordinates of $\sigma^{n_i}\tau$ from which an initial block is deleted of length an integer multiple of the length of the block $t(k-1)$. It follows that the $t(k-1)$-representations of $\sigma^{n_i}\tau$ and $\sigma^{m_i}\tau$ are the same for almost all i. Consequently, the $t(k-1)$-representations of x and y are the same as well. This holds for all $k \in \mathbb{N}$. Now there are two possibilities:
   (a) For every $k \in \mathbb{N}$ there exists $j \in \mathbb{Z}^+$ such that all separators of the common $t(k-1)$-representation of x and y belong to a block $t(k+j)$ of their $t(k+j)$-representation; in this case $x = y$.
   (b) There exists $k \in \mathbb{N}$ such that some separator of the common $t(k-1)$-representation of x and y does not belong to any block $t(k+j)$ of their $t(k+j)$-representation, i.e., is a separator for every $t(k+j)$-representation with $j \geq 0$. In that case x and y can differ only in that separator.

   Finally, if y' is another point of T such that $\varphi(y') = \varphi(x)$ then, similarly, all $t(n)$-representations of y' are the same as those of x. So there are the same possibilities as for y, with in case (b) the same position of the coordinate where x and y' can be different; in particular, if $y \neq x$ and $y' \neq x$ then $y' = y$.

(5) If the pair (x, y) is proximal in T then the pair $(\varphi(x), \varphi(y))$ is proximal in G under f, hence $\varphi(x) = \varphi(y)$. But then the pair (x, y) is asymptotic.

7.10.

(1) Define sets $A_\delta(X, f) := \{(x, y) \in X \times X : \limsup_{n\to\infty} d(f^n x, f^n y) \geq \delta\}$ and $B_0 := \{(x, y) \in X \times X : \liminf_{n\to\infty} d(f^n x, f^n y) = 0\}$. Then it is clear that $A_\delta(X, f) = \bigcap_{(k,n)\in\mathbb{N}\times\mathbb{N}} \{(x, y) \in X \times X : \exists i \geq n \text{ with } d(f^i x, f^i y) > \delta - 1/k\}$ and that $B_0(X, f) = \bigcap_{(k,n)\in\mathbb{N}\times\mathbb{N}} \{(x, y) \in X \times X : \exists i \geq n \text{ with } d(f^i x, f^i y) < 1/k\}$ are G_δ-sets. Next, note that $LY_\delta(X, f) = A_\delta(X, f) \cap B_0(X, f)$.

(3) First, note that the point (x, y) is recurrent under $f \times f$. Moreover, if $\varepsilon > 0$ and the points $x_0, y_0 \in X$ have $d(x_0, y_0) > \text{diam}(X) - \varepsilon$, then approach the point (x_0, y_0) by points $(f^n(x), f^n(y))$.

(4) Use 1 and 3.

7.11.

(1) Every proximal pair is asymptotic.

(3) Modify the construction in Example (3) in 7.3.1 in such a way that the scrambled set obtained is included in the subshift under consideration.

7.12. Identify a minimal subset of X to a point and apply the first case of Corollary 7.3.7 to the induced system – see 1.5.8.

7.13. If not, then the system is equicontinuous, hence it has no scrambled subsets.

7.14.

(1) Consider the factor map $\varphi: (Z, f) \to (\Omega_2, \sigma)$ proved to exist in Proposition 7.4.3. Let M be a non-trivial minimal set in Ω_2. There is a minimal set N in Z (hence in X) such that $\varphi[N] = M$. Obviously, N is not trivial.

(3) Use that each of the two intervals of a horseshoe includes one of the invariant points. Then the interval including the point $2/3$ must include the interval $[1/2; 1]$ and the other interval must include $[0; 1/2]$.

Chapter 8

8.1. By the mean value theorem, for $x, y \in \mathbb{R}$ there exists a ξ between x and y such that $d_n^f(x, y) = |x^{2^{n-1}} - y^{2^{n-1}}| = 2^{n-1}\xi^{2^{n-1}-1}|x - y|$ with $\xi > 2$ if $x, y > 2$.

8.2.

(1) The argument is similar to that used in Example (5) after Corollary 8.1.9. To estimate $\theta_n(M)$, let $2^{k-1} \leq n < 2^k$ and analyze how a block of length n can be positioned with respect to the standard concatenation of the sequence μ mentioned in 5.6.2 (7). One finds at most $4(n-1) + 2(2^k - n)$ possibilities, which is at most $7n$.

(2) *First proof*: Proceed as in Example (4) after Corollary 8.1.9. In order to compute $\theta_n(X)$, write $\theta_n(X)$ as $\theta_n^o + \theta_n^e + \theta_n^1$, where θ_n^o, θ_n^e and θ_n^1 denote the number of X-present n-blocks ending with an odd number ≥ 1 of 0's, an even number ≥ 2 of 0's or no 0 (i.e., ending with 1), respectively.
Second proof: Use Theorem 8.2.7 and Example (5) just before Proposition 5.4.1.

(3) Use that $p_n(X) \leq \theta_n(X)$).

(4) Let η be an expansive coefficient for f. Then every member of any (n, η)-span – also the (n, η)-span with minimal cardinality – contains at most one periodic point with period n. Hence $p_n(X) \leq \text{minspan}_n(\eta, X.f)$.

8.4. Let K be a compact subinterval of X, let $\varepsilon > 0$ and let $n \in \mathbb{N}$. If S is an (n, ε)-separated subset of K then $\#S \leq 1 + |K|\lambda^n/\varepsilon$.

8.5. First, note that X can be covered by compact subsets $K_1, \ldots, K_m$, each with diameter less than or equal to δ. Then show that

$$\text{minspan}_n(\varepsilon, K_1, f) + \cdots + \text{minspan}_n(\varepsilon, K_m, f) \geq \text{minspan}_n(\varepsilon, X, f)$$

and use Lemma 8.6.4 to show that $h(f) = \max\{h(K_1, f), \ldots, h(K_m, f)\}$.

Literature

And further, my son, by these words be admonished: of making many books there is no end, and much study is a weariness to the flesh.
Ecclesiastes 12:12 (21st Century King James Version)

A few textbooks to which I refer several times are indicated by acronyms:

[Eng] R. Engelking [1977]
[GH] W. H. Gottschalk & G. A. Hedlund [1955]
[deV] J. de Vries [1993]

R. L. Adler, Symbolic dynamics and Markov partitions, *Bull. Amer. Math. Soc. (N. S.)* **35**, No. 1 (1998), 1–56.

R. L. Adler, A. G. Konheim and M. H. McAndrew, Topological entropy, *Trans. Amer. Math. Soc.* **114**, (1965), 309–319.

E. Akin, J. Auslander and K. Berg, When is a transitive map chaotic?, in: V. Bergelson (ed.) et al., *Convergence in ergodic theory and probability*, (Conference Columbus, OH, 1993) (Ohio University Math. Res. Inst. Pub., 5), de Gruyter, Berlin, 25–40, 1996.

E. Akin and S. Glasner, Residual properties and almost equicontinuity, *J. d'Anal. Math.* **84**, (2001), 243–286.

E. Akin, E. Glasner, W. Huang, S. Shao and X. Ye, Sufficient conditions under which a transitive system is chaotic, *Ergodic Theory Dyn. Syst.* **30**, (2010), 1277–1310.

J.-P. Allouche and J. Shallit, The ubiquitous Prouhet–Thue–Morse sequence, in C. Ding *et al.* (ed.), *Sequences and their applications* (Proc. Int. Conf., SETA '89, Singapore), Springer Series in Discr. Math. and Theor. Comp. Sc., Springer, pp. 1–16, 1999.

Ll. Alsed'a, S. Kolyada, J. Llibre and L. Snoha, Entropy and periodic points for transitive maps, *Trans. Amer. Math. Soc.* **351**, (1999), 1551–1573.

J. Auslander, *Minimal flows and their extensions*, North-Holland, Amsterdam, 1988.

J. Auslander and J. A. Yorke, Interval maps, factors of maps, and chaos, *Tôhoku Math. Journ.* **32**, (1980), 177–188.

P. C. Baayen, *Universal Morphisms*, Mathematisch Centrum, Amsterdam, 1964.

M. Barge and J. Martin, Chaos, periodicity and snakelike continua, *Trans. Amer. Math. Soc.* **289**, (1985), 355–365.

N. P. Bhatia and G. P. Szegö, *Stability theory of dynamical systems*, Springer-Verlag, Berlin, etc, 1970.

F. Blanchard, E. Glasner, S. Kolyada and A. Maass, On Li–Yorke pairs, *J. Reine Angew. Math.* **547**, (2002), 51–68.

F. Blanchard, B. Host and A. Maass, Topological complexity, *Ergodic Th. and Dynam. Sys.* **20**, (2000), 641–662.

L. S. Block and E. M. Coven, Topological conjugacy and transitivity for a class of piesewise monotone maps of the interval, *Trans. Amer. Math. Soc.* **300**, No 1, (1987), 297–306.

L. S. Block and W. A. Coppel, *Dynamics in One Dimension*, LNM 1513, Springer-Verlag, Berlin, etc, 1992.

L. Block, J. Guckenheimer, M. Misiurewicz and L. S. Young, Periodic points and topological entropy of one dimensional maps, in: *Global Theory of Dynamical Systems*, LNM 819, Springer-Verlag, Berlin, etc., pp. 18–34, 1980.

A. M. Blokh, On the connection between entropy and transitivity for one-dimensional mappings, *Russian Math. Surveys* **42**, (1987), 165–166.

R. Bowen, Periodic points and measures for Axiom A diffeomorphisms, *Trans. Amer. Math. Soc.* **154**, (1971), 377–397.

R. Bowen, Entropy for group endomorphisms and homogeneous spaces, *Trans. Amer. Math. Soc.* **153**, (1971), 401–414, erratum in *TAMS* **181**, (1973), 509–510.

R. Bowen, *Equilibrium states and the ergodic theory of Anosov diffeomorphisms*, LNM 470, Springer-Verlag, Berlin, etc, 1975.

B. F. Bryant, On expansive homeomorphisms, *Pacific J. Math.* **10**, (1960), 1163–1167.

B. F. Bryant and P. Walters, Asymptotic properties of expansive homeomorphisms, *Math. Systems Theory* **3**, (1969), 60–66.

J. Buescu, *Exotic Attractors,* Birkhäuser, 1997.

G. J. Butler and G. Pianigiani, Periodic points and chaotic functions in the unit interval, *Bull. Austral. Math. Soc.* **18**, No. 2, (1978), 255–265.

C. Conley, *Isolated invariant sets and the Morse index*, Regional Conference Series in Mathematics No 38, Amer. Math. Soc., Providence, 1978.

E. M. Coven and G. A. Hedlund, $\overline{P} = \overline{R}$ for maps of the interval, *Proc. Amer. Math. Soc.* **79**, (1980), 316–318.

E. M. Coven, I. Kan and J. A. Yorke, Pseudo-orbit shadowing in the family of tent maps, *Trans. Amer. Math. Soc.* **308**, No. 1, (1988), 227–241.

E. M. Coven and M. Keane, Every compact metric space that supports a positively expansive homeomorphism is finite, *IMS Lecture Notes, Monograph Series, Dynamics and Stochastics,* **48**, (2006), 304–305.

E. M. Coven and I. Mulvey, Transitivity and the center of maps of the circle, *Ergod. Th. and Dynam. Sys.* **6**, (1986), 1–8.

R. L. Devaney, *An introduction to chaotic dynamical systems*, Addison-Wesley Publishing Company, Redwood City, CA, second edition, 1989.

E. I. Dinaburg, The relation between topological entropy and metric entropy, *Soviet Math. Dokl.* **11**, (1970), 13–16.

M. Dirbák and P. Maličký . On the construction of non-invertible minimal skew products, *J. Math. Anal. Appl.* **375**, (2011), 436–442.

B.-S. Du, A simple proof of Sharkovsky's theorem revisited, *Amer. Math. Monthly* **114**, (2007), 152–155.

R. Ellis, *Lectures on topological dynamics*, W. A. Benjamin, Inc., New York, 1969.

R. Ellis and W. H. Gottschalk, Homomorphisms of transformation groups, *Trans. Amer. Math. Soc.* **94**, (1960), 258–271.

R. Engelking, *General Topology*, PWN, Warszawa, 1977.

A. A. Fraenkel, *Abstract set theory*, North-Holland Publ. Company, Amsterdam, 1953.

J. Franks, A variation of the Poincare-Birkhoff Theorem, in: *Hamiltonian Dynamical Systems*, Contemporary Mathematics, Vol. 81, Amer. Mat. Soc., Providence, pp. 111–117, 1988.

H. Furstenberg, *Recurrence in ergodic theory and combinatorial number theory*, Princeton Univ. Press, Princeton, New Jersey, 1981.

L. Gillman and M. Jerison, *Rings of Continuous Functions*, Van Nostrant, New York, 1960.

E. Glasner, *Ergodic Theory via Joinings*, Surveys and Monographs, vol. 101, Amer. Math. Soc,. Providence, RI, 2003.

S. Glasner and D. Maon, Rigidity in topological dynamics, *Ergod. Theory and Dynam. Sys.* **9**, (1989), 309–320.

E. Glasner and B. Weiss, On the construction of minimal skew products, *Israel J. Math.* **34**, (1979), 321–336.

E. Glasner and B. Weiss, Sensitive dependence on initial conditions, *Nonlinearity* **6**, (1993), 1067–1075.

W. H. Gottschalk and G. A. Hedlund, *Topological dynamics*, AMS Colloquium Publ. XXXVI, Amer. Math. Soc., Providence, 1955.

O. Hajek, Representations of dynamical systems, *Funkcial. Ekvac.* **14**, (1971), 25–34.

G. H. Hardy and W. M. Wright, *An introduction to the theory of numbers*, Oxford University Press, Oxford, England (5th edition), 1979.

G. A. Hedlund, Endomorphisms and automorphisms of shift dynamical systems, *Math. Systems Theory* **3**, (1969), 320–375.

E. Hewitt, Compact monothetic semigroups, *Duke Math. J.* **23**, (1956), 447–457.

W. Huang and X. Ye, Devaney's chaos or 2-scattering implies Li–Yorke chaos, *Topology Appl.* **117**, (2002a), 259–272.

W. Huang and X. Ye, An explicit scattering, non-weakly mixing example and weak disjointness, *Nonlinearity* **15**, (2002b), 1–14.

M. Hurley, Attractors, persistence, and the density of their basins, *Trans. Amer. Math. Soc.* **269**, (1982), 247–271.

M. Hurley, Lyapunov functions and attractors in arbitrary metric spaces, *Proc. Amer. Math. Soc.* **126**, No 1, (1998), 245–256.

A. Iwanik, Independence and scrambled sets for chaotic mappings, in: *The mathematical heritage of C. F. Gauss*, World Sci. Publishing, River Edge, NJ, pp. 372–378, 1991.

K. Janková and J. Smítal, A characterization of chaos, *Bull. Austral. Math. Soc.* **34**, No. 2, (1986), 283–292.

H. B. Keynes and J. B. Robertson, Generators for topological entropy and expansiveness, *Math. Systems Theory* **3**, (1969), 51–59.

B. Kitchens, *Symbolic Dynamics*, Springer-Verlag, Berlin, etc, 1998.

S. Kolyada, L. Snoha and S. Trofimchuk, Noninvertible minimal maps, *Fund. Math.* **168**, No. 2, (2001), 141–163.

R. L. Kraft, Chaos, Cantor sets, and hyperbolicity for the logistic maps, *Amer. Math. Monthly*, **106**, (1999), 400–408.

M. Kuchta and J. Smítal, Two-point scrambled set implies chaos, in: *European Conference on Iteration Theory* (Caldes de Malavella, 1987), World Sci. Publishing, Teaneck, NJ, pp. 427–430, 1987.

C. Kuratowski, *Topologie* I, Academic Press, New York, and PWN-Polish Scientific Publishers, Warszawa, 1966. (There is also a quite recent French edition, published by Éditions Jacques Gabay, Sceaux, in 1992.)

Shihai Li, ω-Chaos and Topological Entropy, *Trans. Amer. Math. Soc.* **339**, No. 1, (1993), 243–249.

T. Y. Li and J. A. Yorke, Period three implies chaos, *Amer. Math. Monthly* **82**, (1975), 985–992.

D. Lind and B. Marcus, *An introduction to symbolic dynamics and coding*, Cambridge University Press, Cambridge, 1995.

J. Llibre & M. Misiurewicz, *Combinatorial Dynamics and Entropy in Dimension One*, World Scientific, Singapore, 1993.

J. L. Locher (ed.), *De werelden van M. C. Escher*, Meulenhof International, Amsterdam, 1971.

N. Markley and G. M. E. Paul, Almost automorphic symbolic minimal sets without unique ergodicity, *Israel J. Math.* **34**, (1979), 259–272.

R. M. May, Simple mathematical models with very complicated dynamics, *Nature* **261**, (1976), 459–467.

J. Milnor, On the concept of attractor: correction and remarks, *Commun. Math. Phys.* **102**, (1985), 517–519.

J. Milnor and W. Thurston, On iterated maps of the interval, in: J. C. Alexander (ed.), *Dynamical Systems, Proceedings, University of Maryland*, 1986–87, LNM 1342, Springer-Verlag, Berlin, etc., pp. 465–563, 1988.

M. Misiurewicz, Horseshoes for mappings of the interval, *Bull. Acad. Polon. Sci. Sér. Sci. Math.*, **27**, No. 2, (1979), 167–169.

M. Misiurewicz and W. Szlenk, Entropy of piecewise monotone mappings, *Studia Math.* **67**, No. 1, (1980), 45–63.

M. Morse and G. A. Hedlund, Unending Chess, Symbolic Dynamics, and a Problem in Semigroups, *Duke Math. J.* **11**, (1944), 1–7.

V. V. Nemytskiĭand V. V. Stepanov, *Qualitative theory of differential equations*, Princeton University Press (translated from the Russian), 1960.

Z. Nitecki, Periodic and limit orbits and the depth of the center for piecewise monotone interval maps, *Proc. Amer. Math. Soc.* **80**, (1980), 511–514.

Y. Oono, Period $\neq 2^n$ implies chaos, *Progr. Theoret. Phys.* **59**, (1978), 1029–1030.

W. Parry Symbolic dynamics and transformations of the unit interval, *Trans. Amer. Math. Soc.* **122**, (1966), 368–378.

K. Petersen and L. Shapiro, Induced flows, *Trans. Amer. Math. Soc.* **177**, (1973), 375–390.

H. Price, *Time's arrow and Archimedes' Point, new directions for the physics of time*, Oxford University Press, New York, Oxford, 1996.

W. L. Reddy, Lifting expansive homeomorphisms to symbolic flows, *Math. Systems Theory* **2**, (1968), 91–92.

H. L. Royden, *Real analysis*, Englewood Cliffs, New Yersey, 1988.

S. Ruette, *Chaos for continuous interval maps – a survey of relationship between the various sorts of chaos*. (Preprint available at http://www.math.u-psud.fr/~ruette/., 2003.

S. Ruette, Transitive sensitive subsystems for interval maps, *Studia Math.* **169**, No. 1, (2005), 81–104.

K. Sakai, Anosov maps on closed topological manifolds, *J. Math. Soc. Japan* **39**, (1987), 505–519.

K. S. Sibirsky, *Introduction to topological dynamics*, Noordhoff International Publishing, Leyden, 1975.

J. Smítal, Chaotic functions with zero topological entropy, *Trans. Amer. Math. Soc.* **297**, No. 1, (1986), 269–282.

L. Snoha, Generic chaos, *Comment. Math. Univ. Carolin.* **31**, No. 4, (1990), 793–810.

W. R. Utz, Unstable homeomorphisms, *Proc. Amer. Math. Soc.* **1**, (1950), 769–774.

M. Vellekoop and R. Berglund, On intervals, transitivity = chaos, *Amer. Math. Monthly* **101**, (1994), 353–355.

P. F. Verhulst, Notice sur la loi que la population suit dans son accroissement, *Corr. Math. et Phys. publ. par A. Quetelet, X*, (1838), 113–121.

J. de Vries, A note on topological linearization of locally compact transformation groups in Hilbert space, *Math. Systems Theory* **6**, (1972), 49–59.

J. de Vries, Linearization of actions of locally compact groups, *Trudy Mat. Inst. Steklov* **154**, (1983), 53–70 (Russian; English translation in: *Proc. Steklov Inst. Math.*, **4**, (1984), 57–74).

J. de Vries, *Elements of topological dynamics*, Kluwer, Dordrecht, 1993.

J. B. Wagoner, Strong shift equivalence theory, in: S. G. Williams (ed.), *Symbolic dynamics and its applications*. Proceedings of Symposia in Applied Mathematics 60; AMS Short Course Lecture Notes, American Mathematical Society, Providence, pp. 121–154, 2004.

P. Walters, *An introduction to ergodic theory*, Springer-Verlag, New York etc, 1982.

P. Walters, On the pseudo-orbit-tracing property and its relationship to stability, in: *The structure of attractors in dynamical systems* (Proc. Conf. North Dakota State Univ., Fargo, 1977), LNM 668, Springer-Verlag, Berlin, pp. 231–244, 1997.

A. Wilansky, *Topology for analysis*,Robert E. Krieger Publishing Company, Inc., Malabar, Florida, 1970.

Index

Index: a table for facilitating reference to topics, names, and the like, in a book, usually giving the page on which a particular word or topic may be found; – usually alphabetical in arrangement, and printed at the end of the volume. Typically found only in non-fiction books.
1913 Webster + PJC

Page numbers in bold refer to definitions. References to examples and applications immediately following a definition are not included: moreover, most occurrences of notions that are so general that they can be found on almost every page (like 'dynamical system' or 'orbit') are not included either.

U

De Gruyter Studies in Mathematics

Volume 58
Lubomir Banas, Zdzislaw Brzezniak, Mikhail Neklyudov, Andreas Prohl
Stochastic Ferromagnetism: Analysis and Numerics, 2014
ISBN 978-3-11-030699-6, e-ISBN 978-3-11-030710-8, Set-ISBN 978-3-11-030711-5

Volume 57
Dimitrii S. Silvestrov
American-Type Options: Stochastic Approximation Methods, Volume 2, 2014
ISBN 978-3-11-0312968-1, e-ISBN 978-3-11-032984-1, Set-ISBN 978-3-11-032985-8

Volume 56
Dimitrii S. Silvestrov
American-Type Options: Stochastic Approximation Methods, Volume 1, 2014
ISBN 978-3-11-0312967-4, e-ISBN 978-3-11-032982-7, Set-ISBN 978-3-11-032983-4

Volume 55
Lucio Boccardo, Gisella Croce
Elliptic Partial Differential Equations, 2013
ISBN 978-3-11-031540-0, e-ISBN 978-3-11-031542-4, Set-ISBN 978-3-11-031543-1

Volume 54
Yasushi Ishikawa
Stochastic Calculus of Variations for Jump Processes, 2013
ISBN 978-3-11-028180-4, e-ISBN 978-3-11-028200-9, Set-ISBN 978-3-11-028201-6

Volume 53
Martin Schlichenmaier
KricheverNovikov Type Algebras: Theory and Applications, 2014
ISBN 978-3-11-026517-0, e-ISBN 978-3-11-027964-1, Set-ISBN 978-3-11-028025-8

Volume 52
Miroslav Pavlovic
Function Classes on the Unit Disc: An Introduction, 2014
ISBN 978-3-11-028123-1, e-ISBN 978-3-11-028190-3, Set-ISBN 978-3-11-028191-0

www.degruyter.com